An Introduction to the Amphibians of Ecuador

An Introduction to the Amphibians of Ecuador is the first of four volumes, which are comprehensive, well-illustrated, and authoritative works, making them invaluable to biologists, conservationists, and others. This initial volume delves into the cultural history of amphibians, encompassing ethnobatrachology and folklore, while summarizing the amphibian iconography found in Ecuadorian archaeology. Moreover, it covers topics such as bioprospecting, sustainable management, and biotrade activities. The history and present state of amphibian biology research are also addressed. Furthermore, it explores in comprehensive detail the rich amphibian diversity of Ecuador, providing a thorough review of biogeography, amphibian declines, and conservation.

Subsequent volumes list the characteristics of each species, define each taxon, and compare them to similar other species. Natural history and reproductive behavior, where known, are described, as are data on vocalizations, larvae, and ontogenetic changes. Amphibian distributions are illustrated with physiographic maps with dots. Each volume addresses the declines, extinctions, and conservation status of each species and provides notations of their occurrence in reserves.

KEY FEATURES

- Summarizes the ethnozoological aspects of amphibians
- Provides a thorough history of research
- Introduction to three volumes providing accounts for each of the 3 orders, 19 families, 78 genera, and over 650 species from Ecuador

An Introduction to the Amphibians of Ecuador

Diversity, Conservation, and Cultural History

Volume I

Luis A. Coloma and William E. Duellman

CRC Press
Taylor & Francis Group
Boca Raton London New York

CRC Press is an imprint of the
Taylor & Francis Group, an **informa** business

Designed front cover image: The inaugural amplectant pair, on 21 October 2023, of the harlequin frog *Atelopus bomolochos*, born in the laboratory at Centro Jambatu de Investigación y Conservación de Anfibios in Quito, Ecuador. This critically endangered species is on the brink of extinction. Photo by Luis A. Coloma.

Designed back cover image: Tadpole of the harlequin frog *Atelopus bomolochos*. These miniature gems, once abundant by thousands in the highland streams of the Andes, have nearly vanished since the 1980s. Photo by Steven Guevara Salvador.

First edition published 2025
by CRC Press
2385 NW Executive Center Drive, Suite 320, Boca Raton FL 33431

and by CRC Press
4 Park Square, Milton Park, Abingdon, Oxon, OX14 4RN

CRC Press is an imprint of Taylor & Francis Group, LLC

ISBN: 978-0-367-65356-9 (hbk)
ISBN: 978-0-367-65355-2 (pbk)
ISBN: 978-1-003-12908-0 (ebk)

DOI: 10.1201/9781003129080

In memory of José Q. Coloma

Contents

About the Authors ... xi
Contributors ... xiii
In Memoriam ... xv
Foreword ... xix
Preface ... xxi
Acknowledgments ... xxv

Chapter 1 Ethnobatrachology ... 1

Indigenous and Vernacular Names ... 1
Tadpole Names ... 7
Standardized Names in Spanish and English ... 9
Amphibian Iconography in Ecuadorian Archaeology ... 10
Late Formative Period (1000–500 BCE) ... 11
Regional Development Period (500 BCE–500 CE) ... 14
Integration Period (500–1500 CE) ... 20
Amphibians and Their Relationship with Rain Rituals ... 27
Crafts ... 34
Mythology, Stories, and Legends ... 35
Mestizo and Spanish Popular Culture ... 44
The Toad Game ... 45
The Water Toad Fountains and Drinkers ... 45
Education ... 45
Films on Ecuadorian Amphibians ... 48
Ecuadorian Amphibians in the News ... 48
Poetry and Music ... 50
Sayings, Superstitions, Jokes, and Graffiti ... 53
Food for Humans ... 54
Native Species ... 54
The Bullfrog Industry ... 55
Traditional Medicine ... 59
Biotrade ... 60
Live Frogs for Research, Education, and the Pet Market ... 60
The Sustainable Bioethical Trade ... 66
References ... 79

Chapter 2 The History and Present of Amphibian Biology ... 87

A Linnaean Preamble ... 87
Cornalia and the Start of Amphibian Taxonomy in Ecuador ... 87
Jiménez de la Espada: An Outstanding Spanish Savant ... 89
European Museums, the British Empire, and the Axis of Collections Growth ... 94
Mountaineering and Collecting: Whymper ... 95
Rosenberg: A Successful Natural History Dealer ... 96
The Scientifically Honored but Poorly Known Buckley ... 96
The Consul: Mr. Fraser ... 97
Simons: A Promising Collector and a Tragedy ... 97
Palmer: The Analytical Chemist ... 98
The Italian Enrico Festa during the Ecuadorian Civil War ... 98
Günther and the Explosion of Collections ... 98
Boulenger: The Tireless and Prolific Taxonomist ... 99
The First Half of the 20th Century, a Period of Wars ... 100
Mario G. Peracca: The Aristocratic Italian Zoologist ... 104
Lars Andersson: The Swedish School Teacher ... 105

Emmett Dunn: The American Herpetologist 106
Edward Taylor: The University of Kansas Grandfather 106
Orcés and the Foundations of Amphibian Collections in Ecuador 106
Peters and the Harlequin Frogs 108
A Rebirth for Amphibian Collections 110
The EPN Museum and a Woman Pioneer 111
Del Pino and the Marsupial Frogs 115
Kenneth Miyata: A Legendary Harvard Herpetologist and the Roots of Ecotourism 117
Vigle: Two Decades of Conservation and Amphibian Surveys in the Old Forest 119
Rivero, Heyer, and Flores: Their Transient Work on Ecuadorian Amphibians 120
Rivero and the Cryptic Dendrobatids 120
Heyer: The Smithsonian Taxonomist behind *Leptodactylus* Frogs 121
Glenn Flores: From "Eleuths" to Pediatrics 121
The Duellman–KU Era: The Big Bang in Amphibian Systematics and Ecology 123
Martha L. Crump: From Tying Tags to Field Biology and Conservation 128
Lynch: "Eleuths" and Glassfrogs 130
Trueb: A Skilled Systematist 132
Cannatella: The Generous Supporter of Ecuadorian Scientists 134
Edwards, Hillis, and Their Pioneering Taxonomic Work 137
Duellman, Trueb, and Their Ecuadorian Graduate Students 139
The 1980s: The National Research Takes Off 141
Amphibians at the QCAZ 141
Read: The Great Naturalist and Survivor of an Airplane Crash and a Kidnapping 159
The DHMECN Collection: From the Friends of Nature to the National Biodiversity Institute 160
Three Decades of Amphibian Research at the Universidad San Francisco de Quito USFQ 170
Touzet, Velasco, and the Vivarium of Quito 177
Centro Jambatu: A Fresh Venture 177
Amphibian Conservation in Southern Ecuador: The Amaru Initiative 183
Filling the gap of amphibian research in southern Ecuador: The Museo de Zoología, Universidad Técnica Particular de Loja 188
Women in Amphibian Biology: An Ongoing Trend of Underrepresentation in Academia 189
The (Slow) Progression of Women's Inclusion in the Taxonomy of Ecuadorian Amphibians 190
Women's Contribution to Evolution, Ecology, and Natural History and Conservation of Amphibians of Ecuador 193
Enhancing Gender Inclusion in Amphibian Research 195
The Diversity and Academic Lives of Current Amphibian Biologists 195
References 204

Chapter 3 Diversity, Extinction, and Conservation 224

Molecular Diversity and Bioprospection in Amphibian Skin 224
Benefit Sharing of Genetic Resources: The Case of Epibatidine and Derivatives 226
In Search of Bioactive Peptides from Ecuadorian Frog Skin Secretions 229
Peptide Families from Ecuadorian Frogs 229
Orphan Peptides 235
Amphibian Skin AMPs as a Source of Inspiration for the Development of Therapeutic Antimicrobial Candidates 235
Cruzioseptin Analogs 236
Picturin Analogs 236
Progress in Amphibian Systematics and Taxonomy in Ecuador 236
Taxonomy and Other Fields 238
Discovery and Rates of Descriptions 239
Species Richness and Patterns of Distribution 242
Understanding the Richness of Ecuadorian Amphibians 242
Patterns of Distribution of Amphibians in Ecuador 245

Undescribed Species and *Incertae Sedis* ... 250
Atelopus spp. (*spumarius* Complex) (Anura: Bufonidae) (Figure 3.24) ... 253
Rhinella spp. (*margaritifera* Complex) (Anura: Bufonidae) (Figures 3.25, 3.26) ... 260
Dendropsophus sp. (*minutus* Complex) (Anura: Hylidae) (Figure 3.27) ... 262
Bolitoglossa spp. (*peruviana* Complex) (Caudata: Plethodontidae) ... 263
Introduced Species ... 264
Extinction and Conservation ... 264
The Red List of Ecuadorian Amphibians ... 266
The Data-Deficient Species ... 280
Which Are the Causes of Amphibian Declines in Ecuador? ... 281
Conservation ... 288
What Are the Solutions? ... 289
Conservation Genetics ... 294
Why Should We Focus More on Populations? ... 294
How Can We Apply Genetics to the Conservation of Amphibians in Ecuador? ... 295
Main Techniques to Assess Genetic Variation in Populations ... 295
Genetic Detection, DNA Barcoding, and Environmental DNA ... 296
Resolving Taxonomic Uncertainties ... 296
Conservation Units ... 297
Genetics of Wild Populations ... 297
Conservation Breeding ... 300
References ... 301

Index ... 319

About the Authors

Luis A. Coloma

Coloma is an amphibian biologist who was born in Guaranda, Provincia Bolívar, Ecuador, in 1962. He received his *licenciatura* in biological sciences from Pontificia Universidad Católica del Ecuador (PUCE) in 1987. In 1991, he was granted an MA in the Department of Systematics and Ecology by the University of Kansas, where he was mentored by William E. Duellman; his thesis was "Ecuadorian Frogs of the Genus *Colostethus* (Anura: Dendrobatidae)." For his PhD, he was under the guidance of Linda Trueb; in 1997, he completed his dissertation, "Morphology, Systematics, and Phylogenetic Relationships among Frogs of the Genus *Atelopus* (Anura: Bufonidae)." For 19 years, from 1991 to 2010, he was professor and Senior Lecturer in charge of vertebrates and in charge of the herpetology section at the Escuela de Ciencias Biológicas at PUCE. He mentored 23 *licenciatura* students. Since 2011, he has been Director and Researcher at Centro Jambatu de Investigación y Conservación de Anfibios in Quito, Ecuador. Coloma has published 69 scientific papers, with 56 of these articles focused on amphibians indexed in Scopus (as of 31 May 2023). Additionally, he has published 32 outreach publications, among the latter two coffee table photo books: *Ecuador Megadiverso* and *Sapos, Ecuador Sapodiverso.*

He has described or co-described 31 new species of frogs, among which 28 are Ecuadorian frogs of the genera *Atelopus* (7), *Hyloxalus* (6), *Gastrotheca* (4), *Hyloscirtus* (4), *Engystomops* (3), *Pristimantis* (2), *Leucostethus* (1), and *Epipedobates* (1). Five species of frogs, one lizard, and one earthworm have been named in his honor by his colleagues. In 2007, he was awarded the Sabin Award for Conservation of Amphibians, presented by the World Conservation Union and Conservation International. In 2008, he received the Saint Louis Zoo Conservation Award in recognition of his extraordinary lifelong dedication to the conservation of Ecuadorian biodiversity. In 2009, he was appointed a member of the Latin American Academy of Sciences. Additional information about Coloma's life appears in Chapter 2 and references therein.

William E. Duellman (1930–2022)

Duellman was a herpetologist that was born in Dayton, Ohio, on 6 September 1930. He passed away on 25 February 2022. He earned three degrees from University of Michigan—a BA (1951) in zoology with a minor in geography, an MS (1952) in zoology with a minor in botany, and a PhD (1956) in zoology with a minor in geology. His doctoral dissertation was on snakes of the genus *Leptodeira*. Since 1959, most of Duellman's academic life was at the University of Kansas, where he was a professor in the Department of Systematics and Ecology and Curator and Curator Emeritus of Herpetology in the Natural History Museum (now Biodiversity Institute). He published 386 titles (among them 12 books). Among these books are *Hylid Frogs of Middle America, An Equatorial Herpetofauna, Biology of Amphibians* (with Linda Trueb); *Cusco Amazónico: The lives of Amphibians and Reptiles in a Tropical Rainforest, Terrestrial Breeding Frogs (Strabomantidae) in Peru* (with Edgar Lehr); and *Marsupial Frogs and Allied Genera.* Duellman served as Major Professor for 12 master's and 34 PhD students, and he mentored seven post-doctoral scholars. During his time at the University of Kansas Natural History Museum, he built up the collection from 59,000 to more than 300,000 specimens through extensive field work by himself and with his students and by acquiring other collections, making it the fourth largest herpetological collection in the United States and by far the most significant collection of the herpetofauna of Latin America. He described or co-described 252 species of amphibians from the Neotropics, among which 93 are from Ecuador. Six species of frogs and five species of reptiles are named in his honor. Additional information about Duellman's life is provided in Chapter 2 and references therein.

Contributors

The authors (first author) and coauthors who have contributed to the writing, review, and editing process of specific sections are listed here in alphabetical order, along with their institutional affiliations. All authors, coauthors, and additional participants are credited alongside their respective contributions. These include: Juan Carlos Santos, who produced time-calibrated maximum likelihood phylogenetic trees illustrating the evolutionary relationships of amphibians in Ecuador; Daniel J. Paluh, who provided microCT images and high-resolution computed tomography reconstructions of the skeletons of Ecuadorian amphibian genera. The acknowledgments section further recognizes other contributions made by these individuals.

AUTHORS

Ana Almendáriz-Cabezas
Instituto de Ciencias Biológicas
Escuela Politécnica Nacional
Quito, Ecuador

Ernesto Arbeláez Ortiz
Amaru Bioparque Cuenca—Zoológico
Cuenca, Ecuador

Sebastián Bermúdez-Puga
Universidad de Sao Paulo
São Paulo, Brasil

Patricia A. Burrowes
Department of Biology
University of Puerto Rico
San Juan, Puerto Rico

Diego F. Cisneros-Heredia
Museo de Zoología & Laboratorio de Zoología Terrestre, Instituto de Diversidad Biológica Tropical
Universidad San Francisco de Quito USFQ
Quito, Ecuador

Martha L. Crump
Northern Arizona University and Utah State University
Arizona and Utah, USA

Stephen R. Edwards
Former IUCN official in conservation
USA

María D. Guarderas
Wikiri
Quito, Ecuador

Juan M. Guayasamin
IBIOTROP, Colegio de Ciencias Biológicas y Ambientales COCIBA
Universidad San Francisco de Quito USFQ
Quito, Ecuador

Carlos E. Montalvo-Puente
Museo de Arte Precolombino Casa del Alabado
Quito, Ecuador

Renato E. Naranjo
Centro Jambatu de Investigación y Conservación de Anfibios
Quito, Ecuador

Mónica I. Páez-Vacas
Centro Jambatu de Investigación y Conservación de Anfibios
Quito, Ecuador

John E. Simmons
Museologica Museum Consulting
Pennsylvania, USA

Gregory O. Vigle
Librarius Study Center, Estación Biológica Jatun Sacha
Provincia Napo, Ecuador

Mario H. Yánez-Muñoz
Unidad de Investigación, Instituto Nacional de Biodiversidad
Quito, Ecuador

COAUTHORS

David Brito-Zapata
Museo de Zoología & Laboratorio de Zoología Terrestre, Instituto de Diversidad Biológica Tropical
Universidad San Francisco de Quito USFQ
Quito, Ecuador

Mateo Dávila-Játiva
Museo de Zoología & Laboratorio de Zoología Terrestre, Instituto de Diversidad Biológica Tropical
Universidad San Francisco de Quito USFQ
Quito, Ecuador

Juan M. Guayasamin
Colegio de Ciencias Biológicas y Ambientales COCIBA
Universidad San Francisco de Quito USFQ
Quito, Ecuador

Andrés León-Reyes
Museo de Zoología & Laboratorio de Zoología Terrestre, Instituto de Diversidad Biológica Tropical
Universidad San Francisco de Quito USFQ
Quito, Ecuador

Emilia Peñaherrera-Romero
Museo de Zoología & Laboratorio de Zoología Terrestre, Instituto de Diversidad Biológica Tropical
Universidad San Francisco de Quito USFQ
Quito, Ecuador

Carolina Proaño-Bolaños
Biomolecules Discovery Group, Laboratory of Molecular Biology and Biochemistry, Universidad Regional Amazónica Ikiam, Tena, Ecuador

Carolina Reyes-Puig
Museo de Zoología & Laboratorio de Zoología Terrestre, Instituto de Diversidad Biológica Tropical
Universidad San Francisco de Quito USFQ
Quito, Ecuador

David Romo
Museo de Zoología & Laboratorio de Zoología Terrestre, Instituto de Diversidad Biológica Tropical
Universidad San Francisco de Quito USFQ
Quito, Ecuador

Camilo Rosero-Gómez
Centro Jambatu de Investigación y Conservación de Anfibios
Quito, Ecuador

Juan Carlos Sánchez-Nivicela
Museo de Zoología & Laboratorio de Zoología Terrestre, Instituto de Diversidad Biológica Tropical
Universidad San Francisco de Quito USFQ
Quito, Ecuador

Fausto Siavichay-Pesántez
Amaru Bioparque Cuenca—Zoológico
Cuenca, Ecuador

Andrea Terán-Valdez
Centro Jambatu de Investigación y Conservación de Anfibios
Quito, Ecuador

María Vizcaíno-Barba
Departamento de Economía, Ambiente y Territorio
Facultad Latinoamericana de Ciencias Sociales
Quito, Ecuador

ADDITIONAL CONTRIBUTORS

Daniel J. Paluh
Department of Natural History, Florida Museum of Natural History
University of Florida
FL, USA

Juan Carlos Santos
Department of Biological Sciences, College of Liberal Arts and Sciences
St. John's University
NY, USA

In Memoriam

A tribute to Bill Duellman from five students at the University of Kansas, who conducted amphibian studies in Ecuador.

Patricia A. Burrowes

When I was interviewing for graduate school at University of Kansas (KU), Bill invited me to join his expedition to the Andes of Ecuador in exchange for a few days in my parents' home in Colombia and the chance to do some field work in the southern part of the country. Despite my interest in herpetology, my knowledge was limited to an introductory course in college, and I was absolutely flabbergasted with the opportunity to work with THE BEST! When we got to the Colombia–Ecuador border, Bill was so impatient to meet John E. Simmons and David Hillis on the other side that he made me walk across the Rumichaca bridge (illegally) while border officers were on their lunch break. At the time, I did not dare to put in doubt Dr. Duellman's word, so I innocently walked into Ecuador only to succumb to the police on my return to Colombia for trying to re-enter a country from which I had never left. But this was a very small price to pay for such an unforgettable trip full of awesome frogs, beautiful mountain sites, and even dangerous adventures like the time we were attacked by drunken truckdrivers while frogging on the curbside. What I learned from Bill and my fellow teammates on that trip is unmeasurable and assured my survival in the KU Herp-Division the following fall. Bill's capacity to organize the trip with perfect timings and stops in the right localities to find the species he wanted to collect was admirable. He was incredibly efficient in the field, and that included taking advange of our strengths and weaknesses. While I was mostly designated to catch *Gastrotheca* frogs, John was in charge of collecting the *Telmatobius*, which led to his nickname, "Johnatobius," and David of catching every herp that seemed impossible to catch, which led him to the amazing *Hyloscirtus pacha*, whose dedication I will always cherish. Although this experience in the Ecuadorian Andes is now tinted with the loss of many species that were abundant at the time, I have Bill to thank for the chance to appreciate their amazing diversity before the Anthropocene and all its stressors started the amphibian collapse that we are experiencing in present times.

Martha L. Crump

After graduation from high school in Pennsylvania, I began my freshman year as a zoology major at the University of Kansas. College was an adventure, and I had wanted to go far from home. Little did I know in 1964 that the Curator of the Herpetology Division at the KU Museum of Natural History would change the trajectory of my life forever.

During my first year at KU, I was told by more than one professor that my career choice—wildlife biology—was not a possibility. Only men could be wildlife biologists. That realization sent me to the Museum of Natural History for a part-time job to discover what I *could* do as a woman in zoology. The Curator of Herpetology, Bill Duellman, had an opening for a student to catalog and tie tags onto specimens. Bill interviewed me and the rest is history. For the next three years I cataloged, tagged, cleared, and stained specimens and made histological preparations. I felt a strong sense of belonging. I listened to stories Bill, his wife Linda, and the graduate students told about fieldwork in exotic places, including a recent trip Bill had made to Santa Cecilia, Ecuador, in 1966—his first visit to Ecuador. Bill speculated that this site in the upper Amazon Basin was the most amphibian-diverse site in the world.

After my senior year, I joined Bill, Linda, and several others from the museum on a summer fieldtrip to survey the amphibians and reptiles of Santa Cecilia. What an experience that was, watching 38-year-old Bill Duellman in his element! Nothing bothered him—not the rice three times a day, cold showers, mosquitoes, too little rain, too much rain, mud, or heat. (I do remember, however, Bill being rather dismayed when I kept finding salamanders and he didn't.) He was like a little boy in a candy shop, pure excitement. He drove himself and expected his field companions to be just as dedicated. Bill mentored me in the art of taking detailed and accurate field notes, helped me develop a search image, and shared his passion for the animals and for the field. Bill was intense, but he was also fun. There were card games, lots of rum and cokes, practical jokes, and stories. His enthusiasm was contagious. How could I not fall in love with herpetology?

One particular incident from that first Ecuadorian field experience reflects Bill's love of jokes and his mentoring style. Bill and I were working a trail together, he on one side, I on the other. I spied a golden-brown frog with a flattened body, wicked-looking bony projections on the sides of her triangular head, and a fleshy proboscis. As I plucked her off the branch, she opened her cavernous mouth and displayed a bright yellow-orange lining and tongue. Bill admired my find, no doubt wishing he'd claimed that side of the trail. I asked, "What would she do if I put my finger in her mouth?"

Bill responded, "Try it and find out!"

The frog snapped her jaws shut on my pinky finger. Bill grinned as he told me this was a *Hemiphractus proboscideus* and that these treefrogs eat large insects, other frogs, and small lizards, which they restrain with the fang-like odontoids that were now gripping my finger. I felt like a fool at the time, but later I interpreted Bill's "Try it and find out!" response as something more than a dare. He was encouraging me to take the next step, to explore.

I returned to Santa Cecilia in 1971 to study anuran modes of reproduction for my PhD dissertation. Bill made two brief visits during my year there to check in on me and provide encouragement. He hadn't wanted me to spend a year in the rainforest alone, so he "found" funds (I've always suspected they were personal) to support a field assistant—John E. Simmons, an undergraduate at the University of Kansas. Bill couldn't have chosen a better field assistant for me; somehow, he sensed that John and I would get along and become like sister and brother.

Bill was a giant among herpetologists. I owe much so him for providing support and guidance throughout my decade at the University of Kansas, for fieldwork opportunities in Ecuador, and most of all for believing in me. Bill's influence on Ecuadorian herpetology was monumental. He described or co-described more than 90 species of frogs in Ecuador that are still currently recognized. He mentored many students who worked in Ecuador, including a number of Ecuadorians. And, of course, he published prolifically on the amphibians of Ecuador. Perhaps his most enduring legacy is his passion for fieldwork and his beloved animals, which he shared with his students, who in turn are passionate about fieldwork, amphibians, and reptiles and are now passing that love on to their students. We miss you, Bill.

Stephen R. Edwards

To put my story about Santa Cecilia in perspective, I arrived at KU in the fall of 1967 intending to pursue research on vertebrate evolution. I was in the process of searching for a suitable advisor/major professor when I enrolled in a herpetology course taught by Bill Duellman. The subject was fascinating, offering many opportunities to explore the evolutionary relationships of frogs, which deeply intrigued me. Bill was not only a compelling lecturer but an excellent teacher. He also had a reputation for being a strong advocate for his graduate students, ensuring that they had assistantships and funding for their research.

It was shortly after he had told me that I had passed the departmental qualifying exam that he invited me to join a team going to Santa Cecilia to collect specimens for the collection and to document the fish, amphibian, and reptile diversity of the area. Santa Cecilia represented a dream come true for a kid who grew up in the Pacific Northwest of the US. Wow! It all began with the preparations before we left Lawrence: the supplies, batteries, preservatives, collecting tags, etc. We packed everything in wooden boxes to accompany us to Ecuador. We spent more time in Quito at a small family-owned pension, where we acquired more supplies and acclimated to the local conditions, foods, and culture.

From Quito we made our way to Shell Mera, a village on the edge of the Amazon. There we caught a ride on a DC3 aircraft that was carrying supplies to a dirt landing strip. This airstrip was used by Texaco Oil and Williams Brothers Pipeline to support their oil exploration efforts and the early stages of planning the construction of a trans-Andean Pipeline to Provincia Esmeraldas on the Pacific Coast. From the airstrip it was a short hike to a place then familiarly called "Muñozlandia" on the banks of the Rio Aguarico, where the compound we called Santa Cecilia was located and so-named because of its proximity to a small Kichwa village in the area by that name. The compound was constructed and staffed by the Muñoz family, who had been engaged by Bill the previous year. Muñozlandia included a "lab" where the specimens we collected were processed, cabins, a dining hall, a shower, and outhouses. A cook prepared our meals, except for Sunday dinners, which was a soda and a culinary delight of prunes impaled with toothpicks holding a small chunk of cheese, with a side dish of popcorn. We loved it. It was the time we would share stories, reminisce about home, our lives before, and maybe a bit about the future. It was during these times that Bill's vision for Santa Cecilia was shared, where we got a glimpse of the importance of our contribution.

The compound that we would call home for the next three months was located at 320 m at 4 degrees south latitude, not far from the Colombian border on a hammock between the Río Aguarico and a much smaller river the Río Conejo. There were no roads other than the track from the airstrip where the Texaco operations were located. It was physically isolated from the outside world; there were no mail, phone, or telegraph services; no stores, pubs, or restaurants. Just us, the staff and the field team, and the incredible Amazon Rainforest surrounding us.

Following Bill's vision, a network of trails was established that radiated out from the compound, along which we constructed a few trench traps. Our routine was to walk the trails twice a day, in the morning and after dark. The specimens collected on each foray were put in plastic or cloth bags and brought back to the lab, where they were preserved, tagged, and packaged for return to the KU MNH. Each of us had small, personalized KU Herp tags, which we attached to each specimen we collected, that were used when they were cataloged into the KU MNH Herp collection. Additionally, we kept a journal in which we recorded the herps collected, occasionally describing them as "a small green frog." We also recorded any notes we may have made on individual specimens, along with the corresponding tag number. I do not know the numbers of specimens cataloged into the KU herp collection, but it had to have been in the thousands that summer alone.

I recall many memorable moments of life in the compound and on those collecting forays. One such incident stands out vividly, involving Bill's determined attempt to catch a very long racer that had slithered into the compound. On that day, Bill spotted the snake which was moving right along and yelled, "Watch this, I'm going to grab the snake by its tail, whip it through my legs, clamp it between my thighs and grab its head," which he proceeded to do. The snake was caught, and he slung it back between his legs while we stood entranced, watching the snake turn its head, open its jaws, while passing between Bill's thighs, turned, and clamped its teeth onto Bill's thigh. With Bill's reaction—a loud yelp—we started laughing. Unwilling to let his prey escape, Bill swiftly executed a very quick hand-over-hand maneuver to reel the snake in to grab it behind its jaw. Once he had control, he turned and walked, albeit with a slight limp, to the "lab," where the snake was bagged for processing the next day.

On another occasion, Bill arranged with the Texaco people to allow us to catch a ride on a company helicopter to a camp site in the Andean foothills along the path the pipeline would take over the Andes. The rationale behind this decision was that the trees that were being cleared for the pipeline route likely held unknown arboreal species, which would be invaluable to Bill's research and our understanding of the scope of the diversity of the herpetofauna.

We embarked on a helicopter journey, ascending the zigzagging slopes of the Andean foothills over a never-ending expanse of green below us and landed in a clearing that appeared to be the size of a postage stamp. I'll never forget the moment when Bill and I sat down on rough planks in the dining area and were served "filet mignon with lobster" for dinner! Amazing! After this sumptuous meal, we got down to business, grabbed our headlamps, and headed out to collect herps. I do not remember how long we had been wandering around, but it was very late when Bill, who was in front of me cutting through the trail, disappeared! I ran up to where he should have been where I heard him yelling loudly from below. Our trail had led right up to a drop-off where Bill had taken one step too far and descended quickly into a deep stream cut. Fortunately, he was not hurt, but it did take a long time for him to find a way out of the stream cut. With Bill back up on the ridge, we continued collecting as we made our way back to the campsite.

The next morning, we returned to explore the stream, which we found to originate from a large opening in the face of a ridge above us. From its origin the stream fell to a pool, around which was a shelf of limestone. As we walked around the pool, we suddenly noticed the movement of what looked like small, black tadpoles slithering along a terrace from which water was seeping. Bill immediately said, "Grab it, that is a larval caecilian, this is incredible." So, there we were, standing in the stream catching larval caecilians. How many I do not recall, but there were several. I believe that moment was a first for both of us. It was truly an amazing experience!

There were many such vignettes over the course of our three months at Santa Cecilia—some good, some not so good, some funny, and others sad. But to this day, 56 years later, there is no doubt that the experience was life changing for me. It all began with Bill's invitation to go to Santa Cecilia, which conveyed his confidence in my abilities, and continued with the fact that Bill was always "on" as a teacher/mentor. He simply could not be anything else. John E. Simmons noted in a memorial that Bill had no hobbies. And literally, that is correct when you think of traditional "hobbying." But I contend that his research, his scholarship, his students, and his pursuit of greater understanding of Neotropical herpetofauna were his hobby. His 1978 publication "The Biology of an Equatorial Herpetofauna" (University of Kansas, Museum of Natural History, Miscellaneous Publication No. 65, 352 pp) provides considerable insight into the vision he had in the mid-1960s and the work that was undertaken at Santa Cecilia by a succession of researchers and field teams, which began in 1966.

He invested in his students every day at Santa Cecilia. He always shared some bit of knowledge or understanding about what was going on around us in that miraculous environment. It was this level of his commitment that led me to ask Bill to be my major professor, which he accepted and subsequently suggested that I review the taxonomy of the genus *Colostethus*. He told me that the genus was relatively well defined, yet it had never been studied as a group. The unique behavior of adults carrying tadpoles during part of their life cycle enabled positive larval identification, and individuals were relatively easy to collect. Over the ensuing years, I served as Bill's curatorial assistant and held several research assistantships funded by grants that Bill had received from the NSF. Bill continued to help and guide me as I pursued my doctoral research and the subsequent writing of my dissertation. His support and encouragement inevitably influenced me throughout my career, for which I am forever thankful.

Juan M. Guayasamin

More than any particular anecdote, I just want to stress how Bill, with his daily routine, taught us what it meant to be a scientist. Bill was always in his office, at the very entrance of the KU Herpetology Division. His door always open, his mind always working, his fingers typing. He would stand up from time to time to check a publication or field journal, slide, specimen, or frog recording. Everything was perfectly organized.

Although the idea of "integrative taxonomy" may have a catchy new air, it has been day-to-day taxonomy for Bill for decades. His field work including carrying bulky tape recorders, camaras, flashes, vials, alcohol (2 types), notebooks, maps, etc. The collections that Bill and his students obtained in the 1970s were so complete that they even included tissues that are now used for molecular studies. I only know two or three people with Bill's work capacity. Leading by example, surrounded by students and colleagues, his legacy changed Neotropical herpetology.

John E. Simmons

Late one afternoon in March of 1984, William E. Duellman, David M. Hillis, Patricia Burrowes, and I stopped at a site 17.6 kilometers southeast of Gualaceo, at 3150 m. After finding a batch of recently metamorphosed *Atelopus* and a clutch of *Pristimantis* eggs, we set up our camp stove on the lee side of a shepherd's hut to prepare dinner—two packages of Maggi soup, served with cheese and crackers, followed by a shot of rum. After dark, I set out alone to follow a fast-flowing stream up a gentle slope. By noodling under the overhanging stream bank, I found six of the frogs I was most interested in—*Telmatobius niger.* It was a clear, cold night on the paramo, with just a sliver of moon. Feeling contented, I turned off my headlamp and sat down to rest and enjoy the solitude. The sky was ablaze with crystalline stars, the only sound a few frogs calling in the distance. I remembered how, as a teenager, my dream had been to go on expeditions to remote and fascinating places, a dream that Bill had made possible. I wrote in my field notes for that night that, after many years of travel and fieldwork, the magic was still there. I did not know it at the time, but this was to be my last field trip with Bill in Ecuador, and the last time I would see jambatos on the paramo, crawling about in abundance. During the more than 50 years that I knew Bill, I got to spend thousands of hours in the field with him throughout South America, and I learned the secret of his eternal optimism—there is nothing as challenging, satisfying, and fulfilling as being in the field, looking for frogs. It is how I remember Bill best.

Foreword

This encyclopedia has been years in the making and represents a highly anticipated volume summarizing the megadiversity of Ecuadorian amphibians. It reviews more than 250 years of published research on more than 650 known species in the country—everyone acknowledges that the true number of extant and recently extinct species of is actually far greater, which portends exciting new discoveries in the future. The contrast of extinction and discovery is an important one in Ecuador. The Andes are at once a hotspot both of amphibian diversity and of amphibian extinction. Despite a long co-existence with humans in the region, a fascinating history in itself exhaustively reviewed here in Chapter 1, the modern onslaught of conservation threats such as mining development have proven to be too much for many species with naturally small geographic ranges and high sensitivity to habitat disruption. Making matters far worse is the reality of the epidemic of the emerging infectious disease chytridiomycosis that devastated many populations of Andean frogs and most certainly eradicated species before they were discovered by science (La Marca *et al*., 2005; Lips *et al*., 2008; Scheele *et al*., 2019; Lötters *et al*., 2023). Co-authors Luis A. Coloma and William E. Duellman exemplified this phenomenon in a paper (Coloma *et al*., 2010) with the bittersweet, and brilliant, title "Five new (extinct?) species of *Atelopus* (Anura: Bufonidae) from Andean Colombia, Ecuador, and Peru." I also know well the depressing phenomenon of announcing the should-be exciting discovery of new species under the cloud of the reality of their extinction before they were named, a phenomenon I termed "Forensic Taxonomy" (Mendelson III, 2011). The geography of the Andes also renders the region highly susceptible to the ravages on ongoing climate shifts, forming a miserable synergy when combined with the omnipresent threats of habitat disruption and the now-endemic chytrid fungal pathogen. Likewise, ocean temperatures play a crucial role in dictating climatic conditions and habitats along Ecuador's Pacific Coastal region. Amazonia, which of course contains a great deal of Ecuador's biodiversity, hopefully will be less impacted, and Ecuador will continue to implement more expansive protected areas there to guard against deforestation for oil, mining, urban development, and agriculture.

Coloma and Duellman are the perfect team to prepare this encyclopedia in the context of these extinctions because, in different ways and eras, they helped lead the documentation of both Ecuador's amphibian diversity and its devastating levels of extinctions and population declines, as historically detailed here in Chapter 3. Duellman and many of his students from the University of Kansas did a lot of field work in Ecuador in the 1970s. This was a time before human populations were quite so large as today, and, therefore, undisturbed or minimally disturbed habitats in Ecuador were far more widespread. Their efforts also pre-dated the widespread epidemic of the non-native chytrid pathogen; the pathogen was present at least as early as 1974 (Manzano Pasquel, 2014). It was these collections, dutifully documented as museum specimens, that made it possible for Coloma, beginning in the late 1980s, to document declines and extinctions among Ecuador's amphibians. Because of Duellman's efforts—as well as, of course, the work of others before him—Coloma could form a reasonably good picture of the diversity and abundance of amphibians in Ecuador and thus demonstrate that this diversity was partly missing and abundances were widely decreased. In the 1990s, at a time when herpetologists were arguing among themselves if the apparent phenomenon of global amphibian extinctions even was real (see the historical overview by Collins and Crump, 2009), the work by Duellman and later by Coloma gave the world some of the most compelling evidence that indeed there was an acute, nearly global, problem. In this sense, one can view their respective careers as representing a glorious era of exciting discovery and a sobering era of documenting catastrophic losses. All of this celebration and mourning is captured in this encyclopedia, making it simultaneously both a crucial contribution to understanding amphibian diversity and distributions as well as informing conservation efforts.

Duellman and his students, which included Luis A. Coloma, played important roles in documenting Ecuador's amphibian biodiversity, but there certainly have been other solid contributors representing both Ecuadorian scientists and foreigners as well. Considered altogether, Ecuador stands as a modern powerhouse of amphibian research with levels of productivity matching those of much larger nations such as Brazil and the USA. Appropriately, virtually all recent research has been led by Ecuadorian herpetologists—many of them with direct or second-generation ties to Duellman and Coloma. Some aspects of this encyclopedia will become outdated over time as new species continue to be discovered, distributions and information on natural histories continue to be refined, and phylogenetic and comparative studies continue to broaden our understanding of this megadiversity. But this encyclopedia will form the basis to greatly inform all subsequent research and conservation efforts. As has been documented in many taxa and regions, when a major compendium is produced, there will follow a glorious explosion of activity and discoveries because the assembled information in the volume grounds and informs the activities of all subsequent researchers. Ecuador's biodiversity and the history of its discovery are rightly celebrated in the volume, but I think the really exciting era is just about to launch as practitioners absorb the wealth of information contained herein.

Joseph R. Mendelson III

Preface

As I conclude the writing of this Encyclopedia (four volumes), I find myself engaged in recurrent, lively, and pivotal discussions with my colleagues. Our conversations center around essential and somehow vanguard and futuristic themes, particularly noteworthy within the context of Ecuador. Recently, we obtained results from an eDNA survey in the highlands of the Ecuadorian Andes, and to our astonishment, we uncovered traces of DNA from a couple of harlequin frog species, specifically *Atelopus*, that we have not observed in more than three decades. What immediate steps should we take to locate and effectively safeguard these animals from extinction? How should we apply the genetic and genomic revolution to the conservation of these species? What measures should be taken to accelerate the development of assisted reproductive technologies for amphibians? What infrastructure and human capacity are needed, and what strategies should be devised for the current and future reintroductions of numerous disappeared amphibian populations into either their natural or restored habitats? This includes five species nearly extinct from Quito and about 15 species of harlequin frogs, among many others. How can we leverage cutting-edge artificial intelligence resources to expedite the documentation and understanding of amphibians? We ponder how to raise awareness among politicians, the general public, and local communities about the incredible amphibian diversity in Ecuador, its value, and the crisis it faces. What are the necessary steps to catalyze a surge in amphibian research, an endeavor that has been nearly blocked during decades in Latin America by bureaucratic obstacles (e.g., Fernández, 2011; Prathapan *et al.*, 2018) and a lack of funds? How do we confront the established order and gradual habituation reflected in the passivity and amnesia of human beings (sensu Dangles *et al.*, 2016) in the face of the mass extinction of amphibians?

Reflecting on the past, it is remarkable that these themes were either non-existent or in their infancy when I embarked on the study of amphibians in the early 1980s. The majority of my academic focus back then was on systematics and a bit of ecology, far removed from the challenges we confront today. Beginning in the mid-1980s, the catastrophic declines and extinction of amphibians unfolded unexpectedly, prompting a radical shift in my perspective of amphibian research and conservation. This compelled me to pivot from my initial passion for systematics—which I did not abandon—to the realms of conservation and education, as I witnessed the urgency of addressing these emerging threats. While unfortunately too late for many species, this encyclopedia is anticipated to contribute somehow to addressing this urgency.

It offers a data-rich compilation and overview of the current knowledge regarding all amphibians in Ecuador, a country that is gaining recognition as the world's foremost "sapodiverse" hotspot in the world. Despite its relatively compact size (e.g., 36 times smaller than USA, 30 times smaller than Brazil), the Ecuadorian territory has played a pivotal role as the crucible for an astonishing array of amphibian species, nurturing their evolution and diversification over time. This remarkable diversity may encompass an astounding estimate of nearly 900 potential species, of which only 655 have been described. Toward this eventual goal of documenting the entire amphibian diversity in Ecuador, it has so far taken more than 264 years since the first published studies and academic attempts to describe this diversity. While we acknowledge that significant knowledge gaps remain, even in fundamental areas such as the completion of descriptions of species, calls, tadpoles, and embryos, we believe the time has come for a comprehensive compendium.

Although resources like Amphibian Species of the World, Bioweb Ecuador, and AmphibiaWeb offer insights into accounts of Ecuadorian amphibians, the pressing need for a consolidated, formal publication is evident. *Amphibians of Ecuador* serves as an essential endeavor, aiming to illuminate not only the vast diversity within Ecuador's amphibian fauna and its profound cultural significance and value but also the looming threats that jeopardize its continued existence.

Available both in print and digitally, this work bridges the divide between the traditional and the contemporary. Although a printed edition can quickly become dated due to constant scientific advancements, like the description of new species, it offers advantages in comprehension and accessibility. Digital readers, however, enjoy portability and swift access. Choosing between the two depends on personal preference and context.

Recognizing the importance of accessibility to the local populations, it is evident that the urgent and next crucial step is to produce a Spanish version. While English dominates the scientific world, the increasing prominence of Spanish in research will widen its reach. Furthermore, the scientific information in Spanish is eagerly needed for various purposes that are not necessarily exclusive to scientists, for instance, field guides for "sapowatching," species management in communities for conservation purposes, biotrade that can be used by communities and authorities, local exhibits, etc.

The origins of this encyclopedia are intricate, interwoven with multiple attempts, collaborations, setbacks, and initiatives over the years. Although the idea of writing this compendium has been challenging and brewing since at least the 1980s, it was only in 2015 that our lives (Coloma and Duellman) converged in many ways, and we decided to concentrate our efforts on its production. Our combined efforts, experience, dedication, and individual circumstances have been instrumental in the creation of this encyclopedia. Tragically, William E. Duellman could not see the encyclopedia printed and online, but Bill's efforts were monumental and shine through the entire piece of work.

Saying when this encyclopedia began is not easy task; thus, I place this in an historical context. It formally started with the agreement between Coloma and Duellman with CRC Press/Taylor & Francis Group in 2015. Almost simultaneously, it also began with the execution of two projects carried out by Coloma at Centro Jambatu of Amphibian Research and Conservation (CJ) (and partially at the Universidad Regional Amazónica Ikiam): the Ministerio de Ambiente de Ecuador PARG project and a National Science Foundation project (PI Lauren A. O'Connell), that partially supported this encyclopedia, adding to the permanent support of Wikiri (a biocommerce enterprise) and the Saint Louis Zoo. However, establishing the origin of this encyclopedia ends up being somewhat more complex, and it remains to be said that there is always a before. It was always in Coloma's agenda, at least since the times of studying Biology at Pontificia Universidad Católica del Ecuador (PUCE), after Miyata's checklist of Ecuadorian amphibians was published in 1982 and also after an influential talk given by Duellman in 1984 at PUCE (http://www.anfibiosecuador.ec/recursos/audio/). Thus, the idea of writing a book about Ecuadorian amphibians began to be discussed in the mid-1980s, along with Giovanni Onore. Subsequently, and based on the publication of the amphibian list of Ecuador (Coloma, 1991), the idea of a printed version of the amphibians of Ecuador gave way to the electronic format encyclopedia, which was driven and collaborated on by Néstor Acosta-Buenaño (see Chapter 2) in the 2000s at PUCE. Other digital media also began to develop at about the same time. David Wake at the University of California, Berkeley, founded AmphibiaWeb and began to provide accounts of all amphibian species of the world, including some occurring in Ecuador. Likewise, since 2002, the ASW web page, edited since its beginnings by Darrel R. Frost, has been a reliable source of information on all Ecuadorian amphibians, especially on the taxonomy of species.

The events at the Museo de Zoología of PUCE (QCAZ) by the end of 2010 disrupted Coloma's and Acosta's plans for the Encyclopedia of Ecuadorian Amphibians (refer to Chapter 2, Amphibians at the QCAZ), forcing them to start almost from scratch to create an online encyclopedia at CJ in the 2010s. Meanwhile, from the 2010s to the 2020s, online accounts of amphibians in Ecuador have continued the legacy of the electronic encyclopedia at QCAZ as Bioweb Ecuador.

In the backdrop of our endeavors lies a world grappling with change; the world and Ecuador are undergoing profound transformations with significant challenges while also witnessing revolutions in various fields that directly impact amphibians. On the one hand, humanity is confronted with daunting issues, including climate change, the sixth mass extinction—amphibians being one the most affected taxa—habitat destruction, pollution, and overpopulation, among other threats, many of which seem beyond our immediate control. On the other hand, the realms of synthetic and evolutionary biology, biotechnology, and informatics are experiencing groundbreaking advancements that hold the potential to mitigate some of these threats.

Furthermore, within a sociological context, it is noteworthy that in an unprecedented move, Ecuador, in 2008, became the first nation to grant nature constitutionally protected rights. This landmark development, coupled with the scientific contributions of CJ, has recently been exemplified in a fascinating case where frogs triumphed over financially lucrative mining operations in the Andes, as documented in the section about CJ (see Chapter 2). The Ecuadorian constitution was put into practice, showcasing the real-world impact of scientific knowledge.

Emerging technologies like AI have begun to shape this encyclopedia, from translations to proofreading. As the world further embraces AI, the compiled information and new data on Ecuadorian amphibians, as well as the species accounts provided herein, are expected to eventually bolster AI databases and computational resources. This will help address gaps and misinformation that pose a threat, thus accelerating the knowledge and conservation efforts related to amphibians.

I wish to emphasize that this encyclopedia is the fruitful combination of dedicated individuals, spanning generations and continents. Most of these individuals largely trace their academic lineages (initiated by Duellman PhD's academic lineage) to a strong commitment to amphibian research and, more recently, its conservation. Those who first ventured into the wilds (e.g., Duellman, Trueb, Lynch, Cannatella, Crump, Simmons, Onore, Coloma, Almendáriz, Read, members of the Tapia family, and a few others) had the privilege of witnessing the magnificent diversity of amphibians but, sadly, also the subsequent decline of that diversity. In contrast, a more recent generation did not experience the same abundance observed by their predecessors, and for them, the stories of the earlier biologists are like fables. Both these generations now play pivotal roles in shaping the present and future of research and conservation of Ecuadorian amphibians. They are faced with at least two exceptionally challenging tasks. First, they must transition into the realms of big science and high technology, which are now almost exclusively in the domains of the most developed countries. "Time to think" became a rare commodity within the current Ecuadorian university academic system (most scientists lack the time and are burdened by courses and administrative duties). Furthermore, research funding is often limited and can be influenced by bureaucratic trends rather than being driven by scientific needs. Second, there is a challenge that is expected to counteract the tragic trends of extinction. It is essential that we transition swiftly from merely documenting the disappearances and population collapses to making significant academic advancements (in the science and social realms) that can effectively mitigate and address the current crisis. This must happen fast within a prevailing socio-political system that is increasingly exposing its inadequacy for the task at hand. In such a system, scientists in a country like Ecuador often find themselves facing the frustration of dealing with political pettiness and missed opportunities, as they are undervalued and mistreated by university and official institutions.

Finally, throughout our (Duellman, Coloma) lifetimes, we have experienced the thrill of discovery, the development of knowledge and concepts, and the associated human and physical capabilities. However, I have also witnessed with dismay the rapid declines and extinctions of amphibians (at least 43 species no longer seen). The latter is a tragic narrative, reminiscent of the broader history of Ecuadorian science and conservation of biodiversity worldwide. I eagerly hope for a rapid and positive transformation in the most "sapodiverse" epicenter of the world.

REFERENCES

Collins, J. P., and M. L. Crump. 2009. *Extinction in our times: Global amphibian decline*. New York: Oxford University Press.

Coloma, L. A. 1991. Anfibios de Ecuador: lista de especies, ubicación altitudinal y referencias bibliográficas. *Reportes Técnicos Ecociencia* 2: 1–47.

Coloma, L. A., W. E. Duellman, A. Almendáriz, S. R. Ron, A. Terán-Valdez, and J. M. Guayasamin. 2010. Five new (extinct?) species of *Atelopus* (Anura: Bufonidae) from Andean Colombia, Ecuador, and Peru. *Zootaxa* 2574: 1–54.

Dangles, O., F. Nowicki, and B. Mena. 2016. *La otra tierra. Léxico ilustrado de un planeta a proteger*. 1st edition. Quito, Ecuador: IRD, Quito Alcaldía, Fundación Museos de la Ciudad.

Fernández, F. 2011. The greatest impediment to the study of biodiversity in Colombia. *Caldasia*: iii–v.

La Marca, E., K. R. Lips, S. Lötters, R. Puschendorf, R. Ibañez, J. V Rueda-Almonacid, R. Schulte, C. Marty, F. Castro, J. Manzanilla-Puppo, J. E. García-Perez, F. Bolaños, G. Chaves, J. A. Pounds, E. Toral, and B. E. Young. 2005. Catastrophic population declines and extinctions in Neotropical harlequin frogs (Bufonidae: *Atelopus*). *Biotropica* 37: 190–201.

Lips, K. R., J. Diffendorfer, J. R. Mendelson III, and M. W. Sears. 2008. Riding the wave: Reconciling the roles of disease and climate change in amphibian declines. *PLoS Biology* 6: 441–454.

Lötters, S., A. Plewnia, A. Catenazzi, K. Neam, A. R. Acosta-Galvis, Y. Alarcón Vela, J. P. Allen, *et al*. 2023. Ongoing harlequin toad declines suggest the amphibian extinction crisis is still an emergency. *Communications Earth & Environment* 4: 412. https://doi.org/10.1038/s43247-023-01069-w.

Manzano Pasquel, A. L. 2014. *Revisiting the enigmatic: Emergence of Bd coincide with amphibian decline in Los Andes of Ecuador*. San Francisco, CA: San Francisco State University.

Mendelson III, J. R. 2011. Shifted baselines, forensic taxonomy, and Rabbs' fringe-limbed treefrog: The changing role of biologists in an era of amphibian declines and extinctions. *Herpetological Review* 42: 21–25.

Prathapan, K. D., R. Pethiyagoda, K. S. Bawa, P. H. Raven, P. D. Rajan, and 172 co-signatories from 35 countries. 2018. When the cure kills—CBD limits biodiversity research. *Science* 360: 1405–1406.

Scheele, B. C., F. Pasmans, L. F. Skerratt, L. Berger, A. Martel, W. Beukema, A. A. Acevedo, *et al*. 2019. Amphibian fungal panzootic causes catastrophic and ongoin loss of biodiversity. *Science* 1463: 1459–1463.

Acknowledgments

While William E. Duellman is no longer with us, his indelible mark on the production of this opus and the near-completion of all species' accounts remains immeasurable. His unique perspective, wealth of experience, knowledge, and unwavering enthusiasm for this project, coupled with meticulous attention to detail, continue to serve as a source of inspiration. Hopefully, his passion for science and amphibians will forever be etched in these pages as a tribute to his titanic contribution to the field of amphibian biology, as well as to the enduring nature of his friendship and partnership that transcends time and space. Similarly, I must express my deepest gratitude to Linda Trueb (his partner), my teacher and mentor at KU, who has been by Bill's side in charting the progress of *Amphibians of Ecuador* since 2015, steadfastly supporting him through challenging times as he embarked on what would ultimately become his final Ecuadorian endeavor. Linda's unwavering commitment also extended to securing part of the funding and seeing this Encyclopedia through to its completion. In January 2023, her wise and practical advice was much appreciated. She said, "I really would urge you to cut off adding additional species at a logical time—e.g., no frogs described in 2023! Otherwise, you'll never see the end—kind of like Sisyphus pushing that stone ball up the side of the mountain."

As detailed in Chapter 2 (under the subtitle "From Anfibios de Ecuador to Amphibiawebecuador and to Anfibios del Ecuador-Bioweb at QCAZ") and briefly mentioned in the preface, this opus is the descendant of earlier endeavors involving a vast collective of individuals and friends, encompassing students, collectors, field companions, scientists, administrators, and more. I extend my deepest gratitude for their invaluable contributions. Also, among them, my immense appreciation for their input while at Pontificia Universidad Católica del Ecuador goes to Giovanni Onore, Néstor Acosta-Buenaño, Ítalo G. Tapia, Elisa Bonaccorso, and Juan M. Guayasamin. Onore has always been a source of inspiration and drive, even in the most difficult times. Tapia (the collection manager at the QCAZ) aided, with great energy and empathy, in the field, in the museum, and with the photography of frogs and their development. Many of the photos in the field and in the lab would have not been possible without his collecting efforts and his ability to raise them. Bonaccorso and Guayasamin collaborated throughout leading the JRS and SENECYT projects, which, in 2008–2010, provided the funding necessary to enhance the electronic version of AmphibiaWebEcuador. Since 2011, all of these individuals have continued to offer their support in numerous ways during my tenure at Centro Jambatu of Amphibian Research and Conservation (CJ), and I am immensely grateful. In particular, Onore welcomed CJ under the auspices of the *Fundación Otonga* until 2019, when CJ became an independent NGO. Acosta-Buenaño has been instrumental in building the CJ electronic Encyclopedia (SapopediaEcuador). He not only developed the conceptual and physical models of the database but also played a key role in its programming, design, and layout. His initiative, serenity, and drive are inspiring. During my time at the QCAZ and the establishment of CJ, I received generous support and encouragement from Mark Wanner (Saint Louis Zoo), Joseph Mendelson III, Ron Gagliardo, and Kevin Zippel of Amphibian Ark, and of Lauren A. O'Connell (Stanford University), and Kim L. Hoke (Colorado State University). Their actions spoke louder than words.

For the realization of *Amphibians of Ecuador*, first and foremost, I extend my gratitude to our contributing authors and coauthors who devoted relevant efforts to write some sections. They are (in alphabetical order): Ana Almendáriz-Cabezas, Ernesto Arbeláez Ortiz, Sebastián Bermúdez-Puga, David Brito-Zapata, Mateo Dávila-Játiva, María D. Guarderas, Juan M. Guayasamin, Andrés León-Reyes, Carlos E. Montalvo-Puente, Mónica I. Páez-Vacas, Emilia Peñaherrera-Romero, Carolina Proaño-Bolaños, Carolina Reyes-Puig, David Romo, Camilo Rosero-Gómez, Juan Carlos Sánchez-Nivicela, Andrea Terán-Valdez, Gregory O. Vigle, María Vizcaíno-Barba, and Mario H. Yánez-Muñoz. Many of these authors (and or coauthors) engaged directly with myself, fostering enlightening and fruitful discussions, especially Andrea Terán-Valdez, Mónica Páez-Vacas, Juan M. Guayasamin, María D. Guarderas, Juan Carlos Santos, Ana Almendáriz-Cabezas, Gregory O. Vigle, Carolina Reyes-Puig, and Carlos E. Montalvo-Puente. My gratitude goes also to Daniel J. Paluh, who enthusiastically (and supported by David C. Blackburn from the Florida Museum of Natural History) embarked on the production of micro CT images of each of the genera of Ecuadorian amphibians with amazing results; Juan Carlos Santos, who, in 2021, performed a phylogenetic analysis and obtained a phylogeny and chronogram for the Ecuadorian amphibians; and to Pedro A. Coloma, Morley Read, and Francisco Velásquez Espín, who enthusiastically analyzed the calls of 13 species: *Pristimantis appendiculatus*, *P. carlosceroni*, *P. ecuadorensis*, *P. eugeniae*, *P. glandulosus*, *P. parvillus*, *P. pycnodermis*, and *P. sirnigeli* (PAC and IGT), *Atelopus balios*, *A. coynei*, *A. exiguus*, *A. ignescens* (MR), and *Ranitomeya ventrimaculata* (FVE, MR), most of which were recorded by Ítalo G. Tapia. Carlos E. Montalvo-Puente introduced me to the amazing world of archeology after a visit to Museo El Alabado.

Some of the authors of subtitles in Volumes I and IV also helped with other aspects of the complete opus in various ways. For example, Terán-Valdez, the CJ Collections Manager, edited and revised all of the accounts; she also helped to write descriptions of the tadpoles of 29 species and of many ontogenetic series. Her input to the Encyclopedia in countless ways has been very valuable. She was always there (including non-working hours and vacations), ready to help with anything. Camilo Rosero-Gómez helped with refining maps and associated data with amazing efficiency. Guarderas helped with many administrative tasks and the acquisition of funds related to the Encyclopedia; Juan Carlos Santos reviewed the section about David C. Cannatella; Almendáriz provided some data, photos, and critically reviewed the sections titled: "Peters and the Harlequin Frogs, a Rebirth for Amphibian Collections" in Chapter 2; Vigle also edited and reviewed the MECN section under the subtitle "The DHMECN Collection: From the Friends of Nature to the National Biodiversity Institute" in Chapter 2; Mónica Páez-Vacas edited part of the section that includes the information about calls of Ecuadorian frogs and also helped with the organization of the Red List table. Guayasamin aided in obtaining information of authors from Scopus.

Additional reviews of sections of the Volumes I–IV were conducted by John E. Simmons and Joseph Mendelson III (Chapters 1 and 2), Giovanni Onore (amphibians at the QCAZ), Andrew Gray (Phyllomedusidae), Santiago Castroviejo-Fisher and his PhD student Lourdes Echevarría (*Hemiphractus*), Ross Maynard (accounts of species from Reserva Manduriacu), and Martha L. Crump (part of "In Memoriam"). John E. Simmons drew my (LAC) attention to Jorge Icaza's mention of frogs in Ecuadorian literature. Darrell R. Frost responded quickly and generously to some concerns. Pedro A. Coloma helped with the organization of the Red List information. Sofía Carvajal-Endara and Gustavo Pazmiño-Otamendi helped with refining the geographic information and the production of many of the distribution maps. Néstor Acosta-Buenaño aided in part of the compilation, organization, and editing of the information about vocalizations. I also appreciate the editorial support from Crumly Chuck, Kara Roberts, Balaji Karuppanan, Shayla McKeown, and the rest of the team at CRC Press and Taylor & Francis Group.

I couldn't have compiled the biographies of many of the amphibian biologists included in Chapter 2 without their positive response to my request for their CVs, photos, and biographical information. I thank them all (in alphabetical order): Ana Almendáriz-Cabezas, Alejandro Arteaga, Diego Batallas-Revelo, Jorge Brito Molina, Marcel Adrián Caminer Rodríguez, Sofía Carvajal-Endara, Glenn Flores, Diego F. Cisneros-Heredia, Jaime Culebras, Juan M. Guayasamin, María José Navarrete, Jhael Ortega, Lauren A. O'Connell, Nadia B. Páez, Mónica I. Páez-Vacas, Carolina Proaño-Bolaños, Juan Carlos Santos, Morley Read, Carolina Reyes-Puig, Juan P. Reyes-Puig, Juan Carlos Sánchez-Nivicela, Diana Székely, Paul Székely, Elicio E. Tapia, Ítalo G. Tapia, Andrea Terán-Valdez, Verónica L. Urgilés, Jorge H. Valencia, and Mario H. Yánez-Muñoz.

Numerous other people have contributed in various ways, and their efforts are greatly appreciated. Joseph Mendelson III wrote an insightful foreword for Volume I with great enthusiasm and generosity. He was the perfect person for such a task, given his deep knowledge and his lifelong commitment to amphibian systematics and conservation in the Neotropics. Moreover, his close bond with Bill Duellman, serving as both mentor and friend, and his consistent support of Coloma in various professional pursuits made him an ideal contributor. Many thanks go to Juan Carlos Santos, Juan M. Guayasamin, and Mónica Páez-Vacas for writing the forewords for Volumes II, III, and IV, respectively. Our long-standing friendship, which began when they were my students at PUCE, has enriched our lives and grown into a productive and successful collaboration in amphibian research. Giovanni Onore has called my attention to ethnozoological aspects of amphibians. Tonnie Woeltjes stood out as a dedicated resource, diligently searching for and providing copies of hard-to-find publications, particularly those pertaining to amphibians in the context of hobbyists. Juan Carlos Santos and John E. Simmons also made significant contributions by actively seeking and sharing copies of numerous publications. Jaime Culebras kindly pointed out some of my misidentifications, particularly within the genus *Pristimantis* (as he also provided photos), and generously shared his expertise derived from extensive field experience. He has gathered rare frogs and specimens, contributing to our ex situ colonies and museum collections. Furthermore, he has been a constant source of lively discussions and passion for frogs. Thanks to Connie Dickinson, who generously organized a visit to a Tsáchila community, where Alfonso, Primitivo, José Jacinto, and Milton Aguavil shared with us their hospitality and knowledge of common names of frogs in Tsáfiqui. Leonardo Negrete assisted in compiling the information related to the calls of centrolenid frogs. Néstor Acosta-Buenaño and Sofía Ballaz contributed to editing the information of calls of Ecuadorian frogs included in the Appendix. Alexander Hirtz's openhandedly provided access for photographing his collection of archeological pieces of amphibians. Erika Ullauri P. granted entry to photograph archeological artifacts of the Museo Nacional del Ecuador. Amanda B. Quezada helped to edit thousands of bibliographic references and numerous other editorial tasks with great efficiency.

Our deepest thanks go to the people that collected specimens in our respective institutions (KU, QCAZ, CJ). All are cited in the databases of the museum collections, and the ones that made the most significant contributions are also mentioned in the history section. At Centro Jambatu, we especially thank to Elicio E. Tapia; Mónica I. Páez-Vacas; Morley Read; Ítalo G. Tapia; Arturo Guasti; Patricio Vargas Mena; Mark Wanner; Jaime Culebras; Jan, Jose, and Thomas Verkade; Juan Carlos Santos; Amadeus Plewnia; Manuel Morales-Mite; Sofía Carvajal-Endara; Amanda B. Quezada; Carolina Proaño-Bolaños; H. Mauricio Ortega-Andrade; and about 150 other collectors who brought specimens and associated data to CJ. Some of them also brought live amphibians to be photographed in the lab. My particular gratitude goes to the Verkade family (Jan, Jose, and Thomas) for their searching for specific species of frogs (under my request), and their tadpoles, recordings of calls, and photography skills (all material deposited at CJ).

Additionally, Thomas found a rare caecilian from southwestern Ecuador. Thanks go to Fernando Vaca and Nanto Huamoni, who enthusiastically supported our CJ scientific work and collections at the Shiripuno Lodge.

Most of the photographs of tadpoles, ontogenetic series, and reproduction data would have not been possible without the dedication to raising them in the lab while at the QCAZ (up to 2010) and at CJ (2011–2023) of the following amphibian keepers, vivaria managers, veterinarians, and several assistants. They are Ítalo G. Tapia (2001–2010 at the QCAZ, 2023 to date at CJ), Diego Almeida Reinoso (2004–2010 at the QCAZ, 2011–2014 at CJ), Freddy Almeida Reinoso (Insectarium, QCAZ 2007–2010, CJ 2011–2014), Gabriela Aguiar (2011–2016), Patricio Vargas Mena (2013–to date), Marco De Rossi (CJ vivaria, 2014–to date), Elena Naydenova (veterinarian, 2015–to date), Diana Carolina Salazar (2016–2019), Gabriela Manzano (2016–2017), Carlos Valle Piñuela (2016–2018), Renata Gavilánez (2018–2019), Leonardo Negrete (2018–2019), Berenice Benavides (2021–2022), Estefanía Erazo (2017–2020), Xavier Cisneros (2019–2021), Johanna Paz (2019–2021), Daniel Romero (2019–to date), Fernanda Gordon Domínguez (2021–2023), Alex Achig (2021–to date), Belén Morales (2022–2023), and Mateo Gómezdelatorre (2022–2023). Among them, Ítalo G. Tapia and Patricio Vargas Mena joined me on innumerable collecting field trips, and I value them as great field companions. Carlos Valle Piñuela generously taught me the use of QGIS for the production of maps. Esther M. Langan, Collection Manager at the National Museum of Natural History of the Smithsonian Institution, graciously provided information in response to several inquiries.

When compiling data included in this Encyclopedia during my time at the QCAZ and CJ, I relied heavily on the invaluable assistance provided by the collection managers, curatorial assistants, volunteers, and their dedicated efforts in managing preserved specimens and the living collection. I am grateful for the support and contributions at the QCAZ of Ítalo G. Tapia (2002–2010) and Diego Almeida Reinoso (2004–2010) and at CJ of Sofía Carvajal-Endara (2011–2013), Diego Almeida Reinoso (2011–2014), Elicio E. Tapia (2011–2016), Manuel Morales-Mite (2015–2018), Morley Read (2017–2018), Sebastián Hernández-Nieto (2018), Andrea Terán-Valdez (2018–to date), and Amanda B. Quezada (2021).

Several aspects of the production of *Amphibians of Ecuador* received partial funding from various grants. One of these grants was the PARG project, which provided support during my temporal association with Universidad Regional Amazónica Ikiam from 2015 to 2017 and subsequently at CJ until 2019. I extend my gratitude to Andrea Coloma Santos and Adriana Paulina Guarderas Valverde for their assistance in crafting the PARG project proposal. In 2019–2020, another grant that contributed to this endeavor was an award to Lauren A. O'Connell (No./FAIN No. 1845651) from the US National Science Foundation. This grant played a vital role in covering an important portion of salaries and printing costs. Lauren's aid and research initiatives are greatly acknowledged from the foundation of CJ from 2011 to date. During 2012–2018, research grants to Kim L. Hoke and her student Molly C. Womack by the US National Science Foundation (NSF IOS–1350346, OISE–1157779, and PRFB–1611752) funded activities specifically related to the documentation of ontogenies of Bufonidae; thus, this project became an important aid to CJ and these books.

Additional financial and miscellaneous support for CJ and this work has been provided by Wikiri, an Ecuadorian biotrade enterprise, and the Saint Louis Zoo. These institutions have consistently offered crucial support to CJ and this project, and I could not have accomplished this without their assistance. I wish to express my profound gratitude to María D. Guarderas, the manager of Wikiri and CJ coordinator, whose inspirational leadership and unwavering dedication have been instrumental in sustaining CJ. She has initiated numerous projects to advance our scientific work and adeptly managed the complex bureaucratic and administrative tasks at CJ. She was joined in recent years by Andrea Terán-Valdez, who plays an instrumental function in scientific coordination. Their support while I have been immersed in this opus, often setting aside many other responsibilities at my institution, did not waver.

I would also like to extend my heartfelt thanks to the Saint Louis Zoo, represented by Jeff Bonner, Eric Miller, Lissa Kelley, Mark Wanner, Lauren Augustine, and more recently by Justin Eiden. Their financial assistance has played a pivotal role in our work. In addition, I want to convey my deepest appreciation to Mark Wanner at the Saint Louis Zoo, with whom I have shared numerous field trips over the years in our pursuit of frogs. Our friendship, which began in the bed of a pickup truck while traveling for a nighttime frog search during a frog meeting in Panama in 2006, has endured and enriched both our personal and professional lives.

Finally, I dedicate this work with my profound and infinite gratitude to my extended family, and especially to Alexandra Quiguango Ubillús (my wife); to our sons, Pedro Antonio and José Miguel; and to my mother, Lida María Román. Their devoted emotional support surpasses any words I can write.

PHOTO AND ILLUSTRATION CREDITS

All photos and illustrations are appropriately credited in their respective figure captions. We have employed abbreviations for the primary authors of the majority of the photos, namely Luis A. Coloma (LAC), William E. Duellman (WED), Diego Acosta López (DAL) and Steven Guevara Salvador (SGS). The two latter worked at CJ focusing on photographic documentation and photo designing in 2016–2020 and 2018 and onwards, respectively. Steven generously and patiently assisted with the detective work of finding the data associated with thousands of photos that had not been completely labeled. Thanks go to Rafe M. Brown and Ana Motta for facilitating access to the KU photographic archive and diligently providing scans. We extend our special thanks to Jaime Culebras, Amanda B. Quezada, Morley Read, Paul Székely, and Mario H. Yánez-Muñoz, who graciously shared numerous photos with us, many of which were crucial to expand the photographic documentation of Ecuadorian amphibians. Additionally, we are grateful to Jorge Brito Molina, Juan Carlos Sánchez-Nivicela, Gregory O. Vigle (for his frog photos and photos of Kenneth Miyata), Carolina Reyes-Puig, Elicio E. Tapia, Fernando Coloma Román, Jan C. Verkade, Diego Batallas-Revelo, Juan P. Reyes-Puig, Ross Maynard (The Biodiversity Group), Amadeus Plewnia, Andreas Nöllert, John E. Simmons, Alejandro Arteaga, Fernando Rojas, Joek Sartore, Martín R. Bustamante, Ted Kahn, Cristian Paul Nieves, Diego Almeida Reinoso, Erwin Patzelt, José Vieira, Mateo Vega-Yánez, Daniela Franco-Mena, Patricia A. Burrowes, Alex Achig, Frank Pichardo, Jazmín Buitrón, Jenna Welch (Division of Amphibians & Reptiles, Smithsonian Institution), Matthijs Hollanders, and Francesca Angiolani also contributed generously by providing sets of photos and/or unique important photos.

1 Ethnobatrachology

Humans have evolved in association with amphibians. In the Americas, they have shared their existence for approximately 20,000 – 25,000 years. Thus, in Ecuador, the history of knowledge of its amphibians dates back to the time of the arrival of the first humans to the territories that today encompass the country, more than 12,000 years ago. The long relationship of both ancient cultures and modern civilization with amphibians has generated an immense amount of cognizance in regions where humans have been active. Thus, this interrelation is manifested in material and psychological culture, archaeological remains, traditional knowledge, multiple expressions of millenary and contemporary origin, and in the development of science since the mid-18th century. However, it is evident that a good part of the ancestral knowledge about amphibians was lost with the passage of time because the available archaeological record is very meager, and in several indigenous cultures, the language, cosmovision, and ancestral knowledge are either extinct or disappearing. For example, just in Ecuador, at least seven languages have disappeared since colonial times, and two languages, Andoan and Zápara (= Záparo, Saparo), are nearly extinct. Currently, of 25 languages spoken in Ecuador, 21 are indigenous, and of these, 13 indigenous languages are in danger of disappearing (Dupuis, 2020). This has been aggravated by the cultural collapse of several native ethnic groups beginning with the conquests by the Inca and Spanish Empires in the 15th century. All that remains of the ancestral knowledge are a few poorly studied archaeological sites and an incipient oral tradition (there were no pre-Columbian written documents in Ecuador), which have not been rescued by Western culture.

From the 15th century through the colonial period and until recently, there was a cultural avalanche (especially Spanish, mestizo, and more recently Anglo-Saxon) that for more than 500 years has privileged the knowledge of incoming cultures and generally ignored the knowledge of the native ethnic groups. On the one hand, efforts to save ethnic cultures have been scarce, although the recent declaration in the Constitution of Ecuador that it is a multinational, multicultural, and multiethnic state is an important step that hopefully will promote the recovery and appreciation of their remaining knowledge. On the other hand, the scientific development of knowledge about Ecuadorian amphibians is relatively recent, and although it started in Europe, with Linnaeus, in the 18th century, it only began to consolidate by the end of the 20th century with the input of foreign researchers and since about the 1960s has started nationally. Despite these advances, scientific knowledge is relatively limited and progresses at a slow pace. Additionally, in some cases, potential knowledge is being lost forever with the catastrophic extinction of many species of Ecuadorian amphibians.

No synthesis of the history of knowledge of Ecuadorian amphibians is available, except for general notes in several popular articles published in 2005 in an Ecuadorian outreach magazine, which dedicated a monographic issue to the amphibians of Ecuador (Terra incognita, 2005); a very brief summary (Coloma, 2014); and the historical synopses for a few particular taxa (e.g., *Colostethus* in Coloma, 1995). Herein, in Chapter 1, we compile information, provide new data, and review and synthesize what is known of ancestral and modern knowledge related to folklore (vernacular names, myths, stories and legends, music, popular poetry, art, crafts, jokes, superstitions, customs, etc.), traditional medicine, and the culinary uses of amphibians. Nonetheless, a comprehensive ethnobatrachology treatise is beyond the scope of our expertise, and we acknowledge that much remains to be discovered and written. In Chapter 2, we describe the history of the scientific development of knowledge about amphibians in Ecuador.

INDIGENOUS AND VERNACULAR NAMES

Ecuadorian amphibian species have names of varied origin and with many nuances. There are names that have surely been lost in the mist of time, some that have an ancestral origin and are preserved intact, others that have been modified, and some recently coined, both by the popular imagination and fantasy and also by academia. The names come mostly from traditional indigenous languages and dialects (the 21 native languages in Ecuador), from Quichuañol (half-language), Spanish, and English. The two latter two are increasingly used in the growing herpetological ecotourism industry (locally named Sapowatching and herpwatching) that has an important flow of both Spanish- and English-speaking visitors. Some Spanish names have been locally coined, such as Diablito, Tintín, or Sapito Guitarrero for *Oophaga sylvatica*; Caucherito for *Epipedobates boulengeri*; Sapitos Bullangueros for *Engystomops* spp.; and Pudridoras for *Caecilia* spp., among others. In abundant species with conspicuous calls, onomatopoeic names are locally common, for example, Wikiri for *Ameerega bilinguis* (Félix-Novoa and Acosta Buenaño, 2006) and napi piripri for *Allobates zaparo* (Coloma, 1986). Within the native languages there are names that are applied to groups of species (classified according to their use, morphological similarity, taxonomic affinity, evolutionary relationships, life habits, etc.) and have a more general use. Unique names for a single species are much less frequent; they are usually applied when the amphibian is very abundant or used for some specific purpose (e.g., medicine, food). There are also a few names in common use and with a wide geographic coverage, although most are restricted to particular geographic regions.

DOI: 10.1201/9781003129080-1

In Ecuador, the abundance of vernacular names for tadpoles is notable. Due to this multiple origin and diversity of criteria in the use of vernacular names, a species can have several names. For example, there are no fewer than 11 common names that have been used for *Gastrotheca riobambae* in the scientific and popular literature and in oral communications: hambatu (de Velasco, 1789), Riobamba pouched frog (Cochran, 1961), marsupial frog (Jones *et al.*, 1973), Sapo Marsupial (del Pino, 1975), egg-brooding frog (del Pino *et al.*, 1986), Riobamba marsupial frog (Frank and Ramus, 1995), Hambato (Fuentes Roldán, 1997), Rana Marsupial Andina (Coloma and Ron, 2001), Sapito (Anonymous, 2014), Gualag (Tapia, 2015), and Gualataco (Coloma, 2016b). In contrast, there are instances in which the same name is applied to different species; for example, the ancestral use of the word Jambatu is frequent for most species of anurans, for tadpoles, or, more recently in popular publications, for various species of the genus *Atelopus* (Arbeláez Ortiz and Vega Toral, 2008; Anonymous, 2009a, 2009b, 2015).

There is a tendency to standardize vernacular names for common use and to avoid confusion (see Chapter 1, Section "Standardized Names in Spanish and English"). However, the literature on vernacular names is scarce, fragmentary, and in some cases obviously inappropriate. For example, in Ecuador, some mestizo people (especially in the countryside) use the Spanish word for frog to refer to the females and that for toad for the males (see also the confusion between frogs and toads in a later paragraph). To illustrate the diversity of the use of vernacular names of amphibians, as well as their origins, modifications, and meanings, some relevant examples from Ecuador are provided next. We also provide a list of 102 vernacular names in indigenous languages, the equivalent scientific taxon, and references in Table 1.1.

TABLE 1.1
Names for Amphibians in Ecuador

Vernacular Name	Language	Taxon	Source
Antza	Tsáfiqui	*Engystomops pustulatus*	Primitivo Aguavil (Coloma, 2016c)
Ashanka, Shanga, Ashanga	Shuar chicham or Swiwiar chicham	*Boana boans*	Coloma (1984b), Morales and Schjellerup (1999)
Atya	Kichwa	*Boana appendiculata*	Kohn (2002)
Bahui	Kichwa	*Leptodactylus wagneri*	Kohn (2002)
Bamburé	African language	*Rhaebo blombergi*	Coloma (1984a)
Boroná	Tsáfiqui	*Boana rosenbergi*	José Jacinto Aguavil (Coloma, 2016c)
Bororó	Tsáfiqui	*Boana rosenbergi*	Primitivo Aguavil (Coloma, 2016c)
Burú	Tsáfiqui	*Lepptodactylus peritoaktites*	Primitivo Aguavil (Coloma, 2016c)
Chirimiu, chimius	Shuar chicham	*Dendropsophus bifurcus, D. rhodopeplus, D. sarayacuensis, D. reticulatus*	Botasso (1977), Coloma (1984b), Morales and Schjellerup (1999), Gallardo (2004)
Coré	Tsáfiqui	*Boana rosenbergi*	José Jacinto Aguavil (Coloma, 2016c)
Cotete	Kichwa	*Pristimantis*?	Coloma (2016b)
Cototo	Tsáfiqui	*Engystomops pustulatus*	José Jacinto Aguavil (Coloma, 2016c)
Cuico	Kichwa	*Pristimantis unistrigatus*	EPN museum catalog
Cutín, cutinte	Kichwa	*Pristimantis* spp.	Coloma (1988)
Dequere, dequiri	Kichwa	*Rhinella* sp. (*margaritifera* complex)	Kohn (2002)
Gualag, hualag, huartag	Kichwa	*Leptodactylus pentadactylus, L. rhodomerus*	Enríquez (1999)
Gualama	Tsáfiqui	*Rhinella horribilis*	Primitivo Aguavil (Coloma, 2016c)
Gualataco	Kichwa	*Gastrotheca riobambae*	José Alquinga (Coloma, 2016a)
Gualeque	Shuar chicham	*Boana boans*	Coloma (1984b)
Gualimana, Gualmana	—	*Rhaebo blombergi*	Coloma (1984a)
Guingui	Kichwa	*Pristimantis kichwarum*	Kohn (2002)
Hambatu, Hampatu, hamppatu, hambatug, hamp'atu, hambato, jambatu, jambato, jampatok, jamp'atu, jampatu, janpatu, jambatiu, jambatio, jambatyu, jambatyug, jambatiug, jambacho, xambatu, xanbatu, ambatiug, ambatyu, ambatio, ambato, ampatu, ambatukuna	Kichwa	*Atelopus ignescens, Gastrotheca riobambae, G. pseustes, Phyllomedusa vaillanti, Leptodactylus pentadactylus, Hyloxalus bocagei*	de Velasco (1789), Jiménez de la Espada (1875), Stark and Muysken (1977), *et al.* (see text).

TABLE 1.1 (*Continued*)
Names for Amphibians in Ecuador

Vernacular Name	Language	Taxon	Source
Hanti wirisam	Shuar chicham or Swiwiar chicham	*Dendropsophus reticulatus*	Coloma (1984b)
Huwin	Kichwa	–	Chimbo Aguinda *et al.* (2007), Ministerio de Educación Ecuador (2019)
Iluli, ilulli, ilulu, ilulo	Kichwa	Gymnophiona	Patzelt (1976), Chimbo Aguinda *et al.* (2007)
Ipiakáchi	Shuar	–	Botasso (1977)
Jante, janta	Shuar chicham	*Scinax ruber*	Botasso (1977), Coloma (1984b), Morales and Schjellerup (1999)
Juátu, juato	Shuar chicham or Swiwiar chicham	–	Botasso (1977), Coloma (1984b)
Juiciu jambatu	Kichwa	*Hyloxalus bocagei*	Kohn (2002)
Jurújur, jurujru	Shuar	–	Botasso (1977), Morales and Schjellerup (1999)
Juwin, juin, uwin	Kichwa	*Leptodactylus pentadactylus*	Enríquez (1999), Dea (2006)
Kaka, káka	Shuar chicham	–	Botasso (1977), Pellizzaro (1996), Morales and Schjellerup (1999)
Karakras, karagras	Shuar	–	Coloma (1984b), Morales and Schjellerup (1999)
Katse	Shuar	–	Botasso (1977), Morales and Schjellerup (1999)
Kayla, kayra, caila, K'ayra, Kcaila	Kichwa	*Telmatobius niger*	Merino-Viteri *et al.* (2005), Chimbo Aguinda *et al.* (2007)
Kiamp	Shuar	–	Bianchi (1988)
Kiki	Tsáfiqui	*Oophaga sylvatica*	Primitivo Aguavil (Coloma, 2016c)
Kiria	Shuar	–	Botasso (1977), Bianchi (1988)
Kirisip´	Shuar	–	Botasso (1977)
Kirúupash, kirupash	Shuar	–	Botasso (1977), Morales and Schjellerup (1999)
Kopa	Tsáfiqui	*Caecilia* spp.	Primitivo Aguavil (Coloma, 2016c)
Kotán, Cotán	Tsáfiqui	*Boana pellucens*	Milton Aguavil (Coloma, 2016c)
Kua	Shuar	*Phyllomedusa* sp.?	Botasso (1977), Morales and Schjellerup (1999), Dea (2006)
Kuartank	Shuar chicham	–	Pellizzaro (1996)
Kuguayu, kuwayu	Kichwa	–	Morales and Schjellerup (1999), Ministerio de Educación Ecuador (2019)
Kunawaru	Kichwa	–	Dea (2006), Ministerio de Educación Ecuador (2019)
Kupáip´, Kuraip	Shuar	–	Botasso (1977), Morales and Schjellerup (1999)
Kuwa	Kichwa	–	Ministerio de Educación Ecuador (2019)
Lolo, Lololo	Tsáfiqui	*Rhinella horribilis*	Milton Aguavil (Coloma, 2016c), José Jacinto Aguavil (Coloma, 2016c)
Lukarya, ukarya	Kichwa	–	Ministerio de Educación Ecuador (2019)
Macana-hambato	Kichwa	–	Jiménez de la Espada (1899)
Maculla	Kichwa	*Boana appendiculata*	Kohn (2002)
Marankit	Shuar chicham	–	Botasso (1977)
Mucunda	Shuar chicham or Swiwiar chicham	*Rhinella marina*	Coloma (1984b)
Murintu, murindu, murundi	Kichwa	–	Morales and Schjellerup (1999), Ministerio de Educación Ecuador (2019)
Mutsui, mutsuin	Kichwa	*Rhinella* sp. (*margaritifera* complex)	Kohn (2002)
Napi piripri	Shuar chicham or Swiwiar chicham	*Allobates zaparo*	(Coloma, 1986)
Pai, Paai	Shuar	–	Botasso (1977), Morales and Schjellerup (1999)
Paik´	Shuar	–	Botasso (1977)
Pakui	Shuar chicham or Swiwiar chicham	*Leptodactylus wagneri*, *Adenomera* sp.	Botasso (1977) (1978), Coloma (1984b), Morales and Schjellerup (1999)
Peut	Shuar	–	Botasso (1978)

(*Continued*)

TABLE 1.1 (*Continued*)
Names for Amphibians in Ecuador

Vernacular Name	Language	Taxon	Source
Picu picu	Kichwa	–	de Velasco (1789)
Piripiri	Kichwa	–	Ministerio de Educación Ecuador (2019)
Puach	Shuar	–	Botasso (1977), Bianchi (1988), Morales and Schjellerup (1999)
Puca sapo	Quichuañol	*Atelopus guanujo*	Coloma (2002)
Puint, puint´, puintiag	Shuar chicham or Swiwiar chicham	*Leptodactylus pentadactylus*	Bianchi (1988), Enríquez (1999), Morales and Schjellerup (1999)
Purushám	Shuar	–	Botasso (1977)
Puy	Kichwa	–	Ministerio de Educación Ecuador (2019)
Rusu mama	Kichwa	*Lithobates palmipes*	Gallardo (2004)
Sacha telele	Kichwa	*Ameerega bilinguis*	Kohn (2002)
Sháank	Shuar	–	Botasso (1977)
Shamba, Shanga	Shuar chicham or Swiwiar chicham	*Phyllomedusa tarsius*	Coloma (1984b)
Sosona	Tsáfiqui	–	Connie Dickinson (Coloma, 2016c)
Súwi	Shuar	–	Botasso (1977)
Suya	Shuar	–	Morales and Schjellerup (1999)
Takash	Shuar chicham or Swiwiar chicham	*Rhinella* sp. (*margaritifera* complex)	Coloma (1984b), Morales and Schjellerup (1999)
Takasha	Shuar chicham or Swiwiar chicham	*Bolitoglossa* sp.	Coloma (1984b)
Tentépu	Shuar	–	Botasso (1977)
Tepém	Shuar	–	Botasso (1977)
Tilele	Kichwa	*Rhaebo ecuadorensis*	Kohn (2002)
Tseat	Shuar	–	Botasso (1977), Bianchi (1988)
Tsonpipi	Tsáfiqui	*Rhaebo haematiticus*	Primitivo Aguavil (Coloma, 2016c)
Tuilte	Kichwa	*Pristimantis* spp.	Coloma (2016b)
Tulpi	Kichwa	–	Andrade-Marín (1937)
Tulte	Kichwa	Metamorphs and small frogs	Carlos Mata (Coloma, 2016b)
Tulumba, tulumpa, tullumpa, tulomba, telili, tulumba, yacu tilimbu, tulumba maqui	Kichwa	*Rhinella marina*	Enríquez (1999), Morales and Schjellerup (1999), Kohn (2002), Dea (2006), Ministerio de Educación Ecuador (2019)
Tundele	Kichwa	*Pritimantis*	Juan Alberto Riera (Coloma, 2016b)
Ty'ungil	Kichwa	–	Stark and Muysken (1977)
Uchi-juat	Shuar	–	Botasso (1977)
Uchu cara	Kichwa	*Rhinella* sp. (*margaritifera* complex)	Kohn (2002)
Ucug, Uco, hugo	Kichwa	*Telmatobius niger*	Merino-Viteri *et al.* (2005), Arbeláez Ortiz and Vega Toral (2008)
Ukraip	Shuar chicham or Swiwiar chicham	*Scinax garbei*	Coloma (1984b)
Unculu	Kichwa	–	Morales and Schjellerup (1999)
Urungu	Tsáfiqui	–	Primitivo Aguavil (Coloma, 2016c)
Walak	Wao Terero	*Leptodactylus pentadactylus*	Tirira and Ríos (2018)
Wanip	Shiwiar	*Phyllomedusa vaillanti*	Tapia-Román (2023)
Wapo	Wao Terero	*Leptodactylus pentadactylus*	Tirira and Ríos (2018)
Wikiri	Wao Terero	*Ameerega bilinguis*	Félix-Novoa and Acosta Buenaño (2006)
Win, uwin	Kichwa	–	Morales and Schjellerup (1999), Dea (2006)
Wirisam	Shuar	—	Botasso (1977)
Yaku telele	Kichwa	*Hyloxalus bocagei*	Kohn (2002)
Yamba	kichwa	–	Kohn (2002)
Yamba la ciega	Quichuañol	–	Kohn (2002)
Yampaqui	Shuar chicham	*Allobates zaparo*	Coloma (1984b)
Yantsaush	Shuar chicham	–	Botasso (1977)

Hambatu, Hampatu, Hamppatu, Hambatug, Hamp'atu, Hambato, Jambatu, Jambato, Jampatok, Jamp'atu, Jampatu, Janpatu, Jambatiu, Jambatio, Jambatyu, Jambatyug, Jambatiug, Jambacho, Xambatu, Xanbatu, Ambatiug, Ambatyu, Ambatio, Ambato, Ampatu, Ambatukuna. An example of perhaps the most known widely used name in ancestral and contemporary use in Ecuador is the Kichwa word Hampatu and its derivatives, although it is also used in Peru, Bolivia, and Argentina. In its origin, it has broad and generic uses to designate any species of anuran (toad or frog), and even tadpoles and catfish. A search in the academic and popular literature reveals that there are several versions of its origin and meaning. Hampatu etymologically comes from words of two languages, Poqomchi and Kaqchikel, from Guatemalan ethnic groups, which are part of the Quichean group of languages, which in turn is part of the Mayan linguistic family. Its original form is Jam-pa-tok and means "cry in the rain" (Moreno Mora, 1967). The origin of the name Jambato—which means frog or toad—is also attributed to a modification (to Spanish) of the Panzaleo language (the southernmost form of the Macro Chibcha filiation), which was spoken by the indigenous people of the south of Provincia Pichincha and the provinces of Cotopaxi and Tungurahua (Anonymous, 2003). Hampatu, Hambatu, or Jambatu would come from Kichwa, Cañari, or a language of the Puruhaes, in what is now Provincia Tungurahua. It could be interpreted as the hill of the frog, toad hill, tadpole, hill of the tadpoles, or freshwater catfish (Saá, 2015). According to Caluña Espin *et al.* (2008), the Kichwa name of the natives of the Hampatu nation derives from Jambatiu (Preñadilla, a species of freshwater catfish). The catfish of the Andean region of the genus *Astroblepus*, also called Preñadillas in Spanish, were known as Challua or Hambato in the language of the Cañaris, which mean the same as Hambato in some of the languages of the Puruhaes (Caluña Espin *et al.*, 2008). Ambato, the capital city of Provincia Tungurahua, apparently inherited this name from the Jambatu indigenous nationality and the river that runs through it. Here it is interesting to note what is mentioned by Saá (2015), who highlights the following text about the preñadilla, which suggests the similarity between the tadpole and the Preñadilla (our English translation):

> And Garcilaso notes: everything seems to be of one species, it has no scale, but skin, the head is wide and flat, like that of the toad, and therefore it has a very wide mouth. It is very tasty to eat, they eat it with the skin, which is so delicate that it should not be removed: they call it challua, which means fish.

Outside of Ecuador, one version of the etymology of the word by Jiménez Sánchez Mariscal (2014) indicates that hamppatu in Aymara means large toad, and Pachamama (Mother Earth) has a marked symbolic relationship with the toad as a symbol of fertility. According to Espinosa (1996), in Chile, the Jamp'atu is part of the rituals to bring rain. In Bolivia, Arias (1999) mentions the following (our translation):

> Jampatu (Toad); during the rainy season they abound in the community, especially at night, when they croak insistently, as if they were crying, they announce the arrival of the rains, if they keep quiet, it will mean that it will not rain.

The first written mention of the name Jambatu in Ecuador was made by the Jesuit priest and historian Juan de Velasco (1789), who defined it as follows (our translation):

> Hambatu is the generic name for all species of toads or toads, they are common and come in various colors and sizes. . . . There are some extremely venomous ones in a mountainous desert in the province of Ibarra, completely green, with glowing eyes and up to six fingers long. . . . In very hot climates there are much larger ones, but harmless, although annoying, even getting inside the houses.

De Velasco's description of the Hambatu from Ibarra allows us to hypothesize that it would correspond to *Gastrotheca riobambae* because it is the only species in that area with the aforementioned characteristics. The Spanish historian and naturalist Jiménez de la Espada mentions the following when describing *Leptodactylus pentadactylus* (under the name *L. goliath*) from the Archidona region and the Guacamayos mountain range (Provincia Napo, Ecuador) (our English translation):

> Give it the name of Hatun-Hambato, which is equivalent to a great frog or toad.
>
> (Jiménez de la Espada, 1875)

The Ecuadorian writer Jorge Icaza (1934) used the name Jambatos in his great novel *Huasipungo* and relates that children play with them, making them run races. This kind of game with frogs has been mentioned by elders from Angamarca, Provincia Cotopaxi, and is associated with *Atelopus ignescens* (Coloma, 2016f).

According to Simmons (2009), because *Atelopus ignescens* were slow moving, they were thought to be dimwitted and clumsy, so Indians sometimes use the word jambato as a nickname or as an insult. Several characters in the works of Jorge Icaza have the nickname Jambato. For instance, Icaza (1942) mentions the following in the novel *Media Vida Deslumbrados*:

> abría a patadas la puerta, li' arrastraba de los cabezones por el suelo, le metía a pulso en las cobijas, pero en seguidita se resbalaba como jambato . . . [Our translation: . . . opened the door, he was stubbornly dragged over the floor, and stuck a hand out of the blankets, but at once slipped like a jambato.]

Santos Ortiz de Villalba (1994) uses Janpatu to refer to toads in the language of the Kichwa people on the banks of the Río Napo. Costales *et al.* (1996), in their revision of the

Catequil myth of the Quitu-Cara, refer to the Kichwa word hambatu as small toad, while its equivalent in Quitu-Cara is potato. Kohn (2002) mentions the names Juiciu Jambatu for the little frog *Hyloxalus bocagei* and Cucha Jambatu (Cucha Ambatu) for *Pristimantis* spp. from the Ávila region (Provincia Napo) in the Ecuadorian Amazon. Gabriel Grefa (Coloma, 2010) indicates that the generic Kichwa name in Archidona (Provincia Napo) for toads is Ambatukuna. Indigenous people of the Puruhá people of the paramos at the foot of the Volcán Chimborazo (Provincia Tungurahua) mention the word Ambatio to designate the green amphibians (presumably *Gastrotheca pseustes*) (Coloma, 1989a). Fuentes Roldán (1997), in his book *Quito Tradiciones*, tells the story of the water toad and mentions (our English translation):

> a small Jambato with bright eyes and greenish skin, native to the area.

This description seems to refer to *Gastrotheca riobambae*, which is the only conspicuous green frog in Quito. *Centrolene buckleyi* is also green but used to be less conspicuous and restricted to streams.

The transcription of the name jambatu from an oral language such as Kichwa has been very variable and also reflects several aboriginal dialects. For example, according to Stark and Muysken (1977), Jambatu is used in Zámbiza (Provincia Pichincha), San Miguel de Salcedo, and Tigua (Provincia Cotopaxi); Jambatyu in Chibuleo San Pedro and Guapante (Provincia Tungurahua) and in Troje (Provincia Chimborazo); Jambatyug in Salasaca (Provincia Tungurahua); and Ambatyu in Casichagua (Provincia Bolívar). The entry for Jambatyug is a line drawing of a marsupial frog (Simmons, 2009), probably *Gastrotheca riobambae*. In recent decades, the use of this word—modified to Jambato especially by the mestizo inhabitants of cities such as Quito—has been confined to designate some harlequin frogs of the genus *Atelopus* (Coloma and Ron, 2001; Valencia *et al.*, 2008; Arbeláez Ortiz and Vega Toral, 2008), although, since 1976 has been more exclusively used to designate the critically endangered *Atelopus ignescens* of the Ecuadorian Andes (Patzelt, 1976; Almendáriz, 1992; Coloma and Lombeida, 1992; Coloma, 1999; Ron *et al.*, 2003, 2009; Simmons, 2009; Coloma, 2016e), which formerly was one of the most abundant and conspicuous anuran species in the region (from Imbabura to Bolívar provinces) until the mid-1980s and was rediscovered in 2016. In this publication, we use the name Jambato exclusively for *A. ignescens*.

Kaila, K'ayra, Kayra, Caila, Cayla. This is a widespread name, at least in Ecuador, Peru, and Bolivia. In Peruvian Quechua, it means frog (Rojas Palomino and Quintero Bendezú, 2014). In Bolivia, the name K'ayra is used for species of *Telmatobius*. In Ecuador, it derives from Kichwa and also signifies frog (Moreno Mora, 1967). According to Costales *et al.* (1996), cayla in Kichwa means large frog, and its equivalent in Quitu-Cara is Cotán, which is the voice of thunder. Etymologically, Kayra and Kaila come from words from two languages, Chol and Kaqchikel, from ethnic groups from Chiapas (Mexico) and Guatemala, respectively, which are part of the Cholano and Quichean groups of languages, which in turn are part of the Mayan linguistic family. Its meaning is "sing and give many voices" (Moreno Mora, 1967).

According to de Velasco (1789), Caila is an edible type of Hambatu. The indigenous and mestizo populations of some localities in the Bolívar and Chimborazo provinces in the central-southern Andes of Ecuador attribute the name kayla to *Telmatobius niger* (Merino-Viteri *et al.*, 2005). In this publication, we use the latter exclusively for *T. niger*.

Hualaj, Huartaj, Ty'ungil, Gualag, Gualataco. According to Stark and Muysken (1977), the three former names mean toad, and their names are used in Ilumán (Provincia Imbabura), Nizag (Provincia Chimborazo), and Cachisagua (Provincia Bolívar), respectively. In Lloa (Provincia Pichincha), the names Gualag and Gualataco would be used to designate either *Gastrotheca riobambae* or *G. pseustes*, while Patzelt (1976) attributed Gualag to *Leptodactylus rhodomerus* (under the name *L. pentadactylus*). In this publication, we exclusively use Gualag for *L. pentadactylus* and Gualataco for *G. riobambae*.

Cutín. This name, apparently onomatopoeic, is used by indigenous people from the Community of Santa Rosa de Totoras, Provincia Bolívar, to designate some species of the genus *Pristimantis*. In recent years, its use has been extended to most craugastorid frog species from Ecuador (Coloma and Ron, 2001; Ron *et al.*, 2009; Coloma and Guayasamin, 2011), which we follow in this publication.

Ucug. Arbeláez Ortiz and Vega Toral (2008) associate this onomatopoeic name with *Telmatobius niger* and provide information on its medicinal uses. In this publication, we use it more generically for the three possibly extinct *Telmatobius* species from Ecuador (*T. niger*, *T. cirrhacelis*, and *T. vellardi*).

Ilulu, Ilulo. Patzelt (1976) mentions the use of the word ilulo, ilulu to designate those apodal amphibians (order Gymnophiona) shaped like snakes or earthworms. In this publication, we adopt an exclusive use of the word Ilulu for *Siphonops annulatus*.

Rana or Sapo (frog or toad)? The distinction between the Spanish words rana and sapo (= frog and toad, respectively) lacks taxonomic or scientific rigor, although the use of these two common names is frequent in both the Spanish and English languages. These two common names have their origin in Europe and are part of the Spanish language, which is why their use has spread to other continents that speak this language. According to the Real Academia Española, Sapo is a pre-Roman voice of onomatopoeic origin due to the noise toads make when falling into a puddle or on wet ground and is defined as: Anuran amphibian with a plump and robust body, bulging eyes, short limbs and warty-looking skin. In contrast, Rana comes from Latin and is: Batrachian of the order of the Anurans, about eight to fifteen centimeters long, with a greenish back stained with dark, green, brown, etc., and a white abdomen, a mouth with teeth and a round or vertical slit pupil. As there are 7713 described species of anurans in the world (up to 13 August 2024; AmphibiaWeb, 2024), there is a continuum of sizes, shapes, and colors that covers the entire range of variation and makes these definitions of rana and sapo not applicable for all species. Therefore, the Spanish language does not have adequate terms to name most species of anurans. In this sense, the initial Spanish European distinction is unsupported, but it has remained in vernacular speech and is applied arbitrarily and sometimes with extreme modifications. For example, the terms Sapo and Rana are sometimes used to distinguish the male (Sapo) from the female (Rana). Other peculiar uses are those of some peasants on the Ecuadorian coast, who use Rana for large anurans (as in Cuba) such as *Rhinella horribilis*, which, according to the Spanish definition, should be a typical toad. On the other hand, in the Sierra in general, the diminutive Sapito is used to designate species such as *Pristimantis* spp. (Arbeláez Ortiz and Vega Toral, 2008), *Atelopus guanujo* (Coloma, 1989b), *Gastrotheca riobambae* (Anonymous, 2014), etc., but the diminutive is also used to designate any small species. For example, *Oophaga sylvatica* is called a guitar Sapito in the Alto Tambo area (Tapia, 2011).

TADPOLE NAMES

The widely used Spanish name (including in educational texts) for tadpoles is "Renacuajo", and a less frequent term is "Gusarapo". The Real Academia Española defines "Gusarapo" as a small animal, shaped like a larva, that is raised in liquid. Both Spanish names are used generically for any tadpole. The second is exclusively used in the coastal lowlands. It has been recorded especially in the Guayas and Manabí provinces. However, there are names more frequently used in Ecuador that belong to Kichwa or other native languages. Coloma (2003) provided a map with the approximate distribution of 19 names (including phonetic and written variants) that Ecuadorians use to designate tadpoles in Ecuador. In Table 1.2, we provide an updated list of 50 names with their languages and source references. We also provide a map (Figure 1.1) showing the approximate distribution of many of them, for which more specific geographic information is available. There have been no fewer than four names (including derivatives) recorded for tadpoles in the coastal region, 12 in the Amazonian lowland region, and 30 in the Andean region. Most of the names

TABLE 1.2
Names for Tadpoles in Ecuador

Name	Language	Informer and/or Source
Tufe	A'ingae (= Cofán)	Martín R. Bustamante; Costales Samaniego (2010)
Nanal	Awap'it	Martín R. Bustamante
Kulo	Kichwa Amazónico	Iván V. Jácome (Coloma, 2003)
Putukulu, putu kulu, putukuku	Kichwa Amazónico	Dea (2006), Chimbo Aguinda *et al.* (2007), Ministerio de Educación Ecuador (2019)
Ñeñepaica, paica	Kichwa Amazónico (Quijos)	Humberto Wapi (Coloma, 2003)
Rusu	Kichwa Amazónico (Quijos)	Gallardo (2001), Coloma (2003), (Costales Samaniego, 2010)
Chimpaluk	Kichwa Sierra	Ministerio de Educación Ecuador (2019)
Chuslin	Kichwa Sierra	Chimbo Aguinda *et al.* (2007)
Güishigüishe, uilli-uilli, willi willi, williwilli	Kichwa Sierra	Icaza (1975), Coloma (2003), Chimbo Aguinda *et al.* (2007), Ministerio de Educación Ecuador (2019)
Pilli-guilli, pilligalli, pilliguille	Kichwa Sierra	Coloma (2003)
Pímbalo, pímbalu, pímpalu, chímbalo	Kichwa Sierra	Coloma (2003), Coloma (2016b), Ministerio de Educación Ecuador (2019)
Pinkulu	Kichwa Sierra	Ministerio de Educación Ecuador (2019)
Shugshi, chugli, chuglia, chuglli, chugshi, chufle	Kichwa Sierra	Coloma (2003), Eduardo Toral, Páez-Vacas (2012)
Timbal, timbul, timbilín	Kichwa Sierra	Coloma (2003)
Tucllín, tugllín	Kichwa Sierra	de Velasco (1789), Coloma (2003)
Tundele	Kichwa Sierra	Juan Alberto Riera (Coloma, 2016b)
Tushil	Kichwa Sierra	Ministerio de Educación Ecuador (2019)
Ultio	Kichwa Sierra	Coloma (2003)
Villico, Williku	Kichwa Sierra	Coloma (2003), Ministerio de Educación Ecuador (2019)
Chimbirico, jimbirico	Palta	Coloma (2003), Costales Samaniego (2010))
Guiltre	Palta	Gustavo Abad (Coloma, 2003) Costales Samaniego (2010)
Táish, taisha	Shuar chicham or Swiwiar chicham	Costales Samaniego (2010)
Wampuch', Wampucha	Shuar chicham or Swiwiar chicham	Coloma (2003)
Gusarapo	Spanish	Coloma (2003)
Temblor	Spanish	Andrés Merino-Viteri (Coloma, 2003)
Kodoka, dodoca	Tsáfiki	Milton Aguavil (Coloma, 2016c); Martín R. Bustamante

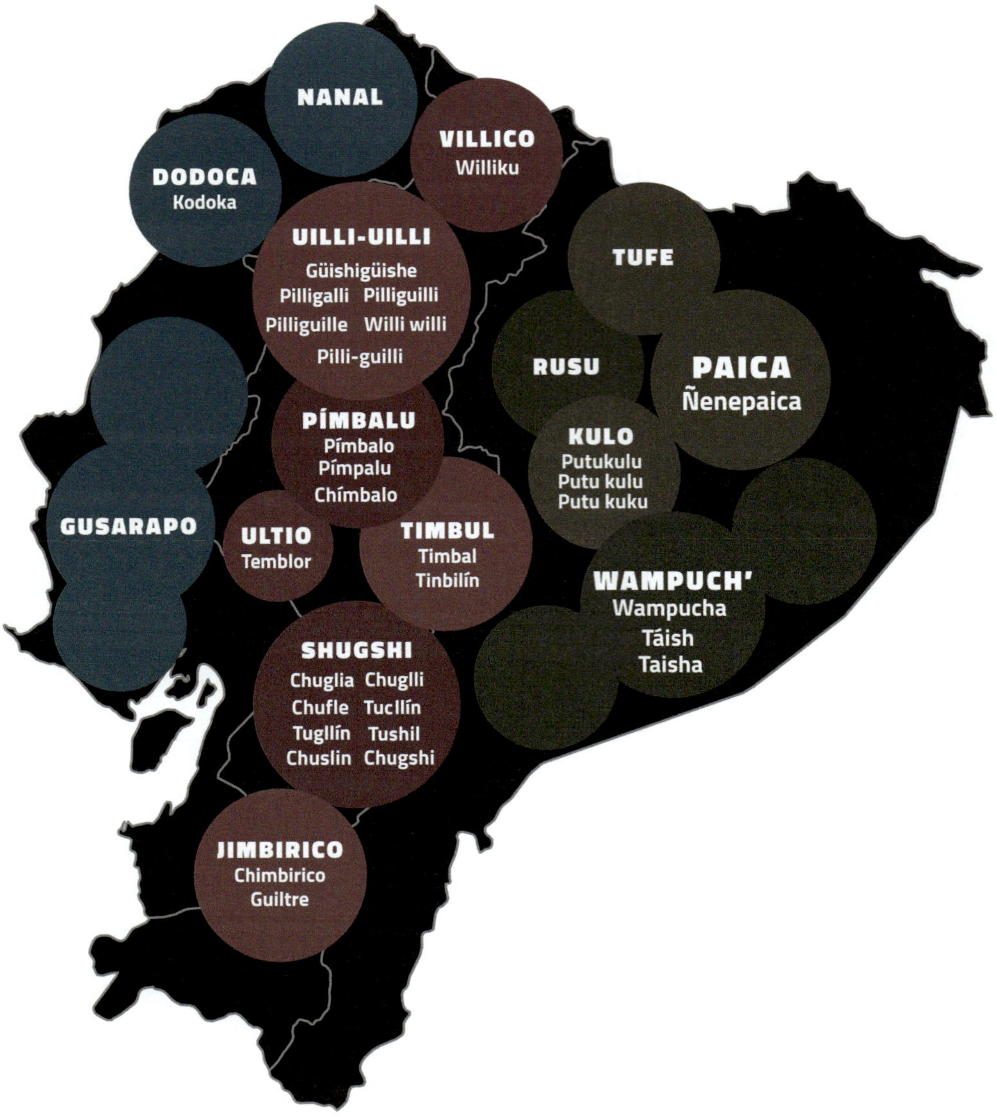

FIGURE 1.1 Approximate distribution of common names for tadpoles in Ecuador. Names are grouped by phonetic similarity and geographic proximity. The ones in capital letters are more commonly used.

used in the Andean region probably originated from the Kichwa language and its dialects, while others belong to the languages of specific indigenous groups inhabiting a given region. Some names are from as yet unknown specific locations (e.g., Chimpaluk), and others are Spanish (e.g., Gusarapo) or a translation from Kichwa to Spanish (e.g., Shugshi = Temblor = earthquake). The use of some of these names has a wide distribution (e.g., Gusarapo in most of the coastal region), while for others it is more restricted. Each of the names applies to tadpoles in a general sense, and thus to any species. However, in the highlands of the Andean region (inter-Andean valleys and paramos) they are more often used for the most abundant species. For example, the names are usually used to refer to tadpoles of several species of *Gastrotheca*, which currently are the most common and abundant local species that have tadpoles, while other species in the region live (or used to live) in streams or rivers (e.g., *Hyloxalus jacobuspetersi*, *Atelopus ignescens*, *A. bomolochos*, *Centrolene buckleyi*, etc.), and are rarely seen, even by scientists looking for them. Herein we provide some information on tadpole names for which more information is available (Table 1.2). Some are grouped according to their phonetic and written similarities and geographic proximity.

Uilli-uilli and variants. These names apply to tadpoles in the Imbabura, Pichincha, and Cotopaxi provinces, especially those of *Gastrotheca riobambae* and *G. pseustes*, which are the most common in swamps in the inter-Andean valleys and paramos. The name uilli-uilli, although it may seem very familiar to the inhabitants of Quito and its surroundings, it is of restricted local use in this city. The indigenous people in the Provincia Cotopaxi especially use the similar words pilliguilli, pilligalli, and pilliguille, and possibly they are the ones that gave rise to the term used in

Quito. According to Alfredo Costales Samaniego (2010), pilliguilli, pilligalli, and pilliguille all have "pi" which means common water, whereas pilligalli means small in the Quitu-Cara language. Thus, presumably the name refers to a small inhabitant of standing waters.

Jorge Icaza, a famous Ecuadorian writer of indigenous novels that aspired to realism, mentions them as güishigüishes in his novel *Huasipungo*—his most popular book, translated into more than 40 languages. The names williku and villico are used in Provincia Carchi mostly for tadpoles of *Gastrotheca espeletia* and *G. riobambae*. They are phonetically similar to uilli and derivatives. According to Costales Samaniego (2010), villico means gentle child.

Ultio. This name probably derives from Kichwa. It is used in Guaranda and surroundings in Provincia Bolívar mostly for tadpoles of *Gastrotheca pseustes*. It is interesting to note that the same epithet is applied in Bolivia for tadpoles. Perhaps this tells us about the Mitimae origin of some of the Kichwa indigenous people inhabiting this province.

Pímbalo and variants. These names are from the Andean Kichwa language. They are used in the Cotopaxi, Tungurahua, Chimborazo, and Bolívar provinces, mostly for tadpoles of *Gastroheca riobambae* and *G. pseustes*. Pi means water. An indigenous community in Simiatug, Provincia Bolívar, is named Pímbalo.

Timbul and variants. These names are of common use for tadpoles of *Gastrotheca riobambae* and *G. pseustes* in Riobamba and surrounding areas of Provincia Chimborazo. According to Costales Samaniego (2010), Timbul derives from the Kichwa word timbu and means to boil. Presumably the word refers to their continuous movements or the movement they cause on the water surface when they are abundant.

Shugshi, Chuglli, and variants. According to Costales Samaniego (2010), these names and phonetic variants are from the Cañari language and mean tremor. The latter presumably refers to their continuous movements or the movement they cause on the water surface when they are abundant. Currently, these names are used in the Cañar, Azuay, and El Oro provinces, mostly for tadpoles of *Gastrotheca cuencana*, *G. litonedis*, *G. pseustes*, and *G. lojana*, although they may also apply for tadpoles of *Hyloxalus vertebralis*, *Telmatobius niger*, and *Ctenophyrne aequatorialis* in the highlands and for *Epipedobates anthonyi* and other frogs in low-elevation sites.

de Velasco (1789) mentions the name Tucllín and describes it like this (our translation):

> It is a unique water tadpole. His original figure is that of a lizard; but as soon as something grows, its tail falls off and, turning into a toad, it comes out of the water and skips through the countryside.

Jimbirico, Chimbirico, and Guiltre. According to Costales Samaniego (2010), these names are variants of the Palta language (a pre-Inca group related to the Amazonian Shuar) and means trembling. Presumably these names are in reference to the tadpole's movement. They are used in Provincia Loja mostly for tadpoles of species of *Gastrotheca* marsupial frogs, which are relatively abundant (*G. elicioi*, *G. lojana*, *G. pseustes*, *G. turnerorum*, and *G. yacuri*).

Dodoka and Kodoka. These names derive from the Tsáfiki (= colorado, tsáchila) language and are used by the Tsáchila indigenous group for tadpoles of any species in Provincia Santo Domingo de los Tsáchilas.

Wampuch' and Wampucha. These names derive from the Shuar chicham or Swiwiar chicham language and are mostly used for any tadpole in the southeastern Amazonia in the Pastaza, Morona Santiago, and Zamora Chinchipe provinces.

Kulu and variants. These names are of generic use for tadpoles in the Amazonian region of Ecuador, especially by Kichwa indigenous groups in the Napo and Pastaza provinces.

STANDARDIZED NAMES IN SPANISH AND ENGLISH

A recurrent problem in naming species is that the vernacular names of amphibians in general do not correspond to the species recognized by science, something that, for example, is not so common with more visible and/or better-known animals such as birds (Diamond, 1966). In amphibians, many vernacular names refer to groups of species or some other non-biological group. Therefore, the scientific nomenclature (a binomial to designate species, usually latinized) is the universal way of recognizing species. Unfortunately, outside the scientific community, the use of scientific names (most of them derived from Greek or Latin) is not widespread, and the use of common nomenclature and vernacular names is increasingly preferred or required, for example in the fields of conservation, biotrade, ecotourism, and outreach. This is articulated by Nicolás Cuvi, an interdisciplinary researcher focusing on socio-environmental issues (Cuvi, 2022) as follows:

> Nevertheless, through the years, I have come to understand that, while this practice is crucial for preventing confusion and easing scientific communication, it can also distance other audiences who might be intimidated by language that appears indecipherable [Our translation].

Therefore, there is a tendency to coin new names for common use. This tendency might complicate matters because names vary according to authors. For example, recently, the frog *Boana picturata*, which is becoming well known as a photographic subject and in the pet trade, is being referred to as the Imbabura treefrog (Frank and Ramus, 1995), Chachi frog (Coloma and Ron, 2001), colorful arboreal frog (Coloma and Guayasamin, 2011), and painted tree frogs (Joel Sartore, Photo Ark). Partial lists of Spanish names are those of Valencia *et al.* (2008), who provide a list for 167 species and that of Valencia and Garzón (2011), who compile names for 113 species. Spanish names are also included on web pages. For example, in a first attempt at standardization, Coloma and Guayasamin (2011) provided an online list of names (most of them in Spanish or a combination with native languages) for Ecuadorian amphibians. Botasso (1977) provided names of some of the species of the Amazon in the Shuar language. Kohn (2002) compiled some Kichwa names of species from the Runa de Ávila community in Loreto (Provincia Napo), while Gallardo (2004) lists the names of some of the edible frogs from Ecuador.

There is no complete and updated printed or online list of standardized names in English for Ecuadorian amphibians. The most complete list of names (not standardized) in English of the species that inhabit Ecuador is that of Frost (2023), who compiled published names in the online accounts, especially those given by Frank and Ramus (1995). Nonetheless, the Frank and Ramus (1995) list has been criticized as inadequate in several cases (Werner, 1996; Caramaschi *et al.*, 2005).

Herein we propose standard Spanish and English names (some of them adopted from native languages) for the amphibian species of Ecuador (Tables 0.1 in Volumens II, III, and IV). The standardization of names in Spanish and English has the potential to facilitate the transmission of information related to a particular species given that Spanish is the most widely used language in Ecuador and spoken by 93% of its population, whereas English is increasingly taught in the school system. We hope that these names (most of them are never, or rarely, used) will become commonly used and thus standardized common names in the future. This, by itself, has a great advantage, especially when the people who communicate in these languages use this information. However, we recognize that some of the proposed names may generate controversy, as is usual with these types of lists (Pinto, 2005 and citations therein), so this list is not immutable, and these names are subject to modifications and additions. For example, the epithets used by Afro-Ecuadorians and Montubios (mestizo people from the coastal region) are mostly not yet known. There also needs to be a compilation of the names that are used in 14 languages and many ancestral dialects that are spoken in Ecuador by the nationalities and indigenous peoples (no fewer than 27).

The English names for families, genera, and species are given in each of the accounts, and Spanish and English names are listed in the Tables 0.1 in Volumens II, III, and IV. Names were chosen for priority use based on the standardization criteria suggested by Caramaschi *et al.* (2005), with some modifications as indicated. These standardization criteria are, in order of priority: (1) Any name that is well established either by its frequent use in publications or by its daily use. Although this list is intended to provide standardized names in Spanish and English, there are exceptions to names or qualifiers incorporated from indigenous languages. The ancestral origin or cultural relevance of the name is also considered, for example, jambato for *Atelopus ignescens*, Kayla for species of *Telmatobius*, and epibatidine poison frog for *Epipedobates anthonyi*. Sometimes, the suggested name in Spanish may differ from the English one, for example, cutín for frogs of the genus *Pristimantis* instead of rainfrogs, and ilulo for species of Gymnophiona instead of caecilian. (2) The etymological intention (explicit or implicit) of the scientists who named the species, for example, small harlequin frog for *A. exiguus* or Onore's harlequin frog for *A. onorei*, which was translated from the original publishing language, usually English, Latin, or Greek. The etymology for each taxon is given in each account. (3) The use of some distinctive aspect (distribution, morphology, behavior, etc.) of the species. A good example of the latter criterion is the common names for genera suggested by Barrio-Amorós (2010) for dendrobatid frogs, which we generally follow, for example, nurse frogs for species of the genus *Allobates*, rocket frogs for *Hyloxalus*, and poison frogs for *Ameerega*, *Epipedobates*, *Oophaga*, and *Ranitomeya*.

AMPHIBIAN ICONOGRAPHY IN ECUADORIAN ARCHAEOLOGY

Carlos E. Montalvo-Puente and Luis A. Coloma

For the pre-Columbian populations of what is now Ecuador, animals held great significance as sources of knowledge and power and were associated with mythical ancestors (Stothert, 2003). In some instances, they were even revered as deities or seen as messengers of life and death (Gutiérrez Usillos, 2002; Stothert, 2003). Animals played an integral role in the cosmology of these populations, woven into their myths and rituals, each serving distinct purposes (Stothert, 2003). Amphibians were no exception to this rich tapestry of beliefs and traditions.

In Ecuador, archeozoological osteological remains of amphibians have been documented as far back as the Early Formative Period (3500 – 1500 BCE) in various archaeological sites (Stahl, 2003). Specifically, anuran animals have been identified in sites such as Loma Alta, Capaperro, Dos Caminos, and San Isidro, located in the coastal Guayas and Manabí provinces (Stahl, 2003). Additionally, Caudata (salamanders) were found in Late Formative contexts at Dos Caminos (Stahl, 2003). This zooarchaeological evidence indicating the presence of amphibians persists over time, as evidenced by these ancient records (Gutiérrez Usillos, 2002).

The iconographic representations of amphibians produced by pre-Columbian populations in Ecuador have been poorly studied; thus, the available literature is limited. The majority of bibliographic references are tied to the general study of specific cultures, such as the Jama Coaque (Gutiérrez Usillos, 2011), the Tolita (Ugalde, 2009), and the Guangala (Uribe Taborda, 2016). There is only one extensive general review on zoomorphic iconography (Gutiérrez Usillos, 2002). As pointed out by María Patricia Ordoñez-Alvarez (2024), who has focused on this issue, there have been no specific efforts or publications to explore iconography in a taxonomic sense by species or higher taxa. One exception is the study of bird iconography conducted by Pablo Quelal (2014) for the Jama Coaque culture.

While animal representations in Ecuadorian archaeology date back to the Valdivia period (4000–1500 BCE), and amphibian bones have been present on archeozoological assemblies since that era, amphibians gained significant representation only during the Chorrera culture (1000–100 BCE) during the Late Formative Period (Cummins, 2003; Gutiérrez Usillos, 2002; Stothert, 2003, 2006). Subsequently, their iconographic presence became a continuous feature in the art and culture of the Pacific coast, Andean highlands, and Amazonian region, persisting until the arrival of Europeans (Dyrdahl and Montalvo-Puente, 2021; Gutiérrez Usillos, 2002; Stothert, 2003, 2006). With this historical context as a reference point, this subchapter seeks to review the literature sources on the iconographic representations of amphibians during pre-Columbian times among the populations that inhabited the present-day territories of Ecuador. We also endeavor to hypothesize the accurate classification of selected anuran representations within various cultures, down to the lowest taxonomic rank possible, while also offering justifications for these assignments. We had the opportunity to photograph and study the artifacts of several collections and museums. Throughout the text we use their acronyms as follows: Museo de Arte Precolombino Casa del Alabado (MAPCA) (collections of Mario Ribadeneira Traversari, MRT), Museo Nacional del Ecuador (MuNa), collections of Alexander Hirtz, materials recovered during archaeological research (collections at Las Orquídeas site, Provincia Imbabura; LO), and Proyecto Arqueológico Huataviro (TH). Unless otherwise indicated, the archaeological pieces from these public and private museums and collections in Ecuador rarely have contextual information, and the localities of provenance are unknown. Chronological and cultural attribution was mainly assigned by style and similarity to archaeological findings during research projects.

This review follows the temporal division of the Ecuadorian archaeology first proposed by Meggers (1966) and further refined by subsequent authors (Stahl, 2003; Ordoñez-Araque *et al.*, 2022). Within each chronological period, the review will progress from north to south, beginning with the cultures of the Pacific coast, moving on to the highland cultures, and concluding with those situated in the Amazon region.

Late Formative Period (1000–500 BCE)

The Chorrera culture (1000–100 BCE) occupied most of the coastal region of modern-day Ecuador, leaving its influence on the highlands and even the Amazonian region. This culture developed distinct local styles for ceramic production (Dyrdahl and Montalvo-Puente, 2022; Lunniss, 2001, 2022; Olsen-Bruhns, 2003; Ugalde *et al.*, 2023; Villalba, 1988; Zeidler, 2003; Zeidler, 2008). This is the first known population that inhabited the region that includes modern Ecuador to portray amphibians. Chorrera ceramics are characterized by a broad repertoire of representation: animals, plants, architectural structures, and people. They are included in a wide type of ceramic vessels, including bowls, pots, lime pots, and whistles (Cummins, 2003; Gutiérrez Usillos, 2002; Montalvo-Puente, 2023; Stothert, 2006). Toads (Figures 1.2, 1.3) and frogs were represented (Figures 1.4, 1.5), following the realistic style of representations that the Chorrera developed for plants and animals (Cummins, 2003). These amphibians were depicted either in the round, known as tutto tondo, or as appliqués. Gutiérrez Usillos (2002) suggested that during the Chorrera era, the focus on poisonous toads and brightly colored frogs may have been related to their toxins, which could have been used for hunting or possibly for hallucinogenic purposes.

In our review, we uncovered artifacts featuring depictions of two frog species, with one of them, *Rhinella horribilis* (see Figure 1.3) (= *Rhinella bella*, see Addendum in Volume IV) likely utilized in traditional medicine, and the other, *Leptodactylus peritoaktites* or *L. rhodomerus* (see Figure 1.5), possibly serving as a food source. Anurans were represented in lime pots (e.g., Figure 1.2), containers used to store lime or ashes of different plants used to chew coca and activate the alkaloid of the plant (Gutiérrez Usillos, 2011), as well as in a vase (see Figure 1.4) for liquids.

The animal depicted in the lime pot (see Figure 1.2) is a robust, medium- to large-sized terrestrial anuran, potentially *Rhinella horribilis.* However, conclusive identification at the species level is challenging due to lack of representation of any sort of skin protuberances in the lime; however, its general morphology fits that of *R. horribilis* well (see Figure 1.3). The connection to *R. horribilis* is bolstered by additional evidence suggesting that the exudates from this particular species were likely utilized for their hallucinogenic and medicinal properties, as has been documented in the Mochica (or Moche) culture (ca 200 – 800 CE) of the north coast of Peru (Schmeda-Hirschmann *et al.*, 2020). According to William E. Duellman (McClelland, 2011), the most frequently depicted frog in the Mochica culture is *R. horribilis* (then named as *R. marina*). This species is also abundant and adapted to live near human habitations. Furthermore, recently, it has been discovered that the compounds found in *R. horribilis* (alkaloids, bufadienolides, and argininyl diacid derivatives) have an anti-cancer effect on human lung cancer cells (Schmeda-Hirschmann *et al.*, 2020).

The frog depicted in the vase (see Figure 1.4) portrays a large, heavy-bodied, terrestrial anuran species discovered in the coastal region of Ecuador. Other noteworthy features include its smooth texture, wide mouth, and distinctive orange–brown coloration. Among the anuran species in coastal Ecuador sharing similar traits, *Leptodactylus peritoaktites*, *L. rhodomerus* (see Figure 1.5), and *Ceratophrys stolzmanni* (see Figure 1.17) stand out. The skin of *C. stolzmanni* is not as smooth and that depicted on the vase, and the size aligns more closely with the dimensions of *Leptodactylus* species rather than those of *Ceratophrys*. The color of the limbs is more similar to *L. peritoaktites*. Consequently, we are inclined to associate this somewhat stylized vase with *L. peritoaktites* inhabiting coastal Ecuador. The wide-open mouth of this depiction may suggest a frog emitting a release call when captured, which is a distinctive feature of species within the *L. pentadactylus* species complex. This call closely resembles the cry of a human child or, as noted

FIGURE 1.2 Anuran (*Rhinella horribilis*) represented in the Chorrera culture (1000 BCE–100 CE). Lime pot, MAPCA (MRT 0410), 65 mm length × 48 width × 32 mm height. Jazmín Buitrón (top) and SGS (bottom).

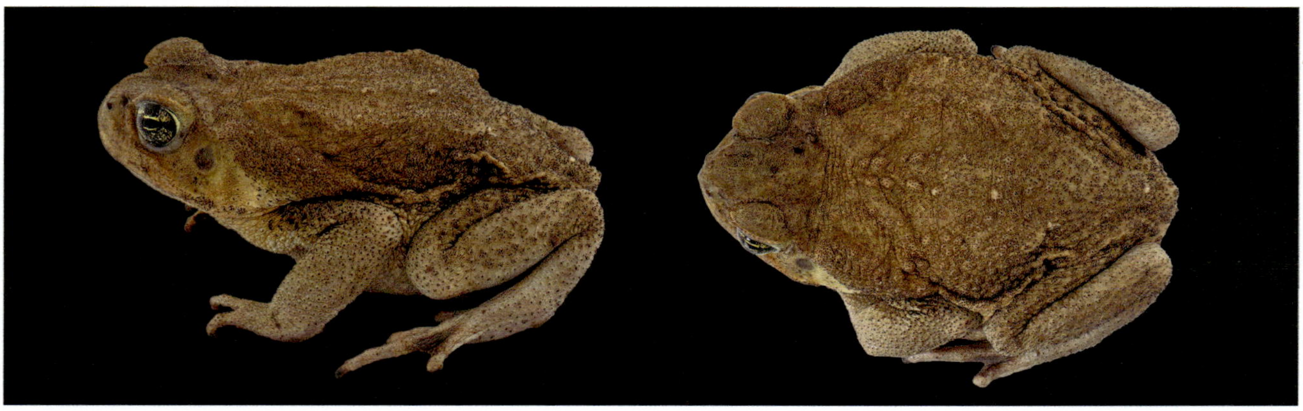

FIGURE 1.3 *Rhinella horribilis* from Bosque Protector Siete Cascadas, Provincia Esmeraldas, 116 mm SVL. SGS.

by Duellman and Trueb (1986: 103): "Upon being seized, these large frogs sometimes emit a loud scream reminiscent of that given by a cat in distress". Notably, *L. peritoaktites* is recognized for its consumption as an edible frog by the Tsáchila culture in modern-day coastal Ecuador. In the Mochica culture, the frog depicted in the artifact known as the botanical frog (McClelland, 2011) is a species of *Leptodactylus* (of the *pentadactylus* complex) as well—although the presence of a species of *Leptodactylus* in northwestern Peru has not yet been documented.

FIGURE 1.4 Anuran (cf. *Leptodactylus peritoaktites*) represented in the Chorrera culture (1000 BCE – 100 CE). Vase, MAPCA (MRT 0273), 154 mm length × 155 mm width × 199 mm height. Museo de Arte Precolombino Casa del Alabado. SGS and Jazmín Buitrón (top left).

FIGURE 1.5 Left: *Leptodactylus peritoaktites*. QCAZ 35265 from Buena Fé, Recinto El Descanso, Dique 2 del embalse del Río Baba, Provincia de los Ríos, Ecuador. LAC. Right: *L. rhodomerus*. CJ 8296, 123 mm SVL, from Reserva Biológica Bilsa, Provincia Esmeraldas, Ecuador. DAL.

Regional Development Period (500 BCE–500 CE)

Subsequent cultures that followed the Chorrera continued the representations of animals with their own styles within the regions of Ecuador where this culture lived. Toads and frogs remained present in the iconographic repertoire during the Regional Development Period across the three continental regions.

The La Tolita–Tumaco culture (600 BCE–300 CE), located on the northern part of the modern Provincia Esmeraldas, Ecuador, and the southern part of Departamento Nariño, continued with the Chorrera tradition of iconographic representations while developing its unique style, which would later be consolidated during the Classic Period (200 BCE–75 CE approx.), departing from the local style known as Tachina (Valdez, 1987; Ugalde, 2009). Toads and frogs were represented, continuing the naturalistic and realistic tradition that had its origins in the preceding period. However, representations of anthropomorphized creatures were relatively uncommon. Ugalde (2009) did not report anthropomorphized representation of toads during her iconographic research on Tolita figurines. However, Gutiérrez Usillos (2002) reported one anthropomorphized representation of a toad. Among Anura amphibians, The Tolita people represented some anurans with a smooth skin (Figure 1.6) and others with a warty and glandular skin (see Figure 1.8).

FIGURE 1.6 Whistle/applique with a representation of a toad (probably *Rhaebo haematiticus*). Tolita culture (600 BCE – 300 CE). MAPCA (MRT 1750), 93 mm length × 113 mm width × 73 mm height. Jazmín Buitrón (top, middle) and SGS (bottom).

We attribute the toad depicted in Figure 1.7 to *Rhaebo haematiticus* (Figure 1.7), primarily due to its skin texture, size, and shape. This species is commonly encountered in the western lowlands of Ecuador. According to Ugalde (2009) and Gutiérrez Usillos (2002), the identifications of the toads shown in Figure 1.8 are supported by the presence of protuberances in their dorsum, which likely replicate the characteristic glands found (parotoid glands, warts) in the skin of bufonids. We agree with their assignment and concur that they might belong to *Rhinella horribilis* (then named as *Bufo*) (see Figure 1.3).

The Jama Coaque culture (350 BCE–1532 CE) was located in the northern part of modern Provincia Manabí, along the Jama and Coaque rivers, as well as the southern part of Provincia Esmeraldas. This culture is renowned for its ceramics. Similar to La Tolita, the Jama Coaque culture evolved from a local tradition of Chorrera ceramics known as Tabuchila, and this tradition persisted until the arrival of the Europeans (Gutiérrez Usillos, 2002, 2011; Zeidler and Pearsall, 1994).

Frogs and toads were a notable feature of the ceramic art of the Jama Coaque culture, primarily appearing as appliqués on various types of ceramic vessels (Figures 1.9, 1.10, 1.11, 1.12) but also on pipes (Figures 1.13, 1.14) and lime pots (Figure 1.15).

FIGURE 1.7 *Rhaebo haematiticus*. CJ 12949, 78 mm SVL, from 5 km northwest of La Florida, Provincia Santo Domingo de los Tsáchilas, Ecuador. SGS.

FIGURE 1.8 Toad (probably *Rhinella horribilis*) representations. Tolita culture (600 BCE–300 CE). Left: MuNA LT-19–50–73, 95 mm length × 85 mm width × 26 mm height. Right: MuNA LT-36–19–70, 130 mm length × 90 mm width × 72 mm height. SGS.

The appliqué in Figure 1.9 seems to feature a representation of a froglet completing metamorphosis in Gosner Stage 45, which is characterized by a tail reduced to stub.

The larger size of the frogs in the appliqués of Figures 1.10, 1.11, and 1.12 is 25 mm SVL. Their body and snout shape resemble *Leptodactylus* spp. (see Figure 1.14), which occur in the Esmeraldas and Manabí provinces. Although the size of the adult animals is larger than the size of the appliqués, we tentatively identify the appliqués as *L. ventrimaculatus* (63 mm maximum SVL) or *L. labrosus* (71 mm maximum SVL) (see Figure 1.14). As for the appliqués in Figure 1.39, they may represent a hylid frog, suggested by the rounded snout; however, identifying the species represented is challenging due to the prevalence of this feature across various frog species. These amphibians were placed in association with plants and people, and amphibian figures

FIGURE 1.9 Lime pot featuring a representation of a froglet completing metamorphosis. Jama Coaque culture (350 BCE–1532 CE). MAPCA (MRT 3725), 60 mm height × 71 mm width. SGS.

FIGURE 1.10 Lime pot or container with anurans (probably *Leptodactylus ventrimaculatus* or *L. labrosus*) as appliqués. Jama Coaque culture (350 BCE–1530 CE). MAPCA (MRT 1878), 113 mm width × 85 mm height. Jazmín Buitrón (left) and SGS (right).

FIGURE 1.11 Bottle or container with anurans (*Leptodactylus ventrimaculatus* or *L. labrosus*?) and plants (*Arachis* sp.) as appliqués. Jama Coaque culture (350 BCE–1530 CE). MAPCA (MRT 1873), 128 mm height × 141 mm width. SGS and Jazmín Buitrón (top right).

especially are associated with rain shamans (see "Amphibians and Their Relationship with Rain Rituals") (see Figure 1.39) and elements tied to agricultural cycles.

The zoomorphic pipe in Figure 1.13 features a representation of an anuran. We attribute this to either *Leptodactylus labrosus* or *Lithobates* sp., primarily due to its shape, skin texture, and size. The protuberance in the snout of this pipe does not fit any species in the region. However, there is one species, *L. labrosus* (see Figure 1.14), that has a spatulate snout, which is not so prominent. This species are commonly encountered in Manabí and Esmeraldas provinces.

The lime pot in Figure 1.15 seems to represent *Rhinella horribilis*, as used in previous cultures. Gutiérrez Usillos (2011) suggests that the poison of these animals, as mentioned previously, may have been consumed for its hallucinogenic properties.

The Bahía culture (500 BCE–650 CE) was located in the southern part of the modern Provincia Manabí, centered around Bahía de Caráquez and Isla de La Plata. Similar to the two cultures described earlier, the Bahía culture evolved from the Chorrera, and the Chorrera style is also known as Engoroy. The Bahía culture continued depicting amphibians, as did the La Tolita and Jama Coaque cultures (Gutiérrez Usillos, 2002). A zoomorphic vase (Figure 1.16) features a representation of a frog, likely *Ceratophrys stolzmanni* (Figure 1.17). Both the representation and the actual frog align in terms of size and shape. The Bahía culture's artistic expressions included a unique representation known as the "Bahía Monster". According to Gutiérrez Usillos (2002), this figure incorporated characteristics of toads and frogs, as well as reptiles and felines, creating a creature that held significant associations with rain rituals.

FIGURE 1.12 Pedestal cup decorated with figurines of rain shamans and frogs (probably *Leptodactylus ventrimaculatus* or *L. labrosus*) as appliqués. Jama Coaque culture (350 BCE–1530 CE). MAPCA (MRT 1869), 165 mm height × 205 mm width. Jazmín Buitrón and SGS (bottom left).

The antropozoomorphic figure shown in Figure 1.18 also seems to have some frog elements such as the rounded digital discs at the fingertips of the hands, a feature present in arboreal tree frogs.

The Guangala culture (500 BCE–800 CE) was located south of the Bahía cultural area, in the modern-day provinces of Manabí (southern part), Santa Elena, and Guayas, occupying the coastline and the Chongón-Colonche hills environment (Gutiérrez Usillos, 2002; Uribe Taborda, 2016). Much like the coastal cultures mentioned previously, the Guangala culture continued the Chorrera tradition of employing iridescent paint on smoked black surfaces, which later evolved into the creation of polychrome fine wares. The culture also continued the practice of crafting anthropomorphic and zoomorphic representations (Gutiérrez Usillos, 2002; Uribe Taborda, 2016). Toads and frogs were depicted in various objects, either as complete representations tutto tondo (e.g., Figure 1.18) or as appliqués incorporated into the overall composition (e.g., Figure 1.19) (Uribe Taborda, 2016). Following Gutiérrez Usillos' proposal, Uribe Taborda (2016) suggested identifying the Guangala depiction of anurans as *Bufo* spp. We concur with their assignment that they might belong to *Rhinella horribilis* (then named as *Bufo*), as in the Tolita examples (Ugalde, 2009). However, there might be a variation in iconographic styles, resulting in a less naturalistic representation of the round glands in the skin. The significance of these representations likely pertained to the potential use of venom for its hallucinogenic properties and its connection with rituals related to rain and agriculture (see "Amphibians and Their Relationship with Rain Rituals") (Uribe Taborda, 2016).

The two lime pots shown in Figure 1.19 appear to depict two species of anurans, which presumably were common in the geographic region of this culture. The one at the top may represent *Ceratophrys stolzmanni* (see Figure 1.17)

FIGURE 1.13 Zoomorphic pipe featuring a representation of an anuran (probably *Lithobates* sp.). Jama Coaque culture (350 BCE–1532 CE). MAPCA (MRT 1946), 63 mm height × 108 mm length × 59 mm width. Jazmín Buitrón (top) and SGS (middle and bottom).

FIGURE 1.14 Left: *Leptodactylus ventrimaculatus*. CJ 7759, 43 mm SVL, from Reserva Biológica Bilsa, Provincia Esmeraldas. Right: *L. labrosus*. CJ 9985, 59 mm SVL, from Puyango, Provincia Loja. DAL.

(currently restricted to protected areas), while the one at the bottom may represent *Rhinella horribilis* (see Figure 1.3) (currently still common). Our identification is based on size, the ornamentation on the dorsum that we interpret to represent glandular areas in the skin, and also the shape of the snout in dorsal view, which differ between the two species.

The frog appliqué on the bowl depicted in Figure 1.20 is more challenging to interpret because it could be either a stylized representation of *Rhinella horribilis* (based on the small points on the dorsum that resemble skin tubercles) or a more slender species of Anura, of which there are many of about that size.

FIGURE 1.15 Lime pot with a representation of an anuran (probably *Rhinella horribilis*). Jama Coaque culture (350 BCE–1530 CE). MAPCA (MRT 1891), 67 mm length × 52 mm width × 43 mm height. SGS.

In the northern Andean highlands, the La Chimba culture (800 BCE–200 CE), located in what is now Provincia Imbabura, began to depict various animals in the Late Formative Period (Athens, 1978; Stahl and Athens, 2001; Olsen-Bruhns, 2003; Zeidler, 2008). At the Las Orquídeas site, zoomorphic representations began to emerge since Phase II (750–650 cal. BCE), likely influenced by contact with Chorreran styles from the northern coast (Dyrdahl, 2017; Dyrdahl and Montalvo-Puente, 2022; Montalvo-Puente, 2016). Amphibians were among the creatures featured in their iconographic repertoire, typically as appliqués adorning the rims of ceramic vessels. However, the schematic nature of these representations makes it challenging to provide identifications at any taxonomic level (Figure 1.21) (Dyrdahl and Montalvo-Puente, 2022; Montalvo-Puente, 2016). This artistic tradition continued over time, extending into the Caranqui Culture during the Integration Period (Athens, 2003; Montalvo-Puente, 2011, 2016).

On the eastern slopes of the Cordillera Oriental of the Andes, in the Amazonian region, the Cosanga-Píllaro culture (1600 BCE–1532 CE) flourished, occupying the modern provinces of Napo, Pastaza, Morona Santiago, Pichincha, Cotopaxi, Tungurahua, and Chimborazo (Ontaneda, 2010). Cosanga ceramics are notable for their thin walls, and throughout their history, they featured anthropomorphic and zoomorphic representations (Bray, 1995; Gutiérrez Usillos, 2002; Ontaneda, 2010). Amphibians were also depicted as appliqués, primarily adorning the rims of vessels, a practice similar to that of La Chimba. An earliest example of this kind of representation was found in the context of Las Orquídeas Phase IV (500–400 calibrated BCE) context (Figure 1.22) (Dyrdahl and Montalvo-Puente, 2022). This tradition of representation, much like that of the Caranqui culture, persisted into the Integration Period (Athens, 1980, 2003; Montalvo-Puente, 2011, 2016).

Integration Period (500–1500 CE)

Throughout the integration period, various traditions of representation persisted and underwent further development in the different geographical areas of modern-day Ecuador. Amphibians continued to be depicted, primarily as appliqués adorning a wide range of ceramic vessels and incised spindle whorls.

The Manteño Huancavilca culture (600–1532 CE) was located in the areas previously occupied by the Bahía and Guangala societies, spanning the provinces of Manabí, Santa Elena, and Guayas, including Isla Puná (Gutiérrez Usillos, 2002; Lunniss, 2018; McEwan and Delgado-Espinoza, 2008; Ontaneda, 2010). Although limited information is available, this culture continued the tradition of representing amphibians from the preceding period. However, they were mostly confined to appliqués and engravings (incised patterns) on spindle whorls (Figures 1.23, 1.24, 1.25, 1.26, 1.27, 1.28, 1.29, 1.30) (Gutiérrez Usillos, 2002; Wilbert, 1974). Many of the amphibians represented in miniature art such as the Ecuadorian spindle whorls from the Manteño Hunacavilca culture generally are stylized and thus difficult to identify, except for some depicted in Figures 1.26, and 1.27, which are anurans; the one at the top later in Figure 1.27 represents an amplectant pair of anurans. Wilbert (1974)

FIGURE 1.16 Zoomorphic vase featuring a representation of a frog (probably *Ceratophrys stolzmanni*). Bahía culture (500 BCE–650 CE). MuNa B 2–19–72, 120 mm length × 115 mm width × 114 mm height. SGS.

suggested that the rows of white dots would represent warts, in which case the creature represented would be rather a warty-skinned amphibian, or they could also represent the characteristic secretions of certain poisonous frogs important in hunting or magical rituals (presumably *Rhinella horribilis*).

The Milagro Quevedo culture (400–1532 BCE) inhabited the inland regions of the coast, specifically the Río Guayas basin, within the modern provinces of Guayas, Los Ríos, and Santo Domingo de los Tsáchilas (Stemper, 1993; Delgado-Espinoza, 2005, 2006; McEwan and Delgado-Espinoza, 2008; Montalvo-Puente *et al.*, 2023).

FIGURE 1.17 *Ceratophrys stolzmanni* from Reserva Ecológica Arenillas, Provincia El Oro. LAC.

Amphibians were prominently featured in the art of this culture, often appearing as appliqués on vessels commonly referred to as "cocina de brujo" (Figure 1.31). These representations can stand alone or in combination with other zoomorphic or anthropomorphic figures (Gutiérrez Usillos, 2002; Salguero-Morán and Ortiz-Quiroz, 2022). Similar to earlier cultures, these depictions likely held significance in connection with propitiatory rain rituals (see "Amphibians and Their Relationship with Rain Rituals") (Salguero-Morán and Ortiz-Quiroz, 2022).

The Pasto culture (700–1500 BCE) thrived in the Andean highlands, in the present-day Provincia Carchi in northern Ecuador, and in Departamento Nariño in southern Colombia (Ontaneda, 2010). The Pastos were renowned for their distinctive polychrome ceramics and metalwork (Granda-Paz, 2010; Lleras-Perez, 2015), and they left a significant artistic legacy. Among their artistic expressions, amphibian representations are discernible in the intricate metalwork pieces (Ontaneda, 2007; Quinatoa Cotacachi, 2013). The figures in the gold discs (Figure 1.32), although stylized, may be loosely interpreted as frogs of the genus *Atelopus*. This interpretation is supported by the pointed snout, a characteristic feature of this genus, which is the only one in the region displaying such a distinctive snout shape. Additionally, these frogs are diurnal and conspicuous due to their vibrant colors and abundance. In historical times, two species were common in the region where the Pasto culture developed: *Atelopus pastuso* (Figure 1.33) and *A. cocha*. The latter only occurred in Departamento Nariño, Colombia.

Continuing south, the Caranqui culture (600–1500 BCE) that also inhabited the northern Andean highlands, specifically in the in the modern-day provinces of Imbabura and Pichincha, maintained the tradition of representations of animals as appliqués, which had been initiated by La Chimba (Bray, 2008; Athens, 1980, 2003; Montalvo-Puente, 2011, 2016). Much like the situation in the La Chimba artifacts, the issue of identifying specific taxa, other than anurans, continues to pose challenges in representations from the early Integration Period (Figure 1.34) (Gutiérrez Usillos, 2002).

The Puruhá (300–1500 BCE) culture occupied the central Ecuadorian Andean highlands encompassing the modern provinces of Cotopaxi, Tungurahua, Bolívar, and Chimborazo (Bray, 2008; Ontaneda, 2010). Although there is a lack of comprehensive archaeological information for this culture during the Late Formative and Regional Development Periods, it is likely that Puruhá ceramics continued the tradition of representing animals, which originated in earlier times (Gutiérrez Usillos, 2002). Amphibians were portrayed as appliqués on ceramic vessels (Figure 1.35). Ugalde (2009) and Gutiérrez Usillos (2002) suggested that the appliqués were *Bufo* spp. There are no species of *Rhinella* (then named *Bufo* spp.) at the altitudes of the Andean region where the Puruhaes lived. Our examination of the pot shown in Figure 1.35 does not fit such an interpretation because the frog represented is very similar in size and shape to the gravid females of marsupial frogs (*Gastrotheca riobambae* and *G. psesutes*) (Figure 1.36). The round protuberances in the dorsum of the frogs fit the shape of the gravid brooding pouch of a marsupial frog well. Both sympatric species are conspicuous inhabitants of the region where the Puruhá culture developed.

To close the review, pre-Columbian populations in various regions of the Amazonian area in present-day Ecuador created representations of amphibians through engraved stones, known as petroglyphs (Quinatoa Cotacachi, 2013; Porras, 1985; Ugalde, 2012; Mosquera-Perugachi, 2014) (see Figure 1.38). These depictions are distinguished by the use of lines made in their artistic representations. Representations of toads and frogs were documented by Pedro Porras (1961, 1985) in the Misahuallí (see Figure 1.37) area, probably linked to the Napo culture (1200–1500 CE). Petroglyphs from Cathazho, Provincia Morona Santiago, were documented by Mosquera-Perugachi (2014).

Porras (1985) describes petroglyphs with frogs as part of the rock art of the Misahuallí Valley (Cotundo, Archidona, Tena and surroundings, Provincia Napo) (Figures 1.37, 1.38). The importance of depictions carved in rock or petroglyphs is a matter of debate. It has been suggested that they were a hobby, but based on the fact that their execution requires considerable time, energy, and artistic skill, in addition to the fact that many petroglyphs have been found far from towns and often in inaccessible areas, they may have had a deeper meaning (Porras, 1985; and publications cited therein). He suggests that they could be the sometimes stylized and almost always naturalistic representation of the Man-Frog, which is known as the pan-Amazonian mythical Man-Animal (a mythological being between man and animal). They were probably carved between 300 BC and 300 CE, as they correspond to the closed contour style, characteristic of the Regional Development Period. According to the same author, the humanization of the raniform features is apparent, as well as the conversion into monkeys and

FIGURE 1.18 Anthropozoomorphic seated figurine with presumed arboreal anuran-like hands with digital discs at the fingertips. Bahía culture (500 BCE–650 CE). MuNa JC. 4–36–75, 120 mm depth × 165 mm width × 190 mm height. SGS.

jaguars—as if the whole is part of a single myth. Frogs are represented in various forms, usually starting from a rhombus or rhomboid shape that represents the body, crowned by a circle or spiral, the head and sides, and the upper and lower limbs, which almost always have three fingers in a trident manner. Sometimes the body is more ovoid or the contour lines remain unconnected in the lower and/or upper section, leaving the head isolated. Some have a more or less long tail,

FIGURE 1.19 Lime pots with a representation of anurans. Guangala culture (500 BCE–800 CE). Top: probably *Ceratophrys stolzmanni*, MAPCA (MRT 1618), 57 mm length × 47 mm width × 33 mm height. Bottom: probably *Rhinella horribilis*, MAPCA (MRT 1619), 48 mm length × 41 mm width × 28 mm height. SGS.

FIGURE 1.20 Small bowl decorated with an unidentified anuran as an appliqué. Guangala culture (500 BCE–800 CE). MAPCA (MRT 1616), 67 mm diameter × 39 mm height. SGS.

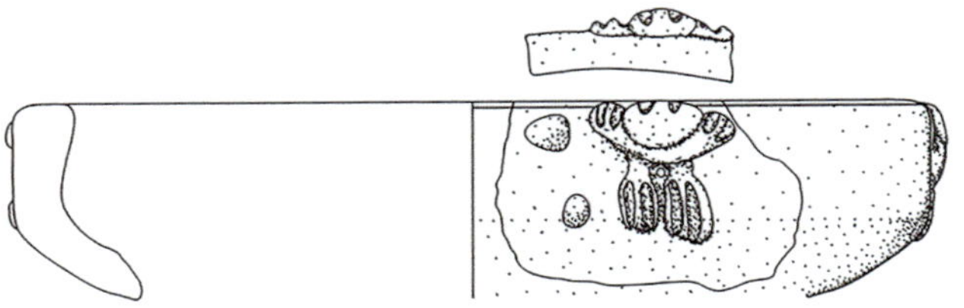

FIGURE 1.21 Plate decorated with probable amphibians as appliqués. La Chimba culture (800 BCE–200 CE). Recovered during the 2014 excavation at Las Orquídeas (LO 270002), Provincia Imbabura. Drawn by Carlos E. Montalvo-Puente.

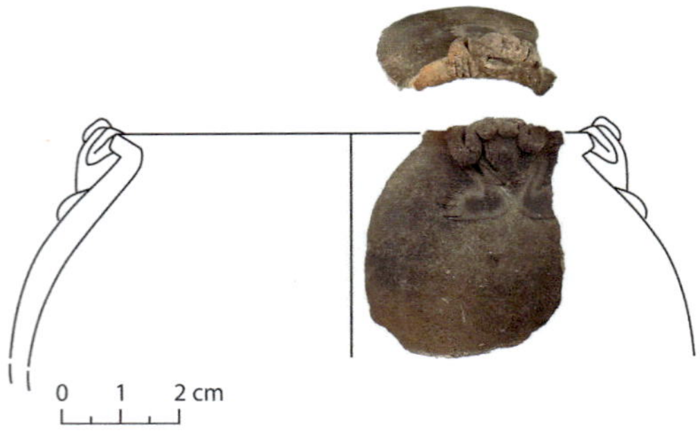

FIGURE 1.22 Pot decorated with anurans as appliqués. Cosanga culture (1600 BCE–1532 CE). Recovered during the 2018 excavation at Las Orquídeas (LO 351001), Provincia Imbabura. Drawn by Carlos E. Montalvo-Puente.

FIGURE 1.23 Miniature art depicting stylized frogs in Ecuadorian spindle whorls (generally 11–15 mm maximum height) from the Manteño Huancavilca culture (600 CE to 1532 CE). Taken from Wilbert (1974) and Shaffer (1985).

FIGURE 1.24 Miniature art depicting a stylized frog as an engraving in an Ecuadorian spindle whorl from the Manteño Huancavilca culture (600 CE to 1532 CE). MAPCA (MRT 4190.1), 28 mm diameter × 22 mm height. SGS.

thus adopting the monkey figure. The head is almost always generally circular, but sometimes within the circle there is a representation of the eyes, mouth, and nose, sometimes of a single eye in the center. Some have something like an aureole on the head, and on the body the indication of a navel or perhaps a vulva (see Figure 1.37). Porras (1985) provides graphic sequences of the possible evolution of the glyphs until they become raniform, either from an angle (circumflex accent), from a circle, or from a relaxed M (bijugal symbol). Among the petroglyphs mentioned, an archaeological piece of enormous importance is the Sapo Rumi (see Figure 1.37), discovered by Julio Rosales, published in 1965 in the Ecuadorian Diario El Día, and later described by Porras (1985). The Sapo Rumi (see Figure 1.37) is in the vicinity of Tena, Provincia Napo, at the confluence of the Yanayaku and Achiyaku rivers. It is a huge rock on the left bank of the Río Achiyaku (of fine-grained granodiorite) 7.5 meters long, 4.5 meters wide, and 1.8 meters high in the shape of a tortoise shell. The petroglyph is estimated to be between 5000 and 8000 years old. On its east side it has seven asexual figures engraved in the shape of frogs: six facing east, the largest with eyes and a mouth, and all have limbs with three "trident" fingers. According to Uzendoski (2009), the stone holds a deep mythology. It is one of the places where the twins (Cuillur and Dusiru) confronted the primordial jaguars. However, the stone represents not only the past but also the present and the future. It holds "kawsay" or vital energy that nourishes the people, the river, and the landscape. Beneath the stone, there is a city of spirits, "yaku warmiguna" (women of the water), and "yaku runaguna" (men of the water) who provide fish and bring dreams to the people. The oral tradition—narrated in January 2024 to Juan P. Reyes-Puig by Federico Kalapucha, Yachak of the Sapo Rumi community—recounts that Wasila and Ishinga are the names of the female frog and one of the males who had incestuous relations. The larger figure represents the female, and others of smaller size represent the males. As a result, they were expelled and imprisoned within the earth alongside a series of other animals, transforming into the hills currently known by those names on the north bank of the Río Hatunyaku. They would remain trapped there until the end of times, when the great final flood known as Yllu pacha arrives. Only then would they be released from the rock, along with a whole series of wild animals. From time immemorial, indigenous peoples have demonstrated an awareness of key biological concepts, such as endogamy, as illustrated in this story.

Amphibians and Their Relationship with Rain Rituals

The primary connections drawn from the existing studies regarding iconography link amphibian representations to rain and water (Gutiérrez Usillos, 2002, 2011). This association is significant for agrarian societies, especially when seasonal rains become a critical factor following the transition from nomadic hunter-gatherer lifestyles to sedentary, agriculture-dependent communities. In Valdivia societies, the emergence of a cult around *Spondylus* and *Strombus* shells reflects the observation of the behavior of the *Spondylus* shell, serving as a bioindicator for predicting the presence or absence of seasonal rains (Marcos, 2005).

It is likely that during the Late Formative Period, particularly with the Chorrera culture, the symbolism and rituals associated with rain and amphibians were integrated. By the early Regional Development Period, and in the subsequent cultures that evolved from it, this relationship became more firmly stablished, as evident in various iconographic representations, with the Jama Coaque being a notable example. In the Jama Coaque culture, numerous objects underscore the connection between anthropomorphic figures adorned with elaborate shell ornaments, often interpreted as rain shamans, and amphibians. These are frequently depicted in conjunction with plants or seeds, highlighting the significance of this relationship (Gutiérrez Usillos, 2011). For

FIGURE 1.25 Miniature art depicting stylized frogs as engravings in Ecuadorian spindle whorls from the Manteño Huancavilca culture (600 CE to 1532 CE). Alexander Hirtz collection. SGS.

instance, the object in Figure 1.39 is a ceremonial table, a plate supported by seven anthropomorphic figures dressed like rain shamans. To convey the idea of abundance, the plate is adorned with a series of frogs—symbols of rain and fertility—and with grains, miraculous seeds that reproduce a thousand times the small grain originally planted (Gutiérrez Usillos, 2011).

Amphibians also have associations with life cycles and fertility. Like the *Spondylus*, they were used as bioindicators to predict the onset of rain. The connection between fertility and water also is closely tied to the development of tadpoles in water during the initial stages, and the warning calls of anurans were believed to be a call for rain (Gutiérrez Usillos, 2002). Representations of anurans in amplexus

FIGURE 1.26 Miniature art depicting stylized frogs as engravings in Ecuadorian spindle whorls from the Manteño Huancavilca culture (600 CE to 1532 CE). Alexander Hirtz collection. SGS.

(their nuptial embrace) are present in Guangala and Jama Coaque iconography (Figure 1.40, see Figure 1.42), including the depiction of their eggs. In these objects, amplectant pairs of *Rhinella horribilis* (Figure 1.41) and *Ceratrophrys stolzamnni* (Figure 1.43) seem to be depicted. This marks the importance of anuran mating in relation to water, where the eggs of many species were typically deposited. In the highlands, the representation of amphibians likely follows similar beliefs. Gutiérrez Usillos (2002) suggests that the representation of toads/frogs or tadpoles is linked to rain

FIGURE 1.27 Miniature art depicting stylized frogs as engravings in Ecuadorian spindle whorls from the Manteño Huancavilca culture (600 CE to 1532 CE). The top spindle whorl represents an amplectant pair of anurans Alexander Hirtz collection. SGS.

FIGURE 1.28 Miniature art depicting stylized frogs as painted engravings in Ecuadorian spindle whorls from the Manteño Huancavilca culture (600 CE to 1532 CE). Alexander Hirtz collection. SGS.

FIGURE 1.29 Miniature art depicting stylized frogs as painted engravings in Ecuadorian spindle whorls from the Manteño Huancavilca culture (600 CE to 1532 CE). Alexander Hirtz collection. SGS.

FIGURE 1.30 Miniature art depicting stylized frogs as painted engravings in Ecuadorian spindle whorls from the Manteño Huancavilca culture (600 CE to 1532 CE). Alexander Hirtz collection. SGS.

FIGURE 1.31 Cocina de brujo pot with representations of anurans, snakes, birds, and anthropomorphic figures as appliqués. Milagro Quevedo culture (600 BCE–300 CE). Collection of Banco Central del Ecuador, Guayaquil. Redrawn by Carlos E. Montalvo-Puente from Holm (1981).

and water provision, and these symbols often adorned vessels associated with liquids.

Wilbert (1974) and Gutiérrez Usillos (2002) propose that the mating calls of anurans is connected to invocations of rain in Ecuadorian pre-Columbian cultures. Both authors draw a parallel to the relationship between the Mayan Chaak and Aztecan Tlaloc, the Mesoamerican rain divinities, and amphibians.

Finally, this first attempt to compile information and explore the amphibian iconography in Ecuador reveals exciting findings. However, it is evident that there are likely hundreds of objects yet to be studied in museum collections and archaeological sites. In-depth research into zoomorphic iconography and identification holds significant potential and could be further developed through interdisciplinary cooperation. Collaborative efforts between archaeologists, biologists, and ethnologists are necessary to advance the study of the role, identification, and importance of the diversity of animals in both the pre-Hispanic and contemporary societies.

CRAFTS

Ecuadorian artisans have incorporated frogs, toads, and tadpoles into objects and jewelry. Crafts representing frogs from the Kichwa in Alto Napo are depicted in Figure 1.44; they are of unknown age. Contemporary pieces are made of tagua (Figures 1.45, 1.46). Tagua—also known as ivory plant, vegetable ivory, corozo, and mococha—is the complex cellulose kernel of the seed of the palm *Phytelephas aequatorialis* (Palmae), which is endemic to the subtropical and tropical forests of the Costa del Ecuador region (0–1500 masl) and is in the category of near threatened according to the IUCN criteria (Montúfar and Pitman,

FIGURE 1.32 Gold discs with embossed anurans (*Atelopus pastuso*, *A. cocha*). Pasto culture (700–1500 CE). Left: MuNa 1–1-57, 79 mm diameter × 7 mm thickness. Right: MuNa 2–1-57, 80 mm diameter × 7 mm thickness. SGS.

FIGURE 1.33 *Atelopus pastuso* MNHG 2258.68, Tulcán, Provincia Carchi. LAC.

2003). The tagua seed is odorless, white, hard, and dense; its texture is smooth; and it acquires a shine when polished. Tagua began to be commercialized in the mid-19th century and currently constitutes an important item of commerce in Ecuador. Most of the anurans carved in tagua crafts usually represent foreign species, although there are some that seem to represent Ecuadorian species.

MYTHOLOGY, STORIES, AND LEGENDS

Wilbert (1974) asserts that the mythology and symbolism surrounding Ecuadorian frogs and toads are intertwined with themes of transformation, metamorphosis, rain, fertility, the moon, and mother earth. These symbolic meanings

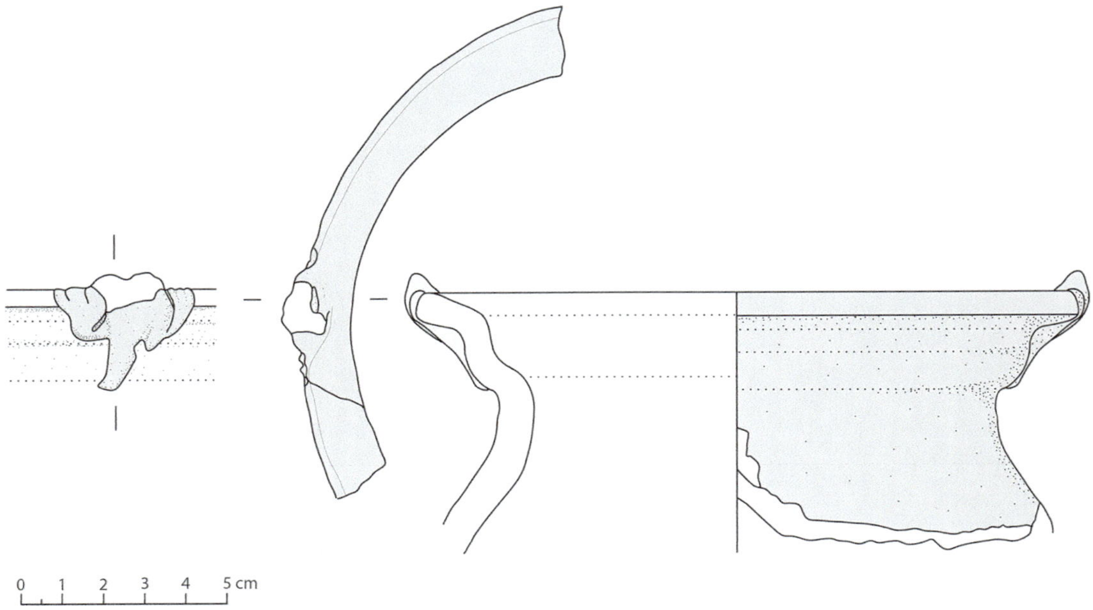

FIGURE 1.34 Vessel decorated with undetermined anurans as appliqués. Caranqui culture (600–1500 CE). Recovered during 2009 excavations at Tola Huataviro (PAH 45N/70E/N7/108/X), Provincia Imbabura. Drawn by Carlos E. Montalvo-Puente.

FIGURE 1.35 Pot decorated with anurans (probably *Gastrotheca riobambae*, *G. pseustes*) as appliqués. Puruhá culture (300–1500 CE). MAPCA (MRT 2670), 196 mm height × 201 mm diameter. Frog in the appliqué 75 mm SVL. SGS and Jazmín Buitrón (top right).

FIGURE 1.36 Gravid females of *Gastrotheca riobambae* (left) from Quito, Provincia Pichincha, and *G. pseustes* (right) from Guaranda, Provincia Bolívar. Approximately 65 mm SVL. LAC.

FIGURE 1.37 Sapo Rumi rock petroglyphs featuring stylized representations of Anura. Napo culture (1200–1500 CE). Top: Federico Kalapucha, the spiritual guide (Yachak) of the local community close to the rock in Pano, Tena, Provincia Napo. Middle: Outline of Sapo Rumi. Bottom: outline of Stone VIII, side Bp., Chuchiyacu, Archidona, Provincia Napo. Outlines redrawn by Carlos E. Montalvo-Puente from Porras (1961).

FIGURE 1.38 Petroglyphs depicting frogs in Amazonian Ecuador. Top: a petroglyph from the Valle Sagrado de Cotundo, near Archidona, Provincia Napo. Bottom: a petroglyph between Limón and San Carlos de Conchay, Provincia Morona Santiago. LAC.

FIGURE 1.39 Ritual table decorated with figurines of rain shamans, frogs/toads (probably hylid tree frogs), and seeds as appliqués. Jama Coaque culture (350 BCE–1530 CE). MAPCA (MRT 0052), 235 mm height × 360 mm width. Jazmín Buitrón.

can vary depending on their context and may even carry ambivalent interpretations. The narratives of indigenous peoples shed light on the everyday realities of communities where the invisible forces of the universe remain an integral part of their lives. According to Paymal and Sosa (1993), in general, in pre-Columbian civilizations, frogs and toads (see their definition in "Indigenous and Vernacular Names") were associated with rain and therefore with germination and the genesis of life. Frogs and toads, due to their undeniable relationship with water, indicated the beginning of the sowing season and the hope for successful harvests. In many cultures, particularly Andean groups, toads and frogs hold a significant place as creatures that influence daily life patterns, with their songs beckoning the rain and their birth

FIGURE 1.40 Lime pot with a representation of an amplectant pair of anurans (probably *Rhinella horribilis*). Guangala culture (500 BCE–800 CE). MAPCA (MRT 4236), 57 mm length × 37 mm width × 28 mm height. Jazmín Buitrón (top, middle) and SGS (bottom).

FIGURE 1.41 Amplectant pair of *Rhinella horribilis* from Reserva Otokiki, Provincia Esmeraldas. Elicio E. Tapia.

from water to inhabit the earth alongside us. The unusual reproductive activities of some species made them natural symbols of fertility.

A short outreach essay on the ancestral vision and some stories in relation to amphibians is provided by Lombeida (2005), which is based on the publications of Botasso (1977), Pellizzaro (1996), and Parra R. (1997), who provide information on the Shuar, Siona, and Kofán ethnic groups. In addition, more than 20 stories and legends from other Ecuadorian ethnic groups have been published, such as those from the Guaraunos, Waoranis, Kichwas de la Sierra, Kichwas Amazónicos, and Salasacas and from Afro-Americans, mestizos, and whites of the Sierra, Costa, and Oriente of Ecuador (Peñaherrera de Costales and Samaniego, 1959; Yuquilema, 1978; de Carvalho-Neto, 1994, 1988; Santos Ortiz de Villalba, 1994; Alarco de Zadra, 1990; Costales *et al.*, 1996; Anhalzer, 2002; Aveiga del Pino, 2003). Herein we provide selected legends and myths from them.

Quitu-Cara myth of the Catequil. Costales *et al.* (1996) describe in detail the myth of the Catequil or the Moon Tiger. In this myth, three species of highland frogs (above 3600 m), Hambatu, Buru, and Kayla, have a broad symbolism and presumably belong to *Atelopus* spp., *Gastrotheca pseustes*, and *Telmatobius niger*, respectively, although the latter does not occur in Provincia Imbabura, but it was abundant from the Bolívar and Chimborazo to Azuay provinces. Jaime Costales Peñaherrera provides an anthropological-psychological interpretation of this myth, in which frogs and toads are some of the animals of darkness in the phase (or station of a four-phase cycle) of the energy that divides and transforms. The myth highlights mythical-socio-astronomical topics and is a communal, warrior-priestly ritual that takes place at midnight, when the moon looks like a jaguar's claw. This ritual or initiation ceremony aims to choose a male (Cari), who has endured the cruel incidents of the great ritual test (cold waters, the onslaught of animals from the darkness, loneliness, darkness). As the day lights up, he will be the driver, teacher, priest, and defender of his people, his land, and his community, a hero who has courageously reached the world of wisdom, light, science, and fortune. The chosen love (Tenu)—for the clarity of his mind and the strength of his heart—will be the one that will find the great-grandparents, mythological beings represented by animals (for example, three species of frogs: Potato or Hambatu, Burú, Cotán or Caila) and other ancestral symbols that populate the

FIGURE 1.42 Seal featuring a representation of an amplectant pair of frogs (probably *Ceratophrys stolzmanni*). Note the presence of eggs and the smaller size of the male frog above. Jama Coaque culture (350 BCE–1530 CE). MAPCA (MRT 1619), 60 mm length × 44 mm width × 46 mm height. Jazmín Buitrón (top) and SGS (middle and bottom).

FIGURE 1.43 Amplectant pair of *Ceratophrys stolzmanni* from Reserva Ecológica Arenillas, Provincia El Oro. LAC.

flowered farm, the indigenous homeland, or land of smiling corn. The "Catequillado" ceremony is carried out in places usually higher than 3600 masl (for example, at 4100 m elevation in the Provincia Chimborazo or in Cochasquí, Provincia Imbabura, etc.) where the Catequil pool (= Pilu) is located. The ritual is performed after the bodies of frogs, destroyed by trampling, are buried under shrubs and herbs (Chamba) and placed on the irregular rim of a ritual vessel (Tasqui), or catequil pool. Costales *et al.* (1996) describe this scene, in relation to frogs, as what was experienced by the Catequil (a water god that takes care of the animals on earth, a god of fortune and values) [Our translation] after drinking a potion extracted from the mature seed of the Huantuc (*Brugmansia arborea*) mixed with the dried blood of the Chucuri Bayo:

> The wise leader (Yachac), raising his arms in an attitude of catching the tiny claw of the moon in the middle of the loneliness of the night, repeatedly shouts:. . .Catequil . . .!Catequil . . .! At the edge of the pool, in a shallow brow or trench, numerous green-black frogs or Hambatus have been arranged, covered with interwoven shrubs. These, in a magical trance, will be crushed, trampled by the sorcer's "Mirucu", while the frogs shed their blood. The Yachac, at full throat, shaking the night, continues shouting: Catequil . . .!Catequil . . .! The moment the frogs shed blood, the moon sets, leaving the deepest darkness. The mythical waters and the darkness cover the Catequil of whom only the shoulders and the head remain without submerging. In absolute solitude, the Catequil will remain until the bursting of the light (Pacarina). The first to emerge from the bottom of the water is the Pini or Llucti, the seed snake that wraps itself around the hero's body. Simultaneously, Caylas, Hambatus and other animals jump into the water. Once the animals that breathe and live underwater have transmitted their powers to him, those that cannot submerge proceed to lick the neck and shoulders of the Catequil, sowing in him their respective qualities [Our translation].

Costales *et al.* (1996) explain the symbolism of three amphibians, the Hambatu, Buru, and Cotán (= Co'tan). The Hambatu frog (= Patata) symbolizes the sun grandfather, the paternal grandfather, the hidden voice that calls, prays, and brings the rain; it is a mythical symbol of the dancer, the rain maker (Tushug). Hambatu is one of the great symbols of social cohesion, the hidden voice that advises, and

FIGURE 1.44 Ancestral crafts representing frogs from the Kichwa in Alto Napo, Provincia Napo. Alexander Hirtz collection. SGS.

his teachings are in a glass that shines like the sun. Two other anurans are also prominent symbols, the Buru and the Cotán (= Cayla, large frog). The Buru would symbolize the ancient people, the social family or kinship, and the ideography of the cruciform vessel. The Buru would transmit to the initiate (Tenu) his powers of rainmaker and teach him the ritual dance. The Cotán is the voice of thunder and brings an ornamental necklace. Therefore, the Cotán as a symbol embellishes necklaces, chest plates, and amulets. She transmits the powerful voice of the one she teaches and expresses that the best adornment, the most effective amulet, is the word. The Cotán represents the ritual blow, the fight, the combat. This frog, as a mythical daughter of the community, is united, is part of it. Those Cotáns are taken and eagerly sought for that ceremony. The Catequil must be caught by them, because they will try to lick the young man subjected to the ritual. This frog becomes a sister and relative of the Catequil.

Shuar mythology. According to Shuar mythology, at one time all living beings were humans, but for some fault they were turned into animals or plants, which is why we must treat all beings with respect, because they are still human (Botasso, 1977). That author and Pellizzaro (1996) provide accounts of four Shuar stories with moral teachings that illustrate this belief [Our translation].

The Yantsaush frog. In this fable, the Yantsaush frog is a woman who was transformed into a frog because she despised the help of Etsa—the god Arútam who comes out of the waters of the river to help the Shuar in hunting—in the garden work. Etsa had established that the tillage implements will work on their own, but Yantsaush, together with Panka, wanted to work herself.

Fabels of kaká. In two fables, the Kaká frogs are symbols of lust and sexual deviations that are punished and corrected by Arútam (a spiritual entity within the life of the Shuar people).

In the story of Manchi and Kaká, the Shuar reveal the origin of the kaká tree frogs. They are all the incestuous offspring of an abusive man named Manchi (grasshopper) and his sister. When the woman became pregnant, there were persistent screams from her belly: kaká kaká kaká!, and so that the relatives would not perceive those voices, the woman wrapped her belly over and over again. Despite the many cloths she put on, she couldn't muffle that endless kaká kaká kaká! When it was her turn to serve chicha (a fermented drink), she did it by squeezing her belly with one hand, so that those she carried in her belly would become quiet. One day she felt a sudden need to urinate and she ran to the edge of the garden. As soon as she crouched down from it, many Kaká frogs began to emerge from her body and they proceeded to venture into the banana trees, the yuca in the garden, and the reeds in the swamps, croaking without interruption. The Shuar, perceiving the continuous kaká, kaká, kaká, went out to see and found out that the kaká frogs had come out of the womb of that woman, who had had incestuous sexual relations with Manchi.

Another tale of the Kaká frogs is about a lustful young man whom the Tsunki (in the form of a manatee)—the god Arútam who heals and gives sexual power—advises to control his lust and violence or he will suffer a poor fortune. Because the young man succumbed to the seductive song of the Kaká frogs, one of them transformed into a beautiful girl who seduced the young man and then turned into a frog again and abandoned him.

The taboo of Kuartank. Pellizzaro (1996) mentioned a Shuar taboo regarding the Kuartank. The latter is a large, green frog that lives in tree branches, eats crickets, and

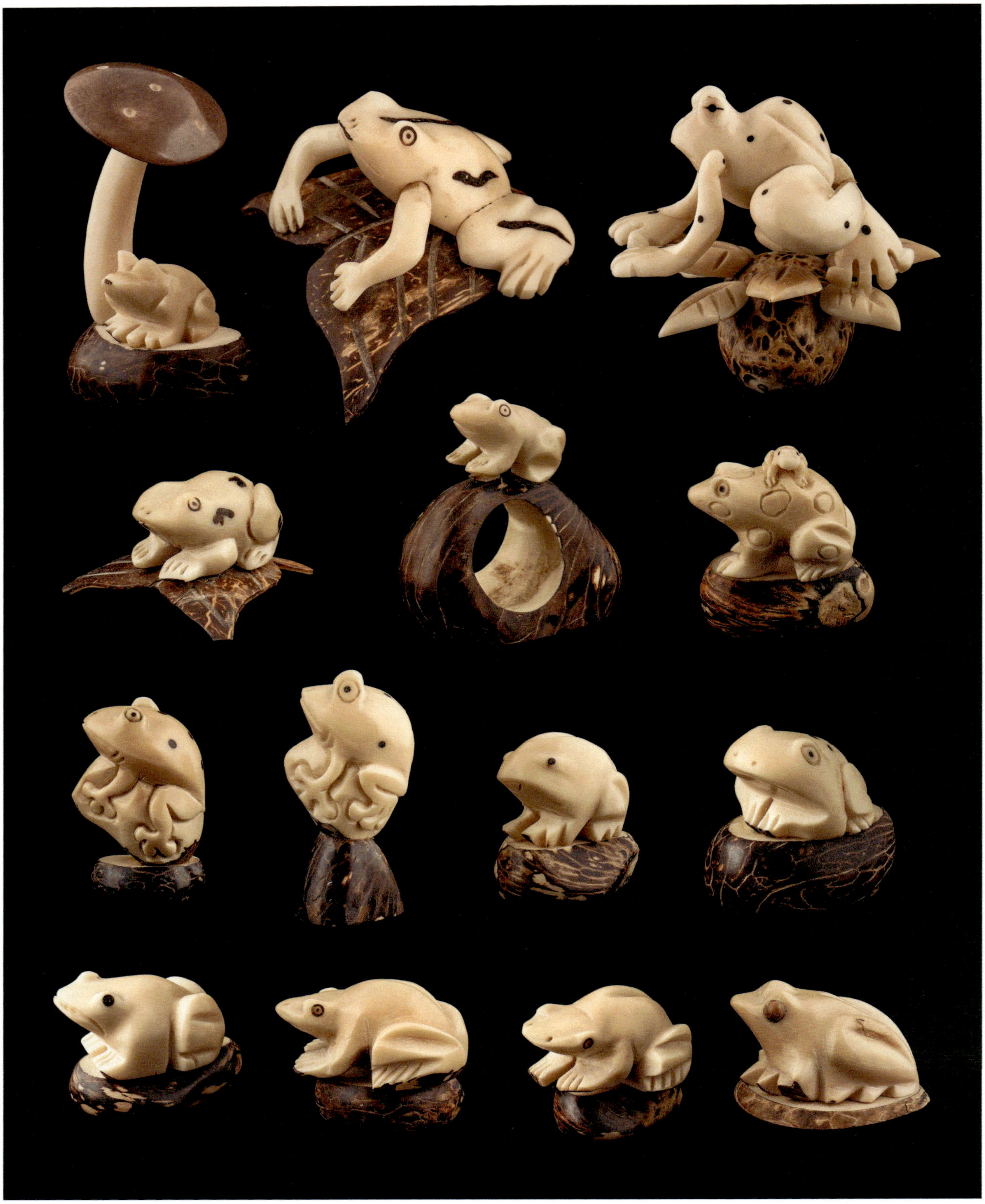

FIGURE 1.45 Contemporary crafts representing frogs carved from tagua, a type of nut found in Ecuador.

is inedible. This frog has a particular song from which it derives its name.

The Kuartank myth tells about a young man that was in the sacred place called Ayámtai, waiting for Arútam, after using tobacco. It was a humid, rainy night, and a kuartank frog kept shouting his kuartanktanktank . . . all the time. The annoyed young man, to make him shut up, yelled at him angrily: Kuatanktanktank! Instantly, an immense

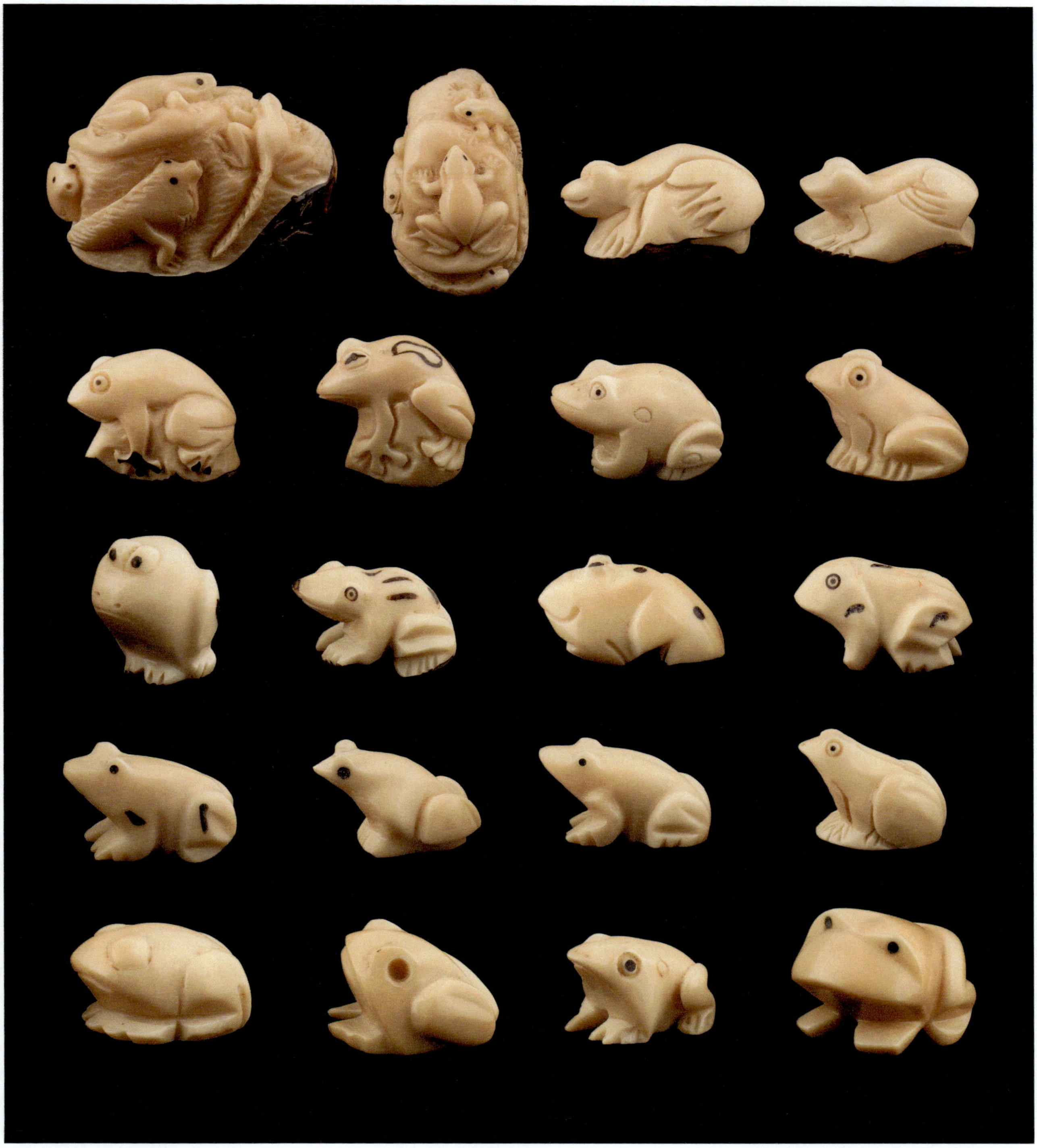

FIGURE 1.46 Contemporary crafts representing frogs carved from tagua, a type of nut found in Ecuador.

jaguar appeared and jumped on him, devouring him. For this reason, the Shuar prohibit their children from imitating the kuartank. Whoever does it will inevitably be devoured by a jaguar.

Siona, Secoyas, and Cofán mythology. Parra R. (1997) provided moral tales from the Siona and Kofán ethnic groups. For the Sionas and Secoyas (inhabitants of northeastern Ecuador) the frog has always represented fertility and the origin of life. Their oral traditions and the use of a hallucinogenic drink made from the magical Yagé plant (*Banisteriopsis caapi*) are the source of these morality lessons. The shaman who drinks the Yagé carries an enormous responsibility because from the interpretation of his visions come guidance on good and evil, the natural world, and on the behavior and destiny of his people.

The man who hunted frogs. In the Siona and Cofán cultures, this story is used to teach people not to kill frogs excessively.

In older times there lived a man of the Siona tribe who couldn't hear frogs singing without wanting to go hunting them. He would bring them back to his house and cook up a big meal for himself. That's how this man lived: wherever there were frogs singing, there he would be to kill them.

As the years passed frogs began to grow scarce. But the man didn't care that he was doing away with them all. So the time came when the singing of frogs was no longer to be heard.

Then one day, all at once the sky clouded over and a fierce wind came up and joined forces with the darkness. It was in this squall that the mother of all the frogs landed. The hunter was sitting in his house without a care in the world when, to his amazement, in hopped Mrs. Frog and sat herself on his right shoulder, and there was no moving her from that spot. From then on, he had to learn to live with the frog and he was in a state of despair because his body and clothes were turning all yellow and smelled rotten. Day and night the poor man wracked his brains to find a way to get rid of her and he even asked the rest of the tribe for advice. But they only replied, "You see what comes from killing so many frogs. Now look what's happened to you!".

For many moons the man could be seen pondering how to free himself from his trouble, of that blessed mother of the frogs. Then, one morning, he set off for the jungle following the bank of a small river in search of a coconut palm. After walking a good while he found one. He asked the frog if she would kindly get down because he was going up for a few coconuts as he was really very hungry. The mother frog took pity on him and she jumped down and sat waiting at the foot of the tree. The man cleaned away the spikes and rough bark with his machete and then began to climb up the palm. He was about twelve feet off the ground when:—SPLASH!—he dived into the water and off he went down stream. The frog, realizing she had been fooled, took the chance to jump into a big nut shell that was passing by in the river, and off she went after him. But the man was far ahead of her and climbed ashore way below and, very glad to be rid of the frog, he went home to change into nice clean clothes. The people couldn't wait to ask him how he had managed to escape and he told them step-by-step, what had happened. Meanwhile, the mother of the frogs, seeing that the man had given her the slip, decided to go on downstream until she reached the big river.

The next day the man was happily relaxing in his house, when again he saw the sky turn black with rain clouds, and suddenly thousands of frogs began to sing. Just then a mighty wind blew up and in the squall the mother of all the frogs returned and up she hopped again onto the man's shoulder and said: "I've come to take you away because now you are going to be my husband." The dark clouds began to disperse, the wind died, and the man who hunted frogs disappeared forever. Since that day frogs can be heard happily singing again.

(Parra, 1997)

MESTIZO AND SPANISH POPULAR CULTURE

Lombeida (2005) provided a review on these topics, representing the only general text written on this subject. Herein, we provide some additional information. The perception that toads and frogs call to bring rain is widespread but is more prominent in rural areas. Because anurans are sometimes seen as harmless little animals to which the power of attracting rain for crops is attributed, they are also known as God's little toads. For example, Giovanni Onore (2017) recounts the following story told to him by a campesino from the paramos of Pujilí, Provincia Cotopaxi: His grandmother told his mother that the Jambatos are the guardians of the farm because they are in charge of eating the worms and insects and that they are the shamans that call the rain when the summer is relentless. According to Carvajal-Endara (2008), in the Sigchos region, Provincia Cotopaxi, frogs and toads cry or scream to attract rain; the campesinos reported that in times of drought, they kill toads with machetes to make it rain. The first appearance of jambatos on the paramo heralded the arrival of the rainy season. It was widely believed that the big jambatos carried the smaller ones around on their backs when they became tired (Simmons, 2009). Similarly, the amplectant pairs were interpreted as "la mamá cargando al guaguito", meaning the mother carrying the baby.

Given the prior abundance of Jambatos (*Atelopus ignescens*) at some localities, they were used as toys by children. For example, elderly campesinos from Pujilí relate (Onore, 2017) that in their youth they used to catch the Jambatos to play with as if they were balls. With bunches of sigsesgrass (*Cortadeira nitida*) they would construct a bat (baseball style), and the "Sapito" would be the ball that they would throw at each other. Lombeida (2005) recalled the time when the beloved Jambatos (*A. ignescens*) were still common and related the following: At that time (the 1970s) this little toad was still quite common. They are not seen anymore and young people do not even remember them. At that time, Jambatos were used from time to time to fill piñatas (Ollas Encantadas, an Ecuadorian traditional game) during celebrations. The diurnal Jambatos were common residents in gardens and parks in Andean cities, and Jambatos were so prominent in popular culture that in March 2002, the Ecuadorian post office issued several jambato postage stamps (see also "Education") (Simmons, 2009).

The common Ecuadorian perceptions about frogs are sometimes negative. For instance, as recounted by Simmons (2009), under some circumstances, the mere presence of jambatos could be a sign of bad luck. A folk belief was that old guinea pigs could turn into toads and then run about grunting. In the Andes, guinea pigs are an important food source and are traditionally kept in the kitchen and fed with fresh grass that is collected daily, usually by the children of the household. Kichwa children were instructed not to collect grass from places that they saw jambatos because it was believed that toad urine transmitted a disease to the guinea pigs that caused their stomachs to swell until they died. According to Simmons, the story of the jambato tells us a lot about how people perceive and understand their environment, both from folkloric and scientific points of view. The belief that toad urine causes fatal disease in guinea pigs probably came about because jambatos favor moist habitats, and wet grass is not healthy for guinea pigs, whose wild relatives graze on dry grass.

Toads are seen as slimy and cold animals, and it is believed that the urine of toads produces warts on the hands. In Quito and other areas, it is said that boys can catch them, while girls should not. Most people in urban areas have lost direct contact with nature, making their encounters with frogs, toads, and tadpoles sporadic. However, the urban populations in large cities maintain indirect contact with a few species of frogs (these species are rarely seen because they are inactive, hiding in shelters during the day) because in winter, the vocalizations (mainly nocturnal) of a few species adapted to the urban habitat are still common. For example, the cuico rainfrog *Pristimantis unistrigatus* in Quito, the five-lined agile frog *Scinax quinquefasciatus* in Guayaquil, and the bifurcated swampfrog (*Dendropsophus bifurcus*) in Tena, Provincia Napo, are common around human habitations.

Frogs are also present in the everyday popular culture in traditions. For example, Quito has two important traditions related to frogs.

The Toad Game

The sapo game (the toad game) consists of throwing small brass disks from about 12 steps away into the different holes in a wooden box where a brass toad with its mouth open sits (Figure 1.47). Points are earned by getting the disks to fall into one of the holes in the top of the box or through the mouth of the toad. The game originated in the Titicaca region as a mystical Inca game. The legend says that the Inca, son of the Sun, and his court threw little gold stones to see if any toad would come out to catch them with his mouth. If they were lucky and the animal took the piece of metal, they were granted a wish and the toad turned into gold.

The Water Toad Fountains and Drinkers

Another tradition is named "the water toad". Fuentes Roldán (1995) and Freire García (2016) provide details of this tradition that symbolizes the unity and ingenuity of the people of Quito. Originally, the toads had a historical context. In 1650, water became scarce in the city, so the city council decided to build an aqueduct of baked brick in Calle Real, in a stream known as La Chorrera, to store water and then distribute it to the inhabitants of the upper and lower neighborhoods of Quito. This distribution of water generated the presence of the "aguateros", indigenous people who worked tirelessly to supply water to those who requested it by hauling it to the houses in large containers, but the residents of the San Sebastián neighborhood looked for another alternative. They built a brick pipe that started at the monastery of Las Clarisas and ended a block from the church. This termination of the pipeline had to be designed so that the water was not wasted and did not fall in gushes. While they were discussing these details, one of the neighbors saw a toad in the puddle formed by the pipe. The presence of this little animal served as an inspiration to the residents, so that they decided that the termination of the pipe should be an effigy of the amphibian. At the beginning of the 20th century, the frog water sources disappeared. However, Marco Chiriboga Villaquirán decided to rescue this tradition, and in 2014 he placed these water faucets in seven places in the Historic Center of Quito (Figure 1.48).

Education

Between the 16th century until near the end of the 18th century, the literature on the amphibians of Ecuador for about 237 years (from the Spanish foundation of Quito in

FIGURE 1.47 The toad game. A popular and traditional game in Calle La Ronda in the Centro Histórico of Quito, Ecuador. Alexandra Quiguango Ubillús and LAC.

FIGURE 1.48 Water toad fountains in the Quito Colonial. Left: Museo Casa de Sucre. Right: a modern water of the toad fountain in Calle La Ronda. LAC.

1534 to 1771) seems to be non-existent. In the 18th century, only two Jesuit monks referred to anurans. The first was the Italian Mario María Cicala—the first geographer of the Kingdom of Quito—who in 1771 described frogs and toads, especially about their abundance, diversity, and uses—his original handwritten manuscripts in Italian are deposited in the Library of the Jesuits in Cotocollao, Quito, Ecuador. Cicala (1994), when referring to frogs, including those of Guayaquil, Ecuador, says the following:

> Of the common frogs, "count the stars if you can". In the cities of Cartagena, Panama and Guayaquil, at the first rains that fall, at the beginning of winter, it is not known where they come from, how they are generated, and how so many frogs of so many different kinds are born within 24 hours, with so many colors and so different sizes. I have seen frogs as large as a nut, but then I have seen them as tiny as a cherry or as a small grape seed and they do not grow any more. From morning to afternoon and from afternoon to morning they do nothing but shout alternately; some shout and others are silent, but immediately the others begin. And all night long, they all scream at the same time. Among these frogs there is a species so small that the whole day passes in silence, but as night falls, that is, the Hail Mary (time of Vespers), all at the same time begin to scream as if four hundred, five hundred or more children were crying, and that cry of the little frogs is so similar to the cry of children that there has been no foreigner or passenger who had not been deceived and the same thing happened to me the first time I spent the winter in Guayaquil and those sad cries of lamentation and weeping were not weak at all, but rather strong and robust, and could be heard more than 300 canas (cana = 1535 m) away and throughout the night until dawn [Our translation].

This description of the songs of the Guayaquil frogs suggests that they would be the chorus of either *Engystomops guayaco*, *E. montubio*, *Scinax quinquefasciatus*, or a combination of them. Cicala also describes toads, which would not be found in cold climates, nor in temperate ones, but only in warm tropical climates. He highlights the monstrous size of the toads and their screams that caused him fear and horror. However, upon arriving in Guayaquil, he says:

> In Panama and in Guayaquil I was not afraid afterwards, only when I went out at night to hear the confessions of the sick, in winter time I was afraid of stepping on them, because some truly deformed ones were seen in the streets some truly misshapen for size. It is an undoubted thing for everyone that toads attract squirrels just by looking at them and having their mouths open, until the squirrel enters the open mouth of the toad an after a thousand turns and twists escapes, just as they do the same with the small birds, the squirrel on the trees, which going around here and there finally gets into the toad's open mouth. Many cases of these have been told to me by subjects worthy of all credit and eyewitness testimony [Our translation].

The toad that Cicala mentions in Guayaquil could be *Rhinella horribilis*, which is one of the largest anurans associated with human habitations. Also, Cicala (1994) describes the medicinal use of toads (see "Traditional Medicine").

Cicala is followed by Father Juan de Velasco, the first historian of the Kingdom of Quito, who in 1789 provided some native names (Hambatu, Picu picu, Caila, Tugllín) and discussed them, especially about their medicinal uses (see also "Indigenous and Vernacular Names and Traditional Medicine").

In the 19th century, the popular literature includes only the references to frogs and toads made by the famous English mountaineer and explorer Edward Whymper (1891) and those of the German scientist Franz Theodor Wolf (1892). In his book *Journeys Through the Majestic Andes of Ecuador*, Whymper (1891) mentions that the only thing they saw belonging to the animal kingdom in the surroundings of the Cotopaxi Volcano in a camp located at 13,455 feet (= 4101.1 masl) was the frog *Phryniscus laevis* (currently *Atelopus ignescens*), which was everywhere. He also indicates that the last specimen in his zoological collection was a *Hylodes* (now *Pristimantis curtipes*) found near Totorillas on the Chimborazo Volcano. A supplementary appendix in the book it says the following about *Atelopus ignescens*, *Pristimantis unistrigatus*, *P. curtipes*, and *Gastrotheca riobambae* and/or *G. pseustes* respectively:

> Only four species of frogs were collected in the interior (refers to areas not lower than 8500 feet except for Ibarra and Guayllabamba), viz. *Phryniscus laevis*, Gthr.; *Hylodes unistrigatus*, Gthr.; *H. whymperi* Bouleng. and *Nototrema marsupiatum* (Dum. & Bibr.). Of these, I think that the one mentioned first is the most widely distributed, and the one mentioned last is the most numerous. In the vicinity of the city of Machachi we saw thousands, and we must have heard hundreds of thousands of frogs, mainly of the latter species which is highly variable in their colors and spot patterns. It should be noted that we get all of our *Hylodes* specimens on the surface of the ground. The species with which Mr. Boulenger has associated my name were observed only in somewhat elevated localities, and their range of altitude seems very restricted; while *Phryniscus laevis* can be observed at almost all altitudes between 7000 and 13,500 feet.
>
> (Whymper, 1891).

According to Boulenger (1891), Whymper says about *Nototrema marsupiatum* (= *Gastrotheca riobambae* and/or *G. pseustes*) that the basic color was bright green: "These frogs were in great numbers in the neighbourhood of Machachi, and in the evening their music was so loud that as almost to interfere with hearing when walking out".

For his part, Theodor Wolf (1892) in his book *Geography and Geology of Ecuador* speaks of the batrachians as follows:

> For zoological geography they have some interest, since they are amphibians, which occur to the Andean region, since the night cry of the toads chase us from the flooded streets of Guayaquil to the paramos of Chimborazo, of course with variations, because mountain toads are different from coastal ones. A small frog, which lives near Totorillas at more than 4100 meters high, and which Mr. Boulenger has called *Hylodes whymperi*, is probably the amphibian, which reaches the highest altitude in Ecuador and perhaps in the entire South American continent.

During most of the 20th century, there was also relatively little public outreach information about amphibians in Ecuador, although this increased notably from the 1990s to the present.

J O R G E   I C A Z A

dejan en huesos a punta de pellizcos con las tijeras que tienen en las manos. Eso es jodido... Esto ca, guagua pantano no más es.

—Y los güishigüishis qui'ay por mishones.

—Acaso hacen nada.

Los aludidos animalucos ponen escalofríos concéntricos en el agua estancada del pantano. Hay millones. El Andrés que se mete en el fango hasta las rodillas los siente que en bombardeo perenne se van robándole las piernas; no sabe si es el frío o es el constante pegarse de aquellos babosos el que ha cercenado la vida de las rodillas para abajo.

FIGURE 1.49 Paragraph showing the original text about tadpoles in the novel *Huasipungo* by Icaza (1934).

Jorge Icaza (1934), in his great opus *Huasipungo*, mentions the abundance of tadpoles when one of the Indians from the troop of "mingueros" asks with curiosity and fear before entering the muddy scrapie (swamp), where tadpoles attach to submerged legs. "—And the 'güishigüishes', which occur by millions?" (Figure 1.49).

Icaza also talks in *Huasipungo* about anurans as follows:

> Era un ruido armónico en la orquesta de los sapos y las alimañas . . . [It was a harmonious noise in the orchestra of toads and vermin . . .]
>
> Y, unos en silencio, otros murmurando en voz baja su mala o buena suerte en la tarea del día, se acurrucaban por los rincones, dejándose arrullar por la música monótona de las goteras, por la orquesta incompleta de los sapos y de los grillos, por el ruido del viento y de la lluvia en el follaje. [And some in silence, others muttering low about their bad luck or good luck in the day's tasks, huddled in the corners of the shed, allowing themselves to be lulled to sleep quickly by the musical monologue of the dripping water, by the skimpy orchestra of the toads and the crickets, by the noise of the wind and the rain on the foliage.]

Beginning in the 20th century, the Ecuadorian educational system in general adopted in its schoolbooks information on amphibians that came from Europe, for example, in the book *El Escolar Ecuatoriano*, first published in 1947 (Carrillo de Landazuri and Arregui de Pazmiño, 1947). However, only in the first half of the 1990s was information on Ecuadorian amphibians added to the book (Carrillo de Landazuri, 1995), after information was provided by Luis A. Coloma to his primary school friend Renán Saltos Chávez, who was in charge of editing the book. The first academic attempt to incorporate Ecuadorian amphibians into schoolbooks was that of Erwin Patzelt, the renowned professor (Dr. hc.) of biological sciences. Of German origin, Patzelt taught at the German High School in Quito during the 1960s and 1970s. Patzelt traveled extensively in Ecuador filming and photographing nature. He took thousands of photographs, many of them of amphibians, including some populations of species that are critically endangered (e.g., *Atelopus ignescens*, *A. bomolochos*). On 9 January 1976, the Ministry of Education and Sports of Ecuador resolved (Oficio # 671) to recommend the publication and the use of the zoological text and didactic material of Professor Erwin Patzelt to be used as consultation tools for schools in Ecuador. Thus, by the end of the 1970s, Patzelt (1978) published the first edition of his book *Fauna del Ecuador*, in which there is a chapter dedicated to amphibians: toads, frogs, ilulus, and salamanders. This first edition was followed by at least one reprint in 1979 and revisions that included important additions (Patzelt, 1989, 2000, 2004).

In the 21th century, printed outreach publications on the amphibians of Ecuador include a complete volume dedicated to Ecuadorian amphibians in the *Terra Incognita* magazine in 2005 and the following amphibian guides: of Parque Nacional Cajas, Provincia Azuay (Arbeláez Ortiz and Vega Toral, 2008); of Ecuador (Valencia *et al.*, 2008) (including short accounts of 163 species); of Quito, Provincia Pichincha (Yanez-Muñoz *et al.*, 2009); of the Chocó, Provincia Esmeraldas (Ortega-Andrade *et al.*, 2010); of the surroundings of the seven stations of the heavy crude oil pipeline (Valencia and Garzón, 2011); of Mindo, Provincia Pichincha (Arteaga *et al.*, 2013); of several protected priority areas for conservation (Morales-Mite, 2013); of La Favorita, Provincia Pichincha (Reyes-Puig *et al.*, 2015); of the Papallacta-El Reventador region, Provincia Napo (Bejarano-Muñoz *et al.*, 2015); of Provincia El Oro (MECN-INB and GADPEO, 2015); of the urban frogs from Cuenca, Provincia Azuay (Siavichay-Pesántez *et al.*, 2016); of Junín, Provincia Imbabura (Freile *et al.*, 2020); and of Alto Río Anzu, Provincia Pastaza (Bentley *et al.*, 2021). Also noteworthy is the publication of coffee table photography books including amphibians from Ecuador by Coloma and Ron (2001), Oxford and Bish (2004), Pérez (2005), and Ron *et al.* (2009). Correos del Ecuador published stamps, stamp series, and philatelic cards depicting Ecuadorian amphibians in 1993, 2001, 2002, 2006, 2010, 2011, 2012, 2015, 2016, 2018, and 2019 (Figures 1.50, 1.51, 1.52).

Numerous websites depicting amphibians, field guides, and information on Ecuadorian amphibians have proliferated in recent years. The quality of their contents varies given that they are not peer reviewed. A pioneer is the website of the Museo de Zoología of Pontificia Universidad Católica del Ecuador (see Chapter 2: "From Anfibios de Ecuador to Amphibiawebecuador and to Anfibios del Ecuador-Bioweb at QCAZ"). Since 2010, the Centro Jambatu of Amphibian Research and Conservation has maintained a website (see Chapter 2). Also, the Field Museum (in Chicago) has 17 photo field guides published online (as of February 2023) (https://fieldguides.fieldmuseum.org/) that include Ecuadorian amphibians from several localities or regions of Ecuador. Other important websites include Amphibian Species of the World (AMNH), (https://amphibiansoftheworld.amnh.org/), AmphibiaWeb (Berkeley, USA) (https://amphibiaweb.org/index.html), FonoZoo.com (https://www.fonozoo.com/eng/index.php), and The IUCN Red List of Threatened Species (https://www.iucnredlist.org/).

Films on Ecuadorian Amphibians

Older films on Ecuadorian amphibians are few in number. The first film on Ecuadorian amphibians dates back to 1957 and is about the biology of the marsupial frog *Gastrotheca riobambae*, created by Emmanuel Ciprian Amoroso in London as a color cinematograph film (see "Biotrade"). In 1986, another film associated with a scientific publication on the biology of *Gastrotheca riobambae* was made in Paris (Aubér-Thomay, 1986). In 1996, Argo films released the documentary "The Last Frog" (Argo, 1996), which was mostly filmed in Ecuador. According to its director, Allison Argo:

> it is a love letter to the frog—and a strange and unnerving murder mystery. Starring our amphibious friends, the drama is supported by a cast of scientific experts and devotees frantically trying to unravel the disturbing puzzle behind the curious decline of frogs.

In 2004, Alfredo Salazar and Juan Diego Pérez produced a documentary video with an overview of the amphibians of Ecuador titled "Chanting and Enchanting Animals" (Salazar and Pérez, 2004). The video was translated into Spanish, English, Italian, and Kichwa, and it was broadcasted to mass audiences on TV in 2005 and 2009. This film helped to improve the public's knowledge of amphibians in Ecuador (Coloma, 2005). William E. Duellman commented: "This film on the amphibians of Ecuador far exceeds any other of its kind. The exquisite photography and excellent narration effectively tell the story of diversity, life and extinctions". The film received some impressive distinctions: Official Selection Latin American Documentaries, Albacete, Spain 2006; Special Jury Mention, Matsalu Film Festival, Estonia 2007; Official Selection Sondrio Film Festival, Italy, 2007. Since those films were released, given the technological advances in video and film production, there has been a steady increase in films depicting Ecuadorian frogs, most of which are compiled by Centro Jambatu (e.g., https://www.youtube.com/user/ArcaDeLosSapos). One of the most recent films is "Heroes" of the series Planet Earth, in which the story of the rediscovery by Jaime Culebras of a harlequin frog (*Atelopus halihelos*) is narrated by David Attenborough. The producer was profoundly impressed by his story, leading to Culebras being showcased in the last episode of Planet Earth III, dedicated to conservation.

Ecuadorian Amphibians in the News

Amphibians of Ecuador are sporadically the focus of journalistic news. The greatest media coverage was related to the Sapari public exhibit that was held in Quito in 2005 and

FIGURE 1.50 Stamps and philatelic card issued in 1993, 2001, 2002, and 2006 depicting Ecuadorian amphibians by Correos del Ecuador.

which attracted the attention of 105,000 visitors during the 90 days that was open to the public (Coloma, 2005). Other events covered issues by local and international media about Ecuadorian amphibians include the discovery of new species (e.g., *Hyloscirtus princecharlesi* and *H. criptico* (Mongabay, 2012), *Hyalinobatrachium mashpi* and *H. nouns* (Posada-Swafford, 2022)), and the recent rediscovery of species thought to be extinct (e.g., *Atelopus bomolochos* (Holland, 2015), *A. ignescens* (Plitt, 2017), *A. longirostris* (Green, 2022)). Since 2020, some Ecuadorian amphibians

FIGURE 1.51 Stamps depicting Ecuadorian amphibians by Correos del Ecuador. Top: *Atelopus ignescens* (2010). Middle: *Boana* spp. (2011, as *Hypsiboas* sp. and *Hypsiboas geographicus*), *Ceratophrys stolzmanni* (2015). Bottom left: *Chimerella mariaelenae* (2018). Bottom right: *Oophaga sylvatica* (2018).

have hit the news (e.g., Montaño, 2020; Graeme, 2021; Kimbrough, 2023; Forrester, 2023) as symbols against the mining industry in the Andes of Ecuador.

Poetry and Music

Although amphibians are not prominent in the poetic and musical expression of Ecuadorians, there are a few examples that deserve attention in some of the better-known writings and songs.

Among the cultured poetry, that is, the poetry of individuals as defined by Ubidia (1983) is "La ranita enamorada" (a little frog in love), a poem written by the well-known Ecuadorian writer and poet Gustavo Alfredo Jácome. This poem is for children, and its figurative language is analyzed in great detail in the unpublished thesis of Bermeo (2015).

FIGURE 1.52 Stamp series issued in 2016 and 2019 by Correos del Ecuador, featuring Ecuadorian amphibians and glassfrogs, respectively.

"La ranita enamorada"

Con un hongo por sombrilla
la Ranita fue a pasear
por el valle y la colina,
por el río y más allá.

Iba toda endomingada,
con un garbo señorial,
y el rumor almionado
de la falda y el fustán.

Los sapitos que miraba
La fachenda sin igual,
Murmuraban:–Qué guapeza!
¡y ese aire al caminar!

Y la Rana, modosita,
simulaba no escuchar;
mas, sentía que de rojo
encendíase su faz.

Esa noche, cuando estuvo
de regreso en el hogar,
se sintió muy enfermita
con un dulce y raro mal.

La Ranita suspiraba,
suspiraba y nada más,
y pasose conaturreando
a la luna un madrigal.

Y la luna sonreía
desde su alto ventanal,
recordando sus amores
cuando estaba en esa edad.

(Jácome, 1956: 77–78)

Alfredo Gangotena Fernández Salvador (1904–1944)—an avant-garde and mystical Ecuadorian poet and representative of French poetry of the first half of the 20th century—wrote in his poem written in French in 1928, "Bebida turbia", "el verde grito del sapo se torna líquido en mi alma".

In popular poetry, "sapos" or "sapitos" are present as well. According to Ubidia (1983), popular poetry is characterized by oral performance, anonymity, a symbiosis with popular music, its adherence to a more or less strict metric, and its functionality.

A popular children's game (common in the mid-20th century) goes like this:

The first child recites:

Sapito sapón
Ponete calzón

The second children jumps and recites:

No puedo ponerme
Porque soy pipón

Cornejo (1970) compiled the following traditional stanzas "pegas, décimas, amorfinos, romances" of toads in Ecuadorian popular poetry from the "montuvio" people in the coastal region. These stanzas usually are transmitted orally from one generation to another.

Dicen que los sapos muerden,
y es mentira de la gente:
yo tuve un sapo en mi casa
y le sobaba la frente.

Allá arriba, en esa loma,
tengo un sapo con calzón:
cada vez que al cerro subo
me regala un chicharrón.

Dos palomitas y un sapo
se fueron a Santa Fe:
las palomitas volando
y el pobre sapito a pie.

Un sapo y una lechuza
se fueron a Santa Clara:
la lechuza llegó atingida
y el sapo como tu cara.

La niña que está bailando
parece vaso floriado,
y el galán que la acompaña
parece sapo rayado.

One of the "décimas" from Afro-Americans in Provincia Esmeraldas compiled by Hidalgo Alzamora (1982) includes a toad. She explains the importance of this kind of poetry.

"El sapo desafiando al ratón"

Una mañana en mi rancho
oí cantando un mongón.
El sapo que se paseaba
y desafiaba al ratón.

Se pusieron en porfía
el zorrín con su macoca
y el ratón con una bola.
El ratón le dío en la boca.
Dijo el zorro: Esto me choca,
no puede haber picardía,
salga uno, salgan dos
y peleamos todo el día.
Y esta porfía vi yo
Una mañana en mi rancho.

Yo no chisto una palabra
por ver la riña hasta el fín.
En esto salió el zorrín
Con una bizarra espada.
La zorra iba arremangada.
El sapo con su bordón,
mas atrá el camaleón
Con revolver en la mano.
Y al pujido de un gusano,
oí cantando un mongón.

Salió el perico orgulloso
con una lanza ofendiendo,
de lejo le grita al oso
que se vaya previniendo.
Y la tortuga en su pozo:
Conmigo tu no te metas
porque te voya fregar:
tus valentías son pocas.
El zorrín con su macoca
y el ratón que se paseaba.

Salió la araña bailando
con su vestio de cola.
Más atrás la lagartija
que esa sí, carga pistola.
El tembán con una bola
de la especie de un cañón.
Se me aflige el corazón,
miraba pa´ todas partes.
El sapo que se paseaba
y desafiaba al ratón.

Mongón = mono
Macoca = machete pequeño
Lejo = lejos
Pa' = para

Since 1962, a song with enormous popularity in Ecuador has been "Sapo cancionero" (https://www.youtube.com/watch?v=mLbe9w7s1K8). This song is a samba—the Argentine national style—with lyrics by Jorge Hugo Chagra and music by Nicolás Toledo. It was popularized especially by the Argentine group Los Chalchaleros and became a standard of the Latin American folk music.

Sapo de la noche, sapo cancionero,
Que vives soñando junto a tu laguna.
Tenor de los charcos, grotesco trovero,
Estás embrujado de amor por la luna.

Yo sé de tu vida sin gloria ninguna;
Sé de las tragedias de tu alma inquieta.

Y esa tu locura de amor a la luna
Es locura eterna de todo poeta

Sapo cancionero:
Canta tu canción,
Que la vida es triste,
Si no la vivimos con una ilusión.

Tú te sabes feo, feo y contrahecho;
Por eso de día tu fealdad ocultas
Y de noche cantas tu melancolía
Y suena tu canto como letanía.
Repican tus voces en franca porfía;
Tus coplas son vanas como son tan bellas.
¿no sabes, acaso, que la luna es fría,
Porque dio su sangre para las estrellas?.

There are several recordings of Ecuadorian folk music featuring the sounds of frogs. The calls of several anurans can be heard in the indigenous music of the Andes, both figuratively and literally (Simmons, 2009). Figuratively, frog calls are represented by the pronounced, high-pitched repetitious pulsing of many tunes (Simmons, 2009). Literally, the calls may be heard in the background of field and festival recordings. The distinctive calls of marsupial frogs, the high peeps of rainfrogs, and other anurans can be heard in various recordings of Enrique Males, the Nanda Mañachi group, and Salasacamanda (Simmons, 2009). For example, the loud "brack-bak-bak-bak" call of the gualataco marsupial frog (*Gastrotheca riobambae*), a chorus of the cuico rainfrog (*Pristimantis unistrigatus*), and the trill of the currently critically endangered Buckley's glass frog (*Centrolene buckleyi*) are identifiable in the traditional song titled "Rafael Taita" (B1.b, 3:30), found on the 1979 vinyl LP directed by Enrique Males titled "Quinchuquimanda Imbabayacuna" (https://www.youtube.com/watch?v=2DugK1wW_Dc). Males is a well-known Ecuadorian Kichwa singer–songwriter from the Imbaya community in Ibarra, Provincia Imbabura, and the recipient of the 2018 national prize Eugenio Espejo. His work is inspired by social memory, aiming to revalue the indigenous worldview, including the natural sounds of his homeland.

A song of Ecuadorian origin is "Venenoso batracio" (https://www.youtube.com/watch?v=5Jg6-dxBSAM). This song was popularized by the singer–songwriter Hugo Idrovo in 1995 in the album "Cuentos del Río Colgado". Born in Guayaquil in 1957, Hugo Idrovo is widely regarded as one of the greatest Ecuadorian musical talents (Wikimedia, 2023). Songs like "Venenoso batracio" influenced the newer generations who were seeking music and lyrics rooted in Ecuador's idiosyncratic culture. The lyrics of this popular song are as follows:

Cual venenoso batracio
tu ponzoñoso corazón
me ahoga en un laberinto
de pecado y de traición

La obscuridad de tu alma
ha perturbado mi razón
Yo que volvía de tan lejos
tu que te ibas con otro amor

Te feriaste mi plata
Me empeñaste la moto
Te has llevado toditas
las joyas de mi madre
Mis amigos me dejan
dormido en la cantina
No tengo mas remedio
que matarte y que matarme
Y que Dios me perdone
Tu no eres mas que una perdida

(Maldita Emelecista)

Yo que te recogí adolescente
y con la piel tan tersa
Compadezco a ese gil
que ahora quiere mantenerte
con esa cara
igualita a la de Elsa
Vieja perversa

Cual venenoso batracio
tu ponzoñoso corazón
me ahoga en un laberinto
de pecado y de maldad

Croac, croac, croac

Some more recent songs that feature Ecuadorian frogs include:

1. "Jambato negro", performed by the group Jacana Jacana and Grecia Albán https://www.youtube.com/watch?v=-Ln3esHRrPw. This song was created to raise awareness about the Jambato harlequin frog *Atelopus ignescens* and the conservation efforts led by the Alianza Jambato.
2. Another song was produced by the US Broadcasting Service for children. This one relates the efforts of the Ecuadorian herpetologist Juan M. Guayasamin in his search for new species in the forests of Ecuador. You can watch it here: https://www.youtube.com/watch?v=F2ZV6SJd0zU.

Sayings, Superstitions, Jokes, and Graffiti

These kinds of Ecuadorian cultural expressions are abundant, and herein, we compile just a few. For example, when speaking of the extinction of amphibians in Ecuador by the end of the 1980s, the inhabitants of each area, mostly indigenous, gave suspicious, fanciful, and ingenious answers to explain the disappearance of the jambato (*Atelopus ignescens*). Some said it was the government's fault. Others indicated that lately the sun was hotter or that the weather had changed. Many attributed their disappearance to the earthquake of 1987 or the visit of the Pope Juan Pablo II to Ecuador in 1985. For some people inhabiting the Central Andes of Ecuador, the reason was that Taita Chimborazo (a volcano) had decided to take them from this world. A child said that a wind came with poison and since then, the "jambatitos" are no longer seen. (Simmons, 2009).

There are many jokes that use the toad (whether in a biological or figurative sense) as the main character. A well-known joke goes like this: What does a toad do when it leaves the garden? Answer: Goes to first grade.

According to the Real Academia Española (in 2023), the word toad (sapo) in Ecuador, Peru, and Bolivia is a colloquial masculine name that means a very alert, lively, and cunning person. Likewise, toads have been the object of the most varied caricatures, whether with political or other overtones. Thus, most people, especially from urban areas, use the word sapo to mock politicians. This connotation (and the image of the anurans) was mischievously picked up by the journalist Simón Espinosa Cordero (2005), who carried out a satirical essay in which he provided a scientific note on the lifestyle of the Ecuadorian toad. He described the toads of Ecuador as divided into five categories: bureaucratic toads, business toads, clerical toads, military toads, and erotic toads. He concludes by saying: "Yo soy el sapito Quiteño. La vida me paso encantado . . ." which means, "I am the Quito toad. I spent my life delighted . . ."

This same negative image of the toads led, in 1960, the former president of Ecuador, Dr. José María Velasco Ibarra (five non-consecutitve presidential terms), to suggest changing the name of the Ecuadorian city formerly known as "La Boca de los Sapos" (the Toad's Mouth). Velasco was well-known as a charismatic leader, famous for his speeches. The origin of the historic name stems from the large number of frogs that reproduced and vocalized in a mouth-shaped cave where the Río Verde joins Río Bulubulu, Provincia Guayas. According to Herrera and Zambrano (2015), the name change was due to the fact that, in the summer of 1960 presidential campaign, Velasco arrived at "La Boca de los Sapos", and his adversaries (outsiders sent by his political enemy Assad Bucaram Elmhalin) intended to boycott his speech and did not stop shouting in a mocking tone, "Los Sapos con Velasco" and "Velasco con los Sapos". Velasco was upset. Thus, in his famous speech he said:

> This name does not suit this town so beautiful, so thriving, so hardworking. I'm going to change the name. I am going to call it "El Triunfo" because this town will give me the triumph. Now we are all going to cheer "El Triunfo con Velasco, El Triunfo con Velasco" [Our translation].

On 8 April 1969, the promise was made effective, and the town's name officially became "El Triunfo".

The very definition of toad and some other sentences are expressed in sayings (or their synonyms such as proverbs, slogans, aphorisms, adages, etc.) extracted from popular wisdom of different origins and usually with moral connotations. Some of the best-known examples used in Ecuador are:

"The toad lives off the fool and the fool, from his work" (Espinosa Cordero, 2005).

"Throwing toads and snakes through the mouth" (RAE, 2023), which means uttering curses, blasphemies, or nonsense.

"What is the fault of the stake if the toad comes and skewers itself" (Jácome, 2006).

"Heal, heal, frog's ass, . . . if you don't heal today, you will heal tomorrow", which is used to soothe pain in children when they become ill or have been injured.

"Toad from another well", which means feeling uncomfortable in an unfamiliar environment.

The word toad also takes shape in the graffiti that Ecuadorians print in various places, especially on walls and vehicles. Graffiti—native to Paris—is a means of popular expression, usually of denunciation and rebellion, in which the word toad has the most diverse connotations. Some of these texts are indicated in Figure 1.52. The one on the top left is a mix of humor and irony and refers to improvisation versus planning. This graffiti could suggest that sometimes it is important to plan and carefully consider our actions instead of simply "jumping" into situations without thinking about the consequences. This graffiti could have a humorous connotation by comparing frogs, which are often associated with intelligence in popular culture, to toads, which may be seen as less intelligent. The phrase could suggest that a lack of planning is a characteristic of toads, in contrast to the smarter frogs. The top right graffiti was depicted in the frog exhibit Sapari held in Quito in 2005 and contrasts poison dart frogs with cunning people (in a bad connotation). The graffiti depicted in the back of the truck is a light and straightforward joke that plays with the double meaning of the word sapo and can be interpreted humorously. It's used in a playful manner to encourage the cars behind to move or act promptly. The graffiti appearing at the bottom right of Figure 1.53 was depicted in a populated area in Quito and was intended to mock a politician, Fabián Alarcón Rivera, a lawyer and Ecuadorian ex-president.

FOOD FOR HUMANS

Native Species

Frogs have been part of the diet of humans since ancient times. The written information about edible frogs in Ecuador is relatively scarce and, to our knowledge, no documentation of eating practices is available. Some authors may have seen those practices; however, in general they do not describe hunting, cooking, or eating activities. Currently and unfortunately, only a little information about four ethnic groups is available, and we lack information for the remaining ones, including the Afro-Americans. The first to mention an Ecuadorian edible frog was de Velasco (1789), who indicated that Caila (a kind of Picu picu frog)—apparently

FIGURE 1.53 Graffiti referring to amphibians in Ecuador. Top left: Baeza, Provincia Napo. Top right: cultural center of Pontificia Universidad Católica del Ecuador, Quito. Bottom left: Road Aloag-Santo Domingo, Provincia Pichincha. Bottom right: Roca Street, Quito, Provincia Pichincha. LAC.

Telmatobius niger from the southern Andes of Ecuador—are given to women (see also "Traditional Medicine"). In the 1960s and 1970s, the Italian Salesian priest Juan Bottasso Boetti edited a series of booklets titled *Mundo Shuar* describing the Shuar culture. In two of them, Botasso (1977, 1978) provided information about native Shuar names for edible frogs. Also, Morales and Schjellerup (1999) provided information on 19 species that are hunted for consumption in large quantities by the Shuar in Mutints (Provincia Morona Santiago). Kohn (2002) revealed that nine species are consumed in Ávila Viejo-Loreto (Provincia Orellana). Gallardo (2004) compiles information from various sources and provides a list with the common names of 45 edible species of amphibians from Ecuador, 12 of them with their scientific identification. Herein, we provide an updated list (Table 1.3). Among the species eaten, 18 species have been identified. It is remarkable that the Shuar indigenous group is an important consumer of many species of frogs (at least 35 species, Table 1.3), in contrast to other groups such as the Waorani that apparently consume only three species of frogs, *Boana boans*, *Leptodactylus pentadactylus*, and *Lithobates palmipes* (Tirira and Ríos, 2018). The Tsáchilas from the western lowlands of Ecuador apparently eat just *Leptodactylus peritoaktites*, whereas the Kichwa people from Amazonia are reported to consume at least 12 species of frogs, and the Kichwa from the Andes of southern Ecuador used to eat one species (*T. niger*) before it evidently became extinct. The largest frogs eaten are *L. pentadactylus*, *L. peritoaktites*, *Rhinella marina*, *Rhaebo ecuadorensis*, *Boana boans*, *L. palmipes*, *Leptodactylus wagneri*, and *T. niger*. However, some small species such as *Dendropsophus bifurcus*, *D. rhodopeplus*, *D. sarayacuensis*, and *D. reticulatus* are also eaten. It should be noted that species such as *R. marina* and *R. ecuadorensis* have toxic compounds in their skin; however, their skin is removed before they are cooked. It is also remarkable that the tadpoles of several species are eaten as well, including *L. palmipes* (Gallardo, 2004) and those of *Boana appendiculata* (Kohn, 2002). The latter author mentions that only the tadpoles but not the adults are eaten by the Kichwa Ávila Runa of Amazonian Ecuador.

The Bullfrog Industry

The bullfrog industry began to develop in Ecuador in the mid-1980s (Gallardo, 2004), and it is steadily growing, especially in Provincia Zamora Chinchipe in southeastern Ecuador. It has been criticized and even prohibited given

TABLE 1.3

Edible Amphibians from Ecuador with the Vernacular Name, the Group That Consumes It, the Scientific Name, and Source References

Atya	Kichwa	*Boana appendiculata*	Kohn (2002)
Bahui	Kichwa	*Leptodactylus wagneri*	Kohn (2002)
Dequere, dequiri	Kichwa	*Rhinella* sp. (*margaritifera* complex)	Kohn (2002)
Gualag	Kichwa	*Leptodactylus pentadactylus*	Enríquez (1999)
Guingui	Kichwa	*Pristimantis kichwarum*	Kohn (2002)
Juwin, juin, uwin	Kichwa	*Leptodactylus pentadactylus*	Enríquez (1999), Dea (2006)
Kayla, kayra, caila, k'ayra, Kaila	Kichwa	*Telmatobius niger*	de Velasco (1789), Coloma (1984c), Merino-Viteri *et al.* (2005), Chimbo Aguinda *et al.* (2007)
Kuguayu, kuwayu	Kichwa	—	Morales and Schjellerup (1999), Ministerio de Educación Ecuador (2019)
Maculla	Kichwa	*Boana appendiculata*	Kohn (2002)
Murintu, murindu, murundi	Kichwa	—	Morales and Schjellerup (1999), Ministerio de Educación Ecuador (2019)
Mutsui, mutsuin	Kichwa	*Rhinella* sp. (*margaritifera* complex)	Kohn (2002)
Rusu mama	Kichwa	*Lithobates palmipes*	Gallardo (2004)
Tilele	Kichwa	*Rhaebo ecuadorensis*	Kohn (2002)
Tulumba, tulumpa, tullumpa, tulomba, telili, tulumba, yacu tilimbu, tulumba maqui	Kichwa	*Rhinella marina*	Enríquez (1999), Morales and Schjellerup (1999), Kohn (2002), Dea (2006), Ministerio de Educación Ecuador (2019)
Uchu cara	Kichwa	*Rhinella* sp. (*margaritifera* complex)	Kohn (2002)
Unculu	Kichwa	—	Morales and Schjellerup (1999)
Win, uwin	Kichwa	—	Morales and Schjellerup (1999), Dea (2006)
Ashanka, shanga, ashanga	Shuar	*Boana boans*	Coloma (1984b), Morales and Schjellerup (1999)
Chirimiu, chimius	Shuar	*Dendropsophus bifurcus*, *D. rhodopeplus*, *D. sarayacuensis*, *D. reticulatus*	Botasso (1977), Coloma (1984b), Morales and Schjellerup (1999), Gallardo (2004)
Ipiakáchi	Shuar	—	Botasso (1977)
Jante, janta	Shuar	*Scinax ruber*	Coloma (1984b), Botasso (1977), Morales and Schjellerup (1999)
Juátu, juato	Shuar	—	Botasso (1977), Coloma (1984b)
Jurújur, jurujru	Shuar	—	Botasso (1977), Morales and Schjellerup (1999)
Kaka, káka	Shuar	—	Botasso (1977), Pellizzaro (1996), Morales and Schjellerup (1999)
Karakras, karagras	Shuar	—	Coloma (1984b), Morales and Schjellerup (1999)
Katse	Shuar	—	Botasso (1977), Morales and Schjellerup (1999)
Kiamp	Shuar	—	(Bianchi, 1988)
Kiria	Shuar	—	Botasso (1977), Bianchi (1988)
Kirisip´	Shuar	—	Botasso (1977)
Kirúupash, kirupash	Shuar	—	Botasso (1977), Morales and Schjellerup (1999)
Kua, kuwa	Shuar	*Phyllomedusa* sp.?	Botasso (1977), Morales and Schjellerup (1999), Dea (2006), Ministerio de Educación Ecuador (2019)
Kupáip´, kuraip	Shuar	—	Botasso (1977), Morales and Schjellerup (1999)
Pai, paai	Shuar	—	Botasso (1977), Morales and Schjellerup (1999)
Paik´	Shuar	—	Botasso (1977)
Pakui	Shuar	*Leptodactylus wagneri*, *Adenomera* sp.	Botasso (1977) (1978), Coloma (1984b), Morales and Schjellerup (1999)
Peut	Shuar	—	Botasso (1978)
Puach	Shuar	—	Botasso (1977), Bianchi (1988), Morales and Schjellerup (1999)
Puint, puint´, puintiag	Shuar	*Leptodactylus pentadactylus*	Bianchi (1988), Enríquez (1999), Morales and Schjellerup (1999)
Purusham	Shuar	—	Botasso (1977)
Sháank	Shuar	—	Botasso (1977)
Súwi	Shuar	—	Botasso (1977)
Suya	Shuar	—	Morales and Schjellerup (1999)
Takash	Shuar	*Rhinella* sp. (*margaritifera* complex)	Coloma (1984b), Morales and Schjellerup (1999)

TABLE 1.3 (*Continued*)
Edible Amphibians from Ecuador with the Vernacular Name, the Group That Consumes It, the Scientific Name, and Source References

Tentépu	Shuar	—	Botasso (1977)
Tepém	Shuar	—	Botasso (1977)
Tseat	Shuar	—	Botasso (1977), Bianchi (1988)
Uchi-juat	Shuar	—	Botasso (1977)
Wirisam	Shuar	—	Botasso (1977)
Burú	Tsáfiqui	*Lepptodactylus peritoaktites*	Primitivo Aguavil (Coloma, 2016c)
—	Waorani	*Boana boans*	Tirira and Ríos (2018)
—	Waorani	*Lithobates palmipes*	Tirira and Ríos (2018)
Walak	Waorani	*Leptodactylus pentadactylus*	Tirira and Ríos (2018)

the dangers of an introduced species to the local ecosystems and native animals (Gallardo, 2004), as has been noted in other countries such as Colombia and Mexico (Rueda-Almonacid, 1999; Lynch, 2005; Casas Andreu *et al.*, 2001). The American bullfrog, *Aquarana catesbeiana*, is also considered an important vector of pathogens (Schloegel *et al.*, 2009).

The bullfrog was introduced to Ecuador in 1985 (Gallardo, 2004; not 1988, as indicated by Cisneros-Heredia, 2004). In 1998, bullfrog farms were established in Guayas and Los Ríos provinces in the southwestern coastal region of Ecuador and in the Napo, Pastaza, Morona Santiago, and Zamora Chinchipe provinces in the Amazonian region (CFN, 1998). According to Cruz-Cordovez *et al.* (2020) and citations therein, during the 1990s, 23 farms were established in Ecuadorian Amazonia (16 in Zamora Chinchipe, 5 in Napo, 2 in Pastaza, and 2 in Morona Santiago) and 6 more farms on the Ecuadorian Coast (4 in Guayas and 2 in Los Ríos). In Amazonia, feral populations of bullfrogs were reported near an abandoned bullfrog farm in Provincia Napo (Cisneros-Heredia, 2004) and in a biological station (CEDAMAZ) located in Provincia Zamora Chinchipe. In the Andes, a feral population of bullfrogs was reported in a recreational park in Loja city. Only a few individuals of this species have been reported on the northern Ecuadorian coast (e.g., at Bahía de Caráquez). These records confirm that bullfrogs are escaping and establishing in natural conditions in Ecuador, even inside a protected Ramsar site, such as Isla Santay, in the outermost part of the Río Guayas basin (Cruz-Cordovez *et al.*, 2020). Herein, we provide an updated map of feral populations with records from museum collections, sightings recorded on iNaturalist, and our personal records (Figure 1.54).

The bullfrog industry had a high peak of exports in 1998. This explosion of farms was followed by the closing of most of them, except in the Amazonian region in Provincia Zamora Chinchipe, where Asopiunank (Asociación de Productores Agropecuarios Piuntza—Nankais) began its activities in 2014 (Figure 1.55). In the global context, Ecuador's bullfrog production, based on data reported by the Food and Agriculture Organization in global aquaculture production, averaged 1.6 million kg/year between 2005 and 2012. It then increased to over 4 million kg in 2013, suggesting that Ecuador is the largest producer and responsible for at least one-fifth of the global market for this species (Sinovas and Price, 2015). The Ecuadorian exportations by Asopiunank are increasing rapidly, especially since 2021. They supply mostly the USA market, but also the local consumption in southeastern Ecuador.

Gallardo (2002, 2004) compiled information about raniculture in Ecuador. Herein we provide her information ordered chronologically and add data up to 2022:

1985. A small breeding stock was imported from Brazil, with which trials began at the Rehabilitation Center of Manabí (Gallardo, 2004). There are unconfirmed records of imports in 1984.

1988. The Ministerio de Agricultura y Ganadería (MAG) granted Franklin Alarcón permit No. 1 for the introduction of 100 adults and 70,000 bullfrog tadpoles to Provincia Guayas, Ecuador (Casares, 1992; Cruz-Cordovez *et al.*, 2020).

1991. The *Fundación Herpetológica Gustavo Orcés* (FHGO) and MAG were alerted by officials from the Ministerio de Medio Ambiente de Colombia about the presence of the bullfrog in Ecuador.

1992 (23 January). the FHGO presented a talk titled “Analysis of Frog Farming in Ecuador”, and they continued to report on the ecological damage of this species in addition to the resolutions taken in Colombia to prohibit the production of this species.

1993. The existence of 20 frog farms in Ecuador was recorded. Most of them were in the Provincia Zamora Chinchipe (CFN, 1998).

1994 (19 January). PUCE alerted the authorities about the ecological risk caused by the bullfrog.

1994. The Instituto Ecuatoriano Forestal, de Áreas Naturales y Vida Silvestre (INEFAN) published a brochure that exposed the negative impacts on the environment and concluded with the prohibition of

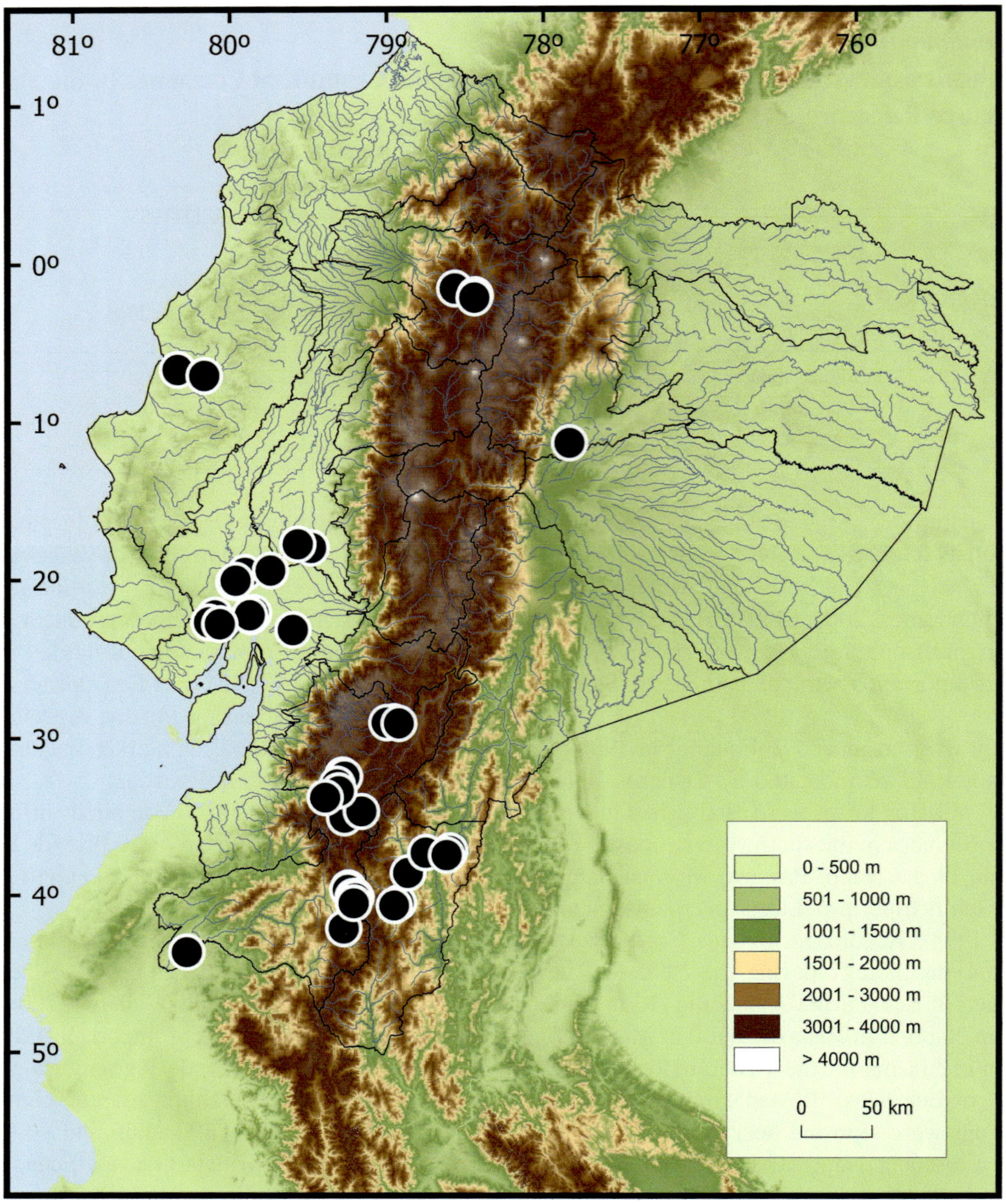

FIGURE 1.54 Occurrences of feral populations of the American bullfrog *Aquarana catesbeiana* in Ecuador. Updated and modified from Cruz-Cordovez *et al.* (2020).

FIGURE 1.55 The American bullfrog *Aquarana catesbeiana* in a frog farm in Provincia Zamora Chinchipe, Ecuador. LAC.

the establishment of frogs in the Ecuadorian Amazonia and in northwestern Ecuador.

1996. INEFAN published information about the illegal establishment of frog farms in Ecuador and indicated the prohibition by this institution of introducing and breeding this species.

1998. The number of frog farms rose to 30, which produced 12.2 tons of frog meat per month, which was mainly exported to the USA, France, Spain, Belgium, the Netherlands, and Germany (CFN, 1998).

1998–2001. A total of 4.1 tons were exported to Chile, Spain, and Argentina (BCE, 2004).

2000–2005. Ecuador was the third largest exporter of live frogs (after Taiwan and Brazil) to the markets in USA (Schloegel *et al.*, 2009).

2006–2009. A total of 51.9 tons were exported to USA (BCE, 2009).

2019–2021. No exportations are recorded.

2022. A total of 156 tons were exported to the USA by Asopiunank (Asociación de Productores Agropecuarios Piuntza—Nankais), Provincia Zamora Chinchipe.

At the end of the 1990s, Gilda G. Gallardo began to study the biology of the native species *Lithobates palmipes* (Rusu mama) from the Amazonian region (Gallardo, 2001). She was looking for ways to develop technologies to raise and breed this native species on an industrial scale in order to replace the species in the farms of the exotic bullfrog *Aquarana catesbeiana*. The cultivation of native species is a desirable and environmentally friendly alternative for local farmers and communities and would eliminate the problems associated with exotic introductions. As reported in Table 1.3, there are at least 50 species of Ecuadorian edible frogs. In 2004, Gallardo published a manual with natural history data and results of her preliminary essays to encourage the farming of native species (Gallardo, 2004). However, these studies and trials did not continue, and farms with these species have not been implemented, whereas the bullfrog industry has continued in Ecuador.

TRADITIONAL MEDICINE

Due to the chemical compounds in their skin, anurans have been part of ancestral medicine and are part of contemporary medicine. In the Ecuadorian popular imagination, toads are used to combat evils of ailments such as "mal agüero, mal de ojo y espanto" (bad omen, the evil eye, and fright) and have been or are used, especially as analgesics (for toothache), antibiotics (to combat diseases such as erysipelas, conjunctivitis, scabies, warts, cataracts), disinfectants, healing agents, etc. They are also currently used in the search for new medicines. The following historical and contemporary accounts provide evidence of this.

The first to mention the medicinal uses of toads (presumably from Guayaquil) is Cicala (1994), in 1771 who said the following:

> Another thing that was not known about toads is the very effective virtue attributed to toad bones, to remove promptly by touch the pain of the teeth no matter how intense it may be and I saw it practiced by many and used even by educated and wise people [Our translation].

Furthermore, de Velasco (1789) confirms this when he says that there are poisonous Hambatus, some listed, which are of medium body size.

> There are some of very active poison in a mountainous desert in the province of Ibarra, entirely green with red eyes, high, six fingers long. These ones are purposely sought for the remedy of cavities in molars, which cannot be removed. A subtle sliver of its bone is gently inserted and covered with cotton, taking care not to touch the tongue, and in a short time the tooth falls out, splintered to its roots [Our translation].

De Velasco (1789) also mentions that the Cailas are given as food to those women raising children whose milk dries up, and then it comes with such abundance that perhaps they need the opposite remedy to contain it.

Simmons (2009) briefly describes the important role of frogs and toads in Andean folk medicine. Perhaps one of the most extensive medicinal (shamanic) uses of the toads is to cure the "mal agüero (mal de ojos) and espanto"—espanto is a form of shock often caused by the consternation of encountering a ghost—and to cure ñahui-nanay (a condition of dry, bloodshot eyes that results from being in the sun without a hat, or working in poor light). To cure ñahui-nanay, the belly of the toad was slit open and rubbed on the patient's neck and forehead.

A woman from Guanujo, Provincia Bolívar, mentioned that the puca-sapo (*Atelopus guanujo*) was traditionally employed to "cleanse children's eyes from snakes and lizards". The latter phrase is a colloquial expression conveying the removal of fears or unsettling thoughts. The animal is rubbed on the child and the toad dies (Coloma, 1989b). Jambato harlequin frogs (*A. ignescens*) have been used to cure "el espanto" (fright) and to treat "chumbes", which are pellets that resemble tacks in sensation when felt against the legs (Coloma, 2016d). This is how Botasso (1977) explains their use among the Shuar in the Amazonian region: "With the marankit frog one cures the 'mal de ojos'".

Traditionally, tadpoles (presumably of *Gastrotheca* spp.) have been used in the Ecuadorian highlands to cure eye tearing and cataracts, although, to our knowledge, no written description of this practice is available. For erysipelas (a skin infection caused by the bacterium *Streptococcus pyogenes*), according to Acero Coral and Pianalto de Dalle Rive (1985), a common remedy in the medicine of the Cacha-Chimborazo Indians is to tie a live toad on the infected part and leave it there until it dies, then remove the toad and throw it away. The toad is tied with maqui (= hand) and yuyu (= nettle) with black sheep wool. To heal wounds, the use of *Rhinella marina* and/or *R. poepigii* has been documented by the Kichwa Indians of Napo in the Misahuallí area. Argo (1996) and Arbeláez Ortiz and Vega

Toral (2008) mention the medicinal uses of ucugs (= kaylas) (*Telmatobius niger*) in the provinces of Cañar and Azuay. The skin of the ucugs was used to make homemade ointments to combat dermatological problems. Live ucugs were rubbed on affected areas of the body to absorb the disease. On other occasions, ucug extracts were made to alleviate illnesses in horses and cows and to improve the quality of their coats. The use of toads such as *R. horribilis* and *R. marina* to cure erysipelas is widespread among Ecuadorians. Erysipelas typically involves the skin's upper layers and can cause redness, swelling, and pain. According to a resident from Provincia Los Ríos (com. pers. to Luis A. Coloma, in 2024) who was affected by this pathology, three toads were rubbed against the affected area on his leg, and then the animals were discarded and died, and he was cured.

BIOTRADE

Live Frogs for Research, Education, and the Pet Market

The Banana Republic Frogs Outflow and Biotrade

Live amphibians have been exported from Ecuador since the mid-20th century, with some accidentally traveling abroad, while others were sent for scientific research and exhibitions in zoos. The majority of exported frogs supply the global amphibian pet trade industry while also fulfilling the needs of the scientific community and zoos. Thus, scientific research, the biotrade industry, and the amphibian hobby have developed in parallel since the 1950s, although each has experienced its own ups and downs based on each country's regulations and diverse circumstances. For about 70 years now, thousands of Ecuadorian frogs have been brought into the international marked sourced from the wild. The live amphibian exports apparently stopped growing when the Ecuadorian wildlife regulations either banned or restricted exports, particularly since 1987, when brightly colored dendrobatids were regulated by the Convention on International Trade in Endangered Species of Wild Fauna and Flora (CITES, Appendix II), and their exportation required permits that were more difficult to obtain. This action was highly controversial and criticized by the scientific community because it affected ongoing research activities. Paradoxically, it is widely known that CITES regulations did not stop the black- and gray-market trafficking of live amphibians, and many wild frogs have been smuggled mainly for the pet trade up through the mid-2010s. In addition, scientific activities involving live frogs were slowed down due to biopiracy fears in the 1990s. An ethical biotrade of live Ecuadorian amphibians began to develop in the 2000s as part of the country's transition to a sustainable bio-economy. Since then, most of the frogs exported are legally lab-raised or farm-raised in Ecuador, serving as commercial stock for companion animals in the pet trade and for education, research, and conservation purposes.

Curiously, the first recorded instances of live Ecuadorian amphibians being sent out of the country were accidental shipments that continued until recently. These incidents began with the postwar banana boom that started in 1948, leading to Ecuador's recognition as a "Banana Republic", with bananas currently being the country's primary non-oil export product. Hylid frogs, such as *Trachycephalus jordani*, *T. quadrangulum*, *Scinax quinquefasciatus*, and *S. tsachila*, which are abundant and well-adapted to live in association with the stalks of banana plants in the western lowlands of Ecuador, were often exported by accident due to inadequate controls and quarantines during the 20th century. They have likely been leaving Ecuador since 1910, when the first banana exports began. However, the first documented specimens that arrived in Europe and the United States as stowaways in banana stalks were shipped from Guayaquil, Ecuador, in the early 1950s (Hartweg, 1955). The frogs (*T. jordani*) survived the long ocean voyages and usually made their appearances in port cities, with some even finding their way to museums. One of these stowaways was received by the UMMZ from a dealer in Detroit. The frog was placed in a terrarium and fed local insects for more than three years, at which time it escaped. Surprisingly, stowaways continue to evade modern controls, occasionally appearing in containers with bananas and other products from Ecuador. Despite increasing regulations and biosecurity measures in the ports of origin, there have been reports of interceptions of Ecuadorian amphibians in New Zealand, Poland, England, and the USA since 1974. Gill *et al.* (2001) reported five interceptions in Auckland, New Zealand, of *T. jordani* (two interceptions) and *S. quinquefasciatus* (three interceptions) in banana shipments coming from Ecuador between 1974 and 2000. The MONZ museum in Wellington, New Zealand, holds five specimens of *T. jordani* and two of *T. quadrangulum* from Ecuador, some of them from Puerto Inca, Provincia Guayas, which were intercepted between 2002 and 2010. More recently, there was a report of a *T. jordani* at SeaWorld San Diego, USA. This animal is named Peter and came into San Diego Harbor as a stowaway in a Dole banana shipment from Ecuador (Sartore, 2021) (Figure 1.56). Another recent record is from December 2019 of a frog (*S. tsachila*) from

FIGURE 1.56 A Jordan's bone-headed tree frog (*Trachycephalus jordani*) at SeaWorld San Diego. Joel Sartore.

Ecuador found in bananas in Northamptom, England, after a 5000-mile trip that began by hopping into a shipping crate (BBC news, 2019).

The Boost of the Dream for Scientists and Frog Lovers in Europe and the USA

The advancement of biology and experimental biology studies in the postwar period of the 20th century generated a surge of interest in organismal biology, leading to the first live frogs being sent abroad from Ecuador for research and zoo purposes to Europe (Britain, Germany, Holland, France) and the USA, starting in 1955. The marsupial frog *Gastrotheca riobambae* (also known as *G. marsupiata* and *G. marsupialis*) was a major focus of the early foreign research attempts with live frogs, which began to be published in 1957 and experienced a significant boost that lasted about three decades, until 1986. During this time, 15 papers on the reproductive and developmental biology of live specimens sent to Europe and the USA were published by researchers such as Walker (1957), Harrison Matthews (1958), Grenard (1958), Oeser (1959), Brenner (1960), Weiss (1960), Deckert (1963), Kästle (1963), Hoogmoed (1967), Spannhof and Spannhof (1971), Jones *et al.* (1973), Ensinck (1974, 1978), Boonman (1985), and Aubér-Thomay and Letellier (1986). They obtained their specimens from Quito, Riobamba, Ibarra, and other locations in Ecuador. Some of these papers were published in scientific journals, while others appeared in terrarium and aquarium magazines. In the meantime, starting in 1975, a significant amount of research on this species was conducted and published locally by the Ecuadorian scientist Eugenia M. del Pino, her students, and collaborators (as discussed in the "Remarks" section of the *G. riobambae* account and the "Del Pino and the Marsupial Frogs" subsection in Chapter 2), making *G. riobambae* the best-known anuran in Andean South America.

The first published report on *Gastrotheca riobambae* was by John Walker in 1957, describing two pairs of frogs that arrived in Britain in October 1955. They were imported from Ecuador by Robert Bustard. Walker published his observations on amplexus and two events of mating and breeding in the *British Journal of Herpetology*. At about the same time, in 1957, Professor Emmanuel Ciprian Amoroso of the Department of Physiology at the Royal Veterinary College in London made a color cinematograph film, which is the first known film on an Ecuadorian frog. The film recorded the biology of *Gastrotheca riobambae*, showing the laying of the eggs, the way they are placed in the female's pouch by the male, and the way the tadpoles are released from it after development. This film was presented by the Director of the Zoological Society of London, Leonard Harrison Matthews, in a conference organized by the Societe Zoologique de France, at which he used it to discuss the evolution of viviparity. The conference proceedings, including Amoroso's photo of a mating pair with the male's hind feet inserted into the female's dorsal pouch, were later published in the *Bulletin de la Societe Zoologique de France* in 1958 (Harrison Matthews, 1958) (Figure 1.57).

In 1957, Steve Grenard in the USA also received a shipment of frogs from Ecuador, which included a gravid female frog sent for identification. Grenard described the reproductive biology, rearing, and maintenance of this frog in his publication. The female frog gave birth to 94 tadpoles, some of which were raised at the New York Zoological Park (Grenard, 1958) (see Figure 1.57). Grenard notes its provenance as follows:

> On November 22, 1957, I received a small shipment of South American anura collected and exported by Robert Copping, in Quito, Ecuador. Included was a female Marsupial Frog, A.M.N.H. 60631, snout to vent, 50, which was sent to Dr. Charles F. Walker, Mus. of Zool., Univ. Mich., for identification.

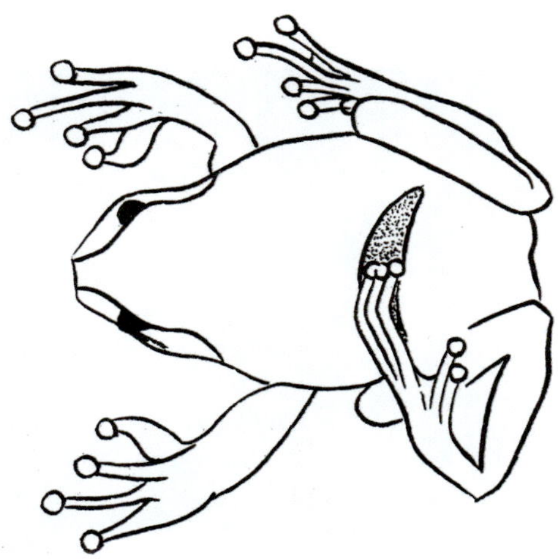

FIGURE 1.57 Left: the first known published photo of an Ecuadorian species taken in 1957 in London by Emmanuel Ciprian Amoroso. It depicts the amplexus of a mating pair of *Gastrotheca riobambae*, where the male is opening the pouch with its hind feet. Taken from Harrison Matthews (1958). Right: female *G. riobambae* opening the pouch with the hindfoot. Taken from Grenard (1958).

In March–April 1958, a researcher in Germany, R. Oeser, received a shipment of "large quantities" of marsupial frogs (*Gastrotheca riobambae*) from Ecuador. Oeser described tadpole growth and aspects of breeding and maintenance, while noting that this species was a dream for frog lovers at the time (Oeser, 1959). Brenner (1960) subsequently described the mating and development of these frogs. On 16 April 1959, Hans Weiss received a pair of marsupial frogs from Herrn Pölzein and described a breeding event and the rearing of tadpoles (Weiss, 1960; Kästle, 1963). Weiss studied and filmed the spawning process, as well as the release of tadpoles from the pouch and their development, while also describing the call and providing photographs taken from the film. Deckert (1963) described in detail the amplexus, mating behavior, and deposition of eggs into the pouch based on three males and four females provided by his colleague Dr. Heinz Wermuth of the Museum of Humboldt University in Berlin. According to Kurt Deckert, these frogs were introduced to Europe a few years earlier.

Researchers in the Netherlands, which later became a bastion of the frog hobby, also studied the marsupial frog *Gastrotheca riobambae*. Nine years after the first shipment to Europe, in early 1964, the Rijksmuseum van Naturlijke Historie in Leiden (RMNHG) received a pair of adults from the vicinity of Quito, Ecuador. These frogs were a gift from W. G. van Oyen. Marinus Steven Hoogmoed (Figure 1.58), a young herpetologist pursuing a master's degree in biology at the University of Leiden and a herpetology assistant at the RMNHG, described aspects of the morphology, mating, and early development based on these animals. He provided quantitative data and information on color changes in young frogs for the first time (Hoogmoed, 1967).

Spannhof and Spannhof (1971) described aspects of the incubation of marsupial frogs and their related behavior, as well as the histology of the pouch, while at the University of Rostok in Germany. They observed not only the vascularization of the pouch but also the formation of egg chambers. They found that the incubation of embryos in the pouch lasts approximately five to six weeks, but the larvae may remain in the pouch for considerably longer periods, depending on environmental factors.

FIGURE 1.58 Marinus Steven Hoogmoed. A herpetologist from the Netherlands who studied *Gastrotheca riobambae* during the 1960s. Taken from Hoogmoed (2019).

Another important early research study based on live Ecuadorian frogs was published by Jones *et al.* (1973) from the Laboratory of Comparative Reproduction of the University of Colorado, USA. They studied *Gastrotheca riobambae* and conducted some experiments using live frogs supplied by Duellman and Thomas H. Fritts. They wrote about their acquisition as follows:

> In late January, 1972, we were fortunate in receiving 13 8–12 week old female *G. riobambae* collected from Ecuador (Chimborazo Province: Riobamba; Pichincha Province: Quito). We also obtained two adult female *G. riobambae* with well-developed brood pouches (from Pichincha Province). We took advantage of this acquisition to study the histology of the formed brood pouch and to explore the hypothesis that the development of this structure is controlled at least in part by estrogen.

At the end of 1968, Frans H. Ensink, a well-known breeder from the Netherlands, received one male and three females via the pet trade, with their provenance marked as "Ecuador?". He successfully bred the species and reported (in Dutch) about its rearing in the journal *Lacerta* (Ensinck, 1974), later providing a German translation in an aquarium magazine (Ensinck, 1978). *Lacerta* was the journal of the Dutch association for herpetology and terrarium science.

In 1985, Boonman, from the Netherlands, described the keeping and rearing of *Gastrotheca riobambae* under terrarium conditions, including its reproduction, larval development, and raising of metamorphs. He noted that the animals were most likely wild-caught from Ecuador and that the species has no extreme environmental demands. The most important factor for successful reproduction, according to Boonman, is a moderate temperature $(16-22^{\circ}C)$ during metamorphosis.

Michèle Aubér-Thomay, a researcher from the Université Pierre et Marie Curie, led relevant and more recent studies of marsupial frogs with live animals in France in the 1980s (Figure 1.59). Aubér-Thomay and Letellier (1986) studied *Gastrotheca riobambae*, and Aubér-Thomay *et al.* (1990) studied *G. plumbea*. A gravid *G. riobambae*, courtesy of Ecuadorian biologist Xavier Silva del Pozo, was collected in 1984 in Ibarra, Provincia Imbabura, and released 124 tadpoles in Paris. The frogs were reared until the third generation, and Aubér-Thomay and Letellier (1986) described the mating behavior and the insertion of the eggs into the pouch and produced a film in collaboration with the Service du film de Recherche Scientifique in Paris, France. Four years later, Aubér-Thomay *et al.* (1990) described some aspects of the reproductive biology of the monophasic species *G. plumbea*, which gives birth to froglets, based on a gravid female from Provincia Bolívar. The study was performed in cooperation with members of the Departamento de Ciencias Biológicas of Pontificia Universidad Católica del Ecuador and included Giovanni Onore, head of the Zoology Museum, and biology student Luis A. Coloma.

FIGURE 1.59 Michèle Aubér-Thomay at Molleturo, Provincia Azuay, Ecuador, on March 1993. LAC.

The German Big Bang of Poison Dart Frogs in Europe

About a decade after the arrival of the marsupial frog *Gastrotheca riobambae* in Europe, another group of Ecuadorian species began to appear in Europe, mostly during the 1960s, 1970s, and 1980s. These were the dendrobatid poison frogs, also known as poison dart frogs. These frogs quickly became popular research animals and some of the most sought-after amphibians for zoos and terrarium keepers, particularly in Europe (Germany, Netherlands, Belgium, Denmark, and Switzerland) and later in the USA, Canada, and Japan, where they also became major players (Lötters *et al.*, 2007; Stuster *et al.*, 2017). They were part of a trend emerging in the late 1960s that focused on the regular importation to Europe of various species of amphibians from Costa Rica, Colombia, and Suriname.

Since the 1960s, Germany has been at the forefront of interest in dendrobatids. Live dendrobatids from Ecuador have been a topic of great scientific interest for decades, leading to a series of publications, starting with Mudrack's 1969 paper and continuing through the 1980s and up to the present day. In 1964, a group of 12 non-aposematic *Hyloxalus awa* individuals (previously identified as *Phyllobates* sp.) were documented as the earliest known dendrobatids from Ecuador to arrive in Europe, specifically in Berlin, Germany. These individuals were collected in the summer of 1964 at about 600 m, 30 km from Pujilí and 20 km from Quevedo, Provincia Cotopaxi. They were acquired by a bus driver who lived in West Berlin, Wolfgang Mudrack, who was a pioneering and well-known herpetoculturist, who kept, bred, and studied them in a terrarium. Mudrack published noteworthy data on their reproductive biology, reporting that:

> Copulation took place mostly in early morning. Four to sixteen eggs were deposited above the water level and guarded by the male and female alternatively. The tadpoles are being carried about by the parents for two to six days, mostly on their back [Our translation].
>
> (Mudrack, 1969)

This observation of both parents taking care of the clutch is unique for a species of the *Hyloxalus* genus, in which only males have been reported to perform parental care. This observation was later confirmed by Correa Monge (1995).

In the 1970s, more Ecuadorian dendrobatids arrived in Europe, starting with the aposematic *Epipedobates anthonyi* from Río Jubones, Provincia Azuay, which was first imported to Germany. Sixteen frogs were part of this shipment (Schmidt, 1981). Hagen Schmidt, who received the frogs, extensively documented the morphology, coloration, variation, tadpoles, behavior, territoriality, and breeding of *E. anthonyi* (formerly known as *Dendrobates tricolor*). He included illustrations and color photographs of this population (Schmidt, 1981). Also, Bech (1982) provided reproductive data for E. anthonyi. Schmidt noted that this species was particularly suitable for terraria due to its ease of care and breeding. The arrival of the import was an exciting moment for Schmidt, as he recounts in his publication (translated from German):

> In the spring of 1976, the Lufthansa cargo department at Hannover Airport informed me about the arrival of a shipment of plants and amphibians. Every animal and plant lover probably knows the tension that such a telephone call causes. There will be postponed appointments, and the daily routine begins to interfere with the unfamiliar hectic pace, and one swings across the highway. Then you finally stand at the airport customs office, take care of the necessary paperwork, and discuss import regulations or restrictions with the officials. Finally, you hold the package in your hands that you have been eagerly anticipating. The tension rises again when unpacking, and the officer may shake his head in disbelief at what he sees as the gray, dried-up plants that can hardly be worth the postage. And then suddenly something moves amongst the *Tillandsia*—a few small, red and yellow striped frogs hop cheerfully in the direction of daylight. Everyone is surprised; no one had ever seen one before. They quickly fall in love with them and from the admiring remarks, one can see again how much the colorfulness of these dendrobatids appeals even to those who otherwise like to distance themselves from these "slippery creatures".
>
> (Schmidt, 1981)

FIGURE 1.60 From left to right: Giovanni Onore, Helmut Zimmermann, Elke Zimmermann, and Luis A. Coloma in March 1986 at the Departamento de Ciencias Biológicas of Pontificia Universidad Católica del Ecuador, in Quito (unknown photographer).

The arrival of Ecuadorian dendrobatids in Europe during the 1970s primarily included *Epipedobates anthonyi*, *E. boulengeri*, *E. espinosai*, and *Oophaga sylvatica* (known at that time as *Phyllobates anthonyi*, *P. tricolor*, *P. boulengeri*, *P. boulengeri-espinosai*, and *Dendrobates histrionicus*, respectively). In the following decade, *Ameerega parvula*, *A. bilinguis*, *Allobates femoralis*, and *A. zaparo* also became available in Europe.

During the 1980s, many of the publications were by two prolific science and outreach writers: Helmut Zimmerman, a German architect, conservationist, and herpetologist, and his daughter Elke Zimmermann (1958–2019), a well-known zoologist, primatologist, and behavioral scientist. Over 15 years, they raised about 5000 individuals of 28 species of Neotropical dendrobatids from several generations of frogs at home. Most of their work was conducted while Elke was a student obtaining her doctorate at the University of Stuttgart-Hohenheim and later working as a research assistant at the Institute of Zoology of the same university. Their detailed observations and studies encompassed various aspects of the acoustics and behavior (Zimmermann and Zimmermann, 1985), taxonomy and zoogeography (Zimmermann and Zimmermann, 1988), conservation (Zimmermann, 1989), and husbandry of dendrobatid poison frogs (Zimmermann and Zimmermann, 1994). The Zimmermanns published many reports on the biology, captive breeding, and conservation of poison dart frogs in the Cologne Zoo journal (*Zeitschrift des Kölner Zoo*). Their studies provided data for the seminal paper on the evolution of parental care by Weygoldt (1987). In 1984, 1986 (Figure 1.60), and 1988, the Zimmermans visited Ecuador to confirm some of their ethological observations, which were previously made in terrarium conditions and in the field, mainly at Centro Científico Río Palenque and around Santo Domingo de los Tsáchilas in the Pacific lowlands of Ecuador. The Zimmermanns' frog research, which was mostly published in aquarium and terrarium magazines, as well as their outreach books (e.g., (Zimmermann, 1979, 1983, 1986) was pioneering and promoted the frog hobby in Germany and elsewhere. Their research had an immense impact on making dendrobatids popular as companion animals in terraria due to their bright coloration, patterns, calls, and unique and interesting parental care behaviors, which were embraced with enthusiasm by the amphibian hobby community. Among the enthusiasts were many members of the Anuran Specialist Group of the German Society for Herpetology and Herpetoculture (Deutsche Gesellschaft fur Herpetologie und Terrarienkunde, DGHT), which was founded to obtain knowledge necessary for breeding and protecting endangered anurans. By 1989, the DGHT had affiliated 4500 professional and amateur herpetologists from 45 nations, and at that time, Helmut Zimmermann was the head of the Anuran Specialist Group. Currently, the DGHT is the largest association of its kind in the world.

One of the members of DGHT, Karl-Heinz Jungfer, was a biology teacher and herpetologist who also studied Ecuadorian poison frogs in terraria. He received a few live Ecuadorian frogs (at that time *Epipedobates parvula* and *E. zaparo*) from Wolfgang Böhme that were sent to the Zoological Research Institute and Alexander Koenig Museum in Bonn in 1985. While observing them in terraria and recording their calls, he discovered that there were more than two taxa among them. As a result, he conducted a taxonomic review (Jungfer, 1989) of the red-backed granulated dendrobatid frogs and described two new species: *Ameerega bilinguis* from Ecuador and *A. ardens* from Peru.

Jungfer also described four species of frogs from Ecuador in the genus *Osteocephalus*, adding information obtained from observations in aquaria and terraria, such as the calls of *O. fuscifascies* and *O. mutabor* and tadpoles of *O. deridens*. These descriptions were published in several papers (Jungfer *et al.*, 2000; Jungfer and Hödl, 2002; Jungfer, 2011).

After these pioneering studies, a large number of publications on poison frogs, tree frogs, and other species kept in terrariums proliferated in the popular literature and specialized magazines from the 1990s onwards. One beautifully illustrated book featuring colorful photos and paintings of dendrobatids is Jerry G. Walls' *Poison Frogs: Biology, Species & Captive Care* (Walls, 1994), which covered 65 species of poison frogs, including many of the species known from Ecuador at that time. A decade later, a series of fully photographically illustrated books (3 volumes, about 1900 photos) focused on *Epipedobates anthonyi*, *E. tricolor*, about a dozen species of *Ranitomeya*, and *Oophaga sylvatica*, all of them occurring in Ecuador, Peru, and Colombia, was published by Christmann (2004). This book describes the routes and localities in which they can be found, their habitats, and information of their breeding in terraria. The epilogue provides the author's insightful and pragmatic reflections on conservation efforts for dendrobatid frogs. In the late 2000s, another significant book was published by a group of specialized biologists and amateur herpetologists (Stefan Lötters, Karl-Heinz Jungfer, Friedrich Wilhelm Henkel, Wolfgang Schmidt, and Frank Mutschmann) (Lötters *et al.*, 2007). This tome compiled most of the scientific and amateur knowledge about dendrobatid poison frogs for the first time, providing the latest information on their systematics, biology, taxonomy, species, and husbandry, including numerous excellent photos. Certainly, this book (versions in German and English) became essential in the library of anyone seriously interested in dendrobatids.

Another compilation book was published a decade later in Dutch by Dendrobatidae Netherlands (Stuster *et al.*, 2017), which covered most of the species of the family Dendrobatidae, including numerous photos depicting variations of Ecuadorian Dendrobatidae and a summary of the Ecuadorian enterprise Wikiri.

During the 1990s, a plethora of publications about poison frogs, tree frogs, and other terrarium-kept species flooded the outreach literature and specialized magazines. This coincided with the emergence of herpetoculture magazines such as *Captive Breeding*, *Reptile & Amphibian Magazine*, *Draco*, *Reptile Hobbyist*, and *DN Magazine*, as well as the formation of organizations like the British Dendrobatid Group, American Society of Herpetoculturists, and Dendrobatidae Netherlands. However, this surge of research was short lived due to various circumstances. Live frog research, which had been thriving in Ecuador, was impeded by CITES restrictions, concerns about biopiracy, and the ambiguous and underdeveloped Ecuadorian and international regulations on genetic resources, which severely limited or even banned the research.

Since the 2010s, legally commercialized lab-bred and farm-reared live frogs, obtained with permits and subject to restrictions established by the Ecuadorian government and CITES authorities, have become increasingly available for scientific research both locally and internationally. Some of the dendrobatids, such as *Oophaga sylvatica*, *Allobates femoralis*, and *A. zaparo*, as well as the Fringe-limbed leaf frog *Cruziohyla craspedopus*, the Gualataco marsupial frog *Gastrotheca riobambae*, and several species of harlequin frogs *Atelopus*, are among the species currently being studied. *Gastrotheca riobambae* and *Atelopus* species are being raised in the lab and studied for conservation–reintroduction trials.

The Bans to Exportation of Ecuadorian Live Amphibians and the Smuggling Waves to Europe and USA

As described previously, live Ecuadorian frogs (mainly *Gastrotheca riobambae* and dendrobatids) had been in Europe and USA since the 1950s. In the USA, early breeding efforts with *Epipedobates anthonyi* (as *E. tricolor*) occurred in 1980, as recounted by Daly *et al.* (2000).

The first strict Ecuadorian regulations on the export of wild live frogs began on 14 August 1969, when the Ecuadorian President, José María Velasco Ibarra, signed a decree prohibiting the exportation of wild live animals for five years. This action was taken because "national and foreign exporters perpetrate abuses by clandestinely, disorderly, and damagingly hunting wild animals for export, circumventing the country's laws, to the detriment of its potential economy" (Comisión Legislativa Permanente, 1969). The decree was accompanied and supported by a regulation through a ministerial agreement of the Ministerio de Agricultura y Ganadería (MAG) in December 1969 (Duarte Valverde and Naranjo Fiallo, 1969). However, the initial regulation lasted only three months of the five planned years. In March 1970, (Espinosa and Durán Arias, 1970) reformed the previous ministerial agreement, responding to claims against it. Thus, the new regulations of March 1970 allowed the exportation of live wild animals, but only after obtaining approval from the MAG. In November of the same year, a new law to protect wildlife (Velasco Ibarra, 1970) transferred the granting of hunting and exportation licenses to the Ministerio de la Producción. This lasted until May 1977, when an official regulation by the MAG

(Vásconez and Peñaherrera Gallardo, 1977) totally prohibited the exportation of wild live amphibians for commercial purposes in its Article 60.

In 1981, revised wildlife laws of the MAG allowed live exports only for scientific purposes, with permits by the MAG (Baca Carbo and Vanegas López, 1981). Two years later, the law was revised again, opening the door to export animals obtained through adequate management procedures under captivity or semi-captivity (Hurtado, 1983). In 1987, the major disruptive change in the outflow and exports of live frogs from Ecuador occurred when brightly colored dendrobatids were included in Appendix II under the regulations by CITES (in response to requests from Suriname and the Netherlands). As a result of this, legal exportation now required fulfilling more complex national and international requirements. This inclusion occurred in spite of the initial objections of some members of the scientific community, represented by Charles Myers (USA), Marinus S. Hoogmoed (Netherlands), Stefan Gorzula (Venezuela), and Stephen R. Edwards (IUCN, Switzerland) (Gorzula, 1996). They foresaw unjustifiable CITES barriers to research (IUCN, 1987). However, after the official inclusion of dendrobatids in CITES, strong reactions were not long in coming, with many criticizing the bureaucratized CITES, including Mrosovsky (1988), who named his essay "The CITES Conservation Circus", published in the journal *Nature*. Others such as Pickett (1987), Bringsøe (1992), and Campbell and Frost (1994) added to the criticism of the bureaucratized CITES. As expected, the CITES restrictions greatly hindered further research on alkaloids from brightly colored dendrobatids, including *Epipedobates anthonyi* (Myers and Daly, 1993; Myers *et al.*, 2000).

On the one hand, Gorzula (1996) examined the legal trade of dendrobatids from 1987 to 1993 under CITES restrictions and found that the reported numbers were significantly lower than those prior to the restrictions. The trade consisted of 82.6% live wild-caught frogs and 13.1% live captive-reared frogs, with only 15,779 frogs reported. Seven legal shipments were made from Ecuador, exporting 70 dendrobatids, of which 52 (*Epipedobates anthonyi*, *E. espinosai*, *E. tricolor*, and *Oophaga sylvatica*) were sent abroad in 1987 without a declared purpose, presumably for the pet trade, and 18 (*E. anthonyi*, then named as *E. tricolor*, and *A. bilinguis*) in 1989 for commercial purposes. Gorzula remarked: "As a very crude illustration, the total number of dendrobatid frogs legally exported from the range countries over a seven-year period would not have filled a large trash can!"

On the other hand, it is widely known that despite national and international regulations under CITES, many live wild frogs were smuggled from the Neotropics for the pet trade between 1987 and about the mid-2010s. The international amphibian pet trade is known to promote smuggled species and morphs of several species through social networks, forums, web pages, fairs, and pet stores (Brown *et al.*, 2011; Yeager *et al.*, 2020, and citations therein). Brown *et al.* (2011) noted that the amphibian pet trade had been tainted by bad practices, such as phantom farming, gray practices, and other illegitimate methods of procuring specimens for the trade, which have caused distortions in the actual prices of some species. Low prices may result from either smuggling or over-collecting large numbers of specimens to meet the demand while avoiding the costly and cumbersome bureaucracy and paperwork required for the legal amphibian trade for commercial purposes under CITES and other regulations. It is worth mentioning that smuggling often leads to high mortality rates because most illegal specimens are exposed to horrendous shipping and living conditions on their way to the market.

Recent international regulations by CITES include those for phyllomedusids and centrolenids. As of 2010, five species of *Agalychnis* (Phylomedusidae) are included in CITES Appendix II, including *Agalychnis spurrelli* found in Ecuador. An attempt to include glassfrogs (Centrolenidae) under CITES in 2019 failed. However, in 2022, a proposal was adopted by consensus to list all glass frog species in CITES Appendix II, put forward by Argentina, Brazil, Costa Rica, Côte d'Ivoire, the Dominican Republic, Ecuador, El Salvador, Gabon, Guinea, Niger, Panama, Peru, Togo, and the United States. The listing was put forward at a family level due to the difficulty of visually differentiating between glass frog species by non-specialists. Nine glass frog species have been found hidden in animal shipments from Central America to Europe and are becoming more popular in the international pet trade. US import records show that imports of glassfrogs increased by approximately 44,000% between 2016 and 2021. The smuggling of glassfrogs from Ecuador has not been reported. While this CITES Appendix II listing is expected to promote increased data collection on the glass frog trade, its impact on smuggling and research remains to be seen.

The Sustainable Bioethical Trade

María D. Guarderas and Luis A. Coloma

From the 1990s to the mid-2010s, the demand for live Ecuadorian frogs in the pet trade market was high, and it was largely met by illegal smuggling and gray-market practices. Gray markets supplied individuals descended from illegal collections, offered by agencies with permits but that were not demonstrably sustainable, or sold under false pretenses, such as claiming to be farmed when actually wild caught (Yeager *et al.*, 2020). In response to market demands, a sustainable and ethical biotrade industry began to emerge as part of a trend in mega-biodiverse developing countries. Biotrade, also known as biocommerce, was born in the context of a transition from extractive economies to sustainable bioeconomies, in which biodiversity became a strategic and central resource in national economies to ensure the political feasibility of sustainable bioeconomic models (Ortega-Pacheco *et al.*, 2018). These concepts were reflected in Ecuadorian development strategies and planning, with the bioeconomy being considered a compromise between economy and nature (Kleemann *et al*., 2022). For example, the economic benefits from the use and management of natural resources were highlighted in the National Biodiversity Strategy 2015–2030 (MAE, 2015). Yeager *et al*. (2020) discussed the problems and benefits related to biotrade using poison frogs (and orchids) as case studies. The authors presented a comprehensive overview of ethical considerations and practical applications related to biotrade and offered recommendations for promoting sustainable practices in the industry. The authors cited the Wikiri enterprise in Ecuador as a successful model for leveraging commercial interests, in which legal commerce contributes to the protection and conservation of species.

The Rise of Amphibian Biotrade: Origins and Global Exports

In the 1990s, the first attempts at developing sustainable biotrade projects for amphibians were made. One of these projects was initiated by a private farm called Fauna Tropical, which obtained provisional permits from the Instituto Ecuatoriano Forestal y de Areas Naturales y de Vida Silvestre (INEFAN) in 1993 to manage and raise *Oophaga sylvatica*, *Epipedobates boulengeri*, and *E. tricolor*, among other species. Fauna Tropical exported 5892 frogs to the United States, including 4210 *O. sylvatica*, 969 *E. tricolor*, and 587 *E. boulengeri*, among others, under the names *Phyllobates* spp., *Dendrobates* spp., *Epipedobates boulengeri*, *E. tricolor*, *Dendrobates histrionicus*, *Oophaga histrionica*, and *O. sylvatica* to the United States (CITES, 2023). However, the reported numbers of frogs imported into the USA were impossible to raise under farm conditions, suggesting that most of them were wild caught. Fauna Tropical's permit was revoked in 1995 after an INEFAN inspection of its facilities near Viche, Provincia Esmeraldas. In 2002, Ecociencia, an Ecuadorian NGO that provides technical advice for the design and implementation of productive initiatives derived from the sustainable use of biodiversity, conducted a pre-feasibility study for the commercialization of ornamental frogs. However, this Ecociencia frog project did not progress further. In the same year, the first legal domestic and international trade of live amphibians began in Ecuador when a private store named Hyla obtained permits (Patente # LM-FAU-OI-DFP, Ministerio de Ambiente de Ecuador) to breed, raise, and commercialize *Gastrotheca riobambae* (Figure 1.61). The first lab-raised live specimens of *G. riobambae* were sold to the School of Biology of Pontificia Universidad Católica del Ecuador (PUCE), and about 100 were exported to Miami, USA. However, Hyla stopped its frog activities in 2004 due to the exacerbating bureaucratic delays in renewing permits.

In 2009, the Zoocriadero Sylvaticus of Pontificia Universidad Católica del Ecuador obtained a permit from the Ministerio de Ambiente del Ecuador (Patente de manejo N° 2009–011-FAU-DPAP-MA) for the management of six species for commercial purposes. In 2010, a private biotrade company, Wikiri, was created as a commercial branch aimed to commercialize frogs produced by the Zoocriadero Sylvaticus and to support research and conservation activities of the Museo de Zoología of PUCE. However, support for the activities of PUCE stopped at the end of 2010, following personnel changes at PUCE. In 2011, Wikiri (Wikiri S.A. and since 2016 Oophaga S.A.) was authorized to carry out frog management for commercial purposes by the Ministerio de Ambiente del Ecuador (Patente de manejo N° 03–2011-FAU-DPAP-MA). Since 2011, Wikiri has been the biocommerce arm of a research and conservation facility, Centro Jambatu de Investigación y Conservación de Anfibios, which boasts state-of-the-art laboratories, a veterinary clinic, and a museum (Guarderas, 2017). The company aims to incorporate science, social, and environmental responsibility into the global amphibian pet trade, and its product, EcuaFrog, was introduced to the market to combat wildlife trafficking and implement new, creative, and multidisciplinary efforts to reverse extinction trends. It incorporates science and social responsibility into the mindset of a business focused on providing high-quality products to the amphibian pet trade in line with BioTrade Principles.

Wikiri's production of amphibians for the pet trade in Ecuador consists of a combination of in situ production in natural locations adapted to maximize breeding and ex situ production in breeding facilities. The company has had strong support from the amphibian pet trade community and Understory Enterprises, a company based in Canada that distributed Ecuadorian frogs to the USA, Europe, and Japan in the first years. Currently, Wikiri has its own distributors in the USA (e.g., Indoor Ecosystems) and Europe.

To access the legal and sustainable pet trade for amphibians, the company has had to address various challenges, including delays in obtaining research and trade permits, fears of biopiracy, the threat of smuggling, and, more recently, concerns about the spread of pathogens.

The long-term growth and success of the Wikiri company will depend mainly on access to and the expansion of this market, which will directly impact its profitability.

Advancements in research for rapid and cost-effective pathogen detection, coupled with procedures for ensuring the biosecure handling of captive animals in trade markets, are enhancing the feasibility of implementing measures to mitigate the risk of disease spread. Consequently, efforts are being made to initiate an industry-led clean-trade certification program (Cavasos *et al*., 2023).

Information and photographs of managed frogs, lab facilities, and Wikiri's farms are provided by Verkade (2012, 2014), Stuster *et al*. (2017), and Guarderas (2017). In Europe, Wikiri (and Centro Jambatu) has been enthusiastically promoted by the Verkade family at frog fairs. They found for the first time in the wild some of the commercial morphs of *Epipedobates tricolor* (Cielito morph) and *Epipedobates darwinwallacei* (Verkade morph) (Figure 1.62).

In August 2022, Wikiri unveiled a new recreational, educational, and touristic enterprise called Wikiri Sapoparque, which is a theme park with the world's largest amphibian

FIGURE 1.61 Materials for the first species of frogs sold by Ecuadorian companies. Top left: booklet on *Gastrotheca riobambae* by Hyla. Top right: Wikiri's export containers in 2011. Bottom: brochure of *Ceratophrys stolzmanni* by Wikiri in 2011.

exhibit, located in Quito. The park features about 60 species of live frogs, including the critically endangered harlequin frogs, and offers visitors first-hand information on science, biotechnology, conservation, and biotrade (see Figure 1.63).

In 2015, Sinovas and Price provided an overview of trade volumes and trends, estimates of the economic value of the trade, and other noteworthy aspects of Ecuadorian amphibians, including CITES-listed species (CL). Their

FIGURE 1.62 José Verkade promoting Wikiri at a frog fair in Nieuwegein, the Netherlands on 4 October 2013. Jan C. Verkade collecting tadpoles south west of Chiriboga, Provincia Pichincha on 14 February 2023.

FIGURE 1.63 The giant Blomberg's toad (*Rhaebo blombergi*) (bottom left), Spurrell's gliding leaf frog (*Agalychnis spurrelli*) (top right), María D. Guarderas (bottom right), Cornelio Williams (top left), and visitors at Wikiri Sapoparque, an amphibian exhibit in Quito, Ecuador. SGS.

report was largely based on trade data provided by Wikiri, the only Ecuadorian private company approved by the government since 2009. According to their report, CL exports comprised five species: *Epipedobates anthonyi* (epibatidine poison frog) (43%), *Oophaga sylvatica* (little devil poison frog, as *O. sylvatica* and *O. histrionica*) (32%), *E. tricolor* (tricolor poison frog) (20%), and *Agalychnis spurrelli* (Spurrell's gliding leaf frog) (5%).

Herein, we provide updated and new data from Wikiri exports up through December 2022. According to Wikiri's records, between September 2010 and December 2022, a total of 6676 live CL (9 species) and non-CL (17 species) frogs, comprising 26 species, were exported from Ecuador (Figure 1.64).

As of 2022, the United States is the main importer of live frogs from Ecuador (41%), followed by the Netherlands

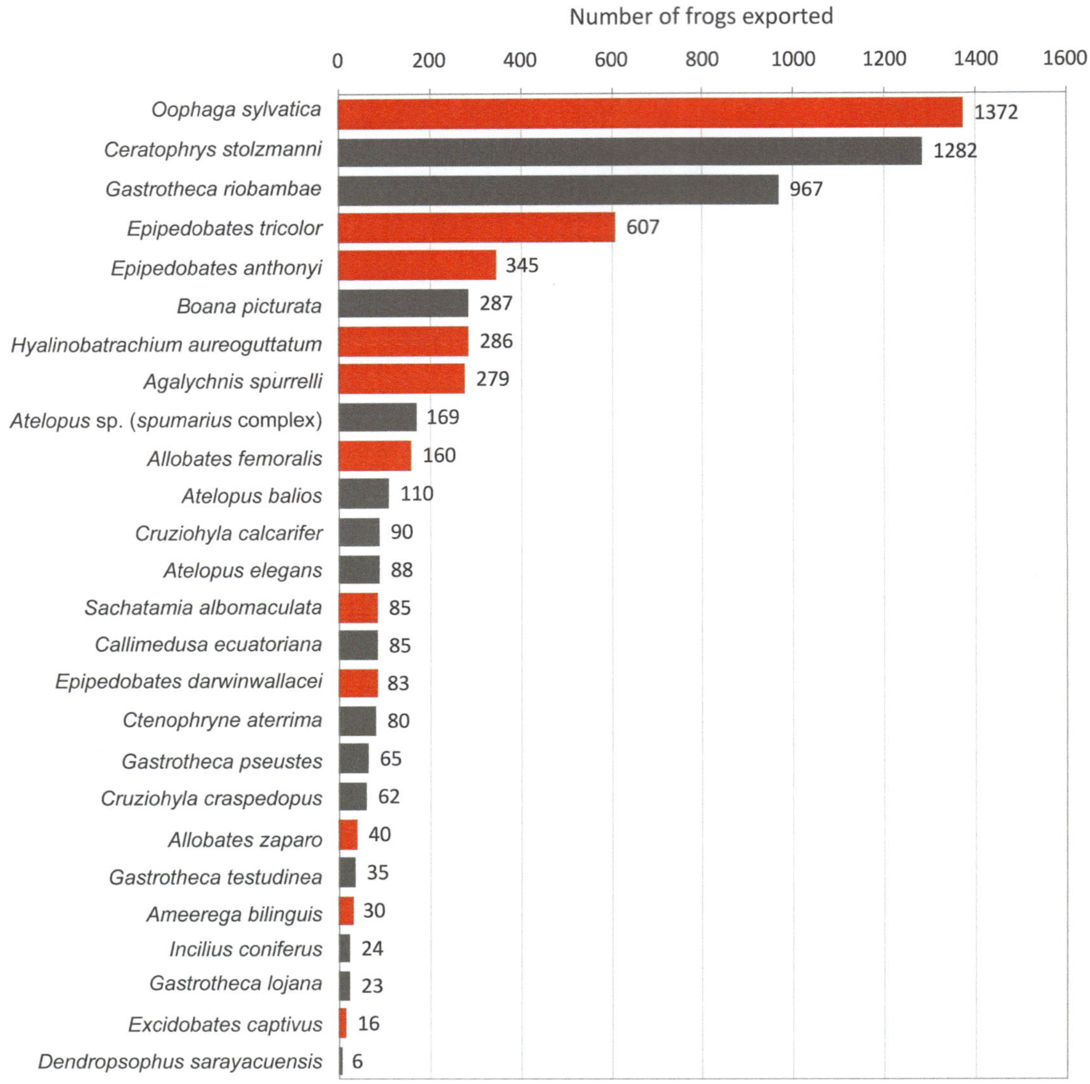

FIGURE 1.64 Exports of live amphibians by species for the period September 2010 to December 2022. CITES-listed (CT) species in red, non-CITES listed (NCL) in gray. Total exported = 6676. Source: Wikiri.

(28%), Canada (11%), Germany (10%), and other countries (10%). Japan stands out as a relatively new country in the trade for frogs from the Neotropics (Figure 1.65). It is important to note that the importing countries are not necessarily the final destination of the frogs, as some of them are ports of entry. For example, the Netherlands and Denmark distribute frogs in other European countries. Canada exported Ecuadorian frogs to the USA, Europe, and Japan as well.

Eight species accounted for 84% of the exports, with *Oophaga sylvatica* (little devil frog) (21%), *Ceratophrys stolzmanni* (Stolzmann's pac-man frog) (19%), *Gastrotheca riobambae* (Gualataco marsupial frog) (14%), *Epipedobates tricolor* (tricolor poison frog) (9%), *E. anthonyi* (epibatidine poison frog) (5%), *Boana picturata* (colorful treefrog) (4%), *Hyalinobatrachium aureoguttatum* (gold-spotted glassfrog) (4%), and *Agalychnis spurrelli* (Spurrell's gliding leaf frog) (4%) being the most highly traded species (Figure 1.66). The other 18 species made up only 16% of the exports, with numbers ranging from 6 to 169 (see Figure 1.65). Wikiri has introduced several new morphs, including the Verkade morph of *E. darwinwallacei* and at least a dozen new species. Among these, three species of harlequin frogs (*Atelopus balios*, *A. elegans*, and *A.* sp. (*spumarius* complex)) (Figure 1.67) have been increasingly requested, with 367 individuals exported.

The data presented do not reveal any clear trends, but some of the species and their numbers are discussed further. Figure 1.68 shows the cumulative number of specimens exported by year, while Figure 1.69 illustrates the marked variation between years from 2010 to 2022, as indicated by the numbers. For example, the numbers ranged from a low of 280 in 2011 to a high of 1044 in 2015. These numbers reflect a gradual increase in the species offered, an increase in production for some species, as well as an increase in exports for some species, but each species has

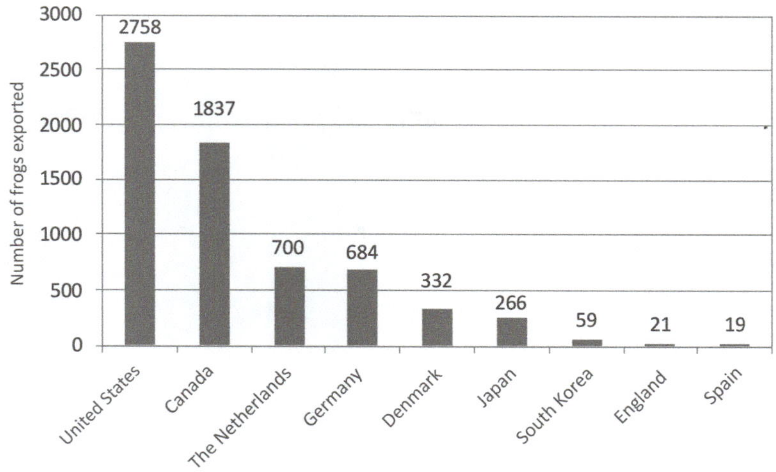

FIGURE 1.65 International trade (countries and number of frogs per importer country) of 6676 live frogs from Ecuador for the period September 2010 to December 2022. Total exported = 6676. Source: Wikiri.

its own unique characteristics. Overall, there were three main declines in the exportation of frogs in 2013, 2016, and 2021. The production drop occurred before 2013 and 2016 due to technological problems. In 2021, the COVID-19 pandemic prevented the air transportation of live frogs.

Each species that is exported has its own unique circumstances that affect its sales and the numbers exported over time. For example, *Ceratophrys stolzmanni*, which accounts for 19% of exports, was only exported from 2010 to 2014, primarily to Canada and Germany. After 2014, the production of this species was discontinued at Wikiri because of technical problems, but assisted-reproductive technologies are currently being used to address this lag. The initial high export numbers for this species can be attributed to several factors, including the popularity of its congeners (e.g., *C. ornata*, *C. calcarata*), which are appreciated for their beauty, ease of maintenance, and relatively low prices. Additionally, support from Understory Enterprises and STC Short Connection, plus the novelty of this species in the pet market explain, its high initial export numbers.

The little devil poison frog *Oophaga sylvatica*, which is the most exported frog with 1372 individuals, has been produced in four different morphs by Wikiri: Paru (orange), Situwa (spring sunshine), Diablito (little devil), and Pata Blanca (white foot) (Figures 1.70, 1.71). These new morphs were introduced to replace the old ones that were smuggled and named after their locality of origin. According to Yeager *et al.* (2020), the use of novel trade names for their offerings helps to distinguish them from illegal lineages. To prevent the laundering of illegal animals with their legal offerings, Wikiri identifies a unique trait for each frog species and chooses a word in the indigenous Kichwa or Spanish language as a trade name. Also, since 2014, each of the exported frogs has a unique code that differentiates legal from illegal specimens. The Paru morph is the only one that is bred in its natural habitat at Reserva Otokiki in Provincia Esmeraldas, using habitat enrichment techniques. The other three morphs are bred at a farm facility called Florida Sapoparque in Provincia Santo Domingo de los Tsáchilas.

Two peaks in the export of *Oophaga sylvatica* occurred in 2012 and 2015, as depicted in Figure 1.72. In both years, the Canadian company Understory Enterprises was the major buyer. However, in 2015, frogs were also sold to the USA, the Netherlands, and Japan. From January 2011 to December 2022, the USA, Canada, and the Netherlands were the main buyers of this species, as illustrated in Figure 1.73.

Domestic Use of Live Amphibians

The practice of keeping live frogs sourced from the wild in Ecuador for research and vivarium exhibits dates back to the 1970s. In 1991, lab-born frogs began to be raised at the Departamento de Ciencias Biológicas of Pontificia Universidad Católica del Ecuador (PUCE). More recently, companies such as Hyla and Wikiri have obtained management permits from the Ministerio de Ambiente of Ecuador to produce lab and farm-raised frogs. While domestically traded frogs by Wikiri are still used for educational and research purposes, they are also contributing to the development of a novel amphibian terraria hobby in Ecuador.

The practice of keeping live wild frogs in terraria for research purposes was first reported in Ecuador in the 1970s by Eugenia M. del Pino. After completing her PhD in Biology at Emory University in 1972 and while working at PUCE, del Pino began to keep the marsupial frog *Gastrotheca riobambae* for embryological studies (del Pino *et al.*, 1975). She described the structure of the pouch and embryos, maternal–embryonic relationships, and changes in the pouch related to reproductive activity. This work is further detailed in "Del Pino and the Marsupial Frogs" in Chapter 2.

FIGURE 1.66 The eight most exported species of frogs from Ecuador. Top left: *Oophaga sylvatica* (little devil poison frog) (SGS). Top right: *Ceratophrys stolzmanni* (Stolzmann's pac-man frog) (LAC). Second row left: *Gastrotheca riobambae* (gualataco marsupial frog) (SGS). Second row right: *Epipedobates tricolor* (tricolor poison frog) (SGS). Third row left: *E. anthonyi* (epibatidine poison frog) (SGS). Third row right: *Boana picturata* (colorful treefrog) (LAC). Bottom left: *Hyalinobatrachium aureoguttatum* (gold-spotted glassfrog) (SGS). Bottom right: *Agalychnis spurrelli* (Spurrell's gliding leaf frog) (SGS).

FIGURE 1.67 Species of harlequin frogs introduced in the markets by Wikiri. Left: *Atelopus* sp. (*spumarius* complex). Middle: *A. elegans*. Right: *A. balios*. (LAC).

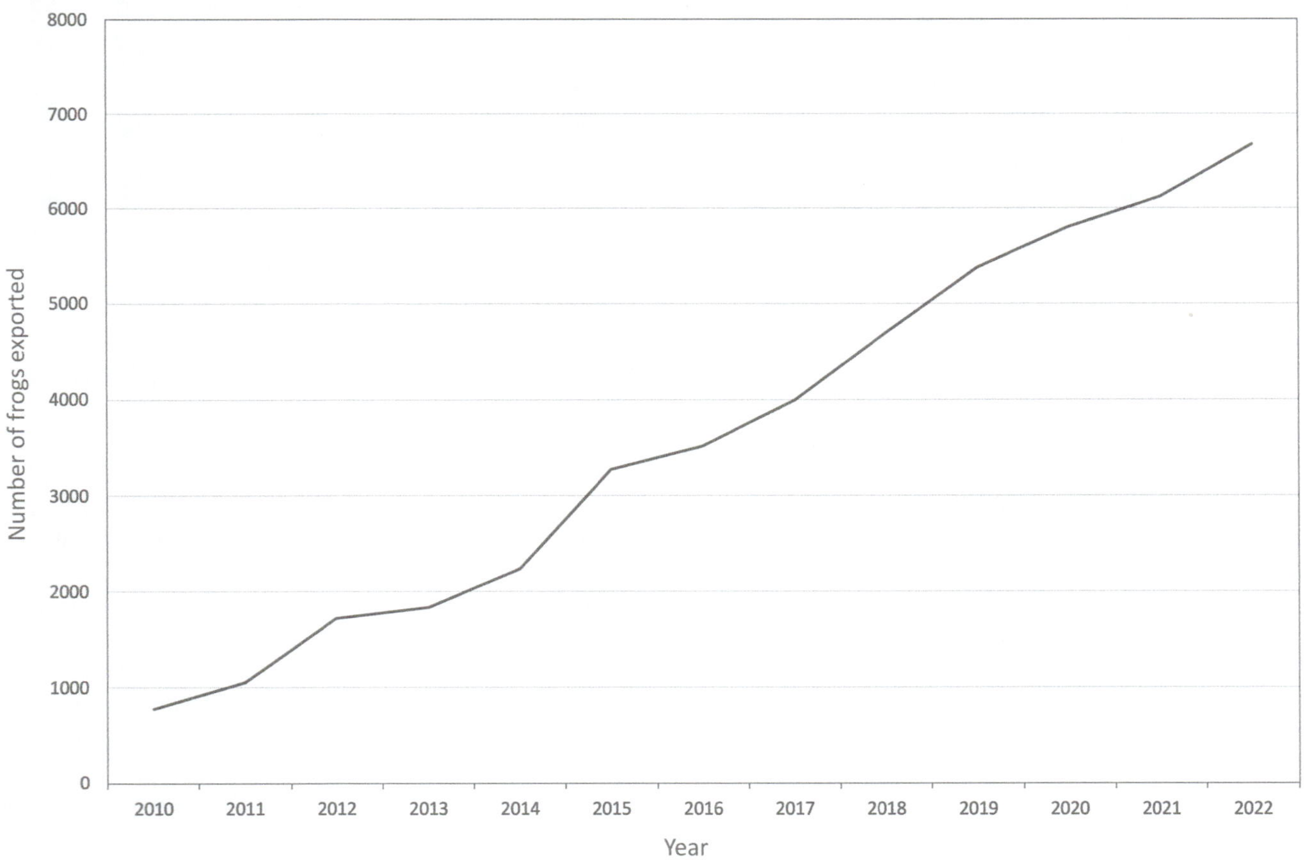

FIGURE 1.68 Cumulative number of Ecuadorian frogs exported by year for the period September 2010 to December 2022. Total exported = 6676. Source: Wikiri.

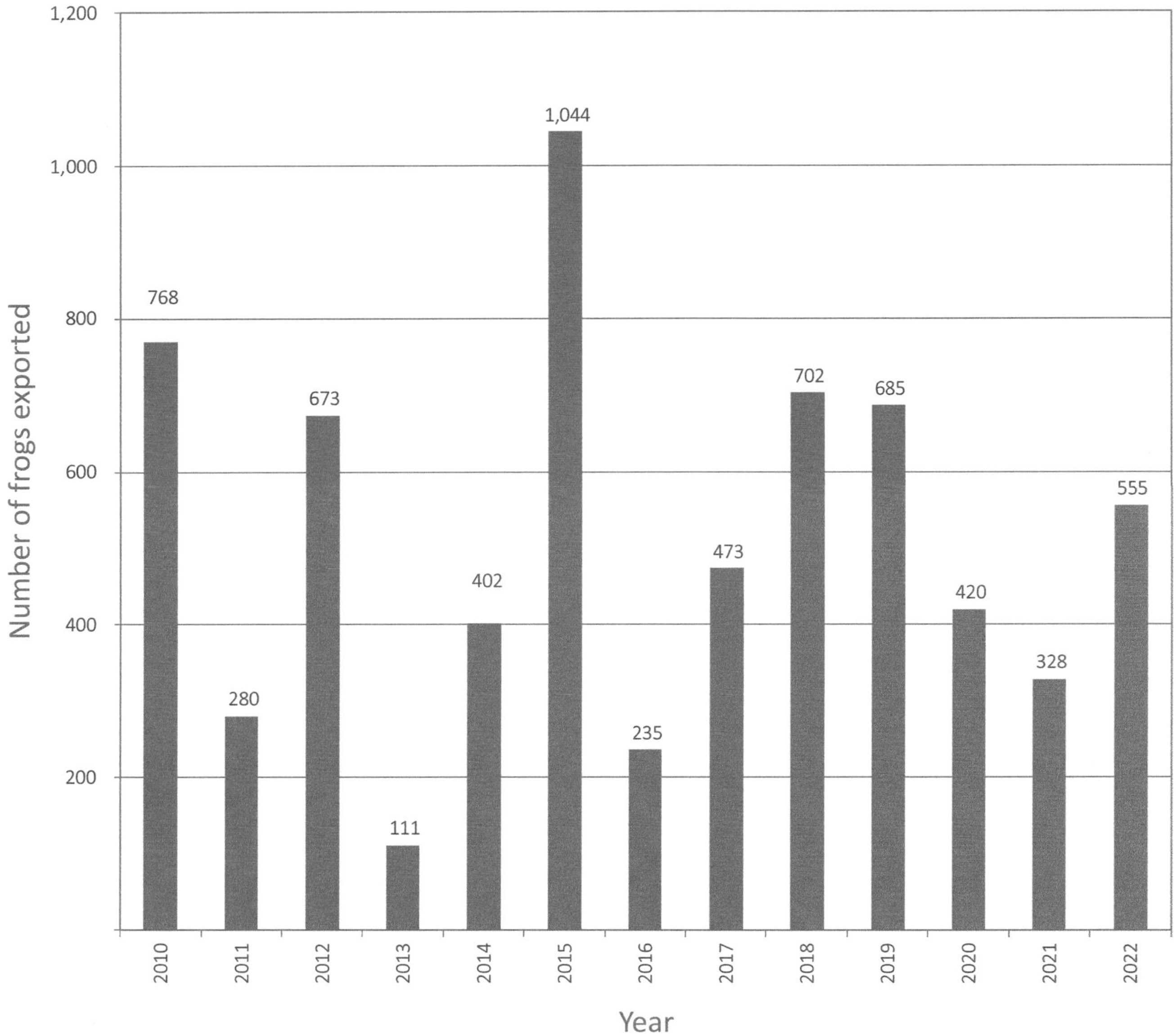

FIGURE 1.69 Number of Ecuadorian frogs exported by year for the period from September 2010 to December 2022. Total exported = 6676. Source: Wikiri.

Gastrotheca riobambae was once common in Quito and could easily be obtained from the gardens of the university campus. Del Pino vividly describes her initial findings on campus in the following passage:

> Adult specimens of *Gastrotheca riobambae* were collected in Quito, Ecuador, on the grounds of Pontificia Universidad Católica del Ecuador, from other localities within the city, and near Machachi, a town located approximately 50 km south of Quito.
>
> (del Pino *et al.*, 1975)

> I decided that I needed to look for an interesting frog. I went to the gardens of the university and found two different frogs. And the most strange was the marsupial frog, which has a pouch and carries the babies inside. The pouch is in the back, like a backpack.
>
> (Mossman, 2007)

In addition to *Gastrotheca riobambae*, *G. plumbea* was also briefly kept live in the embryology lab of PUCE. In the first decade of the 2000s, Del Pino and collaborators were able to use lab-raised frogs from the Balsa de los Sapos program at PUCE for their embryological studies. Several studies were conducted using *Ameerega bilinguis*, *Dendrobates auratus*, *Epipedobates machalilla*, *E. anthonyi*, *E. tricolor*, *Engystomops coloradorum*, and *E. randi* and were published by Del Pino and colleagues in 2007.

In 1991, Coloma initiated several research and conservation projects using live frogs at the Departamento de Ciencias Biológicas of PUCE, which later became known as Balsa de los Sapos in 2005. The focus of the program was on ethological studies and ex situ conservation of endangered amphibians in Ecuador. Among the first species to be housed at the facility were *Atelopus balios*, *Telmatobius*

FIGURE 1.70 Intrapopulation color variation of the Paru (orange) morph of *Oophaga sylvatica* (little devil poison frog) from Provincia Esmeraldas, exported from Ecuador by Wikiri. LAC; SGS, bottom left.

niger, *Hyloxalus awa*, *H. toachi*, *Epipedobates machalilla*, and *Allobates kingsburyi*. The animals of the two first species, which are critically endangered in the wild, were unable to survive in laboratory despite attempts to keep them alive. *Telmatobius niger* is currently considered possibly extinct, while *A. balios* has since been successfully laboratory bred and raised by Centro Jambatu and Wikiri. Dendrobatid species were also kept for ethological studies, resulting in the yet unpublished thesis by Correa Monge (1995) and published papers by Castillo-Trenn and Coloma (2008), Castillo-Trenn (2004), and Quiguango-Ubillús and Coloma (2008). Balsa de los Sapos hosted many other species, and a summary of the program is available in Coloma *et al.* (2010) (See also Chapter 2: "The Life Raft for Frogs at QCAZ").

According to Wikiri's records, a total of 796 live CL (7 species) and NCL (15 species) frogs of 22 species were sold in Ecuador between September 2010 and December 2022 (Figures 1.74, 1.75). This number makes Ecuador one of the largest consumers of frogs, after the USA and Canada. Considering that Canada exports its frogs to the USA, Europe, and Japan, Ecuador may be the second-largest consumer. The number of domestically traded frogs produced by Wikiri has been steadily growing since the 2010s for various reasons, but mostly due to an emergent interest in terraria for hobbyist and cultural purposes. This is exemplified by the terraria and frogs supplied and displayed at the Museo Interactivo de Ciencias, the Museo del Agua, and the INABIO, some of the most visited museums in Quito.

FIGURE 1.71 Three morphs of *Oophaga sylvatica* (little devil poison frog) exported from Ecuador by Wikiri. Top left: Diablito (little devil). Top right: Pata blanca (white footed). Bottom: Situwa (spring sunshine). SGS.

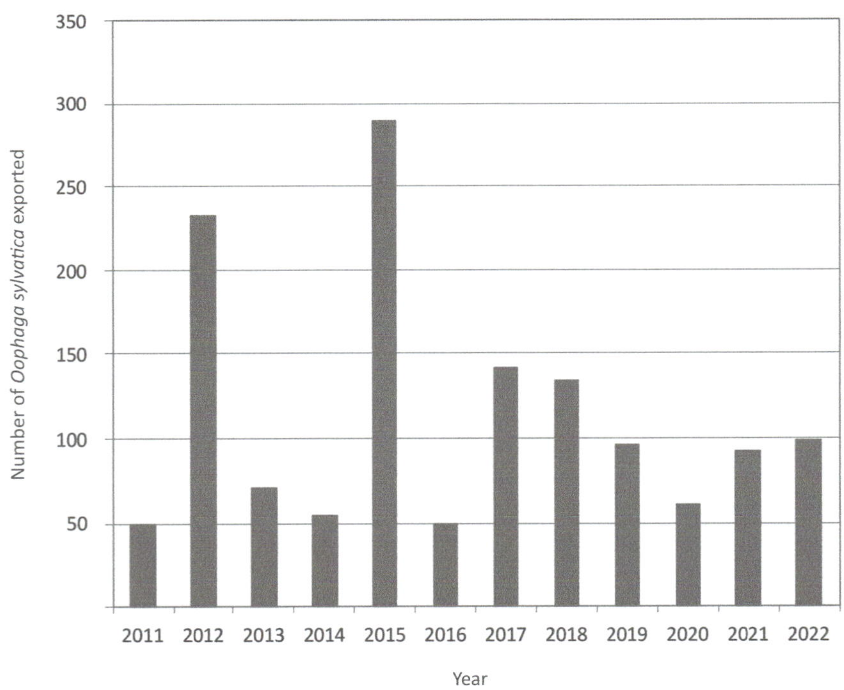

FIGURE 1.72 Number of *Oophaga sylvatica* exported by year for the period from January 2011 to December 2022. Total exported = 1372. Source: Wikiri.

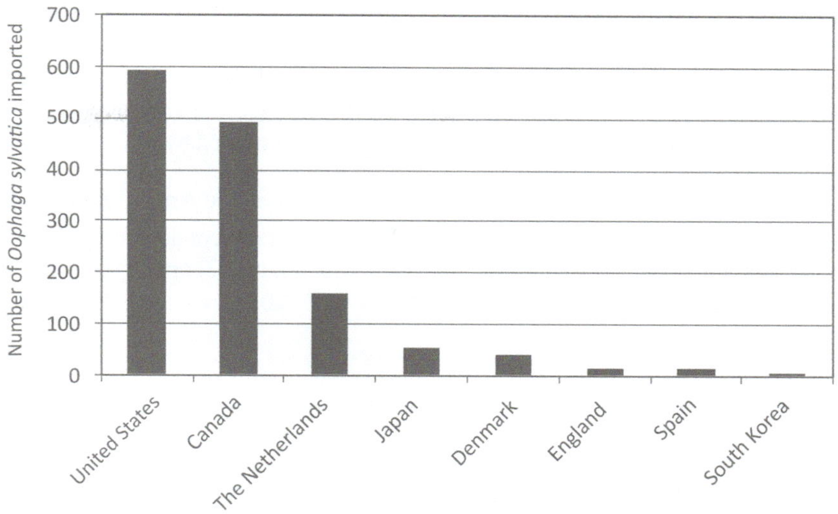

FIGURE 1.73 Number of *Oophaga sylvatica* imported from Ecuador by country for the period from January 2011 to December 2022. Source: Wikiri.

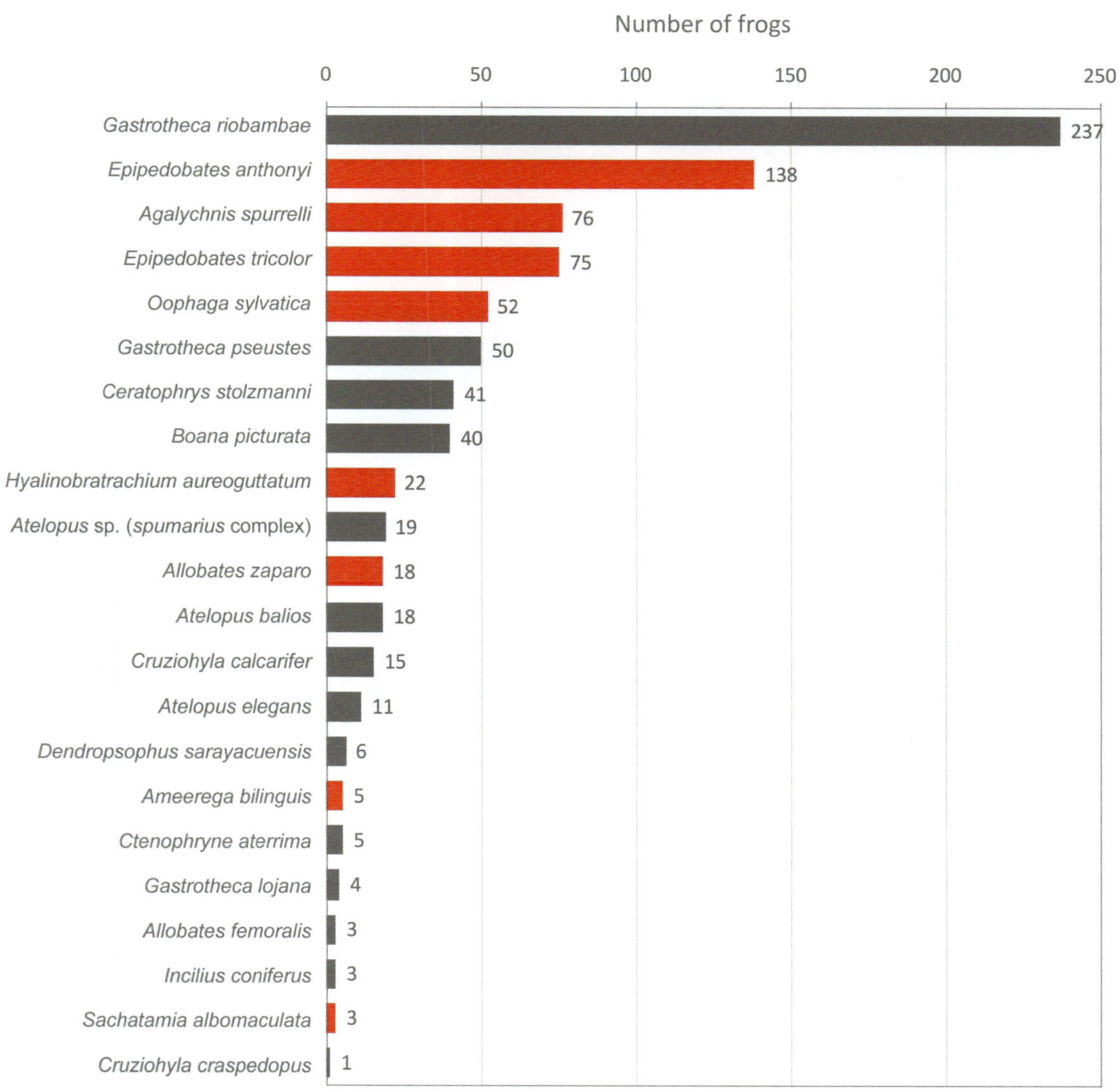

FIGURE 1.74 Domestic trade of 842 live amphibians by species for the period from September 2010 to December 2022. CITES-listed (CT) species in red, non-CITES listed (NCL) in gray. Source: Wikiri.

Gastrotheca riobambae was the most highly traded species, accounting for 30% of sales with 237 individuals, including adults and tadpoles (see Figure 1.74). It was followed by *Epipedobates anthonyi* (epibatidine poison frog) at 15%, *Agalychnis spurrelli* (Spurrell's gliding leaf frog) at 10%, and *Ceratophrys stolzmanni* (Stolzmann's pac-man frog) at 9%. The other 18 species accounted for 36% of sales with lower numbers ranging from 2 to 50 individuals (see Figure 1.75). Comparing the species sold in the international market versus the domestic market, it was found that the most traded species internationally, *Oophaga sylvatica*, was sold in low numbers locally. This is due to the fact that the Ecuadorian hobby is still in its early stages, and this species that is relatively expensive requires specialized care, which is usually provided by advanced hobbyists.

Value of the Live Amphibian Trade

In 2017, a study conducted by the Ministerio del Ambiente del Ecuador and Cooperación Alemana al Desarrollo (MAE and GIZ, 2017) revealed that for every dollar invested in

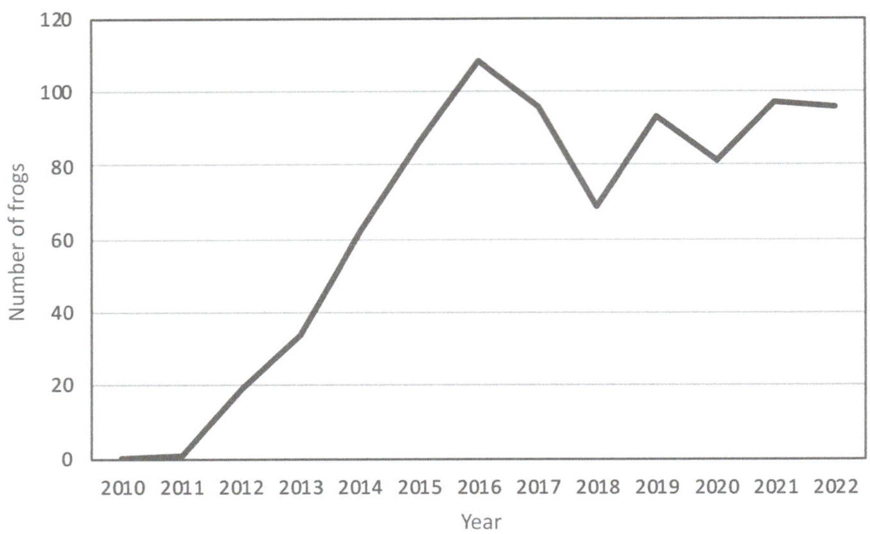

FIGURE 1.75 Number of frogs by year (September 2010–December 2022) traded domestically by Wikiri in Ecuador; 842 live frogs were traded. Source: Wikiri.

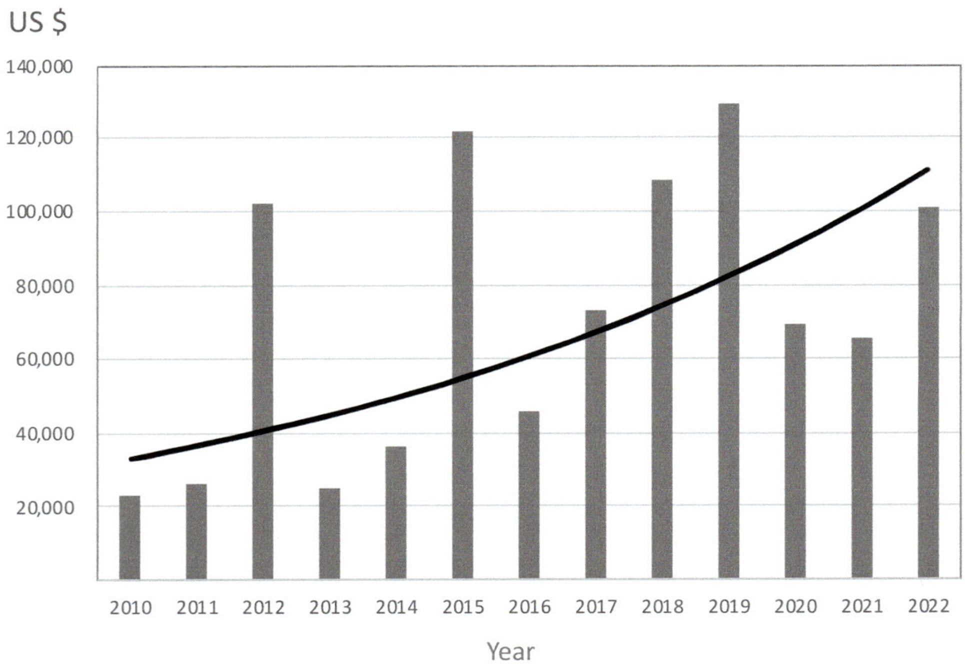

FIGURE 1.76 Total sales in US$ by year of both CITES- and non-CITES-listed live frogs for the period from September 2010 to December 2022. The exponential tendency line is included. Source: Wikiri.

the amphibian trade, US$1.78 of economic benefits were obtained for the country. The internal economic rate of return was calculated to be 44% for this activity, which is higher than the rate assigned as the cost of social opportunity (determined to be 12% for Ecuador). The study concluded that economically, this activity adds value to the country's economy. Therefore, their findings lead to the consideration of amphibian biotrade as a solution that adds value to the national economy while simultaneously contributing to the conservation of natural resources, traditional knowledge, and the distribution of benefits, ultimately helping to prevent the loss of biodiversity. Supporting the amphibian biotrade encourages a long productive chain and the reactivation and diversification of the economy.

According to updated data from Wikiri, the total net sales income for both CL and NCL live frogs for the period from September 2010 to December 2022 was valued at US $912,889, with an average annual value of US $70,222. There was a noticeable steady increase in sales over this period, as shown in (Figure 1.76). Despite accounting for only 18.9% of the commercialized quantity (for the international and domestic trade), *Oophaga sylvatica* accounted for the majority of total sales (50.3%). This high value can be attributed to the species' high unit price, at an average of US $ 322 per unit (with price variation depending on the color morph, e.g., US $ 285). Of the sales of *O. sylvatica*, the Paru morph accounted for the majority (811 frogs, 50.5%).

REFERENCES

Acero Coral, G., and M. A. Pianalto de Dalle Rive. 1985. *Medicina indígena Cacha-Chimborazo*. Quito: Mundo Andino.

Alarco de Zadra, A. 1990. *Brújula para niños: Cuentos, poesías y fábulas*. Cuenca: CIDAP.

Almendáriz, A. 1992. "1991" Anfibios y reptiles. *Revista Politécnica* 16: 89–162.

AmphibiaWeb. 2024. *Information on amphibian biology and conservation*. Berkeley, CA: University of California. http://amphibiaweb.org/.

Andrade-Marín, L. 1937. *Viaje a las misteriosas montañas de Llanganati. Expedición Italo-ecuatoriana. Boschetti-Andrade Marín 933:1934*. Quito: Imprenta Noción.

Anhalzer, J. 2002. *Cuentos del Ecuador*. Guías de Montaña. Quito: Autor.

Anonymous. 2003. La Tierra del "Jambato". *Quincenario Opción*. Ecuador. 44: 14.

Anonymous. 2009a. En la Reserva de Mazán se investiga a los anfibios. *Diario El Comercio* (July 8). https://www.elcomercio.com/actualidad/reserva-mazan-investiga-anfibios-1.html.

Anonymous. 2009b. Científicos descubren nuevas especies animales en Cordillera del Cóndor. *Diario El Comercio* (June 18). https://www.elcomercio.com/actualidad/cientificos-descubren-nuevas-especies-animales.html.

Anonymous. 2014. Los sapitos de agua están en 7 lugares del centro quiteño. *Diario El Comercio* (October 3). https://issuu.com/centrojambatu/docs/los_sapitos_de_agua.

Anonymous. 2015. El redescubrimiento de la rana jambato del Azuay entusiasma a la comunidad científica. *Diario El Comercio* (August 31). https://www.elcomercio.com/tendencias/ciencia/redescubrimiento-ranajambato-azuay-extincion-anfibios.html.

Arbeláez Ortiz, E., and A. Vega Toral. 2008. *Guía de anfibios, reptiles y peces del Parque Nacional Cajas/Cajas National Park amphibian, reptile and fish guide*. Cuenca: Corporación Municipal Parque Nacional Cajas/ETAPA.

Argo, A. 1996. *The last frog*. USA: ArgoFilms.

Arias, C. 1999. Gestión ambiental campesina de agua y las estrategias de acceso al riego en la comunidad de Capellani (Prov. Quillacollo). Tesis de Licenciatura, Universidad Mayor San Simón.

Arteaga, A., L. Bustamante, and J. M. Guyasamin. 2013. *The amphibians and reptiles of Mindo: Life in the cloudforest*. Quito: Scientific Publication Series, Universidad Tecnológica Indoamérica.

Athens, J. S. 1978. Formative period occupations in the highlands of northern Ecuador: A comment on Myers. *American Antiquity* 43: 493–496.

Athens, J. S. 1980. *Proceso evolutivo en las sociedades complejas y la ocupación del período tardío-Cara en los Andes septentrionales del Ecuador*. Otavalo: Biblioteca Hernán Malo González.

Athens, J. S. 2003. *Inventory of earthen mound sites, northern Highland Ecuador. Manuscript on file, National Institute of Cultural Patrimony, Quito*. Quito: National Institute of Cultural Patrimony.

Aubér-Thomay, M. 1986. *Biologie d'une rainette marsupiale (Gastrotheca riobambae). SFRS service du film de recherche scientifique*. SFRS Service du Film de Recherche Scientifique. France. 12 min.

Aubér-Thomay, M., L. Coloma, and G. Onore. 1990. Elevage d'une rainette marsupiale arboricole des forêts nuageuses équatoriennes, *Gastrotheca plumbea* (Boulenger). *Revue Française d'Aquariologie herpétologie* 17: 57–62.

Aubér-Thomay, M., and F. Letellier. 1986. Observations sur le développement de la rainette marsupiale, *Gastrotheca riobambae* (Hylidés). *Revue Française d'Aquariologie* 13: 79–86.

Aveiga del Pino, M. 2003. *Cuentos populares y mitos indígenas del Ecuador*. Vol. 159. Quito: Libri Mundi Ediciones.

Baca Carbo, R., and V. Vanegas López. 1981. *Decreto N° 74. Ley Forestal y de Conservación de Áreas Naturales y Vida Silvestre*. Ecuador: Registro Oficial N° 64, del 24 de Agosto de 1981.

Barrio-Amorós. 2010. A suggestion about common nomenclature for dendrobatids. Poison Dart Frogs. *Dendrobates.org*. https://web.archive.org/web/20160307042838/http://www.dendrobates.org/common_names.html

BBC News. 2019. Frog from Ecuador found in bananas in Northampton Tesco. *BBC News* (December 10). https://www.bbc.com/news/uk-england-northamptonshire-50734441.

BCE (Banco Central del Ecuador). 2004. *Departamento de Comercio Exterior. Exportación de ancas de rana*. Partida Arancelaria N° 020820. Quito.

BCE (Banco Central del Ecuador). 2009. *Departamento de Comercio Exterior. Exportación de ancas de rana*. Partida Arancelaria N° 020890. Quito.

Bech, R. 1982. Pfeilgiftfrösche—Zur haltung und zucht von *Phyllobates vittatus* und *Dendrobates tricolor* im terrarium. *Aquarien -Terrarien, Berlin* 29: 353–356.

Bejarano-Muñoz, P., M. B. Pérez-Lara, and J. Brito. 2015. *Ranas terrestres endémicas y otros anuros emblemáticos de la vía Papallacta-El Reventador Vol. 2*. Quito: Serie de Publicaciones Museo Ecuatoriano de Ciencias Naturales del Instituto Nacional de Biodiversidad (INB-MECN).

Bentley, A. G., J. P. Reyes-Puig, M. H. Yánez-Muñoz, F. Gordón Domínguez, and H. U. Sánchez-Sarango. 2021. *Guía fotográfica herpetofauna del Alto Río Anzu, Ecuador*. Quito: WWF Ecuador.

Bermeo, C. R. 2015. *El lenguaje figurado en la obra "Luz y Cristal" del escritor y poeta Gustavo Alfredo Jácome*. Tesis de Maestría, Universidad Técnica Particular de Loja.

Bianchi, C. 1988. *El Shuar y al ambiente*. Quito: Abya-Yala.

Boonman, J. 1985. *Gastrotheca riobambae*, de buidekikker in het terrarium. *Lacerta* 43: 189–204.

Botasso, J. 1977. Los Animales. In *Mundo Shuar 1. Serie A*. Centro de Documentación e Investigación Cultural Shuar. Sucúa: Imprenta Colegio Técnico Don Bosco.

Botasso, J. 1978. Actividades y Técnicas. In *Mundo Shuar 1. Serie A*. Centro de Documentación e Investigación Cultural Shuar. Sucúa: Imprenta Colegio Técnico Don Bosco.

Boulenger, G. A. 1891. Reptilia & Batrachia. In *Travels amongst the great Andes of the Equator*, ed. E. Whymper, 129–136. London: John Murray.

Bray, T. L. 1995. The Panzaleo Puzzle: Non-local pottery in northern highland Ecuador. *Journal of Field Archaeology* 22: 137–156.

Bray, T. L. 2008. Late pre-Hispanic chiefdoms of highland Ecuador. In *The handbook of South American archaeology*, 527–543. New York: Springer.

Brenner, W. 1960. *Gastrotheca marsupiata*, Ein laubfrosch mit hochinteressanter brutfliege. *Aquarium Und Terrarium Kunde* 7: 210–213.

Bringsøe, H. 1992. The adoption of the poison-arrow frogs of the genera *Dendrobates* and *Phyllobates* in Appendix II of CITES. *Herpetological Review* 23: 16–17.

Brown, J. L., E. Twomey, A. Amézquita, M. B. de Souza, J. P. Caldwell, S. Lötters, R. Von May, *et al*. 2011. A taxonomic revision of the Neotropical poison frog genus *Ranitomeya* (Amphibia: Dendrobatidae). *Zootaxa* 3083: 1–120.

Caluña Espin, N., M. Tisalema Caluña, and T. T. Caluña Til. 2008. *Los Chibuleos: origen, identidad, desarrollo y justicia de un pueblo indígena en los Andes Ecuatorianos*. Quito: Abya-Ayala.

Campbell, J. R., and D. R. Frost. 1994. Conservation biology. *Herpetological Review* 25: 100–102.

Caramaschi, U., R. O. de Sá, and W. R. Heyer. 2005. Common names for the frog genus *Leptodactylus* (Amphibia, Anura, Leptodactylidae). *Herpetological Review* 36: 119–120.

Carrillo de Landazuri, R. 1995. *El Libro del escolar ecuatoriano 4to Grado*. Quito.

Carrillo de Landazuri, R., and F. Arregui de Pazmiño. 1947. *El Libro del escolar ecuatoriano 4to Grado*. Quito: Editorial Voluntad.

Carvajal-Endara, S. 2008. Personal communication by "Campesinos" in the Sigchos Region, Provincia Cotopaxi.

Casares, P. 1992. *Informe sobre el desarrollo del forum "Análisis de la ranicultura en el Ecuador"*. Quito: Fundación Herpetológica Gustavo Orcés.

Casas Andreu, G., X. Aguilar-Miguel, and J. R. Cruz-Aviña. 2001. La introducción y el cultivo de la rana toro (*Rana catesbeiana*): ¿Un atentado a la biodiversidad de México? *Ciencia Ergo Sum* 8: 62–67.

Castillo-Trenn, P. 2004. Description of the tadpole of *Colostethus kingsburyi* (Anura: Dendrobatidae) from Ecuador. *Journal of Herpetology* 38: 600–606.

Castillo-Trenn, P., and L. A. Coloma. 2008. Notes on behavior and reproduction in captive *Allobates kingsburyi* (Anura: Dendrobatidae), with comments on the evolution of reproductive amplexus. *International Zoo Yearbook* 42: 58–70.

Cavasos, K., R. K. Adhikari, N. C. Poudyal, J. L. Brunner, A. R Warwick, and M. J. Gray. 2023. Understanding the demand for and value of pathogen-free amphibians to US pet owners. *Conservation Science and Practice* 5, no. 9: e12995.

CFN (Corporación Financiera Nacional). 1998. *Proyecto cría de rana toro en cautiverio*, 108 pp. Quito, Ecuador: Corporación Financiera Nacional.

Chimbo Aguinda, J. J., J. Saritama, S. Echeverzyría, and P. Vaca. 2007. *Shimiyukkamu Diccionario Kichwa-Español Español -Kichwa*. Sucumbíos: Casa de la Cultura Ecuatoriana Benjamín Carrión.

Christmann, S. P. 2004. *Dendrobatidae—Poison Frogs—A fantastic journey through Ecuador, Peru and Colombia*. Vol. II. Munich: Liebl Druck.

Cicala, M. 1994. *Descripción histórico-topográfica de la provincia de Quito de la Compañía de Jesús*. Vol. 1. Quito: BPR Publishers.

Cisneros-Heredia, D. F. 2004. *Rana catesbeiana* (Bullfrog). Geographic distribution. Ecuador. *Herpetological Review* 35: 406.

CITES. 2023. *CITES trade database*. Compiled by UNEP-WCMC for the CITES Secretariat. https://cites.org/eng.

Cochran, D. M. 1961. *Living amphibians of the world*. London: Hamish Hamilton.

Coloma, L. A. 1984a. Bamburé, Gualmana, Gualimana. Lita-Rocafuerte, Provincia Imbabura. Field notes.

Coloma, L. A. 1984b. Shuar amphibian names. Montalvo, Santiago, Provincia Morona Santiago. Field notes.

Coloma, L. A. 1984c. Kayla. Personal communication by indigenous people. Salinas, Provincia Bolívar.

Coloma, L. A. 1986. El napí pirirpri (*Phyllobates zaparo*). Ranita venenosa del oriente ecuatoriano. *Revista Natura* 68 (October): 8.

Coloma, L. A. 1988. Cutín, cutinte. Personal communication by indigenous people. Cashca Totoras, Provincia Bolívar.

Coloma, L. A. 1989a. Ambato. Personal communication by an indigenous people. Chimborazo Volcano area, Provincia Tungurahua.

Coloma, L. A. 1989b. Sapito. Personal communication by Zoila Ulloa, Guanujo, Provincia Bolívar. Fonoteca Centro Jambatu.

Coloma, L. A. 1995. Ecuadorian frogs of the genus *Colostethus* (Anura: Dendobatidae). *University of Kansas, Natural History Museum. Miscellaneous Publications* 87: 1–75.

Coloma, L. A. 1999. Sobre Orcés y el jambato, su legado. *Revista de La Facultad de Ciencias Exactas, Físicas y Naturales de La PUCE, Nuestra Ciencia* 1: 13–16.

Coloma, L. A. 2002. Two new species of *Atelopus* (Anura: Bufonidae) from Ecuador. *Herpetologica* 58: 229–252.

Coloma, L. A. 2003. Uilli-uillis. *Revista de la Facultad de Ciencias Exactas, Físicas y Naturales de la PUCE, Nuestra Ciencia* 5: 3–6.

Coloma, L. A. 2005. El Sapari, "Imperdonable no visitarlo". *Revista de La Facultad de Ciencias Exactas, Físicas y Naturales de la PUCE, Nuestra Ciencia* 7: 56–57.

Coloma, L. A. 2010. Ambatukuna. Personal communication by Gabriel Grefa, Archidona, Provincia Napo.

Coloma, L. A. 2014. Ecuador sapodiverso. In *El País de la Biodiversidad: Ecuador*, eds. M. García-Saltos, D. Parra-Puente, and P. Mena-Vásconez, 262–267. Quito: Fundación Botánica de los Andes, Ministerio del Ambiente y Fundación EcoFondo.

Coloma, L. A. 2016a. Gualataco. Personal communication by José Alquinga, La Merced, Provincia Pichincha.

Coloma, L. A. 2016b. Pímbalo, tundele. Cotete. Tulte. Personal communication by Juan Alberto Riera and Carlos Mata. Angamarca, Provincia Cotopaxi.

Coloma, L. A. 2016c. Tsáfiki amphibian names. Personal communication by Primitivo Aguavil, José Jacinto Aguavil, Milton Aguavil, Connie Dickinson. Puerto Limón, Provincia Santo Domingo de los Tsáchilas.

Coloma, L. A. 2016d. Chumbes. Personal communication by Zoila Teresa Bazantes, Angamarca, Provincia Cotopaxi.

Coloma, L. A. 2016e. El Jambato negro del páramo, *Atelopus ignescens*, resucitó. *IMCiencia*. https://issuu.com/centrojambatu/docs/coloma_2016_el_jambato_negro_del_p_

Coloma, L. A. 2016f. Field notes in Angamarca, Provincia Cotopaxi.

Coloma, L. A., V. Cano, and M. A. Rodríguez. 2010. A "Life Raft" for amphibians in Ecuador. In *Building a future for wildlife: Zoos and aquariums committed to biodiversity conservation*, eds. G. Dick and M. Gusset, 105–110. Gland: Waza Executive Office.

Coloma, L. A., and J. M. Guayasamin. 2011. 2011–2020. Nombres vernáculos de anfibios de Ecuador, SapoPedia Ecuador. *Fundación Jambatu, Quito, Ecuador*. http://www.anfibiosecuador.ec/index.php?nc,11.

Coloma, L. A., and D. Lombeida. 1992. La extinción desconocida. *Diario El Comercio. Revista Familia* (26 April): 16–17.

Coloma, L. A., and S. Ron. 2001. *Ecuador megadiverso: anfibios, reptiles, aves, mamíferos. Megadiverse Ecuador: Amphibians, reptiles, birds, and mammals*. Quito: Museo de Zoología, Centro de Biodiversidad y Ambiente, Departamento de Ciencias Biológicas, Pontificia Universidad Católica del Ecuador.

Comisión Legislativa Permanente. 1969. *Decreto No. 170. Prohíbese la exportación de animales de nuestra fauna silvestre*. Quito, Ecuador: Registro oficial No. 243 del 14 de Agosto 1969.

Cornejo, J. 1970. *Animales y plantas en la poesía popular ecuatoriana*. Guayaquil: Editorial Cervantes C. A.

Correa Monge, G. M. 1995. Comportamiento, vocalización y notas biológicas de *Colostethus awa* y *C. machalilla* (Anura: Dendrobatidae) en cautiverio. Tesis de Licenciatura, Pontificia Universidad Católica del Ecuador.

Costales, P., A. Costales, and J. Costales-Peñaherrera. 1996. *Mitos Quitu-Cara*. Quito: Abya-Yala.

Costales Samaniego, A. 2010. *Ecuadorian tadpole names*. Quito. Notes.

Cruz-Cordovez, C., I. Herrera, F. Espinoza, K. Rizzo, M. B. Sarmiento, N. Rodas, M. Coello, W. Bravo, and M. Lampo. 2020. New record of a feral population of *Lithobates catesbeianus* Shaw, 1802 in a protected area (Santay Island) in the Ecuadorian coast. *BioInvasions Records* 9: 421–433.

Cummins, T. 2003. Nature as culture's representation: A change of focus in late formative iconography. In *Archaeology of Formative Ecuador*, eds. J. Raymond and R. Burger. Washington, DC: Dumbarton Oaks.

Cuvi, N. 2022. *Historia ambiental y ecología urbana para Quito*. Quito: FLACSO.

Daly, J. W., H. M. Garraffo, T. F. Spande, M. W. Decker, J. P. Sullivan, and M. Williams. 2000. Alkaloids from frog skin: The discovery of epibatidine and the potential for developing novel non-opioid analgesics. *Natural Product Reports* 17: 131–135.

de Carvalho-Neto, P. 1994. *Antología del folklore ecuatoriano*. Quito: Abya-Yala.

de Carvalho-Neto, P. 1988. *Cuentos folklóricos del Ecuador: Costa y Sierra*. Guayaquil: Banco Central del Ecuador.

de la Espada, D. M. J. 1899. Examen descriptivo del grupo de los *Hemiphractus*. *Anales de Historia Natural* 27: 379–410.

de Velasco, J. 1789. *Historia del Reino de Quito en la América Meridional, escrita por el presbítero Dn. Juan de Velasco, nativo del mismo reino. Tomo I y Parte I que contiene la historia natural*. Quito: Imprenta del Gobierno, por Suarez de Valdés, 1841–1844.

Dea, S. 2006. *Diccionario Kichwa—Castellano*. Quito: UNICEF.

Deckert, K. 1963. Die paarung des beutelfrosches (*Gastrotheca marsupiata*). *Zoologische Garten* 28: 12–17.

del Pino, E. M. 1975. Adaptaciones reproductivas para la vida terrestre del sapo marsupial *Gastrotheca riobambae* (Fowler), Anura, Hylidae. *Revista de la Pontificia Universidad Católica del Ecuador* 3: 119–140.

del Pino, E. M., C. M. de A. M. L. Galarza, and A. A. Humphries Jr. 1975. The maternal pouch and development in the marsupial frog *Gastrotheca riobambae* (Fowler). *Biological Bulletin* 149: 480–491.

del Pino, E. M., H. Steinbeisser, A. Hofmann, C. Dreyer, M. Campos, and M. F. Trendelenburg. 1986. Oogenesis in the egg-brooding frog *Gastrotheca riobambae* produces large oocytes with fewer nucleoli and low RNA content in comparison to *Xenopus laevis*. *Differentiation* 32: 24–33.

Delgado-Espinoza, F. 2005. Organización política en la cuenca baja del río Guayas, Ecuador, durante el periodo de integración (ca. 500 DC—Contacto español). *Antropología. Cuadernos de Investigación* 6: 211–228.

Delgado-Espinoza, F. 2006. Organización de la producción de los camellones en la baja cuenca del Guayas durante la ocupación de los Chonos. In *Agricultura ancestral Camellones y Albarradas*, eds. J. Vacher, A. Yépez, J. F. Bouchard, P. Usselmann and F. Valdez, 159–168. Quito: Abya-Yala.

Diamond, J. 1966. Zoological classification system of a primitive people. *Science* 151: 1102–1104.

Duarte Valverde, A., and J. Naranjo Fiallo. 1969. *Decreto N° 800. Reglamento para regular las actividades de caza y exportación de animales vivos de nuestra fauna*. Quito, Ecuador: Registro oficial N° 339, del 31 de Diciembre 1969.

Duellman, W. E., and L. Trueb. 1986. *Biology of amphibians*. New York: McGraw-Hill Co.

Dupuis, C. 2020. Falar é existir: O caso de línguas ameaçadas no Brasil e no Equador. *Revista Amazônia Latitude Review* 1: 70–75.

Dyrdahl, E. 2017. Interregional interaction and craft production at Las Orquídeas, Imbabura, Ecuador, during the Late Formative (800–400 Cal BC). Doctoral dissertation, The Pennsylvania State University.

Dyrdahl, E., and C. Montalvo-Puente. 2021. Insights into a first millennium BC (800–400 Cal BC) social network: Excavations at Las Orquídeas in the northern ecuadorian sierra. *Journal of Global Archaeology*: 1–78.

Dyrdahl, E., and C. Montalvo-Puente. 2022. Late formative (800–400 cal BC). Social networks in northern Ecuador: The evidence from Las Orquídeas, Imbabura. In *87th SAA annual meeting*. Chicago: Society for American Archaeology.

Enríquez, M. S. 1999. *Fauna herpetológica amazónica, especies representativas*. Quito: Abya-Yala.

Ensinck, F. H. 1974. Verslag van het kweken met de buidelkikker *Gastrotheca marsupiata*. *Lacerta* 32: 135–140.

Ensinck, F. H. 1978. Sur zucht des beutelfrosches *Gastrotheca marsupiata*. *Das Aquarium* 12: 401–405.

Espinosa, G. 1996. Lari y Jamp'atu. Ritual de lluvia y simbolismo andino en una escena de arte rupestre de Ariquilda 1. Norte de Chile. *Chungara* 28: 133–157.

Espinosa, R., and J. Durán Arias. 1970. *Decreto No. 099. Reforma al acuerdo ministerial No. 800*. Quito, Ecuador: Registro oficial No. 393, del 18 de Marzo 1970.

Espinosa Cordero, S. 2005. Los Sapos del Ecuador. Ecuador. *Ecuador Terra Incognita*, no. 33 (January–February). https://www.terraecuador.net/revista_33/33_saposdelecuador.htm.

Félix-Novoa, M. C., and N. Acosta Buenaño. 2006. *Manual de cría y manejo de la rana venenosa tricolor del Yasuní (Edición en español). Manual de cría y manejo de la rana Ome Moge tricolor del Yasuní (Edición en español-waoterero)*. Quito: Fundación Yasuní-PUCE.

Forrester, N. 2023. Guardian of Ecuador's diverse—and vanishing—Frog species. *Nature* 615: 960.

Frank, N., and E. Ramus. 1995. *A complete guide to scientific and common names of reptiles and amphibians of the world*. Pottsville: N. G. Publishing.

Freile, J. F., L. A. Coloma, A. Terán-Valdez, D. Acosta López, E. E. Tapia, and G. Pazmiño-Otamendi. 2020. *Anfibios de Junín*. Quito: Centro Jambatu de Investigación y Conservación de Anfibios, Fundación Jambatu.

Freire García, S. 2016. El Sapo de agua quiteño. *Diario Memoria* (January 17).

Frost, D. R. 2023. *Amphibian Species of the World: An online reference*. Version 6.1. American Museum of Natural History, New York, USA. http://research.amnh.org/herpetology/amphibia/index.html.

Fuentes Roldán, A. 1995. *Quito Tradiciones*. Quito: Abya-Yala.

Fuentes Roldán, A. 1997. Cuenta la leyenda que . . . el sapo de agua. *Diario Últimas Noticias* (August 9). https://issuu.com/centrojambatu.

Gallardo, G. G. 2001. Aspectos de la biología reproductiva con énfasis en el crecimiento larval de *Rana palmipes* Spix 1824 (Anura: Ranidae) en la Amazonía ecuatoriana. Tesis de Licenciatura, Pontificia Universidad Católica del Ecuador.

Gallardo, G. G. 2002. *Ranicultura. Ecuador.* Ecuador Terra Incognita, no. 18. Quito: Imprenta Mariscal.

Gallardo, G. G. 2004. *Cría de una rana nativa de la Amazonía ecuatoriana: Rusu Mama; Rana palmipes*. Puyo: Centro Tecnológico de Recursos Amazónicos de la OPIP-Centro Fátima.

Gill, B. J., D. Bejakovtch, and A. H. Whitaker. 2001. Records of foreign reptiles and amphibians accidentally imported to New Zealand. *New Zealand Journal of Zoology* 28: 351–359.

Gorzula, S. 1996. The trade in dendrobatid frogs from 1987 to 1993. *Herpetological Review* 27: 116–123.

Graeme, G. 2021. Frog back from the dead helps fight plans for mine in Ecuador. *The Guardian* (November 23). https://www.theguardian.com/environment/2021/nov/23/frog-back-from-the-dead-helps-fight-mine-plans-in-ecuadors-cloud-forest-aoe.

Granda-Paz, O. 2010. *El sol de los pastos = the sun of the pastos*. Barranquilla: Travesías.

Green, G. 2022. Lost and found: The rediscovery that became "Leonardo DiCaprio's Frog". *The Guardian* (December 29). https://www.theguardian.com/environment/2022/dec/29/leonardo-dicaprio-longnose-harlequin-frog-copper-mining-ecuador.

Grenard, S. 1958. Life history of *Gastrotheca marsupiata* spp. *Herpetologica* 14: 151–152.

Guarderas, L. 2017. EcuaFrog, Wikiri and the amphibian pet trade (Ecuador). In *20 years of BioTrade: Connecting people, the planet and markets*, ed. V. Stone, 33. Geneva: United Nations Publication.

Gutiérrez Usillos, A. 2002. *Dioses, símbolos y alimentación en los Andes: Interrelación hombre-fauna en el Ecuador prehispánico*. Quito: Abya Yala.

Gutiérrez Usillos, A. 2011. *El eje del universo*. Madrid: Ministerio de Cultura. Secretaría Técnica.

Harrison Matthews, L. 1958. "1957". Viviparity in *Gastrotheca* (Amphibia: Anura) and some considerations on the evolution of viviparity. *Bulletin de La Société Zoologique de France* 82: 317–320.

Hartweg, N. 1955. Division of reptiles and amphibians, museum of zoology. In *Report of the director of the Museum of Zoology 1953–1954*, ed. J. S. Rogers, 24–26. Ann Arbor: University of Michigan.

Herrera, I., and R. Zambrano. 2015. *De boca de los sapos al Triunfo*. Ecuador: Irma Herrera Producciones.

Hidalgo-Alzamora, L. 1982. *Décimas esmeraldeñas, recopilación y análisis socioliterario*. Quito: Banco Central Del Ecuador.

Holland, J. 2015. "Extinct" toad rediscovered in Ecuador. *National Geographic, Weird & Wild* (August 31). https://www.nationalgeographic.com/adventure/article/150831-frogs-extinct-rediscovered-science-animals-ecuador.

Holm, O. 1981. *Cultura Milagro-Quevedo*. Guayaquil: Banco Central del Ecuador.

Hoogmoed, M. S. 1967. Mating and early development of *Gastrotheca marsupiata* (Duméril and Bibron) in captivity (Hylidae, Anura, Amphibia). *British Journal of Herpetology* 4: 1–7.

Hoogmoed, M. S. 2019. Unpublished population data of *Dendrobates azureus* Hoogmoed 1969 obtained in 1968 and 1970, and its historical and current taxonomic status. *Amphibian & Reptile Conservation* 13: 95–101.

Hurtado, O. 1983. *El reglamento general de aplicación de la Ley Forestal y de Conservación de Áreas Naturales y Vida Silvestre*. Quito: Ministerio de Agricultura y Ganadería.

Icaza, J. 1934. *Huasipungo*. 1st edition. Quito: Imprenta Nacional.

Icaza, J. 1942. *Media vida deslumbrados*. 1st edition. Quito: Editorial Quito.

Icaza, J. 1975. *Huasipungo*. 1st edition. México: Pepsa editores.

IUCN. 1987. *IUCN analyses of the CITES amendment proposals submitted to the Sixth Meeting of the Conference of the Parties, Ottawa (Canada), 12 to 24 July 1987*. Washington, DC: IUCN/SSC Trade Specialist Group.

Jácome, C. E. 2006. *Sabiduría Popular*. Quito: Ediciones Abya-Yala.

Jácome, G. A. 1956. *Luz y Cristal*. Vol. 2. Quito: Casa de la Cultura Ecuatoriana.

Jiménez de la Espada, M. 1875. *Vertebrados del viaje al Pacífico verificado de 1862 a 1865 por una comisión de naturalistas enviada por el Gobierno Español. Batracios*. Madrid: Miguel Ginesta.

Jiménez Sánchez Mariscal, J. D. 2014. Palimpsesto cultural y palingenesia del pensamiento: Una aportación andina a la filosofía de la religión. Tesis de Licenciatura, Universidad Complutense de Madrid.

Jones, R. E., A. M. Gerrard, and J. J. Roth. 1973. Estrogen and brood pouch formation in the marsupial frog, *Gastrotheca riobambae*. *Journal of Experimental Zoology* 184: 177–184.

Jungfer, K.-H. 1989. Pfeilgiftfrösche der gattung *Epipedobates* mit rot granuliertem rücken aus dem oriente von Ecuador und Peru. *Salamandra* 25: 81–98.

Jungfer, K.-H. 2011. A new tree frog of the genus *Osteocephalus* from high altitudes in the Cordillera del Cóndor, Ecuador (Amphibia: Anura: Hylidae). *The Herpetological Journal* 21: 247–253.

Jungfer, K.-H., and W. Hödl. 2002. A new species of *Osteocephalus* from Ecuador and a redescription of *O. leprieurii* (Duméril & Bibron, 1841) (Anura: Hylidae). *Amphibia-Reptilia* 23: 21–46.

Jungfer, K.-H., S. Ron, R. Seipp, and A. Almendáriz. 2000. Two new species of hylid frogs, genus *Osteocephalus*, from Amazonian Ecuador. *Amphibia-Reptilia* 21: 327–340.
Kästle, W. 1963. Zur brutpflege des beutelfrösches (*Gastrotheca marsupiata*). *Das Aquarium und Terrarium Zeitschrift* 16: 53–56.
Kimbrough, L. 2023. Ecuador court upholds "Rights of Nature," blocks Intag Valley Copper Mine. *Mongabay* (March 31). https://news.mongabay.com/2023/03/Ecuador-Court-Upholds-Rights-of-Nature-Blocks-Intag-Valley-Copper-Mine/.
Kleemann, J., H. Koo, I. Hensen, G. Mendieta-Leiva, B. Kahnt, C. Kurze, D. J. Inclan, *et al.* 2022. Priorities of action and research for the protection of biodiversity and ecosystem services in continental Ecuador. *Biological Conservation* 265: 109404.
Kohn, E. 2002. Natural engagements and ecological aesthetics among the Ávila Runa of Amazonian Ecuador. Doctoral dissertation, University of Wisconsin-Madison.
Lleras-Perez, R. 2015. *Metallurgy in ancient Ecuador: A study of the collection of archaeological metallurgy of the Ministry of Culture, Ecuador*. Oxford: Archaeopress.
Lombeida, D. 2005. Los anfibios en el pensamiento ecuatoriano. *Ecuador Terra Incognita*, no. 33 (January–February). https://www.terraecuador.net/revista_33/33_anfibios_en_pensamiento.htm.
Lötters, S., K. H. Jungfer, F. W. Henkel, and W. Schmidt. 2007. *Poison frogs. Biology, species & captive husbandry*. Frankfurt: Chimaira.
Lunniss, R. 2018. Manteño, geography and culture of. In *Encyclopedia of global archaeology*, ed. C. Smith, 6711–6732. Cham: Springer.
Lunniss, R. 2022. *Cultural identity, transition, and interaction at Salango, coastal Ecuador: A study of pottery from the Early Regional Development funerary precinct*. Oxford: British Archaeological Reports.
Lunniss, R. M. 2001. Archaeology at Salango, Ecuador: An Engoroy ceremonial site on the south coast of Manabí. Doctoral dissertation, University of London.
Lynch, J. D. 2005. An alert concerning a possible threat to the amphibian fauna east of the Andes: Discovery of the American Bullfrog in eastern Colombia. *Revista de La Academia Colombiana de Ciencias Exactas, Físicas y Naturales* 29: 589–591.
MAE (Ministerio del Ambiente del Ecuador). 2015. *Estrategia Nacional de Biodiversidad 2015–2030. Primera Edición*. Quito: Ministerio del Ambiente del Ecuador.
MAE and GIZ (Ministerio del Ambiente del Ecuador, Deutsche Gesellschaft für Internationale Zusammenarbeit). 2017. *Guía de valoración económica de los recursos genéticos de los anfibios en Ecuador. Informe técnico preparado por la firma consultora CTOTAL*. Quito: Ministerio del Ambiente del Ecuador-Deutsche Gesellschaft für Internationale Zusammenarbeit.
Marcos, J. G. 2005. *Los pueblos navegantes del Ecuador prehispánico*. Quito: Abya Yala.
McClelland, D. 2011. The moche botanical frog. *Arqueología Iberoamericana* 3: 30–42.
McEwan, C., and F. Delgado-Espinoza. 2008. Late pre-Hispanic polities of coastal Ecuador. In *The handbook of South American archaeology*, eds. H. Silverman and W. H. Isbell, 505–525. New York: Springer.
MECN-INB, and GADPEO (Museo Ecuatoriano de Ciencias Naturales del Instituto de Nacional de Biodiversidad, Gobierno Autónomo Descentralizado de la Provincia de El Oro). 2015. *Anfibios, Reptiles y Aves de la provincia de El Oro. Una guía para ecosistemas Andino-Costeros* Vol. 7. Quito: Publicaciones Misceláneas MECN-INB.
Meggers, B. 1966. *Ecuador*. New York: Thames & Hudson.
Merino-Viteri, A., L. A. Coloma, and A. Almendáriz. 2005. Los *Telmatobius* de Los Andes de Ecuador y su disminución poblacional. *Monografías de Herpetología* 7: 9–37.
Ministerio de Educación Ecuador. 2019. *Diccionario de Kichwa/Quechua a Español—Traductor de palabras y su significado*. Quito: Ministerio de Educación Ecuador.
Mongabay. 2012. New colorful rainforest frog named after Prince Charles. *Mongabay* (July 4). https://news.mongabay.com/2012/07/new-colorful-rainforest-frog-named-after-prince-charles-pictures/.
Montalvo-Puente, C. 2011. I materiali ceramici dal sitio Huataviro (Imbabura, Ecuador): Metodo di ricerca. Tesis de Maestría, Sapienza Università Di Roma.
Montalvo-Puente, C. 2016. Il sito de Las Orquídeas (Imbabura) nell'ambito del Periodo Formativo Tardo (800–400 A.C.) nel Nord del Ecuador. Sequenza Stratigrafica e Cronologia. Doctoral dissertation, Archeologia. Sapienza Università Di Roma.
Montalvo-Puente, C. E. 2023. Arquitecturas precolombinas: Representar y habitar el pasado. *Index, Revista de Arte Contemporáneo* 15: 131–153.
Montalvo-Puente, C. E., G. Lago, L. Cardarelli, and J. C. Pérez-Molina. 2023. Money or ingots? Metrological research on pre-contact Ecuadorian "Axe-Monies". *Journal of Archaeological Science: Reports* 49: 103976.
Montaño, D. 2020. Las ranas le ganan una batalla judicial a la minería en Ecuador. *Mongabay* (October 13). https://es.mongabay.com/2020/10/Ranas-Le-Ganan-a-La-Mineria-En-Ecuador/.
Montúfar, R., and N. Pitman. 2003. *Phytelephas aequatorialis*. The IUCN Red List of Threatened Species 2003: e.T43981A10836113. https://www.iucnredlist.org/species/43981/10836113 (accessed December 3, 2023).
Morales, P., and I. Schjellerup. 1999. La gente y su cultura. In *La gente y la biodiversidad. Dos estudios en comunidades de las estribaciones de Los Andes en Ecuador*, eds. H. Borgtoft, F. Skov, J. Fjeldsa, I. Schjellerup, and B. Ollgaard, 53–117. Quito: Centro para la investigación de la diversidad cultural y biológica de los bosques pluviales Andinos, Dinamarca/Abya Yala.
Morales-Mite, M. 2013. *Herpetofauna an áreas prioritarias para la conservación*. Quito: Serie de Publicaciones del Museo Ecuatoriano de Ciencias Naturales. Fundación para la Conservación Jocotoco, Fundación Ecominga, Quito, Ecuador.
Moreno Mora, M. 1967. *Diccionario etimológico y comparado del Kichua del Ecuador. Tomo II*. Cuenca, Ecuador: Casa de la Cultura Ecuatoriana. Núcleo del Azuay.
Mosquera-Perugachi, A. 2014. Análisis icónico de los petroglifos de Cathazho—Morona Santiago. Tesis de Licenciatura, Pontificia Universidad Católica del Ecuador.
Mossman, K. 2007. Profile of Eugenia M. del Pino. *Proceedings of the National Academy of Sciences of the United States of America* 104: 17249–17251.
Mrosovsky, N. 1988. The CITES conservation circus. *Nature* 331: 563.
Mudrack, W. 1969. Pflege und zucht eines blattsteigerfrosches der gatung *Phyllobates* aus Ecuador. *Salamandra* 5: 81–84.
Myers, C. W., and J. W. Daly. 1993. Tropical poison frogs: (Letter to the editor). *Science* 262: 1193.

Myers, N., R. A. Mittermeier, C. G. Mittemeier, G. A. B. da Fonseca, and J. Kent. 2000. Biodiversity hotspots for conservation priorities. *Nature* 430: 853–858.

Oeser, R. 1959. Neue frösche und fröschnachzuchten. *Das Aquarium Und Terrarium Zeitschrift* 12: 184–186.

Olsen-Bruhns, K. 2003. Social and cultural development in the Ecuadorean highlands and eastern lowlands during the Formative. In *Archaeology of formative Ecuador*, eds. J. Raymond and R. Burger, 125–176. Washington, DC: Dumbarton Oaks.

Onore, G. 2017. Personal communication by "Campesino". Pujilí, Provincia Cotopaxi.

Ontaneda, L. S. 2007. *Ecuador: Hitos de su pasado precolombino*. Quito: Banco Central del Ecuador.

Ontaneda, L. S. 2010. *Las Antiguas sociedades precolombinas del Ecuador. Un recorrido por la sala de arqueología del Museo Nacional. Catálogo de la sala de arqueología*. Quito: Ministerio de Cultura/Banco Central del Ecuador.

Ordoñez-Alvarez, M. P. 2024. Los Animales y el mundo precolombino. In *Animales y el mundo precolombino*. Catálogo Exhibición. Quito: Museo El Alabado.

Ordoñez-Araque, R., J. Ruales, P. Vargas-Jentzsch, L. Ramos-Guerrero, M. Romero-Bastidas, C. Montalvo-Puente, and S. Serrano-Ayala. 2022. Pre-Hispanic periods and diet analysis of the inhabitants of the Quito Plateau (Ecuador): A review. *Heritage* 5: 3446–3462.

Ortega-Andrade, H. M., P. Meza-Ramos, D. F. Cisneros-Heredia, and M. H. Yánez-Muñoz. 2010. Los anfibios y reptiles del chocó esmeraldeño. In *Serie herpetofauna del Ecuador: El chocó esmeraldeño*, eds. M. Altamirano-Benavides, D. F. Cisneros-Heredia., and P. Meza-Ramos, 95–252. Quito: Museo Ecuatoriano de Ciencias Naturales.

Ortega-Pacheco, D. V., A. Silva, A. López, R. Espinel, D. Inclán, and M. J. Mendoza-Jiménez. 2018. Tropicalizing sustainable bioeconomy: Initial lessons from Ecuador. In *Towards a sustainable bioeconomy: Principles, challenges and perspectives*, eds. W. Filho, D. Pociovălişteanu, P. Borges de Brito, and I. Borges de Lima, 187–204. Cham: Springer.

Oxford, P., and R. Bish. 2004. *Ecuador. Land of frogs*. Ecuador: Cámara ecuatoriana del libro—Núcleo de Pichincha. Palma de Mallorca: Libre Mundi Ediciones.

Páez-Vacas, M. 2012. Chufle. Personal communication by 'campesinos'. El Oro and Azuay provinces.

Parra R., J. H. 1997. *Los cuentos de los abuelos. Tradición oral de los indígenas Siona y Kofán del Putumayo. Our grandparents' stories. Oral tradition of the Siona and Kofán Indians of Putumayo*. Quito: Abya-yala.

Patzelt, E. 1976. *Fauna del Ecuador*. Quito: Imprenta Europa.

Patzelt, E. 1978. *Fauna del Ecuador*. Quito: Imprenta Europa.

Patzelt, E. 1989. *Fauna del Ecuador*. Quito: Banco Central del Ecuador.

Patzelt, E. 2000. *Fauna del Ecuador*. Quito: ImpreFepp.

Patzelt, E. 2004. *Fauna del Ecuador*. Quito: ImpreFepp.

Paymal, N., and C. Sosa. 1993. *Mundos amazónicos. Pueblos y culturas de la amazonía ecuatoriana*. Primera edición. Quito: Sinchi Sacha.

Pellizzaro, S. 1996. *Arutam. Mitología Shuar*. Quito: Abya-yala.

Peñaherrera de Costales, P., and A. Costales Samaniego. 1959. *Coangue o historia cultural y social de los negros del Chota y Salinas*. Quito: Instituto Ecuatoriano de Antropología y Geografía.

Pérez, J. D. 2005. *Sapos, mariposas y orquídeas en la línea equinoccial*. Quito: Trama Ediciones.

Pickett, J. 1987. Pickett's Piece. Crested newts in the Italian Apennines; Poison arrow frogs, CITES, and other interesting matters. *British Herpetological Society Bulletin* 21/22: 57–59.

Pinto, C. M. 2005. Revisión de libros. Nombres de los Mamíferos del Ecuador. Diego Tirira. 2004. Publicación Especial sobre los Mamíferos del Ecuador 5. Ediciones Murciélago Blanco y Museo Ecuatoriano de Ciencias Naturales. Quito, Ecuador, 267 pp. *Mastozoología Neotropical* 12: 289–291.

Plitt, L. 2017. David Jailaca, el niño ecuatoriano que salvó a una especie y se financió sus estudios descubriendo un sapo que se pensaba extinto. *BBC Mundo Ciencia* (July 14). https://www.bbc.com/mundo/noticias-40607553.

Porras, P. 1961. *Contribución al estudio de la arqueología e historia de los valles Quijos y Misaguallí (Alto Napo) en la región oriental del Ecuador*. Quito: Fénix.

Porras, P. 1985. *Arte rupestre del Alto Napo-Valle del Misagualli, Ecuador*. Quito: Artes Gráficas Señal.

Posada-Swafford. 2022. Two new species of see-through frog named in Ecuador. *National Geographic* (March 18). https://www.nationalgeographic.com/animals/article/two-new-species-of-glassfrog-discovered.

Quelal, P. 2014. Representaciones de aves en la iconografía de la cultura Jama-Coaque. *Antropología Cuadernos De Investigación* 13: 27–47.

Quiguango-Ubillús, A., and L. A. Coloma. 2008. Notes on behavior, communication, and reproduction in captive *Hyloxalus toachi* (Anura: Dendrobatidae), an endangered Ecuadorian frog. *International Zoo Yearbook* 42: 78–89.

Quinatoa Cotacachi, E. 2013. *Representaciones ancestrales y colores del cosmos: diseños de los platos del Carchi*. Quito: Ministerio de Cultura y Patrimonio.

RAE. 2023. *Diccionario de la Lengua Española*. https://www.rae.es/.

Reyes-Puig, C., P. A. Meza-Ramos, M. R. Dueñas, P. Bejarano-Muñoz, S. M. Ramírez-Jaramillo, J. P. Reyes-Puig, and M. H. Yánez-Muñoz. 2015. *Guía de identificación de anfibios y reptiles comunes de la estación experimental "La Favorita"*. Quito: Museo Ecuatoriano de Ciencias Naturales del Instituto Nacional de Biodiversidad (INB-MECN).

Rojas Palomino, G., and G. R. Quintero Bendezú. 2014. *Yachakuqkunapa yachakuqkunapa simi qullqa. Ayakuchu chanka Qichwa simipi*. Lima: Perú Suyupi Yachay Kamayuq.

Ron, S. R., M. R. Bustamante, L. A. Coloma, and B. Mena. 2009. *Sapos. Ecuador Sapodiverso*. Quito: Serie de Divulgación No 10. Centro de Biodiversidad y Ambiente, Escuela de Biología, PUCE.

Ron, S. R., W. E. Duellman, L. A. Coloma, and M. R. Bustamante. 2003. Population decline of the Jambato toad (*Atelopus ignescens*, Anura: Bufonidae) in the Andes of Ecuador. *Journal of Herpetology* 37: 116–126.

Rueda-Almonacid, J. V. 1999. Situación actual y problemática generada por la introducción de "Rana toro" a Colombia. *Revista de la Academia Colombiana de Ciencias Exactas, Físicas y Naturales* 23: S367–367.

Saá B., J. L. 2015. *Preñadilla*. https://web.archive.org/web/20211024192636/http://www.trenandino.com/prenadilla.php

Salazar, A., and J. D. Pérez. 2004. *Animales que cantan y encantan*. Quito: Centro de Biodiversidad y Ambiente/Pontificia Universidad Católica del Ecuador.

Salguero-Morán, D., and D. Ortiz-Quiroz. 2022. Iconografía de Las "Cocinas de Brujo" Milagro Quevedo Durante El Periodo de Integración. Un Acercamiento a Su Interpretación Cosmogónica. Tesis de Licenciatura, Escuela Superior Politécnica del Litoral.

Santos Ortiz de Villalba, J. 1994. Sapo Huichu. In *Sacha Pacha. El mundo de la selva*, 176. Quito: Iconografía de las "Cocinas de Brujo" Milagro Quevedo durante el periodo de integración. Un acercamiento a su interpretación cosmogónica. CICAME, Vicariato Apostólico de Aguarico.

Sartore, J. 2021. *ANI110–00458. A Jordan's casque-headed tree frog*. https://www.joelsartore.com/Ani110-00458/ (accessed December 3, 2023).

Schloegel, L. M., A. M. Picco, A. M. Kilpatrick, A. J. Davies, A. D. Hyatt, and P. Daszak. 2009. Magnitude of the US trade in amphibians and presence of *Batrachochytrium dendrobatidis* and ranavirus infection in imported North American bullfrogs (*Rana catesbeiana*). *Biological Conservation* 142: 1420–1426.

Schmeda-Hirschmann, G., J. P. de Andrade, M. R. Soto-Vasquez, P. A. A. Alvarado-García, C. Palominos, S. Fuentes-Retamal, M. Mellado, P. Correa, and F. A. Urra. 2020. The parotoid gland secretion from Peruvian toad *Rhinella horribilis* (Wiegmann, 1833): Chemical composition and effect on the proliferation and migration of lung cancer cells. *Toxins* 12: 608.

Schmidt, H. 1981. *Dendrobates tricolor* (Boulenger 1899) Eine wertvolle bereicherung unserer tropenterrarien. *Herpetofauna* 3: 18–24.

Shaffer, F. W. 1985. *Motivos indígenas del antiguo Ecuador*. Quito: Abya-Yala.

Siavichay-Pesántez, F., G. Maldonado-Cedeño, and D. Mejía-Coronel. 2016. *Anfibios urbanos de Cuenca*. Cuenca: Editorial Don Bosco, Centro Gráfico Salesiano.

Simmons, J. E. 2009. When father Chimborazo took the frogs away. *Leaf Litter* 2: 19–28.

Sinovas, P., and B. Price. 2015. *Ecuador's wildlife trade. UNEP-WCMC*. Ministry of the Environment of Ecuador and the German Development Cooperation (GIZ). UNEP-WCMC. https://www2.unep-wcmc.org//system/comfy/cms/files/files/000/000/808/original/Ecuador%C2%B4s_wildlife_trade.pdf (accessed December 3, 2023).

Spannhof, I., and L. Spannhof. 1971. Beobachtungen zur brutbiologie und larvenentwicklung von *Gastrotheca marsupiata*. *Wissenschaftliche Zeitschrift der Universität Rostock Mathematik Naturwissenschaften* 20: 97–104.

Stahl, P. 2003. Pre-columbian Andean animal domesticates at the edge of empire. *World Archaeology* 34: 470–483.

Stahl, P. W., and J. S. Athens. 2001. A high elevation zooarchaeological assemblage from the northern Andes of Ecuador. *Journal of Field Archaeology* 28: 161–176.

Stark, L. R., and P. C. Muysken. 1977. *Diccionario Español Quichua, Quichua Español*. Guayaquil: Publicaciones de los Museos del Banco Central del Ecuador.

Stemper, D. M. 1993. *The persistence of prehispanic chiefdoms on the Río Daule, coastal Ecuador*. Vol. 7. Pittsburgh: University of Pittsburgh Latin American Archaeology Publications.

Stothert, K. 2003. Expression of ideology in the Formative Period of Ecuador. In *Archaeology of formative Ecuador*, eds. J. Raymond and R. Burger, 337–421. Washington, DC: Dumbarton Oaks.

Stothert, K. 2006. Feeding the living and the dead: Food symbolism in the art of ancient Ecuador. In *Convivium Artium: Food representation in literature, film and the arts*. https://web.archive.org/web/20140505215106/http://flan.utsa.edu/conviviumartium/Stothert.html

Stuster, R., E. H. Poelman, and R. Boesten. 2017. *Gifkikkers*. Dieren: Dendrobatidae Nederland.

Tapia, E. E. 2011. Sapito guitarrero. Personal communication by Manuel Cleotorio Pai, La Unión, Alto Tambo, Provincia Esmeraldas.

Tapia, E. E. 2015. Gualag. Personal communication by "Campesino" from Lloa, Provincia Pichincha. Quito.

Tapia-Román, A. 2023. Wanip. Personal communication to Luis A. Coloma.

Terra incognita. 2005. *Ecuador Terra de Anfibios*. Quito: Ecuador Terra Incognita.

Tirira, D. G., and M. Ríos. 2018. *Monitoreo biológico Yasuní. 8. Uso de flora y fauna por el pueblo Waorani, Amazonía del Ecuador*. Vol. 8. Quito: Ecuambiente Consulting Group.

Ubidia, A. 1983. Ecuador. In *Poesía popular andina*, ed. A. Hernández. Vol. 2. Quito: Instituto Andino de Artes Populares del Convenio "Andrés Bello."

Ugalde, M. F. 2009. *Iconografía de la cultura Tolita: Lecturas del discurso ideológico en las representaciones figurativas del desarrollo regional*. Reichert: DAI-KAAK.

Ugalde, M. F. 2012. Catazho: Arte rupestre en la Amazonía Ecuatoriana. *Zeitschrift Für Archäologie Außereuropäischer Kulturen* 4: 281–310.

Ugalde, M. F., E. Dyrdahl, C. Montalvo-Puente, and H. Prümers. 2023. Novedades desde Machalilla: Aportes al conocimiento de la economía en el Formativo Tardío en la costa ecuatoriana. *STRATA* 1: e8.

Uribe Taborda, S. F. 2016. *La representación zoomorfa en la cultura Guangala: Un análisis pre-iconográfico en el período de desarrollo regional de la costa central ecuatoriana*. Quito: Abya-Yala/Universidad Politécnica Salesiana.

Uzendoski, M. 2009. La textualidad oral Napo Kichwa y las paradojas de la educación bilingüe intercultural en la Amazonía. In *Repensando los movimientos indígenas*, ed. C. Martínez, 147–172. Quito: FLACSO/Ministerios de Cultura del Ecuador.

Valdez, F. 1987. *Proyecto Arqueológico: "La Tolita"*. Quito: Fondo arqueológico del Museo del Banco Central Guillermo Pérez Chiriboga.

Valencia, J., and K. Garzón. 2011. *Guía de anfibios y reptiles en ambientes cercanos a las estaciones del OCP*. Quito: Fundación Herpetológica Gustavo Orcés.

Valencia, J., E. Toral, M. Morales, R. Betancourt, and A. Barahona. 2008. *Guía de campo de anfibios del Ecuador*. Fundación. Quito, Ecuador: Maxigraf S. A.

Vásconez, S. O., and M. Peñaherrera Gallardo. 1977. *Decreto No. 0158. Reglamento para regular la exportación comercial de especímenes y elementos constitutivos de la fauna silvestre*. Ecuador: Registro Oficial N° 329, del 4 de Mayo de 1977.

Velasco Ibarra, J. M. 1970. *Decreto No. 818. Ley de Protección de la Fauna Silvestre y de los Recursos Ictiológicos*. Ecuador: Registro Oficial N° 104, del 20 de Noviembre de 1970.

Verkade, J. 2012. Behoud van bedreigde amfibieën in Ecuador. *DN Magazine* 2: 28–33.

Verkade, J. 2014. Kikkers voor natuurbehoud—Natuurbehoud voor kikkers. *DN Magazine* 3: 34–43.

Villalba, O. 1988. *Cotocollao: Una aldea formativa del Valle de Quito*. Quito: Miscelánea Antropológica Ecuatoriana.

Walker, J. 1957. The breeding habits of the frog *Gastrotheca marsupiatum*. *British Journal of Herpetology* 2: 85–86.

Walls, J. G. 1994. *Jewels of the Rain Forest—Poison frogs of the family Dendrobatidae*. Neptune City: T.F.H. Publications.

Weiss, H. 1960. Die zucht von *Gastrotheca marsupiata*. *Das Aquarium Und Terrarium Zeitschrift* 13: 121–122.

Werner, Y. L. 1996. Review: A complete guide to scientific and common names of reptiles and amphibians of the world by Norman Frank; Erica Ramus. *Herpetological Review*: 1066–1069.

Weygoldt, P. 1987. Evolution of parental care in dart poison frogs (Amphibia: Anura: Dendrobatidae). *Journal of Zoological Systematics and Evolutionary Research* 25: 51–67.

Whymper, E. 1891. *Supplementary appendix to travels amongst the great Andes of the Equator*. London: John Murray.

Wikimedia. 2023. *Hugo Idrovo*. https://es.wikipedia.org/wiki/Hugo_Idrovo (accessed December 3, 2023).

Wilbert, J. 1974. *The thread of life: Symbolism of miniature art from Ecuador*. Washington, DC: Dumbarton Oaks, Trustees for Harvard University.

Wolf, T. 1892. Geografía *y geología del Ecuador*. Leipzig: Tipografía de F. A. Brockhaus.

Yanez-Muñoz, M. H., P. Meza-Ramos, S. M. Ramirez, J. P. Reyes-Puig, and C. L. Oyagata. 2009. Anfibios y reptiles del Distrito Metropolitano de Quito (DMQ). In *Guía de campo de los pequeños vertebrados del Distrito Metropolitano de Quito (DMQ)*, eds. M. H. Yánez-Muñoz, P. Moreno-Cárdenas, P. Mena-Valenzuela, 5: 9–52. Quito: Museo Ecuatoriano de Ciencias Naturales (MECN)/Fondo Ambiental del MDMQ.

Yeager, J., L. Baquero, and A. Zarling. 2020. Mediating ethical considerations in the conservation and sustainable biocommerce of the jewels of the rainforest. *Journal for Nature Conservation* 54: 125803.

Yuquilema, A. 1978. *Ñucanchig llajtamanta parlucuna. Cuentos y leyendas de mi tierra*. Riobamba: Editorial Linguistica.

Zeidler, J. A. 2003. Appendix A: Formative period chronology for the coast and western lowlands of Ecuador. In *Archaeology of formative Ecuador*, eds. J. Scott Raymond and L. Richard Burger, 487–527. Washington, DC: Dumbarton Oaks Research Library and Collection.

Zeidler, J. A. 2008. The Ecuadorian Formative. In *The handbook of South American archaeology*, eds. H. Silverman and W. Isbell, 459–488. New York: Springer.

Zeidler, J. A., and D. M. Pearsall. 1994. *Regional archaeology in northern Manabí, Ecuador, Volume 1: Environment, cultural chronology, and prehistoric subsistence in the Jama River Valley*. Vol. 8. Pittsburgh: University of Pittsburgh Latin American Archaeology Publications.

Zimmermann, E. 1983. Das züchten von terrarientieren. Pflege, verhalten, fortpflanzung. *Reptilia, Testudines*. Stuttgart: Kosmos Verlag.

Zimmermann, E. 1986. *Breeding terrarium animals*. Neptune City: T.F.H. Publications.

Zimmermann, E., and H. Zimmermann. 1994. Reproductive strategies, breeding, and conservation of tropical frogs: Dart-poison frogs and Malagasy poison frogs. In *Captive management and conservation of amphibians and reptiles*, eds. J. B. Murphy, K. Adler, and J. T. Collins, 255–266. Lawrence: SSAR.

Zimmermann, H. 1979. *Tropical frogs*. Neptune City: T.F.H. Publications.

Zimmermann, H. 1989. Conservation studies on the dart-poison frogs (Dendrobatidae) in the field and in captivity. *International Zoo Year Book* 28: 31–44.

Zimmermann, H., and E. Zimmermann. 1985. Brutpflegestrategien bei pfeilgiftfröschen. *verhandlungen der Deutschen Zoologischen Gesellschaft* 78: 220.

Zimmermann, H., and E. Zimmermann. 1988. Etho-Taxonomie und zoogeographische artengruppenbildung bei pfeilgiftfroschen (Anura: Dendrobatidae). *Salamandra* 24: 125–160.

2 The History and Present of Amphibian Biology

In this section, we have compiled and summarized the history of amphibian research, its protagonists, and the most relevant facts in relation to the scientific discoveries and research on Ecuadorian amphibians up to recent times. In doing so, we describe events in nearly chronological order, except in cases where it was necessary to follow up on the work of an individual or an institution during the span of their activities. We draw from published biographies to extract pertinent information while also describing events that have not been published previously, particularly events that occurred in our lifetimes, up to 2022–2023. In describing personal or institutional recent history, we made an effort to include, as contributing authors, several biologists who either wrote (in third person) sections concerning their interests or provided first-hand information; these are contributing authors and are cited under the title of the section. When we and contributing authors deal with our own work, we place an effort to balance self-praise with objectivity and evidence. Nonetheless, some of the adjectives used were chosen by Luis A. Coloma and reviewers, and are of their preference. We acknowledge that there is undoubtedly an inherent bias in the narrative and interpretation of events, given by our own participation in the recent history (from 1966 to the present). The authors who provided information for the biographies are cited in the acknowledgements section. Persons, institutions, and museum names are spelled completely the first time that they are mentioned under a title. Subsequently, last names of persons are used, and institutions and museums are indicated with their abbreviations.

A LINNAEAN PREAMBLE

Prior to 1848, during a period of 90 years (1758–1848), 22 species of widespread Neotropical amphibians, most of them labeled as coming from America, Brazil, and Guyana, were described by European scientists. More than a century later (114 years), beginning in 1872, the first species would be recorded as occurring in Ecuador.

The first binomial names for amphibians appeared in 1758 and were established by the founder of modern binomial taxonomy, Carl Linnaeus (1707–1778) (Linnaeus, 1758). Linnaeus was a Swedish scientist, naturalist, botanist, and zoologist who was a pioneer in the classification of species, to which he assigned binomial names. He named the following five species that subsequently have been found in Ecuador (the type locality, author, and year of publication or first record from Ecuador are in parenthesis): *Rhinella marina* (America; Jiménez de la Espada, 1872), *Ceratophrys cornuta* (unknown locality; Jiménez de la Espada, 1872), *Boana boans* (America; Duellman, 1978), *Pipa pipa* (Suriname; Duellman, 1978), and *Caecilia tentaculata* (America; Taylor and Peters, 1974). With the exception of the caecilian *C. tentaculata*, all of these are common species in the Guyana and Amazon basins. These names were associated with a few specimens and illustrations done by several authors, most in reference to the illustrations published by the Dutch zoologist Albertus Seba (1665–1736) (Seba, 1734) (Figures 2.1, 2.2). In the remainder of the 18th century, over a period of 42 years, two European scientists coined the names for six other common species. Three of them were described by Josephus Nicolaus Laurenti (1735–1805): *Dendropsophus marmoratus* (Suriname), *Scinax ruber* (America), and *Leptodactylus pentadactylus* (Indiis) (Laurenti, 1768). Laurenti was an Austrian naturalist of Italian origin who named 42 species of amphibians and reptiles. Subsequently, in 1799, the German classicist and naturalist Johann Gottlob Theaenus Schneider (1750–1822) provided descriptions of three additional species: *Rhaebo guttatus* (Surinam), *Boana punctata* (Surinam), and *Lithodytes lineatus* (not indicated). During the first half of the 19th century, between 1802 and 1848, 11 abundant species were described from Guyana and Brazil by other European naturalists: Daudin (2), Mikan (1), Spix (5), Wiegmann (1), Tshudi (1), and Troschel (1): *Sphaenorhynchus lacteus* (America), *Siphonops annulatus* (Rio de Janeiro, Brazil), *Hemiphractus scutatus* (Solimões River, Brazil), *Boana cinerascens* (Ega, Tefé, Brazil), *Leptodactylus mystaceus* (Salvador, Estado de Bahía, Brazil), *Lithobates palmipes* (Amazon River, Brazil), *Rhinella proboscidea* (Solimões River, Brazil), *R. horribilis* (Veracruz, Mexico), *R. poeppigi* (Moyobamba, Peru), and *Boana calcarata* (Guyana), respectively. Among these authors, the German Johann Baptist von Spix (1781–1826), who was the first curator of zoology at the Bavarian Academy of Science in Munich, stands out for having participated in an expedition to Brazil in 1817.

CORNALIA AND THE START OF AMPHIBIAN TAXONOMY IN ECUADOR

During the half-century from 1849 to 1900, 108 species were described from Ecuador through the work of 15 European and North American scientists. Italy, Germany, the United States, Spain, and England were at the starting line of the epoch of amphibian research in Ecuador. In 1849, the first amphibian species to be named from Ecuador, *Atelopus ignescens* (Cornalia, 1849), a bufonid, was described in a publication by the Italian scientist Emilio Cornalia (1824–1882) (Figure 2.3). This milestone event happened because in April 1847, another Italian naturalist, Gaetano Osculati (1808–1894), collected several specimens at a

DOI: 10.1201/9781003129080-2

FIGURE 2.1 Types of *Rhinella marina* (top left), *Ceratophrys cornuta* (top right), and syntype of *Boana boans* (bottom) from America. Taken and modified from Seba (1734).

locality described as "Locis humidis circa Latacunga prope Quito" (= wetlands around Latacunga near Quito) and deposited them in the Museum of Natural History in Milan (currently those specimens are missing). Two years later, Cornalia formally named the species and described the external morphology of the endemic *A. ignescens* in a 15-line text account (which was an extensive description for those times). Emilio Cornalia was curator and director of the museum and is well known because he was one of the leading naturalists and scientists of his time; he specialized in comparative anatomy and was one of the founders of the Italian Entomological Society (Conci and Poggi, 1996).

The description of *Atelopus ignescens* marked the beginning of an era of exploration and research by European and North American naturalists, collectors, and researchers who, during more than a century (until 1949), discovered and described 170 additional species of Ecuadorian amphibians. Thirty-six taxonomists worked on the descriptions of these species (we provide information about most of them herein). Among them, Espada and Boulenger stand out because of the number of species they described. Espada excelled in his detailed descriptions, and he is the only one who described specimens that were collected by himself.

In the first half of the 19th century, there were many expeditions to South America, and some of the explorers belonged to or served the German nobility. According to Bustamante (2016), undoubtedly following the example of Humboldt, the exploration of nature became an attractive activity for the aristocracy. In Ecuador, this age of exploration was associated with new species descriptions starting in 1857–1860, when 12 additional species of frogs were collected in Ecuador by the German scientist and explorer Friedrich Johann Carl Moritz Wagner (1813–1887) in an expedition that was funded by the Bavarian King Maximilian II (Wagner *et al.*, 2012). The specimens he collected were deposited at the Zoologische Staatssammlung München (ZSM) museum in Germany, except for one specimen of the lungless salamander *Bolitoglossa palmata*, which was given to the Naturhistorisches Museum Wien (NHMW) in Austria. Some of the specimens served as type material for four species that were described by the German scientist Wilhelm Karl Hartwich Peters (1815–1883), director of the Berlin Zoological Museum, and the Austrian Franz Josef Maria Werner (1867–1939): *Leptodactylus wagneri* (Peters, 1862a), *Hemiphractus fasciatus* (Peters, 1862b), *Pristimantis chalceus* (Peters, 1873), and *Bolitoglossa palmata* (Werner, 1897)—note that *L. wagneri* was named in honor of the collector. Unfortunately,

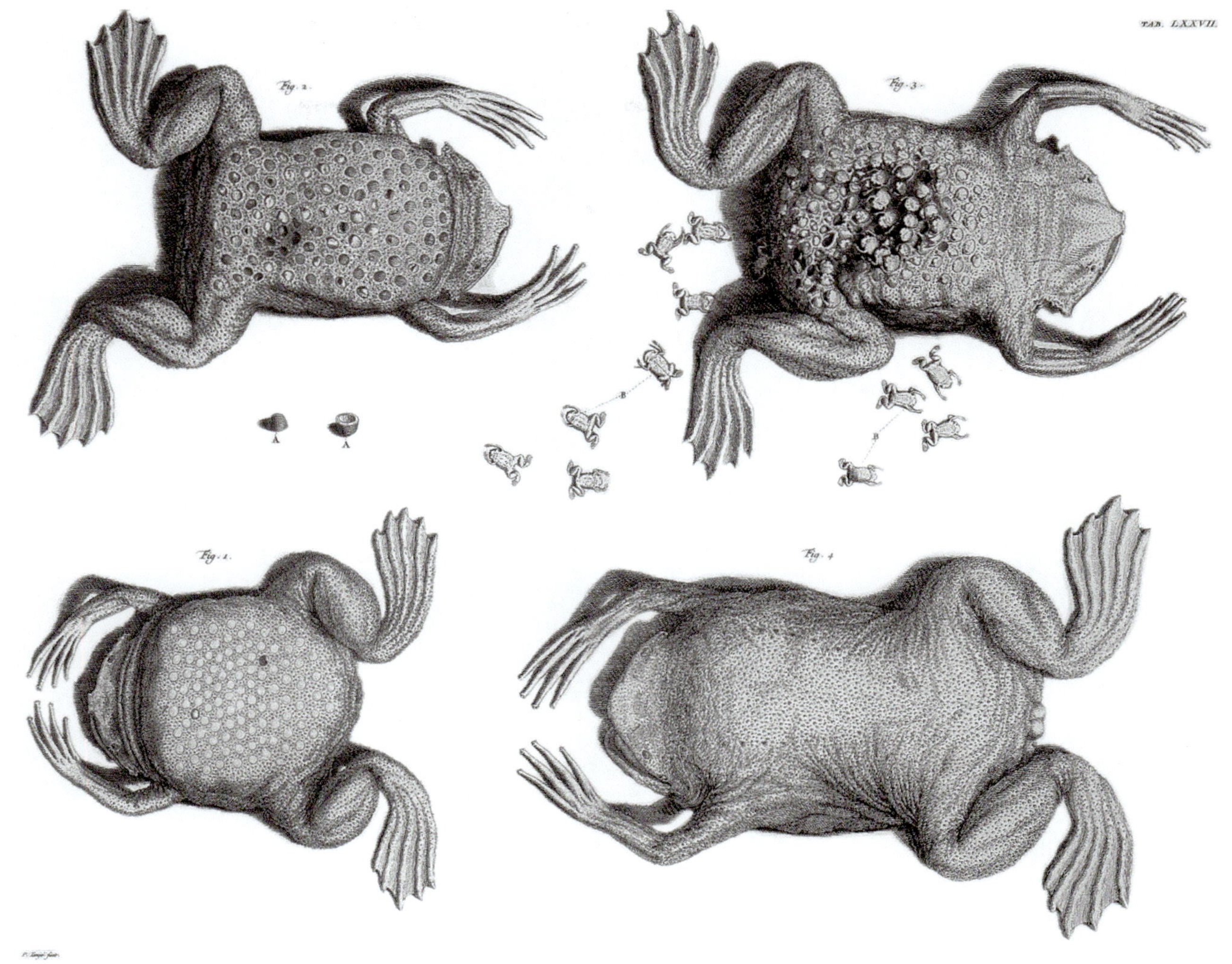

FIGURE 2.2 Syntypes of *Pipa pipa* from Suriname. Taken and modified from Seba (1734).

the material in the ZSM collection was lost due to the bombing during the Second World War. It is possible that some of the specimens are still extant, but this awaits clarification (Wagner *et al.*, 2012).

Another expedition to Ecuador occurred seven years later. As recounted by Cope (1868), an expedition from the United States, for purposes of scientific exploration, was undertaken during the autumn and winter of 1867–1868, under the auspices of the Smithsonian Institution. Professor James Orton of Williams College, Massachusetts, directed the expedition, which was composed mainly of students from the college. Along with many specimens of reptiles, they collected a small black frog on the western slope of Volcán Antisana, 13,000 ft. (3962 m) above sea level. Apparently, this was the only new species of frog they found in Ecuador, and it was described by the prodigious and famous zoologist, paleontologist, and dinosaur expert, Edward Drinker Cope (1840–1897) in 1868. Cope worked during a period of intense fossil-hunting competition now known as the Bone Wars (https://en.wikipedia.org/wiki/Bone_Wars). Cope named the Ecuadorian frog *Atelopus longirostris* (Cope, 1868), mentioning in his publication that it had a black dorsum. However, in the description it is stated that it came from the Valley of Quito, a locality that, according to Peters (1973) is in error. Collections in the 20th century of *A. longirostris* were made in the western slopes of the Cordillera Occidental de los Andes, but the species occurs at a maximum elevation of 1925 m. Cope eventually described 14 species (in 7 publications) as being recorded from Ecuador. Nine of them were from Peru, brought back by Orton, three from Colombia, one from Costa Rica, and one (*A. longirostris*) from Ecuador.

JIMÉNEZ DE LA ESPADA: AN OUTSTANDING SPANISH SAVANT

In 1862, Isabel II of Spain sent an expedition to the Pacific coast of South America. This expedition was a military and political operation, an attempt to help Spain recover its power and influence in the Americas. In this context, and at the last minute, the Spanish Ministry of Development organized the *Comisión Científica del Pacífico*, which included a scientific team of naturalists to undertake the study of the flora and fauna of the former Spanish colonies in America

FIGURE 2.3 Emilio Cornalia. Taken and modified from Smithsonian Institution Archives, Record Unit 95, Box 27A.

and to collect specimens for the museum in Madrid from 1862 to 1865 (Savage, 1978; Miller, 1983; Cabodevilla, 1998; López-Ocón and Ponsati, 2002; González Fernández, 2006). With this expedition, a resurgence in academic interest in America was promoted in Spain, along with the development of Americanism. Although the expedition was an imperial or neo-colonial venture, the "American studies" it initiated laid the foundations for an Ibero-American "federative science" (López-Ocón, 2003). The concern for academic development was also expressed in Ecuador during the government of Gabriel García Moreno, who was interested in science; this period has been called the Conservative Enlightenment (Bustamante, 2016).

The naturalists of the *Comisión* included a prominent pioneer, Marcos Jiménez de la Espada (1831–1898) (Figure 2.4), born in Cartagena de Murcia, Spain. Espada was a well-known and distinguished Americanist, explorer, and scientist (Barreiro, 1928; Colao, 1967; Luer, 1966; Savage, 1978; López-Ocón, 1991; Cabodevilla, 1998; López-Ocón and Pérez-Montes, 2000; González, 2000). His travels, biography, diaries, and writings have been compiled on a web page about the *Comisión* (López-Ocón and Ponsati, 2002). Unfortunately, although the year-long scientific expedition was dedicated to research activities along the entire Pacific coast, from Tierra del Fuego in South America to California in North America, upon reaching Lima, they found that the political situation had seriously deteriorated. An incident in Peru, when half of the Peruvian army declared war on Spain, opened the War of the Pacific (1864–1866). Although Ecuador issued a declaration of war in an act of solidarity, it had no significant participation in the conflict (Bustamante, 2016). Faced with this reality, the scientific mission was suspended in 1864 when Admiral Pinzón arbitrarily ordered the commission members to unload from the ship all their equipment and materials and return to Spain (Savage, 1978). However, Jiménez de la Espada (the most prestigious participant) decided to continue on a different project that was named as the Great Journey, that is, a recreation of the journey that Charles Marie de La Condamine initiated in 1735 from Quito to the Atlantic by way of the Amazon. Espada was an emulator of the great naturalist travelers of the romantic era such as Alexander von Humboldt and Charles Darwin, and his journey through the Andean and Amazonian regions was filled with adventures. Espada arrived in Guayaquil, Ecuador, on 31 October 1864 and collected frogs in Ecuador until about June 1865. His route was from Guayaquil to Quito and then across the Andes to Papallacta, Baeza, Sumaco, Archidona, and the Quijos River. Among his adventures, while descending the Volcán Pichincha, Espada was lost for three days and was near death before being rescued (López-Ocón and Ponsati, 2002). An outline and map of his itinerary is found in Savage (1978), and more detailed information is in the diaries of the expedition members (Jiménez de la Espada *et al.*, 1866; Barreiro, 1928). Sadly, upon their return to Spain, Espada and the other members

FIGURE 2.4 Left: Members of the *Comisión Científica del Pacífico*. Sitting on the foreground is the zoologist Marcos Jiménez de la Espada. Photo by Rafael Castro Ordóñez in Montevideo, Paraguay, in December 1862. Middle: Marcos Jiménez de la Espada in 1864. Photo by Rafael Castro Ordóñez. Right: Marcos Jiménez de la Espada in his later years in 1890–1898. Photo by M. Huerta. © CSIC, ACCHS.

of the *Comisión* did not receive government support, and the indifference and laziness of the museum staff resulted in the deterioration of their collections, which were only recovered recently (González Fernández, 2006). At the peak of his prestige as a naturalist, and after having promoted the Spanish Society of Natural History, of which he was a founder in 1871 along with a handful of colleagues, Espada terminated his research program as a scholar of the Scientific Commission of the Pacific—which was officially terminated in 1872—and redirected his attention to the field of historical geography and American history. He died at the age of 67 (Savage, 1978; López-Ocón and Ponsati, 2002).

Jiménez de la Espada was the first scientist who described Ecuadorian amphibians based on specimens from his own field work. During the expedition, Espada collected 786 specimens of amphibians, most of which were from Ecuador, particularly from the Amazonian side of the Andes, between Baeza and Sumaco, Provincia Napo. A few other Ecuadorian localities were collected with the help of his colleagues on the commission (de La Riva, 2000; González Fernández, 2006). Based on these collections, Espada published descriptions and data for 35 species (34 species and one variant) that belong to 28 currently valid species (Table 2.1). As currently recognized, 18 of the species were new to science, 7 were recorded from Ecuador, and 3 had been reported previously. Furthermore, Espada described six currently valid genera: *Centrolene*, *Edalorhina*, *Engystomops*, *Hyloxalus*, *Oreobates*, and *Pristimantis*. These descriptions were published between 1870 (see Table 2.1 legend) and 1875 (Jiménez de la Espada, 1870, 1872, 1875), among which the Batracios paper of 1875, considered a classic of the zoological literature, is of obligatory use to any amphibian biologist interested in the Neotropics. For four of the new species, Espada only provided illustrations in the 1875 Batracios paper (Table 2.1) because the text was not published due to financial problems. In a paper published in 1899, one year after his death, Espada added extensive morphological descriptions of the horned treefrogs *Hemiphractus* (Jiménez de la Espada, 1899).

Espada worked in close correspondence with renowned herpetologists from other museums in Europe (López-Ocón, 1991). Among them was Wilhelm Peters, to whom he dedicated the specific name of a species, *Engystomops petersi*.

> I dedicate the species to the eminent naturalist Mr. W. Peters, director of the Berlin Zoological Museum, whose wise lessons have contributed so much to direct and illustrate this work of mine.
>
> (Jiménez de la Espada, 1875; our English translation)

The following account of *Hemiphractus* (Jiménez de la Espada, 1899) is based on one of the specimens collected by Wagner in 1857–1860. Herein he also expresses his respect to Peters (our English translation).

> It is not a risk to suppose that the genus *Hemiphractus* would have continued to be proscribed from scientific catalogs without the chance shipment of another specimen found by M. Moritz Wagner (a traveler whose footsteps I have followed through some of the Ecuadorian regions and have left behind in my ascent to Izalco) on the banks of the Pastasa [Pastaza], a river that runs parallel to the Napo and bathes a part of the northern region of the upper Amazon, visited by us. Said copy reached the hands of the wise Director of the Berlin Museum in a state of perfect conservation.

TABLE 2.1
Status of Ecuadorian Amphibian Taxa Described by Marcos Jiménez de la Espada (updated from González Fernández, 2006)

Familiy	Current Identification	Jiménez de la Espada's Names	Years of Publication
Bufonidae	*Atelopus ignescens* (Cornalia, 1849)	*Atelopus ignescens*	1875: 139, Plate 1, figure 8 (as *Phryniscus ignescens*)
Bufonidae	*Atelopus longirostris* Cope, 1868	*Atelopus longirostris*	1875: 155, Plate 2, figures 2, 2a, 2b, 2c
Bufonidae	*Atelopus petersi* (Coloma *et al.*, 2007)	*Phryniscus ignescens* (variet)	1875: Plate 1, figures 7, 7a
Bufonidae	*Atelopus planispina* Jiménez de la Espada 1875	*Atelopus planispina*	1875: 148
Bufonidae	*Atelopus planispina* Jiménez de la Espada 1875	*Atelopus longirostris* (variet)	1875: 155, Plate 2, figures 1, 1a–1d
Bufonidae	*Rhaebo guttatus* (Schneider, 1799)	*Rhaebo leschenaulti*	1875: 167
Bufonidae	*Rhinella marina* (Linnaeus, 1758)	*Bufo marinus napensis*	1875: 201
Bufonidae	*Rhinella* sp. (*margaritifera* complex)	*Oxyrhynchus typhonius*	1875: 171
Bufonidae	*Rhinella* sp. (*margaritifera* complex)	*Oxyrhynchus proboscideus*	1875: 178
Centrolenidae	*Centrolene geckoideum* Jiménez de la Espada 1872*	*Centrolene geckoideum*	1872: 88 (1875: Plate 2, figures 5, 5a–5d)
Ceratophryidae	*Ceratophrys cornuta* (Linnaeus, 1758)	*Ceratophrys megastoma*	1875: 26
Dendrobatidae	*Hyloxalus bocagei* Jiménez de la Espada 1870	*Hyloxalus bocagei*	1870[1]: 59 (1875: Plate 3, figures 2, 2a, 2b)
Dendrobatidae	*Hyloxalus fuliginosus* Jiménez de la Espada 1870*	*Hyloxalus fuliginosus*	1870[1]: 59 (1875: Plate 3, figures 1, 1a, 1b)
Dendrobatidae	*Hyloxalus pulchellus* (Jiménez de la Espada 1875)	*Phyllodromus pulchellum***	1875: Plate 3, figures 3, 3a–3c
Hemiphractidae	*Gastrotheca riobambae* (Fowler, 1913)	*Nototrema marsupiatum****	1875: Plate 4, figure 1
Hemiphractidae	*Gastrotheca testudinea* Jiménez de la Espada 1870	*Nototrema testudineum*	1870: 62 (1875: Plate 4)
Hemiphractidae	*Hemiphractus bubalus* Jiménez de la Espada 1870	*Cerathyla bubalus*	1870: 64 (1875: Plate 5)
Hemiphractidae	*Hemiphractus bubalus* Jiménez de la Espada 1870	*Cerathyla palmarum*	1870: 64 (1875: Plate 6)
Hemiphractidae	*Hemiphractus bubalus* Jiménez de la Espada 1870	*Cerathyla braconnieri*	1870: 64 (1875: Plate 6)
Hemiphractidae	*Hemiphractus proboscideus* Jiménez de la Espada 1870	*Cerathyla proboscidea*	1870: 64 (1875: Plate 5)
Hemiphractidae	*Hemiphractus scutatus* (Spix, 1824)	*Hemiphractus scutatus*	1875: Plate 5, figures 1, 1a
Hylidae	*Dendropsophus reticulatus* Jiménez de la Espada 1870	*Hyla reticulata*	1870: 61 (1875: Plate 3 bis)
Leptodactylidae	*Engystomops petersi* Jiménez de la Espada 1872*	*Engystomops petersi*	1872: 86 (1875: 160, Plate 2, figures 3, 3a–3d)
Leptodactylidae	*Leptodactylus labrosus* Jiménez de la Espada 1875	*Leptodactylus labrosus*	1875: 36
Leptodactylidae	*Leptodactylus pentadactylus* (Laurenti, 1768)	*Leptodactylus goliath*	1875: 57
Leptodactylidae	*Leptodactylus stenodema* (Jiménez de la Espada 1875)	*Leptodactylus stenodema*	1875: 64
Craugastoridae	*Edalorhina perezi* Jiménez de la Espada 1870*	*Edalorhina perezi*	1870: 58
Craugastoridae	*Oreobates quixensis* Jiménez de la Espada 1872*	*Oreobates quixensis*	1872: 87 (1875: Plate 3, figures 5, 5a–5d)
Craugastoridae	*Oreobates quixensis* Jiménez de la Espada 1872	*Hylodes verrucosus*	1875: Plate 3 bis, figure 1, 1a
Craugastoridae	*Oreobates quixensis* Jiménez de la Espada 1872	*Hylodes philippi*	1875: Plate 3 bis, figure 2, 2a
Craugastoridae	*Pristimantis diadematus* Jiménez de la Espada 1875	*Hylodes diadematus***	1875: Plate 3 bis, figure 3, 3a, 3b, 3c
Craugastoridae	*Pristimantis galdi* Jiménez de la Espada 1870*	*Pristimantis galdi*	1870: 61, Plate 3 bis, figure 6, 6a–6f
Craugastoridae	*Pristimantis lacrimosus* Jiménez de la Espada 1875	*Cyclocephalus lacrimosus***	1875: Plate 3 bis, figure 5, 5a, 5b
Craugastoridae	*Pristimantis rubicundus* Jiménez de la Espada 1875	*Hylodes rubicundus***	1875: Plate 3 bis, figure 4, 4a, 4b
Craugastoridae	*Strabomantis cornutus* (Jiménez de la Espada 1870)	*Limnophys cornutus*	1870: 60
Craugastoridae	*Strabomantis cornutus* (Jiménez de la Espada 1870)	*Limnophys napaeus*	1870: 60
Craugastoridae	*Strabomantis cornutus* (Jiménez de la Espada 1870)	*Strabomantis cornutus*	1870: 60 (1872: 85–88; 1875: Plate 3, figures 4, 4a–4c)

*New genus. **Figured only. ***The specimen MNCN 157 (specimen apparently lost) from Guaranda (not published by Espada) and cited by González Fernández (2006) as *Gastrotheca riobambae* may be *G. pseustes*, which is the only similar species occurring in this locality. [1]Several authors (e.g., de La Riva, 2000; González Fernández, 2006) give the publication date as 1871 based on the binding date. However, according to Darrel R. Frost (DRF note 2001, pers. com.) "Zoonomen gives 1870 for this number; 1871 date is a result of the error of considering the whole volume to have come out in the same year as the end-of-year volume cover". It was common for the last numbers of a volume to come out in the following year and then for the binding to reflect the later year even though the earlier numbers (of which Espada's article is) came out in the year that they claimed.

Espada was ahead of his time in providing meticulous and detailed external morphological descriptions (Savage, 1978), and he was the first to publish ample data on osteology, natural history, and diet and to provide illustrations (by F. Díaz Carreño) (Figure 2.5) and ethnozoological information for amphibians occurring in Ecuador. Fragments of his descriptions are interesting to recount herein (translated from Spanish). For example, Espada gives an account (not as detailed as he would have liked; see the following in his own words) of some aspects of the natural history of the once abundant jambato harlequin frog (*Atelopus ignescens*), based on observations made in 1865 at Laguna de la Mica, Antisana, and he included the first known illustration of this species. Currently, *A. ignescens* is almost extinct and was last seen at Laguna de la Mica in 1983 (Mario García Saltos, pers. com.). Espada's encounter with the jambato was during the breeding season, which he describes (our English translation) as follows:

> We saw individuals in the thousands near streams during the months of November, December and January in the grassy and humid meadows, near the streams, ponds or lagoons. On the shores of the one called la Mica, in

FIGURE 2.5 The first illustrations of Ecuadorian frogs. *Edalhorina perezi*, *Atelopus petersi*, and *A. ignescens*. Illustrations by F. Díaz Carreño. Taken and modified from Jiménez de la Espada (1875). © Archivo del Museo Nacional de Ciencias Naturales, Madrid (ACN110B/002/04705).

> Antisana beginning in 1865, I surprised them at the time of their love making, and when the males look for the females to spawn or fertilize their eggs. They chased them through the swamps next to the water with activity and insistence, and so blindly that, struggling to get them, when reaching them, they rolled in groups, tangled with each other. I confess with sorrow that I had then little hesitation in offering myself a more detailed observation, contenting myself with collecting and keeping; for now, I am reduced to mere conjecture about what I could have recorded as a certain fact. Something exceptional happens in the fertilization of *A. ignescens*. In the moment, the four extremities of the females and the forelimbs of the males acquire extraordinary strength; they walk supporting the weight of the male and that of the eggs that swell their ovaries, by which means the act is verified on land. . . . We have six pairs in our collection still tenaciously embracing, just as I surprised them on the spot. . . . Among the loose females there are several that still have very deep impressions of the male embrace. . . . In the stomach of individuals of this species there are abundant remains of coleopterans–many of them elatheroids– of hymenopterans and dipterans–of these there are also larvae and chrysalises, mixed with grains of sand.

Some of the species described by Espada are common frogs in Ecuador; others are rare, and there are some which are possibly extinct. Among the still common ones are *Leptodactylus labrosus* from the Pacific lowlands of Ecuador and *L. pentadactylus* (as *L. goliath*) from the Amazonian region. In a description of the latter, Espada provided the Kichwa name (hatun-hambato), meaning big frog, which might be the first scientifically documented case of the use of the word hambato (see also Chapter 1, under the subtitle "Indigenous and Vernacular Names"). Espada recounts some interesting aspects of these frogs (our English translation):

> [*Leptodactylus labrosus*] inhabits the Guayas or Guayaquil River basin. The two specimens, which along with several other reptiles my partner, Mr. Martínez, owed to the generosity of Mr. Alcides Destruge, a doctor, and a person very knowledgeable about the natural things of the country, come from the banks of the Daule River, a tributary from that to the middle of its course.

> [*Leptodactylus goliath*] inhabits the eastern province of Ecuador, ancient Quijos, where it should not be rare and is found at different altitudes in the plains and mountains. It walks during the day and its habitat is much less aquatic than that of *L. pachypus*, *gracilis*, etc.; it remains hidden underground for a long time and it swelled enormously, like the toads, which the Indians brought along with it, and, in a word, its habits make it a kind of diurnal toad. They give it the name hatun-hambato, which is equivalent to a great frog or toad. Inside its stomach I have found abundant remains of orthopterans and the same coprophagus scarabs that *Bufo agua*, *Oxyrhynchus proboscideus*, etc., usually feed on.

Jiménez de la Espada (1875) also described the rare frog (currently critically endangered according to IUCN criteria) *Leptodactylus stenodema* and wrote (our English translation):

> [*Leptodactylus stenodema*] inhabits the eastern province of Ecuador. The Indians found it together with *L. goliath* and *B. agua*. Inside its stomach I have found the remains of orthopterans and butcher beetles. I assume it has the same customs of the *Leptodactylus* mentioned above.
>
> [*Edalorhina perezi*] lives in the outskirts of the town of Napo, on the bank of the river of that name; I obtained it in the month of May from some Indians employed by us in the collection of natural objects, and despite the many requests that I made to them, I was only able to obtain one specimen. This circumstance indicates that it must be rare.
>
> I dedicated the species to the distinguished professor of the Faculty of Sciences of the University of Madrid, Dr. Laureano Pérez Arcas, my first teacher of Zoology.

As mentioned previously, Espada described species that are possibly extinct, such as *Atelopus planispina* (see its account), and the information he provided is nearly all that is known of the natural history of this species (our English translation).

> [*Atelopus planispina*] inhabits the eastern region of Ecuador; I found it very abundant during the month of May in San José de Moti, at the foot of the Sumaco volcano in the ravines and streams and along the paths of the shady forests; it shows a preference for ants; however, in the stomach of one there was also a curculionid, a larva, and a scorpionid. . . . It must be, like all its congeners, very fecund; it was not spawning at the time in which I observed it: the females showed a depressed body that was as thin as that of the males.

Espada also provided data, in his 1875 paper, for a relatively common toad from the slopes of Volcán Sumaco that still remains undescribed (*Rhinella* sp. of the *margaritifera* complex) (our English translation):

> [*Oxyrhynchus typhonius*] inhabits the eastern Andean region of Ecuador. I have found it abundant in San José de Moti, on the slopes of the extinct volcano Sumaco, a cool place compared to the plains of that region. Its customs are those of a toad. In the stomach of a female, I have found traces of carabids, staphyllinids, curculionids, and lamelicorns, and two flowers; in another, two or three species of ants and snake scales.

EUROPEAN MUSEUMS, THE BRITISH EMPIRE, AND THE AXIS OF COLLECTIONS GROWTH

In the 19th century, European natural history museums emerged and some of them grew exponentially. For example, the British Museum (Natural History) (Natural History Museum, London, since 1992) (BMNH) holds more than 70 million specimens (Simmons, 2016) and about 1280 specimens of amphibians from Ecuador (Natural History Museum, 2023). Simmons discusses more details of European museums becoming the modern museums, which were tied to the rise of research and the so-called the Darwin effect. According to Simmons, in the century after the publication of Linnaeus' work in 1758, and before Darwin's publication on the *Origin of Species* in 1859, natural history collections were mostly typological, meaning that one or two specimens of a species were preserved as sufficient to represent the species. Darwin's theory of natural selection prompted a fundamental revolution in natural sciences and had far-reaching effects on collections, given that his theory laid its foundations on the study of variation within species, and for that purpose large series of specimens of each species throughout its geographic range were needed.

The mid-to-late 19th century and early 20th centuries saw amphibian field surveys in Ecuador being made by several professional naturalists and collectors, mostly European. Among those collectors who stand out are the Englishmen Edward Whymper, William F. H. Rosenberg, Louis Fraser, and Mervin George Palmer; the presumed British Indian (see later) Clarence Buckley; the Italian naturalist Enrico Festa; and the American collector Perry O. Simons. They deposited key specimens at the BMNH and the Museo Zoologico di Torino. These collections were initially studied by the German-born British zoologist Albert Günther, the Belgian-born British zoologist George Boulenger, and the Italian Mario Peracca. Although Günther was from Germany and Boulenger was from Belgium, both worked at the BMNH. Günther worked at the museum from 1857 until 1895 (he was Keeper of Zoology from 1875–1895); Boulenger was what was called a first-class assistant from 1880–1920.

FIGURE 2.6 Edward Whymper, photo by Atelier Boissonnas (vers 1863—vers 1980–1985) taken on 30 September 1909. From Library of Geneva, VG FBB N13x18 Clients 89834A.

Mountaineering and Collecting: Whymper

Edward Whymper (1840–1911) (Figure 2.6) was an English mountaineer, explorer, illustrator, and writer. He is considered the father of world mountaineering and is well known for making the first ascents of several mountains, among them Volcán Chimborazo in Ecuador. In 1880, he organized an expedition to Ecuador, designed primarily to collect data for the study of altitude sickness and the effect of reduced pressure on the human body, but he also took this opportunity to make zoological collections, especially in the highlands. His collection of amphibians was given to George Albert Boulenger at the BMNH in London.

From 1879–1880, Whymper traveled in the northern and Central Andes of Ecuador and made important ascents of the Chimborazo, Sincholagua, Antisana, Cayambe, Sara Urco, and Cotacachi mountains. Whymper arrived in Guayaquil on 9 December 1879 and was on the slopes of Chimborazo by 27 December of that year. He described in detail his journey in the celebrated book, *Travels Amongst the Great Andes of the Equator*, in which he included a Supplementary Appendix with descriptions of many new insects and other invertebrates, amphibians, reptiles, fishes, crustaceans, and also some of the rocks he collected. These descriptions were presented to science by 15 different scientists but all were based on the material he collected. In the preface to the Supplementary Appendix, Whymper further explained the circumstances related to the collections of each of the zoological groups.

Whymper collected numerous specimens of *Gastrotheca riobambae* and/or *G. pseustes* (both species considered to be *Nototrema marsupiatum* at that time) and another 42 amphibian specimens, including two that were new to science: *Paruwrobates whymperi* (described as *Prostherapis whymperi*) and *Atelopus elegans* (described as *Phryniscus elegans*). The collection received compliments from Boulenger, who said that "though containing no striking novelties", the collection was "interesting on account of the care bestowed by its collector in recording the exact locality from which every specimen was obtained". In the case of the new species from Tanti (Provincia Santo Domingo de los Tsáchilas), Whymper provided the following additional data about its collection:

> In Quito, help was rendered by a highly-intelligent Swede (to whom I had been recommended by Baron de Thielmann), Ludwig by name, and through him I acquired the greater part of the collections which come from Milligalli, Tanti, and other places in the West of Ecuador—a region which there was no time to visit.
>
> (Whymper, 1891)

About *Atelopus ignescens*, a species currently nearly extinct in the wild, Whymper says that

> it is one of the most widely distributed, I think the most widely distributed, in the interior of Ecuador. I have seen it almost everywhere from 7000 feet above the sea to 13,500 feet. We could have obtained thousands of specimens.
>
> (Boulenger, 1882a)

It is remarkable that Whymper collected *Atelopus ignescens* (the jambato) in the city of Quito while ascending Panecillo, as this is the only known vouchered record for this locality. Needless to say, this species is now extinct on Panecillo and in the city of Quito. Whymper's solitary visit to this isolated mountain was recounted by him as follows:

> It is indeed obvious that the middle zones of Ecuador have been very imperfectly worked by collectors, even in the localities most frequently visited by them, such as Quito. At the southern outskirts of that city there is a prominent hill called the Panecillo, which is now almost surrounded by houses, and is used as a playground by the youth of the place. I visited this eminence one day alone, to obtain a round of angles, and by beating the dwarf vegetation into my hat secured about thirty species of insects of various orders, without any expectation that a place so frequented would yield interesting results.
>
> (Whymper, 1891)

Whymper mentions the great abundance of *Gastrotheca riobambae* (and probably *G. pseustes*, a cryptic species that occurs in sympatry at that site) (as *Nototrema marsupiatum*) in Machachi by saying:

> the last named is the most numerous. In the vicinity of the town of Machachi we saw thousands, and must have heard hundreds of thousands of frogs, principally of this latter species—which is very variable in its colouring and markings
>
> (Whymper, 1891)

> These frogs were in great numbers at this locality (Machachi); and in the evening their music was so loud as almost to interfere with hearing when walking out.
>
> (Boulenger, 1882a)

Rosenberg: A Successful Natural History Dealer

William Frederick Henry Rosenberg (1868–1957) was an English naturalist, collector, explorer, and natural history specimen dealer. Rosenberg was a Fellow of the Royal Entomological Society of London. He started a business as a natural history dealer in Tring, England, in 1897. He sold specimens from around the world. In South America, he collected or employed collectors of zoological specimens (insects, mammals, birds, amphibians, reptiles, fishes) in Colombia, Ecuador, and Peru, which were subsequently acquired by numerous museums and taxonomists from the United States and Europe, among them the BMNH.

Rosenberg's first expedition was to Colombia in 1894, where he collected insects and birds. In 1896, he went to Ecuador and later returned to Colombia. While in Ecuador, he collected in Cachabé, Paramba, Ibarra, Cayambe, and Chimbo in Western Ecuador and sent the specimens to the BMNH, where George A. Boulenger described the reptiles and batrachians and wrote a paper about them (Boulenger, 1898). Rosenberg provided 215 specimens for Boulenger, so one particularly spectacular amphibian species bears his name; the large hylid gladiator frog *Boana rosenbergi* was named in his honor by Boulenger in 1898. The museum also purchased specimens that Rosenberg had received from Perry O. Simons (Ecuador and Peru), which were reported on by Boulenger (Boulenger, 1899b).

The Scientifically Honored but Poorly Known Buckley

Vane-Wright (1991) and Hämäläinen (2021) provide biographic information for Clarence Buckley (1832–1885 or 1886) (Figure 2.7), the "Prince of Collectors" who was an intrepid and skilled collector of zoological specimens

FIGURE 2.7 Painting depicting Clarence Buckley during his collecting discoveries in Ecuador in c. 1870. It is presumed that the painting was either done by William C. Hewitson or commissioned by him. Taken from Vane-Wright (1991).

(particularly birds and butterflies) in South America. Despite the significance of his zoological collections, his life was nearly a mystery until 2018, when an Australian relative of Clarence Buckley provided on the internet the long-forgotten details of his birth and family connections in India and London, and which were later published by Hämäläinen (2021). Clarence Buckley was born in Baitul in British India. Buckley's wider family was quite wealthy, which apparently helped him finance his extensive expeditions to South America. Clarence Buckley was still recorded as living in London in 1885. Buckley collected in Ecuador, Bolivia, and possibly Colombia during the late 1860s and 1870s. His first Ecuadorian expedition lasted nearly a year during the period July 1868 to June 1869. Setting out from Guayaquil on the 5 July, he had to cross the Andes to reach his objective, the Amazon slopes. The furthest point reached was Santa Rosa de Otas, on the Río Napo. Buckley's second expedition to Ecuador in 1870–1871 began in the autumn of 1870, but it is unclear when he returned to England the following year. It is suggested that another expedition had started in early 1876 and ended before the end of 1879 or the beginning of 1880. During most of this expedition he was headquartered at the village of Sarayacu on the Río Bobonaza, a tributary of the Río Pastaza, from where he made visits to several sites in the surrounding areas. Other Ecuadorian collecting localities on his trips (up until about 1878) included Riobamba, Baños, Santa Inés, Canelos, Curaray, Río Rotuno, Jorge, Sarayacu, the Copataza River, Andoas, Pallatanga, and Intaj (Intag, Intac). Amphibian localities such as Paitanga (= Pallatanga) and Intag have been subject to question and have caused taxonomic confusion. According to Peters (1955), it seems there was an inaccurate labeling of the original material from Buckley. No one knows or has discussed the origin of such mislabeling; it could have been done by Buckley, or by someone at the BMNH. The first possibility is less probable given his recognized perfection. For example, William C. Hewintson, a famous British entomologist and ornithologist, was impressed by Buckley's butterfly collections, and he claimed:

> I do not, of course, compare the collection of Mr Buckley with the . . . collections of Bates and Wallace . . . but I do not hesitate in saying, that during the twenty-five years I have been a student of these things, no such single collection (either for its perfection or extent) has been brought to Europe.

In Ecuador, Buckley collected 125 specimens of amphibians (Natural History Museum, 2023) of about 45 species. Based on these collections, Albert Günther, George A. Boulenger, and Emmett Reid Dunn described nine species (four Gymnophiona, three Craugastoridae, one Hylidae, and one Phyllomedusidae) as new to science. Also, about 90 species have been named in Buckley's honor (Hämäläinen, 2021). Among them are five species of amphibians from Ecuador: *Centrolene buckleyi*, *Osteocephalus buckleyi*, *Agalychnis buckleyi*, *Pristimantis buckleyi*, and *Caecilia buckleyi*.

The Consul: Mr. Fraser

Louis Fraser (1810–1866) was a British zoologist, zookeeper, dealer, and taxidermist. Little is known about Fraser's early life. He was born in 1819 or 1820. In his early years, Fraser was Curator of the Museum of the Zoological Society of London for 14 years. He worked with the anatomist Richard Owen—a famous British comparative anatomist who studied Darwin's fossil collections. In 1850, Fraser was appointed Consul of Quidah, Dahomey (now Benin), West Africa. During 1857–1859, he collected birds, mammals, and amphibians in Ecuador. Gardner (1983) describes his journey in Ecuadorian territory (our English translation):

> After a brief stay in Guayaquil, Fraser proceeded to Cuenca, arriving on 6 October 1857 and remaining there through November. Then he journeyed to Gualaquiza where he worked from December 1857 through February 1858, except for a two-week period in Zamora during January.
>
> Fraser left Gualaquiza for Cuenca on 1 March (arriving 5 March) where he collected through April and May before proceeding to Riobamba. He worked in or near Riobamba during June and again in August on his return from Quito before continuing on to Pallatanga. His collections in the Pallatanga area, with trips to Chillanes, were made from the latter part of August through December 1858.
>
> Leaving Pallatanga in mid-January 1859, Fraser collected on the slopes of Chimborazo north of Riobamba in February on his way to Quito. March, April, May, and part of June were spent in the vicinity of Quito and on the slopes of Pichincha as well as on the western slopes of the Andes northwest of Quito.
>
> Fraser traveled southward from Quito to Babahoyo where he stayed from 10 July through part of September. Then he went on to Guayaquil where, in early October, he left by ship for Esmeraldas. Fraser remained in Esmeraldas until the end of December and then left Ecuador for Guatemala.

Fraser's collection of amphibians (74 specimens) was deposited at the BMNH and was reported on by Boulenger. Among them was the rainfrog *Pristimantis surdus*, which was reported as new to science. Unfortunately, this species and several other frogs collected by Fraser are only labeled as Western Ecuador; thus, the exact locations where they were collected remains unclear.

Simons: A Promising Collector and a Tragedy

Perry O. Simons (1869–1901) was an American scientific collector who predominantly worked in the Neotropics. He spent four years at Stanford University as an electrical engineering major, but his first job in the natural sciences was working for William W. Price as a scientific collector in Mexico from 1896–1897, where he was joined by his brother, Luther. The mammal specimens collected on this 10-month expedition in Mexico were purchased by the BMNH.

The BMNH zoologist Oldfield Thomas was so pleased by the quality of the specimens that he offered Simons a job collecting additional specimens in South America. Luther accompanied Perry for the first two years of the expedition before sailing back to San Francisco. While crossing the Argentinean Andes around the end of December 1901, Simons was murdered by his guide. The motive was said to be robbery, and the guide was captured and imprisoned in Mendoza, Argentina. After Simons' death, Oldfield Thomas called him "the most successful mammal collector that I have ever had to deal with", noting that "we shall not easily find his like again". Simons collected 48 specimens of about 10 species of amphibians, 3 species new to science that were described by Boulenger (Boulenger, 1899b). These are two dendrobatids, *Epipedobates tricolor* and *Hyloxalus vertebralis*, and the craugastorid *Pristimantis crucifer.*

Palmer: The Analytical Chemist

Mervin George Palmer (1882–1954) was an English naturalist and traveler who collected for the BMNH. According to (Beolens *et al.*, 2011), he graduated as an analytical chemist but decided upon a career as a freelance collector and naturalist. He collected in Central and South America. In Ecuador he collected 27 specimens of amphibians of about 10 species. He worked for commercial concerns in Ecuador between 1910 and 1918 before moving to London. *Hyloscirtus palmeri* and *Pristimantis palmeri* are named after him.

The Italian Enrico Festa During The Ecuadorian Civil War

Enrico Luigi Festa (1868–1939) was an Italian scientist and collector who was born into a noble and very wealthy family. He earned a doctoral degree at the University of Turín in 1891 and then worked at the Museum of Zoology and Comparative Anatomy of Turín. He traveled in Asia, Africa, and the Neotropics (including Ecuador), where he collected zoological specimens, some of which were described by him. His travel diaries in Ecuador were detailed in a book (Festa, 1909). Details of his life are given by Rodolfo Pérez Pimentel on a webpage (https://rodolfoperezpimentel.com). Festa visited Ecuador between 1895 and 1898, while a civil war (the liberal revolution) was occurring. He collected numerous specimens of 45 species (at that time) of amphibians, which were deposited at the MZUT (Museo di Zoologia, Instituto di Zoologia e Anatomia Comparata Universitá di Torino) and now housed by the MRSN (Museo Regionale di Scienze Naturali, Torino). The amphibians collected in Ecuador were reported by the Italian Mario G. Peracca in 1904. Peracca honored Festa by using his last name for five species, among which three are currently valid: the morphologically unique long-snouted toad, *Rhinella festae*; the casque-headed frog, *Osteocephalus festae*; and the rainfrog, *Pristimantis festae*. Peracca also provides a synthesis of Festa's route:

In the eastern region: the valley of Gualaquiza, the valley of the Río Zamora, the valley of the Río Santiago, of which the Río Zamora is a tributary, oscillating between 500 and 600 meters above sea level; San José, a village to approximately 1100 m. asl; Papallacta, a village to close to 3100 m asl; Pun, locality to approximately 2500 m. asl., to the border of Ecuador with Colombia, near the sources of the Río Cofanes.

In the Interandean region: Cuenca, capital of the province of Azuay to approximately 2580 m. asl; SigSig, village to east of Cuenca; Cañar a small town at about 3170 m. asl; Tambo, a village north of Cañar, at about 2975 m. asl.; Quito, capital of the Republic; Ibarra, capital of the province of Imbabura at about 2225 m. asl; El Troje, village near Huaca at about 3100 m. asl; Tulcán, capital of the province of Carchi, at about 2977 m. asl. on the border of Ecuador with Colombia; La Concepción, hacienda in the valley of the Río Chota (which in the lower part of its course takes the name of Río Mira) at about 1400 m. asl.

In the Andean region: Paredones, on the road from Cañar to Quito near the hill of Quinza Cruz, at about 4042 m. asl.

In the western region: in the so-called región húmeda de la Costa; Vinces, Balzar and the valley of the Río Peripa, tributary of the Río Daule; in the región seca de la Costa; Puntilla de Santa Elena, narrow peninsula west of Guayaquil.

Günther and the Explosion of Collections

Albert Karl Ludwig Gotthilf Günther (1830–1914) (Figure 2.8) was an eminent German ichthyologist and reptile taxonomist, Keeper of the Zoology Department at the BMNH. Günther worked on many taxa, including amphibians. A biographic summary of him is provided by Adler (2014b) and references therein.

After entering the theological school at the University of Tübingen in Germany, Günther moved to the medical school within the same university due to his interest in science and medicine. At the BMNH, Günther was a great collection builder, remarkably adding about 900,000 specimens from 1875 to 1895 during a major period of expansion of the British overseas empire.

Günther described five anuran species (*Rhaebo caeruleostictus*, *Boana fasciata*, *Dendropsophus rhodopeplus*, *Pristimantis conspicillatus*, *P. unistrigatus*) and one Gymnophiona (*Caecilia pachynema*) that occur in Ecuador. Unfortunately, the type localities for these species were originally vaguely stated as collected either in the Andes von Ecuador, Western Ecuador, or Andes of Western Ecuador by Louis Fraser. The names of the collecting sites refer to general geographic areas; however, according to Caminer and Ron (2014), at least for *B. fasciata*, it is almost certain that the holotype was collected either in Gualaquiza (Provincia Morona Santiago) or Zamora (Provincia Zamora Chinchipe). The latter was inferred from letters and records of the collections of Louis Fraser in Ecuador, which were gleaned by Gardner (1983) and references therein. Certainly, using modern approaches (e.g.,

FIGURE 2.8 Albert Charles Lewis Günther in about 1900. Photo taken from Adler (2014a), courtesy of Kraig Adler.

FIGURE 2.9 George A. Boulenger at about 60 years old. Photo taken from Adler (2014a), courtesy of Kraig Adler.

genome sequencing), it is feasible to solve the geographic enigmas posed by Fraser's vague localities. Because his journeys in Ecuador are known, the sequences of the types could be compared to modern collections from the same localities to determine where the types are situated geographically and phylogenetically.

Günther described *Dendropsophus triangulum* and *Ctenophryne aterrima* coming from Brazil and Costa Rica, respectively, and later both species were recorded as occurring in Ecuador by other authors. Günther is honored in the scientific names of 66 species of reptiles and amphibians, among them two Ecuadorian amphibians, the caecilian *Caecilia guentheri* and the lower-toothed marsupial frog *Gastrotheca guntheri*, which were described by Dunn and Boulenger, respectively. Undoubtedly, his most important contribution to amphibian biology, as remarked by Adler, was his hiring of George A. Boulenger, who became his successor in charge of lower vertebrates at the museum.

Boulenger: The Tireless and Prolific Taxonomist

The person who described most of the species to be collected by British explorers and collectors was George Albert Boulenger (1858–1937) (Figure 2.9). Officially a first-class assistant in the Department of Zoology at the BMNH, Boulenger did not have the title of Curator (the British equivalent then was Keeper); he was a civil servant with the title of "First-Class Assistant" during his entire career there. Boulenger was an honored and leading taxonomist of his era, whose life was dedicated to studying roses (in his free time) and describing the incredible number of 2514 species of fish, amphibians, and reptiles of the world. In all, he described 956 species of amphibians (valid names up to 9 August 2022, from Frost 2022), among them 56 species of amphibians occurring in Ecuador. However, the huge number of species that Boulenger described, many including detailed morphological drawings (see herein), speaks volumes for his undistracted dedication. In the year 1882 alone, he described and named an incredible 393 species of amphibians.

Adler (2014b) summarized Boulenger's biography with an emphasis on his herpetological contributions and provided references about his life. Boulenger was born in Belgium and later became a British citizen. He was hired as assistant in charge of lower vertebrates by Günther in 1882. Between 1882 and 1896, he prepared a new edition of the catalogs of amphibians and reptiles in nine volumes (two volumes of amphibians), in which he described many new taxa. This was a monumental task that covered about 5200 taxa (5170). As remarked by Adler (2014b),

> he had an incredible memory and was a tireless worker with strict self-discipline and great charm. He rarely dissected specimens and even then, he usually made just a small slit through which only gross features could be seen, quite in keeping with the long-standing BM policy at that time that specimens were not to be cut up. Because of this policy,

understandably, he did not make any fundamental anatomical discoveries such as those done by his contemporary in the USA, Edward Cope. Also, Boulenger's habit of not fixing types and of failing to indicate the specific specimens from which descriptions were drawn was custom at the time in Europe. To him scientific names were meant to define and describe, regardless of the legalities involved.

During 38 years, Boulenger described 56 species (13 families) of Anura and Gymnophiona occurring in Ecuador and provided geographic reports of many other species. They were included in 14 publications from 1880 to 1918. His descriptions were based on specimens deposited at the BMNH mostly by Whymper, Buckley, Fraser, Rosenberg, Simons, Palmer, and a few others. Among the species he described are those belonging to Craugastoridae (15), Hylidae (11), Dendrobatidae (10), Hemiphractidae (5), Phyllomedusidae (4), Leptodactylidae (3), Bufonidae (2), Caeciliidae (2), Centrolenidae (1), Eleutherodactylidae (1), Rhinatrematidae (1), and Siphonopidae (1). The first paper on new species from Ecuador was in published in 1880, in which he described *Pristimantis devillei* and *P. glandulosus* from the eastern slopes of the Andes. These descriptions were based on single specimens collected by the Consul of Belgium in Quito, Émile de Ville. Later, in a single year, in 1882, he published the descriptions of 22 currently valid species is his famous catalogs of the amphibians at the BMNH. His major publications that included amphibians from Ecuador (Boulenger, 1882c, 1882a, 1882b, 1898, 1899a) provided good-quality black-and-white illustrations of the adults, as well as of some of their morphological details, which in later years have facilitated their identification (e.g., Figures 2.10, 2.11, 2.12, 2.13).

Boulenger is honored in the specific names of 72 species, among which are two species of frogs from Ecuador: the harlequin frog *Atelopus boulengeri* (Peracca, 1904) and the poison frog *Epipedobates boulengeri* (Barbour, 1909).

THE FIRST HALF OF THE 20TH CENTURY, A PERIOD OF WARS

During the 50 years from 1900 to 1950, despite the effervescence of science in fields such as physics, chemistry, mathematics, medicine, and even biology (in which advances were focused on military applications), the task of describing the diversity of Ecuadorian amphibians was interrupted by the first and second world wars. Only 62 species of Ecuadorian amphibians were described by 20 foreign authors. Among the stand outs are the British zoologists George Albert Boulenger and Hampton W. Parker, the Italian zoologist Mario Peracca, the Swedish workers Lars Andersson and Douglas Melin, and the Americans Thomas Barbour, Emmett Dunn, Gladwyn Kinglsey Noble, Benjamin Shreve, and Edward Taylor.

Some of them (Melin, Barbour, Noble) described species from Peru that later were reported from Ecuador, except for the aquatic *Telmatobius niger* described by Barbour and Noble from Provincia Chimborazo.

Hampton W. Parker (1897–1968) was Keeper of Zoology at the BMNH from 1947 to 1957. He published three papers describing two species from Provincia Loja and adding morphological and distribution data from amphibians in southern Ecuador (Parker, 1932, 1934, 1938). He used the collections made by the Ecuadorian paleontologist and naturalist Clodoveo Carrión Mora (1883–1957) (Figure 2.14). Carrión studied in Europe at the London and Manchester Universities, and upon his return he collected 157 specimens (of about seven species) in his homeland in Loja and Zamora Chinchipe provinces, which he sent to the BM.

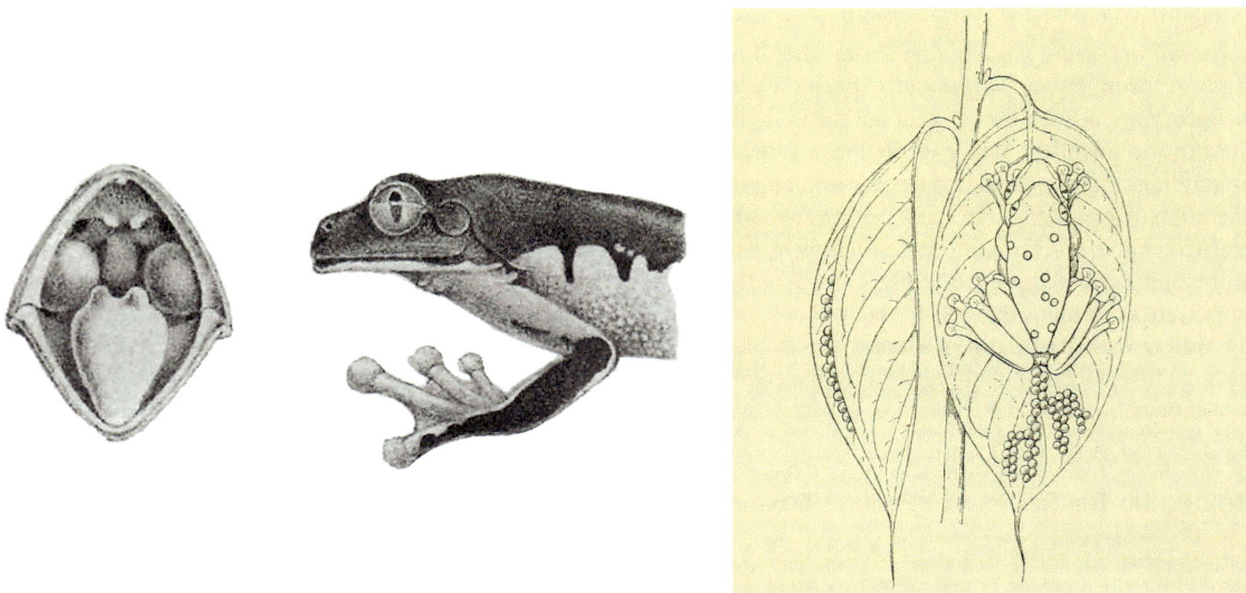

FIGURE 2.10 *Cruziohyla calcarifer.* Left: From Boulenger (1913). Note the tympanum, half the eye's diameter, and the short flank bars, as described by Boulenger, which helped distinguish *C. calcarifer* from the more recently described *C. sylviae.* Right: The first illustration of an Ecuadorian frog showing reproductive behavior: *Agalychnis spurrelli* spawning.

FIGURE 2.11 *Gastrotheca cornuta* showing morphological details. Original plate from Boulenger (1898).

FIGURE 2.12 Illustrations of some anurans described by George A. Boulenger: *Boana rosenbergi* (top left) (taken and modified from Boulenger (1898), *Gastrotheca guntheri* (bottom left), *Rhaebo caerulesotictus* (top center), *Boana appendiculata* (top right), *Tepuihyla tuberculosa* (middle), *Nyctymantis rugiceps* (bottom center), and *Trachycephalus quadrangulum* (bottom right). Taken and modified from Boulenger (1882c).

FIGURE 2.13 Illustrations of some of the Craugastoridae described by George A. Boulenger: *Strabomantis anomalus* (top left) (taken and modified from Boulenger (1898), *Pristimantis surdus* (bottom left), *Pristimantis conspicillatus* (top right), *P. buckleyi* (second right), *P. curtipes* (third right), and *P. unistrigatus* (bottom right). Taken and modified from Boulenger (1882c).

FIGURE 2.14 Left: Hampton W. Parker, Keeper of Zoology at the Natural History Museum in London. Taken from Adler (2014a), courtesy of Kraig Adler. Right: Clodoveo Carrión Mora, the first Ecuadorian academic collector of amphibians. Photographer unknown.

Carrión was the first academic Ecuadorian biologist who collected amphibian specimens for scientific use, and his academic life has been detailed by Loaiza-S (2012). In the introduction of a 1934 paper, Parker acknowledges Carrión:

> British Museum is once again indebted to Prof. Clodoveo Carrión for a fine series of reptiles and amphibians from the province of Loja. In addition to representatives of two species which do not appear to have been previously described, there are examples of several others hitherto known only from single specimens or limited series, and others, again, which add materially to our knowledge of the geographical distribution of the species they represent. The collections were made in three localities, not widely separated but at very different altitudes, and, since our knowledge of the vertical distribution of reptiles and amphibians in the Andes is so meagre, it has been considered worthwhile to give a complete list of all the species collected at each station; but any attempt to generalize on the subject would, at present, be decidedly premature. The altitudes of the different localities are: Loja City, S.E. Ecuador, 7200 ft.; Zamora, 50 kilometers east of Loja, 3250 ft.; 5 kilometers east of Loja, 9200 ft.

Benjamin Shreve (1908–1985) worked directly with specimens collected in the Pastaza and Guayas provinces, Ecuador, and deposited at the Harvard Museum of Comparative Zoology (MCZ). Shreve was an American amateur herpetologist. He was from a wealthy Boston family of jewelers and worked at the MCZ as a volunteer.

Next, we highlight some aspects of four scientists (Peracca, Andersson, Dunn, and Taylor) (Figure 2.15), who boosted the taxonomy of Ecuadorian amphibians and for whom more detailed biographies are provided by Adler (2014b).

Mario G. Peracca: The Aristocratic Italian Zoologist

Adler (2014b) provided a biography of Mario Giacinto Peracca (1861–1923) (see Figure 2.15), the renowned Italian herpetologist, who was born in Turin. Peracca began studying medicine at the University of Turin but soon switched to zoology. While he was studying for his doctoral degree (which he obtained in 1886), he served as an assistant to Michele Lessona, an Italian amphibian specialist. In 1887, Peracca became an assistant at the Zoological Institute of the university and continued his work in herpetology until his passing in 1923. Peracca curated the herpetological collections and added specimens he purchased from collectors around the world or from exchanges with other museums. Adler (2014b) remarked that despite his nobility, Peracca was said to have lived modestly, and he devoted his fortune to zoology and to helping others. Adler added that Peracca's research was clearly in the tradition of George Albert Boulenger, who was a longtime friend. He published 64 herpetological papers (1886–1917) dealing with collections obtained in South America and Africa, including

FIGURE 2.15 Amphibian biologists from the first half of the 20th century that worked with Ecuadorian amphibians: Mario G. Peracca (top left), Lars Gabriel Andersson (top right), Emmett Reid Dunn (bottom left), Edward H. Taylor (bottom right). Photos taken from Adler (2014a), courtesy of Kraig Adler.

Madagascar, but there were also titles on specimens from the Middle East, China, and Australia. Many new genera and 40 valid species of amphibians and reptiles were described, and the collections were eventually donated to the Zoological Institute at Turin. Despite his successes in the field of taxonomy, he remained throughout his life deeply passionate about observing the behavior of reptiles (including the Galápagos tortoises from Ecuador) that he raised in the terraria at his homes.

In 1904, Peracca described the collections made in Ecuador by Enrico Festa in a single paper (Peracca, 1904) that included 45 species of amphibians: 43 Anura, 1 Urodela, and 1 Gymnophiona. Among them there were 6 species new to science (as currently recognized): *Atelopus boulengeri*, *Osornophryne bufoniformis*, *Rhinella festae*, *Osteocephalus festae*, *Ctenophryne aequatorialis*, and *Pristimantis festae*. In that paper, he acknowledged his colleague and friend Boulenger at the BMNH:

> With the kind consent of my good friend Dr. Boulenger, whom I am glad to be able to thank here, I was able to study and compare a great part of the species of the collection at the British Museum of Natural History in London.

Peraccas' contribution not only included the description of new species but also novel geographic and morphological data for species previously described from Ecuador. His descriptions of morphology were relatively detailed and included morphometric data, a tradition pioneered by Jiménez de la Espada. His interpretation of the morphological characters reflected the status quo of taxonomy at that time, in which geographic variation had an important role within species. For example, when describing *Atelopus ignescens*, he mentioned the following (translated from Italian):

> Numerous specimens from El Troje, Tulcán, Paredones, Cañar, Tambo, Sig-Sig, Quito. The specimens from Quito and Cañar recall, due to their small size, the var. *exiguus* Boettger. This species is, moreover, extremely variable both for its size and for its coloration. Thanks to the courtesy of Professor Ehlers, Director of the Zoological-Anatomical Institute of Götingen, with the consent of my friend Dr. F. Werner, I was able to examine the *Atelopus subornatus* Werner and *A. flaviventris* Werner and I can only confirm the opinion expressed to me by Boulenger that these species are identical to the *A. ignescens*. This, moreover, is even better confirmed by the measurements of Werner's types placed in front of the measurements of two specimens of *A. ignescens* from Paredones.

Both Peracca and Boulenger failed to recognize the number of species (as currently defined) that were then considered to be *A. ignescens*. To date, the populations Peracca assigned to *A. ignescens* belong to at least four species: *A. ignescens*, *A. pastuso*, *A. bomolochos*, and *A. subornatus*.

Lars Andersson: The Swedish School Teacher

Lars Gabriel Andersson (1868–1951) (see Figure 2.15) was a Swedish zoologist. He attended Uppsala University, where he graduated as a DPhil in 1909. During most of his life, he was a school teacher in Stockholm, except for the periods of 1894–1895 and 1897–1902, when he was an employee in the vertebrate zoology section at the Royal Natural History Museum in Stockholm. Later, he was a volunteer worker at the museum. Besides several popular zoological works, Andersson published many papers on amphibians and reptiles (1898–1945). According to Adler (2014a), Andersson's contributions were careful and often extensive reviews of collections made by Swedish expeditions to many parts of the world and included descriptions of numerous new species. His special interest was South America, which occupied nearly half of his publications and covered areas from the high Andes to Tierra del Fuego at the southern tip of the continent. His most comprehensive monograph (Andersson, 1945), also his last publication, reviewed the taxonomy of amphibians of eastern Ecuador. It was based on 220 specimens collected in 1937 and 1938 by the American collector William Clarke-Macintyre and the Swedish explorer, naturalist, writer, photographer, and filmmaker Rolf Blomberg. Andersson reported 52 species, 31 of them as new to science. Most of his new species were later synonymized, with only nine of the 31 currently recognized as valid. This is explained as a failure to compare the specimens with taxa previously described by Boulenger.

Emmett Dunn: The American Herpetologist

Emmett Reid Dunn (1894–1956) (see Figure 2.15) was an American herpetologist and specialist on salamanders who is noted for his work in Costa Rica, Panama, and Colombia and for studies in the Eastern United States. His life was summarized by Adler (2014d). Dunn's interest in herpetology first took root in his native Virginia. Dunn received a BA (1915) and MA (1916) from Haverford College, where he became a professor of biology and served as Curator of Reptiles and Amphibians at the Academy of Natural Sciences in Philadelphia. He was editor of *Copeia* from 1923 to 1936. Dunn was a mentor of Jay M. Savage, an American herpetologist best known for his research on the reptiles and amphibians of Central America.

Although Dunn did not visit Ecuador, he reviewed collections made there and described seven species recorded from Ecuador (four anurans and three caecilians) in five papers. Among the species described by him was one of the largest caecilians of the world, *Caecilia guntheri*.

Edward Taylor: The University of Kansas Grandfather

Edward Harrison Taylor (1889–1978) (see Figure 2.15) was an American herpetologist, field collector, and taxonomist who was associated with the University of Kansas and its Museum of Natural History for nearly seven decades, now known as the Taylor era. Taylor traveled throughout much of the world, assembled extraordinary herpetological collections, became a highly respected teacher, initiated a graduate program in herpetology, and rose to international prominence. His extraordinary and enigmatic life has been extensively described specially by William E. Duellman, Kraig Adler, and other authors; the latter are cited in Adler (2014c) and Duellman (2015a).

Taylor spent several weeks in Ecuador, during which he remained entirely in the field. Nevertheless, there is limited information available regarding the specifics of his journey. For his papers on taxonomy, he used specimens of amphibians deposited in museums. His contribution to the knowledge on Ecuadorian amphibians was the description of 14 species that currently are recorded from Ecuador (4 glassfrogs and 10 caecilians) in four papers published between 1949 and 1973, among which his massive book of 1968 about caecilians of the world included 8 new species from Ecuador. Subsequently, Taylor and James A. Peters jointly published a paper in 1974 on the caecilians of Ecuador (Taylor and Peters, 1974). In the introduction to the 1974 publication, they mention that the junior author also did field work in Ecuador: "The senior author [Taylor] visited some weeks in the country, all of which time was spent in the field". The glassfrogs were originally described from Central America and later recorded from Ecuador by other authors.

Orcés and the Foundations of Amphibian Collections in Ecuador

Luis A. Coloma and Ana Almendáriz-Cabezas

In the mid-20th century, in 1949, the amphibian collection of the Escuela Politécnica Nacional (EPN) (museum) was founded. It is the oldest natural history museum in Ecuador and is currently is the fourth largest (10,751 amphibians, as of 29 June 2022). It is also valuable for the historical information it contains, type specimens (10 holotypes, 133 paratypes), specimens of species that are now possibly extinct, and other species that have become very rare.

It all began in 1946 when the President of the Republic of Ecuador, Jose María Velasco Ibarra, reopened the public university called the Escuela Politécnica Nacional (formerly Instituto Politécnico in its first stage, 1869–1876), and the Departamento de Ciencias Biológicas was created. The DCB (EPN) began under the direction of a French zoologist, Robert Julien Hoffstetter (1908–1999), who in a temporary residence in Ecuador led it until 1953. Hoffstetter was a taxonomist and paleontologist, later known for being one of the pioneers of cladistics in France and for his influence on the classification of reptiles and work with the fauna of tertiary mammals of South America. Hoffstetter was also director of the Centre National de la Recherche Scientifique and in 1972 took over as head of the Laboratoire de Paléontologie des Vertébrés et Paléontologie Humaine in Paris, and he taught at Pierre and Marie Curie University. Hoffstetter was part of a French scientific mission in Ecuador and also made expeditions to Peru and Bolivia. Although his stay by Ecuador was relatively short, about eight years, his influence was transcendent. Soon after his arrival in Ecuador, in 1946, he established academic links with the Ecuadorian Gustavo Orcés Villagómez (1902–1999), and together they initiated the formation of the collections of fauna (intended as material for comparison with fossils) of the DCB (EPN), including amphibians, although this was not a particular emphasis, as neither of them was a specialist in that group.

Orcés was the first Ecuadorian vertebrate zoologist. Although he obtained the degree of Professor of Higher Education (1958) at the Universidad Central of Ecuador, his training in zoology was self-taught, and he published 25 scientific articles describing 11 new species of fishes and reptiles (a monumental task for an Ecuadorian of his time) and the taxonomy and distribution of fish, birds, and mammals. His weak point, as he himself acknowledged, was amphibians, about which he only published one paper (issued posthumously), coauthored by his colleague Ana Almendáriz-Cabezas (Alméndariz and Orcés, 2004). He also contributed identifications and information for the classic first published book about the Ecuadorian fauna written by his German friend, biologist, explorer, and writer, Erwin Patzelt (Patzelt, 1978). Patzelt acknowledged the contributions of Orcés by saying:

> My most heartfelt and undying recognition to Professor Gustavo Orcés, an Ecuadorian scientist of international stature,

> who with the modesty of a wise man devotes himself daily and silently to unraveling the mysteries of nature and who always helped me with his limitless generosity.

Nonetheless, Orcés' greatest legacy in the area of amphibians was the formation of the first collections of amphibians (the collection in the EPN (museum) and another personal collection now deposited in the USNM), the creation of the area of herpetology in the DCB (EPN) in 1982, the support of his friend James A. Peters of the Smithsonian Institution in Washington for the study of *Atelopus* harlequin frogs, and his direct influence on some of the Ecuadorian amphibian biologists such as Coloma (his vertebrate zoology student at PUCE) and Almendáriz. About the Orcés inheritance, he expressed of himself, with remarkable modesty [Our translation]:

> I am not a biologist, but I think I have been useful to biologists and someone who has opened trails. I have opened them under very adverse conditions. . . . I leave students who today are good researchers. . . . My main interest was to be able to describe the behavior and the natural history of the animals; but for this it was necessary to identify and classify them taxonomically, and into that task all my life has gone. It was up to me to develop the zoological sciences from its foundations, starting from the formation of bird collections, reptiles and freshwater and marine fish, to acquire books and specialized journals. I would have wanted to deepen the study of some taxon, like zoologists from other countries.
>
> (Báez Tobar, 2017)

These and various aspects of his life have been mentioned and addressed in several publications (Báez, 1983; Ortiz, 1987; Mejía, 1993; Coloma, 1999; Jiménez, 1999; Albuja, 2004; Freile, 2005; Adler, 2007; Báez Tobar, 2017). Orcés was born in Quito to a wealthy family. He was orphaned at an early age, and in his childhood, his mother's relatives sparked his interest in animals by giving him natural history books, although they raised him to be a landlord. In 1946, Orcés was appointed as a research professor of the Department of Zoology of the EPN. After the departure of Hoffstetter in 1953, he served as Director of the DCB (EPN) for 37 years until his retirement in 1990. In addition, Orcés was a zoology professor for more than 40 years at the Universidad Central del Ecuador (beginning in 1940) and PUCE (1970–1974, 1981–1983).

Orcés understood the importance of collections for scientific use and had experience in the formation of collections. In the 1930s and 1940s, he worked with the brothers Olalla (Alfonso, Carlos Rosalino, Ramón, Manuel) and Jorge Olalla (the son of Carlos Rosalino), who made collections for Frank M. Chapman (1864–1945) of the American Museum of Natural History in New York (Figures 2.16 and 2.21). They also provided specimens for the collections of other

FIGURE 2.16 From left to right: Gonzalo Herrera, Ramón Olalla, and natives. Presumably in the Bobonaza region in March 1958. Photo courtesy of Milton Herrera.

museums, including amphibians for the EPN (museum) and the Smithsonian Institution, which were mostly collected by Jorge Olalla. The Olallas—and Clodoveo Carrión, see earlier under the subtitle "The First Half of the 20th Century, a Period of Wars"—emerged as the first Ecuadorian amphibian collectors and were among the most productive vertebrate collectors in the Neotropics for over 50 years, until the 1970s.

In its early years (1946), Orcés activities were in offices and the newly established museum at the EPN facilities located in La Alameda. Orcés assumed the direction of the DCB (EPN) in 1953. In 1960, the EPN campus was built in La Floresta, and the DCB began occupying a laboratory and collections space in the Faculty of Chemical Engineering. Later, it moved to the lower level of the Polytechnic Theater. Using personal funds and later (especially from the 1970s) with the academic support of the EPN Chancellor José Rubén Orellana Ricaurte (1919–1995), he formed vertebrate collections for scientific use, as well as a specialized library, which was also stored on microfilm (scaled-down reproductions of documents used for preservation in libraries) that he printed as photographs. About the books, Orcés commented [Our translation]:

> Then came the time when I started trading toads for books, because I discovered that if I sent them (foreigners) material for their research, they could send me books.
>
> (Ortiz, 1987)

Although Orcés was not a collector, he was a promoter of scientific collections and he acquired and assembled amphibian collections. One part was deposited at the EPN (museum) (772 specimens) and another at the USNM (729 specimens). The former was assembled between 1931 and 1974 and consisted of specimens collected mostly by professional collectors, for example, 518 specimens collected by the Olalla family, 42 specimens from Rodrigo Jiménez (a student of Orcés), and 39 specimens from Antonio Proaño (also a member of the Olalla family). Jiménez collected in the Coastal region, whereas Proaño in the Amazonian, Andean, and Coastal regions. The collections obtained by the Olallas came from several regions of Ecuador but mostly from the eastern region: the area of Ávila, Cotapino, in Provincia Napo and the upper and middle basin of the Pastaza River. The oldest amphibian deposited in the EPN (museum) is a specimen of *Siphonops annulatus* collected in March of 1931 by Ramón Olalla at the Copataza River. César Durán (a taxidermist) was hired by the DCB (EPN) as an assistant of Orcés between 1969–1973 and added 28 amphibians to collection. Orcés was also assisted by Gonzalo Herrera (an exhibit designer), who worked at the DCB (EPN) between 1948 and 1969 and collected 26 specimens. Herrera was hired again from 1974 to 1999. Gonzalo Herrera accompanied the Olalla and Peters on field trips.

In 1970, Orcés sent much of his private amphibian collection (729 specimens) to the USNM. At this time, no EPN resources were allocated either for field trips or for the acquisition of specimens, nor for their maintenance. As a result, Orcés was afraid that his personal collections were going to be lost, and his friendship with Peters prompted him to send his private herpetological collection to the USNM. In 1974, he traveled to the USNM to study faunal collections in that museum. During the 1970s, the specter of the closing of the EPN section of zoology was a latent threat, given that the EPN was focused on developing technical careers and engineering; thus, Orcés had to invest his own resources to keep the museum running. By the end of the 1970s, he had to sell his private bird collections to the MECN. This action caused concern and anger on the part of EPN chancellor Orellana, who offered to Orcés more support that resulted in providing new facilities in the recently built Teatro Politécnico—an iconic architectural work of the 1960s in Quito—and the hiring of personnel.

For his wide-reaching influence on science in Ecuador, Orcés received many awards, including an honorary doctorate from the EPN (1994), the Premio Universidad Central for best publication (1958); the Condecoración de la Orden Nacional al Mérito en la Grado de Oficial, Presidencia de la República (1983); and the Condecoración del la Orden Nacional Honorato Vásquez en el Grado Placa de Oro (1991). Orcés was named Honorary Professor by both the Universidad Central (1983) and the Escuela Politécnica Nacional (1990). In 1989, the Fundación Herpetológica Gustavo Orcés was founded in Quito to promote the conservation of amphibians and reptiles in Ecuador. The natural sciences exhibit, the EPN University's Museum of Natural History, was renamed in his honor and officially created in 2000. Orcés has been honored in the names of about two dozen of species, among them two species of amphibians, the rainfrog *Pristimantis orcesi* (Lynch, 1972) and the now possibly extinct harlequin frog *Atelopus orcesi* (Coloma *et al.*, 2010b).

PETERS AND THE HARLEQUIN FROGS

By 1958, the EPN amphibian collection had grown to include 817 specimens. At that time, the collection began to be an important source of information for researchers abroad. For example, Orcés and Werner C. A. Bokermann (1929–1995), a Brazilian herpetologist, exchanged Ecuadorian specimens (deposited at the Museu de Zoologia of the University of São Paulo) with Brazilian ones (254 specimens from Brazil deposited at the EPN). The collection was also used by the American scientist James A. Peters (1922–1972) (Figure 2.17) of the Smithsonian Institution, who showed particular interest in the herpetofauna of Ecuador.

Peters was as an American herpetologist and zoogeographer whose life has been summarized by Adler (2014b). He was born in Iowa. His main focus of research was Latin America, especially Ecuador. During his 30 years of research in herpetology, he described 17 new species or subspecies, most of them amphibians. Peters died prematurely of liver cancer in 1972 at the age of 50. Peters' specialty was snakes. In 1942, he joined the US Army Air Force and served as a radio operator. Upon his release from the army in 1945, Peters entered the University of Michigan, where he quickly earned three degrees (BS 1948, MS 1950, PhD 1952). During this time, his interest in the Neotropics was initiated when he joined museum expeditions to Mexico in

1949 and 1950, and his publications on the Mexican herpetofauna began in 1950. In 1954, Peters obtained a grant for field work in Ecuador and he continued his work there up to 1969. Ecuador became a major focus of his research, and he published on its herpetofauna the rest of his life. In 1959, he joined the faculty of San Fernando State College, at the northern edge of Los Angeles, and worked there until 1964, when he became Associate Curator of Reptiles and Amphibians at the United States National Museum of Natural History at the Smithsonian. In 1966, Peters became Curator-in-Charge, a position he held until his death in Washington on 18 December 1972. Peters was also a pioneer in the applications of computer technology to questions in systematics, biogeography, and museum curation. Over 100 scientific publications are attributed to Peters. Among them are 5 on Ecuadorian amphibians: a widely used paper on Ecuadorian herpetological type localities (Peters, 1955), another on *Ceratophrys* (Peters, 1967), the taxonomic revisions of *Leptodactylus* (Heyer and Peters, 1971) and caecilians (Taylor and Peters, 1974), and his opus on the harlequin frogs of the genus *Atelopus* (Peters, 1973). The 2 latter papers were published just after his death. His 1955 paper is the first amphibian paper published in Ecuador, curiously in a magazine of entomology and parasitology.

During 1958–1959, Peters was a Fulbright lecturer at the Universidad Central in Quito; this was the second of five trips he made to Ecuador (the others were in 1962, 1965–1966, and 1969). Peters' field trips in Ecuador covered most of the country. They are summarized in Table 2.2. He was pioneer, in the 20th century, of the herpetological exploration of both the Amazonian and Pacific Andean slopes, on which he focused. He describes his sampling efforts as follows:

> In 1958–1959 I made a series of seven transects from the inter-Andean valleys to the Pacific lowlands, and in 1962 my group made five transects on the eastern slopes, usually, although not always, penetrating to the Amazonian lowlands. Such transects perforce follow available routes, and of the total of twelve transects, five were by mule and on foot, six were by road in truck or Landrover, and one was by rail.
> (Peters, 1973)

Peters was particularly interested in collecting reptiles but also placed an emphasis on harlequin frogs. In his seminal paper on the Ecuadorian *Atelopus* (Peters, 1973), he reviewed the taxonomy of all the species and included the descriptions of six new species. He included notes on the ecology and natural history of several species. For example, he made interesting observations of the mass migrations of *Atelopus ignescens*:

> The species engages in mass migrations, apparently at the beginning of the breeding season, and can be seen in the

FIGURE 2.17 James A. Peters with herpetology collections at the National Museum of Natural History of the Smithsonian. 1964. Photo: SIA Acc. 16–126 [MNH-155].

TABLE 2.2
Ecuadorian Provinces Where James A. Peters Collected Amphibians between 1954 and 1966

1954	1958	1959	1962	1965	1966
Imbabura	Esmeraldas	Cotopaxi	Tungurahua	Guayas	Pichincha
Esmeraldas	Cotopaxi	El Oro	Esmeraldas	Pastaza	
Napo	Imbabura	Guayas	Carchi	Manabí	
Pastaza	Pichincha	Imbabura	Napo	Santo Domingo de los Tsáchilas	
Los Ríos	Carchi	Pichincha	Pastaza		
Chimborazo	Napo	Santo Domingo de los Tsáchilas	Cañar		
Cotopaxi	Pastaza		Morona Santiago		
Azuay			Santo Domingo de los Tsáchilas		
Bolívar			Sucumbíos		
Loja					
Pichincha					

> hundreds at such a time. I have see [sic] two such migrations, the first of which was moving across the Pan-American Highway north of Latacunga on December 31, 1958. Large numbers of individuals were smashed on the road over a distance of four or five miles. All living individuals were crossing the highway in the same direction. Some mated pairs were seen. The second such migration was seen in the paramos on the road from Quevedo to Latacunga, on January 22, 1959, at an altitude of 3600 meters. In this case, the movement was again in a single direction, and definitely downhill.

In this *Atelopus* publication, in the acknowledgements section, Peters thanks the Ecuadorians, and especially Orcés:

> I wish particularly to thank Manuel Olalla, Peter Spoecker, Robert Mullen, and Gonzalo Herrera, who have helped in the field. Gustavo Orcés-V. continues as an invaluable friend and critic, and his collections, now at the National Museum of Natural History, Smithsonian Institution, were extremely useful in this study.

Peters has been honored in the names of six Ecuadorian species, including four amphibians, the caecilian *Epichrionops petersi* (Taylor, 1968), the rainfrog *Pristimantis petersi* (Lynch and Duellman, 1980), the Quito rocket frog *Hyloxalus jacobuspetersi* (Rivero, 1991), and the now rare harlequin frog *Atelopus petersi* (Coloma *et al.*, 2007). *Pristimantis petersioides* also bears his last name, given its similarity to *P. petersi*.

A REBIRTH FOR AMPHIBIAN COLLECTIONS

Despite the initial burst in the 1950s, in 1981, the herpetological collection at the EPN (1100 specimens) was almost abandoned due to lack of resources for its care and general lack of interest (except as described earlier in the discussion of the role of Orcés) of the university administration. Orcés (Figure 2.18) focused his efforts on the study of other groups of vertebrates, in particular mammals and fish, for which the EPN hired two renowned Ecuadorian zoologists, Luis Albuja Viteri (in 1974) and Ramiro Barriga (in 1981), who helped him and were in charge of mammals and fish, respectively. Early in 1981, Orcés accepted a young volunteer (Luis A. Coloma), who in 1979 had done a study of the autoecology of *Atelopus ignescens* as part of his high school research thesis, and who was beginning studies in biology at PUCE. Coloma was put in charge of reorganizing and identifying the moribund herpetological collection. This relatively small collection (although at that time the largest in Ecuador) was preserved in formaldehyde and stored on a dusty floor in a shelf-less space where there were also many jars of fish specimens. Coloma started working with the collection to organize it. Specimens that were nearly dry were transferred to ethanol, and their data was recovered. This was a unique and valuable experience for Coloma because in addition to this first herpetological work, discussions with Orcés, Albuja, and Barriga introduced him to the world of collections and science. Coloma recounts one of these conversations as follows [Our translation]:

> In one of my conversations with Profesor Orcés, I remember his excitement when he told me about the thousands of jambatos that in his youth he saw wandering during the day in the valley of Los Chillos and in the Chorreras of the volcano Pichincha, next to Quito (a few specimens from Pichincha are today deposited in the collections of the Mejía high school).
>
> (Coloma, 1999)

At that time, Albuja organized a series of weekly discussions on important publications, among which was the reading and discussion of Duellman's recently published book *The South American Herpetofauna: Its Origin, Evolution, and Dispersal.*

FIGURE 2.18 Left: Gustavo Orcés Villagómez at the EPN (museum) in 1983. Photo courtesy of Ramiro Barriga. Right: Luis A. Coloma (right) and Alden Yépez (left) on an amphibian collecting trip at Lita and Cabeceras del Río Baboso, Imbabura and Carchi Provinces, September 1984. Photo by unknown photographer.

In early April 1983, while studying biology, Coloma accompanied Marinus S. Hoogmoed during fieldwork in western and central Ecuador, under the auspices of the EPN and PUCE. Marinus Steven Hoogmoed (born 19 March 1942 in Rotterdam) is a Dutch herpetologist who has lived in Brazil since 2004. His research focus is on the herpetofauna of Suriname, where he collected more than 3000 specimens for the Rijksmuseum van Natuurlijke Historie (RMNH, predecessor of the Naturalis Museum) and the Amazon region in Brazil. In December 1973, Hoogmoed obtained a doctorate from the University of Leiden. He was Curator in the department of herpetology of the RMNH in Leiden from January 1966 until his retirement in April 2004. From 1990 to 2001 he was Head of the vertebrate department of the museum. Since November 2003, he has been a visiting researcher at the Museu Paraense Emílio Goeldi in Belém, Brazil. In April 1983 and 1987, Hoogmoed made collections of amphibians and reptiles in Ecuador. During that time, he described four species of Bufonidae as new to science from Ecuador: *Rhaebo colomai*, *R. olallai*, *Osornophryne antisana*, and *O. guacamayo* (Hoogmoed, 1985, 1987). Hoogmoed was an excellent herpetological mentor to Coloma and Almendáriz (see later), who learned from him the fundamentals of amphibian fieldwork. According to Hoogmoed, Coloma proved to be an enthusiastic and reliable field partner, whose enthusiasm for herpetology resulted in continuous collections (Hoogmoed, 1985). In fact, from 1983 to 1984, Coloma collected about 350 specimens in locations such as Coca, Tiputini (in the Amazonian lowlands), Centro Científico Río Palenque (in the Pacific region), Lita, and the headwaters of the Río Baboso, Provincia Carchi (see Figure 2.18). Most of the specimens were deposited in the EPN (172), whereas the majority of collections from Carchi, made in September 1984, were deposited in the new Museo Ecuatoriano de Ciencias Naturales (currently INABIO). In 1984, mostly with Coloma's collections (102 amphibians, 6 reptiles), the MECN would begin to develop its herpetological collections, and an Ecuadorian biology student at PUCE, Juan Carlos Matheus Pozo, was temporarily in charge of them.

THE EPN MUSEUM AND A WOMAN PIONEER

Ana Almendáriz-Cabezas

At the beginning of the 1980s, a very auspicious event occurred after about three decades of near abandonment of the amphibian collections at the EPN. In 1982, the Ecuadorian biologist Ana Almendáriz-Cabezas was hired at the DCB as a researcher, and under the initial direction of Orcés and Albuja, she began to organize the bird collections (Almendáriz had done her thesis on birds), but Orcés soon changed his mind and suggested that she take charge of the amphibians and reptiles. That year, Almendáriz began a new era with a focus on fieldwork, inventories, taxonomy, and distribution of amphibians and reptiles, which lasted 37 years until her recent retirement.

At the time of Almendáriz's arrival, the amphibian collection at the EPN consisted of 1106 specimens and had increased to 10,694 specimens at her departure on 17 September 2019. Almendáriz, accompanied by her field assistants, research assistants, and collaborators, collected 6070 specimens (58% of the EPN collection) (Figure 2.19).

EPN collections were also enriched by collections deposited by other researchers who collected 3186 specimens (31%): Jorge Brito Molina (777 specimens), Luis Albuja Viteri (328), Paul Hamilton (243), Mario H. Yánez-Muñoz (234), Jorge Vaca-Guerrero (215), Rommel Mena (185), Luis A. Coloma (172), Ramiro Barriga (147), Paul Meza-Ramos (133), Gregory O. Vigle (93), and others (659).

Almendáriz led several inventory projects, particularly in protected areas that were of government interest (e.g., Reserva de Producción Faunística Cuyabeno, Parque Nacional Yasuní), and conducted other research framed in environmental impact studies for the oil, mining, and hydroelectric industries (e.g., Cordillera del Cóndor). During her time, the EPN (museum) infrastructure improved (currently

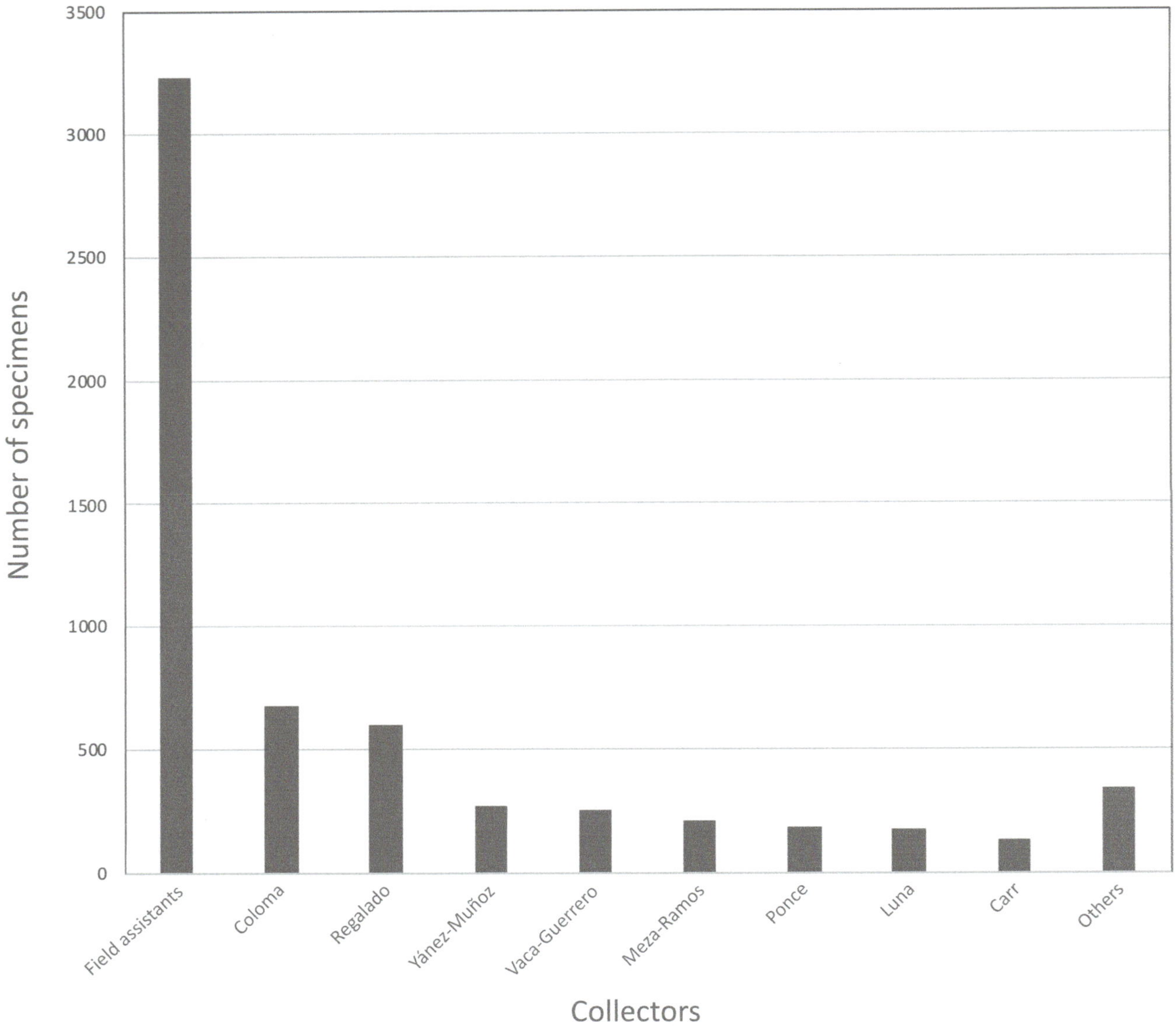

FIGURE 2.19 Number of EPN (museum) amphibian specimens collected by Almendáriz with her field and research assistants and collaborators.

occupying 65 m^2) with funding provided by Banco Interamericano de Desarrollo in 1996. Funds for assistants and field trips were provided by Entrix Inc., a company providing environmental services for the petroleum and mining industries, which signed a contract with the EPN in 1995 that lasted until 2019. She was assisted by students from PUCE and UCE, who collaborated either as field, research, or curatorial assistants. The EPN performed the studies required by the Ecuadorian government for works, projects, and activities of medium or high environmental impact (e.g., on the project OCP, heavy crude oil pipeline).

Almendáriz was one of the first Ecuadorians who studied biology concentrating on research. In fact, she was the fifth student to graduate (second woman) with a *licenciatura* degree in biology in Ecuador, in 1983, from the Departamento de Ciencias Biológicas of PUCE. Soon after Almendáriz began her work at the EPN, an important European herpetologist arrived in Ecuador, which enabled Almendáriz to collaborate, in late April 1983, as a research assistant in the project Herpetofauna of South America. With Hoogmoed in charge, they collected in the Pacific and central Amazonian regions for the RMNH. As a result of the project, 52 specimens were deposited at the EPN (museum). Hoogmoed was the first herpetological fieldwork mentor for Almendáriz, which had a significant influence. Later (1983–1984), Almendáriz continued to make collections in the Montalvo area and other oil exploration locations, often difficult to access, in the Provincia Pastaza, with the idea of filling the gap of knowledge for that region. The northern Amazonian region of Ecuador had already been explored by Duellman and his students. At that time, the central and Amazonian lowlands were still a mostly isolated region in which the indigenous people, military forces, and oil exploration companies were present and interacting. These groups collaborated with Almendáriz and supported her logistically, especially at the oil exploration camps of Corporación Estatal Petrolera Ecuatoriana, to which access was gained by either light

FIGURE 2.20 Left: Luis A. Coloma and Achuar indigenous collectors making herpetological collections at Chuintza, Provincia Pastaza, October 1983. Photo by Ana Almendáriz-Cabezas. Right: Ana Almendáriz-Cabezas (left) at the petroleum exploration site Misión, Provincia Pastaza, 8 March 1984. Photo by LAC.

aircraft or helicopter. Almendáriz took the opportunity to collect specimens in primary forests but also at sites where bulldozers were clearing the forest. She was joined by a couple of undergraduate biology students, Coloma and Alfredo Luna, who actively participated in the project (Figure 2.20). By the end of the project, the group had collected 853 specimens, representing most of the species from the region, as well as tadpole series and recordings of anuran calls, which served as the basis for a review of the amphibians of the Amazonian central region of Ecuador (Almendáriz, 1987). The latter was Almendáriz's first publication and was part of the initial biology series of the Ecuadorian scientific journal *Revista Politécnica.* In subsequent years, Almendáriz would publish nine articles on amphibians in this journal.

In 1986, Almendáriz began a collaboration with John Carr of the University of Illinois and made herpetological collections in the basins of the Cayapas and Santiago Rivers, Provincia Esmeraldas, jointly with John E. Simmons of the University of Kansas and Mark T. Nielsen of the University of Utah. Subsequently, in 1991, Carr and Almendáriz conducted herpetological inventories in forest remnants in the Coastal region and adjacent areas of southwestern Ecuador. Almendáriz's field work resulted in 70 specimens of amphibians and two publications (Almendáriz and Carr, 1992, 2012) (Figure 2.21).

In 1986–1987, the Consejo de Universidades y Escuelas Politécnicas funded a project to increase knowledge of the high Andean Herpetofauna. This project served to make collections along the inter-Andean valleys and adjacent cordilleras. Those amphibian collections are of critical importance, as they were made, by coincidence, at a time when the abundance of frogs was diminishing and catastrophic amphibian declines were occurring in the Andean region. For example, dead specimens of *Atelopus ignescens* were found in the Pita River, and this was reported in the resulting publication made by Almendáriz and Orcés (Almendáriz and Orcés, 2004).

Since 1988, and during her tenure at the EPN, Almendáriz worked abroad with collections (as a visiting researcher) at the Herpetology Division at University of Kansas (KU) (August–October 1988, April–May 1997) (see Figure 2.21), the Department of Herpetology of the MCZ (October 1988), and the Division of Amphibians and Reptiles of the USNM (November 1988, March 1994, June 1997).

Other important surveys made by Almendáriz in protected areas included those carried out in El Pan and Mataje (Reserva Awa), Provincia Esmeraldas, in 1987, jointly with Elizabeth Samec of the Museum of Natural History Vienna. In 1991–1992, inventories were performed in the areas surrounding the Tambococha and Ishpingo oil wells, Provincia Orellana. They are part of the ITT (Ishpingo, Tiputini, Tambococha), the largest oil development project in Ecuador, which is located in one of the most biodiverse areas of the Amazon basin. In the past it, was the subject of an international compensation initiative, launched by former President Rafael Correa Delgado, to leave oil underground, which was finally discarded due to a lack of donors.

In 1991 and 2000, Almendáriz explored the Cuyabeno area (Cofán and Zaparo communities, Iripari and Zancudo lagoons, Concienti and Güepi) as part of the expansion of the Cuyabeno Wildlife Reserve, that in fact (in 1991) was expanded from the initial 254,760 ha to 655,781 ha and later reduced (in 1994) to 603,380 ha.

In 1993, Almendáriz began herpetological studies in the Cordillera del Cóndor and in 2012 in Cerro Plateado, at the highest point of the Cordillera del Cóndor, Refugio de Vida Silvestre El Zarza, and other localities of the mining concession Fruta del Norte. Two young Ecuadorians, Jorge Brito Molina and Jorge Vaca-Guerrero were her main collaborators and worked as research assistants both in the field and at the museum. These were fruitful forays (e.g., nine publications about amphibians of the Cordillera del Cóndor) and led to the finding of five species new to science (Almendáriz *et al.*, 2012, 2014a; Harvey *et al.*, 2013; Brito and Almendáriz, 2018) and also valuable natural history information (Almendáriz and Batallas, 2012a, 2012b, Brito-M. *et al.*, 2014). Diego Batallas-Revelo aided with acoustic analyses for these publications and has been working with the audio library of the EPN. Most of the Cóndor studies were summarized by (Almendáriz *et al.*, 2014b). The new

FIGURE 2.21 Top left: from left to right: John Carr, Gustavo Orcés-Villagómez, and Ana Almendáriz-Cabezas in the EPN (museum). 1986. Photo by John E. Simmons. Top right: part of personnel of the DCB (EPN) in a visit of Erwin Patzelt to the EPN (museum) in 1992. From left to right: Gonzalo Herrera, Gustavo Orcés-Villagómez, Ana Almendáriz-Cabezas, Luis Albuja Viteri, Patricio Mena Valenzuela, Erwin Patzelt, Vilma Guerrero, and Freddy Trujillo. Photo courtesy of Ana Almendáriz-Cabezas. Bottom left: EPN (museum) members on a visit to the amphibian collector Jorge Olalla in Cotapino, Provincia Napo, in 1996. From left to right: Ana Almendáriz-Cabezas, Luis Albuja Viteri, Jorge Olalla, and Ramiro Barriga. Photo courtesy of Ramiro Barriga. Bottom right: from left to right: Linda Trueb, William E. Duellman, John E. Simmons, and Ana Almendáriz-Cabezas in Lawrence, Kansas. 1997. Photo by Ligia Galarza.

species included the poison frog *Excidobates condor*, the stream treefrog *Hyloscirtus condor*, and the snouted frog *Chiasmocleis parkeri*, for which she led the descriptions. These initial inventories—which have been continued by numerous amphibian biologists from several institutions—regularly reveal new species, as well as document the high endemism and fragility of this isolated Amazonian Cordillera. Currently, in the Cordillera del Cóndor, 41,000 ha are protected in the biological reserves El Cóndor, El Quimi, and Cerro Plateado and in the Refugio de Vida Silvestre El Zarza. However, most of the Cordillera is private land, and unfortunately these studies have not served to stop the timber extraction, deforestation, agriculture, and livestock ranching, nor have they served to stop the mining of copper, gold, lead, and silica, which are currently underway and seriously threaten the conservation of the diverse communities of amphibians in this region (Almendáriz *et al.*, 2014b).

In 2004, a survey of the herpetofauna was made by Almendáriz in several locations of the Arenillas Military Reserve (later incorporated to the Ecuadorian system of protected areas) in the southwestern Pacific lowlands with the assistant of Paul Meza-Ramos—Meza has coauthored descriptions of four species new to science. Between 2006 and 2008, collections were carried out in the Waorani community of Guiyero, and in 2008, Almendáriz made amphibian collections in the Llusín River basin, Sangay National Park. Between 2009 and 2010, inventories were carried out jointly with Jorge Brito Molina in the inter-Andean dry forests of the provinces of Imbabura, Pichincha, Azuay, and Loja. Between 2006 and 2012, collections came from the forests of the Provincia Manabí in a project with The Biodiversity Group, an NGO focused on protecting overlooked wildlife (Figure 2.22). In 2014, collections were made along the Curaray and Cononaco Rivers in Yasuní National Park with the assistance of Jorge Vaca-Guerrero. The latter inventories were carried out as part of a project funded by the Secretaría de Ciencia y Tecnología de Ecuador, at a time (2007–2017) when funds for biodiversity research were provided by the government, and significant transformations in science, technology, and innovation were intended from the highest level of decision making but have been discontinued recently (Bonilla *et al.*, 2021).

Between 2008 and 2012, Almendáriz invested a great deal of effort in the writing of the herpetological chapter

FIGURE 2.22 Ana Almendáriz-Cabezas in the collections room at the EPN in 2007. Photo by Vladimir Carvajal.

for a comprehensive book about Ecuadorian vertebrates (Albuja *et al.*, 2012). Currently, Almendáriz, although retired from the EPN, continues to work on pending publications.

Since the second half of 2019, a new era has started at the EPN with an amphibian biologist in charge, Mónica Alexandra Guerra. In 2021, Diego Almeida Reinoso (who has a doctoral degree in biology from UCE) was hired as Collection Manager. Guerra is an Ecuadorian amphibian biologist who studied frogs of the genus *Engystomops* for several years. Guerra has taught at the University of Texas at Austin (USA) and in Ecuador at the Universidad Regional Amazónica Ikiam, the Universidad Técnica Particular de Loja, and currently at the EPN. Her research is focused on amphibians, bioacoustics, and the evolution of animal behavior, for which she incorporates field experiments with molecular genetic techniques.

DEL PINO AND THE MARSUPIAL FROGS

As summarized in Chapter 1 (see "Live Frogs for Research, Education, and the Pet Market"), marsupial frogs were the first Ecuadorian frogs that became sought after for research purposes and for zoos in Europe and the USA, where they had been shipped since 1955. This resulted in an array of foreign studies (15 papers) on their reproductive biology that were published from 1957 to 1986. Nearly a decade and a half after 1955, studies on embryology and amphibian development began in Ecuador in the 1970s, led by the Ecuadorian Eugenia M. del Pino at the Pontificia Universidad Católica del Ecuador (PUCE) (Figure 2.23).

Aspects of del Pino's life and academic achievements have been summarized in a profile published in the *Proceedings of the National Academy of Sciences* (*PNAS*) (Mossman, 2007), an account of the reminiscences of her career (del Pino, 2021), and in outreach publications (e.g., Flores, 2021; Wikipedia, 2023 and citations therein, Serra, 2022). She studied the embryonic development of some marsupial frogs at the cellular and molecular level as well as that of several dendrobatid poison frogs and *Engystomops* and compared them with *Xenopus laevis*. A plethora of her novel findings arose from the realization that marsupial frog development was in some ways (e.g., development that was slow and hormonally controlled) similar to mammalian and bird embryos, as well as the discovery of multinucleate oocytes in *Flectonotus pygmaeus*, the description of a novel developmental pattern in frogs, and the finding that marsupial frog embryos excrete urea instead of ammonia, among other discoveries.

Del Pino established a pioneering laboratory for the comparative analysis of frog development in Ibero-America at PUCE. The value of the research carried out by del Pino and her students is internationally and nationally recognized. Thus, del Pino is considered one of the most outstanding Ecuadorian scientific researchers. In addition, del Pino has been an Ecuadorian pioneer in publishing her research in some of the most prestigious international scientific journals (such as *Scientific American*, *Science* magazine, *Nature*, and

FIGURE 2.23 Left: Eugenia M. del Pino at her office at PUCE. Right: Ligia Galarza (del Pino's first student) in a visit to the developmental biology lab at PUCE on 8 April 1980. Photos by John E. Simmons.

PNAS), as well as being one of the leading scientists to publish in national journals, beginning in 1974 with the *Revista de la Universidad Católica* and in 2005 in the *Revista Ecuatoriana de Medicina y Ciencias Biológicas*.

Del Pino's first steps studying frogs was a study for her PhD in Atlanta, Georgia, at Emory University, mentored by Asa Alan Humphries, Jr., addressing the role of the egg jelly on the fertilization of the aquatic frog *Xenopus laevis* (del Pino, 1973). Soon after her return to Ecuador, in 1972, she wanted an easily accessible frog (*X. laevis* was no longer an option, and she did not like to do field work to collect native frogs) to continue with the analysis of the frog egg jelly and fertilization. However, del Pino did not return to this problem because she found an unusual frog, a marsupial frog, in the gardens of PUCE. At that time, there were only two frog species living naturally in the PUCE gardens, the direct developing rainfrog *Pristimantis unistrigatus* (Nina and del Pino, 1977) and the marsupial frog *Gastrotheca riobambae* (del Pino-Veintimilla, 1975)—marsupial frogs have since disappeared from PUCE and from most parts of Quito. She was amazed by this marsupial frog and decided to combine her attraction for this species with studies of developmental biology. The following are her words:

> Thus, I was very lucky to find a marsupial frog at the doorstep of my laboratory. My first encounter with the marsupial frog *G. riobambae* occurred in 1972 in the gardens of PUCE. The *G. riobambae* female that I saw looked extraordinary to me because this frog incubated her embryos in a dorsal pouch. Of course, this vertebrate looked like a frog, but the incubation of embryos in the body of the mother resembled that of mammals. From the very beginning, the biology of marsupial frogs was appealing. However, I knew that I was in front of a research problem greater than my abilities and resources. I wrote a letter to Dr. Humphries, and I asked him to search in the Emory University library for information on the development of marsupial frogs. He wrote that the development of marsupial frogs was totally unknown, and that I had an important problem in front of me. In addition, he wrote that I could dedicate my scientific career to study the biology and development of marsupial frogs.
>
> (del Pino, 2021)

With basic equipment at PUCE and advice and support from Humphries, including the provision of key reagents, del Pino began her long journey studying marsupial frogs and a few other species. She began to make important developmental discoveries about the marsupial frog *Gastrotheca riobambae*. Her first paper dealt with its reproductive adaptive strategies for terrestrial life and was published in the PUCE magazine (del Pino, 1975). In the same year, she, Humphries, and two of her *licenciatura* students (Ligia Galarza and Carmen Merizalde de Albuja) published a paper (del Pino *et al.*, 1975) revealing that the large eggs and embryos of the marsupial frog *G. riobambae* develop at a very slow pace, like mammalian embryos. They are protected on their mother's back, and their gills, called "bell gills", envelop the embryo in a sac in close proximity to the mother's pouch. The system resembles the mammalian placenta. The embryogenesis of this species was more detailed in two additional papers by del Pino and Escobar (1981) and del Pino and Loor-Vela (1990). Moreover, she and two of her students found that the incubation of embryos by the mother is associated with hormonal control of reproduction, as in mammals (de Albuja *et al.*, 1993; del Pino, 1983; del Pino and Escobar, 1981).

In the second half of the 1970s, del Pino published the discovery of multinucleated oocytes in allies of marsupial frogs (del Pino, 1977; del Pino and Humphries, 1978). In an interview for *PNAS* by Mossman (2007), this relevant discovery is recounted as follows:

> In fact, del Pino discovered that several marsupial frog species have multinucleated oocytes. The most dramatic occurrence is in the Venezuelan marsupial frog, *Flectonotus pygmaeus*, whose oocytes contain more than two thousand nuclei. del Pino showed that these nuclei provide embryos with a large store of ribosomal RNA and provide multiple copies of genes, which most likely allow *Flectonotus* to develop quickly. Eventually, all but one of the nuclei break down, and the cell reabsorbs them. In the common *Xenopus* frog that del Pino first studied at Emory, it is known that oocytes have just one nucleus with two million copies of ribosomal DNA in addition to the chromosomes.

In 1980, del Pino (1980) expanded the descriptions on the morphology of the pouch and incubatory integument for other genera of marsupial frogs and allies. For this study, William E. Duellman, who was working on the taxonomy and evolution of marsupial frogs and allies, enthusiastically promoted del Pino's research on marsupial frogs. Thus, he and his students John D. Lynch and David C. Cannatella provided living frogs of several species for her studies. Additionally, Duellman arranged loans from various museums and arranged her visit to study the world largest collection of hemiphractid frogs at the Natural History Museum of the University of Kansas. In the publication of her reminiscences, del Pino (2021) also acknowledged the support for these studies by the well-known herpetologists Raymond Laurent (Argentina), Scott J. Maness (Venezuela), Paulo E. Vanzolini (Brazil), Pedro M. Ruiz Carranza (Colombia), and John E. Simmons (USA).

In 1983, del Pino and her colleague Richard Elinson (from the University of Toronto, Canada) reported the discovery of a novel development pattern in frogs that included a new mode of gastrulation (according to the timing of notochord elongation) that resulted in the formation of an embryonic disk on the surface of the yolk in marsupial frogs, a process that previously was known to occur only in mammals and birds (del Pino and Elinson, 1983) and that evolved convergently in amphibians as an adaptation to incubation inside the pouch.

In the early 1990s, del Pino, her *licenciatura* students Iliana Alcocer and Ximena Santa Cruz, and German collaborators Karl-Heinz Thierauch (who spent two years teaching at PUCE), Herbert Steinbeisser, and Horst Grunz published an important finding (del Pino and Alcocer, 1994; Alcocer *et al.*, 1992) that is recounted in her interview for *PNAS*:

> Traditional frogs and marsupial frogs also differ in how their embryos excrete waste. Free-swimming tadpoles excrete ammonia, which would be toxic if accumulated in close quarters. del Pino discovered that marsupial frog embryos excrete urea instead of ammonia. This finding allowed her to devise a urea-based medium for the in vitro culture of marsupial frog embryos. Unfortunately, she has never been able to fertilize the eggs artificially.
>
> (Mossman, 2007)

In the decade of the 2010s, the establishment of the ex situ conservation program Balsa de los Sapos, or "Ark for Frogs", at the Museo de Zoología of PUCE produced a welcome turn in del Pino's research, as she became highly interested in comparing the development of other frogs with marsupial frogs. By the mid 2010s, Balsa de los Sapos had become a major breeding program for various species of endangered frogs as well as frogs for research purposes. Thus, Coloma, Ron, and collaborators provided embryos and frogs used in del Pino's lab. An epitome of del Pino's research is the article published in *PNAS* (del Pino *et al.*, 2007). In that year, del Pino and several of her students and colleagues published a comparative analysis of the early development of several frogs mainly based on her studies on *Xenopus laevis*, *Gastrotheca riobambae*, five species of dendrobatids (*Ameerega bilinguis* (as *Epipedobates ingeri*), *Dendrobates auratus*, *Epipedobates machalilla*, *E. anthonyi*, *E. tricolor*), two species of the foam-nesting toads *Engystomops* (*E. coloradorum* and *E. randi*), and the direct developing frog *Eleutherodactylus coqui*. They found variable frog gastrulation patterns and suggested that this variation was related to the frogs' developmental strategies and possibly to their phylogeny rather than to the size of their eggs.

Del Pino has been part of a group of pioneers that established biology as an academic discipline in Ecuador. Since del Pino's retirement in 2013, she has authored and coauthored at least eight scientific articles (as of 2022), while two of her ex-students continued with developmental biology studies at PUCE, Óscar Damián Pérez Vaca and Andrés Romero Carvajal. Only two of her students have continued with frog developmental studies. On the one hand, Pérez was carrying out embryonic developmental studies and amphibian in vitro fertilization assays (Proaño and Pérez, 2017; Browne *et al.*, 2019), but he died prematurely in 2020. On the other hand, Romero Carvajal has wider taxonomic interests in the field of developmental biology, although he recently coauthored a paper on glassfrogs (Salazar-Nicholls *et al.*, 2021) and described the early ontogeny of an *Osornophryne* (Romero-Carvajal *et al.*, 2023).

KENNETH MIYATA: A LEGENDARY HARVARD HERPETOLOGIST AND THE ROOTS OF ECOTOURISM

Gregory O. Vigle

Kenneth Miyata (Figure 2.24) is a pivotal name in the amphibian research and ecotourism of Ecuador for many reasons. His collections were among the largest assembled in Ecuador during his time, and his 1982 checklist and bibliography of Ecuadorian amphibians and reptiles synthesized a chaotic sea of information previously scattered in specialized libraries. Additionally, his pioneering experience as a guide at Río Palenque and his book, *Tropical Nature*, have been inspirational to everyone interested in the Neotropics and ecotourism. Three biographical sketches were published after Miyata's premature death (at age 32) by Collins (1983), Williams (1985), and Lambert (2000); the first and last focus more on Miyata's fly-fishing pursuits, with the lengthy memorial by Williams being the most insightful concerning his life as a biologist. Herein is another biographical sketch of him, as recounted by his friend, Gregory O. Vigle (see biography of Vigle under the title "Vigle, Two Decades of Conservation and Amphibian Surveys in the Old Forest"), written in November 2021:

Kenneth Miyata (1951–1983) was born in Los Angeles, California, and grew up in the nearby suburb of Covina. He developed a keen interest in the natural world (particularly amphibians and reptiles) from an early age, fostered by explorations in the mountains north of LA, and periodic visits to the farm of an aunt and uncle in Idaho. Throughout

FIGURE 2.24 Kenneth Miyata. Top: unknown locality, circa 1980. Photo by B. Wu. Bottom: Centro Científico Río Palenque, Provincia Los Ríos, in 1978. Photo by Gregory O. Vigle.

his youth, he was also fascinated by the intricate sport of fly-fishing, and by the age of 18 was a rising star, able to earn substantial income from selling his own uniquely designed and hand-made fly lures. Later, when in graduate school, he also supplemented his income by writing articles for fishing magazines; by then he had already achieved widespread fame in the close-knit fly-fishing community.

Miyata began studies in biology at the University of California at Berkeley in 1969 and naturally was drawn to the herpetology collections at the Museum of Vertebrate Zoology. He graduated *summa cum laude* in 1973 and began graduate school later that year at Harvard University in the MCZ as a student of Agassiz Professor and Curator of Herpetology, Ernest E. Williams.

Miyata and Williams first met in person at the airport in Miami, Florida, where he joined a group of several of William's students and colleagues on a lengthy field expedition traveling to multiple localities in the Caribbean and Latin America. That journey launched Miyata's interest in tropical biology, and he set his sights on Ecuador, which had always been known for its extraordinary biogeographic diversity and had been the focus of increased herpetological exploration and discovery over the previous decade.

Throughout his graduate school years, Miyata funded his travel expenses and field work by leading Earthwatch and other paying volunteer groups, what is now termed "ecotourism", long before it became popular. Although his main focus was on amphibians and reptiles, he had broad knowledge of many other tropical taxa and ecosystem processes.

Miyata finished his PhD in 1980, with a 787-page dissertation titled "Patterns of Diversity in Tropical Herpetofaunas", with four major chapters: 1) Continental patterns of species diversity in amphibian and reptile communities, 2) Ecological studies on the herpetofauna of the southern Choco of Ecuador, 3) Analysis of a local herpetofauna: Herpetology of the Río Palenque region of northwestern Ecuador, and 4) History and species diversity. The last chapter consisted mostly of six papers describing in total five species of anurans and two lizards, all from Ecuador, some already published or submitted; a complete list of the many subsections within the four major chapters is presented in Miyata, 2013 (also see herein).

Miyata moved to Washington DC the following year for a postdoctoral fellowship at the National Museum of Natural History of the Smithsonian Institution with Roy McDiarmid. He was also coauthoring a book with fellow Harvard graduate and botanist Adrian Forsyth, *Tropical Nature: Life and Death in the Rain Forests of Central and South America* (Forsyth and Miyata, 1984). Written for general audiences, with a forward by Thomas Lovejoy, it consists of 17 essays concerning the ecology and evolutionary biology of tropical forests. The book was widely acclaimed by reviews in *Newsweek*, *Scientific American*, and *Smithsonian Magazine*, and prominent MCZ professor Ernst Mayr wrote that "I know of no better introduction to tropical biology". *Tropical Nature* is unquestionably Ken's greatest legacy in the wider world outside of his specialized work in herpetology, and to this day it is recommended (or required) reading in many universities' tropical biology courses, and countless travel guides and websites.

By the fall of 1983, Miyata had been offered a job with The Nature Conservancy. The position involved plenty of travel, and some of his responsibilities would include the inventory and evaluation of Neotropical forests for conservation assessments. As always, he continued to indulge his passion of fly-fishing, and was working on a book on that subject. In late September, he set out for a three-week fishing trip to Montana (where he had traveled many times), prior to starting his new job. On 14 October, he was fishing alone in the Bighorn River, when he slipped under on the mossy bottom. His heavy rubber waders filled up with water, and he was unable to escape from the rapids. His body was later recovered in a deep pool downriver, with his fishing rod still in his hand.

Although the present volume concerns amphibians, Miyata was a classic old-school herpetologist, equally knowledgeable and skilled with herpetofauna across all taxa, and so no summary of his contributions would be complete without including those concerning reptiles as well. Alone or with coauthors, he described four species of anurans (Lynch and Miyata, 1980; Miyata, 1980a; Vigle and Miyata, 1980) and two lizards (Miyata, 1985a, 1985b), all from Ecuador. His checklist and bibliography of the amphibians and reptiles of Ecuador (Miyata, 1982) was the first single account covering the entire herpetofauna (682 species listed, 343 amphibians), and although now dated, it remains a very useful resource. Additional publications include two brief ecological notes (Miyata, 1980b, 1983) and a massive *Festschrift* volume (with 50 papers by 69 authors) honoring Ernest Williams, which Miyata co-edited with another of Williams' associates (Rhodin and Miyata, 1983).

Although unpublished, the third chapter of his dissertation presented a 292-page account of the Río Palenque region herpetofauna (based on his own field work and collections), with identification keys and detailed treatments of many species; he may have planned to eventually publish that chapter as a field guide, fully illustrated by his excellent photos. Aside from Ken's species descriptions and the inclusion of some of his work in a later paper describing two new *Anolis* (Poe *et al.*, 2009), the only part of his dissertation to see publication to date is an excerpt combining the second and third major sections of his second chapter (Miyata, 2013). Edited by MCZ Curator of Herpetology and *Anolis* specialist Jonathan Losos, it was published in the *MCZ Bulletin* series and includes a supplement with memories from five of his close friends.

In addition to his dissertation and publications, catalog records from the MCZ reveal that Ken contributed about 4200 specimens (80% amphibians) from Ecuador to the herpetology collections on more than 20 lengthy field trips from 1974 to 1979, representing a large percentage of the known fauna of the western lowlands. In addition to the species that he described, he also collected the holotypes (and many associated paratypes) of seven species of anurans and the paratypes of two anurans, one lizard, and one snake, all described later by other workers (and, also from Ecuador, the holotype of one fish!).

His legacy at the MCZ is commemorated today via the Ken Miyata Grants program (funded by the Ken Miyata Fund for Field Research), which offers grants to support herpetological fieldwork by MCZ graduate students "who share Ken's interests as a naturalist, biogeographer, and writer/photographer".

Five species have been named in Miyata's honor, including the anurans *Pristimantis miyatai* (Lynch, 1984), *Dendropsophus miyatai* (Vigle and Goberdhan-Vigle, 1990), and *P. hamiotae* (Flores, 1994; the specific epithet derived from the Latin "of the angler"), one lizard (*Lepidoblepharus miyatai*; Lamar, 1985), and one snake (*Tantilla miyatai*; Wilson, 1987).

A renowned fly-fishing expert who knew and fished with Ken, the late Jack Gartside, said of Miyata's fishing skills, in 2000: "If Ken were alive today, there would be no one who could match him". Perhaps much the same could be said of his potential in herpetology and tropical biology.

VIGLE: TWO DECADES OF CONSERVATION AND AMPHIBIAN SURVEYS IN THE OLD FOREST

Gregory O. Vigle

In the 1970s and 1980s, the first biological stations began to be established in Ecuador. One of them was the Centro Científico Río Palenque (CCRP, founded in 1970) in the Pacific lowlands, and another the Estación Biológica Jatun Sacha (EBJS, founded in 1985), in the eastern Amazonian region. Jatun Sacha is a Kichwa name meaning Old Forest. In the latter, a young US herpetologist, Gregory O. Vigle, began a long-term program to study the diversity and abundance of amphibians (Figure 2.25). Vigle has contributed to the knowledge of taxonomy of Ecuadorian amphibians by collecting specimens (about 2550) and by describing four species, notably the rare and critically endangered rocket frog *Paruwrobates erythromos*, which he discovered at the CCRP.

Vigle developed an interest in herpetology at the early age of nine, seeking out snakes on Boy Scout excursions in Ohio and keeping a variety of amphibians and reptiles as pets. While still in high school, he undertook the first investigations on a now well-known introduced population of European lizards in Cincinnati, Ohio, and presented an oral paper on that population at the SSAR-HL meetings in 1977, where he met the late Ken Miyata (who had already been working in Ecuador for several years).

Vigle then spent the month of January of 1978 in Ecuador, helping Miyata at the CCRP reserve in the western lowlands. That experience provided his initial introduction to tropical biology and herpetology from an ideal mentor, as he learned everything he could from him in the time available. During this time, Vigle also made the first collections of a new dendrobatid species (*Paruwrobates erythromos*), describing the species with Miyata (Vigle and Miyata, 1980).

Eight years later, Vigle again returned to Ecuador, under the sponsorship of Ernest E. Williams and the MCZ at Harvard University. Over two years, from 1986 to 1988, he conducted an extensive initial survey of the amphibians and reptiles at the newly established EBJS in the upper Río Napo region. At the time, Jatun Sacha had only 230 ha of land partially bordering the Napo but included a diverse concentration of both primary forest and river floodplain habitats, forming a microcosm

FIGURE 2.25 Gregory O. Vigle at Jatun Sacha in 1986. Photo by Alejandro Suárez.

representing most of the habitat variation characteristic of the upper Amazon region of Ecuador. The construction of a new road progressing east from Tena passed through part of the reserve in 1987, eventually resulting in increased deforestation in surrounding areas. These processes provided the opportunity to assess the conservation efficacy of a small upper Amazon forest remnant reserve in a fragmented landscape. In addition to his work at Jatun Sacha, a brief field trip in 1988 to another small reserve further downriver on the Napo resulted in the collection and description of a new species of hylid frog that was later named in honor of Kenneth Miyata (*Dendropsophus miyatai*; Vigle and Goberdhan-Vigle, 1990).

Following his 1986–1988 survey (including extensive collections of specimens for the MCZ), Vigle returned to Jatun Sacha in 1998 to again assess the herpetofauna solely via visual-encounter efforts. By then, the land holdings of the reserve had grown to 2334 ha, while the surrounding landscape had continued to suffer ongoing deforestation and fragmentation. At that time, Vigle also acquired an adjoining 14 ha plot of forest on the Río Napo and built a 25-person-capacity facility to host university course groups and researchers studying tropical biology. He named his small reserve the Librarius Study Center, in turn named after *Pristimantis librarius* (Craugastoridae), honoring his late father John, who was a university library director (Flores and Vigle, 1994). The 14 ha forest plot conserves a high concentration of amphibian species in a small area with diverse floodplain habitats. The description of *P. librarius* was written in cooperation with Glenn Flores.

In 2000, Vigle authored the amphibian chapter of an early ecotourist guidebook for Ecuador, which included illustrations based on his photographs (Vigle, 2000). Also in 2000, he led a small group of volunteers on the first survey of the herpetofauna at the Reserva Biológica Bilsa in the Mache Chindul range in northwestern Ecuador (also owned by *Fundación Jatun Sacha*), funded by a grant from the Declining Amphibian Populations Task Force (DAPTF). The first specimens of a new species of dendrobatid frog were collected at that time but only very recently described after the acquisition of additional specimens, in a collaboration including the work of seven authors (*Leucostethus bilsa*; Vigle *et al.*, 2020).

On more than a dozen lengthy trips from 1998 to 2006, Vigle worked to replicate his survey at Jatun Sacha from 1986 to 1988, accumulating approximately the same amount of sampling effort distributed over the same areas and trail routes. Partially supported by a grant from the DAPTF, he also trained and employed Jatun Sacha park guards in the same visual-encounter search protocol, and by using digital cameras to create a permanent record.

Based on broad, coarse-scale comparisons of species richness and abundance, results reported on 20 years of accumulated data suggested that the herpetofauna within the small area surveyed from 1986 to 2006 had not suffered any pronounced declines (Vigle, 2008) and that Jatun Sacha has successfully conserved that fauna as an isolated remnant. The concentration of habitat diversity also revealed a remarkable concentration of species richness. Additional species continue to be recorded, and currently a total of at least 94 species of amphibians (and 90 of reptiles) have been recorded within an area of only about one square kilometer (one of the highest known globally, per unit area).

Over his more than 30-year span of explorations in Ecuador, Vigle also collected about 3400 herpetological specimens (75–80% amphibians, 2550–2720), the majority deposited in the collections of the MCZ, but with substantial numbers housed at several other major museum collections in both Ecuador and the US (in Ecuador, the QCAZ and MECN and in the US, the USNM, AMNH, KU, and UF).

RIVERO, HEYER, AND FLORES: THEIR TRANSIENT WORK ON ECUADORIAN AMPHIBIANS

Rivero and the Cryptic Dendrobatids

Juan Arturo Rivero Quintero (1923–2014) (Figure 2.26) was a prolific Puerto Rican biologist and herpetologist who made major contributions to the amphibian knowledge of Latin America, especially of Puerto Rico, Venezuela, and Colombia. Although his contribution to Ecuadorian amphibians was occasional, it was relevant and focused on the taxonomy of the dendrobatids

Colostethus of his time (currently *Hyloxalus*, *Allobates*, *Epipedobates*, *Leucostethus*).

Rivero's life has been summarized by several contributors to Wikipedia (https://es.wikipedia.org/wiki/Juan_A._Rivero, see also references therein). Rivero obtained a bachelor of science in agronomy from the University of Puerto Rico at Mayagüez in 1945, and in 1951 earned an MS from Harvard University, followed by a PhD from that same institution in 1953. He became a full professor of biology at UPR-Mayagüez in 1958. After founding the UPR-Mayagüez Zoo in 1954, he served as its first director, as well as founding and directing the UPR's Institute of Marine Biology. From 1959 to 1960, he served as director of the Biology Department and from 1962 to 1966 as Dean of Arts and Sciences. Rivero oversaw the work of more than 200 faculty members. In addition to writing 108 scientific papers, he wrote more than 300 outreach publications and 17 books. At least 14 species have been named to honor Rivero's contributions to science, and the zoological park that he founded was named after him in 1998.

Rivero's contribution to the knowledge of the Ecuadorian amphibian fauna was the description of seven dendrobatid frogs of the genus *Hyloxalus* in a single paper (Rivero, 1991), in which he reviewed collections deposited at the USNM, most of which were collected by members of the Olalla family and James A. Peters. Rivero also described *Nymphargus megista* and *Hyloxalus breviquartus* from Colombia, which have now also been recorded in Ecuador. In October 1983, Rivero visited Ecuador, met Almendáriz and Coloma, and jointly embarked on a short trip to the Pifo region, Provincia Pichincha, which resulted in finding a new species (*Pristimantis anae*, currently a junior synonym of *P. curtipes* according to Wiens and Coloma (1992), but see the *P. curtipes* account) and a taxonomic revision of the *P. myersi* subgroup (Rivero, 1986). Although he did not do additional field work in Ecuador, Rivero and Almendáriz began a project that resulted in a semipopular publication on the identification of *Colostethus* (Rivero and Almendáriz, 1991). By the end of the 1980s, at about the same time Coloma was working at the University of Kansas on the taxonomy of Ecuadorian *Colostethus* for his master's thesis, Rivero maintained correspondence with Coloma to avoid overlapping his research. Both of them undertook the difficult task of addressing some aspects of the taxonomy of the genus *Colostethus*, a genus that had been reviewed previously by Stephen R. Edwards for his PhD dissertation (Edwards, 1974a) and most of which remained unpublished, with the exception of one publication that included Ecuadorian taxa (Edwards, 1974b).

Heyer: The Smithsonian Taxonomist behind *Leptodactylus* Frogs

William Ronald Heyer (see Figure 2.26) is a productive American herpetologist whose input to Ecuadorian amphibian knowledge was mostly focused on the genus *Leptodactylus*. He wrote a biographical sketch with his wife (Heyer and Heyer, 2016). He was born on 29 August 1941 in Bellingham, Washington. After completing his master's thesis, he delved into researching the evolution and relationships of the genus *Leptodactylus* for his PhD at the University of Southern California, which he completed in 1968. His passion for zoology was ignited by his undergraduate biology professor, Jens Knudsen, who served as a mentor and inspired him to pursue further studies in graduate school. To honor his professor, Heyer named a species, *L. knudseni* (Heyer, 1972), discovered during two months of fieldwork in Limoncocha, Provincia Sucumbíos, in 1970 and 1971.

In 1973, Heyer joined the Division of Amphibians and Reptiles at the National Museum of Natural History at the Smithsonian Institution as an assistant curator and was promoted to Associate Curator in 1978. Notably, in response to the alarming decline in amphibian populations observed in 1989, he played a pivotal role in the establishment of the Declining Amphibian Populations Task Force in 1990, aiming to investigate the causes behind this ecological disaster and raise public awareness. In 1991–1992, Heyer became its Chairman.

Continuing his research efforts, Heyer described two additional species of the large *Leptodactylus* frogs in the Pacific lowlands of Ecuador in 2005 (Heyer, 2005). Unfortunately, these frogs are currently very rare. Additionally, from the 1990s until 2014, he collaborated with Rafael de Sá at the University of Richmond, extensively studying the phylogeny and relationships of the genus *Leptodactylus*, resulting in a significant publication (de Sá *et al.*, 2014). Throughout his career, Heyer made remarkable scientific contributions, describing 54 species of amphibians and reptiles, and receiving the honor of having eight species of amphibians and reptiles, among them *Noblella heyeri* occurring in Ecuador (Lynch, 1986), as well as one nematode, named after him.

Glenn Flores: From "Eleuths" to Pediatrics

Gregory O. Vigle

Glenn Flores grew up in New York and Connecticut and was interested in zoology from an early age, first keeping tropical fish and then amphibians and reptiles, encouraged by a dedicated high school biology teacher. While still only a sophomore in high school, he passed his advanced placement course in biology; during this period, he also published his first scientific paper, a field study of the hylid frog *Pseudacris crucifer*, published in the *New York Herpetological Society Bulletin* when he was only 16, which garnished him an honorable mention award in the Westinghouse (now Regeneron) Science Talent Search.

After high school, in 1979, he began studies in biology at Harvard University. As a freshman, he approached Ernest E. Williams to ask permission to attend his upper-level herpetology course at the Museum of Comparative Zoology (MCZ). Flores excelled in the course as his interest in herpetology grew and was also employed as a work-study student assistant in the collections at the MCZ Herpetology

FIGURE 2.26 Left: Juan A. Rivero giving a lecture at the Second Latin American Congress of Herpetology in Mérida, Venezuela, 1990. Photographer unknown. Top right: William Ronald Heyer, March 2016. Photo by Ted R. Kahn. Bottom right: Glenn Flores, 2010. Photo courtesy of Glenn Flores.

Department, continuing under the mentorship of Williams, and became friends with then PhD student Kenneth Miyata and Herpetology Collection Manager José Rosado. Along with field studies and collections in the Dominican Republic, Flores developed an interest in the great diversity of "eleuths" (an abbreviation for *Eleutherodactylus* frogs, most of them currently in the genus *Pristimantis*) and began studying the literature and specimens in the MCZ collections.

After the tragic death of Miyata in 1983, Flores identified a number of previously unknown *Pristimantis* species from Ecuador (three collected by Miyata, and one each by Giovanni Onore and Tjitte de Vries), including *P. kirklandi* (Flores, 1985a, named for his high school biology teacher Benjamin Kirkland), *P. ernesti* (Flores, 1987, named for Williams), *P. katoptroides* and *P. rosadoi* (Flores, 1988; the latter named for Rosado), and *P. hamiotae* (Flores, 1993, named for Miyata). He later collaborated on the descriptions of three additional *Pristimantis* species, including *P. librarius* (Flores and Vigle, 1994) and two species from Peru (Flores and Rodríguez, 1997); one of the latter (*P. skydmainos*) has since also been recorded over a wide range in eastern Ecuador. In honor of Flores' contributions, Lynch and Duellman (1997) named *P. floridus* from the Andean slopes of northwestern Ecuador.

Flores also described four species of centrolenid frogs, including two from Ecuador (Flores 1985b, and Flores and McDiarmid 1989). Although both of the Ecuadorian species are now considered synonyms of prior names, his 1987 paper described previously overlooked variations in the nuptial pads and prepollical spines, and those character states have been widely employed in later studies of centrolenids.

Flores graduated *magna cum laude* with a degree in biology from Harvard University in 1984. While he was a top candidate for continuing his herpetological studies in graduate school, he had also been drawn to the field of medicine and the prospect of becoming a physician. Ultimately, he chose the latter path, while continuing to work on the species descriptions cited earlier for more than a decade.

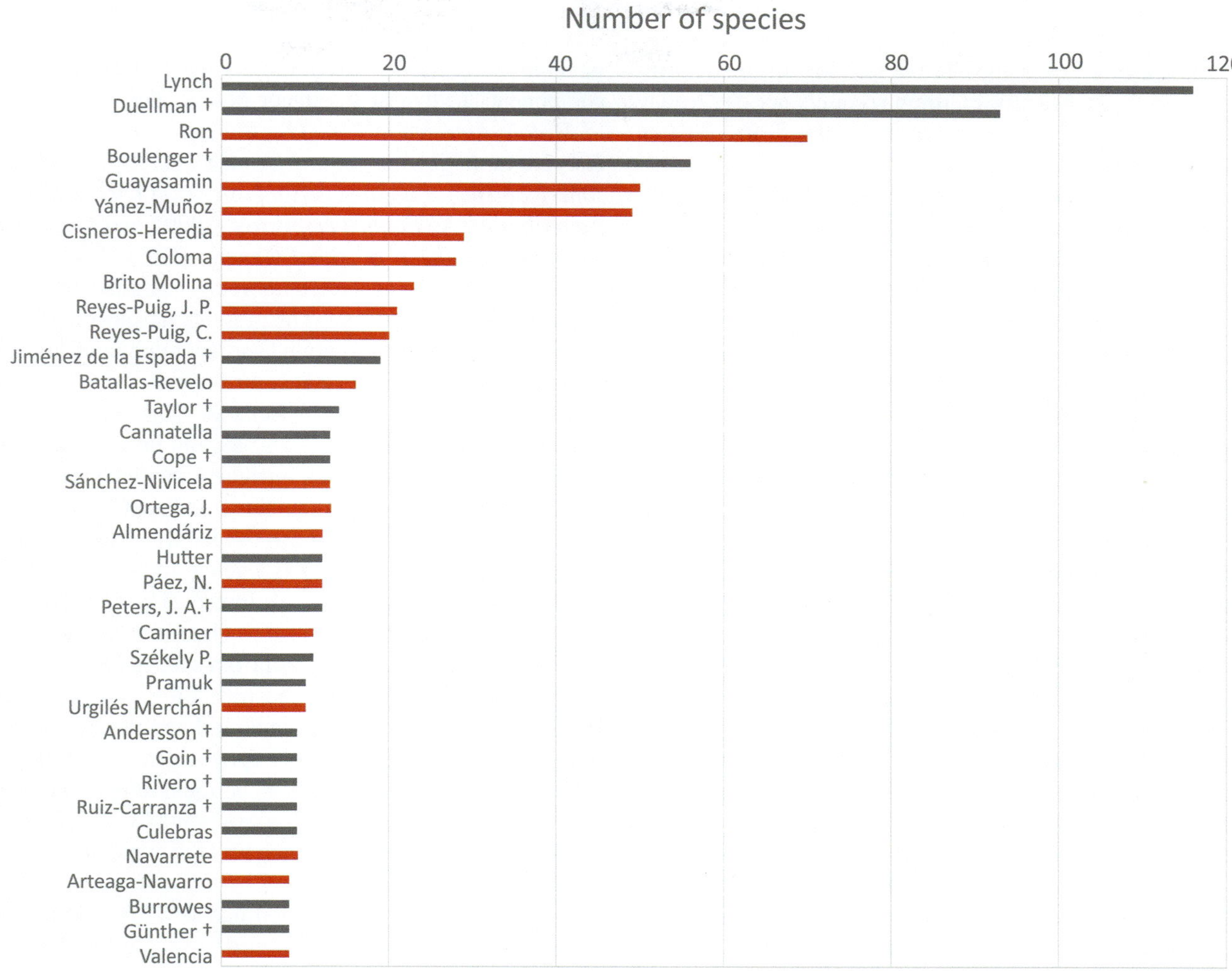

FIGURE 2.27 Thirty-six taxonomists that named most of Ecuadorian amphibians, ordered by the quantity of species they described or co-described up to 2022. Ecuadorian authors in red, foreigners in gray. Cross mark = deceased.

Flores earned his M.D. in pediatric medicine in 1989 at the University of California, San Francisco, School of Medicine. Flores has had a distinguished career as both a practitioner and researcher in pediatric medicine, with a focus on disparities in health care caused by poverty, racism, and lack of access for underserved populations, and has authored more than 250 scholarly publications. He is currently the Chair of Pediatrics, Senior Associate Dean of Child Health, and Professor of Pediatrics and Public Health Services at the University of Miami Miller School of Medicine. And he still enjoys keeping up with literature on the amphibians of Ecuador in his spare time, to the present day.

THE DUELLMAN–KU ERA: THE BIG BANG IN AMPHIBIAN SYSTEMATICS AND ECOLOGY

In 1966, scientists from the University of Kansas (KU)—led by William E. Duellman—began explorations and research in Ecuador. They have formed one of the most important scientific collections (32,171 specimens of amphibians), which are mostly deposited at KU. Duellman and nearly a two dozen of his graduate students have described most of the species of Ecuadorian amphibians (Figure 2.27) known to date and published studies on systematics, ecology, and other fields of biology. Among this team are the outstanding North American John D. Lynch, who has described 116 registered species from Ecuador, as well as the Ecuadorians Coloma, Ron, and Guayasamin, all of whom did graduate studies at KU and, beginning with Coloma in the 1990s, took the lead in amphibian research in Ecuador.

A significant portion of Duellman's academic life, as well as the key events of the Duellman-KU era in Ecuador, has been meticulously documented in Duellman's comprehensive analysis of the centennial history of herpetology at KU (Duellman, 2015a). In this work, he presents biographical profiles of each KU student who conducted research in Ecuador. Recent obituaries authored by Coloma and Guayasamin (2022), Burrowes *et al.* (2022), Mendelson III (2022), and Simmons (2022) succinctly summarize various facets of his life as a tribute to his enduring legacy.

Additional accounts of workers from the Duellman-KU era are available, particularly in the book penned by his student Martha L. Crump (Crump, 2000). In this charming, journal-like narrative, Crump recounts her day-to-day encounters and experiences while engaged in research activities in Ecuador.

This present account aims to highlight Duellman's remarkable undertakings and profound contributions to the study of Ecuadorian amphibians, drawing directly from these published accounts in some instances, either verbatim or in nearly identical form.

Duellman's involvement in Ecuador can be delineated into two significant phases. The initial phase spanned from 1966 to 1990, during which he personally conducted extensive fieldwork in Ecuador alongside his students from the KU. Subsequently, a second phase emerged from 1988 to 2022, during which Duellman, in collaboration with his wife and scientific partner, Linda Trueb, assumed a mentorship role for Ecuadorian students. These students pursued specialization in fields such as systematics, evolutionary biology, and ecology at the University of Kansas. Throughout this period, Duellman and Trueb also maintained their research and publication endeavors centered around Ecuadorian frogs. Duellman's first period began in 1966, when his taxonomic research and resulting publications already included Ecuadorian frogs. For this first paper, on the Neotropical genus *Smilisca*, he examined Ecuadorian specimens of *S. phaeota* deposited at several museums in the US, and it was published by the Museum of Natural History of the University of Kansas (Duellman and Trueb, 1966). Next was a paper that appeared in 1968 in the US journal *Herpetologica*, in which Duellman reviewed the taxonomic status of some American hylid frogs, among them the Ecuadorian *Trachycephalus coriaceus*, based on specimens from Limoncocha (Duellman, 1968). Then, in 1969, three consecutive papers dealt with the Ecuadorian frogs: *Agalychnis buckleyi* (Duellman, 1969a), *Atelopus ignescens*, *A.* sp. (*spumarius* at that time) (Duellman and Lynch, 1969), and *Dendropsophus carnifex* (Duellman, 1969b). The latter was the first new species described by Duellman from Ecuador. These initial studies were followed by an explosion of research and publications about Ecuadorian taxa. Between 1968 and 2020, Duellman published 73 papers that included Ecuadorian frogs among a total of 387 herpetological publications; 67 of the 73 were journal articles, the others were components in five of his major books: *The Hylid Frogs of Middle America* (1970), *The Biology of an Equatorial Herpetofauna in Amazonian Ecuador* (1978), *The South American Herpetofauna* (1979), *Patterns of Distribution of Amphibians: A Global Perspective* (1999), and *Marsupial Frogs: Gastrotheca & Allied Genera* (2015). Also, seven of the journal articles were authored (1) or coauthored (6) by Ecuadorian researchers. The present encyclopedia (published posthumously), compiled in the epilogue of his life, is also added to his important contributions.

Duellman's field work in Ecuador started in November 1966, when he was 36 years old, and lasted until 1984 (Table 2.3). Duellman was intrigued by the idea of an exhaustive study of the amphibians and reptiles at a single site in a tropical rain forest, especially in the Amazon Basin. Essentially nothing was known about herpetological communities in tropical rain forests. The possibility of such a study loomed before him in October 1966, when he received a stimulating phone call from a former KU student, Charles M. Fugler. During the preceding two summers, Fugler had collected amphibians and reptiles in eastern Ecuador (the Oriente) and was impressed by the attitude of Ildefonso Muñoz B., a Colombian political refugee who had a compound near the small village of Santa Cecilia, Provincia Sucumbíos, adjacent to which was the exploration camp of an oil consortium.

On 21 November 1966, Duellman flew from Miami to Quito and then traveled, for the first time, to the Amazonian rainforest, specifically to Santa Cecilia, on the banks of Río Aguarico, in Sucumbíos Province. Soon Duellman was settled into a small, very modestly furnished "guest house". He became acquainted with Muñozlandia and Ildefonso's family—his wife, Blanca, and their four children. The location of Santa Cecilia was ideal—virgin or only slightly disturbed tropical rain forest, 340 m above sea level, and practically on the equator; moreover, it was accessible. Duellman spent two days and nights surveying the area and collecting several species of frogs, lizards, and snakes, many of which were unknown to him. In a

TABLE 2.3
Herpetological Field Work in Ecuador by Personnel from the Natural History Museum, University of Kansas, 1966–1975, 1984

Region	Persons	Person Days	Specimens	Spec/Pers. Day
Pacific	4	35	751	21.5
Western slopes	6	66	251	3.8
High Andes	8	47	929	19.8
Eastern slopes	8	99	1041	10.5
Amazonia	11	2541	4009	1.6
Totals		2788	6981	

prologue for a photography coffee table book of Ecuadorian frogs, he described his first experience at Santa Cecilia as follows:

> As darkness fell, I donned my boots and a headlamp and walked along a trail in the rainforest. Everywhere I looked I saw frogs of all sizes, shapes and colors—big green frogs that made a soft cluck call, large brown frogs with acute snouts that had a laugh-like call, and many kinds of small yellow frogs, some of which had red feet. These were only the tree frogs. By day on the ground there were various kinds of toads and poison-dart frogs. And then there was the frog with a long, fleshy nose and what seemed like horns on its head. Had I died and gone to heaven inhabited by the world's greatest diversity of frogs? No, I was in the Oriente (Amazonia) of Ecuador.
>
> (Duellman, 2009)

With this trip, Duellman began his lifelong journey through Ecuador, toward which he focused a large portion of his professional life. In this first field work of 11 days, he collected 191 specimens of amphibians and reptiles at Santa Cecilia and Limoncocha, adding to his 30,933 field series of amphibians and reptiles he had collected before in the US, Mexico, Guatemala, El Salvador, Honduras, Costa Rica, Nicaragua, Panama, and Venezuela. Duellman would return to conduct field surveys in the Lago Agrio region (Santa Cecilia, Limoncocha, Puerto Libre) in 1967, 1968, 1969, 1971, and 1972. At that, time the oil industry, a Gulf-Texaco consortium in the northern Ecuador Amazonian region, had just begun its activities, and deforestation along with some of the worst oil spills in world history would shortly follow (Kimerling, 1993). Nonetheless, the virgin equatorial rain forest at Lago Agrio and surroundings in Ecuador offered Duellman a unique opportunity to lead and conduct inventories and ecological studies and to document natural histories, sometimes walking behind or stopping bulldozers that were clearing the forests. An exploratory trip would follow next from 28 February to 23 March 1967. In addition to Duellman, three other players would take part in the forthcoming activities—Linda Trueb, who willingly took a break from her work on frog bones, Henry Fitch, who was eager to gain more experience in the tropics, and William G. Saul, a graduate student in ichthyology. Day by day and night by night, they worked the trails and waded in the swamps. After heavy afternoon rains, large numbers of many species of frogs congregated at swamps and ponds, where sometimes the cacophony was earsplitting. Hours were spent recording the calls of individual frogs and always trying to collect the calling male, so that a preserved voucher would be available. Just as many, if not more, hours were spent photographing living individuals, carefully preserving and tagging specimens, plus writing field notes—location, behavior, color in life, etc. By the end of the first week, they had collected 226 specimens of 54 species of amphibians and reptiles. To Duellman's mind, this was incredible. The group continued to work at Santa Cecilia until 24 March. In three weeks, they collected 693 specimens of 94 species of amphibians and reptiles. During a comparable amount of personnel time, a KU field party had obtained 569 specimens of 84 species in the rain forest in Darién, Panama, in 1965. After leaving Santa Cecilia, Fitch and Saul returned to Kansas, while Duellman and Trueb remained in Ecuador for a few days spent in the paramo above Papallacta in the Cordillera Oriental. Subsequently, they spent a week crossing the Cordillera Occidental and collecting in the vicinity of Santo Domingo de los Tsáchilas on the Pacific lowlands. Needless to say, they returned to KU enthusiastic about their recent adventures in the Pacific lowlands and Andean paramos of Ecuador, which they would visit many more times in the coming years, and the rain forest at Santa Cecilia, where they would return to learn more about the diverse herpetofauna. In 1968, Duellman visited Santa Cecilia again, accompanied by a group of biologists. Muñoz arranged for some of the group to visit Puerto Libre, a mining prospectors' camp on the Río Aguarico at the base of the Andes. During the summer two groups flew in small planes that landed on the grass airstrip. Puerto Libre, at 570 m, was slightly higher and seemingly more humid than Santa Cecilia (340 m). Amazingly, the short time at Puerto Libre netted 708 specimens of 68 species, including 102 salamanders and 11 specimens of a new genus and species of microhylid frog that would be named five years later by Charles Frederic Walker (Walker, 1973), Curator of Herpetology at the Museum of Zoology at the University of Michigan. Duellman and Edwards were able to collect at one of the drilling sites northwest of Santa Cecilia, where they obtained 70 specimens of 23 species of amphibians and reptiles in 24 hours. On that trip, a frequent and inquisitive visitor to Muñozlandia was Jim Watt, supervisor of the survey crews for the trans-Andean pipeline to be built by Williams Brothers Construction Company. Jim offered to take a couple of KU people via helicopter to a survey camp in the Andes. Duellman and Edwards eagerly accepted the invitation and spent three days and nights at the well-outfitted tent camp at an elevation of 1150 m on the Cordillera del Dué, an Andean spur overlooking the Río Coca. They accumulated 194 specimens of 30 species, 12 of which were frogs of the genus *Eleutherodactylus* (as they were known then), many of which were new to the KU collection. As they packed up their specimens for departure on 13 August, the field party noted that they had accrued 1095 specimens of amphibians and reptiles from Santa Cecilia representing 124 species, 21 of which had not been collected there previously. Once in Quito, Duellman arranged for a day-long trip to the paramo on Volcán Cotopaxi, where the students were able to investigate the high Andean habitat and find three species of frogs, one of which, the jambato harlequin frog (*Atelopus ignescens*), was so abundant walking around on the ground that they had to be careful not to step on them.

With the aim of composing a comprehensive portrayal of a tropical herpetofaunal community, Duellman recognized the existence of substantial gaps in his understanding of the broader herpetofauna realm. This included the absence of recorded calls from certain frog species and the dearth of information regarding tadpoles from others. Evidently, an additional expedition to Santa Cecilia became imperative, prompting Duellman to once again adeptly secure funding from diverse sources. This time, the expedition team comprised Duellman, Linda Trueb, and a graduate student named Tom Fritts, along with the addition of Charles F. Walker from the University of Michigan (Figure 2.28).

Departing on 2 April 1969, the group journeyed to Quito, proceeding to Santa Cecilia on the following day. A noticeable improvement had been made to Muñozlandia, which now proudly displayed a banner affirming its status as a scientific center. However, amid their appreciation of the enhancements at Muñozlandia, they were disheartened by the discovery that segments of the pristine forest had fallen

FIGURE 2.28 Upper left: from left to right: Linda Trueb, Charles F. Walker, Ildefonso Muñoz B., William E. Duellman, and Thomas H. Fritts at Santa Cecilia, Provincia Sucumbíos, Ecuador, April 1969. Photo by Blanca Muñoz. Right: William E. Duellman recording a treefrog near Tandayapa, Provincia Pichincha, Ecuador, April 1975. Photo by Linda Trueb. Bottom left: William E. Duellman at south slope of Cordillera del Due, Provincia Sucumbíos, August 1968. Photo by Stephen R. Edwards. Bottom right: William E. Duellman and Curatorial Assistant Thomas H. Fritts, identifying recently accessioned specimens from Ecuador. Photo taken from the KU museum magazine *Niobrara* in 1968.

to axes and chainsaws, a consequence of squatters attempting to cultivate crops in the newly cleared areas.

At that time, plans were underway for the construction of a road stretching from Papallacta, situated high in the Andes, to Lago Agrio, the location of a recently established airfield and an oil facility positioned 14 km east of Santa Cecilia. Collecting was reasonably good at Santa Cecilia, where some of the gaps were filled. However, the most interesting finds were at the site of the new airstrip being constructed at Lago Agrio. Workers had cleared an area where helicopters landed with parts of bulldozers, which were then assembled in the middle of the forest. Two large bulldozers were used to clear the forest by knocking down trees and then shoving them aside. There were no facilities at Lago Agrio; personnel were transported back and forth from Santa Cecilia daily. Arrangements were made for KU herpetologists to go to Lago Agrio as space permitted. Tom Fritts went one day and returned nine hours later with 48 specimens of 33 species; a week later the entire KU field team spent three hours at Lago Agrio and obtained 58 specimens of 30 species, 13 of which had not been in the collection made by Fritts the week before. Bulldozer operators were asked to knock some trees into the clearing and not to move them until the herpetologists had an opportunity to examine them. As large trees were knocked down in the clearing, arboreal lizards and frogs leaped out of the branches onto the ground where they were conspicuous and easily picked up. Bulldozers also were scraping the ground and in so doing exposed caecilians and earthworms as large as the caecilians. What a fantastic diversity of amphibians and reptiles! By the end of the trip, the group had expended 251 person-hours collecting 741 specimens of 96 species at Santa Cecilia. They also expended 65 person-hours collecting 364 specimens of 86 species at Lago Agrio. At Santa Cecilia, 2.6 specimens were collected per person-hour, whereas that ratio was 5.6 at Lago Agrio. Bulldozers certainly made a difference. Duellman conducted two brief visits to Santa Cecilia in October 1971 and April 1972. During both occasions, he collaborated briefly with Martha L. Crump and John E. Simmons. In particular, 2 April 1971 stands out as a memorable day. Equipped with their "picnic suppers", Duellman, Crump, and Simmons departed from Muñozlandia and leisurely traversed a trail for approximately one kilometer to a small lake. Shortly before twilight, they settled onto a relatively dry log to enjoy their meal and drinks when an abrupt downpour soaked them for about half an hour. Illuminating their path with headlights, they cautiously retraced their steps back to Muñozlandia. The heavy rain brought out many frogs, so many that they were overwhelmed. When they strode into Muñozlandia at about 1:00 a.m., their headlights were dim, they were wet and tired, but they had collected 56 species of frogs, 14 more than are native to the entire United States east of the Mississippi River. Certainly a night to remember!

Numerous papers were written by Duellman, John D. Lynch, and others describing new species from Santa Cecilia and surroundings, culminating with the publication, in 1978, of Duellman's opus titled *The Biology of an Equatorial Herpetofauna in Amazonian Ecuador.* In this study, he revealed a record of the greatest diversity of reptiles and amphibians for a single site in the world and provided novel information on the ecology and natural history of numerous species. Most of the species' accounts included a photograph of a living individual, but the most notable illustrations were the four plates of watercolor paintings of frogs, lizards, and snakes from the talented hand of Linda Trueb. He also challenged what was then the rule (mostly based on the avian work by Robert MacArthur) at the time in the 1970s in community ecology. Duellman (1978) emphasized: "Herpetological communities in aseasonal tropical forests are not structured in the same way as bird communities. The models generated from bird communities do not generally apply to herpetological communities". This landmark monograph was well received and soon began to be an obligatory reading for anyone interested in herpetology, natural history, and tropical communities. This unique publication on a tropical herpetofauna was well received; although 1200 copies were produced, in four years it was out of print.

This monograph and his impressive book, *Biology of Amphibians* (Duellman and Trueb, 1986), inspired several Ecuadorian herpetologists, among whom was Coloma, who later undertook graduate studies in systematics and ecology at the University of Kansas, with Duellman as his advisor.

Duellman carried out a total of 11 field trips to Ecuador (1966, 1967, 1968, 1969, 1970, 1971, 1972, 1974, 1975, 1984). They varied in length from two days on 30–31 January 1971 to the longest one from 6 March to 18 May 1975 (see Figure 2.28). This field work provided Duellman with the opportunity to travel throughout much of Ecuador, including the lowlands of the Pacific coast, the cloud forest of the Andes (western and eastern slopes), the highlands, and the Amazonian lowlands, and collect 6981 specimens of amphibians and reptiles. Needless to say, he collected for the first time hundreds of species new to science. For example, in just three days on 20, 21, and 23 October 1971, after walking at the margins of the Azuela River in Provincia Napo in the cloud forest of the Amazonian slopes of the Andes, Duellman (with Simmons, Collins, and MacBryde) found 12 new species of frogs (4 centrolenids, 1 bufonid, 2 dendrobatids, and 5 *Pristimantis*). One of the dendrobatids was described by Charles W. Myers and John Daly in 1976, and they remarked the following:

> In 1973, William E. Duellman called our attention to four specimens of a small poison frog that had been collected in Ecuador, by a University of Kansas field party working under his supervision. These specimens represented a previously unknown species, and we therefore visited the locality ourselves to obtain samples of skin toxins for an ongoing biochemical survey of dendrobatid frogs.
>
> (Myers and Daly, 1976)

The numbers of species of amphibians described by Duellman are unreal! He described or co-described 93 from Ecuador (out of 252 he named from the Neotropics), a monumental task for a scientist!

During his field work, Duellman was accompanied mostly by his graduate students and some colleagues: Henry S. Fitch, William G. Saul, Linda Trueb, Martha L. Crump, Stephen R. Edwards, James W. Waddik, Werner C. A. Bokermann, Idelfonso Muñoz B., Thomas H. Fritts, Charles F. Walker, Sandy Echternacht, Bruce MacBryde, Joseph T. Collins, John E. Simmons, Dana T. Duellman (daughter), Alan Savitzky, Patricia A. Burrowes, and David M. Hillis. As part of his and KU's focus on Ecuador, Duellman mentored additional KU students who participated in his fieldwork in Ecuador until 1990. Among them were John D. Lynch, David C. Cannatella, Richard Montanucci, Charles W. Myers, John J. Wiens, and David A. Kizirian, who did collecting trips by themselves.

The KU graduate students mentioned here currently are widely recognized in the herpetological world, and the contributions of most of them to the knowledge of Ecuadorian herpetology have been immense. Their studies were summarized by Duellman (Duellman, 2015a), and herein we briefly highlight the contributions of Crump, Lynch, Cannatella, Edwards, and Hillis.

Martha L. Crump: From Tying Tags to Field Biology and Conservation

At Santa Cecilia, one of Duellman's graduate students, Martha L. Crump (Figure 2.29), undertook an analysis of the ecology and reproductive modes in a tropical anuran community. Nearly five decades later. Duellman (2015a) commented on his student Marty, "What a significant addition (to the field team) that turned out to be!" Also, in 1974, John D. Lynch described a new species of frog as *Eleutherodactylus martiae* (= *Pristimantis martiae*) and remarked that her quest for ecological knowledge of Amazonian frogs has resulted in an outstanding collection of Amazonian eleutherodactylines. During five decades, Crump would work in Ecuador, Brazil, Costa Rica, Argentina, and Chile and would become an active conservationist. Crump, a behavioral ecologist, is now one the main authorities on amphibian awareness and conservation, having written books such as *In Search of the Golden Frog* (2000) and *Extinction in Our Times: Global Amphibian Declines* (Collins and Crump, 2009), coauthored *Herpetology* (four editions, 1997–2015)—one of the most popular modern herpetology textbooks, edited the book *Lost Frogs and Hot Snakes: Herpetologists' Tales From the Field* (Crump, 2024), and published a number of other books for both adults and children.

Crump began her involvement with tropical rainforests and herpetology as an undergraduate student when she was hired to catalog and tie tags onto specimens in the Herpetology Division at the Museum of Natural History KU. In 1967, when she was a 21-year-old senior biology major, she gave a plaintive query to Duellman, "Can't I go, too?", when she heard about his plans to return to Santa Cecilia. In early June 1968, she joined Duellman, ichthyologist Frank Cross, Linda Trueb, and three graduate students, Stephen R. Edwards, James W. Waddick, and F. William Saul, for a summer expedition to eastern Ecuador to continue the inventories Duellman began in the area in 1966. At Santa Cecilia, they were also accompanied by Werner C. A. Bokermann from Sao Paulo, Brazil, who was an inspiration to everyone, especially Marty, who adapted marvelously to life in the "jungle".

She was intrigued by herps, mainly for their ecology and behavior. Her father, a mining geologist, encouraged her as a child to observe and to ask questions, not simply learn the names of the objects of her interest. After obtaining her bachelor's degree, Crump stayed at KU for graduate work, and her fascination led her to become a field biologist. She majored in zoology and received an MA in 1971 and a PhD in 1974. Her master's thesis was titled "Quantitative analysis of the ecological distribution of a tropical herpetofauna" and her PhD dissertation was "Reproductive strategies in a tropical anuran community". She was a postdoctoral researcher at the City University of New York (1974–1976), a professor and researcher at the University of Florida (1976–1992), and adjunct professor at Northern Arizona University in Flagstaff (1992–present) and Utah State University in Logan (2011–present).

Crump spent two months and 10 days at Santa Cecilia during her first trip in 1968 to continue the inventories. Among the tasks assigned to her was collecting and raising tadpoles so that their identities could be determined, since most of the tadpoles were unknown. This was the beginning of knowledge of Ecuadorian tadpoles—a task still incomplete today. Crump was the pioneer in doing this, but for her, the most exciting aspect of the biology of amphibians in Santa Cecilia was the amazing diversity of reproductive modes. She wrote:

> In the United States we are taught that frogs and toads lay eggs in water. The eggs hatch into tadpoles, which develop in the water until they sprout legs, absorb their tails, and climb out onto land. This scenario is indeed true for almost all species in the United States. . . . At Santa Cecilia less than half the species lay eggs in water. . . . At that time I had been amazed by the diversity of the fauna, from the gaudy color patterns of treefrogs to the bizarre shapes of the Suriname toad and the egg-brooding horned treefrog. Even more fascinating, however, had been the wide variety of ways these frogs reproduce: eggs developing out of water, frogs carrying their tadpoles to water, females carrying eggs stuck to their backs.
>
> (Crump, 2000)

Three years later, in 1971, Crump returned to Santa Cecilia and spent 13 months gathering data for her PhD dissertation and trying to answer the question: How can so many species coexist? She speculated that the 81 species coexist, in part, through the many different ways they reproduce. Two years later, Crump published a paper (Crump, 1974) and concluded that the reproductive diversity at Santa

FIGURE 2.29 Martha L. Crump (top left) and Martha and Abraham Goldewicht (top right) at Muñozlandia, Santa Cecilia, Provincia Sucumbíos, 1971. Photos by John E. Simmons. Bottom: Martha L. Crump rearing tadpoles in laboratory at Muñozlandia, June 1968. Photo by Stephen R. Edwards.

Cecilia enables the coexistence of many species through the partitioning of breeding sites. During 1971–1972, Crump's field assistant was John E. Simmons, at that time an undergraduate student working in the Herpetology Division at KU. Simmons later worked as Collection Manager at KU until 2007 and is currently a renowned expert on museum studies, natural history collections, and management, having written more than 150 popular and technical articles and books on museum studies and natural history topics. Simmons subsequently visited Ecuador to participate in field trips with Ana Almendáriz-Cabezas (in 1986) and to teach a course in museum studies at PUCE in 2005.

Crump's field work in Ecuador continued during June–July 1976, September 1977, June–August 1978, May–July 1993, June 1995, and December 1995. During the summer of 1976, she returned to Santa Cecilia and studied the size and caloric content of treefrog eggs. In 1977, Crump made the first inventories, along with a team of scientists led by Dr. David Pearson, in what was to be designated as the Parque Nacional Yasuní. The park was established in 1979 with an area of over 500,000 hectares. Ten days of survey work revealed 33 species of amphibians and, according to Crump, "a mere drop in the bucket of what is likely to occur in the area". During the summer of 1978, she continued to study treefrog eggs at Lago Agrio, about 12 kilometers east of Santa Cecilia (the Ecuadorian army had kicked Ildefonso Muñoz out of Muñozlandia, and he had relocated to Lago Agrio). Another field trip to Yasuní followed in May 1993, with the auspices of EcoCiencia (an Ecuadorian NGO engaged in research applied to conservation, environmental policies, and education), but on her way to Yasuní, she took six days to visit Santa Cecilia and teach a group of 14 Ecuadorian students about field methods for surveying amphibian and reptiles. Two decades after her last visit, she was shocked by the deforestation around Santa Cecilia, which had become a large Ecuadorian army base. She describes and reflects:

> Unbelievably altered, the road to Santa Cecilia is now lined with thatched houses, parcels of *yuca* and *plátano*, and soccer fields instead of rain forest. Colonization within the upper Amazon Basin has spread like a cancer. . . . Muñozlandia was now a thick tangle of second-growth vegetation. I wondered how many other traces of humanity were buried beneath jungle vegetation in the Amazon basin. Perhaps entire civilizations.
>
> (Crump, 2000)

During three days of inventory work at Santa Cecilia, Crump and her students found 22 species of frogs, a small fraction of what actually occurred 25 years earlier. She continued to Parque Nacional Yasuní for an inventory in an area being mapped for a road to oil fields managed by Maxus Energy. At the site she collected 24 species of anurans, despite the lack of support and obstacles to the field crew made by environmental control officers. She experienced first-hand the power of an oil company that showed no interest toward the impact of the road on the amphibians and reptiles, or, in general, biodiversity.

In June–July 1993 Crump made field trips to the Pacific coast, a montane forest southeast of Quito, and to the southern Andes. During her first trip, she accompanied Gloria Correa Monge (Coloma's student at PUCE) in search of the harlequin frog *Atelopus balios* for behavioral studies. After a three-day search effort, they found four frogs at Manta Real, Provincia Azuay, where the frogs had been abundant a decade earlier. She also went to Sierra Azul, Provincia Napo, a cloud forest site in the Amazonian slopes, and then to Podocarpus National Park in southern Ecuador. At Sierra Azul she documented 11 species of frogs during three days. By the end of this trip, she claimed that everywhere she had been, frogs seemed to be declining and disappearing.

In July and December 1995, Crump finished her journey in Ecuador in association with an EcoCiencia conservation project. She conducted workshops, advised on an inventory and monitoring study, and visited field sites in the buffer zone of Reserva Ecológica Cotacachi-Cayapas. The results of the amphibian inventories were summarized by Velasco (1997a).

Lynch: "Eleuths" and Glassfrogs

Lynch's academic career and accomplishments are mostly described by (Duellman, 2015a, Chapters 11, 30). John D. Lynch (see Figure 2.31) significantly expanded the KU collection through his fieldwork in Ecuador, contributing a total of 8410 specimens collected between 1967 and 1978, making him the fourth most productive amphibian collector (Figure 2.30). His remarkable contributions also encompass the description or co-description of an impressive 116 new amphibian species native to Ecuador. This accomplishment positions him as the foremost scientist on the list of taxonomists who have documented Ecuadorian species (see Figure 2.27). His legacy is reflected through the eponymy of nine amphibian species, with two of them originating from Ecuador: *Atelopus lynchi* (Cannatella, 1981) and *Centrolene lynchi* (Duellman, 1980). Additionally, his influence is enshrined in the genus *Lynchius* (Hedges *et al.*, 2008).

Duellman and Lynch's combined efforts resulted in seven titles, including three of the most important and classic studies, one on glassfrogs (Centrolenidae) (Lynch and Duellman, 1973) and two on rainfrogs *Eleutherodactylus* (most of them currently under *Pristimantis*) (Lynch and Duellman, 1980, 1997), in which they jointly described numerous new species to science.

Lynch was born on 30 July 1942 in Collins, Iowa, USA. He received an MS in 1965 with a thesis titled "A review of the frogs of the *mexicanus* and *rhodopis* groups (Genus *Eleutherodactylus*) and the evolution of the genus in northern Central America". He was mentored by famous herpetologist Hobart M. Smith. In 1965, he entered the University of Kansas and worked under the direction of Duellman. In 1969, he completed his PhD with the thesis, "The evolutionary relationships and osteology of the frog family Leptodactylidae". He taught in the Department of Zoology of the University of Nebraska, Lincoln (1969–1997). He then made a major move as Professor and Curator of Amphibians to the Instituto de Ciencias Naturales at the Universidad Nacional de Colombia in Bogotá, where he retired in 2022.

In June and July of 1967, Lynch spent time in Ecuador. While at Santa Cecilia in the Amazon lowlands, he came across numerous species of "eleuths", some of which he had previously observed preserved in Illinois. Returning to

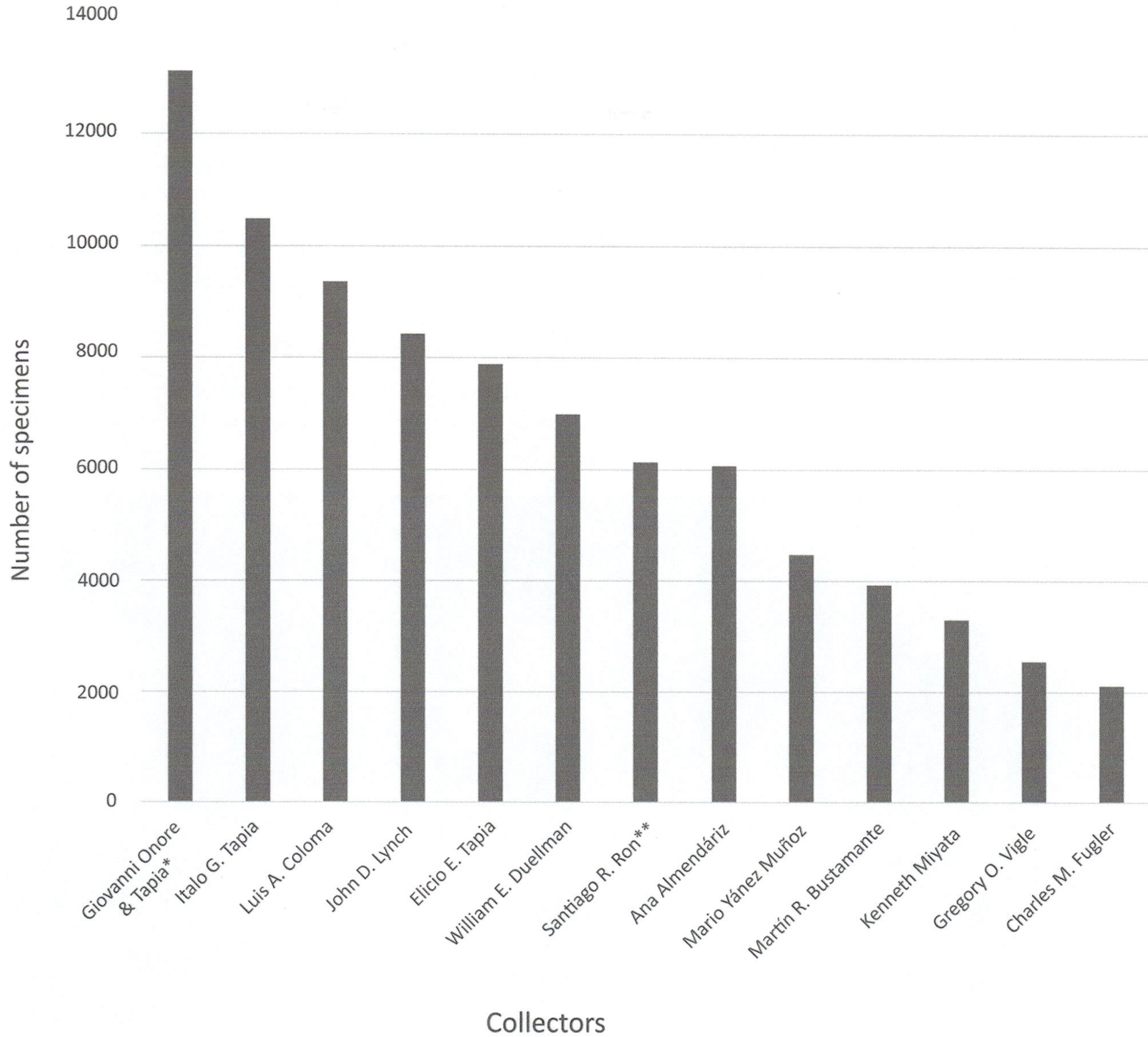

FIGURE 2.30 Twelve primary collectors of Ecuadorian amphibians, regardless of their order in the specimen list of collectors, and the respective numbers of specimens deposited in museums up to December 2022. Data sources from CJ, EPN, KU, MCZ, MECN, MHNG, QCAZ, and UTI databases. *Collections with the Tapia family (see text for details). These specimens are not included in the individual counts of Ítalo and Elicio E. Tapia. **QCAZ data for all collectors of a specimen has not been publicly available since 2011. Therefore, the bar representing Santiago R. Ron's collections may not encompass all his contributions (between 2011 and 2022) unless listed as the first collector.

Ecuador in February 1968, Lynch joined a second KU field expedition led by Duellman. During this trip, John Lynch and Robert Henderson journeyed to the Pacific lowlands of Ecuador, while Duellman and his team explored the Amazonian region. This marked the beginning of Lynch's several trips, lasting until January 1978, when he shifted his focus to Colombia due to the greater variety of "eleuth" species there compared to Ecuador.

During his time in Ecuador, Lynch primarily utilized public transportation for his journeys. Although much of his fieldwork was conducted independently, he was accompanied by Robert Henderson on one occasion and by David C. Cannatella on another.

Following his initial trip to Ecuador, Lynch published descriptions of two new species, *Eleutherodactylus croceoinguinis* and *E. variabilis* (Lynch, 1968). From this point onward until 1980, Lynch authored 31 papers concerning Ecuadorian "eleuths". Among these, he independently described 54 new species, collaborated on one species with Linda Trueb, two with Kenneth Miyata, and 11 with William E. Duellman. These descriptions were supported by substantial series of specimens and featured comprehensive measurements, ratios, and detailed illustrations through drawings and/or photographs.

A significant synthesis of species within the Andes of southern Ecuador (Lynch, 1979) was succeeded by a more comprehensive evaluation of species found on the Amazonian slopes of the Andes (Lynch and Duellman, 1980). Furthermore, Lynch presented an overview of data related to species inhabiting the Andes of northern Ecuador (Lynch, 1981).

During the 1980s, additional descriptions for five species were published, three with Duellman and one with Charles W. Myers. This circumstance left the Pacific slopes and lowlands characterized by numerous specimens simply labeled as "*Eleutherodactylus* sp." While the majority of these specimens were collected by Lynch, Duellman also made significant contributions to the collection. Collaborating over several years, they meticulously reviewed the "eleuths" of western Ecuador, culminating in a comprehensive publication (Lynch and Duellman, 1997). This work unveiled 61 identified species, nine of which were newly designated. Overall, John Lynch described 81 "eleuth" species from Ecuador.

More species continue to be discovered, as indicated by descriptions from Ecuadorian researchers. In the meantime, Lynch has directed his focus toward the taxonomy of "eleuths" in Colombia, where he currently resides. Duellman, on the other hand, shifted his attention to Peruvian "eleuths" and the phylogenetic relationships among these frogs and their relatives (Hedges *et al.*, 2008).

Trueb: A Skilled Systematist

Linda Trueb's academic career has been described in detail by Duellman (2015a, Chapters 8 and 20) in the centennial

FIGURE 2.31 Left: John D. Lynch collecting at Puerto Viejo, Costa Rica, June 1966. Photo by WED. Top right: John D. Lynch and Linda Trueb in the hallway in the Division of Herpetology. Early 1960s. Middle right: John D. Lynch and William E. Duellman working on the manuscript on Ecuadorian *Eleutherodactylus*, September 1995. Photo by Linda Trueb. Bottom right: John D. Lynch at the Universidad de los Andes, Bogotá, Colombia, in June 2014. Photos by John E. Simmons.

history of herpetology in Kansas, in which he remarked on the new dimension that the multitalented Trueb added to the research component of the Division of Herpetology at KU (Figure 2.32). Over the course of her career, Trueb not only maintained an active research agenda but also delivered graduate-level instruction and eventually ascended to administrative roles. Her international recognition in the realm of research, particularly her pioneering exploration of osteological aspects of anurans, including fossilized taxa, is noteworthy. Equally celebrated are her exemplary illustrations, which have consistently accompanied her publications from the outset.

FIGURE 2.32 Linda Trueb at her office in the Division of Herpetology at the University of Kansas and at home. Top left: March 1999. Photo by S. R. Indermauer. Top right: on the first day the Division of Herpetology moved collections into the new fluid wing in May of 1996. Eric R. Wild and Jeffrey Parmelee in the back. Photo by John E. Simmons. Bottom left: in her lab, 1969. Photo by WED. Bottom right: working on illustrations for *Biology of Amphibians* at home, October 1984. Photo by WED.

Trueb's engagement with Ecuadorian amphibian knowledge commenced alongside Duellman with joint fieldwork from 1967 to 1975 (incorporating field trips in 1967, 1968, 1969, 1971, 1974, 1975). Subsequently, a decade of prolific scientific output ensued, culminating in her educational mentorship of Ecuadorian students from 1988 to 2007 at KU.

During the 1970s, Trueb authored a series of comprehensive taxonomic papers featuring Ecuadorian frogs. Noteworthy examples include her seminal examinations of casque-headed tree frogs characterized by co-ossified skulls, spotlighting the detailed cranial morphology of *Osteocephalus jordani* using an Ecuador-sourced specimen (Trueb and Duellman, 1970). In the same year, her collaborative work with Duellman yielded a systematic revision of *O. verruciger* alongside associated natural history insights (Trueb and Duellman, 1970). Sequentially, a taxonomic reassessment of *Osteocephalus* was published (Trueb and Duellman, 1971), followed by a comprehensive phylogenetic analysis of bufonids. Within this, she introduced a novel genus, *Ramphophryne* (now classified under *Rhinella*), inclusive of the Ecuadorian species *Rhinella festae* (Trueb, 1971). In 1974, her monographic endeavor on Neotropical horned frogs of the genus *Hemiphractus* was released (Trueb, 1974), followed by a systematic review of *Nyctimantis rugiceps* in 1976. The zenith of her contribution to Ecuadorian amphibians transpired with a taxonomic revision of Ecuadorian *Telmatobius* in 1979, which featured the unveiling of a new species, *T. cirrhacelis*, which is now possibly extinct.

From 1988 to 2007, Trueb actively participated in collaborative efforts with William E. Duellman to provide support for the Ecuadorian graduate students at KU. For instance, she played a crucial role as an intellectual mentor and the primary advisor for several academic pursuits. She supervised the *Atelopus* PhD thesis of Luis A. Coloma (1991–1997), as well as guiding the research on *Pristimantis* and centrolenids of Juan M. Guayasamin for his MA and PhD theses (2001–2007). From 2007 to 2009, she collaborated with her student Guayasamin in joint publications regarding glassfrogs (Guayasamin and Trueb, 2007; Guayasamin *et al.*, 2008a, 2009). Furthermore, she contributed significantly to the studies on reptiles, overseeing the MA and PhD research endeavors of Omar Torres-Carvajal from 1999 to 2005. It's worth highlighting that her courses on amphibian biology, morphology (including "Truebean osteology"), scientific illustration (an almost obligatory offering for her students), and scientific writing made an indelible impact on her students, becoming an integral component of their studies and subsequent scholarly publications. Three amphibians from Ecuador have been named to honor Trueb, *Pristimantis truebae* (Lynch and Duellman, 1997), *Hyloscirtus lindae* (Duellman and Altig, 1978) and Nymphargus lindae (Guayasamin in Guayasamin et al. 2020).

Cannatella: The Generous Supporter of Ecuadorian Scientists

Duellman introduced David C. Cannatella (Figure 2.33) to the Ecuadorian herpetofauna in the late 1970s and

FIGURE 2.33 Left: David C. Cannatella in the Division of Herpetology at the University of Kansas, 1978. Unknown photographer. Right: David C. Cannatella, Juan Carlos Santos, and Ítalo G. Tapia near Río Negro, Provincia Morona Santiago, Ecuador, on 20 April 2004. LAC.

supervised Cannatella's MS thesis on the systematics of *Phyllomedusa* (currently *Agalychnis*) *buckleyi* group in that country. Since those early years in Cannatella's academic career, he has continued to be a prolific as an author and graduate mentor. As of December 2022, he has published and contributed to the descriptions of 13 species of Ecuadorian frogs. Based at the University of Texas-Austin, Cannatella continues a strong research and academic cooperation programs on Ecuadorian amphibian diversity and especially on dendrobatids and *Engystomops*. Cannatella has engaged in collaborations with US, Ecuadorian, and other international partners, which have extended Duellman's legacy.

Cannatella's input to science in Ecuador has been remarkable and especially important to increase the profile of herpetological research of Ecuadorian amphibians. Cannatella has helped several native Ecuadorians as a graduate mentor and guided them to publish as authors or coauthors in high-impact journals, consequently raising the level of research done in Ecuador (Table 2.4). Beginning with the publication of his master's thesis (Cannatella, 1980) and up through 2022, he has published 29 research articles of Ecuadorian frogs dealing with evolution, systematics, phylogenetics, behavior, and chemical ecology. Most of these publications were published in top science journals of high impact such as *Science*, *PLoS*, *PNAS*, *Systematic Biology*, and the *Proceedings of the Royal Society*, among others. Ecuadorians have participated in 17 publications with Cannatella: Juan Carlos Santos (10), Luis A. Coloma (7), Santiago R. Ron (5), Mónica A. Guerra (2), Pablo A. Menéndez-Guerrero (1), and H. Mauricio Ortega-Andrade (1). Some of these papers are among the most cited research papers on Ecuadorian amphibians. For example, the paper on dendrobatid biogeography published in 2009 jointly with one of his brightest and most hard-working students, Juan Carlos Santos, and other authors (Santos *et al.*, 2009) reached 368 citations through 27 January 2023 and is the fifth most cited paper in the history of amphibian research of Ecuador. Another paper, on aposematism evolution in poison frogs published in 2003 (Santos *et al.*, 2003), is the sixth most cited, with 317 citations.

Cannatella was born in on 25 December 1954 in Opelousas, Louisiana, USA. At KU, he obtained an MA in 1979 with the thesis "A review of the *Phyllomedusa buckleyi* group (Anura: Hylidae)" under the supervision of William E. Duellman. He was then mentored during his PhD by Linda Trueb, with a PhD dissertation on the phylogeny of primitive frogs (Archaeobatrachians). He was a postdoctoral fellow at the University of California at Berkeley (1986–1988) under the mentorship of David and Marvalee Wake. His first academic position was as a professor and curator at Louisiana State University (LSU; 1988–1990), and then he accepted a position as a professor at the University of Texas at Austin and as Curator of the Texas Memorial Museum (1990–present). He has been honored with one Ecuadorian species, *Osteocephalus cannatellai*, from Amazonia (Ron *et al.*, 2012).

Cannatella's connection with Ecuadorian herpetology can be dated to the time when numerical taxonomy was replaced by phylogenetic systematics. He and other fellow KU colleagues are considered founders of the new systematics, and Cannatella's contribution as editor in the 1990s of *Systematic Biology* is considered fundamental to molecular systematics. Cannatella visited Ecuador for the first time in January 1977. During this formative trip, he accompanied William E. Duellman and went to the localities of Chiriboga in Provincia Pichincha (with Eugenia M. del Pino) and Santa Cecilia in Provincia Sucumbíos. Then, in the same year, he returned to Ecuador with John D. Lynch in June–July traveled across the Andes, and collected amphibians in Cotopaxi, Imbabura, and Carchi provinces. Cannatella made another trip to Ecuador in April 1979, this time collecting near Limoncocha in Provincia Francisco de Orellana. In late 1990s, as a faculty member of University of Texas, he engaged in the study of *Engystomops* (jointly with Michael Ryan) and returned to Ecuador to do research on the vocalization and laryngeal morphology of *Engystomops petersi* conducted in Amazonia during 1998 and 2001. He was a co-mentor of Kathryn E. Boul, then a graduate student at the University of Texas (Boul *et al.*, 2007; Boul and Ryan, 2003, 2004) who was assisted in the field by the Ecuadorian herpetologists Juan M. Guayasamin, Fernando Ayala Varela, and Diego Lombeida.

Since the beginnings of his research program in Ecuador, Cannatella placed a fervent emphasis on promoting Ecuadorians for field research and graduate studies abroad, as well as the development of science in Ecuador. Thus, he had the strong support of Coloma (who was in charge of the amphibians at QCAZ) and Laura Arcos Terán (who was the Director of the Escuela de Biología and the Dean of Facultad de Ciencias Exactas y Naturales of PUCE, since its early beginnings) through an institutional agreement. Cannatella visited Ecuador in February 2002 and did field work jointly with Coloma, Santiago R. Ron, and Alisha Holloway in central and southwestern Ecuador with a focus on collecting specimens of *Engystomops* that resulted in finding three new species, *E. montubio*, *E. randi*, and *E. guayaco* (Ron *et al.*, 2004, 2005). Earlier, in January–March 2003, Cannatella traveled with Coloma and Ítalo G. Tapia to Provincia Esmeraldas looking for the northern limits of *Engystomops* in western Ecuador and made general amphibian collections. In February–March 2003, he went to Yasuní National Park and assisted the studies being done by his graduate students, Kathryn E. Boul, Catherine R. Darst, and the Ecuadorian Santiago R. Ron. They were accompanied by Ítalo G. Tapia as a field assistant. Darst did molecular studies of hyloid frogs (Darst and Cannatella, 2004) and of the diet, chemical defenses, and warning signals of poison dart frogs (Darst *et al.*, 2005, 2006). In 2002, following a master's degree from KU, Ron enrolled at the University of Texas for a PhD with Cannatella's mentorship and did studies on the systematics, evolution of calls, and mating preferences of frogs of the genus *Engystomops* (Ron *et al.*, 2004, 2005, 2006) and graduated in 2007. Cannatella returned to Ecuador in

TABLE 2.4

The 22 Most Cited Publications of Ecuadorian Amphibians. R = research, C = commentary, A = publication includes mainly Ecuadorian amphibians, B = publication (with Ecuadorian authors) includes mainly amphibians from abroad, * = Cannatella's research program in Ecuador. Source: Google Scholar and Google Académico (up to 21 August 2023)

Number of Citations	Kind	Citation/Other Ecuadorian Authors	Title	Journal
2390	R/B	Pounds *et al.*, (2006)/Bustamante, Coloma, Ron	Widespread amphibian extinctions from epidemic disease driven by global warming	*Nature*
1702	R/B	Hoffmann *et al.*, (2010)/Ron	The impact of conservation on the status of the world's vertebrates	*Science*
1274	R/A	Duellman (1978)	The biology of an equatorial herpetofauna in Amazonian Ecuador	University of Kansas Museum of Natural History, miscellaneous publications
905	R/B	Scheele *et al.*, (2019)/ Guayasamin	Amphibian fungal panzootic causes catastrophic and ongoing loss of biodiversity	*Science*
804	R/A	Graham *et al.*, (2004)/Ron, Santos	*Integrating phylogenetics and environmental niche models to explore speciation mechanisms in dendrobatid frogs	*Evolution*
649	C/B	Young *et al.*, (2001)/Coloma, Ron	Population declines and priorities for amphibian conservation in Latin America	*Conservation Biology*
627	R/A	Lynch and Duellman (1997)	Frogs of the genus *Eleutherodactylus* (Leptodactylidae) in western Ecuador: systematic, ecology, and biogeography	Natural History Museum, University of Kansas
533	R/B	La Marca *et al.*, (2005)/Toral	Catastrophic population declines and extinctions in Neotropical harlequin frogs (Bufonidae: *Atelopus*)	*Biotropica*
397	R/A	Funk *et al.*, (2011) /Caminer, Ron	*High levels of cryptic species diversity uncovered in Amazonian frogs	*Proceedings of the Royal Society*
387	R/A	Santos *et al.*, (2009)/Coloma	*Amazonian amphibian diversity is primarily derived from late Miocene Andean lineages	*PLoS*
349	C/B	Mendelson III *et al.*, (2006)/ Coloma	Confronting amphibian declines and extinctions	*Science*
333	R/B	Ron (2005)	Predicting the distribution of the amphibian pathogen *Batrachochytrium dendrobatidis* in the New World	*Biotropica*
333	R/A	Santos *et al.*, (2003)/Coloma	*Multiple, recurring origins of aposematism and diet specialization in poison frogs	*PNAS*
278	R/A	Boul *et al.*, (2007)	*Sexual selection drives speciation in an Amazonian frog	*Proceedings of the Royal Society*
275	R/B	Guayasamin *et al.*, (2009)	Phylogenetic systematics of glassfrogs (Amphibia: Centrolenidae) and their sister taxon *Allophryne ruthveni*	*Zootaxa*
251	R/A	Ron *et al.*, (2003)/Coloma, Bustamante	Population decline of the Jambato toad *Atelopus ignescens* (Anura: Bufonidae) in the Andes of Ecuador	*Journal of Herpetology*
209	R/B	Ron (2000)	Biogeographic area relationships of lowland Neotropical rainforest based on raw distributions of vertebrate groups	*Biological Journal of the Linnean Society*
197	R/A	Coloma (1995)	Ecuadorian frogs of the genus *Colostethus* (Anura: Dendrobatidae)	Miscellaneous publications, Museum of Natural History, University of Kansas
195	R/A	Pomerantz *et al.*, (2018)/Peñafiel, Arteaga, L. Bustamante, Coloma, Salazar-Valenzuela	Real-time DNA barcoding in a rainforest using nanopore sequencing: opportunities for rapid biodiversity assessments and local capacity building	*GigaScience*
187	R/A	Darst *et al.*, (2005)/Menéndez-Guerrero, Coloma	*Evolution of dietary specialization and chemical defense in poison frogs (Dendrobatidae): A comparative analysis	*The American Naturalist*
163	R/B	Santos and Cannatella (2011)	*Phenotypic integration emerges from aposematism and scale in poison frogs	*PNAS*
145	R/A	Bustamante *et al.*, (2005a)/ Coloma, Ron	Cambios en la diversidad en siete comunidades de anuros en los Andes de Ecuador	*Biotropica*

2004 to work with his Ecuadorian students and colleagues. During January–February, he participated in a field trip to Yasuní (with Ron, Mónica A. Guerra, and Samael Padilla) and in April to the western lowlands (with Juan Carlos Santos and Martín R. Bustamante) and the Amazonian region (with Santos, Coloma, and Tapia). Throughout his trips he, Coloma, and Ecuadorian students collected 1021 specimens, with associated tissues and recordings. As a result of this cooperation, two additional Ecuadorian students went to the University of Texas in Austin for graduate studies. They were Juan Carlos Santos and Mónica A. Guerra. Santos had worked at PUCE for his *licenciatura* degree (mentored by Coloma) on the molecular phylogeny of two groups of Dendrobatidae (*Epipedobates* and the *Hyloxalus bocagei* complex), so Cannatella foresaw Santos' collections and data as the perfect opportunity to expand to the maximum degree possible the molecular analyses of Dendrobatidae and answer phylogeny-based research questions about taxonomy, toxicity, aposematism, physiology, etc. Mónica A. Guerra studied frogs of the genus *Engystomops* for several years. She completed her *licenciatura* at PUCE in 2005, mentored by Santiago R. Ron and Luis A. Coloma. In 2014, she obtained a PhD in Biology with a major in ecology, evolution, and behavior from the University of Texas at Austin (USA), with a thesis titled "Polymorphic mating signals and female choice in an Amazonian frog". She was mentored by Michael J. Ryan and David C. Cannatella. Cannatella's team in Ecuador also had input from the North American Chris W. Funk, who did research in Ecuador in the 2000s after working (in 1996–1997) as a tourist field guide at la Selva Ecolodge. Since 2000, Funk conducted amphibian monitoring projects in the Andean region, in Cashca Totoras, Provincia Bolívar, and Yanayacu in Provincia Napo (Funk *et al.*, 2003; Guayasamin and Funk, 2009). In 2004, Funk obtained a PhD in organismal biology and ecology at the University of Montana, immediately joined Myke Ryan and Cannatella's team as a postdoctoral scholar (2004–2006) at the University of Texas, and continued investigations of *Engystomops* in Ecuador until 2012 (Funk *et al.*, 2007; Boul *et al.*, 2007).

Through Cannatella's influence and support, two other young woman amphibian neuroscientists from the University of Texas at Austin (Kim Hoke and Lauren A. O'Connell) began amphibian research in Ecuador during the 2010s. They established strong ties with Coloma at Centro Jambatu, the newly developed (in 2010) amphibian institution for amphibian research and conservation. Rebecca Tarvin (University of California Berkeley, UCB) was a PhD student of Cannatella. She had previously worked as a research associate at PUCE and worked on *Atelopus*. During her PhD studies, she worked on the diversity of *Epipedobates* and the evolution of toxin auto-resistance in poison frogs. Much of her field work was done in Ecuador, and Cannatella decisively advocated that she continue her studies in Ecuador. Tarvin has continued her research in Ecuador as an assistant professor at the University of California Berkeley and currently is mentoring the graduate studies of the Ecuadorian María José Navarrate.

Edwards, Hillis, and Their Pioneering Taxonomic Work

Another one of Duellman's students who did fieldwork in Ecuador was Stephen R. Edwards (Figure 2.34), who undertook a phenetic taxonomic revision of the Neotropical

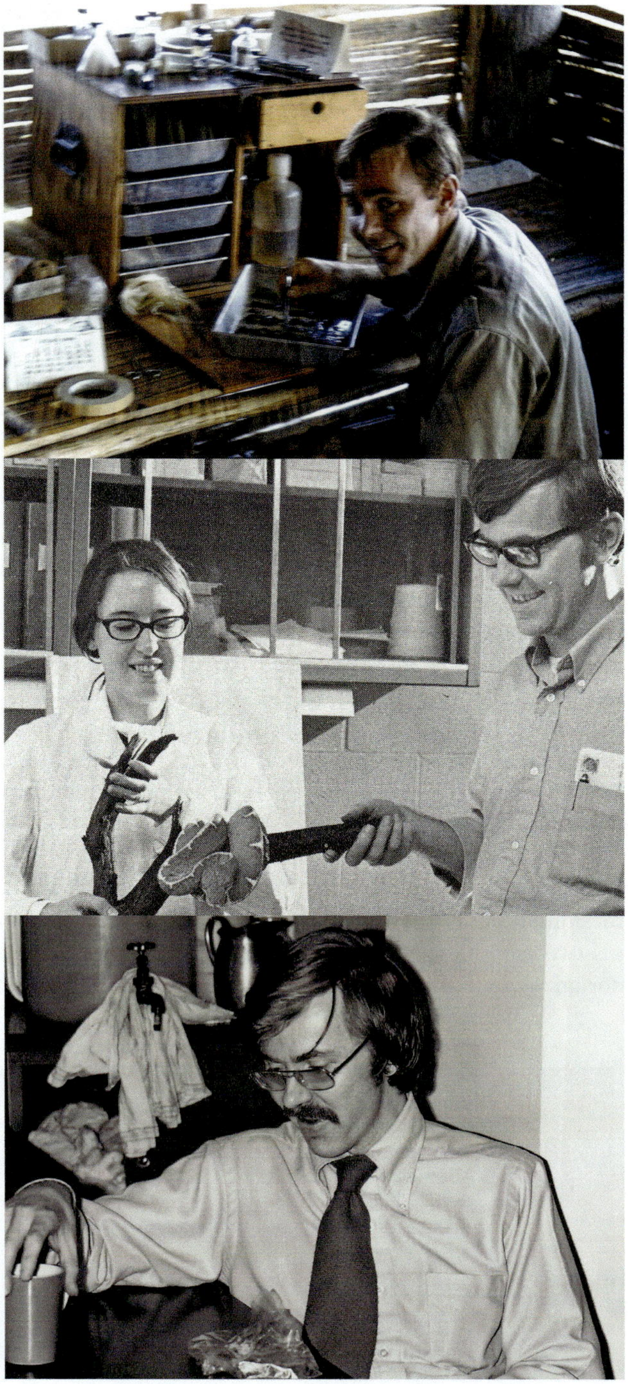

FIGURE 2.34 Top: Stephen R. Edwards in the laboratory at Muñozlandia, June 1968. Unknown photographer. Middle: Martha L. Crump and Stephen R. Edwards in the herpetology laboratory at The University of Kansas in 1968. Photo taken from KU museum magazine *Niobrara*. Bottom: Stephen R. Edwards in the herpetology laboratory at the University of Kansas in February 1976. Photo by John E. Simmons.

genus *Colostethus* (currently under *Allobates*, *Aromobates*, *Colostethus*, *Ectopoglossus*, *Hyloxalus*, *Epipedobates*, *Leucostethus*, *Mannophryne*, *Paruwrobates*, *Rheobates*, and *Silverstoneia*), in which he recognized 19 species as new to science, some of them from Ecuador. In 1968, Edwards spent two months in Ecuador (15 June–16 August) collecting jointly with Duellman and KU students in the Amazonian region and the Andes at Santa Cecilia, Cordillera del Due, and Laguna de Limpiopungo. He began to establish a working foundation of specific taxonomy in the genus *Colostethus*, one of three genera in the family Dendrobatidae at that time, and published two papers on Ecuadorian frogs (Edwards, 1971, 1974a). At the same time (1973–1986), he became involved with the Association of Systematics Collections (1973–1986) and followed an adjacent but different path after earning his PhD in 1974. He was first an administrator at KU, then executive director of an NGO, followed by 25+ years in different roles and leading positions in conservation with IUCN until his retirement in 2007. Unfortunately, most of his dissertation research (Edwards, 1974b) on "*Colostethus*" remained unpublished before Duellman, Lynch, Enrique La Marca, Juan A. Rivero, and Coloma began to undertake a renewed taxonomic revision of *Colostethus*, based on Edwards' extensive work dealing with 45 nominal species at that time. In 1988, Duellman suggested that Coloma review the Ecuadorian *Colostethus* for his master's thesis and enthusiastically supported additional field trips. In the same year, Duellman described (with Lynch) an Ecuadorian species from Cordillera de Kutukú (Duellman and Lynch, 1988). In 2005, when Coloma published his taxonomic revision of *Colostethus* (Coloma, 1995), he acknowledged that Edwards had allowed him to use his PhD dissertation as a reference for the Ecuadorian *Colostethus* deposited at the Natural History Museum of The University of Kansas.

The conclusion of Duellman's journeys to Ecuador and the initiation of a new chapter in Peru was highlighted by his two-month expedition for marsupial frogs in the Ecuadorian Andes during 1984. Accompanying him were David M. Hillis, Patricia A. Burrowes, and John E. Simmons (Figure 2.35). This trip marked the beginning of a new era in systematics and biobanking, in which molecular data began to play a pivotal role. Notably, Hillis was the first graduate student in the herpetology division of the University of Kansas to collect tissues and undertake the complexities of molecular systematics under the direction of Duellman and John Frost. Hillis spent a memorable time with Duellman, collecting tissues, specimens, and studying frogs in the Ecuadorian Andes. After Duellman passed away (in 2022), Hillis would reflect:

> He lived and breathed herpetology. In my mind, I see him with a headlight on, catching frogs.

Duellman and Hillis would publish two seminal systematic papers, one on frogs of the *Hyloscirtus larinopygion* group (Duellman and Hillis, 1990) and the other describing three new species of marsupial frogs, resolving taxonomic problems and their phylogenetic relationships (Duellman and Hillis, 1987). Hillis applied allozyme electrophoresis in solving the systematic problems with these taxa. In addition, Hillis collaborated with Rafael de Sá, a student hailing from Uruguay and studying first at KU and then at the University of Texas under the guidance of Linda Trueb and Hillis, respectively. Together, Hillis and de Sá crafted a comprehensive phylogeny by integrating morphological and biochemical characters of the *Lithobates palmipes* group. This effort resulted in not only a clearer understanding of the group's evolutionary relationships but also the identification of a novel species originating from the Pacific lowlands of southern Ecuador (Hillis and de Sá, 1988).

When Duellman began his journey of discovery of *Gastrotheca* in the Andes in 1967, only two species of biphasic marsupial frogs (*G. riobambae* and *G. lojana*) were recognized from the Ecuadorian Andes, and four monophasic species (*G. guentheri*, *G. plumbea*, *G. testudinea*, and G. *weinlandii*) were known from Ecuador. Currently, 20 species are known, and the task took him about half of a

FIGURE 2.35 Right: David M. Hillis preparing specimens and extracting tissues in Puerto Misahuallí, Provincia Napo, on 30 March 1984. Left: from left to right: David. M. Hillis, Patricia A. Burrowes, and William E. Duellman collecting frogs in the Laguna Cuicocha, Provincia Imbabura, on 28 February 1984. Photos by John E. Simmons.

century. His work on marsupial frogs culminated with the publication of a book (Duellman, 2015b) that was awarded an Honorable Mention in the category of Single Volume Reference Books in Science by the Association of American Publishers in 2016. This was followed by coauthoring a paper on Ecuadorian *Gastrotheca* (Carvajal-Endara *et al.*, 2019), in which they solved pending taxonomic problems of the genus in southern Ecuador. Simmons (2022) recounts the following about Duellman's passion for marsupial frogs:

> Some of the most interesting experiences I had with Bill in the field involved marsupial frogs, from hearing *Gastrotheca* calling at night as snow was falling in a marsh at more than 3300 meters in Ecuador, to the thrill of seeing *Stefania* in the Guiana Highlands of Venezuela, to watching a female *Gastrotheca* use her hind feet to release newly metamorphosed larvae from the pouch on her back. For his book, Bill—ever the orderly maker of lists—reported that he had examined 4,661 preserved adult specimens, 191 lots of tadpoles, 264 young frogs, 95 dried skeletons, and 66 cleared and stained preparations from 78 collections on three continents.

Duellman, Trueb, and Their Ecuadorian Graduate Students

At the end of his *Gastrotheca* and Andean Ecuadorian trip, Duellman presented a somewhat emotional lecture at an event that filled an auditorium in Quito at Pontificia Universidad Católica del Ecuador (PUCE) on 3 April 1984. In his lecture, titled "Los Batracios Andinos: 20 años de estudios y los prospectos para el futuro", he referred to the unparalleled wealth of frogs of Ecuador. He went further and to complain about the difficulties and bureaucratic frustrations of obtaining collecting permits for research—paradoxically, a similar lament still is shared by Ecuadorian scientists. As an epilogue to his talk, he said the following with marked emphasis:

> Interesting and necessary research could be done here in the parks of Quito. And I have a question. Where are the Ecuadorian, Peruvian, and Bolivian herpetologists? The richest region in the world herpetologically, and there are no herpetologists here.
>
> [Translated from Spanish (original audio at: http://www.anfibiosecuador.ec/recursos/audio/DuellmanTalk-PUCE_3April1984.wav. Courtesy of Guillermo Paz-y-Miño)]

In the Ecuadorian nationalistic mood of that time, Duellman's challenge was not necessarily well received, especially by the local bureaucracy and some of the biologists. However, on that day, he did not know that in the auditorium was the seed of the first Ecuadorian herpetologists: Ana Almendáriz-Cabezas and Luis A. Coloma. The former had been in charge of the Escuela Politécnica Nacional herpetology collections since 1983, and the latter had been making collections for Ecuadorian museums since 1982.

Duellman could not have anticipated that, four years after his talk, Coloma would become the first Ecuadorian graduate student at the University of Kansas. Despite Coloma's initial setback of not achieving a high score on the Fulbright English test (a graduate school requirement), Duellman and Trueb (Figure 2.36) remained deeply committed to supporting an Ecuadorian student. They decided to admit Coloma to the graduate school program with the condition that he complete additional English courses at KU. In fact, Coloma would be the first Ecuadorian to obtain a master's and PhD abroad studying systematics and amphibians. Coloma arrived in KU in January 1988 and completed his MA in 1991 and PhD in 1997, after conducting systematic studies on *Colostethus* and *Atelopus*. He was followed by other Ecuadorians: Santiago R. Ron (1996–1998, MA), Omar Torres-Carvajal (1999–2005, MA and PhD), and Juan M. Guayasamin (2001–2007, MA and PhD).

When Coloma arrived in Lawrence, Kansas, in January 1988, he stayed at Duellman and Trueb's house for a few days, where he was warmly welcomed with open arms. He recounts this experience as follows:

> After visiting the amazing herp division facilities and museum and being presented to the herpetology people there I went to his house. As darkness fell, after dinner I walked downstairs to his and Linda's large office. Everywhere I looked I saw hundreds of reprints and books of all sizes, shapes and colors—the big opus of *Biology of Amphibians*, the two volumes of *Hylid frogs of Middle America*, the huge volume of the *Amphibian Species of the World* (promoted by him and edited by his student Darrel R. Frost). There were also large original paintings and illustrations hanging on the wall such as the colorful plates of the frogs and reptiles from Santa Cecilia and the skull drawings of casque-headed frogs (done by Linda Trueb). And then, there were brand new computers, and all ordered to perfection. Had I died and gone to heaven? No, I was in Duellman's and Trueb's empire of knowledge.
>
> (Coloma and Guayasamin, 2022)

Two and a half decades after his last visit to Ecuador, Duellman had these generous words when talking about Ecuadorians:

> The people of Ecuador are fortunate to have such talented scientists who can convey their knowledge so concisely. The people of Ecuador are also fortunate to live in a land of such great frog diversity, but they also are obligated to maintain that diversity for future generations of both frogs and humans.
>
> (Duellman, 2009)

In 1997, Duellman's planned completion of work with Ecuadorian frogs was clearly stated in the influential monograph on the *Eleutherodactylus* (now most are in the genus *Pristimantis*) in western Ecuador:

> Little did we image [sic] three decades ago that the 20th Century would be waning before our endeavors on the *Eleutherodactylus* in western Ecuador would come to

FIGURE 2.36 Left: William E. Duellman at the II Congress of Latin American Herpetology. Universidad de los Andes, Mérida, Venezuela. 1990. Unknown photographer. Right: Linda Trueb with frog skeletons, December 1992. Photo by D. Koch.

> fruition. In the intervening years we have learned much about these frogs and have become increasingly aware that there is so much yet to be known. We leave this challenge to our successors, to whom we bid: *¡Que les vaya bien!*
>
> (Lynch and Duellman, 1997)

Following his paper with Hillis, in 2015, Duellman published his major book about the marsupial frogs on which the Ecuadorian species were included (Duellman, 2015b). He also contributed, along with Ecuadorian researchers, studies on tree frogs (Duellman and Coloma, 1993; Ron *et al.*, 2018a), endangered harlequin frogs (Ron *et al.*, 2003; Coloma *et al.*, 2007, 2010b; Guayasamin *et al.*, 2010), and the description of additional Andean marsupial frogs (Carvajal-Endara *et al.*, 2019).

Duellman retired in 1997 to become probably the most prolific Curator Emeritus in the field of herpetology. Some of his most important publications—a revised edition of *Hylid Frogs of Middle America* (2001); *Cusco Amazónico: the Lives of Amphibians and Reptiles in an Amazonian Rainforest* (2005); *Terrestrial Breeding Frogs (Strabomantidae) in Peru* (2009); *Marsupial Frogs: Gastrotheca & Allied Genera* (2015); *Herpetology at Kansas, A Centennial History* (2015a); and the *Encyclopedia of Ecuadorian Amphibians* (Coloma and Duellman, this publication)—were part of his "retirement" hobbies.

Duellman's influence on Ecuadorian herpetologists also extends to Omar Torres-Carvajal and Santiago R. Ron, both currently working at PUCE on reptiles and amphibians, respectively. Torres-Carvajal is the leading researcher on Ecuadorian Squamata, having made numerous contributions that follow the KU tradition (e.g., Torres-Carvajal, 2000, 2003). Ron has also played a role in amphibian studies, with a focus on the systematics of frogs, especially of the genera *Engystomops* (e.g., Ron *et al.*, 2006) and *Osteocephalus* (e.g., Ron and Pramuk, 1999).

Guayasamin relates the following about Duellman:

> Duellman always was in his office next to the entrance of the Herpetology Division. His door always open, his mind always working. Every printed publication was perfectly organized, as were his field notes, slides, and recordings. Although the idea of "integrative taxonomy" may have a catchy new air, it was clear that for Duellman that was just day-to-day taxonomy. The collections that Duellman and his students obtained in the 1970s were so complete that they even included tissues for allozyme studies. These very tissues were used 40 years later to generate the first

> molecular phylogeny of Ecuadorian harlequin frogs, coauthored by Duellman (Guayasamin *et al.*, 2010). Also, Duellman's work on direct-developing frogs (Lynch and Duellman, 1997) and glassfrogs (Lynch and Duellman, 1973) inspired Guayasamin to pursue these taxa for his MA and PhD theses, transitioning from the classic Truebean osteology (Guayasamin and Bonaccorso, 2004; Guayasamin and Trueb, 2007) to the novelty of molecular systematics in centrolenid frogs (e.g., Guayasamin *et al.*, 2008a, 2009).
>
> (Coloma and Guayasamin, 2022)

Many productive Ecuadorian herpetologists have been inspired by Duellman's work as well. For example, Ana Almendáriz-Cabezas, Martín. R. Bustamante, Alejandro Arteaga, Sofía Carvajal-Endara, Diego F. Cisneros-Heredia, H. Mauricio Ortega-Andrade, Mónica Páez-Vacas, Carolina Reyes-Puig, Juan Carlos Santos, Verónica L. Urgilés, and Mario H. Yánez-Muñoz all visited him at KU or exchanged correspondence for many years.

In an obituary written by Joe Mendelson, one of his most prominent KU graduate students, and Duellman's friend, he states the following:

> However, the first international student arrived in 1962 from Myanmar (then known as Burma), followed later by many students from across Latin America, and a few from other continents as well. Especially well represented was Ecuador and, if you follow both the contributions of Ecuadorian students educated at KU and their academic offspring, it is clear that the top-status of Ecuador globally as a powerhouse of herpetology all leads directly back through the network to Bill.
>
> (Mendelson III, 2022)

Measuring the impact of Duellman on herpetology proves challenging. As Mendelson III (2022) affirms, he held a central position in herpetology during the second half of the 20th century and the first 22 years of the 21st century. His indelible influence in herpetology doubtlessly will extend for centuries. Mendelson submitted that Bill's most significant contributions to herpetology were, unordered, his students, his fieldwork and specimens, and his publications—especially his hundreds of taxonomic efforts and his highly influential regional treatments (e.g., Santa Cecilia; Duellman, 1978), monographs (e.g., Duellman, 1970), and the magnificent textbook on amphibians (Duellman and Trueb, 1986).

As recounted by Simmons (2022) Duellman led an extraordinary life:

> He traveled widely and collected specimens of almost every group of reptiles and amphibians. . . . Bill described 252 currently recognized species, published 386 papers and books, and served as major professor for 34 PhD students. During his time at the University of Kansas Natural History Museum, he built up the collection from 59,000 to more than 300,000 specimens in just 35 years (Ford and Simmons, 1997) through extensive field work by himself and his students and by acquiring other collections, making it the fourth largest herpetological collection in the United States, and with by far the most significant collection of the herpetofauna of Latin America.

Taking a step further, Burrowes *et al.* (2022) asserted that the impact of Bill Duellman's work transcends herpetology with significant contributions to our understanding of tropical community ecology, evolutionary biology, and biogeography. Yet undeniable is Duellman's contribution to Ecuador's recognition as one of Earth's most amphibian species-diverse countries. Ecuadorians had in Duellman and Linda Trueb true mentors, friends, and colleagues. The University of Kansas Natural History Museum was a second home where they listened to endless frog stories, worked in one of the largest collections of Neotropical amphibians, and learned how to study these beautiful animals in the heaven of amphibian academia, built upon the foundation laid by the herpetology titans such as Duellman and Trueb (Coloma and Guayasamin, 2022).

THE 1980s: THE NATIONAL RESEARCH TAKES OFF

The 1980s marked the beginning of amphibian research centered in Ecuador, mostly related to the growth of three museums and the associated academic environment in Quito—EPN, QCAZ, and MECN (currently INABIO). The rise of these institutions was the result of an increasing interest in the natural sciences, which had been consolidating during the 1970s. In addition to the pioneering events at the EPN, two additional events occurred: the opening at PUCE, in 1976, of a new career path in biological sciences that offered a degree of *licenciatura* focused on research (instead of education with a specialization in biology, the career path previously offered) (Arcos Terán, 1998), and the creation of the MECN in 1978 (Moreno Espinosa, 1979). The first biologists graduated from PUCE beginning in December 1980, and the MECN began to assemble the first amphibian collections in 1984.

From the 1980s to the present, Ecuadorian scientists have made significant contributions through the discovery and description of new species and through the generation of information on diverse fields of amphibian biology and conservation. It is worth mentioning that the number of scientific papers is steadily increasing. For example, in 2022, there were no fewer than 35 scientific publications on Ecuadorian amphibians in journals, and most of them included Ecuadorian authors or coauthors.

AMPHIBIANS AT THE QCAZ

After 40 years of amphibian activities, the QCAZ is one of the most prestigious and important centers in Latin America due to its museum, laboratories, high-level academic staff, research, scientific production, and outreach publications. The history of the amphibian division within the Vertebrate Section of QCAZ, from its antecedents in the 1970s and its beginnings in the 1980s up to 2010, were briefly recounted by Coloma (2010). Additional selected information is encountered mainly in Coloma (2008), Coloma *et al.* (2010a), and Lötters (2011). From 2011 to 2022, information is available in Womack *et al.* (2022) and several outreach sources (e.g., Ponce, 2019; https://bioweb.puce.edu.ec/QCAZ/anfibios). So far, the QCAZ holds the largest collection of Ecuadorian amphibians, and it contains the largest number of specimens and species (76,887 holdings as of 30 July 2023). At the same time, the QCAZ built an amphibian genome bank (40,548 tissue samples as of 30 July 2023), in which tissues of most species of amphibians from Ecuador are stored.

Several elements converged to contribute to the QCAZ's prosperity and current state, beginning in 1981 with the dynamism of Giovanni Onore, through two decades of fruitful activities led by Coloma and the subsequent work by Ron. After the 1980s, a new generation of biologists specialized in systematics, evolution, and related sciences returned to Ecuador and were incorporated by PUCE in the 1990s and 2000s. In addition, especially during the 2000s, some investments in science, either through private donations of income tax or funding provided by the government of Rafael Correa Delgado, promoted its growth.

In 2010, a major disruption occurred at the QCAZ due to conceptual disagreements between Coloma and Guayasamin vs. Ron (the latter supported by the Dean, Laura Arcos Terán), entwined with academic and political issues that gave rise to power conflicts and an unfavorable negative academic environment. Thus, Coloma, Guayasamin, and about a dozen of researchers decided to resign, parting ways with PUCE. In the aftermath of their departure, Ron assumed leadership of the amphibian section at the QCAZ, a position he continues to hold.

Onore: An Entomologist at the Roots of Amphibian Research and Conservation

Beginning in 1969, a few vertebrate collections began to accumulate at the Departamento de Ciencias Biológicas (PUCE), which was part of the Instituto de Ciencias of PUCE and currently is in the Facultad de Ciencias Exactas y Naturales. Fernando Ortiz Crespo (1942–2001), a well-known Ecuadorian ornithologist (Freile 2005, and citations therein), was sporadically in charge of the collections up to 1981, and his interest was focused on birds. During this period (1969–1981), the vertebrate specimens were used mainly for didactic purposes, but a few (particularly birds) were used for research.

The first amphibian research collections at PUCE were specimens deposited by Kenneth Miyata (22), and KU researchers (96) (e.g., Duellman, Crump, Lynch, Cannatella). Most of these specimens were given to Eugenia M. del Pino, who also preserved vouchers from the embryological studies she conducted. Later, in 2000, del Pino transferred all of the specimens (about 400) that were under her care to the QCAZ amphibian collections.

In the early 1980s, the arrival in Ecuador of the Italian missionary Giovanni Onore Risso marked the start of a new center of amphibian research at PUCE. Although Onore was mainly an entomologist, his activities covered a broader zoological scope. Onore worked at PUCE for 25 years (1981– 2006) as a professor and researcher. He was in charge of the invertebrate and vertebrate sections of the QCAZ (1982–1991), and from 1991 until 2006, he focused on invertebrates. Subsequently, Onore has been dedicated to conservation and supporting education through the *Fundación Otonga* (FO), an NGO that preserves forests for conservation and helps with public education endeavors.

Onore was the founder of the QCAZ (an acronym derived from Quito, Católica, and Zoología) and the amphibian collections there in 1982. Onore's leadership and dynamism were pivotal for the development of the amphibian scientific collections and, directly or indirectly, for amphibian research and conservation. He revitalized the documentation of the zoological diversity of Ecuador, not only with the formation of collections of many taxa and from many areas of Ecuadorian territory but also with the knowledge and motivation that he transmitted to several generations of biology students. Onore's relationship with the Tapia family (the most productive amphibian collectors in Ecuador, see Figures 2.30, 2.38) and his support of Coloma were significantly influential for the development of the QCAZ up to 2010, the CBA, and later to the startup of CJ.

Onore's scientific, academic, conservationist, and humanist contributions are provided in a detailed biography written by César E. Jácome (Jácome, 2008), and his life has been described as a fiction by some popular media.

While at PUCE, Onore made the greatest contributions to entomology in Ecuador of any single researcher, which have been summarized by Barragán *et al.* (2009). During this period, Onore formed a collection of about one million invertebrates at QCAZ, especially insects, of which about 100,000 were collected and prepared by himself. There are several popular and outreach publications about Onore (e.g., Rodríguez, 2017; Paz Cardona, 2019). From these sources and personal communications, we provide some general aspects of his life and then focus in more detail on his activities and achievements regarding amphibians.

Onore was born in 1941 in Costigliole d'Asti, Italy. His childhood was spent on a family farm where species such as *Bufotes viridis*, *Pelophylax lessonae*, *Rana temporaria*, *Hyla arborea*, *Triturus carnifex*, and *Salamandra salamandra* were common in the surrounding forests and aroused great interest in the "campesino" children. In the village, there was a strong culinary tradition that included *P. lessonae* as an ingredient in the recipes. Onore related about his culinary boyhood:

> In the rush they were simply fried, but the grandmothers knew countless recipes such as: frogs with rice, stewed frogs, grilled frogs, parsley frogs and others. Not far from my home in the town of La Motta every year there was a frog fair and restaurants prepared for diners only dishes based on frogs. A neighbor who had worked as a pawn in the rice fields of the city of Vercelli, the frog capital, had taught us an incredible method of frog farming. Using a piece of my older sister's old nylon stocking as bait, I used it to replace the hook on a fishing rod. At dusk the frogs

were captured by "dancing" the cane among the herbs of the ponds. The frogs, attracted more by the movement of the piece of media than by the bait, grabbed the prey with the maxillary teeth, the ready fisherman lifted the rod with the amphibian at the end of the thread, put it in a basket and continued with the next naive frog. In the house they proceeded to peel them quickly and put them in the frying pan while still jumping. The method was a bit cruel but we considered frogs to be part of the food chain for humans.

In school and later during his higher education his teachers taught him the biology of amphibians and their important ecological role, and since then, in his own words, he "became an apostle of frogs". As a natural science teacher in Rome, Onore took his students to the swamps near the city to observe amphibians. Among his memories, he relates one particularly unforgettable episode.

One day I was lucky enough to find a *Bombina bombina*, known to the Anglo-Saxons as the European fire-bellied toad. I took the opportunity to collect it with my hands and show her curious behavior that consisted in simulating an apparent death by putting her colorful belly in the air. Beware of this little toad—I alerted the students—it has a very poisonous skin secretion! After handling the toad with your hands wash your hands and avoid contact with the mucous membranes! I ended the sentence with a big sneeze. I instinctively rubbed my nose and eyes. A terrible burning forced me to close my eyes and they accompanied me like a blind man to the house where the ophthalmologist helped me regain my sight with abundant ablutions.

Following religious training in high school, Onore served a novitiate year and joined The Society of Mary, also called the Marianists (Latin: Societas Mariae), which is a clerical religious congregation of Pontifical Right for men (brothers and priests) founded for missionary and educational work. He attended the university and did engineering and doctoral studies in agronomic sciences with a specialization in entomology at Università degli Studi di Torino, graduating in 1971. He taught biology at two high schools in Italy and was a university professor of biology, phytopathology, and entomology in Africa in the Republic of the Congo at Brazzaville.

During his decade of work in the Congo, Onore strengthened agricultural production systems by prioritizing work on insect pests. Nonetheless, he also had occasional encounters with frogs, which he claims were not as abundant as in Ecuador in numbers or species. Onore recalled one encounter as follows (Figure 2.37):

FIGURE 2.37 Left: children of the Batékés tribe in a pond from where they collected tadpoles for food. Republic of the Congo. 1978. Photo by Giovanni Onore. Right: Giovanni Onore on a collecting trip to Provincia de Loja, September 1985. Photo by LAC.

> During my entomological expeditions in the town of Mbé, famous for having been the headquarters of King Makoko who had received the explorer Savorgnan from Brazzá and populated by the Batékés tribe, the inhabitants offered me a local dish that pretended to be a "mixed fried" whose ingredients were the fruition of fishing with a sifter in a pond: Odonata larvae, Dytiscidae larvae, Trichoptera larvae, undetermined insects and tadpoles that, before surrendering the spirit, wafted their tails desperately. I confess that I could never associate the tadpoles to the corresponding species. As a diner, I can only certify that I found the dish pleasant, probably because of the hunger caused by the long walk.

Onore arrived in Ecuador from Italy in 1980 when he was 39 years old. Upon his arrival, he helped Italian missioners in Latacunga, Provincia Cotopaxi, where he learned Spanish. Given his passion for insects (and amphibians), Ecuador unveiled a world full of possibilities for research for him. Thus, in 1981, he started teaching invertebrate zoology at PUCE, where his fondness for insects expanded. He became impressed by the abundance of frogs in all the Ecuadorian ecosystems very soon, and began collecting specimens to build a museum at PUCE. Onore was eager to know their names, their biology, and ecology. He recounts an anecdotical talk with an Ecuadorian biologist:

> In the first years of my stay at the biology school of PUCE, it occurred to me to comment to a renowned colleague about my African experience on the palatability of tadpoles. "It is impossible. All tadpoles are poisonous!" she commented. "Well, sorry, I am still alive" was my answer.

Onore's curiosity and observation of details is remarkable, as well as are his didactic abilities, which made his courses the most demanded by biology students. We were told another amphibian anecdote that occurred in a visit to an African palm plantation facility (Palmoriente) in Provincia Orellana:

> I was fascinated by some frogs that cleverly clung to the ceiling of the dining room and acrobatically approached the spotlights in search of insects. When a fly flew in the vicinity of the frog, the amphibian jumped to capture it and, lacking force, resorted to a curious strategy: it released from its cloaca a jet of a liquid that served as a propellant and finally grabbed the prey! On the other hand, my soup was diluted with the propellant. The specimen of the incriminated amphibian ended up in the QCAZ collection and from what I remember belonged to the genus *Osteocephalus*.

Up to 2022, Onore has published extensively, consisting in 63 scientific titles, including three books, primarily focused on Ecuadorian insects, but he also coauthored two research articles on amphibians (Aubér-Thomay *et al.*, 1990; Rödder *et al.*, 2011). Additionally, he has published 11 outreach publications.

Early in Onore's period, the vertebrate section moved to new facilities (in 1984) of approximately 50 m^2, in a new building, which was mostly used by the DCB (PUCE). This new infrastructure was built while Laura Arcos Terán (an Ecuadorian geneticist, who had trained in Germany, who hired Onore) was director of the DCB (Figure 2.38). When Onore proposed to create the museum, an office space, a collection room, and shelves were available, but no budget was allocated for field trips or for collections maintenance. Despite this, Onore began collecting amphibians using personal funds and the assistance of "campesinos", settlers (e.g., the Tapia family), and students (e.g., Coloma). His extensive travels in most of Ecuador took place on weekends, holidays, and vacation periods. In 1985, PUCE began to send specimens to the Muséum d'histoire naturelle de la Ville de Genève (MHNG) under a cooperation agreement (with Volkert Mahnert as curator of herpetology) to collect amphibians and reptiles, which relieved certain field costs and helped to obtain materials for the QCAZ museum. Most specimens were sent to and identified at the MHNG (mainly by the Swiss herpetologist Eugene Kramer), and later a few specimens (representing species that were identified at that time) were returned to the QCAZ. While Onore led the activities of the vertebrate section of the QCAZ (1982–1991), several biology students helped to organize the amphibian collections either as museum assistants or volunteers (e.g., Stella de la Torre, Felipe Campos Yánez, Ana María Velasco, Coloma). Coloma collected about 2100 specimens and deposited them at the QCAZ. In 1986 and 1987, Onore and his student Coloma traveled together in Ecuador and collected amphibians.

Onore's collections during the 1980s have an intangible value, given that they were made at a time when deforestation was rampant and also catastrophic declines in amphibians were occurring in the Andes. Most of the specimens collected by Onore came from areas between San Francisco de las Pampas and Tandapi (in some cases labeled as Santo Domingo de los Tsáchilas and surroundings), Provincia Cotopaxi, in the cloud forests of western Ecuador, and from Coca, Provincia Orellana, in the Amazonian lowlands. In the first location, logging was intensive for colonization purposes, and agriculture and farming were replacing the forests. Additionally, these cloud forest sites were hit hard in the second half of the 1980s by the unexpected and drastic declines in frog populations. Meanwhile, in the Coca region, African palm oil plantations, the petroleum industry, road construction, and colonization were destroying the primary forests. Furthermore, the worst oil spills in world history were occurring in the northern Amazonian region of Ecuador. Onore's inventories in the Andean region during the 1980s provided key data, later used to document catastrophic amphibian extinctions and hypotheses that would explain their causes. Some of the specimens he collected belong to species currently considered possibly extinct. Needless to say, thanks to Onore's specimens deposited at museums (9504 in the MNHG, 2376 in the MCZ, and 1238 in the QCAZ), numerous new species have been named and others continue to be described.

In 1982, a transcendent event occurred that helped to expedite amphibian collections and associated biology data. Onore met the "campesino" family Tapia in a remote locality, San Francisco de Las Pampas, and began a close and long-lasting

FIGURE 2.38 Top left: the Tapia family at their house (currently Reserva Pristirana) in San Francisco de Las Pampas, Provincia Cotopaxi, 1984. Top right: Giovanni Onore with Laura Arcos Terán and Alberto Padilla from PUCE at the paramos of Zumbahua, Provincia Cotopaxi, 1987. Bottom left: Onore and Mario G. Tapia at Reserva Otongachi, 4 April 2010. Bottom right: Onore meets the child (David Jailaca) and his mother Rosa Gancino. David found *Atelopus ignescens* in Angamarca, Provincia Cotopaxi. 21 April 2016. Photos by LAC.

relationship that has resulted in enormous contributions of some of the members of this family (the father César and especially four of his sons, Mario, Elicio, Ítalo, and Queti) to amphibian collections and forest conservation. Onore's encounter and the life of this family (father, mother, and nine sons) is detailed in a book (Pollet, 2003) published by the CBA at PUCE. During the 1980s, Onore and members of the Tapia family collected many specimens from San Francisco de las Pampas and neighboring localities (most currently deposited at the MHNG).

When Onore met the family, he soon realized that the father and the kids had a unique talent for the meticulous observation of details in nature. Thus, he mentored most of them to cultivate these abilities, as well as helping with their education in schools in Quito and later with university studies. In particular, Elicio completed (in 2003) *licenciatura* studies in environmental sciences and development, and Ítalo graduated from high school and has worked at the QCAZ as Collections Manager since 2002. Ítalo and Elicio have amassed thousands of amphibian specimens (see Figure 2.36). In a first period, when they were children, both jointly with their older brother Mario, their father César, and Onore collected more than 13,000 specimens across various locations in Ecuador; they are now deposited at the MHNG and QCAZ (see Figure 2.36, under Onore and Tapia). In the 2000s, they continued collecting for the QCAZ for over a decade, and subsequently, Elicio continued this work in the 2010s when he was employed by Centro Jambatu of Amphibian Research and Conservation (CJ). In this second period, Ítalo collected 10,492 specimens, while Elicio collected 7885 specimens, all of which are housed at the QCAZ, CJ, and Universidad Tecnológica Indoamérica. Both of them have not only collected specimens but also gathered thousands of tissues, recorded hundreds of frog calls, and compiled valuable data on natural history, ex situ biology, and management.

Beginning in 1998, while at PUCE, Onore focused his efforts on in situ conservation through an NGO he founded, *Fundación Otonga*. So far, he has secured ~1756 hectares of forest and its amphibian communities in four reserves: Reserva Integral Otonga (RIO) (1530 ha), Otongachi (66 ha), Otoyacu (~100 ha), and Otokiki (~60 ha). RIO is a subtropical forest in the western slopes and foothills of the Andes of Ecuador, in close proximity to San Francisco de Las Pampas, Provincia Cotopaxi. The care of RIO was entrusted to its co-founders, the Tapia family (see Figure 2.38).

In the 1990s, Onore, César Tapia, and his children began a program of biorestoration of deforested areas (about 250 hectares have been reforested) and, at the same time, they have worked as guides for visitors. The Reserva Otongachi protects cloud forests in the Unión del Toachi town, and a center of environmental education was built there. In this center, Italo, Elicio, and Queti were actively working until recently. The Otoyacu (in the Amazonian foothills) and Otokiki (in the Chocoan foothills) reserves were acquired with Coloma's support, and at these reserves, some amphibian studies have been conducted (Tarvin *et al.*, 2014; Rabeling *et al.*, 2016; Roland *et al.*, 2017).

In 1995, Onore actively participated and supported Coloma in the creation and operation of the Center of Biodiversity and Environment (CBA) with the idea of bringing together all of the QCAZ museum's collections and scientific staff, as well as to begin producing outreach publications—the CBA was coordinated by Diego Lombeida, Gloria Correa Monge, Ruth Boada, and Verónica Cano Molestina and served as support in the management of zoology projects. Although this center never had a formal status within the DCB (PUCE), given the opposition of its administrative authorities to its creation, the CBA produced between 1996 and 2010 a series of scientific and miscellaneous outreach publications (e.g., 17 books, a video documentary, a stamp series, posters) that covered many zoological taxa (e.g., insects, amphibians, lizards, birds, and mammals), most of which were edited by Onore and Coloma. These publications included three on amphibians, the coffee table books *Ecuador Megadiverso* (Coloma and Ron, 2001) and *Sapos* (Ron *et al.*, 2009), and a video about Ecuadorian amphibians (Salazar and Pérez, 2004). From about 2005 to 2010, the CBA also embraced the ex situ amphibian research and conservation program, Balsa de los Sapos, that was led by Coloma (Zippel and Mendelson, 2008; Coloma *et al.*, 2010a). This program had the strong support of Onore at PUCE and, after 2011, of CJ (under the name Arca de los Sapos).

In the 2010s, Onore welcomed the Centro Jambatu (led by Coloma) as part of *Fundación Otonga*, which continued until 2019, when CJ became an NGO by itself. In 2015–2016, Onore proactively participated in the search and rediscovery of the jambato harlequin frog *Atelopus ignescens* that occurred in 2016, a story that traveled around the world given its significance for the world of amphibian conservation (Plitt, 2017) (see Figure 2.38).

In recognition of Onore's achievements, 92 species of animals (89 invertebrates, 3 amphibians) and 5 invertebrate genera (4 of insects, 1 of annelida) bear his name; among the amphibians are the rare caecilian *Cthonerpeton onorei*, the rainfrog *Pristimantis onorei*, and the harlequin frog *Atelopus onorei*. Onore is one of the most eponymically honored scientists, perhaps only Darwin is honored in more scientific names. This recognition is the result of his progressive exchanges of specimens, identifications, and information with taxonomists and important museums worldwide. Onore has been awarded several prices, including the Gambrinus Prize (annually given to him from 1992 to 2007) for the conservation of tropical forests and L'Eroe Mai Cantato (Never Sung Hero) (2009) for conservation and education. Within Ecuador, he received the Gran Collar "Luis Héctor Chislain de Carondelet y Barón de Carondelet" (2021), which is awarded by the Mayor of the city of Quito to honor a foreign citizen promoting science in Ecuador.

Coloma Expands upon the QCAZ Foundation: A Look at the Individual

Luis A. Coloma was closely tied to the QCAZ, initially as a student volunteer helping Giovanni Onore during collecting field trips, beginning in 1982, and from 1991 to 2010 as an assistant professor and later full professor and head of the Vertebrate Section during the 1990s and of the herpetology section during the 2000s. At the QCAZ, he led two fruitful decades of science and education, which resulted in the building of the largest Ecuadorian herpetological collection, the mentoring of 23 biology *licenciatura* students (Table 2.5), the production of scientific and outreach publications, and the founding of an amphibian conservation program.

Duellman (2015a) recounts the academic career of Coloma while at the University of Kansas (KU) (under "Sundry Systematics Studies on Anurans"), his early activities in Ecuador (under "Where Have All the Froggies Gone?") and a biographical sketch (under "Kansas Herpetological Alumni"). Several other aspects of his life have been published in outreach publications (e.g., Duellman, 2009; Crump, 2024), and a compendium from them with up-to-date information is herein provided along with a biographic sketch.

Beginning in 1981, Coloma studied biological sciences at Pontificia Universidad Católica del Ecuador (PUCE) for a *licenciatura* degree that he obtained in 1987. During that period, he was a student of the pioneering zoologists Gustavo Orcés Villagómez (1902–1999), Luis Albuja Viteri, and Onore, and also attended courses taught by Eugenia M. del Pino. Onore gave numerous entomology courses that influenced Coloma's interests, and soon he became captivated by insects; thus, he did a research thesis addressing the diversity and bio-architecture of Ecuadorian stingless bees (Meliponinae). Before he enrolled at PUCE, Coloma studied architecture at Universidad Central del Ecuador for about two years. While at PUCE, he traveled extensively in Ecuador, initially with Almendáriz and later with Onore on collecting trips and also was involved in the conservation of the Bosque Protector Cashca Totoras near his hometown (Figure 2.39).

Coloma pursued his graduate studies at KU from 1988 to 1997. His time there felt like a journey of discovering an entirely new world at the most significant herpetological collection of its time. His mentors, William E. Duellman, Linda Trueb, and John E. Simmons (Collection Manager) were key figures upon his arrival. John E. Simmons, the KU Collection Manager (see "John E. Simmons and Collection Management" in Duellman, 2015a), and his wife,

TABLE 2.5
Students of *Licenciatura* at PUCE Mentored by Coloma for Their Amphibian Theses

Name	Graduation Year	Thesis Title (* = published, ** = partially published)
Gloria Correa Monge	1995	Comportamiento, vocalización, y notas biológicas de *Colostethus awa* y *Colostethus machalilla* (Anura: Dendrobatidae) en cautiverio.
Alexandra Quiguango-Ubillús	1996	Comportamiento social, comunicación y cuidado parental de *Colostethus toachi* (Anura: Dendrobatidae) en cautiverio.*
Diego Lombeida	1996	Secuencias de osificación de *Colostethus bocagei* y *C. machalilla* (Anura: Dendrobatidae).
Marcelo Díaz-Proaño	1998	Comportamiento social, reproducción, cuidado parental y vocalización de *Colostethus infraguttatus* (Anura: Dendrobatidae).
Julio Molineros	1998	Territorialidad, vocalización y notas biológicas de *Colostethus* sp. (*aff. marchesianus*) (Anura: Dendrobatidae) en la Reserva Biológica Yuturi, Amazonía ecuatoriana.
Arturo Paredes-Recalde	1999	Osteología y relaciones filogenéticas del grupo *Hyla larinopygion* (Anura: Hylidae)*
Juan M. Guayasamin	2000	Partición acústica en una comunidad de ranas del género *Eleutherodactylus* (Anura: Leptodactylidae) en un bosque nublado del occidente del Ecuador.**
Gilda Gallardo	2001	Aspectos de la biología reproductiva con énfasis en el crecimiento larval de *Rana palmipes* Spix, 1824 (Anura: Ranidae) en la Amazonía ecuatoriana.*
Néstor Acosta-Buenaño	2002	Vocalización en 9 especies del género *Colostethus* (Anura: Dendrobatidae) del Ecuador.
George Fletcher-Lazo	2002	Taxonomía, distribución e historia natural de los ápodos (Amphibia: Gymnophiona) del occidente ecuatoriano.
Martín R. Bustamante	2002	Monitoreos de poblaciones de anuros en los Andes de Ecuador, con énfasis en el estudio poblacional de las especies del Bosque Protector Cashca Totoras (Provincia Bolívar).*
Juan Carlos Santos	2002	Relaciones filogenéticas de *Colostethus bocagei* y *Epipedobates tricolor* (Anura: Dendrobatidae) inferidas de secuencias de genes mitocondriales 12S, 16S y Citocromo b.*
Patricia Castillo-Trenn	2004	Comportamiento social, reproducción, cuidado parental y vocalización de *Colostethus kingsburyi* (Anura: Dendrobatidae).*
Samael Padilla	2005	Actividad reproductiva de dos comunidades de anuros en el Parque Nacional Yasuní.
Mónica A. Guerra	2005	Selección sexual en *Physalaemus petersi* (Anura: Leptodactylidae), basada en repuestas fonotácticas de las hembras.*
Margarita M. Baquero Rivadeneira	2006	Evolución del canto de anuncio en ranas del clado *tricolor* (Anura: Dendrobatidae).
Cristina Félix-Novoa	2006	Biología reproductiva de *Ameerega bilinguis* (Anura: Dendrobatidae) en el Parque Nacional Yasuní.**
Irene Oña-Alava	2007	Aspectos etológicos de territorialidad, vocalizaciones y reproducción de *Ameerega bilinguis* (Anura:Dendrobatidae) en la Amazonía ecuatoriana.
Verónica Mesías-Tapia	2007	Demografía y aspectos de historia natural de *Engystomops petersi* (Anura: Leiuperidae) en la Amazonía ecuatoriana.
Mónica I. Páez-Vacas	2008	Revisión taxonómica del clado *bocagei* (Anura:Dendrobatidae)*
Sofía Carvajal-Endara	2010	Distribución y patrones de utilización de recursos de *Gastrotheca pseustes* y *Gastrotheca litonedis* (Anura: Hemiphractidae)
Gustavo Pazmiño-Otamendi	2012	Territorialidad, comportamiento social, reproducción y vocalización de *Hyloxalus infraguttatus* (Anura: Dendrobatidae).
Diego Paucar-Guerra	2013	Etología, biología de poblaciones y territorialidad de dos especies de *Hyloxalus* (Anura: Dendrobatidae) de la Amazonía ecuatoriana.

the Ecuadorian Ligia Galarza, were like his family there and soon established a strong emotional bond. Coloma worked as Simmons' curatorial assistant for about a year. Coloma learned a great deal from him, and that experience later helped Coloma in establishing two museums in Ecuador, first, the QCAZ Museum (between 1991 and 2010), and currently, the CJ Life Bank (since 2011).

At the time of Coloma's arrival at KU, amphibian systematics was undergoing a profound revolution, from the traditional phenetic or numeral taxonomy to phylogenetics and from the use of morphological characters to molecular approaches. At KU, Darrel R. Frost, David C. Cannatella, and David M. Hillis were the paladins of applying phylogenetics to amphibian systematics. This era also saw the availability of personal computers with analytical software facilitating the transformation. In fact, part of this transition was reflected in Coloma's master's thesis and PhD dissertation, although he did not perform molecular analyses. However, during a field trip in 1990 to Ecuador, and while at KU, Coloma began collecting amphibian tissues for allozyme and DNA analyses (jointly with the KU students John J. Wiens and David A. Kizirian) following

FIGURE 2.39 From left to right. Top left: Luis A. Coloma on a collecting trip to Santiago, Provincia Morona Santiago, February 1986. Photo by unknown photographer. Top right: Luis A. Coloma, Mario García Saltos, and TJitte de Vries (Director of the Department of Biology of PUCE) working on the plan for the declaration of Cashca Totoras, Provincia Bolívar, as a protected forest, 1987. Photo by Antonio Salazar. Middle left: Felipe Campos Yánez, David A. Kizirian, John J. Wiens, and Stella de la Torre searching for frogs on Bolívar-Chimborazo provinces border, 5 January 1990. Photo by LAC. Middle right: Diego Tirira and Luis A. Coloma (front row), John J. Wiens (second row), Sebastián Valdivieso (third row), in a collecting trip to around Lago Agrio, Provincia Sucumbíos, March 1990. Photographer unknown. Bottom left: Luis A. Coloma revising skeletons in the Division of Herpetology at the University of Kansas, 1991, unknown photographer. Bottom right: Luis A. Coloma, Gustavo Orcés Villagómez, Andrea Encalada, and Gloria Correa Monge at PUCE in February 1993. Photographer unknown.

a trend initiated by William E. Duellman in 1984. This marked a significant step in incorporating molecular approaches into his research endeavors. For example, in the description of *Pristimantis simonbolivari* (Wiens and Coloma, 1992), which was the first species of frog in which an Ecuadorian was involved, John J. Wiens performed the molecular analyses.

For his master's degree, Coloma undertook a systematic review of the non-poisonous frogs of the then dendrobatid genus *Colostethus* in Ecuador, a task that was initiated by Stephen R. Edwards but had remained unpublished. Coloma's studies were based primarily on the large samples in the KU collection, as well as in collections elsewhere in the United States, Ecuador, and Europe. He recognized 31 species, five of which were new, provided detailed osteological descriptions based on cleared and stained specimens of 17 species, and provided call descriptions of six species. The latter were made using a mix of old and vanguard technology–an old vibralizer sound machine was used to print audiospectrograms and brand-new software for Mac computers (SoundEdit) for measuring physical parameters. This was the first time (at least at KU) that computer software was be used for amphibian acoustic analyses. His well-illustrated publication (Coloma, 1995) provided a basis for further studies of this group of frogs in Ecuador and elsewhere. Before leaving Kansas, Coloma completed his PhD requirements except for his dissertation. He engaged in discussions with Duellman and Trueb regarding a proposed doctoral study focusing on the morphology and phylogeny of *Atelopus*, which he intended to do while in Ecuador, where he assembled a vast amount of data on the morphology and osteology of those frogs. In 1991, he and Stefan Lötters (from Germany) did field work in the Ecuadorian Andes and documented dead *Atelopus* in the Provincia Chimborazo (see Figure 2.39). In 1994, Coloma embarked on a lengthy visit to the Instituto de Ciencias Naturales in Bogotá with the valuable support of the renowned Colombian herpetologist Pedro M. Ruiz Carranza (1932–1998). Carranza played a leading role in amphibian studies in Colombia from 1975 to 1998 and described 81 species of frogs from Colombia, eight of which were reported to occur in Ecuador. Upon returning to Kansas, in August 1996, Coloma spent time analyzing data and writing his dissertation, which he defended in December 1997. Coloma used traditional discrete morphological characters (and he added the novelty at that time of continuous and polymorphic characters) and conducted a phylogenetic analysis using PAUP (Phylogenetic Analyses Using Parsimony) software. Coloma continued his systematic work on *Atelopus* and published two major papers in which he and his coauthors described new, possibly extinct, species (Coloma *et al.*, 2007, 2010b).

In 1991, when Coloma returned to Ecuador after completing his MA in systematics and ecology and while enrolled as a PhD candidate at KU, he was rapidly incorporated into PUCE through Giovanni Onore's arrangements and the approval of Laura Arcos Terán (the director of the Department of Biological Sciences). This established the beginnings of Coloma's active research program that initially, in 1991, was on vertebrates and transitioned in 2000 to focus on herpetology, although during nearly two decades, Coloma's priorities were the amphibian collections at the QCAZ as well as the amphibian conservation program. The transition was possible because in 2001, the mammologist Santiago F. Burneo was hired to curate the mammals, birds, and fish, and in 2009, Omar Torres-Carvajal (just returned from completing a PhD at KU and a postdoctoral position at the USNM) was hired to curate the reptiles at the QCAZ.

Coloma was a lecturer and Curator of Vertebrates in the Escuela de Ciencias Biológicas at the Pontificia Universidad Católica del Ecuador. He was well aware of declining populations of Andean anurans, especially frogs of the genus *Atelopus*. While at PUCE, he established La Balsa de Sapos, the "Ark for Frogs" which became a major breeding program for various species of frogs, many of which were declining in their natural habitats. In addition to the breeding program, Coloma popularized frogs in Ecuador by producing colorful postcards, posters, stamps, and attracting media attention. His major production was an amphibian exhibition (Sapari) in the PUCE cultural center in Quito in 2005. These events expanded Coloma's recognition as a leader in conservation in Ecuador. After resigning from his position at PUCE in December 2010, Coloma established the Centro Jambatu de Investigación y Conservación de Anfibios in 2011.

The Dream Team at QCAZ: Two Decades (1991–2010) of Science, Conservation, and Public Awareness

Between 1991 and 2010, the amphibian section of the Museo de Zoología of Pontificia Universidad Católica del Ecuador (QCAZ) had its greatest development, under the direction of Luis A. Coloma. (Figures 2.39, 2.40, 2.41, 2.42, 2.43). In the 1990s, Coloma, jointly with several *licenciatura* biology students, continued the initiative of Giovanni Onore by increasing field work, building the museum's collections (including a genome bank), and performing associated taxonomy research. Coloma encouraged studies of the behavior of amphibians in the laboratory—which were later combined with in situ studies—and began a vigorous amphibian conservation program. All of these activities were closely tied to the production of scientific and outreach publications. Cooperation with researchers from other museums, universities, and institutions at the national and international level was very important for the rapid establishment of the QCAZ as a center for the scientific knowledge of Ecuadorian amphibians. Coloma was joined between 1998 and 2002 and from 2007 to 2010 by Santiago R. Ron and in the second decade (2008–2010) by Juan M. Guayasamin, who did master's and PhD graduate studies in North American universities (see Figure 2.41). At the end of 2010, Guayasamin resigned from the QCAZ, following Coloma's departure.

In the course of his research on Ecuadorian amphibians, Coloma, a group of students and collectors, and other researchers discovered many new species of frogs. Coloma devoted a lot of effort to support taxonomic studies that included field collecting and the description of new species that were deposited in the QCAZ and in other research centers. Specimen loans became a regular task in spite of the burden of obtaining exportation permits. Coloma, students, and researchers (especially Ron and Guayasamin in the 2000s, and Ron *et al.* in the 2010s) were authors and coauthors of the descriptions of dozens of Ecuadorian amphibians, relying on QCAZ specimens. The collection growth at the QCAZ had several peaks of activity (see Figure 2.40), the most significant beginning in 2008 with the funding provided by the Ecuadorian government for a project to conduct inventories and address the genetic and morphological diversity of Ecuadorian amphibians. This was the first time that funds were allocated for collections and systematic and taxonomic research at the QCAZ by a government agency, the Secretaría Nacional de Ciencia y Tecnología (SENACYT).

Cataloging of the QCAZ amphibian collection began in 1984. The first specimen cataloged was a *Gastrotheca guentheri* (a species now apparently extinct) collected by Giovanni Onore and César Tapia. This was the beginning of a saga of collecting throughout continental Ecuador. By the end of 2010, there were 48,714 specimens deposited in QCAZ collections. The oldest specimen with locality data (an *Oophaga sylvatica*) was collected on 10 April 1976 at Centro Científico Río Palenque.

During the 1990s, the amphibian collections continued to grow, and the physical infrastructure was increased and reorganized. The collections of both Ecuadorian and foreign researchers began to be deposited in their entirety at QCAZ, even though at that time the requirement by the Ecuadorian government was that foreign projects were to deposit in an Ecuadorian museum only one voucher specimen of each species collected. The QCAZ departed from this general rule and encouraged foreign researchers to deposit all specimens they collected, and the specimens were then sent for studies abroad under the modality of loan or exchange. Simultaneously, some collectors not working at PUCE began to deposit their specimens at the QCAZ. Among them stands out Morley Read, a British herpetologist resident in Ecuador, who has conducted long-term inventories and monitoring of amphibians in Yasuní National Park and who participated in numerous amphibian projects and the collecting of 1884 specimens of amphibians (829 deposited at the QCAZ and 1055 at CJ).

In 1990, a new era of scientific production at the QCAZ began. Coloma and Onore coauthored a paper published in the *Revue Française d' Aquariologie* that dealt with the maintenance and breeding of the direct-developing marsupial frog *Gastrotheca plumbea* (Aubér-Thomay *et al.*, 1990). The first author was the French scientist Michèle Aubér-Thomay, a friend of Onore who had been working for several years with marsupial frogs (see also Chapter 1, under the subtitle "The Boost of the Dream for Scientists and Frog Lovers in Europe and the USA") at the University Pierre et Marie Curie in París. This article was followed, in the 1990s, by descriptions of new species (e.g., Wiens and Coloma, 1992; Duellman and Coloma, 1993; Kaiser *et al.*, 1994; Coloma, 1995), tadpoles (e.g., Coloma and Lötters, 1996), and taxonomy (e.g., Coloma *et al.*, 2000).

In the 2000s, publications by QCAZ members gradually increased. Between 2005 and 2009, they produced about 10 scientific publications in peer-reviewed journals per year and widened the coverage of amphibian topics to include research on ethology, ecology, and conservation biology. Between 1991 and 2010, Coloma mentored 23 *licenciatura* thesis students who studied amphibians (Table 2.5) and three who studied reptiles. Of the 23 theses, 10 have been published completely or partially, while 13 remain unpublished. It is noteworthy that these theses covered diverse subjects, including osteology, behavior, acoustics, monitoring, natural history, evolution, and reproductive biology; only two focused on taxonomy. One of these theses, by Mónica I. Paéz-Vacas (see her academic biography under the title "The Diversity and Academic Lives of Current Amphibian Biologists"), and a subsequent publication (Páez-Vacas *et al.*, 2010) were among the first works to use integrative approaches in taxonomy in Ecuador.

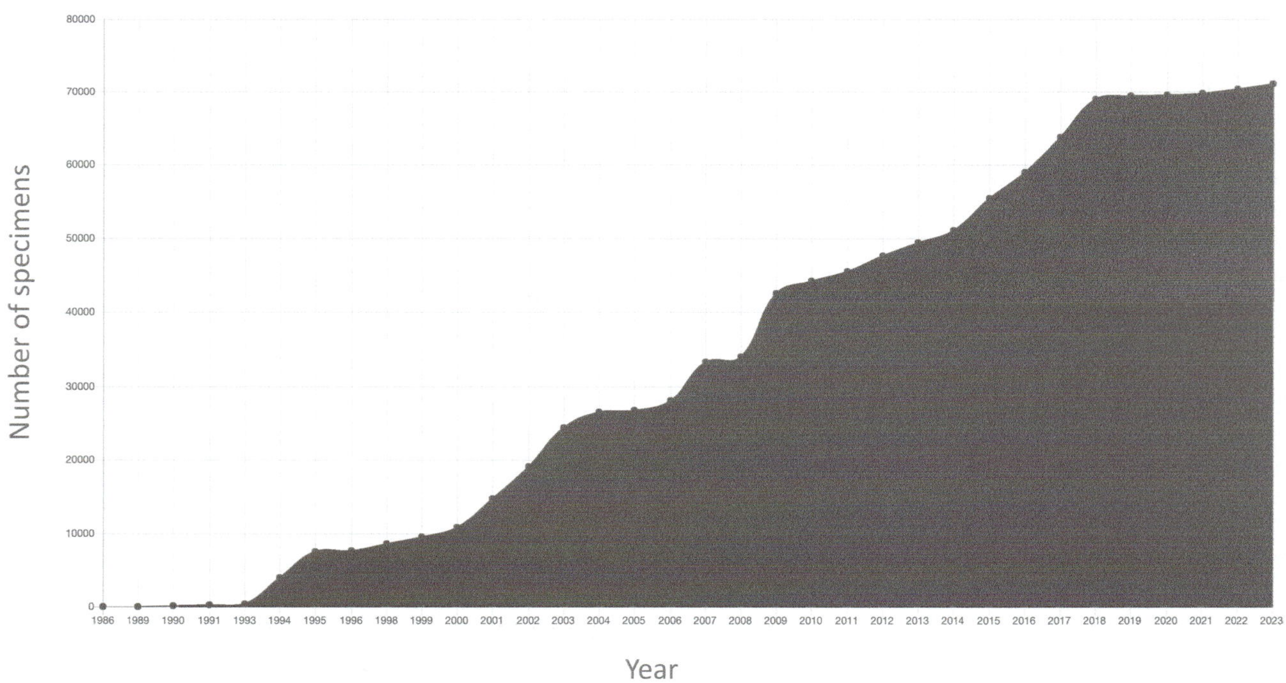

FIGURE 2.40 Amphibian collection growth at QCAZ from 1986 to 2022. Taken from https://bioweb.puce.edu.ec/QCAZ/InformacionGeneralAnfibios.

2001: A Genome Bank and the Return of Brains

The herpetology museums and laboratories were directed between 2001 and 2008 by Coloma (see Figures 2.41, 2.42, 2.43). Between 2001 and 2008, the Vertebrate Section was divided into the subsections of ichthyology, herpetology, ornithology and mammology. During the decade of the 2000s, laboratories and museums of the herpetology sections were consolidated. Progressively, the number of projects grew as well as the collections, the physical

FIGURE 2.41 From left to right. Top left: Stefan Lötters searching for *Atelopus* at the Atillo region, Provincia Chimborazo, 4 August 1991. Photo by LAC. Top right: Martín R. Bustamante, Andrés Merino-Viteri, Juan M. Guayasamin, Santiago R. Ron, and Luis Eduardo López at Laguna de Mojanda, Provincia Imbabura, October 1999. Photo by LAC. Middle left: Mónica I. Páez-Vacas and Diego Almeida Reinoso at Reserva Hola Vida, Provincia Pastaza, 20 March 2007. Photo by Ítalo G. Tapia. Middle right: J. Alan Pounds and Luis A. Coloma at Laguna de Atillo, Provincia Chimborazo, 28 November 2006. Photo by Martín R. Bustamante. Bottom left: Manuel Morales Mitte, Luis A. Coloma, Elicio E. Tapia at Cuellaje, Provincia Imbabura, 24 June 2009. Photo by unknown photographer. Bottom right: standing: Mark Pepper, Allan P. Pessier; sitting: Jackie Pepper, Manuel Sánchez, Ron Gagliardo, Brad Wilson, and Luis A. Coloma on a collecting trip to Alto Tambo, Provincia Esmeraldas, 29 January 2010. Photo by unknown photographer.

FIGURE 2.42 Top left: Martín R. Bustamante at the QCAZ on 8 November 2004. Photo by LAC. Top right: Pablo A. Menéndez-Guerrero at the QCAZ in about 2000–2001. Photo by Martín R. Bustamante. Middle leftt: Chris W. Funk at Washington, D. C., during the Amphibian Conservation Summit, 16 September 2005. Photo by LAC. Middle right: the amphibian exhibit Sapari in the Centro Cultural of Pontificia Universidad Católica del Ecuador. 27 March 2007. Photos by LAC. Bottom left: William E. Duellman showing the new amphibian museum facilities to the Ecuadorians at the KU Biodiversity Center in Lawrence, Kansas, September 2005. From left to right: WED, Juan M. Guayasamin, Fernando Calle (PUCE architect), Luis A. Coloma. Photo by John E. Simmons. Bottom right: part of the QCAZ amphibian team in 11 May 2007. Sitting: Néstor Acosta-Buenaño, Andrés Merino-Viteri, Luis A. Coloma, Mónica A. Guerra, Amaranta Carvajal Campos, Pablo A. Menéndez-Guerrero, and Freddy Almeida Reinoso. Standing: Ítalo G. Tapia, Ramón Zambrano Aveiga, Cristina Félix-Novoa, Mónica I. Páez-Vacas, David Salazar-Valenzuela, Paula Peña Loyola, Verónica Cano Molestina, Miguel Angel Rodríguez, Juan Fernando Dueñas, and Diego Almeida Reinoso. Photo by LAC (self-timer).

space (~700 m^2), the equipment, and the personnel. In 2003, Néstor Acosta-Buenaño (*licenciatura* in biological sciences and master of business administration at PUCE) took charge of bioinformatics, database administration, and other administrative tasks for the herpetology section. Beginning in 2004, Diego Almeida Reinoso (doctorate at the Central University of Ecuador) collaborated as an amphibian keeper and focused on the ex situ management of amphibians in danger of extinction. During this period, Martín R. Bustamante (2000–2008) and Andrés

FIGURE 2.43 From left to right. Top left: Néstor Acosta-Buenaño leading a course for the FaunaWebEcuador at the QCAZ, 8 April 2005. Photo by LAC. Top right: Luis A. Coloma at his office at the QCAZ in 2005. Photo by John E. Simmons. Middle left: Diego Almeida Reinoso at the ranarium of the Balsa de los Sapos Initiative, 14 May 2007. Photo by LAC. Middle right: Margarita M. Baquero Rivadeneira and Juan Carlos Santos at the QCAZ lab working on dendrobatid physiology experiments, 23 March 2005. Photo by LAC. Bottom left: the herp QCAZ team and researchers of the SENACYT project. 13 November 2009. Photographer unknown.

Merino-Viteri (2000–2006) also collaborated on several projects in the herpetology section, carrying out research on amphibian declines.

For the conservation of genetic resources, in 2001, the QCAZ began to strengthen its amphibian genome bank. This reservoir of genetic material is currently one of the largest banks of genomic resources in South America and the largest in Ecuador (40,548 tissue samples as of 30 July 2023, https://bioweb.bio/portal/). This was made possible primarily through a project supported by the Unit of Cooperation and Development of People, an Italian NGO that obtained funds from the European Union. This aid supported the acquisition of an ultra-cold freezer and the storage of tissues collected during a project titled "Inventory and amphibian genome bank of the Cuyabeno Faunal Production Reserve". The resulting collections were pivotal for the launching of the genome bank, which was enthusiastically carried out by several QCAZ research assistants, among them Martín R. Bustamante, Néstor Acosta-Buenaño, and Pablo A. Menéndez-Guerrero, who did field collecting and extracted tissues.

In 2008, the SENACYT project established a molecular laboratory, which was spearheaded by Elisa Bonaccorso and Juan M. Guayasamin, both recently graduated from the University of Kansas, where they specialized in molecular work focused on systematics and evolutionary biology.

By the end of the 1990s, the panorama of progress in Ecuadorian amphibian research was overshadowed by the catastrophic amphibian extinctions that had been occurring since the mid-1980s in the Andes of Ecuador. Traditional conservation strategies (e.g., habitat conservation) were not enough to deal with new threats such as global climate change and lethal pathogens. Thus, Coloma aimed, as one of his research priorities, to understand the causes of amphibian extinctions and understand the magnitude of these loses to the amphibian fauna of Ecuador. This had the enthusiastic support of Martín R. Bustamante and Andrés Merino-Viteri (see Figure 2.42), who were mentored by Coloma and Ron for their *licenciatura* theses. Merino-Viteri initially began studying the natural history of the cuico rainfrog, *Pristimantis unistrigatus*, but quickly became focused on the challenge of discovering the causes of the sudden declines (see more about him later, under the title "Changing of the Guard, Ron and the QCAZ since the 2010s"). In addition, the information generated at the QCAZ served to prepare the first Red List of amphibians in Ecuador and subsequent updates to the Global Amphibian Assessment, beginning in 2004 when Coloma, Ron, and Pablo A. Menéndez-Guerrero prepared the baseline data used for Ecuadorian amphibians.

Bustamante was associated with the QCAZ on several projects when he was a biology student and then as an associate researcher between 2000 and 2008. His contributions to collections and research were pivotal. In 2003–2008, he published 10 scientific papers and 7 outreach publications. He was a prolific collector and participated in the collecting of 3924 specimens of amphibians, many of which resulted in species new to science, and one of them, *Pristimantis gagliardoi*, was described by him and Joseph Mendelson (Bustamante and Mendelson III 2008). The name honored Ron G. Gagliardo (see Figure 2.42), a paladin supporter of the QCAZ amphibian ex situ program (e.g., Gagliardo, 2011). Bustamante became deeply involved in field work and the monitoring of amphibian populations in connection with the amphibian extinction and decline topics (Ron *et al.*, 2003; Funk *et al.*, 2003). His monitoring expertise was mentored and strongly influenced by W. Chris Funk, who initiated amphibian monitoring studies in Ecuador in 2000. In 2005, Bustamante was the leading author of a seminal paper (Bustamante *et al.*, 2005a) that quantitatively demonstrated that the declines and disappearances of frogs were real. From 2000 to 2003, his monitoring of the occurrence and abundance of amphibians in seven historical sites in Ecuador demonstrated that about 50% of the species were missing (he found only 30 species out of 62 occurring in historical times), and revealed a general trend of decline in relative abundance (Bustamante, 2002; Bustamante *et al.*, 2005a). These findings were surprising, as the search efforts during his surveys were more extensive than those in earlier historical periods. Bustamante's field efforts and data were crucial at a time when part of the scientific community was still skeptical about the reality of catastrophic declines, arguing that some of these declines could be explained as part of the natural fluctuations of populations. His publications added to other research and papers by QCAZ authors (e.g., Ron and Merino-Viteri, 2000; Merino-Viteri, 2001; Merino-Viteri *et al.*, 2005) that supported the suspicion that climate change was involved as a proximal cause of declines. At that time, the North American J. Alan Pounds, a resident scientist at the Tropical Science Center's Monteverde Cloud Forest Preserve in Costa Rica, had been gathering data and doing analyses of his work with amphibian declines. In a meeting of the RANA group (Rescate de Anfibios en América Latina) that Bustamante and Andrés Merino-Viteri attended in Panamá in 2004, Pounds invited them, as well as other members of the Ecuadorian QCAZ group and colleagues from other parts of the world, to join as coauthors of his planned paper on amphibians and global warming. In fact, Pounds, Bustamante, Coloma, *et al.* (13 coauthors were ordered in alphabetical order) published, in the journal *Nature* (Pounds *et al.*, 2006), by far the most influential and most cited amphibian paper in which Ecuadorian coauthors have participated (2376 citations up to 12 December 2022). This paper, titled "Widespread amphibian extinctions from epidemic disease driven by global warming", greatly aided the understanding of amphibian declines and after nearly two decades continues to be widely read and cited. This paper was not free of controversy and triggered a response in other high-impact journal publications (e.g., Pounds *et al.*, 2007), disagreements (e.g., Lips *et al.*, 2008), a reply (e.g., Pounds and Coloma, 2008), and discussions (e.g., Alford *et al.*, 2007). Pounds *et al.* (2007) argued that the chytrid fungus was not a lone killer and that climate change, along with pathogens and several other forms of environmental deterioration, were key factors, whereas Lips *et al.* (2008) embraced the paradigm that the chytrid fungus was the main culprit. This discussion is still ongoing (e.g., Scheele *et al.*, 2019; Cohen *et al.*, 2019).

Bustamante not only contributed academically but also to a relaxed, fun-filled atmosphere in the amphibian section, seasoned with soccer and photography (which were his main hobbies). Bustamante used nature photography as a means to educate the public and raise public awareness of nature, and he enthusiastically participated in the publication of at least two photo coffee table books that include amphibians (Bustamante and Cárdenas, 2007; Ron *et al.*, 2009). For the first book, in 2008, Bustamante earned the Enrique Garcés prize, given by the Distrito Metropolitano de Quito. Through the early 2000s, Bustamante had been worried about the poor connection of the university (PUCE) and the QCAZ museum with society, and he encouraged Coloma to take a leadership on the task. Coincidently, in 2000, a new chancellor took office at PUCE, José

Ribadeneira Espinosa S. J., and one of his priorities was to strengthen the public–university connection; for that purpose, he requested Coloma to let the public to know about the work at the QCAZ. In response, Coloma, Bustamante, and Merino Viteri led the QCAZ to produce a large-scale temporary amphibian exhibit held in the cultural center of PUCE in Quito in 2005. The exhibits were in response to the urgent need to educate and increase public awareness about the global amphibian crisis. Thus, the QCAZ held the largest live amphibian exhibit in the world at that time, "Sapari: Venture into a World of Frogs". Sapari is a portmanteau of the words sapo (frog) and safari. This event, unprecedented in Ecuador, was attended by 105,000 people during three months (see Figure 2.42).

In 2007, with the mission to enhance free access to knowledge through the electronic encyclopedia of Ecuadorian amphibians, Bustamante traveled abroad and photographed many type specimens of Ecuadorian amphibians deposited at three museums in the United States, England, and Spain (KU, BMNH, MNCM). These photographs are currently online on the Bioweb site of the QCAZ. Bustamante is now the Director of the Quito Zoo, where he leads efforts to educate the public about the marsupial frog *Gastrotheca riobambae*, a species that was named in 2012 as an emblematic species of fauna of the city of Quito.

The Life Raft for Frogs at QCAZ

The roots of the ex situ research and conservation program at the QCAZ began with Coloma's interest in the behavioral aspects of the Dendrobatidae that led him to initiate ethological studies under laboratory conditions when he arrived at PUCE. Dendrobatid behavior was a side project he took on in addition to his PhD dissertation on *Atelopus.* Thus, in 1992, he mentored Gloria Correa Monge, a *licenciatura* student who conducted studies on *Epipedobates machalilla* and *Hyloxalus awa*. Correa Monge described for the first time many features of both species, including vocalizations, based on laboratory-maintained frogs (Correa Monge, 1995). While working on her thesis, she became interested in the frogs' ex situ conservation after talks with Coloma, who was deeply concerned about the amphibian extinction problems and who had tried to maintain an ex situ colony of *Telmatobius niger* since 1991 without success—currently this species, which was described in 1920 by Barbour and Noble, is probably extinct. Thus, Correa Monge began in 1994 with the maintenance of a small colony of *Atelopus balios* that she collected with Martha L. Crump. Crump had witnessed first-hand the extinction of frogs in Costa Rica (Pounds and Crump, 1994; Crump, 2000). Correa Monge began to develop protocols for terrarium maintenance and insect cultures. She was followed by three of Coloma's *licenciatura* students, who were also studying dendrobatid behavior, María Alexandra Quiguango-Ubillús, who in 1995–1996 studied the behavior of *H. toachi* in terraria (Quiguango-Ubillús and Coloma, 2008); Néstor Acosta-Buenaño, who studied the vocalizations of nine species of dendrobatids recorded in the laboratory and in the field during 1995–1996 (Acosta-Buenaño, 2002); and Patricia Castillo-Trenn, who conducted laboratory and field studies in 2000–2003 on *Allobates kingsburyi* (Castillo-Trenn and Coloma, 2008). In 2000, Ítalo G. Tapia became part of the permanent staff at the herpetology laboratory and took charge of the administration of the herpetological collection (museum specimens and live amphibians); at the same time he began to collect amphibians and reptiles (see also "Onore, an Entomologist at the Roots of Amphibian Research and Conservation"). To feed the live amphibian specimens, Tapia raised insects in large quantities, especially crickets. He adopted aspects of cricket-farming techniques from around the world but refined them for the raising and breeding of a native cricket, *Acheta* sp. (*assimilis* complex), using founders initially collected at Valle de los Chillos. Tapia was responsible (between 2000 and 2010) for the care, maintenance, and breeding of many tadpoles used in the images that document their ontogeny, as well as meticulously helping Coloma in the photography studio to document all the live frogs that arrived at the QCAZ. From 2004 to 2010, Diego Almeida Reinoso, a biologist from Universidad Central del Ecuador, was hired as Amphibian Keeper, and in 2006, his brother, Freddy Almedia Reinoso, began managing the insect colonies.

After those firsts steps, the evolved amphibian ex situ conservation program was consolidated in 2005 after it became clear that the amphibian crisis required new conservation strategies and more far-reaching, coordinated, rapid, and effective actions to prevent the further loss of amphibian biodiversity. That year, parallel to the World Amphibian Summit held in Washington, the strategic plan for research and conservation of amphibians in Ecuador (Balsa de los Sapos) was drawn up, led by Coloma and conceptualized, managed, and coordinated with the support of Verónica Cano Molestina (a PUCE biologist) and Miguel A. Rodríguez (a PUCE pedagogist specializing in biology). Coloma *et al.* (2010a) describe the 2005 plan and its major accomplishments up to 2010. The QCAZ embarked on an ambitious and proactive project, breaking away from the traditional reactive approach. It envisioned a modern-day Noah's ark, with scientists, conservationists, the public, and frogs all on board, aimed at mitigating and preventing the accelerated loss of frogs and toads.

Driven by idealism, and drawing on the knowledge and experience accumulated over three decades by the herpetology laboratory at the QCAZ, Coloma *et al.* developed a strategic plan for the research and conservation of Ecuadorian amphibians. This plan and project were referred as the "Balsa de los Sapos" or "life raft for frogs"—an Ecuadorian local name that emulated Noah's ark.

The strategic plan was crafted to catalyze the conservation efforts for threatened amphibians in Ecuador. Six programs (with numerous projects) guided this ambitious undertaking: 1) research and monitoring of native amphibians; 2) strengthening of local capacities; 3) management and ex situ conservation of species at risk; 4) in situ conservation; 5) public education; and 6) bioinformatics and sharing of information.

As part of the six programs in the plan, in 2006, the infrastructure for the first Research and Conservation Center for Endangered Amphibians in South America (CICA) was established at PUCE, with the support of the Saint Louis Zoo from the USA and Amphibian Ark (AARK)—a partnership between the World Association of Zoos and Aquariums, IUCN/SSC Conservation Breeding Specialist Group, and IUCN/SSC Amphibian Specialist Group. AARK was coordinated by Joseph Mendelson III, Kevin Zippel, and Ron Gagliardo, all of whom provided significant support to Coloma's ex situ program (Zippel and Mendelson, 2008; Gagliardo, 2011). Currently, PUCE still maintains this center for the ex situ management of amphibians at risk of extinction, now directed by Andrés Merino-Viteri.

From Anfibios de Ecuador to AmphibiaWebEcuador and to Anfibios del Ecuador at QCAZ

The history of the online encyclopedia of Anfibios de Ecuador was briefly recounted by Coloma in Renjifo (1999) and later by Ron (see "Agradecimientos" in Ron *et al.* 2022). In the 1990s, the technological advances of the internet and the popularity of the web led to a boom in the dissemination of information through web pages. At that time, Néstor Acosta-Buenaño was knowledgeable about the technological advances and well aware of the importance of these technologies for science public outreach and education. Acosta-Buenaño was a biology student mentored by Coloma who worked on dendrobatid calls for his *licenciatura* thesis. At the end of the 1990s, he proposed displaying a dynamic update of the checklist of the Ecuadorian amphibians (Coloma, 1991), as well as the reptile checklist and other zoological information of the QCAZ, on the PUCE website. Needless to say, his proposal was more than welcome, as it encompassed the importance of having both high-quality and updated check lists and providing information on each species. Thus, the electronic version of Anfibios de Ecuador originated as a technological update of the aforementioned publication. Coloma's printed version of the checklist of Ecuadorian amphibians was the result of a review of the literature related to the description and distribution of each of the 375 Ecuadorian amphibian species known as of 1990. The research and bibliographic compilation work was carried out by Coloma while he was a KU student in the library of the Division of Herpetology at the University of Kansas Museum of Natural History. This compilation was based on the previous works of Miyata (1982) and Frost (1985). Darrel R. Frost and Duellman reviewed the contents of the manuscript before it was published in Ecuador by the environmental NGO *Ecociencia (Fundación Ecuatoriana de Estudios Ecológicos)*, headed by Danilo Silva Chiriboga, who recognized the importance of basic and reliable taxonomic information for conservation biology.

From 2000 to 2010, Coloma and Acosta-Buenaño were highly committed to working on the online encyclopedias on Ecuadorian amphibians that were conceived as part of the online Catalog of the Ecuadorian Fauna (Renjifo, 1999). Coloma wrote and edited its contents, while Acosta-Buenaño developed, programmed, and designed the three layers of the website architecture (databases, administrative, and presentation layers). As the content and informatics aspects of the encyclopedia of Ecuadorian amphibians evolved, several researchers and assistants collaborated (their input was mentioned in the acknowledgements section of each version). Between 2000 and 2010, three versions of the encyclopedia were produced. Version 1 (Coloma and Quiguango-Ubillús, 2000) was an update on the printed list of amphibians of Ecuador by Coloma (1991). In 2005 the list contained about 440 known species and their distribution in altitudinal regions, lists of diversity, and geographic endemism by family, a list of endemic species, and graphs of percentages of species (according to altitude) in natural regions and percentage of endemic species (in ecoregions defined by the WWF) in natural regions. Version 2 appeared between 2005 and 2010 (Coloma, 2005) as the conceptual and physical model of the database of the QCAZ and the encyclopedia was developed. The encyclopedia added species accounts of Bufonidae, Dendrobatidae, and Hemiphractidae, plus bibliographic references for each species description and other scientific and outreach references, as well as photographs of most of Ecuadorian amphibians and distribution maps. Coloma and Pablo A. Menéndez-Guerrero contributed significant efforts into producing the distribution maps and predictive models, with support from Ron (who at that time was studying at the University of Texas). Alexandra Quiguango-Ubillús and Cristina Félix-Novoa helped organize the information in the databases, the bibliographies, and the photo compilations for the summaries of 75 species of Bufonidae and Dendrobatidae. Acosta-Buenaño was responsible for managing the taxonomic backbone of the database. Version 3 was planned in 2008 and began to appear in 2010, with the important advance that the pages of Anfibios del Ecuador were no longer static; a dynamic website was incorporated with interrelated content and associated with the databases.

Between 2003 and 2006, donations from the Ecuadorian government through income taxes made by individuals and private companies supported the QCAZ Life Raft for Frogs Initiative that included the development of the Encyclopedia of Amphibians of Ecuador. Verónica Cano Molestina and Miguel A. Rodríguez worked hard to manage these tax funds and coordinate the initiative. Since 2009, the third version of the encyclopedia has had the financial support from two important sources, the National Secretariat of Science and Technology of Ecuador (SENACYT), which funded an informatic component as part of the project titled "Inventory and genetic and morphological characterization of the diversity of amphibians, reptiles and birds of the Ecuadorian Andes", and the JRS Biodiversity Foundation, which funded the project "Developing a biodiversity information system for Ecuador: Integrating taxonomic expertise, bioinformatics and open access". These projects were fully embraced by Elisa Bonaccorso (JRS) and Juan M. Guayasamin (SENACYT), respectively. They participated with boundless energy and efficiency in their conceptualization, development, fund acquisition, direction, and coordination. Verónica Cano Molestina and Miguel Angel Rodríguez also helped with developing the SENACYT project and María D. Guarderas with administrative tasks. In 2008, data entry assistants, computer experts, and research assistants were incorporated with the production of the electronic encyclopedia, by then named AmphibiaWebEcuador (and part of FaunaWebEcuador), and they contributed to the content published from 2010 onwards. Cristina Félix-Novoa assisted Coloma (in 2007–2008) in taking photographs, organizing photographic databases, performing bibliographic compilation, and digitizing information for summaries of most species. Caty Frenkel helped Coloma and Guayasamin (in 2009–2010) to complete information on species accounts of Bufonidae, Dendrobatidae, Centrolenidae, Ceratophryidae, Craugastoridae, Dendrobatidae, Hemiphractidae, Microhylidae, Ranidae, Caudata, and Gymnophiona. She also edited X-ray images and other figures and compiled biographies of authors of various amphibian descriptions. From January to November 2010, Morley Read worked on the accounts of Hylidae. Pablo A. Menéndez-Guerrero participated with great efficiency in the elaboration of maps, first during 2002–2003 in elaborating the distribution maps for the "Global amphibian assessment" and from 2007–2008 on the preparation of predictive distribution of endemic species in the framework of the project, "Identification of priorities for the conservation of Ecuadorian amphibians through the use of geographic information systems". In 2010, Belén Baus updated current and predictive distribution maps of the majority of species. Ítalo G. Tapia, who worked as the QCAZ Collections Manager (2000–2011), brought creativity, energy, and dedication to the organization of the collections and their data, taking photographs, and especially providing photographic documentation of the ontogenetic variation in numerous species (see Figure 2.43). Beginning in 2008, Elicio E. Tapia and Sofía Carvajal-Endara helped with collecting specimens, activities in the photography studio, and with various miscellaneous tasks. Diego Almeida Reinoso worked from 2004 to 2010 in the care, maintenance, and breeding of many tadpoles used for the images that document their ontogeny. Without his dedication to the meticulous job of raising the tadpoles and frogs, many of the photographs would not have been taken. Matilde Moreano organized and edited the images between January and October 2009, and then Paulina Santiana took over at the end of 2009 up to 2014. Damián Nicolalde and Denisse Estévez helped with the administration of the servers and some programming aspects. In 2011, and after the departure from PUCE of the leading actors of the encyclopedia (Coloma, Acosta-Buenaño, Bonaccorso, and Guayasamin), the encyclopedia (AmphibiaWebEcuador up to 2019) continued by incorporating the work done up to 2010 and by adding new data. In 2011, Ron began to edit the encyclopedia. At present, the encyclopedia is named Anfibios del Ecuador and is part of Bioweb (https://bioweb.bio/). Six versions were published between 2011 and 2019 (Ron *et al.*, 2011, 2013b, 2013c, 2014a, 2014b, 2019). Other contributors are indicated in the acknowledgements section of the current version (Ron *et al.*, 2022). At present, the encyclopedia has real-time access to information on the ~75,000 specimens in the amphibian collection of the QCAZ, an image gallery with more than 200,000 photographs of amphibians, a call database that gives access to recordings of more than 220 species, and pages compatible with mobile devices.

Changing of the Guard: Ron and the QCAZ since the 2010s

After Coloma's departure from PUCE in December 2010, Santiago R. Ron (Figure 2.44) took the lead of the QCAZ, and in 2011, Andrés Merino-Viteri (see Figure 2.44) became the Director of the Balsa de los Sapos ex situ program.

Ron joined the herpetology section of the QCAZ in 1998 after finishing his master's studies at KU and then left in 2002 for graduate studies and returned in 2007. Ron is part of Duellman's academic lineage (Adler, 2012), having completed graduate studies at the University of Kansas and University of Texas, and was mentored by Duellman,

FIGURE 2.44 Left: Santiago R. Ron and Andrés Merino-Viteri at QCAZ in 3 January 2005. Right: Santiago R. Ron at Reserva de Producción de Fauna Cuyabeno in 3 February 2008. Photos by LAC.

Trueb, and Cannatella. Ron's academic career was briefly recounted by Duellman (2015a) (under "More Frogs, Salamanders and Even Caecilians"), who also provided a biographical sketch (under "Kansas Herpetological Alumni"). Several other aspects of Ron's life have been described in outreach publications (e.g., Molineros, 2016, Balarezo, 2023), and extracts from them and additional information is herein provided, along with an academic biography.

Ron became fascinated with frogs in his infancy when a marsupial frog caught his attention in Quito. He described his inclination to the study of amphibians in a popular article as follows:

> "I was a completely city-raised kid, and I remember vividly the first time I found a little frog in Quito. It was the rainfrog of Quito, and it seemed to me like the most incredible little creature in the world. In fact, the rainfrog of Quito is a common species in the city's gardens and surroundings, where its singing can be heard during both day and night, especially when it rains." "If someone comes across this little frog, they'd probably find it ugly, but to me, it's adorable." I never hesitated to pursue a degree in Biology, and during my first semester, I confirmed that my true calling was studying frogs.
>
> (Our English translation from Molineros, 2016)

During his time at the QCAZ between 1998 and 2002, Ron developed a keen interest in the systematics of *Osteocephalus* frogs. At that time, these frogs posed significant challenges in terms of accurate identification. They were found in abundance at several fieldwork sites, particularly in Parque Nacional Yasuní, and of course there were many unsolved taxonomic problems and also cryptic species awaiting descriptions. Ron completed his first description of a new species of amphibian (Ron and Pramuk, 1999) and a year later described a couple of new species from Jatun Sacha jointly with the German Karl Heinz Jungfer (Jungfer *et al.*, 2000) (see more about him under "The German Big Bang of Poison Dart Frogs in Europe"). This was followed by detailed phylogenetic analyses and taxonomic revisions with the description of three other species of *Osteocephalus* in 2012 and 2020 (Ron *et al.*, 2012; Chasiluisa *et al.*, 2020).

During his graduate studies at the University of Texas, Ron became interested in the systematics and sexual selection of foam nest frogs of the genus *Engystomops* under the mentorship of David C. Cannatella. His passion for these frogs and research led to several publications (Ron *et al.*, 2004, 2005, 2006, 2014c). The *Engystomops* research also marked the beginning of describing cryptic species using the topologies of molecular phylogenies for their discovery. Over the past decade, the incorporation of the latter has become almost mandatory in QCAZ systematic publications authored by Ron and his *licenciatura* students (e.g., Ron *et al.*, 2012, 2013a, 2015, 2016, 2018a, 2018b, 2020, 2022; Caminer and Ron, 2014, 2020; dos Santos *et al.*, 2015; Caminer *et al.*, 2017; Rivadeneira *et al.*, 2018; Páez and Ron, 2019; Chasiluisa *et al.*, 2020; Zumel *et al.*, 2021; Ortega *et al.*, 2022). Among Ron's most significant students, Marcel Adrián Caminer Rodríguez, Nadia B. Páez, Jhael Ortega, and María José Navarrete not only prepared several relevant taxonomic studies and new species descriptions but continued professionally, engaged either in graduate studies abroad (Caminer, Páez, Navarrete) or continued involvement with the QCAZ (Ortega). Thus, the 2010s was highly productive in the description of species new to science. Ron participated in the description of at least 59 species between 2011 and 2022 for sum a total of 70 species from Ecuador that he has described either as author or coauthor since 1999. This makes him the third most prolific author (after Lynch and Duellman) in terms of the number of new species described (see Figure 2.27). Ron and QCAZ personnel were involved in the organization of the XI Latin American Congress of Herpetology, which was held in July 2017 at PUCE in Quito.

Andrés Merino-Viteri began as a student at the School of Biological Sciences at PUCE in 1994 and since 2001 has been a professor. At the QCAZ, he conducted research on the causes of amphibian declines in the Andes and collaborated in the amphibian exhibit Sapari from 2004 to 2005. He permanently joined the QCAZ in 2011 as Director of the Balsa de Sapos (Merino-Viteri, 2015), for which Freddy Almeida Reinoso is the Collection Manager. In 2018, Merino-Viteri obtained his PhD in Tropical Ecology at James Cook University in Australia, where he specialized in the potential effects of climate change on amphibians through the study and analysis of physiological parameters and the microenvironmental conditions in which organisms live. An overview of the project Balsa de Sapos is given by Ron (Ponce, 2019).

Currently, the QCAZ maintains the largest collection of Ecuadorian amphibians, tissues, and associated data and is actively involved in research. A summary of QCAZ advancements was provided by Ron for the online portal Anfibios del Ecuador Bioweb and in Womack *et al.* (2022).

> With 656 amphibian species (as of November 2021), Ecuador is the third most species-rich country in the world. In 2018, the zoology museum of the Catholic University of Ecuador (QCAZ) launched the website Anfibios del Ecuador (https://bioweb.bio/faunaweb/amphibiaweb) to give access to comprehensive information about all amphibian species of Ecuador, which represent nearly 8% of the amphibian species worldwide. Anfibios del Ecuador was inspired by AmphibiaWeb and has a similar scope but at a country level. Anfibios del Ecuador replaced the previous portal "AmphibiaWebEcuador," which operated between 2010 and 2017. The website is currently part of a larger Ecuadorian biodiversity database called BIOWEB.
>
> Anfibios del Ecuador provides detailed species accounts, which include an extensive photo gallery with 206,785 images. Species phenotypic variation is well documented with an average number of photographs per species of 323 (maximum 8746 for *Pristimantis achatinus*) and a median of 144. Species distribution is documented with a database of over 50,000 geographic records from

the literature and the specimen database of the QCAZ museum. Those records are used to build species distribution models under current and future environmental conditions using the MAXENT algorithm . . .

The website also presents overviews of the biogeography, species richness, and the conservation status of the Ecuadorian amphibians. Overviews of the natural regions, geography and climate of Ecuador are also provided. Anfibios del Ecuador gives access to: (1) the database of the amphibian QCAZ collection, the largest for Ecuadorian amphibians (76,500 specimens), and (2) the QCAZ sound collection with 1297 recordings for 222 species.

Crucially, Anfibios del Ecuador is in Spanish and helps to provide scientific data in languages other than English. For most species, Anfibios del Ecuador presents the only comprehensive species accounts available online in Spanish. Since 2018, Anfibios del Ecuador has been connected with AmphibiaWeb through a web API that allows visualizing the Spanish-language species accounts and photographs from Anfibios del Ecuador within AmphibiaWeb.

Anfibios del Ecuador has high visitor traffic (data from Google Analytics). Between June and November 2021, for example, it received approximately 16,000 monthly visitors from 106 countries or sovereignties. 51% of visitors shared demographic data, of which 72% were women and 28% men. Younger age groups are the most frequent visitors: 18– 24 years old represented 37% of the visits, followed by 25–34 years (23%).

(Womack *et al.*, 2022)

Read: The Great Naturalist and Survivor of an Airplane Crash and a Kidnapping

Morley Read (Figure 2.45), a freelance researcher, began his Ecuadorian amphibian journey in 1986 when he participated as a herpetologist in a British expedition to develop a management plan for the Reserva Río Mazán, west of Cuenca, Provincia Azuay. For that project, he conducted inventories and monitoring of amphibians (Read, 1986). At that time, he was able to document in Mazán the presence and relative abundance of nine species of amphibians that soon would begin to collapse. Four of the species he recorded have not been seen again in the Mazán Valley, in spite of intensive search efforts. They are: *Hyloxalus vertebralis*, *H. anthracinus*, *Centrolene buckleyi*, and *Telmatobius niger*, while others became very rare (*Atelopus exiguus*, *Gastrotheca cuencana*, and *G. pseustes*).

Read was the pioneer in Ecuador in producing a series of CDs with the songs of Ecuadorian amphibian species and has formed a valuable audio bank (Read, 2000) that has been deposited at the QCAZ and CJ. He has done much the same with photographs and videos of the amphibians of Ecuador. These images are distributed worldwide, and the quality of his work is high. He has filmed sequences for David Attenborough's award-winning BBC series *Life on Earth* and *The Living Planet*. Some of his archives have immense historical value because he documented species during the 1980s and 1990s that are currently probably

FIGURE 2.45 Left: Morley Read in Trinidad in 1984. Photo courtesy of Morley Read. Right: Morley Read and Nanto Huamoni (Waorani leader) at Shiripuno River, Provincia Orellana, Ecuador, May 2017. Photo by LAC.

extinct. His photos are deposited at the Science Photo Library, Getty Images, Alamy, Shutterstock, CJ, and QCAZ photo archives, among others. Most of his audio recordings are deposited in the CJ and QCAZ audio libraries.

Read is one of the most knowledgeable amphibian field biologists in Ecuador. His field recording experience is unique and is the accumulation of a life passionately dedicated to this activity. Perhaps the most important contributions of Read have been his inventories in areas where the petroleum industry has been active in the Amazonian region. One of his long-term projects is the monitoring of amphibians in Parque Nacional Yasuní (PNY), Provincia Orellana, which continues to date. Read (2018) summarized the monitoring from 1994 to 2012 at the PNY, and he provided color photos of most species. At that time, he had documented the presence of 113 species out of 135 currently recorded from the PNY (Ron *et al.*, 2022), a world record compared to similar sites. No declines were noticed for most of the species, in contrast to declines that have occurred at highland sites.

Read has been highly committed to environmental causes, especially conservation projects and the fight against the petroleum and mining industries, using amphibians as indicator species. Among other projects, he was involved in establishing the nature reserve of Los Cedros on the western slopes of the Andes (he was director from 1991 to 1993). Recently, Los Cedros was in the news because of a clear victory for the science produced over the years and the rights of nature. To the satisfaction of the Supreme Court of Ecuador, mining has practically been excluded from the reserve, along with lumbering and other extractive activities. From 1997 to 2004, Read was involved in a project for the biorestoration of Ilaló, a mountain ecological reserve in Provincia Pichincha. While doing the amphibian work, Read was able to participate with indigenous communities in the conservation of the area, as was the case in PNY, in which he was directly involved in the field delimitation work of the park, walking hundreds of kilometers and opening trails through the jungle with the Waorani (e.g., Nanto Huamoni) and Kichwa people. A few mishaps and misadventures have been also part of his intense amphibian work. One that caught the attention of the British news media (e.g., *The Guardian*, *The Mail*, *The Times*) was his kidnapping, which occurred in February 1997 in the Amazonian region by the Achuar indigenous community in southeastern Ecuador, while he had been hired by Walsh Environmental Scientists and Engineers for CGC Energy to conduct an environmental impact survey. In a publication in *BBC Wildlife* he recounted his experience (Figure 2.46).

The DHMECN Collection: From the Friends of Nature to the National Biodiversity Institute

Mario H. Yánez-Muñoz

Forty-three years have passed (from 1979 to June 2022) since the amphibian collection of the DHMECN (transferred in 2017 to the INABIO) was initiated. It contains 13,411 amphibians from most of the Ecuadorian territory (Figure 2.47), holds nearly 80% of the amphibian species of Ecuador, and includes 39 holotypes (38 anura, one caudata) and 173 paratypes. It is the second largest collection of preserved amphibians in Ecuador and is increasing at an average rate of about 750 specimens per year. The main collectors are indicated in Figure 2.48. Standing out among them is Mario H. Yánez-Muñoz, who has been working with the collection since 1999 and who officially took a position as a public servant to accomplish curator duties in charge of the herpetology collection in 2004 and later, in 2017, as Aggregate Researcher of the INABIO. Under the leadership of Yánez-Muñoz, a prolific taxonomic and systematic research program closely tied to the DHMENC collections and biology students, particularly from UC, has been developed and has recently been evolving from traditional phenetic taxonomy (e.g., Yánez-Muñoz *et al.*, 2011) to phylogenetic and integrative approaches (e.g., Yánez-Muñoz *et al.*, 2021). Inspired by the titanic historical legacy initiated by William E. Duellman (Coloma and Guayasamin, 2022; Burrowes *et al.*, 2022) and motivated by the Ecuadorian researchers who laid the foundations for the country's amphibian biology research, the DHMECN maintains a strong commitment to collection-based amphibian systematic research. This research is closely connected to the activities of NGOs (e.g., Jocotoco, Ecominga) dedicated to land acquisition and in situ conservation activities (MECN, 2013; Reyes-Puig *et al.*, 2022a).

As recounted in the first publication of the MECN (Moreno Espinosa, 1979), by the end of the 1960s, the idea of having an Ecuadorian national museum began to be explored, and for this purpose, a group of naturalist promoters met in 1967 and in October 1969 and formed a group named Amigos de la Naturaleza (Friends of Nature), which, in 1973 (according to statutes approved by the Ecuadorian government through the Ministerio de Agricultura y Ganadería on 29 July 1974), founded the Sociedad Ecuatoriana Francisco Campos de Amigos de la Naturaleza. Among these promoters, the most active were Miguel Moreno Espinosa (1918–2004, a civil engineer and naturalist), Samuel Valarezo Delgado (1911–1980, a pioneering Ecuadorian ornithologist), Fernando Ortiz Crespo (1942–2001, well-known Ecuadorian ornithologist), Gustavo Orcés Villagómez (the Ecuadorian pioneer zoologist), Manuel Guerra Leiva, and Rosario de Lafabre. One of the aims of this society was to develop a project to create a national museum. This proposal was supported by PUCE, the EPN, the Instituto Geográfico Militar, the WWF, and *Fundación Natura*, among others. The efforts paid off during the military dictatorship (led by Alfredo Poveda Burbano), with the creation of the Museo Ecuatoriano de Ciencias Naturales (MECN) on 24 May 1978 as part of the Casa de la Cultura Ecuatoriana under the leadership of Moreno Espinosa (executive director), Valarezo Delgado (cataloger), and Juan Carlos Matheus Pozo (specimen preparator). In 1979, the MECN hired Ortiz Crespo

TALES FROM THE BUSH

Morley Read thought he'd had his share of bad luck when his plane crashed in Ecuador's rainforest. But, two weeks later, while on the same environmental survey, he got himself kidnapped.

The distant sound of a small plane broke the rainforest peace. Our hopes lifted – perhaps rescue had come at last. We ran to the walkie-talkie to see if someone had come to give us news. But we couldn't get through, and the plane disappeared into the distance.

From our initial group of five, only Mark (a geologist and the project leader) and I – a zoologist – were still being held captive by Achuar Indians in the rainforest village of Shaimi, deep in the Ecuadorian Amazon. The botanist and the anthropologist had both been released because they were ill, and the representative of the company carrying out the environmental survey had left a few days ago to go to Puyo (the capital of Pastaza Province in which Shaimi lies) to negotiate our release. Until he managed to do so, we were stuck.

Escape was not an option. Though we had the freedom of the village, our maps, satellite photographs and GPS (geographical positioning system) showed us that the nearest road was 33km away and the terrain hilly, densely forested and without paths. There was nothing we could do but wait.

It was eight days since our environmental study had been brought to an abrupt end, when Luis Vargas – an Indian politician – and two members of FINAE (the organisation which represents the Achuar communities in the Ecuadorian Amazon) had suddenly arrived at Shaimi, where we were camped, in a chartered Cessna and kidnapped us.

Luis, dressed in military paraphernalia from the recent war with Peru, had told us, "You came here without permission. We do not allow oil companies in our territory, and you are forbidden to leave until compensation of $1 million is paid."

At the time, he had looked rather uncomfortable, because he and I knew each other well, having worked together for a Channel 4 production *Flames in the Forest*, broadcast in the UK in 1991, which highlights the devastating environmental problems that petroleum exploitation has caused in the Ecuadorian Amazon.

Though our team had, indeed, been contracted by an oil company, our report was to have been lodged with the government, which requires environmental studies to be carried out before a company is allowed into the area. We had been surveying a 200,000ha oil concession granted to an Argentinean company. I was documenting the fauna and was especially interested in finding rare and endemic species – in particular reptiles and amphibians, my specialities. The more I found, the more reason there would be that the oil company should keep out of the area. Others in my group were studying the plant life and taking soil and water samples.

The Achuar Indians, who have lived in this area for centuries and have been granted legal title to their land, are very concerned at any threat to their territory. What I did not realise at the time was that the company that had contracted me had not obtained permission in the proper manner from the Indians to enter their territory.

Mark and I had no idea how long we would be held hostage. We whiled away the time trying to work out how long the runway would take to dry out sufficiently to allow a landing, and we were already familiar with every rut and puddle.

Later that day, heavy rainclouds gathered. There would be no more planes flying over today. We left off pacing the runway and went to the village schoolroom. Luis was waiting for us, having called a meeting of the villagers. "I will defend Achuar territory until I die," he announced, though this was probably the longest time that he had spent in the community since his childhood. His comment was aimed as much at the inhabitants of Shaimi as at us, because it was apparent that they were becoming embarrassed at our presence. They were hoping to attract tourists to their village – the house where we were held captive had, in fact, been built with this in mind. But who would want to come to stay with Indians who had a reputation as kidnappers?

We spent the rest of the day reading some of last year's newspapers, which the botanist had brought to press his plants.

That night, it didn't stop raining. When the sun rose the next day, we went straight to examine the runway. It was sodden. There would be no planes landing today. As the day drew on, the weather cleared a little, but the river continued to rise. Branches – even entire trees – were swept downstream in the flood. By the afternoon, the water level was only one metre from the runway surface.

Then, suddenly, we heard a motor. It was a helicopter, which would be able to land where planes couldn't. The vice-president of FINAE climbed down with the authorisation of our release. We scrambled aboard, and took off almost immediately, immensely relieved that our captivity was over.

◆ *The oil company paid a ransom of £1,000 to FINAE for Morley's and Mark's release.*

Crash course. Jungle aviation is perilous, as the author *(right)* found out when, shortly before the kidnapping, a downdraft forced the plane he was in onto a runway prematurely. Though the plane hit a log and flipped over, nobody was seriously hurt. He now knew, however, that a flimsy 5-seater plane wouldn't be able to land on a muddy runway and rescue them.

Reluctant guests, reluctant hosts. The Indians in Shaimi (about 10km from the village of Achuar, *below*, where the crash occurred) were as much innocent players in the kidnapping as the author.

106 May 1997 BBC Wildlife

FIGURE 2.46 *BBC Wildlife* reportage recounting Morley Read's kidnapping in Ecuador, May 1997.

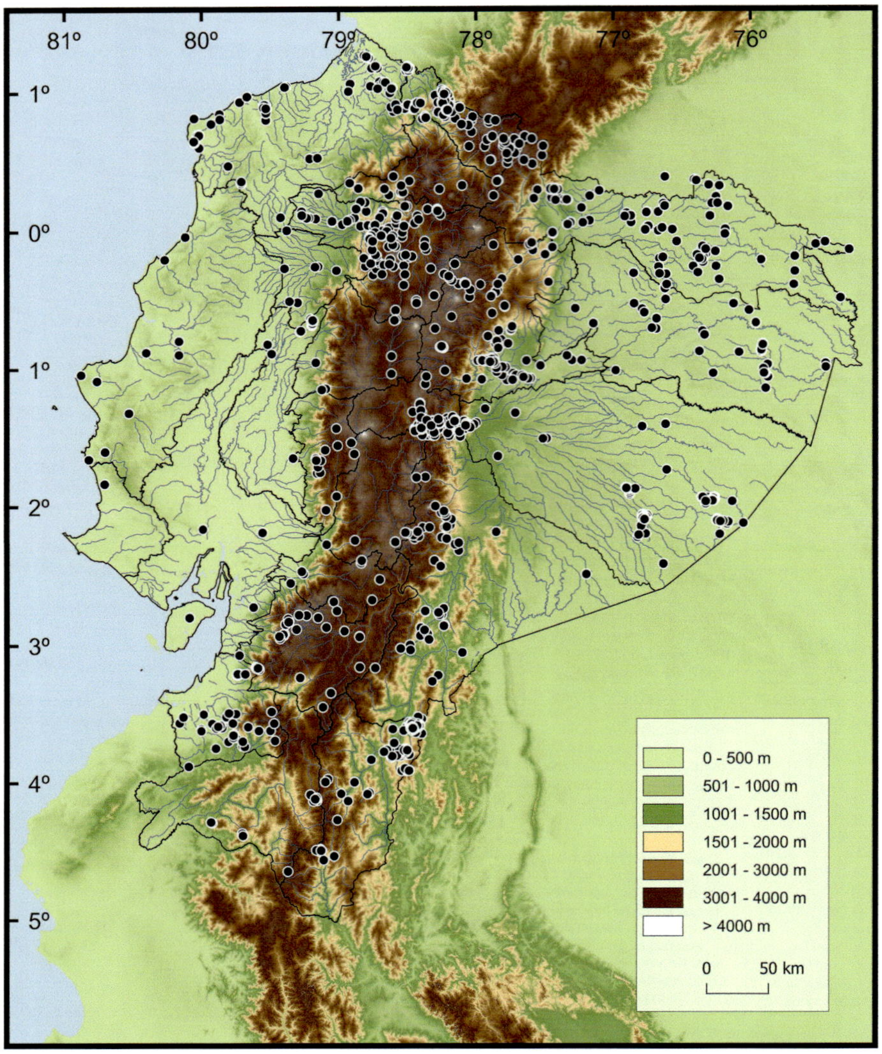

FIGURE 2.47 Ecuadorian collecting localities of the amphibian specimens deposited in the DHMECN of the INABIO. Data extracted from the national database of biodiversity by the INABIO (https://bndb.sisbioecuador.bio/bndb/index.php, downloaded on 10 July 2022).

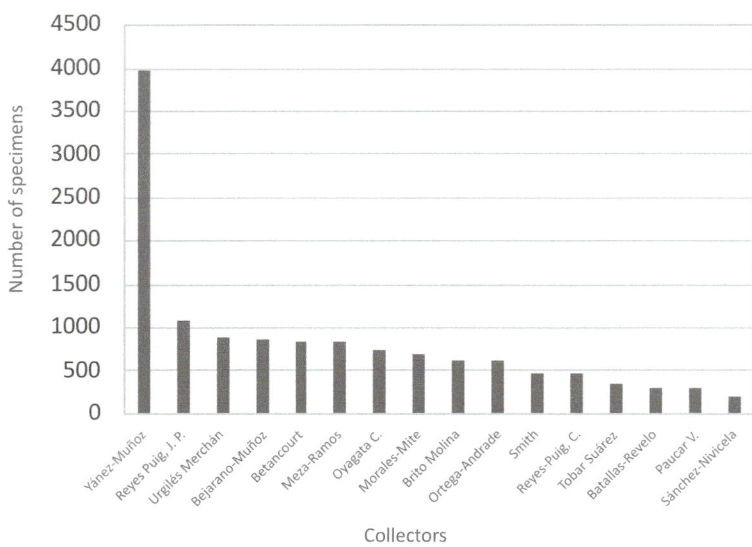

FIGURE 2.48 Fifteen main collectors (either first or secondary collector) and numbers of specimens that each deposited at the DHMECN. Data extracted from the national database of biodiversity by the INABIO (https://bndb.sisbioecuador.bio/bndb/index.php, downloaded on 10 July 2022).

FIGURE 2.49 Top left: personnel of the Museo Ecuatoriano de Ciencias Naturales in 1982. From left to right: standing: Guido Román Jiménez, Hernando Román Jiménez, Fausto Sarmiento Rodríguez, Miguel Moreno Espinosa, Fernando Ortiz Crespo, Juan Carlos Matheus Pozo, Diego Bastidas Andrade, and Angel Román Carrión. Sitting: Leticia Avilés Chevalier, Clemencia Vela, Mercedes Ron, Bethy Valencia Rojas, Amparo Avilés (photographer unknown). Top right: Gustavo Orcés Villagómez at the ornithological collection at the EPN in 1980. Photo by Fernando Ortiz Crespo. Bottom left: Miguel Moreno and his assistant Roberto Reece Dousdebés at the MECN inspecting the recently acquired dermestid colony for skeletal preparations. Bottom right: Samuel Valarezo Delgado with the ornithological collection at the MECN. 25 April 1980. Photos by John E. Simmons.

(cataloger), and Roberto Reece Dousdebés (laboratory assistant, an entomologist) (Figure 2.49).

At that time, the amphibian DHMECN collections began with 11 specimens collected between 1964 and March 1979 in the western and eastern lowlands of Ecuador by Valarezo Delgado and 130 specimens collected by Duellman's students David C. Cannatella and Thomas J. Berger at Limoncocha in April 1979. Two bufonids (*Rhinella alata* and *Rhaebo haematiticus*) collected by Samuel Valarezo in Puerto Quito, Provincia Pichincha, in February 1964 are the oldest specimens in the DHMECN collection.

After 12 years (up to 1992), the collection had grown to 644 amphibian specimens (MECN 1992) and was supervised by Orcés Villagómez and Matheus Pozo (a biology student at PUCE). Their approach allowed the museum to sustain a reference collection, represented by several families of Ecuadorian amphibians representing about 10% of the species reported at that time. The initial collections were greatly enriched by the amphibian specimens deposited by Eduardo Azanza and Leticia Avilés (45 specimens collected between 1979 and 1984, mostly in Reserva de Producción Faunística Cuyabeno, Provincia Sucumbíos), Jean Marc Touzet (119 specimens collected in 1983–1987, most from San Pablo de Kantesya, Provincia Sucumbíos), and Luis A. Coloma (104 specimens collected in 1984–1985, most from Cabeceras Río Baboso, Provincia Carchi, and other northwestern and Andean localities in Ecuador), among others. The amphibian specimens deposited at the DHMECN were mainly used for research conducted by Hoogmoed (1985) on *Andinophryne* (now *Rhaebo*), Coloma (1995) on *Colostethus* (currently under *Allobates*, *Colostethus*, *Ectopoglossus*, *Hyloxalus*, *Epipedobates*, *Leucostethus*, and *Paruwrobates*), and Lynch and Duellman (1997) on *Eleutherodactylus* (most now *Pristimantis*). The research interests of Orcés Villagómez and Matheus Pozo focused on reptiles, and their only project with amphibians did not come to fruition; their planned paper reporting *Phyllomedusa psylopigion* (now *Agalychnis psylopigion*) for the first time from Ecuador was never published.

Between 1992 and 1999, the herpetological collection increased to 1123 amphibians, most from opportunistic collections; some were deposited in the MECN by researchers as duplicates of their collections deposited elsewhere. An exception to these occasional collections was the amphibian collection made by Gregory O. Vigle (137 specimens deposited in 1995, 136 from Estación Biológica Jatun Sacha, Provincia Napo) for his amphibian surveys and conservation activities, which continued for two decades. During this time, Fabián Navas, Igor Castro R., and Juan Rivadeneira R. (today a professor at UCE) were the catalogers of the collection–their work focused strictly on the care and maintenance of the specimens; their research interests were focused on fish and terrestrial micromammals, not amphibians.

The New Millennium: Biologists from the Universidad Central and Taxonomy at the DHMECN

While working for almost five years (from 1998 to 2002) as a curatorial and field assistant for Ana Almendáriz-Cabezas at the EPN museum, Mario H. Yánez-Muñoz, a student at UCE who was mentored by Jean Marc Touzet—a well-known French herpetoculturist and founder of the Vivarium and *Fundación Herpetológica Gustavo Orcés*—learned about the existence of the MECN herpetological collection (now the DHMECN) in 1998. Almendáriz asked Yánez-Muñoz to return type material used for the description of the lizard *Anolis vanzolini* (Williams *et al.*, 1996) to the director of the institution, Moreno Espinosa. This was the opportunity for Yánez-Muñoz to meet and befriend the cataloger of the collection, Igor Castro, who a year later (1999) invited him to review the collection and organize the information associated with the historical catalogs of specimens (Yánez-Muñoz, 2002).

From 1999 to 2001, as a volunteer, Yánez-Muñoz took on the task of unifying four separate catalogs of the 1123 specimens that made up the MECN amphibian collection into a single catalog. Assisted by Marco M. Reyes-Puig and Cecilia Tobar Suárez (biology students at UCE) and in coordination with Juan F. Rivadeneira (cataloger), the first digital version of the collection's catalog was created, with a manual paper-based backup. This information was incorporated into the first global amphibian assessment led by Stuart *et al.* (2004).

The second group of collections from Yánez-Muñoz's *licenciatura* thesis are the specimens that symbolically began the new stage of the MECN's herpetological collection (the first one was deposited at the EPN, 234 specimens). For his research, Yánez-Muñoz recruited as collaborators Paúl A. Meza-Ramos, Juan P. Reyes-Puig, Flor Ortiz, Marco M. Reyes-Puig, and David Veintimilla Yánez (students at UCE). This enabled the compilation of collections from several priority points in the Andes and Chocó (Yánez-Muñoz, 2005). These collections supported the future degree research of Meza-Ramos on the population ecology of *Hyloxalus delatorreae* and Juan P. Reyes-Puig's work on the anuran diversity of the Tungurahua volcano. UCE students received guidance and mentorship from Nelson Gallo, an Ecuadorian biologist who held the position of Professor of Zoology at the Departamento de Biología of the Universidad Central del Ecuador and has been teaching numerous UC biologists since 1988. He has been honored in the name of the rainfrog *Pristimantis nelsongalloi* (Valencia *et al.*, 2019).

With the strong support of Francisco Sornoza-Molina (a member of *Fundación Jocotoco* in 2002) and other colleagues who funded 11 field trips between November 2001 and August 2004, Yánez-Muñoz and collaborators added new distribution records for several species of *Pristimantis*, *Agalychnis psilopygion*, and *Hyloscirtus tapichalaca* and threatened populations of *Atelopus* sp. (*spumarius* complex), *Centrolene buckleyi*, and *Hyloxalus delatorreae* (Meza-Ramos and Yánez-Muñoz, 2003, and references in MECN, 2013; Yánez-Muñoz *et al.*, 2020). These efforts were enhanced by the collections made by H. Mauricio Ortega-Andrade (a student at UCE and Ecociencia Foundation scholarship holder) in the Chocó region. Ortega-Andrade became a pioneer in producing ecological and natural history data on the amphibian and reptile communities of the Chocó (Ortega-Andrade *et al.*, 2010). His line of research promoted the undergraduate work of Cecilia Tobar Suárez, Mónica Arellano, and Christian R. Paucar Veintimilla.

After the 2001 retirement of Miguel Moreno Espinosa as executive director of the MECN and several transitions and failed administrative performances by Sonia Sandoval and Marco *Altamirano-Benavides*, in 2004, the MECN hired Yánez-Muñoz as Curator-in-Charge of the herpetological collection. Up until 2009, Yánez-Muñoz made several collections (and associated publications) in the Yasuní and Cuyabeno national parks (Yánez-Muñoz and Venegas, 2008); the western and eastern foothills of provinces of Imbabura, Carchi, and Sucumbíos (Yánez-Muñoz and Chimbo, 2007; Yánez-Muñoz and Mueses-Cisneros, 2009); the Chocó in Provincia Esmeraldas; the semi-deciduous forests in the Provincia El Oro (Yánez-Muñoz *et al.*, 2009c); and the forests of the Distrito Metropolitano de Quito (DMQ) (Yánez-Muñoz *et al.*, 2009b) (Figure 2.50).

Since the mid-2010s, Yánez has collaborated with other Ecuadorian amphibian biologists from diverse institutions, including Diego F. Cisneros-Heredia, Juan M. Guayasamin, Martín R. Bustamante, and H. Mauricio Ortega-Andrade, to produce the first reports from Ecuador of centrolenid glassfrogs and the descriptions of new species (*Teratohyla amelie* and *Nymphargus buenaventura*). The holotypes of the latter were the first holotypes in the DHMECN amphibian collection, and by 2009, the collection had increased to 6938 amphibian specimens (Guayasamin *et al.*, 2006a; Bustamante *et al.*, 2007; Cisneros-Heredia and Meza-Ramos, 2007; Cisneros-Heredia and Yánez-Muñoz, 2007a, 2007b; Yánez-Muñoz and Cisneros-Heredia, 2008; Yánez-Muñoz *et al.*, 2009a). At the same time, the monitoring efforts for threatened species were strengthened (Yánez-Muñoz and Meza-Ramos, 2006), and inventories were conducted in the southern Andes of Ecuador, the Pastaza River basin (Ortega-Andrade, 2009; Ortega-Andrade and Valencia, 2010), and around the Napo River. Meza-Ramos, Juan P. Reyes-Puig, Eric N. Smith, Carlos C. Martínez, Andrew Willworth, and Ortega-Andrade collaborated on these inventories and monitoring work through 10 research projects that were funded and supported

FIGURE 2.50 Some of the MECN and INABIO outreach publications that include Ecuadorian amphibians. Available at http://inabio.biodiversidad.gob.ec/nuestros-libros/.

by numerous NGOs and institutions. At this stage, Mauricio Herrera Madrid, Miguel A. Urgilés Merchán, Christian R. Paucar Veintimilla, Carolina Reyes-Puig (a herpetologist and mammologist currently at ZSFQ and the sister of Marco and Juan P. Reyes-Puig), María Pérez Lara, Luis A. Oyagata, and Patricia Bejarano-Muñoz (all from UCE) joined as collaborators. They assisted in the management and maintenance of the database, as well as on the expeditions to the DMQ and the northern extreme of Ecuador. Urgilés Merchán, Pérez Lara, Oyagata, and Bejarano-Muñoz did their undergraduate thesis research on the amphibian and reptile diversity of the DMQ and the upper Pastaza River basin. (Figures 2.51, 2.52).

The visit of Eric N. Smith (herpetologist at the University of Texas at Arlington) and Hussam Zaher (herpetologist from the University of Sao Paulo) in 2008 radically influenced the DHMECN's procedure in relation to protocols, collection management, and information associated with the specimens (extraction and management of tissues and photographic archives). Smith traveled with the DHMECN team to nearly 50 localities in the provinces of Esmeraldas, Imbabura, Pichincha, Tungurahua, and Morona Santiago, which resulted in finding *Atelopus palmatus*, a critically endangered species that had not been recorded for 20 years (Yánez-Muñoz *et al.*, 2011d), as well as the discovery of at least five new species of amphibians (among a dozen species of amphibians and reptiles).

An Explosion of New Species

After five years of intensive analysis of the collections in the Andes of Ecuador, Yánez-Muñoz, motivated by Cisneros-Heredia, produced a flood of new amphibian species in a special issue of the USFQ journal. Nine new species of frogs

FIGURE 2.51 MECN researchers and associates during field expeditions. Upper left: from left to right, Francisco Sornoza-Molina, Patricia Bejarano-Muñoz, Mario H. Yánez-Muñoz, María Pérez Lara, Luis Oyagata C., and Christian Castro in Yunguilla, Provincia Azuay, in 2010. Photo by Mario H. Yánez-Muñoz. Upper right: Salomón Ramírez J. (left) and Miguel A. Urgilés Merchán (right) at Reserva Tapichalaca, Provincia Zamora Chinchipe, 2008. Photo by Juan Pablo Reyes-Puig. Middle left: Mario H. Yánez-Muñoz (left) and H. Mauricio Ortega-Andrade (right) at Reserva Colonso-Chalupas, Provincia Napo, 2021. Photo by Miguel A. Urgilés Merchán. Middle right: Paúl A. Meza-Ramos at Morán, Provincia Carchi, 2006. Photo by Mario H. Yánez-Muñoz. Bottom left: Carolina Reyes-Puig at Reserva Narupa, Provincia Napo, 2012. Photo by Manuel Morales-Mite. Bottom right: Jorge Brito Molina at Reserva Guango Lodge, Provincia Napo, 2015. Photo by Mario H. Yánez-Muñoz.

(*Osornophryne angel*—a junior synonym of *O. bufoniformis* sensu in this encyclopedia, *O. occidentalis*, *Epipedobates darwinwallacei*, *Pristimantis yumbo*, *P. romanorum*, *P. sirnigeli*, *P. loujosti*, *P. tungurahua*, and *P. rufoviridis*), were described for the international year of biodiversity in the journal *Avances en Ciencias e Ingenierías* (Yánez-Muñoz *et al.*, 2011; Cisneros-Heredia and Gluesenkamp, 2011; Cisneros-Heredia and Yánez-Muñoz, 2011; Reyes-Puig *et al.*, 2011; Valencia *et al.*, 2011) (the publication date is December 8, 2010, for both the online and printed versions, but the online version was made available on the web only in 2011, according to Darrell R. Frost (DRF note 2024, pers. com.). The printed version was distributed in Ecuador starting in December 2010 and globally from 2011 onward (Diego F. Cisneros-Heredia, pers. com.).

FIGURE 2.52 MECN-associated researchers. From left to right and top to bottom: Eric N. Smith (Photo by Sven Smith, 2017); Marco M. Reyes-Puig (Photo by MYM, 2007); Miguel A. Urgilés Merchán (Photo by MYM, 2021); Luis Oyagata C. (Photo courtesy of Luis Oyagata C., 2021); Paúl A. Meza-Ramos (Photo by Sandra Navas, 2012); Christian R. Paucar Veintimilla (Photo by MYM, 2021), Cecilia Tobar Suárez (Photo courtesy of Cecilia Tobar Suárez, 2017); Raquel Betancourt (Photo by Raquel Betancourt, 2021); E. Patricia Bejarano-Muñoz (Photo by MYM, 2015). MYM = Mario H. Yánez-Muñoz.

Therefore, wide distribution occurred only in 2011). In 2010, *Osornophryne cofanorum* was described in the journal *Papéis Avulsos de Zoología* (Mueses-Cisneros *et al.*, 2010). John Jairo Mueses Cisneros is a Colombian amphibian biologist who studied mainly the amphibians from Departamento Nariño on the Colombian border with Ecuador. This description was followed by the publication of a monographic book, *Serie Herpetofauna de Ecuador: El Chocó Esmeraldeño* (Ortega-Andrade *et al.*, 2009). The latter publication was the first identification guide and ecological summary for that region after the combined field efforts during five years of Ortega-Andrade and Yánez-Muñoz in more than a dozen localities of the Provincia Esmeraldas. John D. Lynch commented in the book's preface:

> This contribution is evidence of the development of herpetology by Ecuadorian scientists and represents a sign of its maturity. Each of the authors has participated in the documentation of Ecuadorian herpetofauna and in this contribution they are collaborating to produce a document that is more than a scientific study, a study that offers diverse users evidence of the biological diversity of an important sector of Ecuador.

At the beginning of 2010, inspired by the work of Luis A. Coloma and his collaborators who consolidated the amphibian and reptile sections of the QCAZ and a leading amphibian research center in South America, Yánez-Muñoz joined the master's program in conservation biology at PUCE. This academic opportunity supported the amphibian systematic and taxonomic research activities at the DHMECN. This was an excellent opportunity for him to benefit from the knowledge of the Ecuadorian specialists Coloma, Juan M. Guayasamin, Elisa Bonaccorso, Santiago R. Ron, Pablo A. Menéndez-Guerrero, and Omar Torres-Carvajal.

During the next two years, Yánez-Muñoz deepened his study of the terrestrial frogs *Pristimantis* to complete his master's thesis under the direction of Ron. During this time, Manuel Morales-Mite (a biologist from southern Ecuador with an extensive trajectory) assumed the post of Curator for the herpetological collection of the DHMECN. Between 2011 and 2012, Yánez-Muñoz curated of the collections of *Pristimantis* at the QCAZ and edited the content related to the taxonomic group Terrarana on the online site AmphibiaWebEcuador. During these years, *Pristimantis bellae* and *Osornophryne simpsoni* were described, two species that would initiate a close and long-lasting collaboration with *Fundación Ecominga* in the upper Pastaza River basin (Reyes-Puig and Yánez-Muñoz, 2012; Páez-Moscoso *et al.*, 2011). Upon their return to the MECN, Yánez-Muñoz and Morales-Mite concentrated their efforts on writing and editing the book *Herpetofauna en áreas prioritarias para la conservación: El sistema de reservas Jocotoco y Ecominga* (MECN, 2013), promoted by their friends Francisco Sornoza, Zoltan Waliczky, Rocío Merino, Nigel Simpson, Lou Jost, and Javier Robayo. The famous British broadcaster and naturalist, David Attenborough, commented in the book's foreword:

> In some parts of the world making such a fauna list, while time-consuming, does not present too great a challenge for competent naturalist. In the English Midlands where I grew up, there were eleven species of amphibians and reptiles. But in Ecuador there are 822. The numbers alone are daunting. When you add to that the wildness and inaccessibility of the places where the researchers had to work, from the bleak misty slopes of the paramo down to the warm humid rainforests of the Amazon basin, the magnitude of the task they faced becomes truly alarming.

While at the MECN, Morales-Mite collaborated as coauthor in the detailed revision of Ecuadorian *Hyloscirtus* that included a description of two new species from Ecuador (Coloma *et al.*, 2012), and in 2013, he ended his temporary position at the DHMECN and became the Collection Manager of the museum at CJ.

In 2013, the academic and administrative environment at the MECN was tense. Inappropriate decisions by the administrative authorities led to the imminent closure of the institution. At that time, Yánez-Muñoz was appointed Executive Director with the objective of stabilizing and restructuring the institution. He served in this capacity until 2017, when the MECN became part of the new Instituto Nacional de Biodiversidad (INABIO), a novel initiative of the government of Rafael Correa Delgado.

In 2013, Juan Carlos Sanchez-Nivicela (a biologist from Universidad del Azuay in the southern Andes of Ecuador) became associated with the MECN until 2017 and then with the INABIO. He has worked on several projects. From 2013 to 2015, he led research in the Provincia El Oro that, with the support of Pérez Lara, Oyagata C., Meza-Ramos, Verónica L. Urgilés, Valentina Posse, and Bruno Timpe, gathered information on 50 species of amphibians, including about half a dozen new species (Sánchez-Nivicela *et al.*, 2018, 2019, 2021, 2022).

The friendship of Yánez-Muñoz with Sánchez-Nivicela and Urgilés Peñafiel (biologists from Universidad del Azuay) generated a publication with the description of *Pristimantis gualacenio* and *P. tinajillas* (Urgilés *et al.*, 2014).

From 2014 to 2017, C. Reyes-Puig was hired with the duties of herpetological curator of the DHMECN and was involved in the description of seven new species (*Pristimantis allpapuyu, P. hamptusami, P. kuri, P. marcoreyesi, P. pinchaque, P. puruscafeum*, and *P. punzan*) using specimens and data collected in the upper Pastaza River basin, in the foothills of the Pichincha volcano (Yánez-Muñoz *et al.*, 2015), and in the montane forest of Provincia El Oro (Yánez-Muñoz *et al.*, 2016). One of the species described was in named honor of her brother Marco M. Reyes-Puig (see review in Reyes-Puig *et al.*, 2014, 2015), who passed away early in his life (at 35 years old) and who was influential to the DHMECN collections and research. The Reyes-Puig siblings (Carolina, Juan Pablo, and Marco) have described or co-described 30 species of Ecuadorian amphibians to date; 27 of them are *Pristimantis.*

In 2015, Jorge Brito Molina (a productive mammologist and herpetologist), became associated with the DHMECN as a MECN researcher and, in 2017, as an INABIO researcher. His input led to the description of four new species from the

Sangay National Park River basin (*Pristimantis albujai*, *P. saturninoi*, *P. sambalan* and *P. roni*) in collaboration with Diego Batallas-Revelo (an Ecuadorian amphibian biologist and expert in bioacoustics) and Bejarano-Muñoz (Yánez-Muñoz *et al.*, 2014; Brito *et al.*, 2017). Bejarano-Muñoz (research associate at DHMECN) received a research grant (supported by Ecociencia) to study the distribution and richness of terrestrial *Pristimantis* frogs in an altitudinal gradient of the upper Napo River basin, generating outreach publications (Bejarano-Muñoz *et al.*, 2015) and the discovery of at least two new species (Yánez-Muñoz *et al.*, 2014, 2015). During this time, Raquel Betancourt (an Ecuadorian herpetologist) deposited important collections of amphibians and reptiles from the Cordillera del Cóndor, which consolidated the collection of this priority area in Ecuador. The synergy of all DHMECN collaborators increased the collection to 9210 amphibian specimens by 2017.

The INABIO and the Coining of Species Names for Amphibian Conservation

At the end of 2017, Yánez-Muñoz returned to his position as principal investigator of the herpetological collection incorporated into the Instituto Nacional de Biodiversidad (INABIO) (Figure 2.53) under executive director Diego Inclán. Maintaining the DHMECN acronym, Miguel A. Urgilés Merchán assumed the duties of collection management in 2017.

This new period as the INABIO has been characterized by an increased level of collaboration with various amphibian collections in Ecuador, particularly the EPN, QCAZ, UTI, ZSFQ, FHGO, and ZFMK. These synergies have fostered an increase in the annual rate of description of new amphibians, especially of the most diverse group of Neotropical vertebrates, Terrarana. Of note are the descriptions of *Pristimantis andinogigas*, *P. afrox*, *P. cisnerosi*, *P. barrigai*, *P. chocoensis*, *P. erythros*, and *Noblella naturetrekii* in Craugastoridae (Sánchez-Nivicela *et al.*, 2018; Yánez-Muñoz *et al.*, 2019; Reyes-Puig *et al.*, 2019a, 2019b, 2020a); *Elachistocleis araios* in Microhylidae (Sánchez-Nivicela *et al.*, 2020); and the plethodontid salamanders *Oedipina villamizariorum* and *O. ecuatoriana* (Reyes-Puig *et al.*, 2020b). Most of this work was carried out along the lines of integrative taxonomy, with the collaboration of the ZSFQ (Guayasamin, Cisneros-Heredia, and C. Reyes-Puig), QCAZ (Ron, Jhael Ortega, María José Navarrete), ZFMK (Claudia Koch), and Ikiam (Ortega-Andrade). As of 31 December 2022, the DHMECN collection included the holotypes of 37 species: *Osornophryne cofanorum*, *O. occidentalis*, *Nymphargus buenaventura*, *Teratohyla amelie*, *Epipedobates darwinwallacei*, *Dendropsophus shiwiarum*, *Hyloscirtus conscientia*, *Noblella naturetrekii*, *P. afrox*, *P. albujai*, *P. allpapuyu*, *P. andinogigas*, *P. bellae*, *P. burtoniorum*, *P. chocoensis*, *P. churuwiai*, *P. cisnerosi*, *P. erythros*, *P. gualacenio*, *P. hampatusami*, *P. kuri*, *P. loujosti*, *P. marcoreyesi*, *P. maryanneae*, *P. matidiktyo*, *P. pichincha*, *P. pinchaque*, *P. punzan*, *P. romanorum*, *P. roni*, *P. sambalan*, *P. saturninoi*, *P. sirnigeli*, *P. tinajillas*, *P. tungurahua*, *P. yumbo*, and *Oedipina villamizariorum*.

The strong and long-standing association of Yánez-Muñoz with Ecominga (Juan P. Reyes-Puig, Lou Jost, and Javier Robayo) has allowed the MECN–INABIO to develop a taxonomic-systematic research program connected to land acquisition and in situ conservation. After 10 years of collaboration with the DHMECN, Ecominga has funded the exploration of new sites in priority ecoregions of Ecuador. As a result, 13 new species have been described in the Llangantes–Sangay Ecological Corridor (*Osornophryne simpsoni*, *Noblella naturetrekii*, *Pristimantis tungurahua*, *P. loujosti*, *P. bellae*, *P. ardyae*, *P. marcoreyesi*, *P. puruscafeum*, *P. pinchaque*, *P. sacharuna*, *P. malli*, *P. burtoniorum*, and *P. maryanneae*). Remarkably, several names assigned to new species, enabled the securing of funds to conserve their habitats (Reyes-Puig *et al.*, 2022a). The use of scientific names for the recognition of conservation enthusiasts has resulted in raising funds to conserve the habitats of new species: *O. simpsoni*, *P. bellae*, *P. puruscafeum*, *P. ardyae*, *P. burtoniorum*, and *P. maryanne*. These species were named in recognition of donors who have contributed substantially to the acquisition of 5000 hectares for the conservation and connectivity of the montane forests

FIGURE 2.53 Left: Juan P. Reyes-Puig reviewing specimens at Universidad del Valle, in Cali, Colombia, 2017. Photo by Mario H. Yánez-Muñoz. Right: Mario H. Yánez-Muñoz at INABIO, in Quito, Ecuador, 2023. Photo by Miguel A. Urgilés Merchán.

of Llanganates National Park and Sangay National Park (Reyes-Puig *et al.*, 2022b).

The same initiative was replicated in northwestern Ecuador in the montane forests of the San Juan River basin, Provincia Carchi. *Hyloscirtus conscientia* was part of a naming designation. The geographic range of participants in the coining of this name (627 entries from 36 counties in South America, North America, Africa, and Europe) indicates that engaging the public can be an effective method to educate those outside the scientific community, especially young people, about the importance of environmental issues and to motivate communities both locally and globally to support conservation (Yánez-Muñoz *et al.*, 2021). The efforts of the INABIO have brought the attention of the local authorities to the formal recognition of 1000 hectares of forest of the Reserva Drácula of *Fundación Ecominga* as part of the subsystem of protected areas of Tulcán, Provincia Carchi.

In 2018, the information stored in the DHMECN database was made available for the assessment of the conservation status of Ecuadorian amphibians and the production of the Red List. This pivotal work, led by H. Mauricio Ortega-Andrade, brought together the efforts of 33 Ecuadorian herpetologists (Ortega-Andrade *et al.*, 2021).

Three Decades of Amphibian Research at the Universidad San Francisco de Quito USFQ

Diego F. Cisneros-Heredia, Carolina Reyes-Puig, Emilia Peñaherrera-Romero, Juan M. Guayasamin, David Brito-Zapata, Mateo Dávila-Játiva, Andrés León-Reyes, Juan Carlos Sánchez-Nivicela, and David Romo

The Universidad San Francisco de Quito USFQ (USFQ) is a private university based in Quito, Ecuador. Its mission is to cultivate freethinking, innovative, creative, and enterprising individuals within the framework of liberal arts. Since its foundation in 1988, the USFQ has been intertwined with the study of amphibians, with a strong research program focused on taxonomy, biogeography, and conservation. The university has a valuable collection of amphibians deposited at its *Museo de Zoología* (ZSFQ) and a recently established biobank. The museum currently houses around 3500 specimens of amphibians, including the type specimens of 25 species and a collection of more than 1000 frozen samples. USFQ scientists have participated in describing 54 species of amphibians new to science (as of December 2022), and the USFQ is a major contributor to amphibian research in Ecuador. This section chronicles the amphibian research, collections, and people through more than 30 years of history at USFQ.

The Initial Steps (1988–1996)

On 1 September 1988, the USFQ opened its doors in an early Republican house in Quito, Ecuador. The first fully self-financed university in Ecuador, the USFQ was an initiative of Santiago Gangotena González (1945–2023), an Ecuadorian nuclear physicist and multifaceted entrepreneur, with the support and contributions of Carlos Montúfar, Bruce Hoeneissen, Doris Cashmore de Mantilla, Miguel Vallejo, Emilio Gangotena, Conto Patiño, 12 founding professors, and 122 founding students. Despite its young age, by 1989, the USFQ had signed its first agreement for international student exchange with a US university: the University of North Carolina at Wilmington (UNCW). This agreement was achieved thanks to the friendship and trust between Gangotena and US herpetologist Charles M. Fugler (1929–1999). Fugler and Gangotena met while studying at the University of Auburn in Alabama, USA, where they were roommates.

Fugler was a former KU student and professor of biology at UNCW who did extensive biological collections in several Neotropical countries (he collected amphibians in Mexico, Honduras, Bolivia, and Ecuador) and conducted herpetological and ethnological research in Ecuador during the 1960s and 1970s (Fugler, 1971, 1966; Fugler and Schwaner, 1968; Fugler and Walls, 1979a, 1978, 1979b; Williams and Fugler, 1968). Fugler's biological collections from Ecuador have been the basis for many studies of animal taxa, among them his amphibian collections in the provinces of Cotopaxi, Sucumbíos, and Morona Santiago. He collected 2113 specimens and deposited them at the MCZ; thus, he is one of the most prolific collectors of Ecuadorian amphibians. His first collections were of *Atelopus ignescens* in the paramos west of Pujilí in April 1963, at a time when this species was still abundant. Then he moved to the Amazonian region, where he collected at Limoncocha, Santa Cecilia, and surroundings in 1965 and 1966, at a time when the oil industry was beginning its activities in the Ecuadorian Amazonian lowlands. In 1975, he collected in the Provincia Morona Santiago. Fugler's amphibian collections were initially studied by Lynch (1968) and Duellman (1978), and he published a paper on Ecuadorian amphibians (Fugler and Walls, 1979a). Another remarkable contribution by Fugler was his phone call in October 1966 to William E. Duellman, in which he encouraged him to study the amphibians and reptiles at Santa Cecilia (Duellman, 2015b), and that resulted in the landmark work on the herpetofauna of Santa Cecilia (Duellman, 1978).

In 1989 and 1990, Fugler's support was important for the creation of the USFQ unit today known as the College of Biological and Environmental Sciences, which was founded by Gangotena and Gunther Reck (as Founding Dean), together with several biologists and conservationists, including Kelly Swing, Carlos Fabara, Luis Suárez, and Enrique Lasso. During its early years, the USFQ began to acquire small biological collections to use in its educational activities. These collections were first housed in a "cabinet of curiosities" style at the second campus of the university at Monteserrín. Eventually, they were transferred to the first biology laboratory that the university built on its campus in Cumbayá, in the suburbs of Quito, where it moved in 1994.

The First Researchers and Collections (1997–2008)

In 1996, the *Instituto de Ecología Aplicada* (ECOLAP) of the USFQ, created by Gunther Reck, developed the project "Design and formulation of the national strategy for the protection and sustainable use of wildlife." Three consultants were involved in overseeing the analysis of amphibian data for the project: Felipe Campos Yánez, Jean-Marc Touzet, and Ana María Velasco (ECOLAP, 1998a, 1998b; Valle *et al.*, 1999). Touzet, a French herpetologist and conservationist, and Velasco, an Ecuadorian biologist and conservationist, founded the *Fundación Herpetológica Gustavo Orcés* (FHGO). The FHGO, thanks to the efforts and amazing field skills of Touzet and his collaborators, had one of the best scientific collections of amphibians and reptiles at that time in Ecuador and the first library of amphibians, reptiles, and conservation literature open to the public in the country.

Jean-Marc Touzet had worked at the *Vivarium de Lausanne* before moving to Ecuador in 1977. Upon arriving in the country, he did field work that resulted in taxonomic papers (e.g., Schätti and Kramer, 1993), natural history, and parasitology of amphibians (e.g., Ben and Durette-Desset, 1995), reptiles, and other vertebrates—as a research correspondent of the *Muséum d'Histoire Naturelle de Genève*, Switzerland, venomous snakes and snakebite, and ex situ management of herpetofauna (Valle *et al.*, 1999; ECOLAP, 1998a), collaborating with different national and international institutions, including the *Instituto Francés de Estudios Andinos* (IFEA), the *Office de la recherche scientifique et technique outre-mer* (ORSTOM) of France, and the *Museo Ecuatoriano de Ciencias Naturales* (MECN; nowadays the *Instituto Nacional de Biodiversidad*, INABIO).

Ana María Velasco obtained a BS in biology at PUCE and a master's in wetland management at the *Université d'Angers* (France). Throughout her career, she has coordinated several herpetological projects in Ecuador, including the herpetological component of the project "Sustainable Use of Biological Resources" (SUBIR) at the Cotachi-Cayapas and Cayambe-Coca National Parks and the "Tropical Amphibian Monitoring" project in the Jatun Sacha Reserve, Amazonian Ecuador. She has also worked extensively as a consultant for Ecuadorian and international organizations, including the IUCN, GEF, GRET, *Fondation Ensemble*, Probona, Ministry of Environment of Ecuador, among others (Pearman *et al.*, 1995; Velasco, 1997b; Valle *et al.*, 1999).

In 1997, Reck, as dean of the USFQ College of Environmental Sciences, invited Touzet and Velasco to join the USFQ faculty as professors, and they became the first to offer courses in herpetology and ex situ conservation. Also in 1997, the USFQ and FHGO signed an agreement that allowed for the FHGO's herpetological collection and library to be housed in the USFQ's Cumbayá campus (in the Galileo building until 2002, afterward at new facilities specifically built to house the collections in the Newton Plaza building, which was later renamed the Darwin building). Based on the interinstitutional agreement, a joint research laboratory was created, the *Laboratorio de Anfibios & Reptiles FHGO-USFQ* (Lab FHGO-USFQ), making it possible to increase the number of USFQ personnel involved in the collection, to reinforce research projects, and to actively use the collections and library in the USFQ's educational and research programs (Cisneros-Heredia, 2003b).

In 1997, David Romo (Figure 2.54) joined the USFQ as the On-Site Manager for the Tiputini Biodiversity Station (TBS), while still studying for his PhD. The TBS is a research station founded in 1994 by the USFQ, covering 744 hectares of undisturbed Amazonian lowland rainforest on the northern bank of the Tiputini River. It is within the Yasuní Biosphere Reserve, and in the land of the "El Edén" Kichwa community in Ecuador. The site has been documented as one of the regions with the highest species richness of amphibians in the world (Bass *et al.*, 2010; Romo *et al.*, 2017). For his PhD in conservation biology at the University of Minnesota, Romo proposed a novel method using computerized sound analysis for monitoring frog diversity, aiming at the development of a "tool that can be used for assessing environmental quality in areas influenced by the oil industry in the Amazon basin of Ecuador determining at least changes in the geographical distribution of frogs," hoping that the "early detection of changes in these distributions will help herpetologists concentrate their monitoring efforts on the species at high risk" (Romo, 2001). Romo became a full-time professor at the USFQ and was the co-director of the Tiputini Biodiversity Station until 2021. Part of his data collection experiments with frog calls was conducted with the participation of his USFQ students at the Jatun Sacha Reserve and the Tiputini Biodiversity Station. Romo was also part of the team of Latin American scientists who developed regional workshops to discuss the catastrophic amphibian declines that had just started to be understood (Young *et al.*, 2001).

In 1998, David Romo met a student, Diego F. Cisneros-Heredia (see Figure 2.54), who was in the first semester of his first year at university, during a USFQ colloquium and invited him to visit the Tiputini Biodiversity Station during the New Year's break. A few weeks later, Jean-Marc Touzet (see Figure 2.54) opened a search for a teaching and research assistant to help him in his USFQ courses and herpetological studies and selected the same first-year student, Cisneros-Heredia, for the position—these opportunities would change Cisneros-Heredia's life forever. After his trip to Tiputini and working with Romo, Touzet, and Velasco (see Figure 2.54) for several years, Cisneros-Heredia became entranced by amphibians and reptiles (Touzet and Cisneros-Heredia, 1998a, 1998b; Cisneros-Heredia, 2003a, 2004b).

In 1999, Andrés León-Reyes joined the team of the Lab FHGO-USFQ, working as a collection's assistant until 2001. León-Reyes and Cisneros-Heredia conducted many herpetological expeditions across Ecuador, including the Cuyabeno Wildlife Reserve, Yasuní National Park, the Río Guajalito Reserve, and the Sumaco Ecoroute

FIGURE 2.54 Top left: Jean-Marc Touzet and Ana María Velasco at the Vivarium in Quito in 1992. Photographer unknown. Top right: David Romo (left) and students on a field trip to the Tiputini Biodiversity Station in 2009. Photo by Leo Zurita. Bottom left: Diego F. Cisneros-Heredia, William W. Lamar, and Roy W. McDiarmid at the 2004 meeting of the Association of Ichthyologists and Herpetologists (ASIH) in Norman, Oklahoma. Photo by John E. Simmons. Bottom right: Juan M. Guayasamin and a centrolenid frog (*Ikakogi tayrona*) in Sierra Nevada de Santa Marta, Colombia, in August 2022. Photo by Pedro Peloso.

(Cisneros-Heredia, 2004b, 2005a; Cisneros-Heredia *et al.*, 2005). León-Reyes completed his senior thesis for his bachelor's in ecology at the USFQ, studying the diversity of amphibians and reptiles of the Río Guajalito Reserve in northwestern Ecuador (León-Reyes, 2008).

About 2000 specimens collected at Tiputini, Cuyabeno, Río Guajalito, Machalilla National Park, and other localities by Cisneros-Heredia, León-Reyes, and other collaborators were deposited at the Lab FHGO-USFQ and were used for several publications (Bustamante *et al.*, 2005b, 2007; Cisneros-Heredia, 2005b, 2005c, 2006c). The Lab FHGO-USFQ came to house nearly 6000 specimens and the library more than 5000 titles (Cisneros-Heredia, 2003b). Several national and international researchers deposited their specimens at the laboratory, e.g., Felipe Campos Yánez, Fernando Nogales, and Oxford University's AndinoHerps 2000 expedition (Schultz *et al.*, 2000). The curators of the laboratory were Touzet (1997–1999), Velasco (1999–2000), and Cisneros-Heredia (2000–2003).

In 2003, the FHGO and USFQ terminated their agreement, and the collections were split between the two organizations. The USFQ collections remained in the care of Cisneros-Heredia for several years, being cited under the abbreviation "DFCH-USFQ" and used in many publications (Cisneros-Heredia and Touzet, 2004; Mueses-Cisneros *et al.*, 2012; Elmer *et al.*, 2013; Faivovich *et al.*, 2005; Guayasamin *et al.*, 2006b; López-Rojas and Cisneros-Heredia, 2012; Cisneros-Heredia and Reynolds, 2007; McCracken and Forstner, 2006). Due to the limited available space, most DFCH-USFQ herpetological scientific collections were kept in the guest room of Cisneros-Heredia's house, while the collections used for educational activities were held at the university's biology laboratory. Educational collections included amphibians obtained during field activities organized as part of courses such as Introduction to Biology, Zoology, Herpetology, and Tropical Ecology by different professors, including Romo, Kelly Swing, Nelson Zabala, Vlastimil Zak, Arturo Paredes-Recalde, and Cisneros-Heredia.

Cisneros-Heredia at the Starting Point

Cisneros-Heredia's life and achievements have been recounted in popular articles, books, and press multimedia (e.g., Sierra Praeli, 2021; Montgomery and Lechuga, 2021), and his academic biography can be found under the title "The Diversity and Academic Lives of Current Amphibian Biologists." Here, we highlight some related aspects of his trajectory at the USFQ. His first professional activity with amphibians and reptiles was in 1996, when he volunteered at the Vivarium of Quito. Later, he became associated with it and the *Fundación Herpetológica Gustavo Orcés* (1998–2002). Between 1998 and 2002, Diego F. Cisneros-Heredia conducted a four-year research project on the diversity of amphibians and reptiles of the Tiputini Biodiversity Station as his senior thesis for his bachelor's degree in ecology at the USFQ (Cisneros-Heredia, 2003a), which was part of the foundational information for the first comprehensive synthesis of biodiversity data for the Yasuní National Park (Bass *et al.*, 2010). Cisneros-Heredia has been captivated by amphibians since the age of three, when he met some toads (*Rhinella horribilis*), which he remembers as "gigantic and wonderful," during a beach trip in northern Ecuador. Later in life, Cisneros-Heredia became fascinated by glassfrogs after he collected his first specimens of Centrolenidae (*Chimerella mariaelenae* and *Hyalinobatrachium yaku*) in 1998 with Touzet (Cisneros-Heredia and McDiarmid, 2007; Guayasamin *et al.*, 2017a).

In 2002, Cisneros-Heredia was accepted in the Research Training Program of the USNM in Washington D. C., where he worked under the mentorship of the renowned herpetologist Roy W. McDiarmid, curator of amphibians and reptiles for the US Geological Survey, Patuxent Wildlife Research Center, at the USNM. McDiarmid has worked for most of his career on the systematics, biogeography, ecology, and behavior of amphibians and reptiles of tropical America and is a world expert on amphibian monitoring and tadpole biology (McDiarmid, 1971; Heyer *et al.*, 1975; McDiarmid and Altig, 1999; Altig and McDiarmid, 2007; Heyer Ronald *et al.*, 2014; Lovich *et al.*, 2012).

McDiarmid has contributed significantly to the understanding of the evolution, natural history, reproductive biology, and behavior of glassfrogs, which he studied in the field in Costa Rica, Ecuador, Peru, and Mexico (McDiarmid and Altig, 1999; Altig and McDiarmid, 2007; McDiarmid and Adler, 1974; Flores and McDiarmid, 1989; Cadle and McDiarmid, 1990; Schwalm and Starrett, 1977; McDiarmid and Savage, 1984; McDiarmid, 1975, 1978, 1983; Delia *et al.*, 2017; Cisneros-Heredia and McDiarmid, 2006a, 2007; Luna *et al.*, 2018). McDiarmid explored several localities across Ecuador in the late 1970s, accompanied by the late Kenneth Miyata and other colleagues, and conducted extensive fieldwork at the Río Palenque Scientific Center, Provincia Los Ríos, one of the last remnants of tropical semideciduous forest in western Ecuador (McDiarmid and Altig, 1990; Cisneros-Heredia and McDiarmid, 2007). McDiarmid marveled at Palenque's diverse glassfrog community and wanted to study it in detail. When he first met Cisneros-Heredia in the summer of 2022, he offered to teach him and do research together on Palenque's glassfrogs. It was the beginning of productive collaborative efforts involving glassfrogs and other amphibians (Cisneros-Heredia and McDiarmid, 2003, 2004, 2005, 2006a, 2006b, 2006c, 2007, 2016a, 2016b; Mueses-Cisneros *et al.*, 2012; Guayasamin *et al.*, 2020). One of their landmark scientific contributions was the revision of the taxonomic characters of the glassfrogs, family Centrolenidae (Cisneros-Heredia and McDiarmid, 2007), published when many characters important for understanding the diversity of glassfrogs were still confused, poorly defined, or little explored. McDiarmid took Cisneros-Heredia to his first international scientific congress, the 2002 Joint Meeting of Ichthyologists and Herpetologists in Kansas City, where he introduced him to some of the most renowned herpetologists who have worked on the herpetofauna of Ecuador, including Duellman, John D. Lynch, and David Wake. Since then, Cisneros-Heredia has published studies on the taxonomy and systematics of glassfrogs (Cisneros-Heredia and McDiarmid, 2006a; Cisneros-Heredia, 2007a; Cisneros-Heredia and Meza-Ramos, 2007; Torres-Gastello *et al.*, 2007; Cisneros-Heredia and Guayasamin, 2014; Cisneros-Heredia *et al.*, 2023), their nomenclatural history (Cisneros-Heredia, 2013, 2009; Guayasamin *et al.*, 2008b), reproduction and natural history (Delia *et al.*, 2010; Cisneros-Heredia *et al.*, 2008a, 2011), distribution (Cisneros-Heredia and McDiarmid, 2005, 2006b; Cisneros-Heredia and Yánez-Muñoz 2007a; Toledo *et al.*, 2009; Yánez-Muñoz *et al.*, 2009a; Jaramillo-Martinez *et al.*, 2015), and conservation (Cisneros-Heredia and McDiarmid, 2006c; Cisneros-Heredia *et al.*, 2010). In 2008, Cisneros-Heredia and collaborators dedicated the specific name of a glassfrog, *Rulyrana mcdiarmidi*, in honor of Roy W. McDiarmid (see Figure 2.54) in recognition of his contributions to the understanding of Neotropical herpetology and for "his constant support, friendship, and guidance," which Cisneros-Heredia will always recognize as invaluable.

Cisneros-Heredia's scientific contributions to amphibians include more than 70 papers addressing a wide variety of topics concerning amphibians in tropical ecosystems, including diversity, biogeography, and the natural history of frogs of the families Bufonidae (Cisneros-Heredia and Gluesenkamp, 2011; Yánez-Muñoz *et al.*, 2011d; Mueses-Cisneros *et al.*, 2012), Dendrobatidae (Cisneros-Heredia, 2016; Cisneros-Heredia and Yánez-Muñoz, 2010; Cisneros-Heredia and Flechas, 2016), Hemiphractidae (Almendáriz and Cisneros-Heredia, 2005; Urgilés *et al.*, 2017), Hylidae (Cisneros-Heredia, 2005d, 2007b; Reyes-Puig *et al.*, 2022b; Yánez-Muñoz *et al.*, 2021; Sánchez-Nivicela *et al.*, 2023), Leptodactylidae (Cisneros-Heredia, 2006a), Microhylidae (López-Rojas and Cisneros-Heredia, 2012), and Craugastoridae (Yánez-Muñoz *et al.*, 2019; Brito-Zapata *et al.*, 2021; Sánchez-Nivicela *et al.*, 2018); salamanders of the family Plethodontidae (Cisneros-Heredia, 2006c; Reyes-Puig *et al.*, 2020b); caecilians of the families Caeciilidae (Cisneros-Heredia and Mosquera, 2011) and Rhinatrematidae (Cisneros-Heredia, 2004b); amphibian communities (Cisneros-Heredia, 2006b; Armijos-Ojeda *et al.*, 2021); anuran communication (Peñaherrera-Romero and Cisneros-Heredia, 2022); conservation and extinction risks of amphibians (Ortega-Andrade *et al.*, 2021; Cox *et al.*, 2022; Cisneros-Heredia, 2008a, 2008b; Reyes-Puig *et al.*, 2019a; Cisneros-Heredia *et al.*, 2004; Cisneros-Heredia and Jungfer, 2004; Sánchez-Carvajal *et al.*, 2021); anuran invasive species (Cisneros-Heredia, 2004a, 2018; Valarezo-Aguilar *et al.*, 2016); and anuran historical taxonomy and cartography

(Cisneros-Heredia, 2013, 2017; Urgilés *et al.*, 2017; Orton, 2022). Among several management tasks and positions, at present Cisneros-Heredia, is a full-time professor at the USFQ and Director of the *Laboratorio de Zoología Terrestre* and ZSFQ.

Beginning in the 2000s, a strong partnership developed between the USFQ and the MECN, thanks to the active collaborations with Mario H. Yánez-Muñoz. Yánez-Muñoz and Cisneros-Heredia first met and became friends in a course on herpetology taught by Touzet in 1999. Since then, Yánez-Muñoz and Cisneros-Heredia have collaborated on the study of many new species of amphibians (Table 2.6), (Cisneros-Heredia and Yánez-Muñoz, 2007b; Cisneros-Heredia *et al.*, 2008b, 2009, 2011, 2023; Reyes-Puig *et al.*, 2011, 2022b; Yánez-Muñoz *et al.*, 2009a, 2011b, 2011c, 2011a, 2019, 2021). In 2004, Cisneros-Heredia was one of the founders of the *Red Ecuatoriana de Herpetología*, which ran an active news email service until 2015. In addition, he has collaborated on several projects studying amphibians across Ecuador and the tropical Andean countries and has been a research associate at The Natural History Museum of London (2015–2020), IUCN Red List Authority Regional Coordinator for Ecuador, and a member of the IUCN-SSC Amphibian Specialist Group (2008–present). Cisneros-Heredia has been involved in the editing of three books and the production of or participation in about 50 instances of outreach publications through multimedia communication, including radio, newspapers, TV, and new media (as of December 2022).

Consolidation of the Museum and Laboratories (2014–Present)

Guayasamin and Reyes-Puig join USFQ

Given the growth of the collections and their importance at a national and international level, in 2014, Cisneros-Heredia founded the *Instituto de Zoología Terrestre* at the USFQ to develop scientific research on the fauna of Ecuador and promote the maintenance and growth of the USFQ zoological collections. Together with the institute, he founded two centers: the *Museo de Zoología* (ZSFQ = Zoology San Francisco de Quito) to consolidate all scientific and educational animal collections and the *Laboratorio de Zoología Terrestre* to develop research programs and social outreach initiatives. The new museum was created based on the herpetological collections obtained by Cisneros-Heredia and his collaborators, plus the USFQ educational collections. The collections were housed in a newly created laboratory in the Darwin building of the USFQ's Cumbayá campus. The curator of all scientific collections was Cisneros-Heredia, while the administrators of the educational collections were Pablo Riera and Jaime Guerra.

Given the growing need to expand the collections and intending to create a specialized research center, the USFQ built new facilities in 2018 thanks to the support of Gangontena, Montúfar, Diego Quiroga, Jorge Gómez Tejada, and Stella de la Torre, the TUERI Wildlife Hospital, and the *Instituto de Zoología Terrestre* became the *Instituto de Biodiversidad Tropical* (IBIOTROP). The ZSFQ, the *Laboratorio de Zoología Terrestre*, the TUERI Wildlife Hospital, and a newly created USFQ Quito Biobank are part of the IBIOTROP, which is a research institute integrated under the College of Biological and Environmental Sciences and the School of Veterinary Medicine of the USFQ.

In 2013, Emilia Peñaherrera-Romero (Figure 2.55) joined Cisneros-Heredia's laboratory as an undergraduate student studying communication and behavior in treefrogs. She completed her senior thesis for her bachelor's degree in biology at the USFQ with a study of the visual signals of the treefrog *Dendropsophus carnifex* (Peñaherrera-Romero and Cisneros-Heredia, 2022). After graduating, she began working as a research assistant with Cisneros-Heredia, learning about taxonomy, biogeography, geography, and environmental modeling. In 2017, she started to work as a curatorial assistant at the ZSFQ, helping to consolidate the museum databases and supporting the growth of the collection. Emilia has expanded her research into several taxa and topics but continues conducting research with amphibians (Cisneros-Heredia and Peñaherrera-Romero, 2020; Peñaherrera-Romero *et al.*, 2022). She obtained her master's in tropical ecology and conservation at the USFQ under the mentorship of Cisneros-Heredia and Susana Cardenas with a thesis evaluating of the significance of important bird and biodiversity areas in the tropical Andes for amphibian conservation.

In 2016, Juan M. Guayasamin (see Figure 2.55) joined the USFQ and co-founded, with Elisa Bonaccorso, the *Laboratorio de Biología Evolutiva*. Guayasamin's life and academic achievements have been described by Duellman (2015a), in outreach publications (Vélez Zuazo, 2019), and in a song by the US Public Broadcasting Service for children (https://www.youtube.com/watch?v=F2ZV6SJd0zU). In this book, some of his academic activities are mentioned in this chapter under the subtitles: "Duellman, Trueb, and Their Ecuadorian Graduate Students," "The Dream Team at QCAZ, Two Decades (1991–2010) of Science, Conservation, and Public Awareness," and "The Diversity and Academic Lives of Current Amphibian Biologists." In the field of amphibian biology, the most significant contributions by Guayasamin have been his work on glassfrogs (family Centrolenidae). This group was the focus of his PhD dissertation at the University of Kansas, under the direction of Linda Trueb. Since starting his research on glassfrogs, he has led studies on their evolutionary relationships (Guayasamin *et al.*, 2008a, 2020), systematics (Guayasamin *et al.*, 2009), osteology (Guayasamin and Trueb, 2007), and ecology and conservation (Guayasamin *et al.*, 2020). He has described about 10% of the currently recognized species of the family (Table 2.6) and participated in studies on their biogeography (Hutter *et al.*, 2013; Castroviejo-Fisher *et al.*, 2014, 2020), parental behavior (Salgado and Guayasamin, 2018), visual displays (Brunner and Guayasamin, 2020), call evolution (Mendoza-Henao *et al.*, 2023), and interaction with exotic species (Martín-Torrijos *et al.*, 2016). Guayasamin has made several contributions to the documentation of other amphibian groups, leading or collaborating on the

TABLE 2.6
Species of Amphibians Described by Researchers from the Universidad San Francisco de Quito USFQ, in Chronological Order

Number	Species	Year	Family	References
1	*Chimerella mariaelenae*	2006	Centrolenidae	Cisneros-Heredia and McDiarmid (2006b)
2	*Boana nympha*	2006	Hylidae	Faivovich *et al.* (2006)
3	*Pristimantis aureolineatus*	2006	Craugastoridae	Guayasamin *et al.* (2006a)
4	*Cochranella erminae*	2007	Centrolenidae	Torres-Gastello *et al.* (2007)
5	*Espadarana durrellorum*	2007	Centrolenidae	Cisneros-Heredia (2007a)
6	*Nymphargus buenaventura*	2007	Centrolenidae	Cisneros-Heredia and Yánez-Muñoz (2007b)
7	*Nymphargus laurae*	2007	Centrolenidae	Cisneros-Heredia and McDiarmid (2007)
8	*Teratohyla amelie*	2007	Centrolenidae	Cisneros-Heredia and Meza-Ramos (2007)
9	*Rulyrana mcdiarmidi*	2008	Centrolenidae	Cisneros-Heredia *et al.* (2008b)
10	*Centrolene condor*	2008	Centrolenidae	Cisneros-Heredia and Morales-Mite (2008)
11	*Pristimantis loujosti*	2010	Craugastoridae	Yánez-Muñoz *et al.* (2011b)
12	*Pristimantis romanorum*	2010	Craugastoridae	Yánez-Muñoz *et al.* (2011c)
13	*Pristimantis sirnigeli*	2010	Craugastoridae	Yánez-Muñoz *et al.* (2011c)
14	*Pristimantis tungurahua*	2010	Craugastoridae	Reyes-Puig *et al.* (2011)
15	*Pristimantis yumbo*	2010	Craugastoridae	Yánez-Muñoz *et al.* (2011c)
16	*Osornophryne occidentalis*	2010	Bufonidae	Cisneros-Heredia and Gluesenkamp (2011)
17	*Epipedobates darwinwallacei*	2010	Dendrobatidae	Cisneros-Heredia and Yánez-Muñoz (2011)
18	*Rhaebo ecuadorensis*	2012	Bufonidae	Mueses-Cisneros *et al.* (2012)
19	*Hyloscirtus criptico*	2012	Hylidae	Coloma *et al.* (2012)
20	*Hyloscirtus princecharlesi*	2012	Hylidae	Coloma *et al.* (2012)
21	*Pristimantis ecuadorensis*	2017	Craugastoridae	Guayasamin *et al.* (2017b)
22	*Hyalinobatrachium yaku*	2017	Centrolenidae	Guayasamin *et al.* (2017a)
23	*Pristimantis erythros*	2018	Craugastoridae	Sánchez-Nivicela *et al.* (2018)
24	*Pristimantis gralarias*	2018	Craugastoridae	Guayasamin *et al.* (2018)
25	*Pristimantis caniari*	2018	Craugastoridae	Ramírez-Jaramillo *et al.* (2018)
26	*Pristimantis andinogigas*	2019	Craugastoridae	Yánez-Muñoz *et al.* (2019)
27	*Nymphargus manduriacu*	2019	Centrolenidae	Guayasamin *et al.* (2019a)
28	*Gastrotheca cuencana*	2019	Hemiphractidae	Carvajal-Endara *et al.* (2019)
29	*Gastrotheca elicioi*	2019	Hemiphractidae	Carvajal-Endara *et al.* (2019)
30	*Gastrotheca turnerorum*	2019	Hemiphractidae	Carvajal-Endara *et al.* (2019)
31	*Gastrotheca yacuri*	2019	Hemiphractidae	Carvajal-Endara *et al.* (2019)
32	*Noblella naturetrekii*	2019	Craugastoridae	Reyes-Puig *et al.* (2019)
33	*Hyalinobatrachium adespinosai*	2019	Centrolenidae	Guayasamin *et al.* (2019b)
34	*Pristimantis mallii*	2019	Craugastoridae	Reyes-Puig *et al.* (2019b)
35	*Nymphargus colomai*	2020	Centrolenidae	Guayasamin *et al.* (2020)
36	*Nymphargus humboldti*	2020	Centrolenidae	Guayasamin *et al.* (2020)
37	*Nymphargus lindae*	2020	Centrolenidae	Guayasamin *et al.* (2020)
38	*Pristimantos afrox*	2020	Craugastoridae	Reyes-Puig *et al.* (2020c)
39	*Pristimantis chocoensis*	2020	Craugastoridae	Reyes-Puig *et al.* (2020c)
40	*Pristimantis cisnerosi*	2020	Craugastoridae	Reyes-Puig *et al.* (2020c)
41	*Oedipina villamizariorum*	2020	Plethodontidae	Reyes-Puig *et al.* (2020b)
42	*Oedipina ecuatoriana*	2020	Plethodontidae	Reyes-Puig *et al.* (2020b)
43	*Noblella worleyae*	2020	Craugastoridae	Reyes-Puig *et al.* 2020a)
44	*Pristimantis andinodiabolus*	2021	Craugastoridae	Sánchez-Nivicela *et al.* (2021)
45	*Pristimantis ledzeppelin*	2021	Craugastoridae	Brito-Zapata and Reyes-Puig (2021)
46	*Hyloscirtus conscientia*	2021	Hylidae	Yánez-Muñoz *et al.* (2021)
47	*Noblella mindo*	2021	Craugastoridae	Reyes-Puig *et al.* (2021)
48	*Pristimantis daquilemai*	2021	Craugastoridae	Brito-Zapata *et al.* (2021)

(*Continued*)

TABLE 2.6 (*Continued*)
Species of Amphibians Described by Researchers from the Universidad San Francisco de Quito (USFQ), in Chronological Order

Number	Species	Year	Family	References
49	*Hyalinobatrachium mashpi*	2022	Centrolenidae	Guayasamin *et al.* (2022)
50	*Hyalinobatrachium nouns*	2022	Centrolenidae	Guayasamin *et al.* (2022)
51	*Pristimantis maryanneae*	2022	Craugastoridae	Reyes-Puig *et al.* (2022a)
52	*Pristimantis burtoniorum*	2022	Craugastoridae	Reyes-Puig *et al.* (2022a)
53	*Hyloscirtus sethmacfarlanei*	2022	Hylidae	Reyes-Puig *et al.* (2022b)
54	*Pristimantis morlaco*	2022	Craugastoridae	Sánchez-Nivicela *et al.* (2022)
55	*Pristimantis donnelsoni*	2023	Craugastoridae	Reyes-Puig *et al.* (2023)
56	*Pristimantis kayi*	2023	Craugastoridae	Guayasamin *et al.* in Reyes-Puig *et al.* (2023)

FIGURE 2.55 USFQ amphibian biologists in 2022–2023. Left: Juan M. Guayasamin in an amphibian conservation workshop as part of the Atelopus Survival Initiative at the Sierra Nevada de Santa Marta, Colombia, in August 2022. Photo by Pedro Peloso. Right: members of the *Laboratorio de Zoología Terrestre* and *Museo de Zoología* of the USFQ. From left to right: Mateo Dávila-Játiva, Diego F. Cisneros-Heredia, Carolina Reyes-Puig, Emilia Peñaherrera-Romero, and David Brito-Zapata in 2023. Photo by Margarita López.

study of bufonids (Coloma *et al.*, 2010b; Guayasamin *et al.*, 2010; Pereyra *et al.*, 2016; Jaynes *et al.*, 2022), marsupial frogs (Carvajal-Endara *et al.*, 2019), rainfrogs (Arteaga *et al.*, 2016; Guayasamin *et al.*, 2015a; Hutter *et al.*, 2016), torrent treefrogs (Coloma *et al.*, 2012; Yánez-Muñoz *et al.*, 2021; Reyes-Puig *et al.*, 2022a; Guayasamin *et al.*, 2015b), and salamanders (Jaramillo-Martinez *et al.*, 2020), among others. While at the USFQ, Guayasamin has led and collaborated on 59 scientific articles, addressing topics as broad as the drivers of tropical diversity (Gill *et al.*, 2016; Polato *et al.*, 2018), amphibian decline and diseases (Guayasamin *et al.*, 2014; Scheele *et al.*, 2019), the rights of nature (Guayasamin *et al.*, 2021), and the current threats to the Amazon basin (Albert *et al.*, 2023). Today, the Laboratory of Evolutionary Biology counts on the invaluable and active research of Gabriela Gavilanes and Daniela Franco-Mena, as well as undergraduate and graduate students.

In 2017, Carolina Reyes-Puig (see Figure 2.55) joined the College of Biological and Environmental Sciences at the USFQ as a professor, IBIOTROP researcher, and curator of the vertebrate collection of the ZSFQ (see her academic biography under the title "The Diversity and Academic Lives of Current Amphibian Biologists"). Her research has focused on the taxonomy, systematics, ecology, and conservation of amphibians in Ecuador, mainly on the Andean slopes and in the lowlands of Ecuador, especially in the Llanganates Sangay Ecological Corridor (Reyes-Puig *et al.*, 2014; Alvear and Reyes-Puig, 2016; Reyes-Puig *et al.*, 2019b, 2019c, 2022a). Her most significant contributions have been in taxonomy. She has described 25 new species of amphibians (up to 2023), including two species of salamanders of the rare genus *Oedipina* and 20 of *Pristimantis* (Table 2.6). Her work has also included analyzing trends associated with describing new species of *Pristimantis* (Reyes-Puig and Mancero, 2022; Reyes-Puig *et al.*, 2020c).

USFQ's collections, a core role in education

The growth of the ZSFQ and the *Laboratorio de Zoología Terrestre* has attracted new researchers who have contributed to increasing collections and developing new research programs. For example, in 2013, Mateo Dávila-Játiva started as a field assistant on projects dealing with amphibians and reptiles in mainland Ecuador and the Galápagos islands, eventually becoming dedicated to reptile ecology and now working as the IBIOTROP's operative coordinator. In 2017, David Brito-Zapata, a young Ecuadorian herpetologist, started collaborating as a research assistant on different projects run by IBIOTROP-USFQ, and in 2023, he obtained his master's in tropical ecology and conservation at the USFQ under the mentorship of Carolina Reyes-Puig and Cisneros-Heredia. Brito-Zapata has deposited specimens of amphibians from environmental consulting and research projects at the ZSFQ, and he has participated in the description of new species and published amphibian ecological notes (Brito-Zapata *et al.*, 2021; Brito-Zapata, 2021; Brito-Zapata and Reyes-Puig, 2021; Reyes-Puig *et al.*, 2021). In 2018, Juan Carlos Sánchez-Nivicela joined the *Laboratorio de Zoología Terrestre* as an associated researcher and has helped to increase the collections of the ZSFQ, especially with specimens from southern Ecuador (Sánchez-Nivicela *et al.*, 2021, 2022, 2023; Urgilés *et al.*, 2019).

In 2022, the ZSFQ opened its first public exhibit at the USFQ, displaying live and preserved amphibians, among other animals, and received more than 800 visitors in the first six months. The *Laboratorio de Zoología Terrestre* develops research programs on the taxonomy, evolution, histology, biogeography, natural history, population ecology, urban ecology, conservation, and management of amphibians across mainland Ecuador and the Galápagos. It currently admits about six Ecuadorian and international undergraduate and graduate students per year to collaborate as research assistants in projects involving amphibians. Since 2014, 28 students have conducted their senior theses on topics related to amphibians under the direction of C. Reyes-Puig and Cisneros-Heredia, and four students under the direction of Guayasamin.

In summary, the USFQ, through collaborative efforts, has contributed to the discovery and description of more than 50 species of amphibians (Table 2.6) and laid a solid foundation for long-term amphibian studies with the development of a diverse array of research programs focused on their study and conservation across the Neotropics. As a university, creating educational and training programs for new generations of biologists and conservationists is its core mission and will further expand the USFQ's impact on amphibian research and conservation for many more years to come.

Touzet, Velasco, and the Vivarium of Quito

The *Fundación Herpetológica Gustavo Orcés* (FHGO) is a non-profit organization dedicated to research and education on the conservation of amphibians and reptiles in Ecuador. This organization maintains a relatively small amphibian collection for scientific use and through its vivarium works on the conservation of the marsupial frog *Gastrotheca riobambae* and showcases some amphibians.

FHGO was founded on 5 June 1989 by Jean-Marc Touzet and Ana María Velasco (see also about them in the USFQ history section), together with their friends and colleagues Juan José Espinoza, Rocío Paz y Miño, Xavier Silva del Pozo, Miguel Samaniego, Jorge Anhalzer, and Inés María Jaramillo. The foundation's emblematic project was the *Vivarium de Quito*, an Ecuadorian center specializing in the ex situ management of reptiles and amphibians and public exhibits, and which began its operation on 1 December 1989.

Although Jean-Marc Touzet and Ana María Velasco primarily focused on education about reptiles, they made several contributions to knowledge of amphibians. The FHGO created an important scientific collection of amphibians in Ecuador and the first library on amphibians open to the public in Ecuador. Initially, this was done as part of the Lab FHGO-USFQ headed by Touzet (1997–1999), Velasco (1999–2000), and Cisneros-Heredia (2000–2003) at the USFQ. Among their collections are the type material of several species of amphibians (e.g., *Atelopus orcesi* (Coloma *et al.*, 2010b), *Osornophryne guacamayo* (Hoogmoed, 1987), *Boana nympha* (Faivovich *et al.*, 2006), *Osteocephalus fuscifacies* (Jungfer *et al.*, 2000), *O. mutabor* (Jungfer and Hödl, 2002), *Pristimantis colomai* (Lynch and Duellman, 1997), *P. degener* (Lynch and Duellman, 1997), *P. esmeraldas* (Guayasamin, 2004), and *Hyalinobatrachium yaku* (Guayasamin *et al.*, 2017a)). Touzet also collected some amphibians between 1984 and 1988, which were subsequently deposited at the MHNG (135 specimens) and KU (109 specimens). The Vivarium began to display a few amphibians, among them the marsupial frog *Gastrotheca riobambae*, a program that was continued by their colleague, María Elena Barragán, who became the executive director of the Vivarium in 2013. In 1990, she began to work at the Vivarium as a volunteer, six months after the FHGO was created, while she was studying biology. In 1999, Jorge H. Valencia, a student from the Universidad Central del Ecuador, began to work at the Vivarium as a volunteer and from 2000 worked as research assistant, curatorial assistant, and curator until 2013. In 2008, the FHGO produced a field guide that included 167 species of Ecuadorian amphibians (Valencia *et al.*, 2008), which was followed, in 2009, by three other books for which Valencia and Raquel Betancourt were the lead authors (Valencia *et al.*, 2009a, 2009b; Betancourt *et al.*, 2009). Currently, the Vivarium maintains a few amphibians on display and has an education program.

Centro Jambatu: A Fresh Venture

Centro Jambatu for amphibian research and conservation began its activities in January 2011 under the direction of Coloma. As director of the center, Coloma oversees publications, educational programs, and the website, Anfibios-Ecuador. Coloma is closely involved in an ex situ breeding program to keep assurance colonies for more than 70 species of Ecuadorian frogs.

The foundation of CJ was enthusiastically supported by many people, including a team of biologists and personnel who previously worked at PUCE. The team included María D. Guarderas, Juan M. Guayasamin, Néstor Acosta-Buenaño, Sofía Carvajal-Endara, Manuel Morales-Mite, Elicio E. Tapia, Diego Almeida Reinoso, Freddy Almeida Reinoso, and Tatiana Dávila, among others (see the academic biographies of Guayasamin, Carvajal, Morales-Mite, Tapia under the title "The Diversity and Academic Lives of Current Amphibian Biologists"). Seven of them were hired to work at CJ. The center had the support and strong commitment of Giovanni Onore, president of *Fundación Otonga*, and functioned under its legal status for about eight years. In June 2019, CJ acquired its own legal status as part of a new NGO, *Fundación Jambatu*.

Since its beginnings, CJ was sponsored and partially funded by a private company, Wikiri (an allied biocommerce enterprise. See also Chapter 1 under the subtitle "The Sustainable Bioethical Trade") headed by María D. Guarderas. The international cooperation has been pivotal for the continuous support of the Saint Louis Zoo and funding for several projects (see herein).

Throughout 12 years, CJ has focused its efforts on integrative actions to save amphibians from extinction based on a sound research program that includes in situ and ex situ actions and the formation of a biobank (biorepository of specimens, tissues, skin, sperm, photos, call recordings, and other associated data), an ex situ facility for live amphibians, and the creation of public exhibits (under the name Wikiri Sapoparque, the educational branch of CJ).

The CJ biobank and ex situ program received support from 2015 to 2021 from the Ecuadorian government and throughout its Ministerio de Ambiente (with most funds from the Global Environment Facility) throughout a project titled "Conservation of Ecuadorian Amphibian Diversity and Sustainable Use of its Genetic Resources" (PARG project), which aimed to safeguard the globally significant biodiversity of Ecuador through building capacity for access to genetic resources, using amphibians as a pilot case. Currently (up to 5 October 2023), CJ stores 13,343 specimens of about 443 species and associated cryopreserved tissue samples of nearly all of the species, which are fixed and stored in ethanol. Tadpole series are fixed and stored in formalin. Six types and 30 paratypes of six species are part of these collections and result from the descriptions of seven species of amphibians new to science of the genera *Hyloscirtus* (2), *Gastrotheca* (4), and *Leucostethus* (1). Specimens are fixed and stored in ethanol (except eggs and larvae, which are fixed in formaldehyde), while tissues and sperm are cryopreserved and skin extracts are lyophilized. CJ holds the third largest collection of preserved amphibian specimens in Ecuador and the largest collection of lyophilized skin extracts for peptide analyses (1692 samples of 381 species). Carolina Proaño-Bolaños (Figure 2.56) has been analyzing these extracts, finding many new molecules (e.g., Proaño-Bolaños *et al.*, 2016, 2017; Cuesta *et al.*, 2021) with potential biomedical uses (Mendes *et al.*, 2020; Valdivieso-Rivera *et al.*, 2022). Carolina Proaño-Bolaños is an Ecuadorian expert

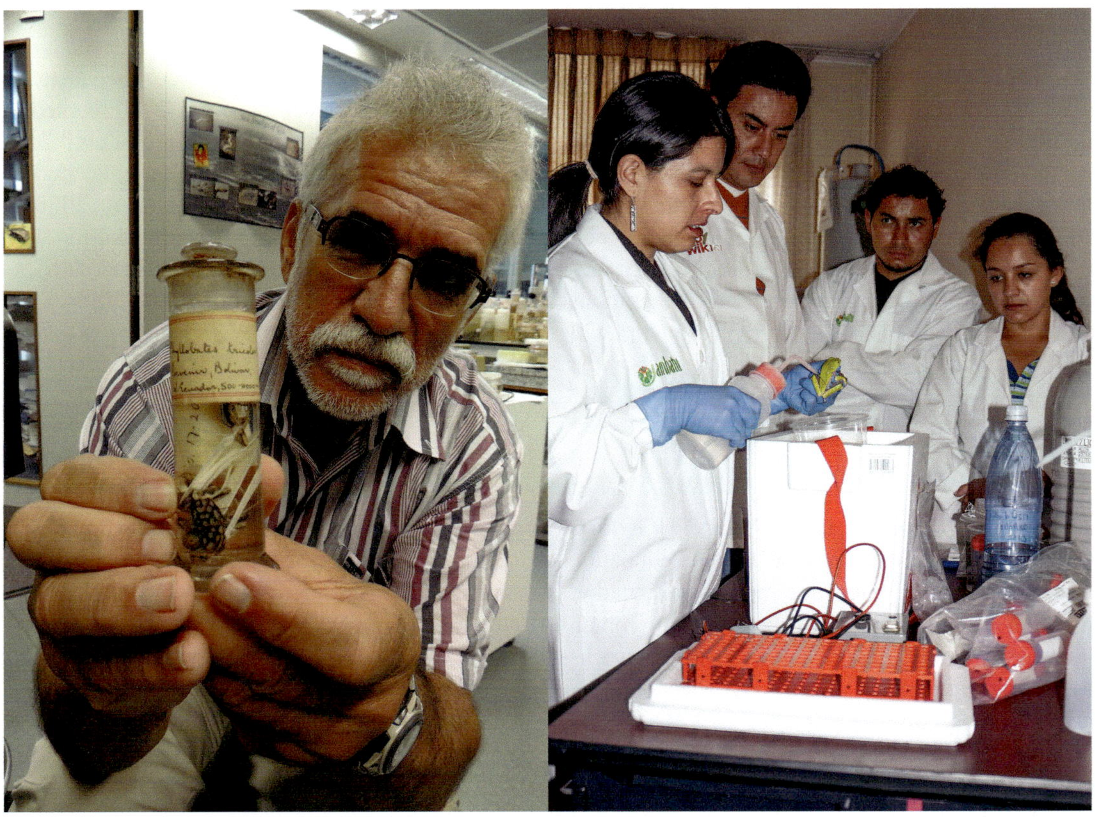

FIGURE 2.56 Left: Jan C. Verkade at The Natural History Museum, London, examining the type of *Epipedobates tricolor*, 7 July 2012. Right: Carolina Proaño-Bolaños teaching procedures for the extraction of peptides at the CJ lab; from left to right: Manuel Morales-MIte, Patricio Vargas Mena, Gabriela Aguiar, 29 October 2013. Photos by LAC.

on molecular aspects of amphibian skin secretions (see her academic biography under the title "The Diversity and Academic Lives of Current Amphibian Biologists"), who studied at Queen's University in Belfast, UK, and in 2016 was hired by Universidad Regional Amazónica Ikiam. Ikiam established a molecular biology lab to accomplish some of the goals of the PARG project.

Resurrecting the Jambato Harlequin Frog

The ex situ program of CJ has the largest live amphibian collection in the world and is working tirelessly to ensure many amphibians are not lost to extinction. Currently (up to 5 October 2023), the center safeguards assurance colonies of about 2600 individuals that represent 76 species of amphibians, which are mainly used for conservation and research purposes. Backup colonies of some species are being maintained at the Amaru Zoo, the Saint Louis Zoo, and some private breeders. The Amaru Zoo (directed by Ernesto Arbeláez Ortiz) also has mounted an amphibian rescue center in Cuenca, Provincia Azuay, and works in cooperation with CJ.

Among the species maintained at CJ are harlequin frogs *Atelopus* (14 species), marsupial frogs (8 species), glassfrogs (9 species), and others (36 species). One of the most iconic species maintained is *Atelopus ignescens*, the jambato harlequin frog. One relict population was discovered in Angamarca, Provincia Cotopaxi, in northwestern Ecuador in 21 April 2016, after the species had not been seen for 28 years, despite intensive searching for this formerly common diurnal frog inhabiting the northern Andes of Ecuador (Coloma, 2016). The discovery attracted media attention at the local and international level (Plitt, 2017). Individuals from the population were collected in 2016 to form the basis of an ex situ breeding program to conserve the likely last remaining individuals of the species. Nine out of 35 individuals tested positive for the pathogen that causes amphibian chytridiomycosis (*Batrachochytrium dendrobatidis*, *Bd*). In 2017, CJ bred the jambato for the first time. A brief field survey of the site conducted in November 2019 by Sara W. Fitzpatrick and her PhD graduate student Kyle Jaynes from Michigan State University resulted in the initiation of a genetic rescue program for the species in cooperation with CJ and other Ecuadorian researchers. The first genomic data for the contemporary rediscovered population was obtained (Jaynes *et al.*, 2022), as well as the genomic signatures of multiple species of recently rediscovered harlequin frogs that were nearly extinct in in Ecuador. Currently, the Fitzpatrick lab, in cooperation with CJ and the Jambato Alliance, continues to carry out genomic studies and harness genomic tools to help reverse the decline of *A. ignescens*. In June 2021 an in situ monitoring project began simultaneously with the formation of the Jambato Alliance. The Jambato Alliance includes at least 30 members of numerous private and public organizations and seeks to unite efforts to position the jambato, *A. ignescens*, as an emblematic species and save it from extinction through both in situ and ex situ research and conservation efforts involving the local community (Alianza Jambato, 2023).

A Victory for Nature's Rights over Mining: Cooperative Work, Two Frogs, and the Science Crusade

Another species that has been successfully bred at CJ is the longnose harlequin frog, *Atelopus longirostris*, a species that has become a symbol of the fight for the rights of nature in Ecuador—a campaign in which Andrea Terán-Valdez, Collection Manager at CJ (Figure 2.57), has been actively participating by joining the historical efforts of environmental organizations and members of local communities to resist mining.

The story of the longnose harlequin frog as an activist emblem began in the Intag Valley, a highly conflicted region due to the presence of large deposits of copper and other minerals that is located in the Intag-Toisán corridor, a key biodiversity area (KBA) within the Tropical Andes Biodiversity Hotspot, characterized by an immense and unique biodiversity. The discovery of metal ore in the area began in 1980 with a Belgian expedition, but exploration intensified in the early 1990s with the arrival of Bishimetals, one of the most important mining companies at that time. The presence of this company triggered the creation of DECOIN, led by Carlos Zorrilla, an environmental organization that has sustained the longest battle of resistance against mining in Ecuador. Since that time, Bishimetals and other companies have been unsuccessful in exploration of the area because of the resistance of the local communities. The situation changed in 2014, when a partnership between the Chilean enterprise Codelco and the Ecuadorian company ENAMI aggressively moved into the Junín area, aided by government forces, to perform advanced exploration activities. The resistance did not stop, and DECOIN, in an attempt to defend the territory, contacted CJ to carry out a rapid amphibian inventory with the aim of enhancing knowledge of the biodiversity in the area (Freile *et al.*, 2020). And so, the longnose harlequin frog, *Atelopus longirostris*, was rediscovered. The species was rediscovered by Elicio E. Tapia, a researcher at CJ, on 28–31 March 2016, 27 years after it was last seen, in 1989 (Tapia *et al.*, 2017), reversing the speculation that it was extinct. In 2004, it had been categorized as extinct in the IUCN Red List. The species was found in a new locality where the species had not been recorded in the past, in Junín, in the Intag Valley. Following the rediscovery, CJ led a rescue effort to bring four of the frogs into their ex situ conservation program with the hope of successfully breeding them in the laboratory and one day reintroducing this species in the wild. The mining activities at the headwaters of the Chalguayacu and Junín rivers have a high environmental impact; thus, they threaten this last-known, tiny population of *A. longirostris*, as well as the biodiversity and local communities. CJ bred this species for the first time on 26 July 2019 (using hormone stimulation) and developed technological protocols to successfully breed the species—as well of six other species of harlequin frogs.

FIGURE 2.57 The CJ team: Andrea Terán-Valdez, Collection Manager, 23 November 2022. Photo by Jaime Culebras. Top right: Elena Naydenova, Veterinarian, 28 June 2017. Photo by LAC. Bottom left: Marco De Rossi, Bioteria Keeper, 27 June 2017. Photo by LAC. Bottom right: Manuel Morales Mitte and Luis A. Coloma at the CJ museum, 17 April 2017. Photo CJ archive.

However, the founder colony is still too small, consisting of just 16 field parentals and about 180 laboratory-born frogs, given that after nine months of population monitoring in the area, done by CJ researchers from August 2019 to March 2020, only four adults and two groups of tadpoles were sighted—30 tadpoles were collected for the ex situ breeding program.

Although the ex situ efforts were relatively successful, the in situ conditions for the species were of great concern due to the risk from mining activities for this fragile population, especially after a scientific analysis of the environmental impact assessments (EIA) (approved by the Ministry of Environment of Ecuador) revealed serious methodological problems and gaps in the reported information (e.g., the longnose harlequin frog was not included). With the species having been lost for 27 years, the luxury of squandering a singular chance to rescue it from extinction was simply unacceptable. Therefore, in 2020, Andrea Terán-Valdez from CJ joined environmental organizations and local communities to launch a legal battle to defend the longnose harlequin frog and all the biodiversity threatened by mining. In August 2019, a new species of rocket frog, *Ectopoglossus* sp., was found by CJ researchers (e.g., Morley Read) at the Cascadas Gemelas in the Intag Valley. It has been provisionally named (its scientific description is still pending) as the Intag resistance rocket frog, which added an additional powerful reason to protect this unique site. The discourse of "responsible mining" employed by the government soon became unsustainable, given that the environmental license was approved with a poor EIA that did not include appropriate mitigation or protection measures to reduce and avoid the impact on the unique biodiversity of the area. A "precautionary measure" was demanded by the court of Cotacachi, Provincia Imbabura, to stop the mining activities due to their imminent impact on the survival of the longnose harlequin frog and the resistance frog, using as background the protection of the rights of nature, recognized by the Ecuadorian Constitution since 2008 (Guayasamin *et al.*, 2021). In the trial, other scientists supported the cause, including Juan M. Guayasamin and Jaime Culebras, by providing scientific information on the possible impacts of mining on amphibian populations and other threatened biodiversity present in the area. Despite the compelling evidence, the anti-mining case prevailed in the first trial (Cotacachi, municipal level), only to falter during the subsequent appeal (Imbabura, province level). A second lawsuit was initiated in 2022 by environmental organizations (e.g., CEDENMA, DECOIN) and members of the local communities (e.g., Junín, Chalguayacu, Cerro Pelado), citing a lack of consultation with the local communities, as well as infringement of the rights of nature. CJ participated by providing scientific information to the court that could help resolve the case. Nevertheless, the judges ruled against the communities and nature in the first trial (municipal level). It is important to understand the economic and political pressures that Ecuadorian judges face when working on these cases, and a clear example was that the President of Ecuador sent his lawyer to the trial to represent him as an interested party in the case. However, the frogs helped raise the flag of resistance, and the case gained support from the international conservation community. For example, the actor and activist Leonardo Di Caprio actively participated in the case with support messages on social media for CJ and to protect Intag's biodiversity (Graeme, 2022). On 29 March 2023, in a historical decision, the Imbabura Provincial Court ruled that the Chilean copper producer Codelco and Ecuador's Empresa Nacional Minera (ENAMI EP) had violated communities' constitutional right to consultation as well as the rights of nature, thus canceling the environmental licenses that allowed them to work in the territory. The decision represents a milestone for communities in the Intag Valley, which have been living in limbo for many years, dealing with threats and conflicts between families and community members due to the presence of mining companies. The court's decision represents a triumph of science supporting the conservation efforts of communities to conserve their frogs and biological diversity. Environmental organizations such as Re: Wild and media outlets have spotlighted this achievement, accompanied by images of the longnose harlequin frog *A. longirostris* and the resistance frog *Ectopoglossus* sp. (Montaño, 2020; Graeme, 2021; Kimbrough, 2023; Forrester, 2023). The former has begun to be known as De Caprio's longnose harlequin frog.

The Foreign Input to Research at CJ

Research at CJ has been conducted by its staff and in cooperation with several local and foreign researchers covering a diverse array of topics from systematics (including the description of amphibians new to science) to morphology, neurophysiology, ecophysiology, reproductive biology, and ex situ management, among others. Cooperative research with Lauren O'Connell (on dendrobatid evolution of behavior, physiology) and Kim Hoke (on Bufonidae ear evolution) from the United States has resulted in several scientific publications. So far, CJ has been involved in 26 peer-reviewed publications and 13 outreach publications stemming from such international collaborations.

Since its founding, international support has been critical for the development of CJ and its activities, and numerous researchers and institutions have worked in cooperation with CJ. A permanent and pivotal institutional collaboration is that of the WildCare Institute of the Saint Louis Zoo, which has provided funds for the CJ project titled: "Providing sustainability to the ex-situ conservation program (Arca de los Sapos) of amphibians in risk of extinction". This collaboration is effective thanks to its CEO, Jeff Bonner, the director of its WildCare Institute, Eric Miller; Mark Wanner, the Zoological Manager of Herpetology and Aquatics; and more recently Lauren Augustine with Mark Wanner. Wanner has been in charge of a cooperation program for the ex situ conservation of amphibians in Ecuador. The cooperative work started in February 2006, when Coloma and Wanner attended an ex situ workshop in Panama to participate in the elaboration of the Global Amphibian Conservation Action Plan leaded by IUCN/Species Survival Commission/ Amphibian Specialist Group and established a friendship. For four years, the Saint Louis Zoo supported the ex situ program titled Balsa de los Sapos while Coloma was at PUCE. When Coloma left PUCE, in December 2010, Wanner encouraged the Saint Louis authorities to continue supporting Coloma in his new institution, CJ. Wanner participated in several collecting field trips jointly with Coloma to the Amazonian region, the Andes, and the Pacific Coast, on which they found and collected several species new to science (not yet described) and collected founders for ex situ assurance colonies (e.g., a rare individual of *Atelopus longirostris*) and numerous other specimens maintained live at CJ and for the preserved collections. In 2022, Wanner accepted a new position at the Brookfield Zoo in Chicago.

Jan C. Verkade (see Figure 2.56) and José Verkade (his wife) from Germany have been collaborating with Coloma since 2008 by collecting specimens and associated data such as call recordings and photographs. They did monthly field trips to many regions of Ecuador in 2008, 2009, 2010, 2012, 2014, 2015–2020, and 2022. In 2020 and 2022, they were joined by their son, Thomas Verkade. The amphibian specimens they collected were deposited at the QCAZ until 2010 (292 preserved specimens, plus live specimens) and at CJ since 2011 (397 preserved specimens of 75 species, plus live specimens). During this time, they produced a series of outreach publications in *Dendrobatidae Netherland Magazine*, a specialized publication for amphibian breeders and the general public. They published several articles (Verkade and Verkade, 2009, 2011; Verkade, 2013, 2016) and enthusiastically promoted CJ and its allied enterprise, Wikiri (Verkade, 2012, 2014). Additionally, Jan C. Verkade collaborated on the scientific description of the tadpoles of *Ameerega parvula* and *A. bilinguis* (Poelman *et al.*, 2010). Coloma joined Jan and José Verkade in 2012 in London at the Natural History Museum to gather historical data and photographs of some type specimens of *Epipedobates tricolor* (Verkade, 2013) and *Gastrotheca*, and at present they are continuing their collaboration with CJ to fulfill specific collecting needs to increase the parental stock of endangered species, to solve taxonomic issues, and to perform photographic documentation of the development of many species of frogs depicted in this encyclopedia.

As part of the Amphibian Ark strategy to coin names from famous people in an effort to call attention to the amphibian crisis, one discovery was named *Hyloscirtus princecharlesi* in recognition of H. R. H. Prince Charles' remarkable efforts to protect rainforests around the world (Coloma *et al.*, 2012). Personnel at Amphibian Ark coordinated a special event at Prince Charles' Highgrove House residence in England in

July 2012, at which time Prince Charles thanked Coloma for the eponymic honor (Figure 2.58). Another species, a marsupial frog, was named *Gastroheca turnerorum* to honor the Turner family and highlight their exceptional, unwavering and ever-growing commitment to conservation of the world natural resources (Carvajal-Endara *et al.*, 2019).

Two international collaborators stand out for their highly productive research programs: Lauren A. O'Connell and her research on the evolution of dendrobatid poison frogs, and Kim Hoke addressing the evolution of the amphibian ear (Figure 2.59).

O'Connell is a neuroscientist of the highest caliber specializing in cellular and molecular biology who graduated at the University of Texas at Austin (see her academic biography under the title "The Diversity and Academic Lives of Current Amphibian Biologists"). She began to work with dendrobatid poison frogs in collaboration with Coloma shortly after the founding of CJ in 2011. She studies behavior, physiology, and evolution, trying to understand how genetic and environmental factors contribute to biological diversity and adaptation. Her research program is deeply rooted in the concepts of community outreach, the decolonization of field-based science, and the inclusion of groups historically excluded from conducting research with the goal that the scientific enterprise will transform into something truly beneficial for all members of the society (O'Connell, 2020). This conceptual framework led her to support not only research that included Ecuadorian researchers (e.g., Coloma, Juan Carlos Santos, David Donoso, Elicio E. Tapia, Alejandro Marcillo Lara) but several related CJ and Wikiri activities in Ecuador such as the mounting of exhibits (Wikiri Sapoparque), the growth of the CJ biobank, and the production of this encyclopedia. In a reciprocal manner, O'Connell has greatly benefited from access to field facilities (La Florida Sapoparque, Otokiki), access to live lab-raised frogs provided by Wikiri, Ecuadorian contributions to study design, grants that funded research and field work, and the writing of publications. In an interview published in the *Journal of Experimental Biology*, Lauren remarked on her work at the Otokiki Reserve (part of *Fundación Otonga*) in Provincia Esmeraldas, Ecuador, as follows:

> All of our field work in Ecuador is done in collaboration with local scientists at Centro Jambatu de Investigación y Conservación de Anfibios. They knew the frogs in this particular location are quite variable as each frog has a different colour or pattern. This is confusing because most poison frog populations agree on a specific colour or pattern and then they all use that colour or pattern to advertise that they are unpalatable to predators and so, in theory, you don't want to vary that pattern, because predators might not learn to avoid you. This population was a mystery to us and so we began a collaboration to try to figure out what was going on. I believe it is critical to collaborate with local scientists in international field work.
>
> (O'Connell, 2017)

FIGURE 2.58 King Charles III (formerly known as The Prince of Wales) of Great Britain greeting Luis A. Coloma and Kevin Zippel (Amphibian Ark Director) at Highgrove House, England, 6 July 2012. Photo by Arthur Edwards.

So far, O'Connell's investigations have resulted in 10 publications in high-impact, peer-reviewed journals that have included Ecuadorian authors or coauthors (Rabeling *et al.*, 2016; McGugan *et al.*, 2016a, 2016b; Roland *et al.*, 2017; Santos *et al.*, 2018; Caty *et al.*, 2019; Fischer *et al.*, 2019; Moskowitz *et al.*, 2020; Moskowitz *et al.*, 2022; Pasukonis *et al.*, 2022). The knowledge acquired in a decade of O'Connell's dendrobatid research is remarkable, and the results up to 2020 were discussed in a review article titled "Lessons from poison frogs on ecological drivers of behavioral diversification" (O'Connell, 2020).

Kim Hoke, an exceptional neuroscientist, graduated from the Stanford University School of Medicine and began collaboration with CJ in 2012 to study frog ear evolution and to unveil the flexibility, constraints, and selection on repeated ear loss and regain in toads. Hoke participated in the field, collecting jointly with Coloma and Elicio E. Tapia to collect frogs that later were bred and raised to obtain ontogenetic series for CT scanning. Her graduate student, Molly Womack, performed experimental work at the facilities at CJ as well, which resulted in five published papers (Pereyra *et al.*, 2016; Womack *et al.*, 2017, 2018; Stynoski *et al.*, 2020).

Amphibian Conservation in Southern Ecuador: The Amaru Initiative

Ernesto Arbeláez Ortiz and Fausto Siavichay-Pesántez

Since its inception in 2001, the Centro de Conservación de Anfibios Amaru (CCA-Amaru) of Bioparque Amaru has played a pivotal role as a private initiative dedicated to amphibian conservation and education in southern Ecuador. This remarkable endeavor has been spearheaded by biologists Ernesto Arbeláez Ortiz (the founder of Amaru and its president), along with co-founders Fausto Siavichay-Pesántez (coordinator of CCA-Amaru) and Amanda Vega Toral (Figure 2.60). The CCA-Amaru has not only made substantial contributions to the discovery of rare amphibians but has also actively engaged public institutions such as the Cuenca municipal government in critical efforts to monitor and rescue endangered frogs. Moreover, the center is presently deeply involved in in situ amphibian conservation through the acquisition of land. Over the years, Amaru has cultivated vital partnerships with researchers, herpetology groups, environmental organizations, political entities, and academic institutions. These collaborative efforts have resulted in the initiation of numerous conservation projects,

FIGURE 2.59 Top left: Kim Hoke, Sara Amstrong, Elicio E. Tapia, and Mark Wanner on a collecting trip to Nueve de Octubre, Provincia Morona Santiago, 9 February 2014. Photo by LAC. Top right: Molly Womack at Centro Jambatu lab conducting audio experiments, 30 November 2015. Photo by LAC. Bottom: Lauren A. O'Connell with David Donoso (EPN) and Aurora Alvarez-Buylla (Stanford University) studying *Oophaga sylvatica* at Wikiri facilities in Sapoparque La Florida, Provincia Santo Domingo de los Tsáchilas, April 2017. Photos courtesy of Lauren A. O'Connell.

FIGURE 2.60 Top left: from left to right: Ernesto Arbeláez-Ortiz, Amanda Vega Toral, and Fausto Siavichay-Pesántez at the Bioparque Amaru on 20 September 2023. Photo courtesy of Bioparque Amaru. Top right: the Reserva Quitahuaycu in Cordillera Occiental of the Andes, Provincia Azuay. Bottom left: the marsupial frog *Gastrotheca cuencana* in Bioparque Amaru, where a wild population still occurs. Bottom right: *Pristimatis erythros* at CCA-Amaru, the only species of *Pristimantis* (the genus with the highest number of threatened species in Ecuador) in an ex situ conservation program. Photos by Jaime Culebras.

the establishment of facilities dedicated to amphibian rescue and research, and the creation of a valuable collection comprising rare amphibian species in collaboration with the QCAZ until 2010, the Philadelphia Zoo since 2009, and CJ since 2011. Amaru currently holds a valuable collection of 24 species in its conservation program, including colonies of five critically endangered harlequin frog species native to the southern Andes of Ecuador.

The Birth of Amaru in the New Century

Construction of the Amaru Zoo began toward the end of 2001, and it was officially inaugurated as a private initiative recognized by the Ecuadorian government on 29 October 2002. Its primary focus has been to engage in activities related to the care, management, and ex situ breeding of fish, amphibians, and reptile species, alongside initiatives aimed at raising awareness, providing education, conducting research, and protecting wildlife and its ecosystems.

During its first ten years, Amaru's operations were based in downtown Cuenca, within the heritage zone of the Tomebamba River ravine. Amaru Zoo exhibits began on the ground floor of the house of the Arbeláez family. In this relatively small space of just 180 m^2, the initial infrastructure became the birthplace of significant activities, experiences, facilities, and talent development. These efforts were all geared toward fostering the study, care, and interest in the conservation of a wide range of fish, amphibians, and reptiles that occurred both within the City of Cuenca and throughout the southern region of the country. This ex situ wildlife conservation center soon became a specialized and central hub for knowledge exchange and a valuable resource for consultations on herpetology and ichthyofauna-related topics.

In its early days, the Amaru Zoo managed colonies of more than 35 species of a wide range of families of native amphibians, and a few exotic ones; among the latter were the axolotl *Ambystoma mexicanum* and the African clawed frog *Xenopus laevis*. The living amphibian collection was key for training in amphibian care and health, as well as education and outreach. Additionally, Amaru built and maintained a valuable preserved herpetological research collection of about 200 specimens (more than a half them being amphibians). In 2006, this scientific collection was donated to Universidad del Azuay, which helped to establish its emerging zoology museum at the Escuela de Biología.

During its first decade (up to 2012) of operation and service to the community and tourists, more than 165,000 visitors passed through Amaru's facilities, where they were educated through guided tours about the importance, diversity, and conservation status of their animals, including the amphibians. In this first decade, Amaru also served as a platform for a group of students from the southern region of Ecuador who were studying biology at the Universidad del Azuay, enabling them to get involved in the study and work with amphibians in the region. As a result, significant synergies and spaces were created for studies, projects, initiatives, and actions focused on the conservation of amphibians in southern Ecuador and the development of the amphibian community in Cuenca.

Beginning in 2004, Amaru initiated the first care-in-laboratory trials for harlequin frogs of the genus *Atelopus* and began its amphibian ex situ conservation program with an undescribed but endangered harlequin frog, *Atelopus* sp. (*spumarius* complex), using a parental stock that had been rescued from degraded and threatened areas in the Cordillera del Cóndor region, Provincia Zamora Chinchipe. In the mid- to late 2000s, Amaru launched a population-monitoring project at this site, funded by the Disney Worldwide Conservation Fund. Additionally, Amaru collaborated with the team at the QCAZ of Pontificia Universidad Católica del Ecuador (Luis A. Coloma; Andrés Merino-Viteri, David Salazar-Valenzuela, Paula Peña Loyola, Diego Acosta López, Verónica Cano, and Miguel A. Rodríguez), who was conducting studies on natural history and working on in situ conservation of the same species at Quebrada Napinaza, Provincia Morona Santiago. This collaboration marked the beginnings of a long-lasting cooperation with Coloma, initially at the QCAZ and later at CJ. Joint efforts have been carried out since 2004 to investigate, document, and showcase the diversity of amphibians in the southern part of the country. Fascinating field expeditions took place in the montane forests of the Andes, the chilly paramos, and the tropical Amazonian forests, leading to the discovery of new species and the addition of geographic records that were crucial for understanding the conservation status and distribution of some species. Amaru personnel benefited by attending workshops, as well as personalized training opportunities at the QCAZ and CJ focused on ex situ amphibian practices. Coloma was one of the best mentors that Amaru has had, sharing his invaluable experience and knowledge, instilling in us a passion for the study and protection of amphibians, and connecting us with key actors and researchers in the international community.

The Amphibian Conservation Program began in 2006 with initial management trials of threatened species such as *Atelopus* sp. (*spumarius* complex), the vertebral rocket frog *Hyloxalus vertebralis*, and the equatorial snouted frog *Ctenophryne equatorialis*, both of the latter being rare inhabitants of the City of Cuenca. They were housed in a small room at the Amaru Zoo that used a setup made from recycled drinking water bottles as terraria. In 2007, through a strategic alliance with ETAPA EC of the municipal government of Cuenca, the Amphibian Conservation Center for Threatened Species was established inside the forest of Bosque Protector de Mazán, adjacent to Parque Nacional Cajas. Facilities were constructed in situ to care for threatened amphibian species. The species housed in this facility were the harlequin frogs *A. nanay* and *A. exiguus*, the rocket frog *H. vertebralis*, and the marsupial frog *Gastrotheca pseustes*, all of which have become very rare in both protected areas. The objectives of this center were: 1) to advance the understanding of the causes of the decline in amphibian populations; 2) to prevent the extinction of these species through in situ and ex situ research; 3) to protect these species with the assistance of local scientists, students, and people from local rural communities; and 4) to equip local stakeholders with the necessary tools to protect amphibians and the ecosystems they inhabit. This initiative achieved significant success in the care and breeding of the critically endangered sad harlequin frog *A. nanay*, which was bred for the first time on two occasions. We discovered and documented the mode of amplexus, egg laying, egg development, tadpole care, and growth and subsequently recorded the metamorphosis and growth of the froglets (Siavichay *et al.*, 2011). We successfully established a backup population of approximately 30 healthy individuals of this species

In 2008, Ernesto Arbeláez Ortiz and Amanda Vega Toral published the first guide focused on the natural history, distribution, and conservation status of amphibians in Parque Nacional Cajas (Arbeláez Ortiz and Vega Toral, 2008). This popular guide featured several color photographs of most species and maps describing their potential distributions. The guide not only promoted knowledge of these animals among park officials but provided a starting point for new generations of biologists interested in amphibians to conduct more in-depth and detailed research. Specialists from Amphibian Ark visited the facilities of the Amaru Zoo and the Amphibian Conservation Center in the Mazán Forest in 2009; ten members of our technical team received training during their visit.

In 2009, a strategic alliance was established with the Philadelphia Zoo to support the work of the Amphibian Conservation Center. The Philadelphia Zoo, through the conservation biologist Carlos Martínez Rivera, joined CCA-Amaru to assist endangered amphibians in the region, specifically amphibians from the five southernmost provinces of Ecuador (Azuay, Loja, El Oro, Morona Santiago, and Zamora Chinchipe). Resources were concentrated on continuing the program to protect endangered amphibians in Cajas National Park and its surroundings in the Provincia Azuay. Both institutions conducted important master's and PhD studies with international students, which allowed us to understand the prevalence of the pathogenic fungus *Saprolegnia* in the water bodies in the Bosque Protector de Mazán and the chytrid fungus in the populations of harlequin and marsupial frogs in Cajas National (Pérez, 2010). We organized field expeditions to remote areas of Cajas

National Park that had never been assessed before, aiming to locate amphibian populations, identify threats, and study their conservation status.

We conducted rescues of species at risk, which became part of our ex situ backup population in the center's laboratories. We promoted ongoing training for our technical team and the Cajas park rangers and developed methodologies to ensure the proper care and breeding of species ex situ. Grants from the Mohamed Bin Zayed Foundation, the Disney Conservation Worldwide Fund, and Amphibian Ark supported the improvement of infrastructure, equipment, awareness campaigns, and outreach material for neighboring rural communities. These funds also supported the establishment of monitoring for threatened amphibian species in new areas in the southern part of the country. In 2011, the Amphibian Conservation Center for Threatened Species in Bosque Protector Mazán closed due to operational difficulties caused by the scarcity of resources from the public company ETAPA EC. Therefore, in the middle of 2011, all live amphibian specimens and the mobile infrastructure of the laboratories were relocated to the new facilities at Bioparque Amaru in the city of Cuenca.

2011–2023: A New Chapter in Our History

In November 2011, CCA-Amaru began its activities as part of the Bioparque Amaru, an eight-hectare property owned by Amaru and located in the city of Cuenca. In this new location, quarantine laboratories, a vivarium, a veterinary clinic, four climate-controlled rooms, and an education and exhibit room were constructed in an area of about 1500 m^2. This new facility focused on the dedicated care and study of endemic endangered species from southern Ecuador (24 species as of September 2023, Table 2.7) and continues to achieve positive results in the care of species from Parque Nacional Cajas. These species, along with an additional eight (Table 2.7), contribute to raising awareness and providing environmental education about amphibians, particularly for tourists and residents of Azuay, Guayas, and other provinces in the southern region of Ecuador.

Several collaborations have enhanced CCA-Amaru's development over the last decade. In 2013, the environmental section of the Cuenca municipal government supported the first monitoring and relocation of a population of the dendrobatid rocket frog *Hyloxalus vertebralis*, a threatened amphibian species living in unused lots in the urban area of Cuenca. During this period, our collaborative efforts with CJ focused on the joint management of backup populations of the harlequin frogs *Atelopus* sp. (*spumarius* complex), *A. nanay*, *A. bomolochos*, and *A. exiguus*; the equatorial snouted frog *Ctenophryne aequatorialis*; and the Cuenca marsupial frog *Gastrotheca cuencana*, for which we have exchanged specimens and training.

Between 2015 and 2020, the Ecuadorian Ministry of Environment and Water (MAAE) funded the PARG project carried out by CJ, the Universidad Regional Amazónica Ikiam, and CCA-Amaru. The PARG project played a pivotal role in improving our infrastructure and equipment while also providing vital support for veterinary and personnel training, field work, and outreach activities. Within the framework of this project, Siavichay-Pesántez took the lead in monitoring critically endangered harlequin frog species such as *Atelopus balios* from southwestern Ecuador and *A. coynei* and *A.* sp. (*aff. longirostris*) from Chinambí, Provincia Carchi, and he also contributed significantly to the legal protection of their natural habitats. Additionally, Siavichay-Pesántez and colleagues produced a book recounting their experiences of amphibian monitoring (MAAE, 2020). In 2021, the monitoring of the two species of *Atelopus* from Chinambí was continued through a joint project with CJ and supported by Re: Wild.

Amaru has fostered a cooperative partnership with Naturaleza y Cultura Internacional. In 2016, under this collaboration, we developed a traveling amphibian and reptile exhibit in the city of Loja, conducted herpetofauna monitoring in the dry forests of southwestern Ecuador, and participated in workshops, the development of the management plan, and the creation of the Sangay–Podocarpus Connectivity Corridor. We are part of the Herpetofauna Group of the corridor, and since 2020, we have been collaborating to establish a new conservation area, the Reserva Quitahuayco—the first privately protected 807-hectare protected forest owned by Amaru. In addition, our associated researcher Juan Carlos Sánchez-Nivicela published information on the amphibians of this corridor (Sánchez-Nivicela, 2022).

Since 2017, through the Andean Condor Foundation, of which we are founding members, we have been promoting the protection of the natural habitat of various amphibian species. Since 2020, Siavichay-Pesántez has been a part of the herpetofauna subcommittee of the Latin American

TABLE 2.7

Amphibians Currently Maintained at CCA-Amaru for the Ex Situ Conservation Program and Education. The asterisk (*) indicates species used only for education purposes

Atelopus bomolochos	*Atelopus* sp. 1 (*spumarius* complex)
Atelopus exiguus	*Atelopus* sp. 2 (*spumarius* complex)
Atelopus nanay	*Atelopus* sp. (*halihelos* complex)
Rhaebo ceratophrys	*Rhaebo ecuadorensis*
Rulyrana mcdiarmidi	*Espadarana audax*
Cochranella sp.	*Centrolene buckleyi*
**Nymphargus cariticommatus*	*Pristimantis* sp. (aff. *ruidus)*
Pristimantis erythros	**Pristimantis morlaco*
**Hyloxalus vertebralis*	*Epipedobates anthony*
**Epipedobates tricolor*	*Gastrotheca cuencana*
Gastrotheca pseustes	*Gastrotheca riobambae*
Gastrotheca lojana	*Hyloscirtus phyllognatus*
Hyloscirtus tapichalaca	**Phyllomedusa tarsius*
**Hyloxalus* sp. (*aff. elachyhistus*)	*Bolitoglossa* sp. (*peruviana* complex)
**Callimedusa tomopterna*	*Ctenopryne aequatorialis*
**Agalychnis hulli*	

Association of Zoos and Aquariums (ALPZA). Its primary objectives include training and preparing zoo personnel to promote effective actions for the care of threatened amphibian colonies in the region. In 2021, we began to collaborate with faculty researchers and students of the Universidad Católica de Cuenca on various projects related to assisted reproduction, microbiota analysis, and topics related to the ecology and biology of native amphibians.

CCA-Amaru has been involved in several conservation actions related to amphibians in Provincia Azuay. In 2018, an analysis of the amphibian community and the use of different microhabitats in the Minas River basin was carried out with the support of the Sígsig municipal government. Since 2021, the rescue, monitoring, and management of amphibians have been conducted on the premises where the wastewater treatment plant in Guangarcucho, which is being constructed by ETAPA EP, will be installed. In 2022, the rescue and recovery of endemic fauna affected in the influence zone of the Escombrera Ucubamba Alto in the City of Cuenca was carried out by the Empresa Municipal de Aseo de Cuenca (EMAC EP). During the course of this project, several dozen amphibians of three species were rescued.

CCA-Amaru has benefited from the work of a team of biologists during several periods. They are: Chelsea Korfel (2009–2012), Diego Alvarado Montesdeoca (2009–2012), José Cáceres Andrade (2011–2012), Juan Fajardo (2014–2018), Xavier Buestan (2016–2018), Nohemí Torres (2016–2019), Juan Diego Carvajal (2019), Nataly Aguilar (2021–present), Jackeline Arpi (2021–present), and Juan Carlos Sánchez-Nivicela (2022–present).

Several members of the team, as well as ourselves, have done their undergraduate, master's, or doctoral theses on amphibians and coauthored several scientific publications (Lehr *et al.*, 2007; Siavichay, 2008; Arbeláez Ortiz and Vega Toral, 2014; Ortega-Andrade *et al.*, 2021; Arpi and Ocampos, 2022; Aguilar, 2022; Jaynes *et al.*, 2022; Lojano *et al.*, 2023). Amaru has published two books to raise public awareness about the City of Cuenca's amphibians (Siavichay-Pesántez *et al.*, 2016; Salazar *et al.*, 2022). An "Action Plan for Urban Amphibian Conservation in Cuenca" was developed (Siavichay *et al.*, 2016) under the auspices of the municipal government of Cuenca. This plan includes criteria, strategies, and technical recommendations for future urban amphibian protection in the city, spanning the next two decades. Important outreach publications include an educational tool for children (Siavichay *et al.*, 2016) that was designed with the support of the Comisión de Gestión Ambiental of Cuenca.

CCA-Amaru: Inspiring Students and Communities through Education

The Bioparque Amaru annually welcomes approximately 85,000 visitors to its facilities dedicated to wildlife contemplation and education about biodiversity. In the 11 years since its opening, Amaru has received around 900,000 visitors. Forty-five percent of these were children or minors; the remainder were adults and senior citizens. We receive visitors from all social strata, economic levels, and regions of the country. Approximately 12% of the total visitors are foreign tourists. Through effective visitor management, we promote various activities every day of the year to ensure that our visitors, especially young people and children, can marvel at the faunal richness of our country and our region. All of these visitors have passed through the educational and awareness areas developed by CCA-Amaru to promote knowledge and appreciation of amphibians.

Amaru collaborates each year with teachers and educational authorities to establish cooperation agreements based on the curriculum to provide training and education to their students. In this way, Amaru has become a specialized pedagogical platform for promoting biodiversity knowledge. Due to our institutional mission and history, Amaru always places special emphasis on promoting education and training related to amphibians and reptiles.

For more than two decades, we have been actively involved in educating professionals on topics related to amphibians, their care, and conservation. In 2011, as part of our commitment to educating young university students in fields such as biology, veterinary medicine, environmental engineering, and ecotourism, we initiated voluntary and pre-professional internships. These internships resulted in the participation of approximately 760 students. We consistently promote the organization of various educational events, including workshops, lectures, symposia, working meetings, training courses, and conferences. Notable among these events are:

1. The International Symposium on Amphibian Conservation, held in collaboration with the PARG project, MAATE, CJ, and the ETAPA enterprise in 2019 in the City of Cuenca.
2. The Herpetofauna Photography Course, conducted in 2019, with the valuable support of Photo Wildlife Tours.
3. The First Workshop on Inventory and Monitoring Techniques for Amphibians and Reptiles in the Sangay–Podocarpus Connectivity Corridor.
4. In 2022, a course titled "Amphibians: In-Situ Monitoring and Ex-Situ Management" in Cuenca, specifically for students of the Universidad Estatal Península de Santa Elena.
5. The inaugural "Symposium on Biodiversity Conservation in Southern Ecuador" in 2022.

Our commitment to educating and engaging with students and professionals remains unwavering as we strive to advance the cause of amphibian conservation.

Protecting More than 100 Amphibian Species in Natural Reserves

CCA-Amaru works on several strategies aimed at preventing the loss of natural habitat, monitoring wild populations, restoring original habitat, and protecting species in imminent danger. Our organization has been constantly evolving and since 2020, we have been under the umbrella of an NGO, the Amaru Foundation, a non-profit legal entity that seeks to develop management strategies to generate and implement more biodiversity conservation projects, especially for threatened species, and to support existing initiatives such as CCA-Amaru.

Together with some local stakeholders, Amaru owns two natural reserves, Reserva Quitahuayco and Reserva Natural Selva Eterna. These reserves protect approximately 1080 hectares of western Andean forests, paramos, and Amazon rainforest. Both house more than a hundred native amphibian species. Reserva Quitahuayco (see Figure 2.60), located in the cold Andean foothills of Molleturo, Provincia Azuay, spans altitudes from 2300 to 4170 meters on about 807 hectares. Through three expeditions carried out with herpetologist collaborators, we have recorded a dozen species to date. We maintain an area with a small residual population of *A. nanay*, and we have recorded new species to science, which are being described elsewhere. We hope that in some corner of these pristine forests and paramo habitats with watercourses that are difficult to access, we may find *A. onorei* or perhaps the mythical *Telmatobius niger*, two species last recorded in the 1990s. Reserva Natural Selva Eterna is located at 200 meters above sea level in the Amazon region of southern Ecuador, in the town of Morona, Provincia Morona Santiago. In the 273 hectares of flooded and upland Amazon tropical forests that we strive to protect, we have recorded more than 70 amphibian species to date. Some of these species are potentially new to science, and we have obtained significant new distribution records in this underexplored area. The CCA-Amaru team is challenged to continue studying the amphibian communities in these two natural reserves and to disseminate the importance of these areas through scientific publications, with the aim of generating greater interest in the protection and study of likely more than a hundred native and endemic amphibian species.

Amaru's efforts and initiatives have earned recognition both locally and internationally. Notably, in 2012, the Cuenca municipal government honored us for our dedicated work in support of urban amphibians. In 2019, during the Annual Congress of the Latin American Association of Zoological Parks and Aquariums (ALPZA) held in Santiago, Chile, CCA-Amaru received certification as an ALPZA Conservation Project, underlining our commitment to conservation. In 2017, Fausto Siavichay-Pezántez, the coordinator of CCA-Amaru, was honored with the prestigious Global Conservation Award by the Philadelphia Zoo in the USA. This award recognized his outstanding contributions to the conservation and research of amphibians in southern Ecuador.

Filling the Gap in Amphibian Research in Southern Ecuador: The Museo de Zoología, Universidad Técnica Particular de Loja

In 2014, the Universidad Técnica Particular de Loja (UTPL), a private university located in Loja, southern Ecuador, embarked on an exciting journey in the field of amphibian research. This newfound interest in amphibians was sparked by a unique occurrence: the arrival to Ecuador of Paul Székely and his wife Diana Székely from Romania (Figure 2.61). The Székelys' research efforts at the UTPL served as the cornerstone for the university's active involvement in amphibian research, which has flourished over the past decade. Their dedication and expertise have paved the way for ongoing advancements in the ecology, systematics, and conservation of amphibians of the region.

Paul arrived as a research fellow at the UTPL through the Prometeo Project. This project, initiated by the Ecuadorian government of Rafael Correa Delgado, aimed to enhance research and academic capabilities in Ecuadorian institutions by bringing world-class experts with PhDs from foreign countries to Ecuador. These experts were invited to reside in Ecuador for a period of two to 12 months, enabling them to engage in research activities.

Paul Székely, along with Diana Székely, played a crucial role in the early stages of this transformative initiative. They dedicated their time to studying the ecology, distribution, and bioacoustics of amphibians in southern Ecuador. Their contributions were so significant that they decided to

FIGURE 2.61 Left: Diana Székely, Paul Székely, and Diego Armijos Ojeda at Abra de Zamora, 2022. Right: Paul Székely in the Museo de Zoología, Universidad Técnica Particular de Loja, 2023. Photos by Diego Armijos Ojeda.

take up permanent residence in Ecuador and continue their research endeavors, particularly focusing on the amphibians of the southern Ecuadorian dry forest. So far, their research in these forests has resulted in ten publications (Székely *et al.*, 2016, 2017, 2018a, 2018b, 2018c, 2019, 2020a, 2020b; Bardier *et al.*, 2020; Armijos-Ojeda *et al.*, 2021). About their residence in Ecuador, Paul remarked:

> For a long time now (since my undergraduate years as a student), I've dreamed of working in frog systematics. However, this wasn't possible in Europe (Romania, a country almost the same size as Ecuador that has only 20 species of amphibians). So, it was an incredible opportunity for us to work here. And luckily, Diana felt the same way.

Between 2014 and 2017, as part of one of the most productive research groups from the UTPL, the Laboratorio de Ecología Tropical y Servicios Ecosistémicos (EcoSs-Lab), the Székelys studied the amphibians and laid the foundations of the FUTPL sound library (Fonoteca UTPL), which currently holds around 500 amphibian sound recordings.

The Museo de Zoología of UTPL (MUTPL) was first created in 2002 as an entomological collection under the name of Colección de Invertebrados del Sur del Ecuador (CISEC). Under Paul's leadership, the vertebrate collection was formally created in 2018, so the zoology museum now includes fish, amphibian, reptile, bird, and mammal collections. The museum has benefited from the involvement of various UTPL researchers, and in the case of the amphibian collection, especially from the important work of Diana Székely and Diego Armijos Ojeda (see Figure 2.57). The latter is an engineer in management and conservation, master in zoology and animal ecology, and has a PhD degree in the conservation of natural resources at the Universidad Rey Juan Carlos (Spain). He has been involved in biological diversity research in Ecuador since 2007 and has contributed to 15 amphibian scientific publications and books. Since January 2015, he has been affiliated with the UTPL as an assistant professor in the Department of Biological and Agricultural Sciences.

By 2023, the MUTPL amphibian collection had grown to more than 1700 specimens, representing about 25% of the Ecuadorian amphibian species, mostly from southern Ecuador. The collection includes 9 holotypes and 84 paratypes of amphibian species described between 2018 and 2023 and an impressive number of undescribed species of anurans, salamanders, and caecilians.

Currently, at the UTPL, the amphibian section of the Museum of Zoology leads the research in the field of systematic zoology in southern Ecuador. It has embraced the current practice of integrative taxonomy, especially by incorporating molecular characters to build phylogenies and additional lines of evidence such as sound recordings to delimit species (e.g., Székely *et al.*, 2018d, 2020c, 2023a, 2023b; Carvajal-Endara *et al.*, 2019),

> The MUTPL is involved in the continuous effort to promote amphibian conservation. In an alliance with the environmental NGO Naturaleza y Cultura Internacional, it supported the increased protection of habitats and established monitoring programs for endangered amphibian species, with a focus on emblematic sites such as Fierro Urco, Reserva Numbala and Parque Nacional Podocarpus. A particular emphasis has been given to the global biodiversity hotspot, the "Abra de Zamora", with the museum's researchers engaging in several projects to protect and restore this critical ecosystem. The site is the source of most of the water for the City of Loja, stores vast reserves of carbon, is home to at least 12 endemic amphibian species, and it is the type locality for 14 species of frogs.
>
> (Székely *et al.*, 2020c)

WOMEN IN AMPHIBIAN BIOLOGY: AN ONGOING TREND OF UNDERREPRESENTATION IN ACADEMIA

Mónica I. Páez-Vacas, Andrea Terán-Valdez, and María Vizcaíno-Barba

Gender differences in academia have been extensively studied across various disciplines and countries, with documented disparities in the number of authors, citations, recognition, funding, and salaries (Kozlowski *et al.*, 2022; Larivière *et al.*, 2013; Maas *et al.*, 2021; Ross *et al.*, 2022; Shen, 2013). Multiple factors, such as family responsibilities, career length, peer review process, stereotypes, and work climate, have been proposed as explanations for these gender differences (Huang *et al.*, 2020; Morgan *et al.*, 2021; Salerno *et al.*, 2019). In some disciplines, such as the biological sciences, female students comprise nearly 50% of the total enrollment. However, this percentage gradually drops in higher career stages (Farr *et al.*, 2017; Frances *et al.*, 2020). Despite the increase in women's representation in academia over the last five decades, women only account for 30 – 35% of total authorships in STEM (Campbell *et al.*, 2013; Frances *et al.*, 2020; Huang *et al.*, 2020). Nonetheless, when dropout rates are considered, women's productivity is equivalent to that of men (Huang *et al.*, 2020). Moreover, gender parity has been shown to lead to higher-quality papers and greater scientific novelty (Campbell *et al.*, 2013; Hofstra *et al.*, 2020).

The global trend indicates that nearly 94% of countries represented in the Web of Science database are still far from achieving gender parity in terms of published papers (Larivière *et al.*, 2013). However, South American and Eastern European countries stand out for displaying relatively higher levels of gender equality in this regard (Larivière *et al.*, 2013). For example, in the fields of ecology and zoology, the proportion of female authors is slightly higher in Latin America compared to other regions. Nevertheless, this is primarily driven by specific countries such as Argentina, Venezuela, Brazil, and Colombia, which have higher female representation than other countries (Salerno *et al.*, 2019). However, the increased presence of female authors in certain Latin American countries may be attributed to a form of gender segregation. Female principal investigators in the field of ecology and zoology often tend to publish more frequently with their female colleagues, given a trend of male-biased connections that can exclude women and perpetuate the gender gap both regionally and globally (Grosso *et al.*, 2021; Salerno *et al.*, 2019).

Herpetology, as a field, has traditionally been predominantly male dominated (Perry *et al.*, 2020; Rock *et al.*, 2021). Women accounted for only 5% of authors on manuscripts focusing on amphibians and reptiles in the United States from 1937 to 2006 in the context of wildlife research (Nicholson *et al.*, 2008) and less than 15% in three herpetological journals by the 1990s (Wilson, 1998). Recent studies also indicate that women's representation in herpetology journals ranges from 0% to 35% (Grosso *et al.*, 2021; Salerno *et al.*, 2019). Additionally, in the field of herpetology, women make up less than 20% of distinguished speakers at ecology conferences (Farr *et al.*, 2017). In Ecuador, for example, in amphibian research on systematics of the diverse genus *Pristimantis* in recent decades, only 20% of the contributions include women, and a mere 2% have women as principal authors (Reyes-Puig and Mancero, 2022). Nonetheless, it is worth highlighting that the percentage of female authors has shown an upward trend in herpetology worldwide, from 8% in the 1970s to 15% in the 1990s, further rising to 27.7% in 2010, and reaching 34.5% in 2019 (Rock *et al.*, 2021; Wilson, 1998).

Herein, we aimed to investigate women's representation in amphibian biology in Ecuador throughout the years. For this, we evaluated women's contributions to (1) the description of the 655 species of amphibians in Ecuador up to December 2022 and (2) studies in the fields of the evolution, ecology and natural history, and conservation of amphibians in Ecuador between 2010 and 2022 (see Methods in Box 1).

The (Slow) Progression of Women's Inclusion in the Taxonomy of Ecuadorian Amphibians

Early amphibian taxonomy in Ecuador was primarily established by male scientists from North America and Europe. It was not until 1952, 173 years after the first description of an amphibian found in Ecuador, that Bertha Lutz, a distinguished Brazilian naturalist, zoologist, professor, and feminist pioneer, described two new species from Brazil, which are present in Ecuador (*Pristimantis carvalhoi* and *Vitreorana ritae*). Over the next four decades, scattered contributions were made by female herpetologists in studies led by male herpetologists, and mainly from North America. Doris M. Cochran, from the Smithsonian Institution, described five species from Colombia that are currently known from Ecuador: *Pristimantis myersi* in 1963 and *Boana rubracyla*,

BOX 1 METHODS

To analyze women's contributions to the description of the 655 species of amphibians in Ecuador, we built a taxonomy database with the authorship of each species description. We focused on the individual species descriptions rather than the papers themselves, as a single paper may contain descriptions of multiple species. To evaluate women's contributions to the fields of (1) evolution, (2) ecology and natural history, and (3) conservation of amphibians in Ecuador, we built three databases, one for each field. We performed literature searches in Google Scholar from 2010 to 2022, since 2010 marks the beginning of a consistent upward trend in female participation, using the following search term combinations for each of the three databases: Evolution: *("amphibians" OR "anfibios") AND ("Ecuador") AND ("evolution" OR "evolutionary" OR "evolución" OR "evolutivo" OR "evolutiva")*; Ecology: *("amphibians" OR "anfibios") AND ("Ecuador") AND ("ecology" OR "ecological" OR "ecología" OR "ecológico" OR "ecológica" OR "natural history" OR "historia natural")*; and Conservation: *("amphibians" OR "anfibios") AND ("Ecuador") AND ("conservation" OR "conservación")*. The results retrieved in each search were further filtered for inclusion: only published manuscripts in peer-reviewed journals and only manuscripts with data on amphibians from Ecuador were kept. We also excluded manuscripts that were exclusively species descriptions, as these were already considered in the taxonomy database. However, manuscripts with species descriptions that also included evolutionary, ecological/natural history, and/or conservation contributions were maintained in the corresponding database. Studies that contributed to more than one of the three fields of study were kept in each of the corresponding lists. For each article (or species described in taxonomy database), we recorded the (1) year; (2) number of authors; (3) gender of the first author; (4) whether the first author was Ecuadorian; (5) gender of the last author; (6) whether the last author was Ecuadorian; (7) proportion of female authors; (8) proportion of Ecuadorian authors; and (9) proportion of Ecuadorian female authors. An author was considered Ecuadorian by birth or if his/her professional training started in Ecuador, for example, undergraduate studies. To assign gender, we used the given name and online images of researchers. We acknowledge that the binary method employed to determine gender may inadvertently exclude certain authors. We used generalized linear models (GLMs) with quasibinomial error distribution to evaluate the proportion of coauthors through time with the R package *glm* (R Core Team, 2023). We ran additional models with single-authored papers or species descriptions omitted, but our general findings did not differ from the full datasets.

Centrolene medemi, *Nymphargus grandisonae*, and *Osteocephalus cabrerai* in 1970. In 1974, Martha L. Crump, a University of Kansas (KU) student, was the first women to describe an amphibian from Ecuador, *Dendropsophus brevifrons*, while Linda Trueb, also from KU, described *Telmatobius cirrhacelis* in 1979 and *P. quinquagesimus* in 1980. During the same year, Rebecca A. Pyles from KU described *Gastrotheca orophylax*. It is noteworthy that Linda Trueb's description in 1979 was the only solo contribution, while the others were collaborative efforts with male colleagues.

Between 1989 and 1998, Patricia A. Burrowes and Cristina Ardila Robayo, two Colombian herpetologists, participated in the description of eight and two species, respectively, from Colombia, which were later reported from Ecuador. In 1999, Jennifer B. Pramuk contributed to the description of nine species originally described from Peru and one from Ecuador. Notably, in 2000, Ana Almendáriz-Cabezas, from the Escuela Politécnica Nacional of Ecuador (EPN), became the first Ecuadorian female herpetologist to contribute to the description of amphibian species known from Ecuador, *Osteocephalus deridens* and *O. fuscifacies* (see also under the subtitle "The EPN Museum and a Woman Pioneer"). During the 2000s, Pramuk and Fatemeh Kadivar (one species), Elisa Bonaccorso (one species), Karen Siu-Ting (two species), and Kathryn Elmer (two species), none of them Ecuadorian, also contributed to the taxonomy of amphibians of Ecuador (Figure 2.62 left).

The final years of the 2000s saw the emergence of female Ecuadorian herpetologists in amphibian taxonomy (Figure 2.62 right). The first contributions of this period were of Andrea Terán-Valdez (2009) and Mónica I. Páez-Vacas (2010), both at the Pontificia Universidad Católica del Ecuador (PUCE), along with Raquel Betancourt from *Fundación Herpetológica Gustavo Orcés*, Ana Almendáriz-Cabezas (EPN), and Miryan Rivera (PUCE) in 2010. Notably, Mónica I. Páez-Vacas became the first female Ecuadorian herpetologist to lead the description of new species from Ecuador. Her work was a pioneering study in the integrative taxonomy of amphibians in Ecuador, utilizing multiple lines of evidence that included molecular phylogeny, adult and tadpole morphology, morphometrics, and acoustic data. This research represents the beginning of a new era in amphibian systematics in Ecuador.

From 2012 to 2022, there was an increased contribution by female herpetologists to taxonomy (see Figure 2.62 left; Table 2.8). The average representation of women during this period was 20%, with the majority being from Ecuador (18%). Noteworthy female Ecuadorian herpetologists that have played a significant role in systematics and taxonomy are Carolina Reyes-Puig, who has described more than 20 species (since 2014), and Ana Almendáriz-Cabezas, who has described 12 species (since 2000). Other female Ecuadorian herpetologists who have made important contributions to this field include Jhael Ortega (12 species), Nadia B. Páez (10 species), Verónica L. Urgilés (10 species), María José Navarrete (8 species), Andrea Terán-Valdez (7 species), Sofía Carvajal-Endara (6 species), Diana Székely (6 species), Patricia Bejarano Muñoz (5 species), Claudia Koch (5 species), María Pérez Lara (5 species), Daniela Franco-Mena (4 species), Andrea Varela-Jaramillo (4 species), and Yerka Sagredo (3 species).

In our analyses, we found that only 45 out of 269 authors who participated in species descriptions were female, accounting for just 8% of the total number of species described. Overall female representation between 1990 and 2022 was 17%. Notably, only 17 women have

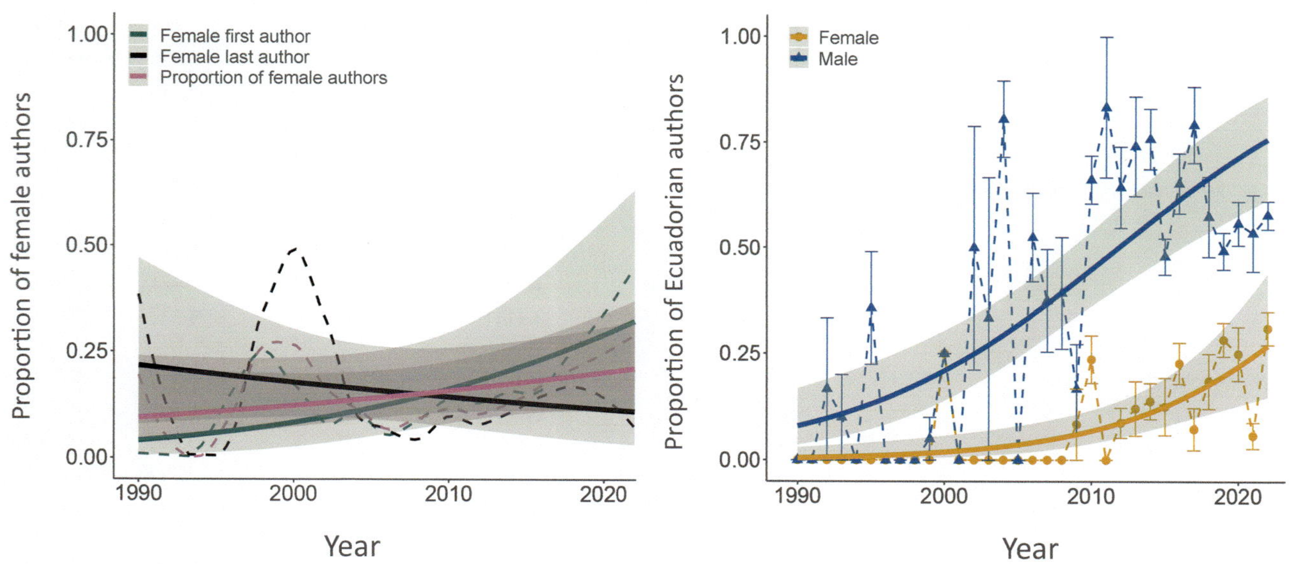

FIGURE 2.62 Proportion of authors in the description of species of amphibians in Ecuador. Left: proportion of female first, last, and overall authorship per species per year. Right: proportion of Ecuadorian authors per species per year. Values are mean proportions per year for every species described from 1990. Note that the first contribution of an Ecuadorian biologist in a species description was in 1992 by Luis A. Coloma. The first Ecuadorian female biologist to contribute to the description of a species was Ana Almendáriz-Cabezas in 2000. Gray areas represent 95% confidence intervals.

TABLE 2.8
Results of Quasibinomial GLMs Describing the Proportion of Female Authors in Studies on the Taxonomy, Evolution, Ecology and Natural History, and Conservation of Amphibians of Ecuador

Response Variable	Intercept Estimate	Predictors	Estimate	Standard Error	t-Value	p-Value
		Taxonomy				
Proportion of female authors	—79.45276	Year	0.0387	0.01	3.723	0.0002*
Proportion of female last authors	70.22713	Year	—0.0358	0.01654	—2.166	> 0.03*
Proportion of Ecuadorian authors	—274.71385	Year	0.13686	0.01349	10.14	< 0.0001*
Proportion of female Ecuadorian authors	—292.77362	Year	0. 14436	0.01766	8.172	< 0.0001*
Proportion of female authors	—1.9741	Last author female	1.7422	0.1962	8.882	< 0.0001*
Proportion of female authors	—2.0170	First author female	1.6310	0.1914	8.52	< 0.0001*
		Evolution, Ecology and Natural History, and Conservation				
Proportion of female authors	—2.9575	Year	0.0004	0.0183	0.021	0.9833
		Field: Ecology	0.4339	0.2280	1.903	0.0581
		Field: Evolution	0.3253	0.1878	1.732	0.0845
		Last author female	1.5471	0.1428	10.836	< 0.0001*
		First author female	1.1914	0.1450	8.217	< 0.0001*
Proportion of female first authors	—52.3250	Year	0.0253	0.0374	0.676	0.4996
		Field: Ecology	—0.2550	0.48367	—0.527	0.5985
		Field: Evolution	0.2422	0.3724	0.650	0.5160
		Last author female	1.0959	0.2980	3.677	0.0003*
Proportion of female Ecuadorian authors	—48.2532	Year	0.0215	0.0255	0.842	0.4008
		Field: Ecology	—0.0200	0.3214	—0.062	0.9504
		Field: Evolution	—0.1788	0.2744	—0.652	0.5151
		Last author female	—0.4095	0.4390	—0.933	0.35189
		First author female	—1.2633	0.4817	—2.623	0.0093*
		Last author Ecuadorian female	2.7704	0.4583	6.044	< 0.0001*
		First author Ecuadorian female	2.9667	0.5169	5.740	< 0.0001*
First author Ecuadorian female	—96.6512	Year	0.0467	0.0504	0.927	0.3550
		Field: Ecology	0.7653	0.6627	1.155	0.2492
		Field: Evolution	0.4248	0.5858	0.725	0.4690
		Last author female	—1.7506	1.0386	—1.686	0.0931
		Last author Ecuadorian female	2.5144	1.1055	2.274	0.0238*
Last author Ecuadorian female	4.300e+01	Year	—3.263e-02	3.720e-02	—0.877	0.3813
		Field: Ecology	1.017e+00	4.284e-01	2.373	0.0184*
		Field: Evolution	—9.949e-03	3.575e-01	—0.028	0.9778
		Last author female	2.192e+01	1.055e+03	0.021	0.9834
		First author Ecuadorian female	2.748e+00	6.145e-01	4.472	< 0.0001*
Proportion of male Ecuadorian authors	44.7217	Year	—0.0220	0.0224	—0.980	0.3281
		Field: Ecology	—0.1234	0.2733	—0.452	0.6519
		Field: Evolution	—0.8307	0.2218	—3.746	0.0002
		Last author female	—1.1209	0.2614	—4.289	< 0.0001*
		Last author Ecuadorian female	0.7835	0.3570	2.194	0.0291*
		First author Ecuadorian female	—0.5595	0.2866	—1.952	0.0520

For Taxonomy, Year affects the proportion of female Ecuadorian authors, showing change over time. In Evolution, Ecology and Natural History, and Conservation, Year does not predict female representation. However, female last authorship predicts female representation.

* Significant

ever been first authors, representing 18% of the total number of species described where women are involved. However, beginning in 2010, a consistent upward trend in female participation can be observed. We documented a rise in female representation in the field over the last two decades, increasing from 27% in 1990 to 36% in 2022 (see Figure 2.62 left; Table 2.8). In terms of coauthors, 23 women have been last authors, accounting for 15% of species described. Interestingly, this percentage was 28% between 1990 and 2000, and then decreased and remained relatively consistent over the last two decades, with 11.6% between 2000 and 2010 and 12% in the most recent decade (see Figure 2.62 left; Table 2.8). However, although the last author is often considered one of the primary contributors to a publication in many STEM fields (Bendels *et al.*, 2018), caution must be exercised in this regard since this may not apply or may only recently apply in amphibian taxonomy in Ecuador (Luis A. Coloma, *pers. comm.*). Single-authored papers have been very rare in recent years in biology, and from all 348 species described by single authors, only three were described by a woman: *Pristimantis carvalhoi* and *Vitreorana ritae* (Lutz, in Lutz and Kloss, 1952), and *Telmatobius cirrhacelis* (Trueb, 1979). Female representation as sole authors is consistently underrepresented in all amphibian research worldwide compared to the overall proportion of female authors (Rock *et al.*, 2021).

Despite the existing gender disparities and the relatively low percentages of female representation, there has been a positive trend in the inclusion of Ecuadorian women in amphibian taxonomy over the past three decades. An overall increase in male and female Ecuadorian researchers in amphibian taxonomy from 11% between 1990 and 2000 to 79% in the last decade has translated into an increase in female Ecuadorian representation from 0.7% to 18.6% in the same period (see Figure 2.62 right; Table 2.8). In line with trends observed in herpetology globally, women involved in amphibian taxonomy in Ecuador tend to be overrepresented as first authors but underrepresented as last authors, compared to their overall representation (see Figure 2.62 left; Rock *et al.*, 2021). Most female contributions as first authors often come from students or early career professionals working within a team led by a male author (Duffy, 2017). Nonetheless, it is worth noting the significant role of mentorship in fostering the rise of female authors, particularly through the guidance of researchers such as Luis A. Coloma, Juan M. Guayasamin, Santiago R. Ron, Diego F. Cisneros-Heredia, Ana Almendáriz-Cabezas, Mario H. Yánez-Muñoz, and Jorge Brito Molina (Reyes-Puig and Mancero, 2022). Although we are aware that last author may not always correspond to the leading author in amphibian taxonomy in Ecuador, our analyses indicate that species descriptions with a female last author tend to have more female coauthors (Table 2.8). Thus, it is crucial to acknowledge that achieving true and significant female representation requires an increase in the number of women assuming leading roles in the coming years.

Women's Contribution to the Evolution, Ecology, and Natural History and Conservation of Amphibians of Ecuador

Until recently, the fields of amphibian evolution, ecology, and conservation in Ecuador were largely dominated by male researchers. However, there have been significant contributions from female scientists dating back to the late 1970s, when Eugenia M. del Pino made groundbreaking discoveries in the development of marsupial frogs and other Neotropical taxa. Del Pino's research involved several female students beginning in the 1980s, and she is still considered one of the most influential Ecuadorian scientists of all time (see also under the subtitle "Del Pino and the Marsupial Frogs"). Another notable female contributor to Ecuadorian amphibian fauna is Ana Almendáriz-Cabezas, who had been active since the late 1980s. Martha L. Crump contributed to the conservation of amphibians in Ecuador with a study in 2002 (Toral *et al.*, 2002), while Elisa Bonaccorso collaborated on an ecological study published in 2004 (Guayasamin *et al.*, 2004). Patricia Castillo-Trenn, in 2004, provided one of the most complete descriptions of tadpoles to date (Castillo-Trenn, 2004). In 2006, Diana Freire and Carla Fernández, two undergraduate students mentored by Fernando Nogales at the Universidad Técnica Particular de Loja, published the first analysis of *Batrachochytrium dendrobatidis* (Bd) in southern Ecuador (Freire *et al.*, 2006). In 2008, Castillo Trenn and María Alexandra Quiguango Ubillús published detailed studies on dendrobatid behavior (Castillo-Trenn and Coloma, 2008; Quiguango-Ubillús and Coloma, 2008). Since 2009, Miryan Rivera has contributed to our understanding of amphibian cytogenetics and skin secretions (e.g., Cipriani and Rivera, 2009). Kathryn Elmer's doctoral dissertation, in the late 2000s, resulted in the first comprehensive study of the evolution of Neotropical amphibians from Ecuador, focusing on a group within *Pristimantis* (Elmer *et al.*, 2006, 2007a, 2007b).

Since 2010, there has been an increase in amphibian research in Ecuador, with 159 articles published on evolution, 50 on ecology, and 51 on conservation (Figure 2.63 top left). Some female researchers who stand out for their contributions to amphibian evolution, ecology, and conservation in Ecuador are Eugenia del Pino, Ana Alméndariz-Cabezas, Elisa Bonaccorso, Kathryn Elmer, Catherine Graham, Miryan Rivera, Lauren A. O'Connell, Janeth Lessmann, Kim Hoke, Molly Womack, Eva K. Fischer, Diana Székely, Carolina Proaño-Bolaños, Rebecca Tarvin, and Mónica I. Páez-Vacas. Other researchers who have led at least one study that advanced amphibian knowledge in Ecuador in these fields include Rebecca Brunner, Teresa Camacho-Badani, Sofía Carvajal-Endara, Sueny dos Santos, Katherine Krynak, Laura Martín-Torrijos, Nora Moskowitz, María José Navarrete, Belén Proaño, Carolina Reyes-Puig, María José Salazar-Nicholls, María José Sánchez-Carvajal, Florina Stănescu, Andrea Terán-Valdez, Cecilia Tobar Suárez, Verónica L. Urgilés, Anyelet Valencia-Aguilar, and María Vizcaíno-Barba.

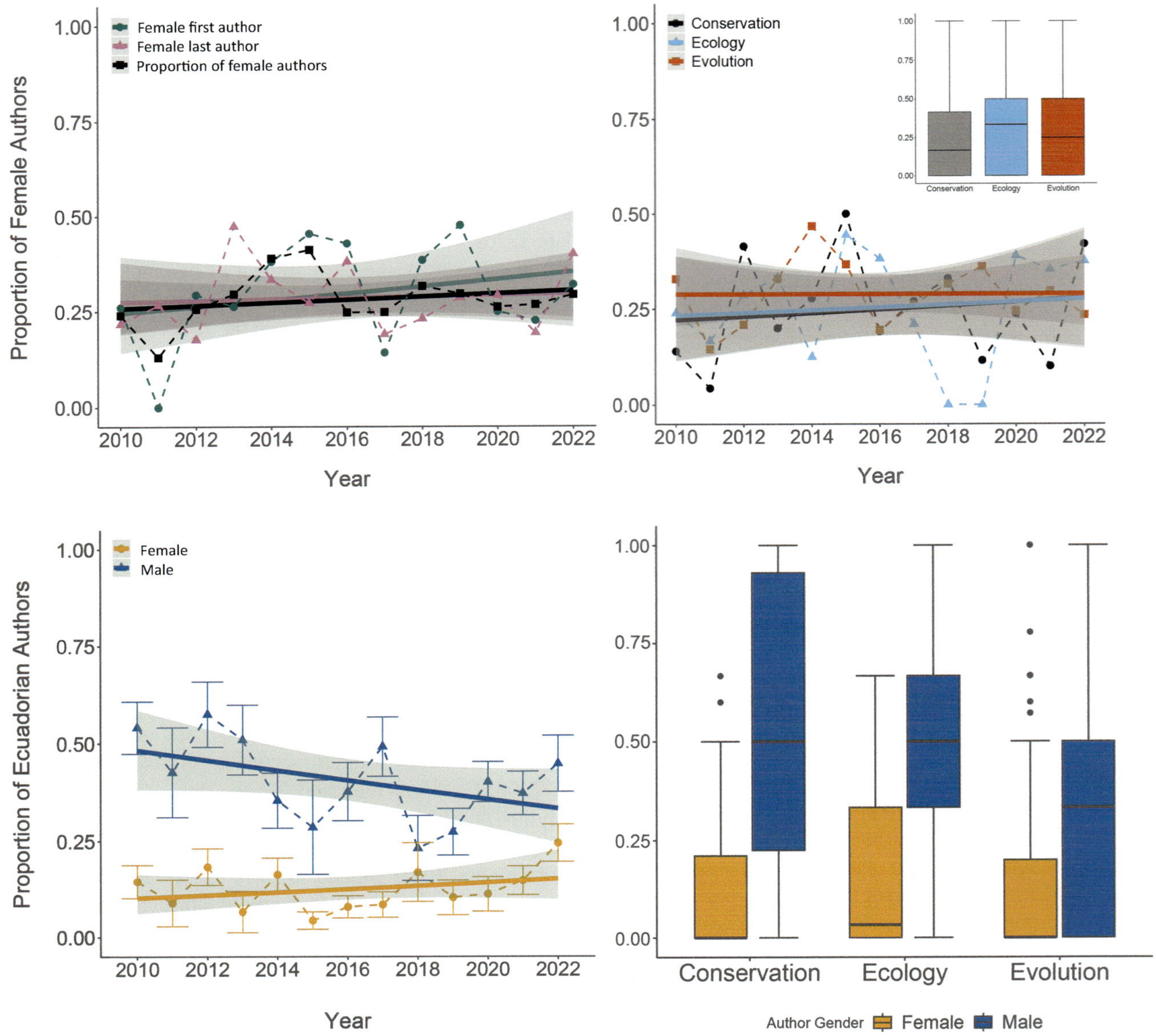

FIGURE 2.63 Proportion of authors in studies on the conservation, ecology, and evolution of amphibians of Ecuador. Top left: proportion of female first, last, and overall authorship per year. Top right: proportion of female authors in the three disciplines per year and overall (inset). Bottom left: proportion of female and male Ecuadorian authors per year. Bottom right: proportion of female and male Ecuadorian authors for the three disciplines. Boxplots show median, first and third quartiles, minimum, and maximum.

Despite the contributions of these female researchers, women's representation has not increased since 2010 (see Figure 2.63 top left; Table 2.8). On average, women represent 27.9% of authors in publications across the three fields, with ecology/natural history and evolution having the highest female representation with 29.1% each, while conservation has 23% (Figure 2.63 top right). Both first and last female authors correspond to 29% each, which has not changed through time either (see Figure 2.63 top left). When the last author is a female, the proportion of female coauthors and first female authors increases (Table 2.8). This strong bias in female authorship has been identified in several studies, and it is widespread in related and other disciplines in STEM and across regions (e.g., Grosso *et al.*, 2021; Rock *et al.*, 2021).

In terms of Ecuadorian authors, overall representation of Ecuadorian female researchers remains at only 13%, fluctuating between 4 and 24%, and has not increased since 2010 (Figure 2.63 bottom left; Table 2.8). As of 2022, Ecuadorian female representation has finally surpassed 20%. In contrast, male Ecuadorian researchers' participation corresponds to 54%, ranging from 33 to 76%, with a slight decrease over time (see Figure 2.63 bottom left; Table 2.8). First and last female Ecuadorian authors account for less than 12% each, and these proportions have not changed significantly between 2010 and 2022, nor do they differ across disciplines (Table 2.8). Among Ecuadorian female researchers, ecology/natural history has the highest representation at 18%, followed by evolution and conservation at 11.9 and 11.2%, respectively (Figure 2.63 bottom right).

These values contrast with the higher participation rate of male Ecuadorian researchers in evolution (47%), ecology (70%), and conservation studies (60.5%; see Figure 2.63 bottom right). These values have remained stable over the years, except for evolution, in which we see a decrease in the participation of male Ecuadorian authors (see Figure 2.63 bottom left; Table 2.8). Although we acknowledge that the last author may not always be the leading author, our analysis also shows that the proportion of female Ecuadorian authors increases when the last author is a female and particularly when the last author is female and Ecuadorian (Table 2.8). Additionally, the proportion of female Ecuadorian authors also increases when the first author is female, and even more so if the first author is an Ecuadorian female researcher (Table 2.8).

Enhancing Gender Inclusion in Amphibian Research

There is still a concerningly low representation of women in the taxonomy, evolution, ecology, and conservation of amphibians in Ecuador. While there has been an increase in Ecuadorian authors in the field of taxonomy over the last decade, this increase has not been balanced between male and female researchers. Moreover, the same progress has not been observed in research related to evolution, ecology, and conservation. It is worth mentioning that work in conservation often remains unpublished or is only recorded in reports and outreach publications; hence, the contribution of female researchers—or conservation biologists as a whole—may have been underestimated in this sample from peer-reviewed literature. Therefore, it is also important to encourage all contributors and teams working in amphibian conservation to increase the academic publication of their results.

We acknowledge the positive impact of inclusive mentorship where predominantly male researchers have welcomed young female researchers to their teams. Such initiatives have the potential to be transformative in fostering diversity and inclusion. However, our results also underscore the importance of female role models and the urgent need for more women to advance in their careers and assume leadership positions in academia. It is crucial to create an environment that supports and empowers women to excel in their fields, knowing the challenges and barriers that women face in Ecuador. Notably, our data indicate that when Ecuadorian female authors take on leading roles, they have the greatest impact on improving the representation of females in Ecuadorian amphibian biology. This further highlights the significance of supporting and promoting women in leadership positions because they can serve as influential role models and catalysts for change in increasing gender diversity in amphibian research in Ecuador. Academia cannot disregard the intellectual capabilities of half of its workforce. Consequently, it is imperative to explore alternative mechanisms that promote the inclusion of women in science.

THE DIVERSITY AND ACADEMIC LIVES OF CURRENT AMPHIBIAN BIOLOGISTS

The scientists who are currently studying amphibians in Ecuador are a group with diverse nationalities, genders, academic levels, backgrounds, and interests. There is a prevalence of Ecuadorians but some foreigners as well. Most have pursued formal education at the university level. Overall, this team of scientists is a relatively small group compared to those in other countries or regions with less amphibian diversity (e.g., USA, Europe), but it is a large group compared with other countries in Latin America.

In terms of academic level, there are some with PhD titles obtained abroad (e.g., 12 out of the 30 selected herein), others with master's degrees acquired either at international or national universities (10 out of 30), and some with *licenciatura* degrees (8 out of 30) obtained in Ecuador—half of the latter are currently enrolled in graduate studies. One notable case is that of Ítalo G. Tapia (see Figures 2.41, 2.64) who completed high school but has gained practical knowledge through fieldwork and research experience mostly working as a collector (he is the most prominent collector) and the Collection Manager at the QCAZ (see also this chapter, under the subtitle "The Life Raft for Frogs at QCAZ"). In Ecuador, most scientists are currently associated with universities (e.g., Diego F. Cisneros-Heredia, Juan M. Guayasamin, Carolina Reyes-Puig, Mónica I. Páez-Vacas, Santiago R. Ron, Paul and Diana Székely), but others work for NGOs (e.g., Luis A. Coloma, Juan P. Reyes-Puig) or government organizations (e.g., Mario Yánez-Muñoz, Jorge Brito Molina), a few work independently (e.g., Morley Read, Alejandro Arteaga, Jaime Culebras), and a few are retired but still publishing (e.g., Ana Almendáriz-Cabezas, Eugenia M. del Pino).

Most amphibian biologists live in Ecuador, although some reside abroad, either working at universities (e.g., Juan Carlos Santos, Lauren A. O'Connell) or studying (e.g., Jorge H. Valencia, Marcel Caminer, Nadia B. Páez, María José Navarrete). The majority are working in systematics and taxonomic fields, but some are focused on other areas of biology (e.g., Eugenia M. del Pino, developmental biology; Diana Székely, ecology; Juan Carlos Santos, molecular biology; Lauren A. O'Connell, physiology, behavior). In addition to their research and academic pursuits, some of the amphibian biologists are involved in public outreach and education efforts (e.g., Luis A. Coloma, Jaime Culebras, Diego F. Cisneros-Heredia). In terms of gender, women amphibian researchers began to play a major role in the 1970s, and their participation is also analyzed in this chapter under the subtitle, "Women in Amphibian Biology: An Ongoing Trend of Underrepresentation in Academia".

In addition to the information provided previously for many of the amphibian biologists, herein, we include brief academic biographical sketches of 30 selected amphibian biologists who are currently contributing to Ecuadorian amphibian knowledge. They were selected based on their scientific productivity, as reflected by their publications or publication impact (e.g., as provided by Scopus, an indexed publications citation database of peer-reviewed literature including scientific journals, books, and conference proceedings), with a few exceptions. Fifteen of them are working on systematics and have described or co-described eight or more species of amphibians reported from Ecuador as of 30 June 2022 (Figures 2.64, 2.65, 2.66). This number (eight species) was arbitrarily chosen to represent some of the most prolific authors in terms of naming species,

FIGURE 2.64 Biologists studying amphibians in Ecuador. From upper left to lower right: Eugenia M. del Pino, John D. Lynch*, David C. Cannatella*, Ana Almendáriz-Cabezas, Morley Read, Luis A. Coloma, Santiago R. Ron, Paul Székely, Juan M. Guayasamin, Elicio E. Tapia, Juan Carlos Santos, Ítalo G. Tapia (collector)*, Mario H. Yánez-Muñoz, Jorge H. Valencia, and Diego F. Cisneros-Heredia. Photos (from up down, left to right): 1 (Martin Blum); 3, 5, 7, 9, 10, 11, 12 (LAC); 2 (John E. Simmons); 6 (Diego Acosta López); 4 (Vladimir Carvajal), 8, 13, 14, 15 (courtesy of the depicted biologists). *Biography not included but mentioned in other sections of Chapter 2.

although we acknowledge that there are other taxonomic authors who described fewer species but have made relevant taxonomic contributions (e.g., Glenn Flores, Manuel Morales-Mite, H. Mauricio Ortega-Andrade, Karl-Heinz Hungfer, and Gregory O. Vigle, among others). Some of their activities are mentioned in other sections of this encyclopedia. Biographical sketches of foreign-living non-residents who have described eight or more species from Ecuador (Lynch, Cannatella, Hutter, Pramuk, and Burrowes) are given by Duellman (2015a), and their contributions are highlighted through the text in the history chapter. Among the 30 selected amphibian biologists, we also include biographical sketches for 10 authors who are making relevant contributions to several aspects of amphibian biology and/or this encyclopedia. They are: Sofía Carvajal-Endara, Eugenia M. del Pino, Mónica I. Páez-Vacas, Carolina Proaño-Bolaños, Lauren A. O'Connell, Morley Read, Elicio E. Tapia, Juan Carlos Santos, Diana Székély, and Andrea Terán-Valdez. Some other authors who have made few but relevant contributions in any field are mentioned elsewhere. The biographies that follow are arranged chronologically by date of birth and were updated as of 31 December 2022, unless indicated otherwise.

Eugenia M. del Pino (see Figure 2.64) is a developmental biologist who works with amphibians. She was born in Quito, Provincia Pichincha, Ecuador, in 1945. From 1963 to 1967, she studied for a *licenciatura* degree in secondary education from Pontificia Universidad Católica del Ecuador (PUCE). In 1969 she received an MS degree from Vassar College in New York, where she studied protozoans. In 1972, she obtained a PhD from Emory University, where she was mentored by Asa Alan Humphries, Jr., with a thesis dealing with eggs of *Xenopus laevis*. She worked at PUCE for 41 years. She was Head of Biological Sciences (1972–1974), Professor (1972–2013), and now Professor Emerita (since 2013). Del Pino has been involved in almost a hundred publications, including 66 scientific articles focused on amphibians, with 43 of these articles indexed in Scopus (as of 31 May 2023). She wrote a textbook on embryonic early development in animals. She is widely recognized for her studies on marsupial frogs, especially *Gastrotheca riobambae*, in which she made important discoveries. Her 17 national and international awards include the L'Oreal-Unesco Prize for Women in Science for Latin America (2000); the most prestigious Ecuadorian award for natural sciences, the Premio Nacional Eugenio Espejo (2012); and the Latin American Society Prize for Developmental Biology for her strong contributions to research in Ecuador (2019). In 2006, she became the first Ecuadorian citizen to be elected to the United States National Academy of Sciences. In 2010–2013, del Pino participated in the foundation of the Ecuadorian Academy of Sciences. The rainfrog *Pristimantis eugeniae* was named in her honor (Lynch and Duellman, 1997).

Ana Almendáriz-Cabezas (see Figure 2.64) is a herpetologist who was born in Quito, Provincia Pichincha, Ecuador, in 1956. In January 1983, she received her *licenciatura* degree in biological sciences from PUCE after conducting a thesis on the altitudinal distribution of finches and related vegetation in the Galápagos islands. Nearly two decades later, in 2002, she earned a master's degree in conservation and management of the natural environment from Universidad Internacional de Andalucía in Spain after completing a thesis on herpetological monitoring in the Andes. For 34 years, from June 1982 until May 2016, she was in charge of the Herpetology section of the Instituto de Ciencias Biológicas (currently Departamento de Biología) of the Escuela Politécnica Nacional, initially as an assistant researcher and later (beginning in 2005) as a researcher. After her formal retirement in 2016, she continued collaborating with the EPN as an honorary researcher until September 2019. Almendáriz has been involved in 53 publications, including 34 scientific articles focused on amphibians, and the herpetological components of two books, with nine of these articles indexed in Scopus (as of 31 May 2023). Almendáriz is one of the most productive collectors of Ecuadorian amphibians (see Figure 2.36). Ana has described or co-described 20 species of amphibians and reptiles, of which 14 are amphibians (12 from Ecuador and two from Colombia). *Pristimantis almendariz* (Brito-M and Pozo-Zamora, 2013) and *Boana almendarizae* (Caminer and Ron, 2014) were named in her honor.

Morley Read (see Figure 2.64) is a British herpetologist, Onychophora specialist, and nature photographer who was born in Fowley, England, in 1958. In 1981, he graduated with a BS degree in zoology from University College of North Wales, Bangor, where he also obtained a PhD degree in zoology in 1985, with a dissertation on the ecology and taxonomy of Onychophora velvet worms. While studying for his doctorate, he did fieldwork in Trinidad, where he began recording frog calls. Since 1986, Read has worked independently on several projects studying Ecuadorian amphibians and been associated with NGOs, particularly the QCAZ and CJ, where he has deposited most of his amphibian collections (1895 specimens) and the largest collection of frog audio recordings. Read worked as an assistant researcher for the QCAZ in 2010 and for CJ from 2017 to 2018. He participated as cameraman for Survival Anglia, filmed amphibian reproduction sequences in Yasuní for the documentary "Frogs the Movie", and received the Semi-Finalist Award for an outstanding film on animal behavior at the Animal Behavior Society Film Festival of 2000. As of 31 May 2023, Read has been involved in 17 publications, with six amphibian articles indexed in Scopus, plus one report, two CDs with frog calls, and one book on Ecuadorian amphibians. Read described or co-described five species of amphibians, among which three are from Ecuador of the genera *Osteocephalus* (2) and *Leucostethus* (1) and two are from Trinidad.

Luis A. Coloma (see Figure 2.64). Refer to biographies of authors in the introductory pages and Chapter 2 about "The History and Present of Amphibian Biology".

Santiago R. Ron (see Figure 2.64) is an Ecuadorian herpetologist who was born in Quito, Provincia Pichincha, in 1970. In 1995, he earned a *licenciatura* degree in biological sciences from Pontificia Universidad Católica del Ecuador after completing a thesis on crocodilian awareness. In 1998, he received an MA in systematics and ecology from the University of Kansas, where he was mentored by William E. Duellman; his master's thesis was titled "Biogeographic area relationships of lowland Neotropical rain forests based on cladistic analyses of anurans". In 2007, he obtained a PhD in behavior, ecology, and evolution from the University of Texas at Austin, where he was mentored by David C. Cannatella, with the dissertation "Systematics and the evolution of calls and mating preferences in frogs of the genus *Engystomops*". He is a professor at the Pontificia Universidad Católica del Ecuador (2000–2002, 2007–present) and Curator of Amphibians at the QCAZ (2010–present). Ron has been involved in about 150 publications, including 130 scientific articles focused on amphibians, with 96 of these articles indexed in Scopus (as of 31 May 2023). He participated in two photographic coffee table books including amphibians. He is one of the most productive collectors of Ecuadorian amphibians (see Figure 2.30). He has described or co-described 62 species of Ecuadorian frogs of the genera *Pristimantis* (31), *Osteocephalus* (6), *Boana* (6), *Engystomops* (4), *Hyloscirtus* (2), *Atelopus* (3), *Bolitoglossa* (2), *Lynchius* (1), *Noblella* (1), *Chiasmocleis* (1), *Elachistocleis* (1), *Scinax* (1), *Dendropsophus* (1), *Amazophrynella* (1), *and Excidobates* (1). *Pristimantis roni* (Yánez-Muñoz *et al*., 2014) was named in his honor.

Paul Székely (see Figure 2.64) is a Romanian (of Hungarian ethnicity) herpetologist working mainly on the systematics, evolutionary ecology, and conservation of amphibians. He was born in Bicaz, Romania, in 1974. He completed his undergraduate and master studies in 2004 and defended his PhD dissertation in biology in 2010 at Babeș-Bolyai University, Romania. His previous work focused on the ecology of the European spadefoot toads (*Pelobates fuscus* and *P. balcanicus*). Since 2014, he has worked in southern Ecuador, and since 2017, he has been an associate professor at the Universidad Técnica Particular de Loja, Ecuador. Paul has been the Director of the Museo de Zoología, Universidad Técnica Particular de Loja (MUTPL) and the Curator of the vertebrate collection since 2018. He has mentored nine undergraduate and one graduate (MS) students. Székely has been involved in 42 publications, including 38 scientific articles focused on amphibians, with 32 (21 of Ecuadorian amphibians) of these articles indexed in Scopus (as of 31 May 2023). He has described or co-described 15 species of frogs from Ecuador of the genera *Pristimantis* (10), *Centrolene* (1), and *Gastrotheca* (4) (up to July 2023). Currently, he is continuing his work, along with his wife Diana Székely, on the herpetofauna of Southern Ecuador.

Juan M. Guayasamin (see Figure 2.64) is an Ecuadorian-Bolivian herpetologist born in 1974 in Quito, Provincia Pichincha, Ecuador. In 2000, he earned a *licenciatura* degree in biological sciences from PUCE after completing a thesis titled "Acoustic partition in a frog community (Anura: Leptodactylidae) of the cloud forests of western Ecuador"; he was mentored by Luis A. Coloma. In 2003, he received a MA in ecology and evolutionary biology at the University of Kansas; his thesis, "The *Eleutherodactylus orcesi* group (Anura: Leptodactylidae): Comparative osteology and new species" was mentored by Linda Trueb. For his PhD, he was also mentored by Trueb and did a dissertation on "Phylogeny, taxonomy, character evolution, and biogeography of glassfrogs (Amphibia: Centrolenidae)", and he graduated in 2007. He was a professor and researcher at PUCE (2008–2010) and Universidad Tecnológica Indoamérica (2011–2016) and has been on the faculty of the Universidad San Francisco de Quito since 2016. He has mentored 12 graduate (MS) and six *licenciatura* students. Guayasamin has been involved in 118 indexed, peer-reviewed publications, with 90 of these articles indexed in Scopus (as of 31 May 2023). His contributions include co-authoring four books and authoring four articles for the general public. He has described seven glassfrog genera (*Celsiella*, *Chimerella*, *Espadarana*, *Ikakogi*, *Rulyrana*, *Sachatamia*, and *Vitreorana*) and has described or co-described 61 new species of amphibians from Ecuador, Peru, Colombia, Venezuela, and Brazil. Guayasamin is a member of the Ecuadorian Academy of Sciences, the Atelopus Survival Initiative, the IUCN SSC Amphibian Specialist Group, Alianza Jambato and the Sciencia Panel for the Amazon.

Elicio E. Tapia (see Figure 2.64) is an Ecuadorian naturalist focused in herpetology and arachnology who was born in Cutzualó, Cantón Sigchos, Provincia Cotopaxi, in 1975. In 2008, he obtained his *licenciatura* degree in environmental sciences and development from Pontificia Universidad Católica del Ecuador, campus Ibarra. Tapia has worked as a herpetologist on several projects at the Museo de Zoología of Pontificia Universidad Católica del Ecuador (2008–2011) and at Centro Jambatu (2011–2017), where he worked as a collection manager and field researcher. In 2019, Elicio co-founded, along with Nadine Dupérré (his wife), the Reserva Biológica Pristirana, a private preserve dedicated to the protection of the type locality of *Pristimantis ecuadorensis* in the Andean Chocoan forests of Ecuador. Tapia has been involved in nearly 30 scientific articles, with 13 of these focused on amphibians and indexed in Scopus (as of 31 May 2023). He is one of the most productive collectors of Ecuadorian amphibians (see Figure 2.30). The marsupial frog *Gastrotheca elicioi* has been described in his honor.

Juan Carlos Santos (see Figure 2.64) is a molecular ecologist, evolutionary biologist, and herpetologist mostly working with amphibians and lizards. He was born in Quito, Provincia Pichincha, Ecuador, in 1977. In 2002, he earned a *licenciatura* degree in biological sciences from Pontificia Universidad Católica del Ecuador after completing a thesis on the molecular phylogenetic relationships of the complex of species related to *Hyloxalus bocagei* and *Epipedobates tricolor*; he was mentored by Luis A. Coloma. In 2009, he received an MS in statistics and a PhD in ecology, evolution, and behavior at the University of Texas, Austin. For his PhD, he was mentored by David C. Cannatella with a dissertation titled "Phylogeography and the evolution of correlated traits under multiple origins of aposematism in the poison frog family". He has been a postdoctoral fellow at the National Evolutionary Synthesis Center (2009–2012) at Duke University, USA; at the University of British Columbia (2014–2017), Canada; and at Brigham Young University, USA (2014–2017). Since 2017, he has been an assistant professor in biological sciences at St. John's University, USA. He has mentored 16 graduate students. Santos has been involved in 41 peer-reviewed publications, including 26 scientific articles focused on amphibians and indexed in Scopus (as of 31 May 2023). He has co-described 12 species of dendrobatids (3 from Ecuador, 9 from Venezuela).

Mario H. Yánez-Muñoz (see Figure 2.64) is a herpetologist who was born in Quito, Provincia Pichincha, Ecuador, in 1978. In 2005, he graduated with a *licenciatura* degree in biological sciences from the Universidad Central del Ecuador, with a thesis on the diversity and structure of 11 amphibian and reptile communities in Ecuador. In 2014, he obtained a master's degree in conservation biology from Pontificia Universidad Católica del Ecuador with a thesis that includes descriptions and phylogenetic relationships of two new species of crested *Pristimantis* from the northern Andes of Ecuador; he was mentored by Santiago R. Ron. Since 1998, Yánez-Muñoz has worked on several projects studying amphibians and has been associated with diverse institutions: the EPN (1998–2002, curatorial and field assistant), MECN (1999–2017, volunteer/curatorial assistant/curator/executive director), INABIO (2017–present, researcher), and PUCE (2011–2018, associate taxonomist). Yánez-Muñoz has been involved in about 100 publications, including 64 scientific articles and six books focused on amphibians, with 25 of these articles indexed in Scopus (as of 31 May 2023). He has described or co-described 40 new species of amphibians (out of 58 new species of amphibians, reptiles, and invertebrates) from Ecuador of the genera *Pristimantis* (30), *Osornophryne* (2), *Oedipina* (2), *Noblella* (1), *Lynchius* (1), *Elachistocleis* (1), *Hyloscirtus* (1), *Nymphargus* (1), and *Epipedobates* (1). He also is one of the 10 most prolific amphibian collectors (3823 specimens) (Figure 2.30). The rainfrog *Pristimantis yanezi* has been described on his honor.

Jorge H. Valencia (see Figure 2.64) is a herpetologist born in Ibarra, Provincia Imbabura, Ecuador, in 1979. He received his degree in biological sciences from the Universidad Central del Ecuador with a thesis titled "Monitoreo de herpetofauna en bosque húmedo tropical, estudio de caso: los bosques de las comunidades kichwas Chuyuyacu y Pandenuque, cantón Arajuno, Provincia de Pastaza". He was mentored by Nelson Gallo. He received a master's degree in sciences, biology network, and vertebrate conservation by studying the species diversity, distribution, and conservation status of amphibians and reptiles of the Cordillera del Cóndor, a region between Ecuador and Peru. Currently, he is pursuing a doctoral degree in sciences at the Instituto de Ecología, Xalapa, Veracruz, México, with a dissertation on coral snakes. As a herpetologist, Valencia has worked independently with amphibians and on numerous inventories, monitoring, rapid assessments, and environmental impact studies, especially for the petroleum, mining, and hydroelectric power industries. He has been associated with diverse institutions, including the Vivarium of *Fundación Herpetológica Gustavo Orcés* (1999–2013), where he began as a volunteer and later worked as a researcher and curator. In 2001–2002, he conducted amphibian studies for Ecociencia. Currently, he is a professor of taxonomy and ecology of amphibians and reptiles for the Asociación Veterinaria de Anfibios y Reptiles de México. Valencia has been involved in about 50 publications, including 40 scientific articles (12 on amphibians) and five books including amphibians, with seven of these articles indexed in Scopus (as of 31 May 2023). He has described or co-described 8 species of the genus *Pristimantis*.

Diego F. Cisneros-Heredia (see Figure 2.64) is a naturalist, zoologist, and herpetologist who was born in Quito, Provincia Pichincha, Ecuador, in 1980. He earned his undergraduate degree in applied ecology at the Universidad San Francisco de Quito USFQ, Ecuador, in 2006, with a dissertation on the "Herpetofauna of the Tiputini Biodiversity Station, Ecuadorian Amazonia" under the mentorship of David Romo. In 2008, he obtained a master's degree in environmental modeling, monitoring, and management from King's College London in the United Kingdom with a thesis titled "Habitat loss and climate change impacts on Neotropical anurans" under the mentorship of Mark Mulligan. In 2019, he received a PhD with a dissertation on the "Spatial Patterns and Impact of Habitat Change on the Vertebrate Diversity of North-Western South America". Since 2009, Cisneros-Heredia has worked as a professor at the USFQ and has held several management positions, including being the Founder and Director of the ZSFQ (2014–present). He has worked on several projects studying amphibians and has been associated with the MECN (INABIO since 2017). Cisneros-Heredia has published 165 titles that include 17 book chapters and 5 books. Among these publications, there are 72 peer-reviewed scientific publications, including 37 scientific articles focused on amphibians and indexed in Scopus (as of 31 May 2023). He has described or co-described a new genus (*Nymphargus*); 3 species of amphibians from Colombia, Peru, and Bolivia; and 29 new species of amphibians from Ecuador of the genera *Pristimantis* (10), *Nymphargus* (4), *Hyloscirtus* (2), *Oedipina* (2), *Osornophryne* (1), *Rhaebo* (1), *Noblella* (1), *Boana* (1), *Centrolene* (1), *Chimerella* (1), *Cochranella* (1), *Espadarana* (1), *Hyalinobatrachium* (1), *Rulyrana* (1), *Teratohyla* (1), and *Epipedobates* (1). The rainfrog *Pristimantis cisnerosi* was named in his honor.

Carolina Proaño-Bolaños (see Figure 2.65) is a biochemist and microbiologist who studies amphibian secretions. She was born in Quito, Provincia Pichincha, Ecuador, in 1980. In 2004, she completed a thesis and received her *licenciatura* degree in biology at Pontificia Universidad Católica del Ecuador. She studied the occurrence of antimicrobial compounds in 21 species of Ecuadorian amphibians and partially purified five peptides of *Agalychnis spurrelli*. She attended

FIGURE 2.65 Biologists studying amphibians in Ecuador. Continued. From upper left to lower right: Carolina Proaño-Bolaños, Diana Székely, Mónica I. Páez-Vacas, Sofía Carvajal-Endara, Juan Carlos Sánchez-Nivicela, Jorge Brito Molina, Juan P. Reyes-Puig, Lauren A. O'Connell, Diego Batallas-Revelo, Andrea Terán-Valdez, Marcel Adrián Caminer Rodríguez, Carolina Reyes-Puig, Jaime Culebras, Verónica L. Urgilés, and Alejandro Arteaga. Photos (from up down, left to right): 2, 4–6, 7, 8, 10–15 (courtesy of the depicted biologists); 1, 4 (LAC); 10 (Jaime Culebras).

Universidad San Pablo, CEU, Spain and received a master's degree in 2007 in management and biodiversity conservation in the tropics after doing a thesis on the genetic structure of a declining population of *Atelopus* sp. in Ecuador. Subsequently, she attended the Universidad San Francisco de Quito, where, in 2012, she completed a master's degree in microbiology specialized in micro- and molecular biology. She next attended the School of Pharmacy at Queen's University, Belfast, UK, where she obtained a PhD in pharmacy in 2016 with a dissertation titled "Unravelling skin secretion peptidomes of *Cruziohyla calcarifer, Agalychnis spurrelli*, and *Hypsiboas picturatus* (Hylidae)" under the mentorship of Chris Shaw. Since July 2016, she has worked at the Universidad Regional Amazonica Ikiam as a professor and researcher. Proaño-Bolaños has been involved in 23 publications, including 14 scientific articles focused on amphibians and indexed in Scopus (as of 31 May 2023). Her publications include the remarkable finding and description of novel peptides (including cruzioseptins, picturins, pictuseptins, new families) and related molecules in the skin of phyllomedusids and hylids, which show promising biomedical uses.

Diana Székely (see Figure 2.65) is a Romanian herpetologist focused on the behavioral adaptations that allow amphibians to survive in extreme environments. She was born in 1981 in Constanţa, Romania. Diana obtained a bachelor's degree in biology and a master's degree in protected areas management from Babeş-Bolyai University, Romania. Under the guidance of Prof. Dan Cogălniceanu from Ovidius University, Romania, and Prof. Mathieu Denoël from the University of Liège, Belgium, she finished her PhD studying the "Life-history and behavioral adaptations to xeric environments in fossorial amphibians". Her study methods include field monitoring and experiments under controlled conditions using diverse laboratory techniques and new integrative computing approaches. Her main contribution to the study of Ecuadorian amphibians is her work on the biology and ecology of the poorly known Pacific horned frog, *Ceratophrys stolzmanni*, and the advancement of our understanding of the complexity of life-history adaptations and behavioral strategies that allow fossorial anurans to survive in extremely dry or drastically seasonal environments. She has authored and co-authored 32 publications, including 29 scientific articles focused on amphibians and indexed in Scopus (as of 31 May 2023). Out of the latter publications, 18 concern Ecuadorian species. She also participated in the description of 11 species of *Pristimantis*, all from Ecuador.

Mónica I. Páez-Vacas (see Figure 2.65) is an amphibian biologist born in Quito, Provincia Pichincha, Ecuador, in 1982. She earned the degree of *licenciatura* in Biological Sciences at Pontificia Universidad Católica del Ecuador in 2008 after completing a thesis titled "Taxonomic revision of the *bocagei* clade (Anura: Dendrobatidae)" and was mentored by Luis A. Coloma. In 2016, she earned a PhD in ecology at Colorado State University, USA, with the dissertation "Mechanisms of population divergence along elevational gradients in the Tropics". Her major advisor was W. Chris Funk. Between 2016 and 2022, Páez-Vacas was an assistant professor at Universidad Tecnológica Indoamérica in Quito, Ecuador. In 2023, she taught at the Universidad Central del Ecuador. She has mentored three graduate (MS) and four undergraduate students. She has participated in 11 peer-reviewed publications, including five scientific articles focused on amphibians and indexed in Scopus (as of 31 May 2023). Páez-Vacas has described two species of the genus *Hyloxalus*. She is a member of the Atelopus Survival Initiative, and Alianza Jambato.

Sofía Carvajal-Endara (see Figure 2.65) is an amphibian biologist born in Cuenca, Ecuador, in 1983. She earned a degree of *licenciatura* in biological sciences at Pontificia Universidad Católica del Ecuador in 2010 after completing a thesis on the effect of environmental factors and interspecific competition on the distribution of *Gastrotheca riobambae* and *G. pseustes* (Anura: Hemiphractidae) and was mentored by Luis A. Coloma. She worked as a research assistant at PUCE (2010) and Centro Jambatu (2011–2013) and in 2019 earned a PhD in biology at McGill University, Canada. From 2021 to 2022, Carvajal taught at the Universidad Tecnológica Indoamérica in Quito, Ecuador. In 2023, she was associated with the Universidad San Francisco de Quito. She has participated in eight peer-reviewed publications, including five scientific articles focused on amphibians and indexed in Scopus (as of 31 May 2023). She described or co-described six species of the genera *Hyloscirtus* (2) and *Gastrotheca* (4).

Juan Carlos Sánchez-Nivicela (see Figure 2.65) is a herpetologist born in Piñas, Provincia El Oro, in southwestern Ecuador, in 1983. He earned a biology degree at the Universidad del Azuay in 2013 after completing an undergraduate thesis on the diversity and distribution of amphibians in a montane humid forest in Zamora, Ecuador. In 2019, he enrolled in a master's program in biology at the Universidad Nacional de Colombia with the thesis "Functional ecological patterns associated with dermal macroglands in directly developing frogs of the genus *Pristimantis* (Anura: Strabomantidae) in the high Andes of northern South America" under the tutelage of Yaneth Muñoz Saba, professor and member of the Instituto de Ciencias Naturales of UNAL, and Diego F. Cisneros-Heredia from the USFQ. Since 2008, Sánchez-Nivicela has worked as a herpetologist, primarily on monitoring projects, and has contributed to the Museo de Zoología of Universidad del Azuay (2012–2018), MECN (2013–2017), INABIO (2017–present), Universidad San Francisco de Quito (2019–present), and Instituto de Ciencias Naturales of the Universidad Nacional de Colombia (2019). Sánchez-Nivicela has been part of the field team that helped develop the Río Negro–Sopladora National Park. He is the author or coauthor of four amphibian guides. He has participated in 15 peer-reviewed publications, including 11 scientific articles focused on amphibians and indexed in Scopus (as of 31 May 2023). Sánchez-Nivicela has described or co-described 12 species of endangered amphibians endemic to southern Andean Ecuador, 10 of the genus *Pristimantis* and one of each *Lynchius* and *Elachistocleis*.

Jorge Brito Molina (see Figure 2.65) is a mammalogist and herpetologist who was born in Gualaquiza, Provincia Morona Santiago, Ecuador, in 1983. He earned his *licenciatura* degree in biology and environmental sciences from the Universidad Central del Ecuador in 2013 after completing a thesis on the diversity and abundance of small mammals in Reserva Ecológica el Angel. Brito Molina worked as a volunteer/assistant researcher at the EPN (2008–2014), mentored by Ana Almendáriz-Cabezas. At the EPN, he contributed to field work and the resulting publications, particularly in the inventories done in the Cordillera del Cóndor that revealed several species new to science. He is a researcher at the Mammalogy Division of the MECN (2015–2017) and INABIO (2017–present). He has made important contributions to the knowledge of amphibians of the Amazonian montane forests of the Upano River and Sangay National Park. Brito Molina has authored or coauthored about 90 titles, including four book chapters, two books, and nine guides and checklists. Among these publications, 56 are peer-reviewed, including 11 scientific articles focused on amphibians and indexed in Scopus (as of 31 May 2023). He has described or co-described 31 species of vertebrates, most of which (16) are amphibians of the genera *Pristimantis* (11), *Noblella* (2), *Chiasmocleis* (1), *Excidobates* (1), and *Hyloscirtus* (1).

Juan P. Reyes-Puig (see Figure 2.65) is a herpetologist born in Quito, Provincia Pichincha, Ecuador, in 1984. In 2008, he received a *licenciatura* degree in biological sciences from the Universidad Central del Ecuador with a thesis about patterns of diversity and the conservation status of the frog assemblages (*Eleutherodacytlus*) occurring in the northern versant of the Tungurahua volcano. Since 2004, Juan P. Reyes-Puig has worked on several projects studying amphibians and has been associated with several institutions: the EPN (2004–2006, intern/field assistant), MECN (INABIO since 2017) (2002–2021, associate researcher), *Fundación Oscar Efrén Reyes* (2005–present, technical director), and *Fundación Ecominga* (2008–2023, administrator, executive director). In the latter role, he is active in land acquisition and the creation of reserves to protect amphibians. Reyes-Puig has participated in 14 peer-reviewed publications, including 9 scientific articles focused on amphibians and indexed in Scopus (as of 31 May 2023). He has described or co-described 18 species of craugastorid frogs (17 *Pristimantis*, 1 *Noblella*).

Lauren A. O'Connell (see Figure 2.65) is a cellular and molecular biologist born in rural Texas, USA, in 1984. In 2006, she received a BS degree in biological sciences from Cornell University. In 2011, she obtained a PhD in cellular and molecular biology at the University of Texas at Austin with a dissertation titled "Evolution of neuroendocrine mechanisms regulating adaptive behavior" and was mentored by Hans Hofmann. From 2012 to 2017, she was in an independent postdoctoral position as a Bauer Fellow at Harvard University and since 2017 has been an assistant professor at Stanford University. Since 2011, she has worked on several projects studying poison frogs in situ and ex situ. She has mentored, at Harvard and Stanford universities, 12 postdoctoral scholars, eight graduate students (including the Ecuadorian Daniela Pareja Mejía), and dozens of high school and undergraduate students. O'Connell has participated in 72 peer-reviewed publications, including 35 scientific articles focused on amphibians and indexed in Scopus (as of 31 May 2023). She has received many awards and prizes, including the L'Oréal-USA For Women in Science fellowship in 2014, and she was recognized for her outreach work by L'Oréal USA in 2016.

Diego Batallas-Revelo (see Figure 2.65) is an amphibian biologist born in Tulcán, Provincia Carchi, Ecuador, in 1984. He earned a degree of *licenciatura* in biological and environmental sciences at the Universidad Central del Ecuador in 2014 after completing a thesis titled "Bioacoustic analysis of anuran calls from Cormoran Lagoon, Sardinayacu lacustrine complex, Sangay National Park, Ecuador". In 2018, he earned an MS in zoology at the Universidad Complutense de Madrid, Spain, with the thesis "Bioacoustic analysis of the songs of *Pristimantis* species of the *myersi* group from the montane and high montane forests of the province of Carchi-Ecuador" and is currently a doctoral candidate at the same institution. From 2010 to 2013, Batallas-Revelo worked as a research assistant at the Escuela Politécnica Nacional, mentored by Ana Almendáriz-Cabezas, where he organized audio recordings and began publishing about frog vocalizations. Up to 2014, he headed the *Fundación Naturaleza Kakaram*, an NGO dedicated to research and education. From 2016 to 2018, he was a research assistant at the MECN and then at the Instituto Nacional de Biodiversidad. Batallas-Revelo has authored or coauthored 22 publications on amphibians, among which 13 publications are peer reviewed, including 12 scientific articles focused on amphibians and indexed in Scopus (as of 31 May 2023). He has described or co-described 16 species of the genera *Pristimantis* (12), *Hyloscirtus* (2), *Noblella* (1), and *Chiasmocleis* (1).

Andrea Terán-Valdez (see Figure 2.65) is an amphibian biologist who was born in Quito, Provincia Pichincha, Ecuador, in 1984. She earned the degree of *licenciatura* in biological sciences at Pontificia Universidad Católica del Ecuador in 2008 after completing a thesis titled "Sexual Selection in *Engystomops puyango*. Experimental research in laboratory to test the Sexual Selection hypothesis with frog calls". She was mentored by Santiago R. Ron. In 2013, she received a master's degree in technology and resources management in the tropics and subtropics at the University of Applied Sciences of Cologne, Cologne (Germany) and the Universidad Autónoma de San Luis Potosí (México). She worked with amphibians as a research assistant at PUCE (2009–2011) and as a collections manager and research coordinator at Centro Jambatu (2018 to present). During 2015–2016, she was a teacher and researcher at Universidad Estatal Amazónica. Since 2018, Terán-Valdez has worked as Collection Manager at Centro Jambatu. She has participated in 16 publications, among which 9 were peer-reviewed, including 7 scientific articles focused on amphibians and indexed in Scopus (as of 31 May 2023). She has described and co-described 9 species (7 from Ecuador) of the genera *Atelopus* (5), *Pristimantis* (2), *Noblella* (1), and *Engystomops* (1).

Marcel Adrián Caminer Rodríguez (see Figure 2.65) is an Ecuadorian-German amphibian biologist born in Berlín, in 1985. He earned a degree of *licenciatura* in biology at Pontificia Universidad Católica del Ecuador in 2010 after completing a thesis titled "Systematics of the species complex *Hypsiboas fasciatus* (Günther, 1858) and *Hypsiboas* *calcaratus* (Troschel, 1848) (Anura: Hylidae)". He was mentored by Santiago R. Ron. In 2014, he earned a MS in Evolutionary Biology at Universidad Complutense de Madrid, Spain with the thesis "Phylogenetic relationships, taxonomic revision and phylogeography of the species complex *Dendropsophus leucophyllatus* (Beireis, 1783) and *Dendropsophus triangulum* (Gunther, 1869) (Anura Hylidae)". He is currently a PhD candidate in the Gene Regulation in Evolution Program of the Institute of Molecular Biology, Johannes Gutenberg University of Mainz, Germany. Caminer has participated in 10 publications about amphibians; eight of them are indexed in Scopus (as of 31 May 2023). He has described or co-described 11 species in the genera *Boana* (6), *Pristimantis* (2), *Scinax* (1), *Hyloscirtus* (1), and *Osteocephalus* (1).

Jaime Culebras (see Figure 2.65) is a herpetologist and wildlife photographer born in Cáceres, Spain, in 1985. He earned the degree of *licenciatura* in biological sciences at Extremadura University, Spain, in 2010. He graduated with a master's degree in teaching and learning of experimental, social, and mathematical sciences from Extremadura University, Spain, in 2011 and a master's degree in biodiversity and conservation of tropical areas from CSIC and Menendez Pelayo University, Spain, in 2012. Since then, he has been involved in many projects concerning science, photography, and conservation with an emphasis on amphibians. His amphibian photography has been published in international books and magazines such as *National Geographic* and *BBC Wildlife Magazine*, and he has been awarded several international photography prizes, including Wildlife Photographer of the Year, World Press Photo, Montphoto, and the Big Picture Photo Competition. In 2017, he co-founded Photo Wildlife Tours, an ecotourism company that organizes, supports, and promotes tours and research in several countries with a mission to conserve wildlife. Culebras has participated in 19 publications, including 10 scientific articles focused on amphibians and indexed in Scopus (as of 31 May 2023). He has co-described 9 species in the genera *Pristimantis* (4), *Hyloscirtus* (1), *Noblella* (1), and *Hyalinobatrachium* (3).

Carolina Reyes-Puig (see Figure 2.65) is a herpetologist and mammalogist born in Quito, Provincia Pichincha, Ecuador, in 1988. In 2012, she received a *licenciatura* degree in biological and environmental sciences from the Universidad Central del Ecuador. In 2015, she was certified in basic biostatistics mediated with the R software platform from Universidad Nacional de Córdoba, Argentina, and received a master's degree in conservation biology from Pontificia Universidad Católica del Ecuador. She is currently a PhD candidate in biodiversity, genetics, and evolution at University of Porto, Portugal, and works at the USFQ. As a herpetologist, and since 2008, Carolina Reyes-Puig has worked on several projects studying amphibians and has been associated with several institutions: the MECN (2008–2010, 2012, 2015–2017), *Fundación Oscar Efrén Reyes* (2009–present), Ecuambiente Consulting Group (2010– 2011), Ecociencia (2013), *Fundación Herpetológica Gustavo Orcés* (2014), QCAZ (2014–2015), and USFQ (2017–present). In the latter two, her position is Curator of Herpetology. Carolina Reyes-Puig has participated in 39 publications, one field guide, and two books. Among these publications, 24 are peer-reviewed, including 12 scientific articles focused on amphibians and indexed in Scopus (as of 31 May 2023). She has described or co-described 21 species of craugastorid frogs (18 *Pristimantis*, 3 *Noblella*) and two salamanders of the genus *Oedipina*.

Verónica L. Urgilés (see Figure 2.65) is an amphibian biologist who was born in Cuenca, Provincia Azuay, Ecuador, in 1990. In 2014, she graduated as a biologist specializing on ecology and management from the Universidad del Azuay, Ecuador, with a thesis titled "Composition and structure of amphibians in an altitudinal gradient of the south eastern Cordillera of the Ecuadorian Andes". In 2020, Urgilés earned an MS in biology from the University of Central Florida, USA, where she is currently working on her integrative biology PhD. While studying in Florida, she has engaged in amphibian research in collaboration with Ecuadorian institutions, including the INABIO and Universidad San Francisco de Quito. Urgilés has participated in 10 peer-reviewed publications focused on amphibians and indexed in Scopus (as of 31 May 2023). In half of them she is the senior author, including the description of 10 species of *Pristimantis* (8), *Lynchius* (1), and *Elachystocleis* (1).

Alejandro Arteaga (see Figure 2.65) is an Ecuadorian-Venezuelan herpetologist and tourism entrepreneur born in 1991 in Caracas, Venezuela. In 2015, he obtained his *licenciatura* degree in biology from Pontificia Universidad Católica del Ecuador. He is the President and Research Director of Tropical Herping, an initiative he co-founded in 2009 to preserve tropical reptiles and amphibians through tourism, photography, education, and scientific research. Arteaga's research has focused on the systematics and biogeography of tropical amphibians and reptiles. He is the author of two books (*Reptiles of the Galápagos* and *The Amphibians and Reptiles of Mindo*) and 19 peer-reviewed publications, including 9 scientific articles focused on amphibians and indexed in Scopus (as of 31 May 2023). Arteaga has described or co-described 20 species; among them, seven are of the genus *Pristimantis* and one of *Hyloscirtus*. He has been involved in land acquisition for conservation through *Fundación Jocotoco*. His photographic work has been featured in National Geographic and the Discovery Channel. In 2015, Arteaga was awarded in the Big Picture Natural World Photography competition.

Nadia B. Páez (see Figure 2.66) is an amphibian biologist who was born in Quito, Ecuador, in 1992. In 2014, she earned a *licenciatura* degree in biological sciences from Pontificia Universidad Católica del Ecuador after completing a thesis titled "Systematics of the *Pristimantis phoxocephalus* species complex (Anura, Craugastoridae) in Ecuador". She was mentored by Santiago R. Ron. She worked as an amphibian researcher at PUCE until 2017. In 2018, Páez Rosales entered a zoology PhD program at the University of British Columbia in Canada, studying

FIGURE 2.66 Biologists studying amphibians in Ecuador. Continued. From left to right: Nadia B. Páez, Jhael Ortega, and María José Navarrete. Photos courtesy of the depicted biologists.

invertebrate ecology. Páez has participated in three peer-reviewed publications focused on amphibians and indexed in Scopus (as of 31 May 2023). These publications include the description of 13 species in the genera *Pristimantis* (12) and *Elachistocleis* (1).

Jhael Ortega (see Figure 2.66) is an amphibian biologist who was born in Quito, Provincia Pichincha, Ecuador, in 1993. In 2020, she earned a *licenciatura* degree in biological sciences from Pontificia Universidad Católica del Ecuador after completing a thesis titled "Two new species of *Pristimantis* (Anura: Strabomantidae) from Llanganates National Park with comments on the community composition at two localities". She was mentored by Santiago R. Ron. While doing her undergraduate studies, she was a volunteer field and research assistant at the QCAZ. Currently, she is an assistant professor at PUCE. Ortega has participated in four peer-reviewed publications focused on amphibians and indexed in Scopus (as of 31 May 2023). They include the description of 13 species in the genera *Pristimantis* (12) and *Noblella* (1).

María José Navarrete (see Figure 2.66) is an amphibian biologist who was born in Ibarra, Provincia Imbabura, Ecuador, in 1994. In 2017, she earned a *licenciatura* degree in biological sciences from Pontificia Universidad Católica del Ecuador after completing a thesis titled "Systematics of the *Pristimantis chloronotus* species group (Anura, Craugastoridae) with insights into their historic biogeography". She was mentored by Santiago R. Ron. While completing her undergraduate studies, she was a volunteer research assistant and curatorial assistant at the QCAZ. Since 2020, Navarrete has been enrolled in a PhD program in integrative biology at the University of California, Berkeley, USA, studying the evolution of tetrodotoxins in amphibians under the mentorship of Rebecca Tarvin. Navarrete has participated in five publications on amphibians, among which three are and indexed in Scopus (as of 31 May 2023). They include the description of nine species of the genera *Pristimantis* (8) and *Lynchius* (1).

REFERENCES

Acosta-Buenaño, N. A. 2002. Vocalización de 9 especies del género *Colostethus* (Anura: Dendrobatidae) del Ecuador. Tesis de Licenciatura, Pontificia Universidad Católica del Ecuador.

Adler, K. 2007. Orcés V., Gustavo (1903–1999). In *Contributions to the history of herpetology*, ed. K. Adler, 394. Dexter: Society for the Study of Amphibians and Reptiles.

Adler, K. 2012. *Contributions to the history of herpetology*. Vol. 3. Issued to Commemorate the 7th World Congress of Herpetology, Vancouver.

Adler, K. 2014a. Andersson, Lars Gabriel (1868–1951). In *Contributions to the history of herpetology. Volume I (revised and expanded)*, ed. K. Adler, 68. Ithaca: Society for the Study of Amphibians and Reptiles.

Adler, K. 2014b. *Contributions to the history of herpetology. Volume 1 (revised and expanded)*. Ithaca: Society for the Study of Amphibians and Reptiles.

Adler, K. 2014c. Dunn, Emmett Reid (1894–1956). In *Contributions to the history of herpetology. Volume I (revised and expanded)*, ed. K, Adler, 92–93. Ithaca: Society for the Study of Amphibians and Reptiles.

Adler, K. 2014d. Peters, James A. (1922–1972). In *Contributions to the history of herpetology. Volume I (revised and expanded)*, ed. K. Adler, 129–130. Ithaca: Society for the Study of Amphibians and Reptiles.

Aguilar, N. 2022. Influencia de la dieta en el desarrollo de renacuajos de la rana cohete de Cuenca *Hyloxalus vertebralis* para el manejo bajo cuidado humano. Tesis de Licenciatura, Universidad Estatal Península de Santa Elena.

Albert, J. S., A. C. Carnaval, G. A. Suzette, L. G. Lohmann, C. C. Ribas, R. Douglas, J. D. Carrillo, *et al.* 2023. Human impacts outpace natural processes in the Amazon. *Science* 379: eabo5003.

Albuja, L., A. Almendáriz, R. Barriga, and P. Mena Valenzuela. 1993. Inventarios de los vertebrados del Ecuador. In *La investigación para la conservación de la diversidad biológica en el Ecuador: Memorias del Simposio llevado a cabo del 10 al 12 de junio de 1992*, eds. P. Mena and L. Suárez, 83–103. Quito: EcoCiencia, Fundación Ecuatoriana de Estudios Ecológicos.

Albuja, L., A. Almendáriz, R. Barriga, L. Montalvo, F. Cáceres, and J. Román. 2012. *Fauna de vertebrados del Ecuador*. Quito: Instituto de Ciencias Biológicas. Escuela Politécnica Nacional.

Albuja, V. L. 2004. Dr. Gustavo Orcés Villagómez (1903–1999). *Politécnica* 25: 7–10.

Alcocer, I., X. Santacruz, K. H. Thierauch, and H. Steinbeisser. 1992. Ureotelism as the prevailing mode of nitrogen excretion in larvae of the marsupial frog *Gastrotheca riobambae* (Fowler) (Anura, Hylidae). *Comparative Biochemistry and Physiology* 101: 29–231.

Alford, R. A., K. S. Bradfield, and S. J. Richards. 2007. Global warming and amphibian losses. *Nature* 447: E3–4.

Alianza Jambato. 2023. *Alianza jambato*. https://alianzajambato.org/.

Almendáriz, A. 1987. Contribución al conocimiento de la herpetofauna centroriental Ecuatoriana. *Politécnica* 12: 77–133.

Almendáriz, A., and D. Batallas. 2012a. Ecología y distribución geográfica de *Hyloxalus mystax* Duellman y Simmons, 1988 (Amphibia: Anura: Dendrobatidae). *Revista Politécnica* 30: 54–61.

Almendáriz, A., and D. Batallas. 2012b. Nuevos datos sobre la distribución, historia natural y el canto de *Centrolene condor* Cisneros-Heredia y Morales-Mite 2008 (Amphibia: Anura: Centrolenidae). *Politécnica* 30: 42–53.

Almendáriz, A., J. Brito, D. Batallas, and S. Ron. 2014a. Una especie nueva de rana arbórea del género *Hyloscirtus* (Amphibia: Anura: Hylidae) de la Cordillera del Cóndor. *Papéis Avulsos de Zoologia (São Paulo)* 54: 33–49.

Almendáriz, A., and J. Carr. 1992. Herpetofauna of Bilsa area. In *Status of forest remnants in the Cordillera de la Costa and adjacent areas of southwestern Ecuador*, eds. T. Parker and J. Carr, 1–28. Washington, DC: Conservation International.

Almendáriz, A., and J. Carr. 2012. Lista actualizada de los anfibios y reptiles registrados en los remanentes de bosque de la Cordillera de la Costa y áreas adyacentes del suroeste del Ecuador. *Revista Politécnica* 30: 184–194.

Almendáriz, A., and D. F. Cisneros-Heredia. 2005. Nuevos datos sobre la distribución e historia natural de *Gastrotheca longipes* (Boulenger, 1882), una rana marsupial amazónica poco conocida (Amphibia: Anura: Hylidae). *Politécnica, Biología* 26: 20–27.

Alméndariz, A., and G. Orcés. 2004. Distribución de algunas especies de la herpetofauna de los pisos: altoandino, temperado y subtropical. *Politécnica* 25: 97–150.

Almendáriz, A., J. Simmons, J. Brito, and J. Vaca-Guerrero. 2014b. Overview of the herpetofauna of the unexplored Cordillera del Cóndor of Ecuador. *Amphibian & Reptile Conservation* 8: 45–64.

Almendáriz, C. A., S. R. Ron and J. Brito-M. 2012. Una especie nueva de rana venenosa de altura del género *Excidobates* (Dendrobatoidea: Dendrobatidae) de la Cordillera del Cóndor. *Papéis Avulsos de Zoologia. São Paulo* 52: 387–399.

Altig, R., and R. W. McDiarmid. 2007. Morphological diversity and evolution of egg and clutch structure in amphibians. *Herpetological Monographs* 21: 1–32.

Alvear, G., and C. Reyes-Puig. 2016. Reptilia, Sauria, Sphaerodactylidae, *Lepidoblepharis conolepis* Avila-Pires, 2001: Distribution Extension in Northern Ecuador. Check List 12: 1983.

Andersson, L. G. 1945. Batrachians from east Ecuador collected 1937, 1938 by Wm Clarke-MacIntyre and Rolf Blomberg. *Arkiv för Zoologi* 37: 1–88.

Arbeláez Ortiz, E., and A. Vega Toral. 2008. *Guía de anfibios, reptiles y peces del Parque Nacional Cajas*. Cuenca: ETAPA/Corporación Municipal Parque Nacional Cajas.

Arbeláez Ortiz, E., and A. Vega Toral. 2014. Aspectos del desarrollo larval de *Gastrotheca litonedis*, 1987 (Anura: Hemiphractidae) en cautiverio. Tesis de Licenciatura, Universidad del Azuay.

Arcos Terán, L. 1998. Orígenes, actividad y proyección de la Facultad de Ciencias Exactas y Naturales. *Prospecto General*. Pontificia Universidad Católica del Ecuador. *Facultad de Ciencias Exactas y Naturales* 12–25.

Armijos-Ojeda, D., D. Székely, P. Székely, D. Cogălniceanu, D. F. Cisneros-Heredia, L. Ordóñez-Delgado, A. Escudero, and C. I. Espinosa. 2021. Amphibians of the equatorial seasonally dry forests of Ecuador and Peru. *ZooKeys* 1063: 23–48.

Arpi, J., and E. Ocampos 2022. Prevalencia de parásitos gastrointestinales, *Batrachochytrium dendrobatidis* y Ranavirus en tres especies de anuros en las zonas urbanas y periurbanas de la ciudad de Cuenca. Tesis de Licenciatura, Universidad del Azuay.

Arteaga, A., P. Alexander, N. Peñafiel, P. Romero-Barreto, J. Culebras, L. Bustamante, M. H. Yánez-Muñoz, and J. M. Guayasamin. 2016. Comparative phylogeography reveals cryptic diversity and repeated patterns of cladogenesis for amphibians and reptiles in northwestern Ecuador. *PloS One* 11: e0151746.

Aubér-Thomay, M., L. Coloma, and G. Onore. 1990. Elevage d'une rainette marsupiale arboricole des forêts nuageuses équatoriennes, *Gastrotheca plumbea* (Boulenger). *Revue Française d'Aquariologie* 17: 57–62.

Báez, O. 1983. Gustavo Orcés Villagómez, primer zoólogo ecuatoriano. Suplemento Cultural del Diario El Comercio.

Báez Tobar, O. 2017. Semblanza de Gustavo Orcés Villagómez, el primer zoólogo ecuatoriano. *Periódico Opción* (January 23). https://periodicoopcion.com/semblanza-de-gutavo-orces-villagomez-el-primer-zoologo-ecuatoriano/.

Balarezo, G. 2023. Santiago Ron mantiene viva su curiosidad. *Revista Familia* (February 13). https://www.revistafamilia.ec/vida-hoy/santiago-ron-viva-curiosidad-biologia.html.

Barbour, T. 1909. Some new South American cold-blooded vertebrates. *Proceedings of the New England Zoological Club* 4: 47–52.

Bardier, C., D. Székely, G. Augusto-Alves, N. Matínez-Latorraca, B. R. Schmidt, and S. S. Cruickshank. 2020. Performance of visual vs. software-assisted photo-identification in mark-recapture studies: A case study examining different life stages of the pacific horned frog (*Ceratophrys stolzmanni*). *Amphibia-Reptilia* 42: 17–28.

Barragán, A. R., O. Dangles, R. E. Cárdenas, and G. Onore. 2009. The history of entomology in Ecuador. *Annales de la Société entomologique de France* 45: 410–423.

Barreiro, A. J. 1928. *Biografía de D. Marcos Jiménez de la Espada (1831–1898)*. Madrid: Establecimiento Tipográfico de Rafael G.-Menor.–Toledo.

Bass, M. S., M. Finer, C. N. Jenkins, H. Kreft, D. F. Cisneros-Heredia, S. F. McCracken, N. C. A. Pitman, *et al.* 2010. Global conservation significance of Ecuador's Yasuní National Park. *PloS One* 5: e8767.

Bejarano-Muñoz, P., M. B. Pérez-Lara, and J. Brito. 2015. *Ranas terrestres endémicas y otros anuros emblemáticos de la vía Papallacta-El Reventador*. Vol. 2. Quito: Serie de Publicaciones Museo Ecuatoriano de Ciencias Naturales del Instituto Nacional de Biodiversidad (INB-MECN).

Ben, B., and M.-C. Durette-Desset. 1995. *Oswaldocruzia* (Nematoda, Trichostrongylina, Molineoidea) Parasitic in Brazilian and Ecuadorian amphibians, with redefinition of the type species *O. subauricularis* (Rudolphi, 1819) and O. Mazzai Travassos, 1935. *Revue Suisse de Zoologie* 102: 635–653.

Bendels, M. H. K., R. Müller, D. Brueggmann, and D. A. Groneberg. 2018. Gender disparities in high-quality research revealed by Nature Index journals. *PloS One* 13: e0189136.

Benslimane, B., and M. C. Durettedesset. 1995. *Oswaldocruzia* (Nematoda, Trichostrongylina, Molineoidea) parasitic in Brazilian and Ecuadorian amphibians, with redefinition of the type species *O. subauricularis* (Rudolphi, 1819) and *O. mazzai* Travassos, 1935. *Revue Suisse de Zoologie* 102: 635–653.

Beolens, B., M. Watkins, and M. Grayson. 2011. *The eponym dictionary of reptiles*. Baltimore: Johns Hopkins University Press.

Betancourt, R. M., J. H. Valencia, P. Yánez, F. Arauz, M. Cueva, P. Farinango, L. Manitio, and E. Ramos. 2009. *Anfibios y reptiles en ambientes cercanos a Borja, Papallacta y Nono*. Quito: Fundación Herpetológica Gustavo Orcés.

Bonilla, K., M. Serafim, and E. Bámaca-López. 2021. Science diplomacy in Ecuador: Political discourse and practices between 2007 and 2017. *Frontiers in Research Metrics and Analytics* 6: 1–17.

Boul, K. E., W. C. Funk, C. R. Darst, D. C. Cannatella, and M. J. Ryan. 2007. Sexual selection drives speciation in an Amazonian frog. *Proceedings of the Royal Society B-Biological Sciences* 274: 399–406.

Boul, K. E., and M. J. Ryan. 2003. Call variation and correlated vocal production mechanisms: Intraspecific and interspecific comparisons from the *Physalaemus pustulosus* species group. PhD thesis, The University of Texas at Austin.

Boul, K. E., and M. J. Ryan. 2004. Population variation of complex advertisement calls in *Physalaemus petersi* and comparative laryngeal morphology. *Copeia* 2004: 624–631.

Boulenger, G. A. 1882a. Account of the reptiles and batrachians collected by Mr. Edward Whymper in Ecuador in 1879–80. *Annals and Magazine of Natural History* 5: 457–467.

Boulenger, G. A. 1882b. *Catalogue of the Batrachia Gradientia s. Caudata and Batrachia Apoda in the collection of the British Museum*. 2nd edition. London: British Museum of Natural History.

Boulenger, G. A. 1882c. *Catalogue of the Batrachia Salientia s. Ecaudata in the Collection of the British Museum*. 2nd edition. London: Taylor and Francis.

Boulenger, G. A. 1898. An account of the reptiles and batrachians collected by Mr. W. F. H. Rosenberg in western Ecuador. *Proceedings of the Zoological Society of London* 1898: 107–126.

Boulenger, G. A. 1899a. Descriptions of new batrachians in the collection of the British Museum (Natural History). *Journal of Natural History* 3: 273–277.

Boulenger, G. A. 1899b. Descriptions of new reptiles and batrachians collected by Mr. P. O. Simons in the Andes of Ecuador. *Annals and Magazine of Natural History* 4: 454–457.

Boulenger, G. A. 1913. On a collection of batrachians and reptiles made by Dr. H. G. F. Spurrell, F. Z. S., in the Choco, Colombia. *Proceedings of the Zoological Society of London* 1913: 1019–1038.

Brito, J., and A. Almendáriz. 2018. Una especie nueva de rana *Pristimantis* (Amphibia: Strabomantidae) de ojos rojos de la Cordillera de Cóndor, Ecuador. *Cuadernos de Herpetología* 32: 31–40.

Brito, J., D. Batallas, and M. H. Yánez-Muñoz. 2017. Ranas terrestres *Pristimantis* (Anura: Craugastoridae) de los bosques montanos del río Upano, Ecuador: Lista anotada, patrones de diversidad y descripción de cuatro especies nuevas. *Neotropical Biodiversity* 3: 125–156.

Brito-M, J., A. Almendariz, and D. Batallas. 2014. *Phyllomedusa ecuatoriana* Cannatella (Amphibia: Hylidae): variación, descripción del renacuajo, vocalización y anotaciones sobre la historia natural. *Papéis Avulsos de Zoologia* 54: 419–433.

Brito-M, J., and G. Pozo-Zamora. 2013. Una nueva especie de rana terrestre del género *Pristimantis* (Amphibia: Craugastoridae), de la Cordillera de Kutukú, Ecuador. *Papéis Avulsos de Zoologia* 53: 315–325.

Brito-Zapata, D. 2021. First report of partial leucism in the poison frog *Epipedobates anthonyi* (Anura: Dendrobatidae) in El Oro, Ecuador. *Neotropical Biodiversity* 7: 1–4.

Brito-Zapata, D., and C. Reyes-Puig. 2021. A new species of terrestrial-breeding frog, *Pristimantis* (Anura: Strabomantidae), from the Cordillera del Cóndor, Zamora-Chinchipe, Ecuador. *Neotropical Biodiversity* 7: 213–222.

Brito-Zapata, D., C. Reyes-Puig, D. Cisneros-Heredia, D. Zumel, and S. R. Ron. 2021. Description of a new minute frog of the genus *Pristimantis* (Anura: Strabomantidae) from Cordillera del Cóndor, Ecuador. *Zootaxa* 5072: 351–372.

Browne, R. K., A. J. Silla, R. Upton, G. Della-Togna, R. Marcec-Greaves, N. V. Shishova, V. K. Uteshev, *et al*. 2019. Sperm collection and storage for the sustainable management of amphibian biodiversity. *Theriogenology* 133: 187–200.

Brunner, R. M., and J. M. Guayasamin. 2020. Nocturnal visual displays and call description of the cascade specialist glassfrog, *Sachatamia orejuela*. *Behaviour* 157: 1257–1268.

Burrowes, P. A., J. R. Mendelson III, and I. de la Riva. 2022. The passing of a titan: William E. Duellman. *Herpetological Review* 53: 368–370.

Bustamante, M. R. 2002. Monitoreos de poblaciones de anuros en los Andes de Ecuador, con énfasis en el estudio poblacional de las especies del Bosque Protector Cashca Totoras (Provincia Bolívar). Tesis de Licenciatura, Pontificia Universidad Católica del Ecuador.

Bustamante, M. R., and R. E. Cárdenas. 2007. *Bichos, una mirada a la macrodiversidad tropical*. Quito: Trama/Escuela de Biología, Pontificia Universidad Católica del Ecuador.

Bustamante, M. R., D. F. Cisneros-Heredia, M. H. Yánez-Muñoz, H. M. Ortega-Andrade, and J. M. Guayasamin. 2007. Amphibia, Centrolenidae, *Cochranella pulverata*, *Hyalinobatrachium aureoguttatum*: distribution extension, Ecuador. *Check List* 3: 271–276.

Bustamante, M. R., and J. R. Mendelson III. 2008. A new frog species (Strabomantidae: *Pristimantis*) from the high Andes of southeastern Ecuador. *Zootaxa* 1820: 49–59.

Bustamante, M. R., P. A. Menéndez-Guerrero, and D. F. Cisneros-Heredia. 2005b. *Bufo guttatus*. *Herpetological Review* 36: 331.

Bustamante, M. R., S. R. Ron, and L. A. Coloma. 2005a. Cambios en la diversidad en siete comunidades de anuros en los Andes de Ecuador. *Biotropica* 37: 180–189.

Bustamante, T. 2016. *Historia de la conservación ambiental en Ecuador. Volcanes, tortugas, geólogos y políticos*. Quito: FLACSO Ecuador.

Cabodevilla, M. A. 1998. Ecuador (1864–1865) y la Comisión Científica del Pacífico. In *El gran viaje*, eds. M. Jiménez de la Espada, F. de P. Martínez, M. Almagro, and J. Isern, 250 Quito: Abya-Yala.

Cadle, J. E., and R. W. McDiarmid. 1990. Two new species of *Centrolenella* (Anura: Centrolenidae) from northwestern Peru. *Proceedings of the Biological Society of Washington* 103: 746–768.

Caminer, M. A., B. Milá, M. Jansen, A. Fouquet, P. J. Venegas, G. Chávez, S. C. Lougheed, and S. R. Ron. 2017. Systematics of the *Dendropsophus leucophyllatus* species complex (Anura: Hylidae): Cryptic diversity and the description of two new species. *PloS One* 12: e0171785

Caminer, M. A., and S. R. Ron. 2014. Systematics of treefrogs of the *Hypsiboas calcaratus* and *Hypsiboas fasciatus* species complex (Anura, Hylidae) with the description of four new species. *ZooKeys* 370: 1–68.

Caminer, M. A., and S. R. Ron. 2020. Systematics of the *Boana semilineata* species group (Anura: Hylidae), with a description of two new species from Amazonian Ecuador. *Zoological Journal of the Linnean Society* 190: 149–180.

Campbell, L. G., S. Mehtani, M. E. Dozier, and J. Rinehart. 2013. Gender-heterogeneous working groups produce higher quality science. *PloS One* 8: e79147.

Cannatella, D. C. 1980. A review of the *Phyllomedusa buckleyi* group (Anura: Hylidae). *Occasional Papers of the Museum of Natural History University of Kansas* 87: 1–40.

Cannatella, D. C. 1981. A new *Atelopus* from Ecuador and Colombia. *Journal of Herpetology* 15: 133–138.

Carvajal-Endara, S., L. A. Coloma, M. A. Morales-Mite, J. M. Guayasamin, P. Székely, and W. E. Duellman. 2019. Phylogenetic systematics, ecology, and conservation of marsupial frogs (Anura: Hemiphractidae) from the Andes of southern Ecuador, with descriptions of four new biphasic species. *Zootaxa* 4562: 1–102.

Castillo-Trenn, P. 2004. Description of the tadpole of *Colostethus kingsburyi* (Anura: Dendrobatidae) from Ecuador. *Journal of Herpetology* 38: 600–606.

Castillo-Trenn, P., and L. A. Coloma. 2008. Notes on behavior and reproduction in captive *Allobates kingsburyi* (Anura: Dendrobatidae), with comments on evolution of reproductive amplexus. *International Zoo Yearbook* 42: 58–70.

Castroviejo-Fisher, S., J. M. Guayasamin, A. Gonzalez-Voyer, and C. Vilà. 2014. Neotropical diversification seen through glassfrogs. *Journal of Biogeography* 41: 66–80.

Caty, S. N., A. Alvarez-Buylla, G. D. Byrd, C. Vidoudez, A. B. Roland, E. E. Tapia, B. Budnik, S. A. Trauger, L. A. Coloma, and L. A. O'Connell. 2019. Molecular physiology of chemical defenses in a poison frog. *Journal of Experimental Biology* 222: 1–12.

Chasiluisa, V. D., M. A. Caminer, A. Varela-Jaramillo, and S. R. Ron. 2020. Description and phylogenetic relationships of a new species of treefrog of the *Osteocephalus buckleyi* species group (Anura: Hylidae). *Neotropical Biodiversity* 6: 21–36.

Cipriani, I., and M. Rivera. 2009. Detección de alcaloides en la piel de cuatro especies de anfibios ecuatorianos (Anura: Dendrobatidae). *Revista Ecuatoriana de Medicina y Ciencias Biológicas* 30: 42–49.

Cisneros-Heredia, D. F. 2003a. Herpetofauna de la Estación de Biodiversidad Tiputini, Amazonía Ecuatoriana. In *Memorias del I Congreso de Ecología y Ambiente, Ecuador país megadiverso*, eds. de la Torre and G. Reck, 21. Quito: Universidad San Francisco de Quito USFQ.

Cisneros-Heredia, D. F. 2003b. Laboratorio de anfibios & reptiles USFQ, una nueva herramienta de investigación y conservación en Ecuador. In *Resúmenes VI Congreso Latinoamericano de Herpetología*. Lima: Museo de Historia Natural Universidad Nacional Mayor de San Marcos/Centro de Conservación, Investigación/Manejo de Áreas Naturales CIMA-Cordillera Azul.

Cisneros-Heredia, D. F. 2004a. Geographic distribution: *Rana catesbeiana*. *Herpetological Review* 35: 406.

Cisneros-Heredia, D. F. 2004b. New distributional records of amphibians in the Andes of Ecuador. *Herpetozoa* 17: 194–195.

Cisneros-Heredia, D. F. 2005a. Amphibia, *Epipedobates bilinguis*, *Agalychnis calcarifer*, *Eleutherodactylus croceoinguinis*: Distribution extensions. *Check List* 1: 14–15.

Cisneros-Heredia, D. F. 2005b. *Barycholos pulcher*. *Herpetological Review* 36: 331.

Cisneros-Heredia, D. F. 2005c. *Eleutherodactylus malkini*. *Herpetological Review* 16: 74.

Cisneros-Heredia, D. F. 2005d. On the distribution and natural history of *Hyla miyatai* Vigle & Golberghan-Vigle, 1990 in Amazonian Ecuador. *Herpetozoa* 19: 71–72.

Cisneros-Heredia, D. F. 2006a. Distribution and ecology of the western Ecuador frog *Leptodactylus labrosus* (Amphibia: Anura: Leptodactylidae). *Zoological Research* 27: 225–234.

Cisneros-Heredia, D. F. 2006b. Amphibians, Machalilla National Park, province of Manabí, Western Ecuador. *Check List* 2: 45–54.

Cisneros-Heredia, D. F. 2006c. Amphibia, Caudata, Plethodontidae: *Bolitoglossa equatoriana* and *Bolitoglossa biseriata*: Range extensions, new provincial records from Ecuador, and Natural History. *Check List* 2: 64.

Cisneros-Heredia, D. F. 2007a. A new species of glassfrog of the genus *Centrolene* from the foothills of Cordillera Oriental of Ecuador. *Herpetozoa* 20: 27–34.

Cisneros-Heredia, D. F. 2007b. Notes on the natural history of the casque-headed treefrog *Trachycephalus jordani* (Stejneger & Test, 1891). *Herpetozoa* 20: 92–94.

Cisneros-Heredia, D. F. 2008a. Habitat loss and climate change impacts on Neotropical anurans: Implications for in-situ conservation. A case study with glassfrogs from eastern Ecuador (Amphibia: Anura: Centrolenidae). MS dissertation, King's College London.

Cisneros-Heredia, D. F. 2008b. Seven new species of the Ecuadorian glassfrog discovered: Uncertain future. *Froglog (Newsletter of the IUCN/SCC Amphibian Specialist Group)* 89: 7–10.

Cisneros-Heredia, D. F. 2009. Amphibia, Anura, Centrolenidae, *Chimerella mariaelenae* (Cisneros-Heredia and McDiarmid, 2006), *Rulyrana flavopunctata* (Lynch and Duellman, 1973), *Teratohyla pulverata* (Peters, 1873), and *Teratohyla spinosa* (Taylor, 1949): Historical records, distribution extension and new provincial record in Ecuador. *Check List* 5: 912–916.

Cisneros-Heredia, D. F. 2013. *Centrolene ritae* Lutz es un sinónimo senior de *Cochranella oyampiensis* Lescure y *Cochranella ametarsia* Flores (Anura: Centrolenidae). *Avances en Ciencias e Ingenierías* 5: 1–4.

Cisneros-Heredia, D. F. 2016. Machalilla poison frog *Epipedobates machalilla* (Coloma, 1995). In *Aposematic poison frogs (Dendrobatidae) of the Andean countries: Bolivia, Colombia, Ecuador, Perú and Venezuela*, eds. T. R. Kahn, E. La Marca, S. Lötters, J. L. Brown, E. Twomey, and A. Amézquita, 337–339. Tropical Field Guide Series. Arlington: Conservation International.

Cisneros-Heredia, D. F. 2017. The type localities of *Anolis aequatorialis* Werner, 1894 (Sauria: Iguania: Dactyloidae) and *Pristimantis appendiculatus* (Werner, 1894) (Amphibia: Anura: Craugastoridae). *Zootaxa* 4216: 190–196.

Cisneros-Heredia, D. F. 2018. The hitchhiker wave: Non-native small terrestrial vertebrates in the Galapagos. In *Understanding invasive species in the Galapagos Islands*, eds. M. de L. Torres and C. F. Mena, 95–139. Cham: Social and Ecological Interactions in the Galapagos Islands/Springer.

Cisneros-Heredia, D. F., A. Almendáriz, M. H. Yánez-Muñoz, L. A. Coloma, and S. R. Ron. 2004. Chimbo rocket frog, *Hyloxalus infraguttatus*: E.T55095A11241073. The IUCN Red List of Threatened Species 2004: e.T55095A11241073. https://doi.org/10.2305/IUCN.UK.2004.RLTS.T55095A11241073.en (accessed April 2, 2022).

Cisneros-Heredia, D. F., J. Delia, M. H. Yánez-Muñoz, and H. M. Ortega-Andrade. 2008a. Natural history and intraspecific variation of the Ecuadorian blue glassfrog *Cochranella mache* Guayasamin & Bonaccorso, 2004. *Herpetozoa* 21: 57–66.

Cisneros-Heredia, D. F., J. Delia, M. H. Yánez-Muñoz, and H. M. Ortega-Andrade. 2010. Endemic Ecuadorian glassfrog *Cochranella mache* is critically endangered because of habitat loss. *Oryx* 44: 114–117.

Cisneros-Heredia, D. F., and S. V. Flechas. 2016. Ecuador poison frog *Ameerega billinguis* (Jungfer, 1898). In *Aposematic poison frogs (Dendrobatidae) of the Andean countries: Bolivia, Colombia, Ecuador, Perú and Venezuela*, eds. T. R. Kahn, E. La Marca, S. Lötters, J. L. Brown, E. Twomey, and A. Amézquita, 127–129. Tropical Field Guide Series. Arlington: Conservation International.

Cisneros-Heredia, D., and A. G. Gluesenkamp. 2011. "2010". A new Andean toad of the genus *Osornophryne* (Amphibia: Anura: Bufonidae) from northwestern Ecuador, with taxonomic remarks on the genus. *Avances en Ciencias e Ingenierías* 2: 64–73.

Cisneros-Heredia, D. F., and J. M. Guayasamin. 2014. Notes on the taxonomy of some glassfrogs from the Andes of Peru and Ecuador (Amphibia: Centrolenidae). *Papéis Avulsos de Zoologia (São Paulo)* 54: 161–168.

Cisneros-Heredia, D. F., and K.-H. Jungfer. 2004. *Dendropsophus carnifex*. The IUCN Red List of Threatened Species 2004: e.T55433A11310850. https://doi.org/10.2305/IUCN.UK.2004.RLTS.T55433A11310850.en (accessed April 3, 2022).

Cisneros-Heredia, D. F., A. León-Reyes, and S. Seger. 2005. *Boa constrictor* predation on a Titi Monkey, *Callicebus discolor*. *Neotropical Primates* 13: 11–12.

Cisneros-Heredia, D. F., and R. W. McDiarmid. 2016a. Notes on some glass frogs (Anura: Centrolenidae) from western Ecuador: Taxonomy, natural history, and conservation status, with the description of a new species. *PeerJ Preprints* 4: e2159v2.

Cisneros-Heredia, D. F., and R. W. McDiarmid. 2016b. The glass frogs (Centrolenidae) from the 'Río Palenque' Science Center, Western Ecuador, with the description of a new species. *PeerJ Preprints* 4: e2159v1.

Cisneros-Heredia, D. F., and R. W. McDiarmid. 2003. Ecuadorian glass frogs: Current state of knowledge, new research trends, and conservation. *PeerJ Preprints* 4: e2164v.

Cisneros-Heredia, D. F., and R. W. McDiarmid. 2004. Observations on the taxonomy, distribution, and conservation status of the centrolenids from Ecuador (Anura: Centrolenidae). In *Abstracts 2004 joint meeting of ichthyologists and herpetologists*. Oklahoma: ASIH, SSAR, HL, AES.

Cisneros-Heredia, D. F., and R. W. McDiarmid. 2005. Amphibia, Centrolenidae, *Centrolene peristictum*, *Centrolene prosoblepon*, *Cochranella cochranae*, *Cochranella midas*, *Cochranella resplendens*, *Cochranella spinosa*, *Hyalinobatrachium munozorum*: Range extensions and new provincial records. *Check List* 1: 18–22.

Cisneros-Heredia, D. F., and R. W. McDiarmid. 2006a. Review of the taxonomy and conservation status of the Ecuadorian glassfrog *Centrolenella puyoensis* Flores & McDiarmid (Amphibia: Anura: Centrolenidae). *Zootaxa* 1361: 21.

Cisneros-Heredia, D. F., and R. W. McDiarmid. 2006b. A new species of the genus *Centrolene* (Amphibia: Anura: Centrolenidae) from Ecuador with comments on the taxonomy and biogeography of glassfrogs. *Zootaxa* 1244: 132.

Cisneros-Heredia, D. F., and R. W. McDiarmid. 2006c. Primer registro de *Hyalinobatrachium ruedai* (Amphibia: Centrolenidae) en Ecuador, con notas sobre otras especies congénericas. *Herpetotropicos* 3: 21–28.

Cisneros-Heredia, D. F., and R. W. McDiarmid. 2007. Revision of the characters of Centrolenidae (Amphibia: Anura: Athesphatanura), with comments on its taxonomy and the description of new taxa of glassfrogs. *Zootaxa* 1572: 1–82.

Cisneros-Heredia, D. F., and P. Meza-Ramos. 2007. An enigmatic new species of glassfrog (Amphibia: Anura: Centrolenidae) from the Amazonian Andean slopes of Ecuador. *Zootaxa* 1485: 33–41.

Cisneros-Heredia, D. F., and M. Morales-Mite. 2008. A new species of glassfrog from elfin forests of the Cordillera del Cóndor, southeastern Ecuador. *Herpetozoa* 21: 49–56.

Cisneros-Heredia, D. F., and D. Mosquera. 2011. "2010". Primer registro de un cánido (*Atelocynus microtis*) depredando a un caecilian amphibian. *Avances en Ciencias e Ingenierías* 2: 5–6.

Cisneros-Heredia, D. F., and E. Peñaherrera-Romero. 2020. Invasion history of *Harmonia axyridis* (Pallas, 1773) (Coleoptera: Coccinellidae) in Ecuador. *PeerJ* 8: e10461.

Cisneros-Heredia, D. F., and R. P. Reynolds. 2007. New records of *Phyllonastes* Heyer 1977 (Anura, Brachycephalidae) from Ecuador and Peru. *Herpetozoa* 19: 184–186.

Cisneros-Heredia, D. F., and J.-M. Touzet. 2004. Distribution and conservation status of *Bothrops asper* (Garman, 1884) in Ecuador—2004. *Herpetozoa* 17: 135–141.

Cisneros-Heredia, D. F., P. J. Venegas, M. Rada, and R. Schulte. 2008b. A new species of glassfrog (Anura: Centrolenidae) from the foothill Andean forests of Ecuador and Peru. *Herpetologica* 64: 341–353.

Cisneros-Heredia, D. F., and M. H. Yánez-Muñoz. 2007a. Notes on geographic distribution. Amphibia, Centrolenidae, *Centrolene balionotum*, *Centrolene geckoideum* and *Cochranella caritocommata:* Distribution extensions, new provincial records, Ecuador. *Check List* 3: 30–42.

Cisneros-Heredia, D. F., and M. H. Yánez-Muñoz. 2007b. A new species of glassfrog (*Centrolenidae*) from the southern Andean foothills in the west Ecuadorian region. *South American Journal of Herpetology* 2: 1–10.

Cisneros-Heredia, D. F., and M. H. Yánez-Muñoz. 2011. "2010". A new poison frog of the genus *Epipedobates* (Dendrobatoidea: Dendrobatidae) from the north-western Andes of Ecuador. *Avances En Ciencias e Ingenierías* 2: 83–86.

Cisneros-Heredia, D. F., M. H. Yánez-Muñoz, and M. H. Ortega-Andrade. 2009. Description of a new species of *Teratohyla* Taylor (Amphibia: Athesphatanura: Centrolenidae) from north-western Ecuador. *Zootaxa* 2227: 53–62.

Cisneros-Heredia, D. F., M. H. Yánez-Muñoz, and M. H. Ortega-Andrade. 2011. "2010". *Teratohyla sornozai* Cisneros-Heredia, Yánez-Muñoz y Ortega-Andrade es un sinónimo junior de *Rulyrana orejuela* Duellman y Burrowes (Amphibia, Anura, Centrolenidae). *Avances en Ciencias e Ingenierías* 2: 3–4.

Cisneros-Heredia, D. F., M. H. Yánez-Muñoz, J. C. Sánchez-Nivicela, and S. R. Ron. 2023. Two new syntopic species of glassfrogs (Amphibia, Centrolenidae, *Centrolene*) from the southwestern Andes of Ecuador. *PeerJ* 11: e15195.

Cohen, J. M., D. J. Civitello, M. D. Venesky, T. A. McMahon, and J. R. Rohr. 2019. An interaction between climate change and infectious disease drove widespread amphibian declines. *Global Change Biology* 25: 927–937.

Colao, A. 1967. *Jiménez de la Espada. Aventura de un científico hispanista*. Cartagena: Athenas Ediciones.

Collins, D. 1983. Ken Miyata: A legend is dead. *The Washington Post* (October 23). https://www.washingtonpost.com/archive/sports/1983/10/25/ken-miyata-a-legend-is-dead/8eb9fc16-d111-4272-94ad-cccbdfe33d8e/.

Collins, J. P., and M. L. Crump. 2009. *Extinction in our times: Global amphibian decline*. New York: Oxford University Press.

Coloma, L. A. 1991. Anfibios de Ecuador: Lista de especies, ubicación altitudinal y referencias bibliográficas. *Reportes Técnicos Ecociencia* 2: 1–47.

Coloma, L. A. 1995. Ecuadorian frogs of the genus *Colostethus* (Anura: Dendrobatidae). *University of Kansas, Natural History Museum. Miscellaneous Publications* 87: 1–75.

Coloma, L. A. 1999. Sobre Orcés y el Jambato, su legado. *Revista Nuestra Ciencia* 1–4.

Coloma, L. A. 2005. *Anfibios de Ecuador*. Museo de Zoología, Pontificia Universidad Católica del Ecuador.Ver. 2.0. Wayback

Machine. http://web.archive.org/web/20110721135416/http:/www.puce.edu.ec/zoologia/vertebrados/amphibiawebec/anfibiosecuador/index.html (accessed December 3, 2023).

Coloma, L. A. 2008. Pontificia Universidad Católica in Quito, Ecuador. *Herpetological Review* 39: 26–27.

Coloma, L. A. 2010. 40 años de historia del área de vertebrados del QCAZ. *Nuestra Ciencia* 12: 22–26. https://issuu.com/centrojambatu/docs/coloma2010.

Coloma, L. A. 2016. El Jambato negro del páramo, *Atelopus ignescens*, resucitó. *IMCiencia*. https://issuu.com/centrojambatu/docs/coloma_2016_el_jambato_negro_del_p_

Coloma, L. A., V. Cano, and M. A. Rodríguez. 2010a. A "life raft" for amphibians in Ecuador. In *Building a future for wildlife: Zoos and aquariums committed to biodiversity conservation*, eds. G. Dick and M. Gusset, 105–110. Gland: Waza Executive Office.

Coloma, L. A., S. Carvajal-Endara, J. F. Dueñas, A. Paredes-Recalde, M. Morales-Mite, D. Almeida-Reinoso, E. E. Tapia, C. R. Hutter, E. Toral, and J. M. Guayasamin. 2012. Molecular phylogenetics of stream treefrogs of the *Hyloscirtus larinopygion* group (Anura: Hylidae), and description of two new species from Ecuador. *Zootaxa* 78: 1–78.

Coloma, L. A., W. E. Duellman, A. Almendáriz, S. R. Ron, A. Terán-Valdez, and J. M. Guayasamin. 2010b. Five new (extinct?) species of *Atelopus* (Anura: Bufonidae) from Andean Colombia, Ecuador, and Peru. *Zootaxa* 54: 1–54.

Coloma, L. A., and J. M. Guayasamin. 2022. In Memoriam: William E. Duellman, 1930–2022. Studying the world's greatest diversity of frogs. *Phyllomedusa* 21: 103–111.

Coloma, L. A., and S. Lötters. 1996. The tadpole of *Atelopus balios* (Anura: Bufonidae) from the Pacific lowlands of Ecuador. *Herpetologica* 52: 66–70.

Coloma, L. A., S. Lötters, W. E. Duellman, and A. Miranda-Leiva. 2007. A taxonomic revision of *Atelopus pachydermus*, and description of two new (extinct?) species of *Atelopus* from Ecuador (Anura: Bufonidae). *Zootaxa* 32: 1–32.

Coloma, L. A., S. Lötters, and A. W. Salas. 2000. Taxonomy of the *Atelopus ignescens* complex (Anura: Bufonidae): Designation of a neotype of *Atelopus ignescens* and recognition of *Atelopus exiguus*. *Herpetologica* 56: 303–324.

Coloma, L. A., and A. Quiguango-Ubillús. 2000. *Anfibios de Ecuador: lista de especies y distribución*. Museo de Zoología, Centro de Biodiversidad y Ambiente, Pontificia Universidad Católica del Ecuador. Ver. 1.2. Wayback Machine. https://web.archive.org/web/20000229032729/http://www.puce.edu.ec:80/ (accessed December 3, 2023).

Coloma, L. A., and S. R. Ron. 2001. *Ecuador Megadiverso: Anfibios, reptiles, aves, mamíferos*. Quito: Centro de Biodiversidad y Ambiente, Pontificia Universidad Católica del Ecuador.

Conci, C., and R. Poggi. 1996. Iconografía de entomólogos italianos, con datos biográficos esenciales. *Memorie della Società entomologica Italiana* 75: 159–382.

Cope, E. D. 1868. An examination of the Reptilia and Batrachia obtained by the Orton Expedition to Equador and the Upper Amazon, with notes on other species. *Proceedings of the Academy of Natural Sciences of Philadelphia* 20: 96–140.

Cornalia, E. 1849. *Vertebratorum synopsis in Musaeo Mediolanense extantium quae per novum Orbem Cajetanas Osculati collegit Annis 1846–47–48. Speciebus novis vel minus cognitis adjectis, nec non Descriptionibus atque Iconibus illustratis*. Corbetta: Mediolani.

Correa Monge, G. M. 1995. Comportamiento, vocalización y notas biológicas de *Colostethus awa* y *C. machalilla* (Anura: Dendrobatidae) en cautiverio. Tesis de Licenciatura, Pontificia Universidad Católica del Ecuador.

Cox, N., B. E. Young, P. Bowles, M. Fernandez, J. Marin, G. Rapacciuolo, M. Böhm, *et al.* 2022. A Global reptile assessment highlights shared conservation needs of tetrapods. *Nature* 605: 285–290.

Crump, M. 2000. *In search of the Golden Frog*. Chicago: The University of Chicago Press.

Crump, M. L. 1974. Reproductive strategies in a tropical anuran community. *Miscellaneous Publications of the Museum of Natural History*. University of Kansas 61: 1–68.

Crump, M. L. 2024. *Lost frogs and hot snakes: Herpetologists' Tales from the field*. Ithaca: Cornell University Press.

Cuesta, S. A., C. Reinoso, F. Morales, F. Pilaquinga, G. Morán-Marcillo, C. Proaño-Bolaños, A. Blasco-Zúñiga, M. Rivera, and L, Meneses. 2021. Novel antimicrobial cruzioseptin peptides extracted from the splendid leaf frog, *Cruziohyla calcarifer*. *Amino Acids* 53: 853–868.

Darst, C. R., and D. C. Cannatella. 2004. Novel relationships among hyloid frogs inferred from 12S and 16S mitochondrial DNA sequences. *Molecular Phylogenetics and Evolution* 31: 462–475.

Darst, C. R., M. E. Cummings, and D. C. Cannatella. 2006. A mechanism for diversity in warning signals: Conspicuousness versus toxicity in poison frogs. *Proceedings of the National Academy of Sciences of the United States of America* 103: 5852–5857.

Darst, C. R., P. A. Menéndez-Guerrero, L. A. Coloma, and D. C. Cannatella. 2005. Evolution of dietary specialization and chemical defense in poison frogs (Dendrobatidae): A comparative analysis. *American Naturalist* 165: 56–69.

de Albuja, C. M., Campos, M., and E. M. del Pino. 1983. Role of progesterone on oocyte maturation in the egg-brooding hylid frog *Gastrotheca riobambae* (Fowler). *The Journal of Morphology* 227: 271–276.

de la Riva, I. 2000. La obra herpetológica de Jiménez de la Espada: su relevancia y validez después de un siglo. In *Tras la senda de un explorador Marcos Jiménez de la Espada (1831–1898)*, eds. L. López-Ocón and C. M. Pérez-Montes, 76–90. Madrid: CSIC.

del Pino, E. M. 1973. Interactions between gametes and environment in the toad *Xenopus laevis* (Daudin) and their relationship to fertilization. *Journal of Experimental Zoology* 185: 121–132.

del Pino, E. M. 1975. Adaptaciones reproductivas para la vida terrestre del sapo marsupial *Gastrotheca riobambae* (Fowler), Anura, Hylidae. *Revista de la Pontificia Universidad Católica del Ecuador* 3: 119–140.

del Pino, E. M. 1977. Núcleos múltiples en el oocito de algunos sapos marsupiales (Hylidae). *Revista de la Pontificia Universidad Católica del Ecuador* 5: 21–30.

del Pino, E. M. 1980. Morphology of the pouch and incubatory integument in marsupial frogs (Hylidae). *Copeia* 1980: 10–17.

del Pino, E. M. 1983. Progesterone induces incubatory changes in the brooding pouch of the frog *Gastrotheca riobambae* (Fowler). *Journal of Experimental Zoology* 227: 159–163.

del Pino, E. M. 2021. Experiences with the marsupial frogs: Reminiscences of a developmental biologist. *International Journal of Developmental Biology* 65: 103–110.

del Pino, E. M., and I. Alcocer. 1994. Urea is necessary for the culture of embryos of the marsupial frog *Gastrotheca riobambae* and is tolerated by embryos of the aquatic frog *Xenopus laevis*. *Development, Growth & Differentiation* 36: 73–80.

del Pino, E. M., and R. P. Elinson. 1983. A novel development pattern for frogs: Gastrulation produces an embryonic disk. *Nature* 306: 589–591.

del Pino, E. M., and B. Escobar., B. 1981. Embryonic stages of *Gastrotheca riobambae* (Fowler) during maternal incubation and comparison of development with that of other egg-brooding hylid frogs. *Journal of Morphology* 167: 277–295.

del Pino, E. M., M. L. Galarza, C. M. de A., and A. A. Humphries, Jr. 1975. The maternal pouch and development in the marsupial frog *Gastrotheca riobambae* (Fowler). *Biological Bulletin* 149: 480–491.

del Pino, E. M., and A. A. Humphries, Jr. 1978. Multiple nuclei during early oogenesis in *Flectonotus pygmaeus* and other marsupial frogs. *Biological Bulletin* 154: 198–212.

del Pino, E. M., and S. Loor-Vela. 1990. The pattern of early cleavage of the marsupial frog *Gastrotheca riobambae*. *Development* 110: 781–787.

del Pino, E. M., M. Venegas-Ferrín, A. Romero-Carvajal, P. Montenegro-Larrea, N. Sáenz-Ponce, I. M. Moya, I. Alarcón, N. Sudou, S. Yamamoto, and M. Taira. 2007. A comparative analysis of frog early development. *Proceedings of the National Academy of Sciences of the United States of America* 104: 11882–11888.

del Pino-Veintimilla, E. M. 1975. Adaptaciones reproductivas para la vida terrestre del sapo marsupial *Gastrotheca riobambae* (Fowler). *Revista de la Pontificia Universidad Católica del Ecuador* 3: 119–140.

Delia, J., L. Bravo-Valencia, and K. M. Warkentin. 2017. Patterns of parental care in Neotropical glassfrogs: Fieldwork alters hypotheses of sex-role evolution. *Journal of Evolutionary Biology* 30: 898–914.

Delia, J., D. F. Cisneros-Heredia, J. Whitney, and R. Murrieta-Galindo. 2010. Observations on the reproductive behavior of a Neotropical glassfrog, *Hyalinobatrachium fleischmanni* (Anura: Centrolenidae). *South American Journal of Herpetology* 5: 1–12.

dos Santos, S. P., R. Ibáñez, and S. R. Ron. 2015. Systematics of the *Rhinella margaritifera* complex (Anura, Bufonidae) from western Ecuador and Panama with insights in the biogeography of *Rhinella alata*. *ZooKeys* 145: 109–145.

Duellman, W. E. 1968. The taxonomic status of some American hylid frogs. *Herpetologica* 24: 194–209.

Duellman, W. E. 1969a. *Phyllomedusa buckleyi* Boulenger: Variation, distribution and synonymy. *Herpetologica* 25: 134–140.

Duellman, W. E. 1969b. A new species of frog in the *Hyla parviceps* group from Ecuador. *Herpetologica* 25: 241–247.

Duellman, W. E. 1970. *The Hylid frogs of Middle America. Monograph of the Museum of Natural History*. Lawrence: The University of Kansas.

Duellman, W. E. 1978. *The biology of an equatorial herpetofauna in Amazonian Ecuador*. University of Kansas Museum of Natural History, Miscellaneous Publications 65. Lawrence: University of Kansas.

Duellman, W. E. 1980. The identity of *Centrolenella grandisonae* Cochran and Goin (Anura: Centrolenidae). *Transactions of the Kansas Academy of Science* 83: 26–32.

Duellman, W. E. 2009. Prólogo. Foreword. In *Sapos*, eds. S. R. Ron, M. Bustamante, L. A. Coloma, and B. Mena, 20. Quito: Museo de Zoología, Pontificia Universidad Católica del Ecuador.

Duellman, W. E. 2015a. *Herpetology at Kansas: A centennial history*. Ithaca: Society for the Study of Amphibians and Reptiles.

Duellman, W. E. 2015b. *Marsupial frogs: Gastrotheca and Allied Genera*. Baltimore, MD: JHU Press.

Duellman, W. E., and R. Altig. 1978. New species of tree frogs (Hylidae) from the Andes of Colombia and Ecuador. *Herpetologica* 34: 177–185.

Duellman, W. E., and L. A. Coloma. 1993. *Hyla staufferorum*, a new species of treefrog in the *Hyla larinopygion* group from the cloud forests of Ecuador. *Occasional Papers of the Museum of Natural History* 161: 1–11.

Duellman, W. E., and D. M. Hillis. 1987. Marsupial frogs (Anura: Hylidae: *Gastrotheca*) of the Ecuadorian Andes: Resolution of taxonomic problems and phylogenetic relationships. *Herpetologica* 43: 141–173.

Duellman, W. E., and D. M. Hillis. 1990. Systematics of frogs of the *Hyla larinopygion* group. *Occasional Papers of the Museum of Natural History University of Kansas* 134: 1–23.

Duellman, W. E., and J. D. Lynch. 1969. Descriptions of *Atelopus* tadpoles and their relevance to atelopodid classification. *Herpetologica* 25: 231–240.

Duellman, W. E., and J. D. Lynch. 1988. Anuran amphibians from the Cordillera de Cutucú, Ecuador. *Proceedings of the Academy of Natural Sciences of Philadelphia* 140: 125–142.

Duellman, W. E., and L. Trueb. 1966. Neotropical hylid frogs, genus *Smilisca*. *University of Kansas Publications Museum of Natural History* 17: 281–375.

Duellman, W. E., and L. Trueb. 1986. *Biology of the amphibians*. New York: JHU Press.

Duffy, M. A. 2017. Last and corresponding authorship practices in ecology. *Ecology and Evolution* 7: 8876–8887.

ECOLAP. 1998a. *Proyecto INEFAN/GEF, Actividad 20: El manejo para la protección y el uso sostenible de la vida silvestre en el Ecuador, diagnóstico de la situación actual*. Tomos I y II. Quito: Instituto de Ecología Aplicada, Universidad San Francisco de Quito.

ECOLAP. 1998b. *Proyecto INEFAN/GEF, Actividad 20: Propuesta de políticas para la protección y el uso sustentable de la vida silvestre en el Ecuador*. Quito: Instituto de Ecología Aplicada, Universidad San Francisco de Quito.

Edwards, S. R. 1971. Taxonomic notes on South American *Colostethus* with descriptions of two new species (Amphibia, Dendrobatidae). *Proceedings of the Biological Society of Washington* 84: 147–162.

Edwards, S. R. 1974a. Taxonomic notes on South American dendrobatid frogs of the genus *Colostethus*. *University of Kansas, Natural History Museum, Occasional Papers* 30: 1–14.

Edwards, S. R. 1974b. A phenetic analysis of the genus *Colostethus* (Anura: Dendrobatidae). PhD thesis, The University of Kansas.

Elmer, K. R., R. M. Bonett, D. B. Wake, and S. C. Lougheed. 2013. Early Miocene origin and cryptic diversification of South American salamanders. *BMC Evolutionary Biology* 13: 59.

Elmer, K. R., J. A. Dávila, and S. C. Lougheed. 2006. Isolation of simple and compound polymorphic tetranucleotide microsatellites for the neotropical leaflitter frog *Eleutherodactylus ockendeni* (Leptodactylidae). *Molecular Ecology Notes* 6: 891–893.

Elmer, K. R., J. A. Dávila, and S. C. Lougheed. 2007a. Applying new inter-individual approaches to assess fine-scale population genetic diversity in a Neotropical frog, *Eleutherodactylus ockendeni*. *Heredity* 99: 505–515.

Elmer, K. R., J. A. Dávila, and S. C. Lougheed. 2007b. Cryptic diversity and deep divergence in an upper Amazonian leaflitter frog, *Eleutherodactylus ockendeni*. *BMC Evolutionary Biology* 7: 1–14.

Faivovich, J., C. F. B. Haddad, P. C. A. Garcia, D. R. Frost, J. A. Campbell, and W. C. Wheeler. 2005. Systematic review of the frog family Hylidae, with special reference to Hylinae: Phylogenetic analysis and taxonomic revision. *Bulletin of the American Museum of Natural* History 294: 1–240.

Faivovich, J., J. Moravec, D. F. Cisneros-Heredia, and J. Köhler. 2006. A new species of the *Hypsiboas benitezi* group from the Western Amazon Basin (Amphibia: Anura: Hylidae). *Herpetologica* 62: 96–108.

Farr, C. M., S. P. Bombaci, T. Gallo, A. M. Mangan, H. L. Riedl., L. T. Stinson, K. Wilkins, D. E. Bennett, T. Nogeire-McRae, and L. Pejchar. 2017. Addressing the gender gap in distinguished speakers at professional ecology conferences. *BioScience* 67, no. 5: 464–468.

Festa, E. 1909. *Nel Darien e nell'Ecuador: Diario di viaggio di un naturalista*. Torino: Unione tip.-editrice torinese.

Fischer, E. K., A. B. Roland, N. A. Moskowitz, C. Vidoudez, N. Ranaivorazo, E. E. Tapia, S. A. Trauger, M. Vences, L. A. Coloma, and L. A. O'Connell. 2019. Mechanisms of convergent egg provisioning in poison frogs. *Current Biology* 29: 4145–4151.

Flores, G. 1985a. A new *Eleutherodactylus* from the Amazonian slopes of the Ecuadorian Andes. *Herpetologica* 41: 447–450.

Flores, G. 1985b. A new *Centrolenella* (Anura) from Ecuador, with comments on nuptial pads and prepollical spines in *Centrolenella*. *Journal of Herpetology* 19: 313–320.

Flores, G. 1987. A new *Eleutherodactylus* from Volcan Sumaco, Ecuador. *Herpetologica* 43: 90–95.

Flores, G. 1988. Two new species of Ecuadorian *Eleutherodactylus* (Leptodactylidae) of the *E. crucifer* assembly. *Journal of Herpetology* 22: 34–41.

Flores, G. 1993. A new species of earless *Eleutherodactylus* (Anura: Leptodactylidae) from the Pacific slopes of the Ecuadorian Andes, with comments on the *Eleutherodactylus surdus* assembly. *Herpetologica* 49: 399–420.

Flores, G. 2021. Eugenia del Pino: El azar también está presente en la ciencia. *Diario El Comercio* (June 14). https://www.elcomercio.com/tendencias/cultura/eugenia-pino-entrevista-azar-ciencia.html.

Flores, G., and R. W. McDiarmid. 1989. Two new species of South American *Centrolenella* (Anura: Centrolenidae) related to *C. mariae*. *Herpetologica* 45: 401–411.

Flores, G., and L. O. Rodríguez. 1997. Two new species of the *Eleutherodactylus conspicillatus* group (Anura: Leptodactylidae) from Peru. *Copeia* 1997: 388–394.

Flores, G., and G. O. Vigle. 1994. A new species of *Eleutherodactylus* (Anura: Leptodactylidae) from the lowland rainforests of Amazonian Ecuador, with notes on the *Eleutherodactylus frater* assembly. *Journal of Herpetology* 28: 416–424.

Ford, L. S., and J. E. Simmons. 1997. The diffusion of knowledge: Agassiz (1807–1873), Ruthven (1882–1971), and the growth of herpetological collections. *Collection Building in Ichthyology and Herpetology*: 577–593.

Forrester, N. 2023. Guardian of Ecuador's diverse—and vanishing—Frog species. *Nature* 615, no. 7954: 960.

Forsyth, A., and K. Miyata. 1984. *Tropical nature: Life and death in the tropical rain forests of Central and South America*. Touchstone: Charles Scribners Sons/Macmillan Publishing Company.

Frances, D. N., C. R. Fitzpatrick, J. Koprivnikar, and S. J. McCauley. 2020. Effects of inferred gender on patterns of co-authorship in ecology and evolutionary biology publications. *The Bulletin of the Ecological Society of America* 101: e01705.

Freile, J. F. 2005. Gustavo Orcés, Fernando Ortiz y el desarrollo de la ornitología hecha en Ecuador. *Ornitología Neotropical* 16: 321–336.

Freile, J. F., L. A. Coloma, A. Terán-Valdez., D. Acosta López, E. E. Tapia, and G. Pazmiño-Otamendi. 2020. *Anfibios de Junín*. Quito: Centro Jambatu de Investigación y Conservación de Anfibios, Fundación Jambatu.

Freire, D., C. Fernández, and Nogales, F. 2006. Analysis of the presence of the fungus *Batrachochytrium dendrobatidis Bd.* in the amphibian populations of 8 localities in the southern Ecuadorian Andes. *Lyonia* 9: 7–18.

Frost, D. R. 1985. *Amphibian Species of the World. A Taxonomic and Geographical Reference*. D. R. Frost, ed. Lawrence, KS: Allen Press and the Association of Systematics Collections.

Fugler, C. M. 1966. *Lepidodactylus lugubris* Duméril and Bibron in western south America. *Journal of the Ohio Herpetological Society* 5: 162–162.

Fugler, C. M. 1971. Biological and ethnological observations on the Cofán, Secoya, and Awishiri Indians of eastern tropical Ecuador. *Proceedings of the Oklahoma Academy of Science* 51: 106–119.

Fugler, C. M., and T. D. Schwaner. 1968. Sexual dichromatism in the genus *Uracentron* Kaup in eastern Ecuador. *Herpetologica* 24: 253–255.

Fugler, C. M., and A. B. Walls. 1978. Snakes of the Upano Valley of Amazonian Ecuador. *Journal of the Tennessee Academy of Science* 53: 81–87.

Fugler, C. M., and A. B. Walls. 1979a. Individual variation in the Neotropical salamander *Bolitoglossa peruviana* (Boulenger, 1883) (Amphibia: Plethodontidae) in Eastern tropical Ecuador. *Revista de Biología Tropical* 27: 1–5.

Fugler, C. M., and A. B. Walls. 1979b. The lizards of the Upano Valley of southeastern Ecuador. *Journal of the Tennessee Academy of Science* 54: 120–121.

Funk, W. C., D. Almeida-Reinoso, F. Nogalas-Sornosa, and M. R. Bustamante. 2003. Monitoring population trends of *Eleutherodactylus* frogs. *Journal of Herpetology* 37: 245–256.

Funk, W. C., J. P. Caldwell, C. E. Peden, J. M. Padial, I. de la Riva, and D. C. Cannatella. 2007. Tests of biogeographic hypotheses for diversification in the Amazonian Forest frog, *Physalaemus petersi*. *Molecular Phylogenetics and Evolution* 44: 825–837.

Funk, W. C., M. Caminer, and S. R. Ron. 2011. High levels of cryptic species diversity uncovered in Amazonian frogs. *Proceedings of the Royal Society B: Biological Sciences* 279: 1806–1814.

Gagliardo, R. 2011. Veterinarios de ranas al rescate! El programa Amphibian Veterinary Outreach Program continúa su trabajo en Ecuador. *Boletín Informativo Amphibian Ark* 15: 21–22.

Gardner, A. L. 1983. *Proechimys semispinosus* (Rodentia: Echimydae): Distribution, type locality, and taxonomic history. *Proceedings of the Biological Society of Washington* 96: 134–144.

Gill, B. A., B. C. Kondratieff, K. L. Casner, A. C. Encalada, A. S. Flecker, D. G. Gannon, C. K. Ghalambor, *et al.* 2016. Cryptic species diversity reveals biogeographic support for the 'Mountain Passes Are Higher in the Tropics' hypothesis. *Proceedings of the Royal Society B: Biological Sciences* 283: 20160553.

González, J. E. 2000. Marcos Jiménez de la Espada, "El Coleccionista". In *Tras la senda de un explorador Marcos Jiménez de la Espada (1831–1898)*, eds. L. López-Ocón and C. M. Pérez-Montes, 72–75. Madrid: Consejo Superior de Investigaciones Científicas.

González Fernández, J. E. 2006. Anfibios colectados por la Comisión Científica del Pacífico (entre 1862 y 1865) conservados en el Museo Nacional de Ciencias Naturales de Madrid. *Graellsia* 62: 111–158.

Graeme, G. 2021. Frog back from the dead helps fight plans for mine in Ecuador. *The Guardian* (November 23). https://www.theguardian.com/environment/2021/nov/23/

frog-back-from-the-dead-helps-fight-mine-plans-in-ecuadors-cloud-forest-aoe.

Graeme, G. 2022. Lost and found: The rediscovery that became "Leonardo DiCaprio's frog". *The Guardian* (December 29). https://www.theguardian.com/environment/2022/dec/29/leonardo-dicaprio-longnose-harlequin-frog-copper-mining-ecuador.

Graham, C. H., S. R. Ron, C. Moritz, J. C. Santos, and C. J. Schneider. 2004. Integrating phylogenetics and environmental niche models to explore speciation mechanisms in dendrobatid frogs. *Evolution* 58: 1781–1793.

Grosso, J., J. Fratani, G. Fontanarrosa, M. Chuliver, A. S. Duport-Bru, R. G. Schneider, M. D. Casagranda, D. P. Ferraro, N. Vicente, and M. J. Salica. 2021. Male homophily in South American herpetology: One of the major processes underlying the gender gap in publications. *Amphibia-Reptilia* 42: 407–418.

Guayasamin, J. M. 2004. A new species of *Eleutherodactylus* (Anura: Leptodactylidae) from the northwestern lowlands of Ecuador. *Herpetologica* 60: 103–116.

Guayasamin, J. M., A. Arteaga, and C. R Hutter. 2018. A new (singleton) rainfrog of the *Pristimantis myersi* group (Amphibia: Craugastoridae) from the northern Andes of Ecuador. *Zootaxa* 4527: 323–334.

Guayasamin, J. M., and E. Bonaccorso. 2004. A new species of glass frog (Centrolenidae: *Cochranella*) from the lowlands of northwestern Ecuador, with comments on the *Cochranella granulosa* group. *Herpetologica* 60: 485–494.

Guayasamin, J. M., E. Bonaccorso, W. E. Duellman, and L. A. Coloma. 2010. Genetic differentiation in the nearly extinct harlequin frogs (Bufonidae: *Atelopus*), with emphasis on the Andean *Atelopus ignescens* and *A. bomolochos* species complexes. *Zootaxa* 2574: 55–68.

Guayasamin, J. M., E. Bonaccorso, P. A. Menéndez, and M. R. Bustamante. 2004. Morphological variation, diet, and vocalization of *Eleutherodactylus eugeniae* (Anura: Leptodactylidae) with notes on its reproduction and ecology. *Herpetological Review* 35: 17–23.

Guayasamin, J. M., R. M. Brunner, A. Valencia-Aguilar, D. Franco-Mena, E. Ringler, A. Medina Armijos, C. Morochz, L. Bustamante, R. J. Maynard, and J. Culebras. 2022. Two new glassfrogs (Centrolenidae: *Hyalinobatrachium*) from Ecuador, with comments on the endangered biodiversity of the Andes. *PeerJ* 10: e13109.

Guayasamin, J. M., S. Castroviejo-Fisher, J. Ayarzagüena, L. Trueb, and C. Vilà. 2008a. Phylogenetic relationships of glassfrogs (Centrolenidae) based on mitochondrial and nuclear genes. *Molecular Phylogenetics and Evolution* 48: 574–595.

Guayasamin, J. M., S. Castroviejo-Fisher, L. Trueb, J. Ayarzagüena, M. Rada, and C. Vilà. 2009. Phylogenetic systematics of glassfrogs (Amphibia: Centrolenidae) and their sister taxon *Allophryne ruthveni*. *Zootaxa* 2100: 1–97.

Guayasamin, J. M., D. F. Cisneros-Heredia, and S. Castroviejo-Fisher. 2008b. Taxonomic identity of *Cochranella petersi* Goin, 1961 and *Centrolenella ametarsia* Flores, 1987. *Zootaxa* 1815: 25–34.

Guayasamin, J. M., D. F. Cisneros-Heredia, R. J. Maynard, R. L. Lynch, J. Culebras, and P. S. Hamilton. 2017a. A marvelous new glassfrog (Centrolenidae, *Hyalinobatrachium*) from Amazonian Ecuador. *ZooKeys* 673: 1–20.

Guayasamin, J. M., D. F. Cisneros-Heredia, R. W. McDiarmid, P. Peña, and C. R. Hutter. 2020. Glassfrogs of Ecuador: Diversity, evolution, and conservation. *Diversity* 12: 222.

Guayasamin, J. M., D. F. Cisneros-Heredia, J. Vieira, S. Kohn, G. Gavilanes, R. L. Lynch, P. S. Hamilton, and R. J. Maynard. 2019a. A new glassfrog (Centrolenidae) from the Chocó-Andean Río Manduriacu Reserve, Ecuador, endangered by mining. *PeerJ* 7: e6400.

Guayasamin, J. M., D. F. Cisneros-Heredia., M. Yánez-Muñoz, and M. Bustamante. 2006b. Amphibia, Centrolenidae, *Centrolene ilex*, *Centrolene litorale*, *Centrolene medemi*, *Cochranella albomaculata*, *Cochranella ametarsia:* Range extensions and new country records. *Check List* 2: 70–75.

Guayasamin, J. M., and W. C. Funk. 2009. The amphibian community at Yanayacu Biological Station, Ecuador, with a comparison of vertical microhabitat use among *Pristimantis species* and the description of a new species of the *Pristimantis myersi* group. *Zootaxa* 2220: 41–66.

Guayasamin, J. M., C. R. Hutter, E. E. Tapia, J. Culebras, N. Peñafiel, R. A. Pyron, C. Morochz, W. C. Funk, and A. Arteaga. 2017b. Diversification of the rainfrog *Pristimantis ornatissimus* in the lowlands and Andean foothills of Ecuador. *PloS One* 12: e0172615.

Guayasamin, J. M., T. Krynak, K. Krynak, J. Culebras, and C. R. Hutter. 2015a. Phenotypic plasticity raises questions for taxonomically important traits: A remarkable new Andean rainfrog (*Pristimantis*) with the ability to change skin texture: Phenotypic plasticity in Andean rainfrog. *Zoological Journal of the Linnean Society* 173: 913–928.

Guayasamin, J. M., A. M. Mendoza, A. V. Longo, K. R. Zamudio, and E. Bonaccorso. 2014. High prevalence of *Batrachochytrium dendrobatidis* in an Andean frog community (Reserva Las Gralarias, Ecuador). *Amphibian & Reptile Conservation* 8: 33–44.

Guayasamin, J. M., M. Rivera-Correa, A. Arteaga, J. Culebras, L. Bustamante, R. A. Pyron, N. Peñafiel, C. Morochz, and C. R. Hutter. 2015b. Molecular phylogeny of stream treefrogs (Hylidae: *Hyloscirtus bogotensis* group), with a new species from the Andes of Ecuador. *Neotropical Biodiversity* 1: 2–21.

Guayasamin, J. M., S. R. Ron, D. F. Cisneros-Heredia, W. Lamar, and S. F. McCracken. 2006a. A new species of frog of the *Eleutherodactylus lacrimosus* assemblage (Leptodactylidae) from the western Amazon Basin, with comments on the utility of canopy surveys in lowland rainforest. *Herpetologica* 62: 191–202.

Guayasamin, J. M., and L. Trueb. 2007. A new species of glass frog (Anura: Centrolenidae) from the lowlands of northwestern Ecuador, with comments on centrolenid osteology. *Zootaxa* 1447: 27–45.

Guayasamin, J. M., R. Vandegrift, T. Policha, A. C. Encalada, N. Greene, B. Ríos-Touma, L. Endara, *et al*. 2021. Biodiversity conservation: Local and global consequences of the application of 'Rights of nature' by Ecuador. *Neotropical Biodiversity* 7: 541–545.

Guayasamin, J. M., J. Vieira, R. E. Glor, and C. R Hutter. 2019b. A new glassfrog (Centrolenidae: *Hyalinobatrachium*) from the Topo River Basin, Amazonian slopes of the Andes of Ecuador. *Amphibian & Reptile Conservation* 13: 133–144.

Hämäläinen, M. 2021. The marvellous polythorid damselfly *Euthore mirabilis* and other new Odonata species discovered by Clarence Buckley, with notes on the zoological legacy of this poorly known late 19th century collector in Ecuador and Bolivia. *Agrion* 25: 54–73.

Harvey, M. B., A. Almendáriz, J. Brito-M, and D. Batallas-R. 2013. A new species of *Noblella* (Anura: Craugastoridae) from the Amazonian slopes of the Ecuadorian Andes with comments on *Noblella lochites* (Lynch). *Zootaxa* 3635: 1–14.

Hedges, S. B., W. E. Duellman, and M. P. Heinicke. 2008. New World direct-developing frogs (Anura: Terrana): Molecular phylogeny, classification, biogeography, and conservation. *Zootaxa* 1737: 1–181.

Heyer, W. R. 1972. The status of *Leptodactylus pumilio* Boulenger (Amphibia, Leptodactylidae) and the description of a new species of *Leptodactylus* from Ecuador. *Contributions in Science, Natural History Museum, Los Angeles County* 231: 1–8.

Heyer, W. R. 2005. Variation and taxonomic clarification of the large species of the *Leptodactylus pentadactylus* species group (Amphibia: Leptodactylidae) from Middle America, northern South America, and Amazonia. *Arquivos de Zoologia. São Paulo* 37: 269–348.

Heyer, W. R., M. A. Donnelly, M. Foster, and R. W. McDiarmid. 2014. *Measuring and monitoring biological diversity: Standard methods for amphibians*. Washington, DC: Smithsonian Institution.

Heyer, W. R., and M. M. Heyer. 2016. Biographical sketch and bibliography of W. Ronald Heyer. *Smithsonian Hepetological Infomation Service* 147: 1–18.

Heyer, W. R., R. W. McDiarmid, and D. L. Weigmann. 1975. Tadpoles, predation and pond habitats in the Tropics. *Biotropica* 7: 100–111.

Heyer, W. R., and J. Peters. 1971. The frog genus *Leptodactylus* in Ecuador. *Proceedings of the Biological Society of Washington* 84: 163–170.

Hillis, D. M., and R. de Sá. 1988. Phylogeny and taxonomy of the *Rana palmipes* group (Salientia: Ranidae). *Herpetological Monographs* 2: 1–26.

Hoffmann, M., C. Hilton-Taylor, A. Angulo, M. Böhm, T. M. Brooks, S. H. M. Butchart, K. E. Carpenter, *et al*. 2010. The impact of conservation on the status of the world's vertebrates. *Science* 330: 1503–1509.

Hofstra, B., V. V. Kulkarni, S. Munoz-Najar Galvez, B. He, D. Jurafsky, and D. A. McFarland. 2020. The diversity–innovation paradox in science. *Proceedings of the National Academy of Sciences* 117: 9284–9291.

Hoogmoed, M. S. 1985. A new genus of toads (Amphibia: Anura: Bufonidae) from the Pacific slopes of the Andes in northern Ecuador and southern Colombia, with the description of two new species. *Zoologische Mededelingen* 59: 251–273.

Hoogmoed, M. S. 1987. New *Osornophryne* (Amphibia: Anura: Bufonidae) from the Atlantic versant of the Andes in Ecuador. *Zoologische Mededelingen* 61: 209–242.

Huang, J., A. J. Gates, R. Sinatra, and A. L. Barabási. 2020. Historical comparison of gender inequality in scientific careers across countries and disciplines. *Proceedings of the National Academy of Sciences* 117: 4609–4616.

Hutter, C. R., S. Escobar-Lasso, J. A. Rojas-Morales, P. D. A. Gutiérrez-Cárdenas, H. Imba, and J. M. Guayasamin. 2013. The territoriality, vocalizations, and aggressive interactions of the red-spotted glassfrog, *Nymphargus grandisonae*, Cochran and Goin, 1970 (Anura: Centrolenidae). *Journal of Natural History* 47: 3011–3032.

Hutter, C. R., V. Liu, T. Kell, J. A. Lyons, and J. M. Guayasamin. 2016. The natural history, distribution, and conservation of Lonely rainfrogs, *Pristimantis eremitus*. *Herpetologica* 72: 13–22.

Jácome, C. 2008. *Vida de Giovanni Onore. El héroe nunca cantado*. Quito: Cámara Ecuatoriana del Libro, Núcleo de Pichincha.

Jaramillo-Martinez, A. F., I. de La Riva, J. M. Guayasamin, J. C. Chaparro, G. Gagliardi-Urrutia, R. C. Gutiérrez, I. Brcko, C. Vilà, and S. Castroviejo-Fisher. 2020. Vastly underestimated species richness of Amazonian salamanders (Plethodontidae: *Bolitoglossa*) and implications about Plethodontid diversification. *Molecular Phylogenetics and Evolution* 149: 106841.

Jaramillo-Martinez, A. F., A. Valencia-Z., V. E. Cardona, F. Castro-Herrera, and D. F. Cisneros-Heredia. 2015. Range extension of *Cochranella mache* Guayasamin and Bonaccorso, 2004 (Anura Centrolenidae). *Herpetology Notes* 8: 161–163.

Jaynes, K. E., M. I. Páez-Vacas, D. Salazar-Valenzuela, J. M. Guayasamin, A. Terán-Valdez, F. R. Siavichay, S. W. Fitzpatrick, and L. A. Coloma. 2022. Harlequin frog rediscoveries provide insights into species persistence in the face of drastic amphibian declines. *Biological Conservation* 276: 109784.

Jiménez, P. 1999. Gustavo Orcés: científico y maestro. *Ecuador Terra Incognita* (August). https://www.terraecuador.net/revista_4/4_orces.htm.

Jiménez de la Espada, M. 1870. "1871." Faunae neotropicalis species quaedam nondum cognitae. *Jornal de Sciencias Mathematicas Physicas e Naturaes* III: 57–65.

Jiménez de la Espada, M. 1872. Nuevos batrácios americanos. *Anales de la Sociedad Española de Historia Natural. Madrid* 1: 84–88.

Jiménez de la Espada, M. 1875. *Vertebrados del viaje al Pacífico verificado de 1862 a 1865 por una Comisión de Naturalistas enviada por el Gobierno Español*. Batracios. Madrid: Miguel Ginesta.

Jiménez de la Espada, M. 1899. Examen descriptivo del grupo de los *Hemiphractus*. *Anales de Historia Natural* 27: 379–410.

Jiménez de la Espada, M., F. de P. Martínez, M. Almagro, and J. Isern. 1866. *"1998". El gran viaje*. M. A. Cabodevilla, ed. Quito, Ecuador: Ediciones Abya-Yala.

Jungfer, K.-H., and W. Hödl. 2002. A new species of *Osteocephalus* from Ecuador and a redescription of *O. leprieurii* (Duméril & Bibron, 1841) (Anura: Hylidae). *Amphibia-Reptilia* 23: 21–46.

Jungfer, K.-H., S. Ron., R. Seipp, and A. Almendáriz. 2000. Two new species of hylid frogs, genus *Osteocephalus*, from Amazonian Ecuador. *Amphibia-Reptilia* 21: 327–340.

Kaiser, H., L. A. Coloma, and H. M. Gray. 1994. A new species of *Colostethus* (Anura, Dendrobatidae) from Martinique, French-Antilles. *Herpetologica* 50: 23–32.

Kimbrough, L. 2023. Ecuador court upholds "rights of nature," blocks Intag Valley copper mine. *Mongabay* (March 31). https://news.mongabay.com/2023/03/ecuador-court-upholds-rights-of-nature-blocks-intag-valley-copper-mine/?fbclid=IwAR3u4mrqLSguiJmmznF4b0Q3FZJiyCmU3gghCuIrcQJpDBywaThs6q7irqs.

Kimerling, J. 1993. *Crudo Amazónico*. Quito: Abya-Yala.

Kozlowski, D., V. Larivière, C. R. Sugimoto, and T. Monroe-White. 2022. Intersectional inequalities in science. *Proceedings of the National Academy of Sciences* 119: e2113067119.

La Marca, E., K. R. Lips, S. Lötters, R. Puschendorf, R. Ibañez, J. V. Rueda-Almonacid, R. Schulte, *et al*. 2005. Catastrophic population declines and extinctions in neotropical harlequin frogs (Bufonidae: *Atelopus*). *Biotropica* 37: 190–201.

Lamar, W. W. 1985. A new *Lepidoblepharis* (Sauria: Gekkonidae) from the north coast of Colombia. *Herpetologica* 41: 128–132.

Lambert, C. 2000. Ken Miyata: Brief life of a legendary fly fisherman: 1951–1983. *Harvard Magazine* (July 1). https://www.harvardmagazine.com/2000/07/ken-miyata-html.

Larivière, V., C. Ni, Y. Gingras, B Cronin, and C. R. Sugimoto. 2013. Bibliometrics: Global gender disparities in science. *Nature* 504: 211–213.

Laurenti, J. N. 1768. *Specimen Medicum, Exhibens Synopsin Reptilium emendatum cum experimentis circa venena at antidota reptilium austriacorum*. Viena: Joan Thomae.

Lehr, E., L. Trueb, P. J. Venegas, and E. Arbeláez. 2007. Descriptions of the tadpoles of two Neotropical microhylid frogs, *Melanophryne carpish* and *Nelsonophryne aequatorialis* (Anura: Microhylidae). *Journal of Herpetology* 41: 581–589.

León-Reyes, A. 2008. Inventario de la herpetofauna del Bosque Protector Río Guajalito, un bosque nublado montano del occidente de la Provincia de Pichincha, Ecuador. Tesis de Licenciatura, Universidad San Francisco de Quito.

Linnaeus, C. 1758. *Systema naturae per regna tria naturae, secundum classes, ordines, genera, species, cum characteribus, differentiis, synonymis, locis*. 10th edition. Estocolmo: Impensis Direct. Laurentii Salvdii.

Lips, K. R., J. Diffendorfer, J. R. Mendelson III, and M. W. Sears. 2008. Riding the wave: Reconciling the roles of disease and climate change in amphibian declines. *PloS Biology* 6: 441–454.

Loaiza-S, C. R. 2012. *Clodoveo Carrión Mora (1883–1957). Científico insigne de la Provincia de Loja*. Loja: Universidad Técnica Particular de Loja.

Lojano, J. D. A., J. C. Sánchez-Nivicela, F. Siavichay, and N. Aguilar. 2023. Descripción del renacuajo y variación ontogenética de *Rulyrana mcdiarmidi* (Anura: Centrolenidae) del sur de Ecuador. *Revista Latinoamericana de Herpetología* 6: e682–621.

López-Ocón, L. 1991. *De viajero naturalista a historiador: las actividades americanistas del científico español Marcos Jiménez de la Espada (1831–1898)*. Madrid: Editorial de la Universidad Complutense de Madrid.

López-Ocón, L. 2003. La comisión científica del Pacífico: de la ciencia imperial a la ciencia federativa*. *Bulletin de l'Institut français d'études andines* 32: 479–515.

López-Ocón, L., and C. M. Pérez-Montes. 2000. *Marcos Jiménez de la Espada (1831–1898): tras la senda de un explorador*. Madrid: Consejo Superior de Investigaciones Científicas, Instituto de Historia.

López-Ocón, L., and A. Ponsati. 2002. *Comisión Científica del Pacífico*. CSIC. http://www.pacifico.csic.es.

López-Rojas, J. J., and D. F. Cisneros-Heredia. 2012. *Synapturanus rabus* Pyburn, 1977 in Peru (Amphibia: Anura: Microhylidae): Filling Gap. *Check List* 8: 274–275.

Lötters, S. 2011. "Balsa de Sapos" am Ende? *Elaphe* 1: 38–39.

Lovich, J. E., N. J. Scott, R. B. Bury, C. Kenneth, and R. W. McDiarmid. 2012. A history of herpetologists and herpetology in the U.S. Department of the Interior. *Herpetological Conservation and Biology* 7: 1–45.

Luer, H. G. 1966. La Comisión Científica del Pacífico (1862–1865). *Noticiario Mensual, Museo Nacional de Historia Natural de Chile* 10: 1–4.

Luna, M. C., R. W. McDiarmid, and J. Faivovich. 2018. From erotic excrescences to pheromone shots: structure and diversity of nuptial pads in anurans. *Biological Journal of the Linnean Society* 124: 403–446.

Lutz, B., and G. R. Kloss. 1952. Anfíbios anuros do alto Solimões e Rio Negro. Apontamento sôbre algumas formas e suas vicariantes. *Memórias do Instituto Oswaldo Cruz, Rio de Janeiro* 50: 625–678.

Lynch, J. D. 1968. Two new frogs of the genus *Eleutherodactylus* from eastern Ecuador (Amphibia: Leptodactylidae). *Journal of Herpetology* 2: 129–135.

Lynch, J. D. 1972. Two new species of frogs (*Eleutherodactylus:* Leptodactylidae) from the páramos of northern Ecuador. *Herpetologica*: 141–147.

Lynch, J. D. 1979. Leptodactylid frogs of the genus *Eleutherodactylus* from the Andes of southern Ecuador. *The University of Kansas, Museum of Natural History, Miscellaneous Publications* 66: 1–62.

Lynch, J. D. 1981. Leptodactylid frogs of the genus *Eleutherodactylus* in the Andes of northern Ecuador and adjacent Colombia. *Miscellaneous Publication, University of Kansas Museum of Natural History* 72: 1–46.

Lynch, J. D. 1984. New frogs (Leptodactylidae: *Eleutherodactylus*) from cloud forests of the northern Cordillera Oriental, Colombia. *Milwaukee Public Museum Contributions in Biology and Geology* 60: 1–22.

Lynch, J. D. 1986. New species of minute leptodactylid frogs from the Andes of Ecuador and Peru. *Journal of Herpetology* 20: 423–431.

Lynch, J. D., and W. E. Duellman. 1973. A review of the centrolenid frogs of Ecuador, with descriptions of new species. *Occasional Papers of the Museum of Natural History, University of Kansas* 16: 1–66.

Lynch, J. D., and W. E. Duellman. 1980. The *Eleutherodactylus* of the Amazonian slopes of the Ecuadorian Andes (Anura: Leptodactylidae). *Miscellaneous Publication University of Kansas Museum of Natural History* 69: 1–86.

Lynch, J. D., and W. E. Duellman. 1997. Frogs of the genus *Eleutherodactylus* in western Ecuador: Systematics ecology and biogeography. *Special Publication. Natural History Museum, University of Kansas* 23: 1–236.

Lynch, J. D., and K. Miyata. 1980. Two new species of *Eleutherodactylus* (Amphibia: Leptodactylidae) from the lowlands and lower cloud forests of western Ecuador. *Breviora* 457: 1–12.

MAAE (Ministerio del Ambiente y Agua). 2020 December. *Experiencias del monitoreo biológico de especies focales PARG y Centro de Conservación de anfibios AMARU. Proyecto Conservación de la Biodiversidad de anfibios ecuatorianos y Uso sostenible de sus recursos genéticos PARG*. Quito: Ministerio del Ambiente y Agua de Ecuador.

Maas, B., R. J. Pakeman, L. Godet, L. Smith, V. Devictor, and R. Primack. 2021. Women and global south strikingly underrepresented among top-publishing ecologists. *Conservation Letters* 14: e12797.

Martín-Torrijos, L., J. Vladimir Sandoval-Sierra, J. Muñoz, J. Diéguez-Uribeondo, J. Bosch, and J. M. Guayasamin. 2016. Rainbow trout (*Oncorhynchus mykiss*) threaten Andean amphibians. *Neotropical Biodiversity* 2: 26–36.

McCracken, S. F., and M. R. J. Forstner. 2006. Reproductive ecology and behavior of *Eleutherodactylus aureolineatus* (Anura, Brachycephalidae) in the canopy of the upper Amazon Basin, Ecuador. *Phyllomedusa: Journal of Herpetology* 5: 135.

McDiarmid, R. W. 1971. Comparative morphology and evolution of frogs of the Neotropical genera *Atelopus, Dendrophryniscus, Melanophryniscus,* and *Oreophrynella*. *Bulletin of the Los Angeles County Museum of Natural History*: 1–66.

McDiarmid, R. W. 1975. Glassfrog romance along a tropical stream. *Terra, The Natural History Museum of Los Angeles County* 13: 14–18.

McDiarmid, R. W. 1978. Evolution of parental care in frogs. In *The development of behavior: Comparative and evolutionary aspects*, eds. G. M. Burghardt and M. Bekkof, 127–147. New York: Garland STPM Press.

McDiarmid, R. W. 1983. *Centrolenella fleischmanni* (Ranita de Vidrio, Glass Frog). In *Costa Rican natural history*, ed. D. H. Hanzen, 389–390. Chicago: The University of Chicago Press.

McDiarmid, R. W., and K. Adler. 1974. Notes on territorial and vocal behavior of Neotropical frogs of the genus *Centrolenella. Herpetologica* 30: 75–78.

McDiarmid, R. W., and R. Altig. 1990. Description of a bufonid and two hylid tadpoles from western Ecuador. *Alytes* 8: 51–60.

McDiarmid, R. W., and R. Altig. 1999. *Tadpoles: The biology of anuran larvae*. Chicago: The University of Chicago Press.

McDiarmid, R. W., and J. M. Savage. 1984. Taxonomic status of the frog genus *Centrolenella* Noble (Anura: Centrolenidae). *Journal of Herpetology* 18: 213–215.

McGugan, J. R., G. D. Byrd, A. B. Roland, S. N. Caty, N. Kabir, E. E. Tapia, S. A. Trauger, L. A. Coloma, and L. A. O'Connell. 2016a. Response to Heethoff, Norton, and Raspotnig: Ant and mite diversity drives toxin variation in the little devil poison frog and erratum. *Journal of Chemical Ecology* 42: 845–848.

McGugan, J. R., G. D. Byrd, A. B. Roland, S. N. Caty, N. Kabir, E. E. Tapia, S. A. Trauger, L. A. Coloma, and L. A. O'Connell. 2016b. Ant and mite diversity drives toxin variation in the little devil poison frog. *Journal of Chemical Ecology* 42: 537–551.

MECN (Museo Ecuatoriano de Ciencias Naturales). 2013. *Herpetofauna en áreas prioritarias para la conservación: El sistema de Reservas Jocotoco y Ecominga*. Monografía 6. Quito: Museo Ecuatoriano de Ciencias Naturales, Fundación para la Conservación Jocotoco, Fundación Ecominga.

Mejía, S. 1993. Gustavo Orcés, científico ecuatoriano. Historia de un abre trochas. *Diario El Comercio* (July 18).

Mendelson III, J. R. 2022. William E. Duellman (1930–2022). *Ichthyology & Herpetology* 110: 623–629.

Mendelson III, J. R., K. R. Lips, R. W. Gagliardo, G. B. Rabb, J. P. Collins, J. E. Diffendorfer, *et al*. 2006. Confronting amphibian declines and extinctions. *Science* 313 (July): 48–48.

Mendes, B., C. Proaño-Bolaños, F. R. Gadelha, J. R. Almeida, and D. C. Miguel. 2020. Cruzioseptins, antibacterial peptides from *Cruziohyla calcarifer* skin, as promising leishmanicidal agents. *Pathogens and Disease* 78: 1–8.

Mendoza-Henao, A., R. Zamudio, J. M Guayasamin, M. Escalona, and G. Parra-Olea. 2023. Environment rather than character displacement explains call evolution in glassfrogs. *Evolution* 77: 355–369.

Merino-Viteri, A. 2001. Análisis de posibles causas de las disminuciones de poblaciones de anfibios en los Andes del Ecuador. Tesis de Licenciatura Pontificia Universidad Católica del Ecuador.

Merino-Viteri, A. 2015. Andrés Merino-Viteri, Balsa de los Sapos, Ecuador. *AARK Newsletter*, no. 30 (March). http://www.amphibianark.org/Newsletters/AArk-newsletter-30-Spanish.pdf.

Merino-Viteri, A., L. A. Coloma, and A. Almendáriz. 2005. Los *Telmatobius* de los Andes de Ecuador y su disminución poblacional. *Monografías de Herpetología* 7: 9–37.

Meza-Ramos, P., and M. Yánez-Muñoz. 2003. Nuevos registros de distribución y adición para la herpetofauna de Ecuador. Sociedad Ecuatoriana de Biología. In *Resúmenes de las XXVII Jornadas Ecuatorianas de Biología "Pedro Núñez Lucio". Sociedad Ecuatoriana de Biología*. Quito: Escuela de Biología de la Universidad Central del Ecuador.

Miller, R. R. 1983. *Por la ciencia y la gloria nacional. La expedición científica española a América (1862–1866)*. Barcelona: Ediciones del Serbal.

Miyata, K. 1980a. A new species of *Atelopus* (Anura: Bufonidae) from the cloud forests of northwestern Ecuador. *Breviora* 458: 1–10.

Miyata, K. 1980b. Notes on the occurrence of *Eleutherodactylus appendiculatus* in Ecuador. *Journal of Herpetology* 14: 85–86.

Miyata, K. 1982. A check list of the amphibians and reptiles of Ecuador with a bibliography of Ecuadorian herpetology. *Smithsonian Herpetological Information Service* 1982: 1–70.

Miyata, K. 1983. Notes on the *Phenacosaurus heterodermus* in the Sabana de Bogotá, Colombia. *Journal of Herpetology* 17: 102–105.

Miyata, K. 1985a. A new *Anolis* of the *lionotus* group from northwestern Ecuador and southwestern Colombia (Sauria: Iguanidae). *Breviora* 481: 1–13.

Miyata, K. 1985b. A new *Lepidoblepharis* from the Pacific slope of the Ecuadorian Andes (Sauria: Gekkonidae). *Herpetologica* 41: 121–127.

Miyata, K. I. 2013. Studies on the ecology and population biology of little known Ecuadorian Anoles. Online Supplement "Ken Miyata Some Remembrances". *Bulletin of the Museum of Comparative Zoology* 161: 45–78.

Molineros, D. 2016. Santiago Ron conoce a los anfibios mejor que a los humanos. *Diario El Telégrafo* (July 6). https://www.eltelegrafo.com.ec/noticias/septimo/1/santiago-ron-conoce-a-los-anfibios-mejor-que-a-los-humanos.

Montaño, D. 2020. Las ranas le ganan una batalla judicial a la minería en Ecuador. *Mongabay* (October 13). https://es.mongabay.com/2020/10/ranas-le-ganan-a-la-mineria-en-ecuador/.

Montgomery, H. L., and M. Lechuga. 2021. *What's in your pocket? collecting nature's treasures*. Watertown: Charlesbridge.

Moreno Espinosa, M. 1979. Creación del Museo Ecuatoriano de Ciencias Naturales—Antecedentes y datos históricos. *Publicaciones del Museo Ecuatoriano de Ciencias Naturales*: 5–9.

Morgan, A. C., S. F. Way, M. J. Hoefer, D. B. Larremore, M. Galesic, and A. Clauset. 2021. The unequal impact of parenthood in academia. *Science Advances* 7: eabd1996.

Moskowitz, N. A., A. Alvarez-Buylla, C. R. Morrison, A. Chamba, J. Rentería, E. E. Tapia, L. A. Coloma, D. A. Donoso, and L. A. O'Connell. 2022. Poison frog diet and chemical defense are influenced by availability and selectivity for ants. *bioRxi* 2022: 495949.

Moskowitz, N. A., B. Dorritie, T. Fay, O. C. Nieves, C. Vidoudez, 2017 Biology Class, C. R., *et al*. 2020. Land use impacts poison frog chemical defenses through changes in leaf litter ant communities. *Neotropical Biodiversity* 6: 75–87.

Mossman, K. 2007. Profile of Eugenia M. del Pino. *Proceedings of the National Academy of Sciences of the United States of America* 104: 17249–17251.

Mueses-Cisneros, J. J., D. Cisneros-Heredia, and R. W. McDiarmid. 2012. A new Amazonian species of *Rhaebo* (Anura: Bufonidae) with comments on *Rhaebo glaberrimus* (Günther, 1869) and *Rhaebo guttatus* (Schneider, 1799). *Zootaxa* 3447: 22–40.

Mueses-Cisneros, J. J., M. H. Yánez-Muñoz, and J. M. Guayasamin. 2010. Una nueva especie de sapo del género *Osornophryne* (Anura: Bufonidae) de las estribaciones Amazónicas de los Andes de Ecuador. *Papéis Avulsos de Zoologia* 50: 269–279.

Myers, C. W., and J. W. Daly. 1976. A new species of poison frog (*Dendrobates*) from Andean Ecuador, including an analysis of its skin toxins. *Occasional Papers of the Museum of Natural History, The University of Kansas, Lawrence, Kansas* 59: 1–12.

Natural History Museum 2023. Data portal query on 1 resource created at 2023-07-26 20:34:45.663709 PID. https://doi.org/10.5519/qd.isg77b4z.

Nicholson, K. L., Krausman, P. R., and Merkle, J. A. 2008. Hypatia and the Leopold standard: Women in the wildlife profession 1937–2006. *Wildlife Biology in Practice* 4: 57–72.

Nina, H. L. V., and del Pino, E. M. 1977. Estructura histológica del ovario del sapo *Eleutherodactylus unistrigatus* y observaciones sobre el desarrollo embrionario. *Revista de la Universidad Católica Ecuador* 5: 31–41.

O'Connell, L. 2017. Early career researchers: An interview with Lauren O'Connell. *Journal of Experimental Biology* 220: 2303–2305.

O'Connell, L. A. 2020. Frank Beach Award Winner: Lessons from poison frogs on ecological drivers of behavioral diversification. *Hormones and Behavior* 126: 104869.

Ortega, J. A., J. Brito, and S. R. Ron. 2022. Six new species of *Pristimantis* (Anura: Strabomantidae) from Llanganates National Park and Sangay National Park in Amazonian cloud forests of Ecuador. *PeerJ* 10: 1–66.

Ortega-Andrade, H. M. 2009. Amphibia, Anura, Eleutherodactylidae, *Adelophryne adiastola* Hoogmoed and Lescure, 1984: First countries records and distribution extension from Ecuador and Brazil. *Check List* 5: 139.

Ortega-Andrade, H. M., P. Meza-Ramos, D. F. Cisneros-Heredia, and M. Yánez-Muñoz. 2010. Los anfibios y reptiles del Chocó esmeraldeño. In *Serie herpetofauna del Ecuador: El Chocó Esmeraldeño*, eds. M. Altamirano-Benavides, D. F. Cisneros-Heredia, and P. Meza-Ramos, 95–252. Quito: Museo Ecuatoriano de Ciencias Naturales.

Ortega-Andrade, H. M., M. Rodes Blanco, D. F. Cisneros-Heredia, N. Guerra Arévalo, K. G. López de Vargas-Machuca, J. C. Sánchez-Nivicela, D. Armijos-Ojeda, *et al.* 2021. Red List assessment of amphibian species of Ecuador: A multidimensional approach for their conservation. *PloS One* 16: e0251027.

Ortega-Andrade, H. M., and J. H. Valencia. 2010. First country records of *Pristimantis luscombei* (Duellman and Mendelson) and *Syncope tridactyla* (Duellman and Mendelson) in eastern lowlands of Ecuador (Amphibia: Anura: Strabomantidae, Microhylidae). *Herpetology Notes* 3: 251–256.

Ortiz, G. 1987. El padre de la zoología del país. *Diario Hoy* (August 15).

Orton, J. 2022. *Los Andes y el Amazonas o notas de un viaje desde Guayaquil a Pará.* Quito: USFQ PRESS.

Páez, N. B., and S. R. Ron. 2019. Systematics of *Huicundomantis*, a new subgenus of *Pristimantis* (Anura, Strabomantidae) with extraordinary cryptic diversity and eleven new species. *ZooKeys* 868: 1–112.

Páez-Moscoso, D. J., J. M. Guayasamin, and M. Yánez-Muñoz. 2011. A new species of Andean toad (Bufonidae, *Osornophryne*) discovered using molecular and morphological data, with a taxonomic key for the genus. *ZooKeys* 108: 73–97.

Páez-Vacas, M., L. A. Coloma, and J. C. Santos. 2010. Systematics of the *Hyloxalus bocagei* complex (Anura: Dendrobatidae), description of two new cryptic species, and recognition of *H. maculosus*. Zootaxa 2711: 1–75.

Parker, H. W. 1932. Some new or rare reptiles and amphibians from southern Ecuador. *The Annals and Magazine of Natural History* 9: 21–26.

Parker, H. W. 1934. Reptiles and amphibians from Southern Ecuador. *Annals and Magazine of Natural History* 14: 264–273.

Parker, H. W. 1938. The vertical distribution of some reptiles and amphibians in southern Ecuador. *Annals and Magazine of Natural History Series* 11: 438–450.

Pašukonis, A., S. J. Serrano-Rojas, M. T. Fischer, M. C. Loretto, D. A. Shaykevich, B. Rojas, M. Ringler, *et al.* 2022. Contrasting parental roles shape sex differences in poison frog space use but not navigational performance. *eLife* 11: 1–31.

Patzelt, E. 1978. *Fauna del Ecuador*. Quito: Imprenta Europa.

Paz Cardona, A. J. 2019. En Ecuador todo es para estudiar, somos tierra virgen. *Mongabay* (March 28). https://es.mongabay.com/2019/03/insectos-poco-estudio-aun-en-ecuador/.

Pearman, P. B., A. M. Velasco, and A. López. 1995. Tropical amphibian monitoring: A comparison of methods for detecting inter-site variation in species composition. *Herpetologica* 51: 325–337.

Peñaherrera-Romero, E., and D. F. Cisneros-Heredia. 2022. First report of visual displays in the Neotropical treefrog *Dendropsophus carnifex*. *Acta Ethologica* 26: 75–78

Peñaherrera-Romero, E., S. Espinoza, D. de la Torre, D. Espinoza, and D. F. Cisneros-Heredia. 2022. Integrando la ciencia ciudadana y la educación para fomentar los vínculos entre las personas y la naturaleza en áreas urbanas. *Esferas* 3: 112–133.

Peracca, M. G. 1904. Viaggio del Dr. Enrico Festa nell' Ecuador e regioni vicine. Rettili ed anfibi. *Bollettino dei Musei di Zoologia e Anatomia Comparata della R. Universita di Torino* 19: 1–41.

Pereyra, M. O., M. C. Womack, J. S. Barrionuevo, B. L. Blotto, D. Baldo, M. Targino, J. J. Ospina-Sarria, J. M. Guayasamin, L. A. Coloma, K. L. Hoke, T. Grant, and J. Faivovich. 2016. The complex evolutionary history of the tympanic middle ear in frogs and toads (Anura). *Scientific Reports* 6: 1–9.

Pérez, A. 2010. Saprolegnia de anfibios: una enfermedad emergente en puestas de *Atelopus nanay* en el Parque Nacional Cajas. Tesis de licenciatura, Universidad Internacional Menéndez Pelayo.

Perry, G., E. Muths, M. Crump, C. Distel, R. Hansen, M. Harvey, J. Rowe, S. Walls, and S. Woodley. 2020. Increasing diversity and inclusivity in herpetology: Small but critical steps from the editors of herpetological journals. *Herpetological Review* 51: 765–766.

Peters, J. A. 1955. Herpetological type localities in Ecuador. *Revista Ecuatoriana de Entomología y Parasitología* 2: 335–352.

Peters, J. A. 1967. The generic allocation of the frog *Ceratophrys stolzmanni* Steindachner, with the description of a new subspecies from Ecuador. *Proceedings of the Biological Society of Washington* 80: 105–112.

Peters, J. A. 1973. The frog genus *Atelopus* in Ecuador (Anura: Bufonidae). *Smithsonian Contributions to Zoology* 145: 1–49.

Peters, W. C. H. 1862a. Eine neue gattung von laubfröschen, *Plectromantis*, aus Ecuador. *Monatsberichte der Königlich Akademie der Wissenschaften zu Berlin* 1862: 232–233.

Peters, W. C. H. 1862b. Über die Batrachier-Gattung *Hemiphractus. Monatsberichte der königlich Akademie der Wissenschaften zu Berlin* 1862: 144–152.

Peters, W. C. H. 1873. Über eine neue Schildkrötenart, *Cinosternon effeldtii* und einige andere neue oder weniger bekannte Amphibien. *Monatsberichte der königlich Akademie der Wissenschaften zu Berlin* 1873: 603–618.

Plitt, L. 2017. David Jailaca, el niño ecuatoriano que salvó a una especie y se financió sus estudios descubriendo un sapo que se pensaba extinto. *BBC Mundo Ciencia* (July 14). https://www.bbc.com/mundo/noticias-40607553.

Poe, S., J. Velasco, K. Miyata, and E. E. Williams. 2009. Descriptions of two nomen nudem species of *Anolis* lizards from northwestern South America. *Breviora* 516: 1–16.

Poelman, E. H., J. C. Verkade, R. P. A. van Wijngaarden, and C. Félix-Novoa. 2010. Descriptions of the tadpoles of two poison frogs, *Ameerega parvula* and *Ameerega bilinguis* (Anura: Dendrobatidae) from Ecuador. *Journal of Herpetology* 44: 409–417.

Polato, N. R., B. A. Gill, A. A. Shah, M. M. Gray, K. L. Casner, A. Barthelet, P. W. Messer, *et al.* 2018. Narrow thermal tolerance and low dispersal drive higher speciation in tropical mountains. *Proceedings of the National Academy of Sciences* 115: 12471–12476.

Pollet, D. 2003. *A la sombra de Otonga*. Serie de divulgación. Quito: Centro de Biodiversidad y Ambiente, Pontificia Universidad Católica del Ecuador.

Pomerantz, A., N. Peñafiel, A. Arteaga, L. Bustamante, F. Pichardo, L. A. Coloma, C. L. Barrio-Amorós, D. Salazar-Valenzuela, and S. Prost. 2018. Real-time DNA barcoding in a rainforest using nanopore sequencing: Opportunities for rapid biodiversity assessments and local capacity building. *GigaScience* 7: 1–14.

Ponce, I. 2019. La Balsa de los Sapòs: el proyecto científico que protege más de 1500 anfibios en Ecuador. *Mongabay* (June 19). https://es.mongabay.com/2019/06/ecuador-anfibios-la-balsa-proyecto/.

Pounds, J. A., M. R. Bustamante, L. A. Coloma, J. A. Consuegra, M. P. L. Fogden, P. N. Foster, *et al.* 2006. Widespread amphibian extinctions from epidemic disease driven by global warming. *Nature* 439: 161–167.

Pounds, J. A., M. R. Bustamante, L. A. Coloma, J. A. Consuegra, M. P. L. Fogden, P. N. Foster, *et al.* 2007. Global warming and amphibian losses; The proximate cause of frog declines? (Reply). *Nature* 447: E5–6.

Pounds, J. A., and L. A. Coloma. 2008. Beware the lone killer: Why are harlequin frogs disappearing across the American tropics? A resifting of the evidence backs up the conclusion that global warming is a key conspirator in the losses. *Nature Reports Climate Change* 57–59.

Pounds, J. A., and M. L. Crump. 1994. Amphibian declines and climate disturbance—the case of the golden toad and the harlequin frog. *Conservation Biology* 8: 72–85.

Proaño-Bolaños, C., R. Li, M. Zhou, L. Wang, X. Xi, E. E. Tapia, L. A. Coloma, T. Chen, and C. Shaw. 2017. Novel Kazal-type proteinase inhibitors from the skin secretion of the Splendid leaf frog, *Cruziohyla calcarifer*. *EuPA Open Proteomics* 15: 1–13.

Proaño, B., and O. D. Pérez. 2017. In vitro fertilizations with cryopreserved sperm of *Rhinella marina* (Anura: Bufonidae) in Ecuador. *Amphibian and Reptile Conservation* 11: 1–6.

Proaño-Bolaños, C., M. Zhou, L. Wang, L. A. Coloma, T. Chen, and C. Shaw. 2016. Peptidomic approach identifies cruzioseptins, a new family of potent antimicrobial peptides in the splendid leaf frog, *Cruziohyla calcarifer. Journal of Proteomics* 146: 1–13.

Quiguango-Ubillús, A., and L. A. Coloma. 2008. Notes on behavior, communication and reproduction in captive *Hyloxalus toachi* (Anura: Dendrobatidae), an endangered Ecuadorian frog. *International Zoo Yearbook* 42: 78–89.

R Core Team. 2020. *R: A language and environment for statistical computing*. R Foundation for Statistical Computing. R Foundation for Statistical Computing, Vienna, Austria. http://www.r-project.org/index.html

Rabeling, C., J. Sosa-Calvo, L. A. O'Connell., L. A. Coloma, and F. Fernández. 2016. *Lenomyrmex hoelldobleri:* A new ant species discovered in the stomach of the dendrobatid poison frog, *Oophaga sylvatica* (Funkhouser). *ZooKeys* 2016: 79–95.

Ramírez-Jaramillo, S., C. Reyes-Puig, D. Batallas, and M. H. Yánez-Muñoz. 2018. Ranas terrestres en los ecosistemas surandinos de Ecuador IV: Una nueva especie de *Pristimantis* (Anura: Strabomantidae) de los páramos sur del Parque Nacional Sangay. *Avances en Ciencias e Ingenierías* 10: 1–17.

Read, M. 1986. *Reptiles and amphibians*. In *Río Mazán Project 1986 Report*, ed. F. Robinson. Oxford: Department of Plant Sciences, Oxford University.

Read, M. 2000. *Frogs of the Ecuadorian Amazon: A guide to their calls*. Compact Disk. Cornwall: Morley Read Productions.

Read, M. 2018. *Monitoreo biológico Yasuní. Anfibios y reptiles*. Quito: Ecuambiente Consulting Group.

Renjifo, A. 1999. Fauna del Ecuador y catálogo de la fauna ecuatoriana: un proyecto de largo alcance. *Revista Nuestra Ciencia*: 20–24.

Reyes-Puig, C., G. B. Bittencourt-Silva, M. Torres-Sánchez, M. Wilkinson, J. W. Streicher, S. T. Maddock, R. Kotharambath, *et al.* 2019a. Rediscovery of the endangered Carchi Andean toad, *Rhaebo colomai* (Hoogmoed, 1985), in Ecuador, with comments on its conservation status and extinction risk. *Check List* 15: 415–419.

Reyes-Puig, C., J. M. Guayasamin, C. Kohn, D. Brito-Zapata, M. Holland, M. Costales, and D. Cisneros-Heredia. 2021. A new species of the genus *Noblella* (Amphibia: Strabomantidae) from Ecuador, with new information for *Noblella worleyae*. *Acta Herpetologica* 16: 63–87.

Reyes-Puig, C., and E. Mancero. 2022. Beyond the species name: An analysis of publication trends and biases in taxonomic descriptions of rainfrogs (Amphibia, Strabomantidae, *Pristimantis*). *ZooKeys* 1134: 73–100.

Reyes-Puig, C., R. J. Maynard, S. J. Trageser, J. Vieira, P. S. Hamilton, R. Lynch, J. Culebras, S. Kohn, J. Brito, and J. M. Guayasamin. 2020a. A new species of *Noblella* (Amphibia: Strabomantidae) from the Río Manduriacu Reserve on the Pacific slopes of the Ecuadorian Andes. *Neotropical Biodiversity* 6: 162–171.

Reyes-Puig, C., J. P. Reyes-Puig, D. A. Velarde-Garcéz, N. Dávalos, E. Mancero, M. J. Navarrete, M. H. Yánez-Muñoz, D. F. Cisneros-Heredia, and S. R. Ron. 2019b. A new species of terrestrial frog *Pristimantis* (Strabomantidae) from the upper basin of the Pastaza River, Ecuador. *ZooKeys* 832: 113–133.

Reyes-Puig, C., D. B. Wake, R. Kotharambath, J. W. Streicher, C. Koch, D. F. Cisneros-Heredia, M. H. Yánez-Muñoz, and S. Ron. 2020b. Two extremely rare new species of fossorial salamanders of the genus *Oedipina* (Plethodontidae) from northwestern Ecuador. *PeerJ* 8: e9934.

Reyes-Puig, C., M. Yánez-Muñoz, J. A. Ortega, and S. R. Ron. 2020c. Relaciones filogenéticas del subgénero *Hypodictyon* (Anura: Strabomantidae: *Pristimantis*) con la descripción de tres especies nuevas de la región del Chocó. *Revista Mexicana de Biodiversidad* 91: 18.

Reyes-Puig, J. P., D. Recalde, F. Recalde, C. Koch, J. M. Guayasamin, D. F. Cisneros-Heredia, L. Jost, and M. H. Yánez-Muñoz. 2022b. A spectacular new species of *Hyloscirtus* (Anura: Hylidae) from the Cordillera de Los Llanganates in the eastern Andes of Ecuador. *PeerJ* 10: 1–34.

Reyes-Puig, J. P., C. Reyes-Puig, D. Franco-Mena, L. Jost, and M. H. Yánez-Muñoz. 2022a. Strong differentiation between amphibian communities on two adjacent mountains in the Upper Rio Pastaza watershed of Ecuador, with descriptions of two new species of terrestrial frogs. *ZooKeys* 1081: 35–87.

Reyes-Puig, J. P., C. Reyes-Puig, M. Pérez, and M. Yánez-Muñoz. 2015. Dos nuevas especies de ranas *Pristimantis* (Craugastoridae) de la cordillera de los Sacha Llanganatis, vertiente oriental de los Andes de Ecuador. *Avances en Ciencias e Ingenierías* 7: 61–74.

Reyes-Puig, J. P., C. Reyes-Puig, J. S. Ramírez, M. Pérez-Lara, and M. H. Yánez-Muñoz. 2014. Tres nuevas especies de ranas terrestres *Pristimantis* (Anura: Craugastoridae) de la cuenca alta del Río Pastaza, Ecuador. *Avances en Ciencias e Ingenierías* 6: 51–62.

Reyes-Puig, J. P., C. Reyes-Puig, S. Ron, J. A. Ortega, J. M. Guayasamin, M. Goodrum, F. Recalde, J. Vieira, C. Koch, M. H. Yánez-Muñoz. 2019c. A new species of terrestrial frog of the genus *Noblella* Barbour, 1930 (Amphibia: Strabomantidae) from the Llanganates-Sangay Ecological Corridor, Tungurahua, Ecuador. *PeerJ* 7: e7405.

Reyes-Puig, J. P., M. Urgilés-Merchán, D. Franco-Mena, J. M. Guayasamin, D. Batallas-Revelo, and C. Reyes-Puig. 2023. Two new species of terrestrial frogs of the *Pristimantis gladiator* complex (Anura, Strabomantidae) from the Ecuadorian Andes, with insights on their biogeography and skull morphology. *ZooKeys* 1180: 257–293.

Reyes-Puig, J. P., and M. H. Yánez-Muñoz. 2012. Una nueva especie de *Pristimantis* (Anura: Craugastoridae) del corredor ecológico Llanganates-Sangay, Andes de Ecuador. *Papéis Avulsos de Zoologia* 52: 81–91.

Reyes-Puig, J. P., M. H. Yánez-Muñoz, D. F. Cisneros-Heredia, and S. Ramírez. 2011. "2010". Una nueva especie de rana *Pristimantis* (Terrarana: Strabomantidae) de los bosques nublados de la cuenca alta del río Pastaza, Ecuador. *Avances en Ciencias e Ingenierías* 2: 78–82.

Rhodin, A. G. J., and K. Miyata. 1983. *Advances in Herpetology and Evolutionary Biology: Essays in Honor of Ernest E. Williams*. Cambridge: Museum of Vertebrate Zoology.

Rivadeneira, C. D., P. J. Venegas, and S. R. Ron. 2018. Species limits within the widespread Amazonian treefrog *Dendropsophus parviceps* with descriptions of two new species (Anura, Hylidae). *ZooKeys* 726: 25–77.

Rivero, J. A. 1986. Descripcion de una nueva especie de *Eleutherodactylus* (Amphibia: Leptodactylidae) del subgrupo *myersi*, con anotaciones sobre las relaciones y el origen de ese grupo de especies. *Caribbean Journal of Science* 22: 107–114.

Rivero, J. A. 1991. New Ecuadorian *Colostethus* (Amphibia, Dendrobatidae) in the collection of the National Museum of Natural History, Smithsonian Institution. *Caribbean Journal of Science* 27: 1–22.

Rivero, J. A., and A. Almendáriz. 1991. La identificación de los *Colostethus* (Amphibia: Dendrobatidae) de Ecuador. *Politécnica* XVI: 99–152.

Rock, K. N., I. N. Barnes, M. S. Deyski, K. A. Glynn, B. N. Milstead, M. E. Rottenborn, N. S. Andre, A. Dekhtyar, O. Dekhtyar, and E. N. Taylor. 2021. Quantifying the gender gap in authorship in herpetology. *Herpetologica* 77: 1–13.

Rödder, D., A. Schmitz, G. Onore, and W. Böhme. 2011. New records of the endangered *Pristimantis eugeniae* (Lynch and Duellman, 1997) (Anura: Strabomantidae) from Ecuador. *Herpetology Notes* 4: 247–248.

Rodríguez, A. 2017. Giovanni Onore, el padre adoptivo de los insectos. *Diario El Telégrafo* (January 15). https://www.eltelegrafo.com.ec/noticias/septimo/1/giovanni-onore-el-padre-adoptivo-de-los-insectos.

Roland, A. B., C. Santos, B. C. Carriker, S. N. Caty, E. E. Tapia, L. A. Coloma, and L. A. O'Connell. 2017. Radiation of the polymorphic little devil poison frog (*Oophaga sylvatica*) in Ecuador. *Ecology and Evolution* 7: 9750–9762.

Romero-Carvajal, L. Negrete, M. J. Salazar-Nicholls, K. Vizuete, A. Debut, P. H. Dias, and F. Vera Candioti. 2023. Direct development or endotrophic tadpole? Morphological aspects of the early ontogeny of the plump toad *Osornophryne occidentalis* (Anura: Bufonidae). *Journal of Morphology* 284: e21582.

Romo, D. 2001. A computerized sound analysis method for monitoring anuran diversity in the Amazon Basin of Ecuador. PhD thesis, The University of Minnesota.

Romo, D., D. Mosquera, K. Swing, A. di Fiore, J. G. Blake, B. A. Loiselle, T. B. Ryder, *et al.* 2017. *Los secretos del Yasuní: avances en investigación en la Estación de Biodiversidad Tiputini, Universidad San Francisco de Quito USFQ*. Quito: USFQ PRESS.

Ron, S., and J. B. Pramuk. 1999. A new species of *Osteocephalus* (Anura: Hylidae) from Amazonian Ecuador and Peru. *Herpetologica* 55: 3–446.

Ron, S. R. 2000. Area relationships of Neotropical lowland rainforests based on cladistic analysis of vertebrate groups. *The Biological Journal of the Linnean Society* 71: 379–402.

Ron, S. R. 2005. Predicting the distribution of the amphibian pathogen *Batrachochytrium dendrobatidis* in the New World. *Biotropica* 37: 209–221.

Ron, S. R., A. Almendariz, and D. C. Cannatella. 2013a. The *Phyllomedusa perinesos* group (Anura: Hylidae) is derived from a Miocene Amazonian lineage. *Zootaxa* 3741: 289–294.

Ron, S. R., M. R. Bustamante, L. A. Coloma, and B. Mena. 2009. *Sapos. Ecuador Sapodiverso*. Serie de Divulgación No. 10. Quito: Centro de Biodiversidad y Ambiente, Escuela de Biología, PUCE.

Ron, S. R., M. A. Caminer, A. Varela-Jaramillo, and D. Almeida-Reinoso. 2018b. A new treefrog from Cordillera del Cóndor with comments on the biogeographic affinity between Cordillera del Cóndor and the Guianan Tepuis. *ZooKeys* 124: 97–124.

Ron, S. R., D. C. Cannatella, and L. A. Coloma. 2004. Two new species of *Physalaemus* (Anura: Leptodactylidae) from western Ecuador. *Herpetologica* 60: 261–275.

Ron, S. R., J. Carrión, M. A. Caminer, Y. Sagredo, M. J. Navarrete, J. A. Ortega, A. Varela-Jaramillo, G. A. Maldonado-Castro, and C. Terán. 2020. Three new species of frogs of the genus *Pristimantis* (Anura, Strabomantidae) with a redefinition of the *P. lacrimosus* species group. *ZooKeys* 2020: 121–155.

Ron, S. R., L. A. Coloma, and D. C. Cannatella. 2005. A new, cryptic species of *Physalaemus* (Anura: Leptodactylidae) from western Ecuador with comments on the call structure of the *P. pustulosus* species group. *Herpetologica* 61: 178–198.

Ron, S. R., L. A. Coloma, J. M. Guayasamin, and M. H. Yánez-Muñoz. 2011. *AmphibiaWebEcuador*. Versión 2011. Museo de Zoología, Pontificia Universidad Católica del Ecuador. http://zoologia.puce. edu.ec/Vertebrados/anfibios/AnfibiosEcuador.

Ron, S. R., W. E. Duellman, M. A. Caminer, and D. Pazmiño. 2018a. Advertisement calls and DNA sequences reveal a new species of *Scinax* (Anura: Hylidae) on the Pacific lowlands of Ecuador. *PloS One* 13: 1–26.

Ron, S. R., W. E. Duellman, L. A. Coloma, and M. R. Bustamante. 2003. Population decline of the Jambato toad *Atelopus ignescens* (Anura: Bufonidae) in the Andes of Ecuador. *Journal of Herpetology* 37: 116–126.

Ron, S. R., J. M. Guayasamin, and M. H. Yánez-Muñoz. 2013b. *AmphibiaWebEcuador*. Version 2013.0. Museo de Zoología, Pontificia Universidad Católica del Ecuador. http://zoologia.puce.edu.ec/Vertebrados/anfibios/AnfibiosEcuador.

Ron, S. R., J. M. Guayasamin, M. H. Yánez-Muñoz, and A. Merino-Viteri. 2013c. *AmphibiaWebEcuador*. Version 2013.1. Museo de Zoología, Pontificia Universidad Católica del Ecuador.

Ron, S. R., J. M. Guayasamin, M. Yánez-Muñoz, A. Merino-Viteri, and D. A. Ortiz. 2014a. *AmphibiaWebEcuador*. Version 2014.0. Museo de Zoología, Pontificia Universidad Católica del Ecuador. http://zoologia.puce.edu.ec/Vertebrados/anfibios/.

Ron, S. R., J. M. Guayasamin, M. H. Yánez-Muñoz, A. Merino-Viteri, D. A. Ortiz, and D. A. Nicolalde. 2014b. *AmphibiaWebEcuador*. Version 2014.0. Museo de Zoología, Pontificia Universidad Católica del Ecuador. Quito, Ecuador. http://zoologia.puce.edu.ec/Vertebrados/anfibios/.

Ron, S. R., and A. Merino-Viteri. 2000. Amphibian declines in Ecuador: Overview and first report of chytridiomycosis from South America. *Froglog* 42: 2–3.

Ron, S. R., A. Merino-Viteri, and D. A. Ortiz. 2019. *AmphibiaWeb-Ecuador*. Version 2019.0. Museo de Zoología, Pontificia Universidad Católica del Ecuador. Quito, Ecuador. http://zoologia.puce.edu.ec/Vertebrados/anfibios.

Ron, S. R., A. Merino-Viteri, and D. A. Ortiz. 2022. *Anfibios del Ecuador*. Version 2022.0. Museo de Zoología, Pontificia Universidad Católica del Ecuador. Quito, Ecuador. https://bioweb.bio/faunaweb/amphibiaweb.

Ron, S. R., J. J. Mueses-Cisneros, P. D. A. Gutiérrez-Cárdenas, A. Rojas-Rivera, R. L. Lynch, C. F. Duarte Rocha, and G. Galarza. 2015. Systematics of the endangered toad genus *Andinophryne* (Anura: Bufonidae): Phylogenetic position and synonymy under the genus *Rhaebo*. *Zootaxa* 3947: 347–366.

Ron, S. R., A. Narváez, and G. E. Romero. 2014c. Reproduction and spawning behavior in the frog, *Engystomops pustulatus* (Shreve 1941). *Amphibian & Reptile Conservation* 8: 25–32.

Ron, S. R., J. C. Santos, and D. C. Cannatella. 2006. Phylogeny of the tungara frog genus *Engystomops* (= *Physalaemus pustulosus* species group; Anura: Leptodactylidae). *Molecular Phylogenetics and Evolution* 39: 392–403.

Ron, S. R., P. J. Venegas, H. M. Ortega-Andrade, G. Gagliardi-Urrutia, and P. E. Salerno. 2016. Systematics of *Ecnomiohyla tuberculosa* with the description of a new species and comments on the taxonomy of *Trachycephalus typhonius* (Anura, Hylidae). *ZooKeys* 2016: 115–154.

Ron, S. R., P. J. Venegas, E. Toral, M. Read, D. A. Ortiz, and A. L. Manzano. 2012. Systematics of the *Osteocephalus buckleyi* species complex (Anura, Hylidae) from Ecuador and Peru. *ZooKeys* 229: 1–52.

Ross, M. B., B. M. Glennon, R. Murciano-Goroff, E. G. Berkes, B. A. Weinberg, and J. I. Lane. 2022. Women are credited less in science than men. *Nature* 608: 135–145.

de Sá, R. O., T. Grant, A. Camargo, W. R. Heyer, M. L. Ponssa, and E. Stanley. 2014. Systematics of the Neotropical genus *Leptodactylus* Fitzinger, 1826 (Anura: Leptodactylidae): Phylogeny, the relevance of non-molecular evidence, and species accounts. *South American Journal of Herpetology* 9: 1–128.

Salazar, A., and J. D. Pérez. 2004. *Animales que cantan y encantan*. Quito: Centro de Biodiversidad y Ambiente, Escuela de Biología, PUCE.

Salazar, J., F. Siavichay, and G. Maldonado. 2022. *Haciendo visible lo invisible: Fauna urbana de la Universidad Católica de Cuenca*. Cuenca: Editorial Don Bosco.

Salazar-Nicholls, M. J., F. Hervas, S. I. Muñoz-Tobar, A. B. Carrillo, H. Ricaurte, S. R. Ron, and A. Romero-Carvajal. 2021. A polymorphism in oocyte pigmentation in natural populations of the glass frog *Espadarana prosoblepon* (Centrolenidae). *International Journal of Developmental Biology* 65: 333–344.

Salerno, P. E., M. Páez-Vacas, J. M. Guayasamin, and J. L. Stynoski. 2019. Male principal investigators (almost) don't publish with women in ecology and zoology. *PloS One* 14: e0218598.

Salgado, A. L., and J. M. Guayasamin. 2018. Parental care and reproductive behavior of the minute dappled glassfrog (*Centrolenidae: Centrolene peristictum*). *South American Journal of Herpetology* 13: 211–219.

Sánchez-Carvajal, M. J., G. C. Reyes-Ortega, D. F. Cisneros-Heredia, and H. M. Ortega-Andrade. 2021. Rediscovery of Laura's glassfrog *Nymphargus laurae* (Anura: Centrolenidae) with new data on its morphology, coloration, phylogenetic position and conservation in Ecuador. *PeerJ* 9: e12644.

Sánchez-Nivicela, J. C. 2022. *Anfibios y Reptiles del Corredor de Conectividad Sangay Podocarpus*. Cuenca: Naturaleza y Cultura Internacional.

Sánchez-Nivicela, J. C., E. Celi-Piedra, V. Posse-Sarmiento, V. L. Urgilés, M. H. Yánez-Muñoz, D. F. Cisneros-Heredia. 2018. A new species of *Pristimantis* (Anura, Craugastoridae) from the Cajas Massif, southern Ecuador. *ZooKeys* 751: 113–128.

Sánchez-Nivicela, J. C., J. M. Falcón-Reibán, and D. F. Cisneros-Heredia. 2023. A new stream treefrog of the genus *Hyloscirtus* (Amphibia, Hylidae) from the Río Negro-Sopladora National Park, Ecuador. *ZooKeys* 1141: 75–92.

Sánchez-Nivicela, J. C., P. L. V. Peloso, V. L. Urgiles, M. H. Yánez-Muñoz, Y. Sagredo, N. Páez, and S. Ron. 2020. Description and phylogenetic relationships of a new trans-Andean species of *Elachistocleis* Parker 1927 (Amphibia, Anura, Microhylidae). *Zootaxa* 4779: 323–340.

Sánchez-Nivicela, J. C., E. Toral-Contreras, and V. L. Urgiles. 2022. Una nueva especie de *Pristimantis* (Anura: Strabomantidae) de la provincia del Azuay, Andes sur de Ecuador. *Neotropical Biodiversity* 8: 305–322.

Sánchez-Nivicela, J. C., V. L. Urgiles, J. Cedeño-Palacios, H. Abad-Peñafiel, and J. M. Guayasamin. 2021. Una fantástica nueva especie del grupo *Pristimantis orcesi* de los Andes sur de Ecuador. *Neotropical Biodiversity* 6: 224–237.

Sánchez-Nivicela, J. C., V. L. Urgilés, M. J. Navarrete, M. H. Yánez-Muñoz, and S. Ron. 2019. A bizarre new species of *Lynchius* (Amphibia, Anura, Strabomantidae) from the Andes of Ecuador and first report of *Lynchius parkeri* in Ecuador. *Zootaxa* 4567: 1–24.

Santos, J. C., and D. C. Cannatella. 2011. Phenotypic integration emerges from aposematism and scale in poison frogs. *Proceedings of the National Academy of Sciences* 108: 6175–6180.

Santos, J. C., L. A. Coloma, and D. C. Cannatella. 2003. Multiple, recurring origins of aposematism and diet specialization in poison frogs. *Proceedings of the National Academy of Sciences* 100: 12792–12797.

Santos, J. C., L. A. Coloma, K. Summers, J. P. Caldwell, R. Ree, and D. C. Cannatella. 2009. Amazonian amphibian diversity is primarily derived from late Miocene Andean lineages. *PLoS Biology* 7: 0448–0461.

Santos, J. C., R. D. Tarvin, L. A. O'Connell, D. C. Blackburn, and L. A. Coloma. 2018. Diversity within diversity: Parasite species richness in poison frogs assessed by transcriptomics. *Molecular Phylogenetics and Evolution* 125: 40–50.

Savage, J. M. 1978. Marcos Jiménez de la Espada, naturalist explorer of the Andes and Upper Amazon basin. In *Vertebrados del Viaje al Pacífico*, ed. K. Adler. Athens: SSAR.

Schätti, B., and E. Kramer. 1993. Ecuadorian pitvipers of the genera *Bothriechis*, *Bothrops* and *Porthidium*. *Revue Suisse de Zoologie* 100: 235–278.

Schätti, B., E. Kramer, and J.-M. Touzet. 1990. Systematic remarks on a rare crotalid snake from Ecuador, *Bothriechis albocarinata* (Shreve), with some comments on the generic arrangement of arboreal Neotropical pitvipers. *Revue Suisse de Zoologie* 97: 877–885.

Scheele, B. C., F. Pasmans, L. F. Skerratt, L. Berger, A. Martel, W. Beukema, A. A. Acevedo, P. A. Burrowes, T. Carvalho, A. Catenazzi, I. De Riva, M. C. Fisher, and S. V. Flechas. 2019. Amphibian fungal panzootic causes catastrophic and ongoing loss of biodiversity. *Science* 1463: 1459–1463.

Schultz, N., H. Forsey, B. McCormick, J. Mather-Hillon, and L. Frolich. 2000. *Report of the Oxford University Expedition: AndinoHerps 2000, a herpetological survey of the High Andes of northern Ecuador, July to September 2000.* Oxford: University of Oxford.

Schwalm, P. A., and P. H. Starrett. 1977. Infrared reflectance in leaf-sitting Neotropical frogs. *Science* 196: 1225–1227.

Seba, A. 1734. *Locupletissimi rerum naturalium thesauri accurata descriptio, et iconibus artificiosissimus expressio, per Universam Physices Historium. Opus, cui in hoc Rerum Genere, Nullum par Exstitit. Volume 1*. Amstelaedami: Apud J. Wetstenium, & Gul. Smith, & Janssonio-Waesbergios.

Serra, C. 2022. Building a career through Ecuadorian frogs. The Society for Developmental Biology awards TWAS Fellow Eugenia del Pino Veintimilla for her lifetime contribution to developmental biology. *TWAS UNESCO* (February 8). https://twas.org/article/building-career-through-ecuadorian-frogs.

Shen, H. 2013. Inequality quantified: Mind the gender gap. *Nature News* 495: 22.

Siavichay, F. 2008. Diagnóstico de la comunidad de anfibios para el manejo y gestión del Área de Conservación Municipal y Uso Sustentable Cordillera Oriental. Del cantón Sígsig, Azuay. Tesis de Maestría, Universidad del Azuay.

Siavichay, F., D. Alvarado, E. Arbeláez Ortiz, and C. C. Martínez. 2011. Amplexus, desove y eclosión de renacuajos del jambato negro del Cajas en el Centro de Conservación de Anfibios del Bosque de Mazán en Ecuador. (versión español, inglés). *AARK Newsletter*, no. 15 (June). http://www.amphibianark.org/Newsletters/boletin_informativo_de_amphibian_ark_15.pdf.

Siavichay, F., G. Maldonado, D. Mejía, J. F. Webster, N. Torres, and K. Costa. 2016. *Plan de manejo para la conservación de los anfibios urbanos de Cuenca*. Cuenca: I. Municipalidad de Cuenca.

Siavichay-Pesántez, F., G. Maldonado-Cedeño, and D. Mejía-Coronel. 2016. *Anfibios urbanos de Cuenca*. Cuenca: Editorial Don Bosco, Centro Gráfico Salesiano.

Sierra Praeli, Y. 2021. "La situación de conservación y el riesgo de extinción de algunas especies ha empeorado": Diego Cisneros-Heredia. *Mongabay* (June 10). https://Es.Mongabay.Com/2021/06/Conservacion-y-Riesgo-de-Extincion-de-Especies-Ha-Empeorado-Entrevista/.

Simmons, J. E. 2016. *Museums. A history*. London: Rowman & Littlefield.

Simmons, J. E. 2022. William E. Duellman-a Remembrance. *Revista Latinoamericana de Herpetología* 5: 132–144.

Stuart, S. N., J. S. Chanson, N. A. Cox, B. E. Young, A. S. L. Rodrigues, D. L. Fischman, and R. W. Waller. 2004. Status and trends of amphibian declines and extinctions worldwide. *Science* 306: 1783–1786.

Stynoski, J. L., M. C. Womack, F. A. Trama, L. A. Coloma, and K. L. Hoke. 2020. Whispers from vestigial nubbins: Arrested development provokes trait loss in toads. *Evolution and Development* 23: 5–18.

Székely, D., D. Cogălniceanu, P. Székely, D. Armijos-Ojeda, V. Espinosa-Mogrovejo, and M. Denoël. 2020a. How to recover from a bad start: Size at metamorphosis affects growth and survival in a tropical amphibian. *BMC Ecology* 20: 1–8.

Székely, D., D. Cogălniceanu, P. Székely, and M. Denoël. 2018a. Dryness affects burrowing depth in a semi-fossorial amphibian. *Journal of Arid Environments* 155: 79–81.

Székely, D., D. Cogălniceanu, P. Székely, and M. Denoël. 2020b. Adult—Juvenile interactions and temporal niche partitioning between life-stages in a tropical amphibian. *PloS One* 15: e0238949.

Székely, D., M. Denoël, P. Székely, and D. Cogălniceanu. 2017. Pond drying cues and their effects on growth and metamorphosis in a fast developing amphibian. *Journal of Zoology* 303: 129–135.

Székely, D., F. P. Gaona, P. Székely, and D. Cogălniceanu. 2019. What does a Pacman eat? Macrophagy and necrophagy in a generalist predator (*Ceratophrys stolzmanni*). *PeerJ* 7: 1–15.

Székely, D., P. Székely., M. Denoël, and D. Cogălniceanu. 2018b. Random size-assortative mating despite size-dependent fecundity in a Neotropical amphibian with explosive reproduction. *Ethology*: 1–9.

Székely, D., P. Székely, F. Stănescu, D. Cogălniceanu, and U. Sinsch. 2018c. Breed fast, die young: Demography of a poorly known fossorial frog from the xeric Neotropics. *Salamandra* 54: 37–44.

Székely, P., M. Córdova-Díaz, D. Hualpa-Vega, S. Hualpa-Vega, and D. Székely. 2023a. A new glassfrog species of the genus *Centrolene* (Amphibia, Anura, Centrolenidae) from Cordillera del Cóndor, southern Ecuador. *ZooKeys* 1149: 53–84.

Székely, P., J. S. Eguiguren, L. Ordóñez-Delgado, D. Armijos-Ojeda, and D. Székely. 2020c. Fifty years after: A taxonomic revision of the amphibian species from the Ecuadorian biodiversity hotspot Abra de Zamora, with description of two new *Pristimantis* species. *PloS One* 15: 1–59.

Székely, P., J. S. Eguiguren, D. Székely, L. Ordóñez-Delgado, D. Armijos-Ojeda, M. L. Riofrío-Guamán, and D. Cogălniceanu. 2018d. A new minute *Pristimantis* (Amphibia: Anura: Strabomantidae) from the Andes of southern Ecuador. *PloS One* 13: e0202332.

Székely, P., D. Székely, D. Armijos-Ojeda, S. Hualpa-Vega, and J. Vörös. 2023b. Molecular and morphological assessment of rain frogs in the *Pristimantis orestes* species group (Amphibia: Anura: Strabomantidae) with the description of three new cryptic species from southern Ecuador. *Herpetological Monographs* 37: 41–69.

Székely, P., D. Székely, D. Armijos-Ojeda, A. Jara-Guerrero, and D. Cogălniceanu. 2016. Anfibios de un bosque seco tropical: Reserva Ecológica Arenillas, Ecuador. *Revista Ecosistemas* 25: 24–34.

Tapia, E. E., L. A. Coloma, G. Pazmiño-Otamendi, and N. Peñafiel. 2017. Rediscovery of the nearly extinct longnose harlequin frog *Atelopus longirostris* (Bufonidae) in Junín, Imbabura, Ecuador. *Neotropical Biodiversity* 3: 157–167.

Tarvin, R. D., P. Peña, and S. R. Ron. 2014. Changes in population size and survival in *Atelopus spumarius* (Anura: Bufonidae) are not correlated with chytrid prevalence. *Journal of Herpetology* 48: 291–297.

Taylor, E. H. 1968. *The caecilians of the world: A taxonomic review*. Lawrence: University of Kansas Press.

Taylor, E. H., and J. A. Peters. 1974. The caecilians of Ecuador. *University of Kansas Science Bulletin* 50: 333–346.

Toledo, L. F., O. G. S. Araújo, R. W. Ávila, R. A. Kawashita-Ribeiro, D. H. Morais, and D. F. Cisneros-Heredia. 2009. Amphibia, Anura, Centrolenidae, *Cochranella adenocheira*: Distribution and range extension, Brazil. *Check List* 5: 380–382.

Toral, E., P. Feinsinger, and M. L. Crump. 2002. Frogs and a cloud-forest edge in Ecuador. *Conservation Biology* 16: 735–744.

Torres-Carvajal, O. 2000. Ecuadorian Lizards of the genus *Stenocercus* (Squamata: Tropiduridae). *Scientific Papers Natural History Museum, The University of Kansas* 15: 1–38.

Torres-Carvajal, O. 2003. Cranial osteology of the Andean lizard *Stenocercus guentheri* (Squamata: Tropiduridae) and its postembryonic development. *Journal of Morphology* 255: 94–113.

Torres-Gastello, C. P., J. Suárez-Segovia, and D. F. Cisneros-Heredia. 2007. *Cochranella erminea*, a new species of Centrolenidae (Amphibia: Anura: Athesphatanura) from Amazonian Peru. *Journal of the National Museum (Prague), Natural History Series* 176: 1–12.

Touzet, J.-M., and D. F. Cisneros-Heredia. 1998a. Lianas con dientes o serpiente palo? *Reptinotas* 1.

Touzet, J.-M., and D. F. Cisneros-Heredia. 1998b. Las serpientes del género *Rhinobothryum bovallii* en Ecuador con notas sobre su manejo en cautiverio. *Reptinotas* 4: 3.

Trueb, L. 1971. Phylogenetic relationships of certain Neotropical toads with the description of a new genus (Anura: Bufonidae). *Los Angeles County Museum Contributions in Science* 216: 1–40.

Trueb, L. 1974. Systematic relationships of Neotropical horned frogs, genus *Hemiphractus* (Anura: Hylidae). *Occasional papers of the Museum of Natural History* 29: 1–60.

Trueb, L. 1979. Leptodactylid frogs of the genus *Telmatobius* in Ecuador with the description of a new species. *Copeia* 1979: 714–733.

Trueb, L., and W. E. Duellman. 1970. The systematic status and life history of *Hyla verrucigera* Werner. *Copeia* 1970: 601–610.

Trueb, L., and W. E. Duellman. 1971. A synopsis of Neotropical hylid frogs, genus *Osteocephalus*. *Occasional papers of the Museum of Natural History* 1: 1–47.

Urgilés, V. L., J. C. Sánchez-Nivicela, and D. F. Cisneros-Heredia. 2017. Notes on the distribution of *Gastrotheca testudinea* (Jiménez de la Espada, 1870) in Ecuador. *Check List* 13: 121–125.

Urgilés, V. L., J. C. Sánchez-Nivicela, C. Nieves, and M. H. Yánez-Muñoz. 2014. Ranas terrestres en los ecosistemas surandinos de Ecuador I: Dos nuevas especies de *Pristimantis* (Anura: Craugastoridae) de la ladera oriental. *Avances en Ciencias e Ingenierías* 6: 1–59.

Urgilés, V. L., P. Székely, D. Székely, N. Christodoulides, J. C. Sánchez-Nivicela, and A. E. Savage. 2019. Genetic delimitation of *Pristimantis orestes* (Lynch, 1979) and *P. saturninoi* Brito, *et al.* 2017 and description of two new terrestrial frogs from the *Pristimantis orestes* species group (Anura, Strabomantidae). *ZooKeys* 864: 111–146.

Valarezo-Aguilar, K., D. F. Cisneros-Heredia, and O. Ordonez-Gutiérrez. 2016. A new distribution record for the invasive american bullfrog, *Lithobates catesbeianus* (Shaw 1802) (Anura: Ranidae), from eastern Ecuador. *IRCF Reptiles & Amphibians* 23: 147–149.

Valdivieso-Rivera, F., S. Bermúdez-Puga, C. Proaño-Bolaños, and J. R. Almeida. 2022. Deciphering the limitations and antibacterial mechanism of Cruzioseptins. *International Journal of Peptide Research and Therapeutics* 28.

Valencia, J., E. Toral, M. Morales, R. Betancourt, and A. Barahona. 2008. *Guía de Campo de Anfibios del Ecuador*. Quito: Fundación. Herpetológica Gustavo Orcés, Simbioe.

Valencia, J. H., R. M. Betancourt, P. Yánez, P. González, J. Olmedo, and R. Flores. 2009a. *Anfibios y reptiles en ambientes cercanos a Nueva Loja y Lumbaqui*. Quito: Fundación Herpetológica Gustavo Orcés.

Valencia, J. H., R. M. Betancourt, P. Yánez, J. P. Hidalgo, and J. Campaña. 2009b. *Anfibios y reptiles en ambientes cercanos a Pedro Vicente Maldonado y Balao*. Quito: Fundación Herpetológica Gustavo Orcés.

Valencia, J. H., F. Valladares-Suntasig, L. Tipantiza-Tuguminago, and M. R. Dueñas. 2019. A new species of terrestrial-breeding frog of the genus *Pristimantis* (Anura: Terrarana: Craugastoridae) from the eastern Andean slopes of the southern Ecuador. *Zootaxa* 4658: 509–525.

Valencia, J. H., M. H. Yánez-Muñoz, R. Betancourt-Yépez, A. Terán-Valdez, and J. M. Guayasamin. 2011. "2010". Una llamativa nueva especie de *Pristimantis* (Anura: Terrana: Strabomantidae) de las estribaciones noroccidentales de los Andes de Ecuador. *Avances en Ciencias e Ingenierías* 2: 41–45.

Valle, C., V. Arias, F. Campos, R. Hidalgo, P. Ponce, R. Proaño, G. Reck, F. Rodríguez, M. Sáenz, J.-M. Touzet, H. Valdebenito, and A. M. Velasco. 1999. *Estrategia nacional para la protección y el uso sustentable de la vida silvestre en Ecuador*. Quito: Universidad San Francisco de Quito.

Vane-Wright, R. I. 1991. A portrait of Clarence Buckley, Zoologist. *The Linnean* 7: 30–33.

Velasco, A. M. 1997a. Diversidad y abundancia relativa de la herpetofauna en dos localidades en San Miguel, Reserva Ecológica Cotacachi-Cayapas, Esmeraldas, Ecuador. In *Estudios biológicos para la conservación, diversidad, ecología y etnobiología*, eds. P. A. Mena, A. Soldi, R. Alarcón, C. Chiriboga, and L. Suárez, 109–114. Quito: EcoCiencia.

Velasco, A. M. 1997b. Resumen comparativo de los inventarios herpetológicos en las reservas ecológica Cayambe-Coca y Cotacachi-Cayapas estudiadas dentro del convenio EcoCiencia-SUBIR entre 1993 y 1994. In *Estudios biológicos para la conservación, diversidad, ecología y etnobiología*, eds. P. A. Mena, A. Soldi, C. Alarcón, C. Chiriboga, and L. Suárez, 243–247. Quito: EcoCiencia.

Vélez Zuazo, A. 2019. Juan Manuel Guayasamin: el herpetólogo ecuatoriano que ha descubierto 50 especies de anfibios. *Mongabay* (May 19). https://es.mongabay.com/2019/05/Ecuador-Anfibios-Herpetologo-Conservacion/.

Verkade, J. 2012. Behoud van bedreigde amfibieën in Ecuador. *DN Magazine* 2: 28–33.

Verkade, J. 2013. Herontdekking van de typelocatie van Epipedobates tricolor. *DN Magazine* 2: 42–43.

Verkade, J. 2014. Kikkers voor natuurbehoud—Natuurbehoud voor kikkers. *DN Magazine* 3: 34–43.

Verkade, J. 2016. Mijn ervaringen met *Epipedobates tricolor* "cielito". *DN Magazine* 1: 10–15.

Verkade, J., and Verkade, J. 2009. *Epipedobates tricolor*. *DN Magazine* 3: 28–39.

Verkade, J., and Verkade, J. 2011. *Epipedobates anthonyi*. *DN Magazine* 2: 14–23.

Vigle, G. O. 2000. Amphibians (Chapter 5). In *The ecotravellers' wildlife guide to Ecuador and its Galapagos Islands*, eds. D. L. Pearson and L. Beletsky, 46–63. San Diego: Academic Press.

Vigle, G. O. 2008. The amphibians and reptiles of the Estación Biológica Jatun Sacha in the lowland rainforest of Amazonian Ecuador: A 20-year record. *Breviora* 514: 1–30.

Vigle, G. O., L. A. Coloma, J. C. Santos, S. Hernández-Nieto, H. Mauricio Ortega-Andrade, D. J. Paluh, and M. Read. 2020. A new species of *Leucostethus* (Anura: Dendrobatidae) from the Cordillera Mache-Chindul in northwestern Ecuador, with comments on similar *Colostethus* and *Hyloxalus*. *Zootaxa* 4896: 342–372.

Vigle, G. O., and D. Goberdhan-Vigle. 1990. A new species of small colorful *Hyla* from the lowland rainforest of Amazonian Ecuador. *Herpetologica* 46: 467–473.

Vigle, G. O., and K. Miyata. 1980. A new species of *Dendrobates* (Anura: Dendrobatidae) from the lowland rain forests of western Ecuador. *Breviora* 459: 1–7.

Wagner, P., A. M. Bauer, and W. Böhme. 2012. Amphibians and reptiles collected by Moritz Wagner, with a focus on the ZFMK collection. *Bonn Zoological Bulletin* 61: 216–240.

Walker, C. F. 1973. A new genus and species of Microhylid frog from Ecuador. *Occasional Papers of the Museum of Natural History, The University of Kansas* 20: 1–7.

Werner, F. 1897. Über einige noch unbeschriebene Reptilien und Batrachier. *Zoologischer Anzeiger* 20: 261–267.

Whymper, E. 1891. *Supplementary appendix to travels amongst the great Andes of the Equator*. London: John Murray.

Wiens, J. J., and L. A. Coloma. 1992. A new species of the *Eleutherodactylus myersi* (Anura: Leptodactylidae) assembly from Ecuador. *Journal of Herpetology* 26: 196–207.

Wikipedia. La enciclopedia libre 2023. *Eugenia del Pino*. Wikipedia, La enciclopedia libre. https://es.wikipedia.org/w/index.php?title=Eugenia_del_Pino&oldid=147540900 (accessed December 3, 2023).

Williams, E. E. 1985. In Memoriam (Kenneth Ichiro Miyata). *Herpetological Review* 16: 34–35.

Williams, E. E., V. Orcés-V., J. C. Matheus, and R. Bleiweiss. 1996. A new giant phenacosaur from Ecuador. *Breviora* 505: 1–32.

Williams, K. L., and C. M. Fugler. 1968. An additional record of *Pseudoeryx plicatilis ecuadorensis* from Ecuador. *Journal of Herpetology* 1: 104–105.

Wilson, D. S. 1998. Patterns in publishing in three north American herpetological journals: Gender biases. *Herpetologica* 54: S35–42.

Wilson, L. D. 1987. A résumé of the colubrid snakes of the genus *Tantilla* of South America. *Milwaukee Public Museum Contributions in Biology and Geology* 68: 1–35.

Womack, M. C., J. Christensen-Dalsgaard, L. A. Coloma, J. C. Chaparro, and K. L. Hoke. 2017. Earless toads sense low frequencies but miss the high notes. *Proceedings of the Royal Society B: Biological Sciences* 284: 1–9.

Womack, M. C., E. Steigerwald, D. C. Blackburn, D. C. Cannatella, A. Catenazzi, J. Che, M. S. Koo, J. A. McGuire, S. R. Ron, and C. L. Spencer. 2022. State of the Amphibia 2020: A review of five years of amphibian research and existing resources. *Ichthyology & Herpetology* 110: 638–661.

Womack, M. C., J. L. Stynoski, M. K. Voyles, L. A. Coloma, and K. L. Hoke. 2018. Prolonged middle ear development in *Rhinella horribilis*. *Journal of Morphology* 279: 1518–1523.

Yánez-Muñoz, M. 2002. *Importancia de las colecciones científicas de anfibios y reptiles. Memorias: Taller de introducción a la educación ambiental*. Quito: Museo Ecuatoriano de Ciencias Naturales. Casa de la Cultura.

Yánez-Muñoz, M. 2005. Diversidad y estructura de once comunidades de anfibios y reptiles en los Andes de Ecuador. Tesis de Licenciatura, Universidad Central del Ecuador.

Yánez-Muñoz, M. H., M. Altamirano-Benavides, D. F. Cisneros-Heredia, and A. G. Gluesenkamp. 2011a. "2010". Nueva especie de Sapo Andino del género *Osornophryne* (Amphibia: Bufonidae) del norte de Ecuador, con notas sobre la diversidad del género en Colombia. *Avances en Ciencias e Ingenierías* 2: 46–53.

Yánez-Muñoz, M. H., D. Batallas, D. Franco-Mena, P. A. Meza-Ramos, L. A. Oyagata, D. Padilla, C. Paucar, *et al*. 2020. *Anfibios en los Ecosistemas Andino-Tropicales de la Provincia del Carchi*. Publicación Miscelánea 14: 1–340. Quito: Serie de Publicaciones del Gobierno Autónomo Descentralizado Provincial del Carchi y el Instituto Nacional de Biodiversidad.

Yánez-Muñoz, M. H., P. Bejarano-Muñoz, J. Brito-M., and D. R. Batallas. 2014. Ranas terrestres de los Andes surorientales de Ecuador II: Una nueva especie de *Pristimantis* verde espinosa de los bosques montanos del Parque Nacional Sangay (Anura: Craugastoridae). *Avances en Ciencias e Ingenierías* 6: 63–77.

Yánez-Muñoz, M. H., and A. Chimbo. 2007. Anfibios y Reptiles. In *Rapid Biological Inventories. Ecuador: Territorio Cofán Dureno*, eds. R. Borman, C. Vriesendorp, W. S. Alverson, D. K. Moskovits, and A. del Campo, 96–99. Chicago: The Field Museum.

Yánez-Muñoz, M. H., and D. F. Cisneros-Heredia. 2008. Amphibia, Anura, Centrolenidae: *Cochranella orejuela*, first country records from Ecuador. *Check List* 4: 50–54.

Yánez-Muñoz, M. H., D. F. Cisneros-Heredia, and J. P. Reyes-Puig. 2011b. "2010". Una nueva especie de rana terrestre *Pristimantis* (Anura: Terrarana: Strabomantidae) de la cuenca alta del Río Pastaza, Ecuador. *Avances en Ciencias e Ingenierías* 2: 28–32.

Yánez-Muñoz, M. H., and P. A. Meza-Ramos. 2006. Generando información para conservar a *Colostethus delatorreae* (Anura: Dendrobatidae): una especie de rana nodriza críticamente amenazada en los Andes ecuatorianos. *Boletín Especies Amenazadas UICN* 6: 1–3.

Yánez-Muñoz, M. H., P. A. Meza-Ramos, D. F. Cisneros-Heredia, and J. P. Reyes-Puig. 2011c. "2010". Descripción de tres nuevas especies de ranas del género *Pristimantis* (Anura: Terrarana: Strabomantidae) de los bosques nublados del Distrito Metropolitano de Quito, Ecuador. *Avances en Ciencias e Ingenierías* 2: 16–27.

Yánez-Muñoz, M. H., P. A. Meza-Ramos, C. C. Martínez, and M. Reyes-Puig. 2009c. Anfibios y reptiles del sur occidente de Ecuador. *Rapid Color Guide*, no. 237 (January). https://fieldguides.fieldmuseum.org/sites/default/files/rapid-color-guides-pdfs/237_Anfibios_y_Reptiles_SW_Ecuador_04_0.pdf.

Yánez-Muñoz, M. H., P. A. Meza-Ramos, S. M. Ramírez, J. C. Reyes-Puig, and L. A. Oyagata. 2009b. Anfibios y reptiles del Distrito Metropolitano de Quito (DMQ). In *Guía de Campo de Los Pequeños Vertebrados del Distrito Metropolitano de Quito (DMQ)*, eds. M. Yánez-Muñoz, P. Moreno-Cárdenas, P. Mena-Valenzuela, 5: 9–52. Quito: Museo Ecuatoriano de Ciencias Naturales (MECN)/Fondo Ambiental del MDMQ.

Yánez-Muñoz, M. H., and J. J. Mueses-Cisneros. 2009. Amphibians and Reptiles. In *Ecuador: Cabeceras Cofanes-Chingual. Rapid biological and social inventories*, eds. C. Vriesendorf, W. S. Alverson, A. del Campo, D. F. Stotz, D. K. Moskowitz, B. Fuentes-Cáceres, B. Coronel-Tapia, and E. P. Anderson, 191–199. Chicago: The Field Museum.

Yánez-Muñoz, M. H., P. Pérez-Peña, and D. Cisneros-Heredia. 2009a. New country records of *Hyalinobatrachium iaspidiense* (Amphibia, Anura, Centrolenidae) from the Amazonian lowlands of Ecuador and Peru. *Herpetology Notes* 2: 49–52.

Yánez-Muñoz, M. H., C. Reyes-Puig, P. Bejarano-Muñoz, and S. R. Ron. 2015. Otra nueva especie de rana *Pristimantis* (Anura: Terrarana) de las estribaciones occidentales del Volcán Pichincha, Ecuador. *Avances en Ciencias e Ingenierías* 7: 52–60.

Yánez-Muñoz, M. H., J. P. Reyes-Puig, D. Batallas-Revelo, C. Broaddus, M. Urgilés-Merchán, D. F. Cisneros-Heredia, and J. M. Guayasamin. 2021. A new Andean treefrog (Amphibia: *Hyloscirtus bogotensis* group) from Ecuador: An example of community involvement for conservation. *PeerJ* 9: e11914.

Yánez-Muñoz, M. H., J. C. Sánchez-Nivicela, and C. Reyes-Puig. 2016. Tres nuevas especies de ranas terrestres *Pristimantis* (Anura: Craugastoridae) de la Provincia de El Oro, Ecuador. *Avances en Ciencias e Ingenierías* 8: 5–25.

Yánez-Muñoz, M. H., D. A. Veintimilla-Yánez, D. Batallas, and D. F. Cisneros-Heredia. 2019. A new giant *Pristimantis* (Anura, Craugastoridae) from the paramos of the Podocarpus National Park, southern Ecuador. *ZooKeys* 852: 137–156.

Yánez-Muñoz, M. H., D. A. Veintimilla-Yánez, E. N. Smith, and M. Altamirano-Benavides. 2011d. "2010". Descubrimiento de dos poblaciones sobrevivientes de sapos arlequín (Amphibia: Bufonidae: *Atelopus*) en los Andes de Ecuador. *Avances en Ciencias e Ingenierías* 2: 74–77.

Yánez-Muñoz, M. H., and P. Venegas. 2008. Ecuador-Perú: Cuyabeno Güeppí. In *Rapid biological and social inventories Report 20*, eds. W. S. Alverson, C. Vriesendorp, A. del Campo, D. K. Moscovits, D. F. Stottz, M. García D, and L. A. Borbor, 90–96. Chicago: The Field Museum.

Young, B. E., K. R. Lips, J. K. Reaser, R. Ibáñez, A. W. Salas, J. Rogelio Cedeño, L. A. Coloma, *et al.* 2001. Population declines and priorities for amphibian conservation in Latin America. *Conservation Biology* 15: 1213–1223.

Zippel, K. C., and J. R. Mendelson. 2008. The amphibian extinction crisis: A call to action. *Herpetological Review* 39: 23–24.

Zumel, D., D. Buckley, and S. R. Ron. 2021. The *Pristimantis trachyblepharis* species group, a clade of miniaturized frogs: Description of four new species and insights into the evolution of body size in the genus. *Zoological Journal of the Linnean Society* 195: 315–354.

3 Diversity, Extinction, and Conservation

MOLECULAR DIVERSITY AND BIOPROSPECTION IN AMPHIBIAN SKIN

Sebastián Bermúdez-Puga, Luis A. Coloma, and Carolina Proaño-Bolaños

Amphibian skin is a morphologically, biochemically, and physiologically complex organ essential for amphibian survival. Its functions include temperature control, reproduction, camouflage, respiration, anti-predatory activity, and defense against pathogens (Clarke, 1997). The granular glands are responsible for the synthesis of the chemical compounds that provide protection against potential predators. These include biogenic amines, bufadienolides, bufogenines, steroids, alkaloids, proteins, and peptides (Clarke, 1997; Costa-Neto, 2005; Wang *et al.*, 2012). In recent decades, amphibian skin secretions have been a subject of interest because of their compounds with unique biological properties and potential biotechnological applications (van Zoggel *et al.*, 2012; Xu and Lai, 2015; Wang *et al.*, 2015).

Amphibian-produced biomolecules have been extensively studied due to their multifunctional activities, including analgesic, antioxidant, immunomodulatory, anti-inflammatory, anti-hypertensive, antimicrobial, antiviral, anticancer, and antiparasitic properties (Wang, 2020). These compounds have been identified in the Alytidae, Bombinatoridae, Hylidae, Hyperoliidae, Leiopelmatidae, Leptodactylidae, Myobatrachidae, Phyllomedusidae, Pipidae, and Ranidae families (Conlon, 2011; Conlon and Mechkarska, 2014).

Several frogs and toads, including endemic species, belonging to the families mentioned earlier are also found in Ecuador. In Ecuador, there are 655 described species, of which 319 are endemic. As mentioned in Chapter 1, the toxins and other substances from the skin of some frogs have been used since ancient times by indigenous communities in Ecuador in their diets, rituals, and also in traditional medicine. These compounds have a high potential for application in the therapeutics, cosmetics, pharmaceutical, biomedicine, and other industries. Therefore, the skin secretions of Ecuadorian frogs could represent an untapped source of lead drugs to combat global health problems such as multidrug-resistant bacterial infections, cancer, and pain, among many others.

Since the 1970s, a series of exploratory and groundbreaking studies of alkaloids in frogs from Ecuador was conducted by John W. Daly at the National Institutes of Health (USA). These studies triggered a series of countless studies focused on epibatidine (see under the subtitle "The Fight for Sovereignty over Genetic Resources: The Case of Epibatidine and Derivatives") and on other alkaloids as well, including some more recent studies by American and Ecuadorian researchers (e.g., Tarvin *et al.*, 2016; McGugan *et al.*, 2016; review by Santos *et al.*, 2016; Tarvin *et al.*, 2017; Moskowitz *et al.*, 2022). Furthermore, researchers mentored by Miryan Rivera of Pontificia Universidad Católica de Ecuador (PUCE) since the 2000s and by Carolina Proaño-Bolaños since 2016 at Universidad Regional Amazónica Ikiam are conducting studies to determine the biological activity of the skin secretions of Ecuadorian frogs, as well as the molecules associated with them. Proaño-Bolaños, an Ecuadorian biologist, studied the bioactive properties of chemicals extracted from the skin of amphibians through a National Secretariat of Science and Technology (SENESCYT) scholarship at one of the world's most important research centers for bioprospecting the skin secretions of amphibians (Queen's University Belfast, Northern Ireland, UK). An additional and crucial step was taken by the Ecuadorian government through its Ministerio de Ambiente in 2016 when it supported a project conceived by researchers at Centro Jambatu of Amphibian Research and Conservation (CJ), titled "Conservation of Ecuadorian Amphibian Diversity and Sustainable Use of its Genetic Resources". Through this project, a biobank was established at CJ, as well as a bioprospecting facility led by Proaño-Bolaños at Ikiam.

The PUCE and Ikiam studies are focused on the evaluation of the antimicrobial and anticancer activity of crude skin secretions (Proaño-Bolaños, 2004; Caicedo, 2007; Torres, 2009; Chuang, 2012; Vargas Hurtado, 2012; Cilveti, 2013; Vallejo Rodríguez, 2016; Díaz Escobar, 2017; Llumiquinga, 2017; Viteri Dávila, 2020). In a pioneering study, Proaño-Bolaños (2004) found antimicrobial activity against *Escherichia coli*, *Shigella* sp., *Bacillus subtilis*, and *Streptococcus* sp. in 13 of the 21 analyzed species that belong to four amphibian families (Dendrobatidae, Hylidae, Phyllomedusidae, and Ranidae) (Table 3.1). Further analysis of the skin secretion of *Agalychnis spurrelli* identified additional antiproliferative, antifungal, and antibacterial activity against resistant pathogenic strains (Caicedo, 2007; Torres, 2009; Vargas Hurtado, 2012; Chuang, 2012).

Later, Cilveti (2013) determined the antibacterial activity of the skin secretion of the bufonid frog *Atelopus elegans* against multidrug-resistant clinical isolates of *Serratia marcescens*, *S. rubidaea*, *Klebsiella pneumoniae*, *K. oxytoca*, *Escherichia coli*, *Enterobacter cloacae*, *E. aerogenes*, *Pseudomonas aeruginosa*, and *Morganella morganii*. In 2016, a study by Vallejo Rodríguez determined the antibacterial and anticancer effects of 15 frog skin secretions belonging to species of Bufonidae, Hylidae, and Leptodactylidae. *Staphylococcus aureus* was susceptible to skin secretions of *Boana geographica*, *B. lanciformis*, *Callimedusa tomopterna*, *Phyllomedusa tarsius*, and *P. vaillantii*, while *E. coli* was susceptible to *C. tomopterna*, *P. tarsius*, and *P. vaillantii*. Furthermore, the cytotoxic properties were evaluated using acute lymphoblastic leukemia cells. The cutaneous secretion of the

DOI: 10.1201/9781003129080-3

TABLE 3.1

Exploratory Analysis of the Bioactivity of Frog Skin Secretions from Ecuador. + indicates that frog skin secretion has bioactivity. ABA = antibacterial activity, AFA = antifungal activity, APA = anti-proliferative activity, ACS = analysis of crude secretions, AFS = analysis of fractionated secretion

Family	Species	ABA	AFA	APA	ACS	AFS	References
Bufonidae	*Atelopus elegans*	+			+		Cilveti (2013)
	Atelopus nanay	+	+			+	López-Tacoaman (2022)
	Atelopus sp.	−			+		Llumiquinga (2017)
Dendrobatidae	*Allobates fratisenescus* (under *A. insperatus*)	−			+		Proaño-Bolaños (2004)
	Allobates sp.	−			+		Proaño-Bolaños (2004)
	Allobates zaparo	−			+		Proaño-Bolaños (2004)
	Ameerega parvula	−			+		Proaño-Bolaños (2004)
	Epipedobates anthonyi	−			+		Proaño-Bolaños (2004)
	Oophaga sylvatica	+			+		Proaño-Bolaños (2004), Llumiquinga (2017)
Hemiphractidae	*Gastrotheca* sp.	−			+		Llumiquinga (2017)
Hylidae	*Boana boans*	+			+		Vallejo Rodríguez (2016)
	Boana calcarata	+		+	+		Vallejo Rodríguez (2016)
	Boana cinerascens	+		+	+		Proaño-Bolaños (2004), Vallejo Rodríguez (2016)
	Boana geographica	+			+		Proaño-Bolaños (2004), Vallejo Rodríguez (2016)
	Boana lanciformis	−		+	+		Vallejo Rodríguez (2016)
	Boana pellucens	−			+		Llumiquinga (2017)
	Boana picturata	+			+		Proaño-Bolaños (2004)
	Boana rosenbergi	+			+		Llumiquinga (2017), Díaz Escobar (2017)
	Dendropsophus bifurcus	−			+		Proaño-Bolaños (2004)
	Hylocirtus pantostictus	+			+		Viteri Dávila (2020)
	Hyloscirtus lindae	+			+		Viteri Dávila (2020)
	Osteocephalus taurinus	−			+		Proaño-Bolaños (2004)
	Smilisca phaeota	−			+		Llumiquinga (2017)
	Trachycephalus jordani	−			+		Llumiquinga (2017)
	Trachycephalus cunauaru (under *T. resinifictrix*)	−			+		Proaño-Bolaños (2004)
	Leptodactylus mystaceus	+			+		Vallejo Rodríguez (2016)
Leptodactylidae	*Leptodactylus wagneri*	+			+		Vallejo Rodríguez (2016)
	Lithodytes lineatus	+	−			+	Yambay (2022)
	Agalychnis buckleyi	+			+		Proaño-Bolaños (2004)
Phyllomedusidae	*Agalychnis hulli*	+		+	+		Vallejo Rodríguez (2016)
	Agalychnis spurrelli	+	+		+	+	Proaño-Bolaños (2004), Caicedo (2007), Torres, (2009), Ramirez (2012), Po Hao (2012), Cajiao Garcés (2018)
	Callimedusa ecuatoriana	+	−			+	Bonilla-Jiménez (2021)
	Callimedusa tomopterna	+		+	+		Proaño-Bolaños (2004), Cajiao Garcés (2018), Vallejo Rodríguez (2016)
	Cruziohyla calcarifer	+			+		Proaño-Bolaños, 2004
	Phyllomedusa tarsius	+		+	+		Proaño-Bolaños (2004), Vallejo Rodríguez (2016)
	Phyllomedusa vaillantii	+		+	+		Proaño-Bolaños (2004), Vallejo Rodríguez (2016)
	Rana bwana	+			+		Proaño-Bolaños (2004)
Ranidae	*Lithobates palmipes*	+			+		Proaño-Bolaños (2004)
	Lithobates sp.	+			+		Proaño-Bolaños (2004)
	Lithobates sp.	+			+		Llumiquinga (2017), Díaz Escobar (2017)

species *Agalychnis hulli*, *B. calcarata*, *B. cinerascens*, *B. lanciformis*, *C. tomopterna*, *P. tarsius*, and *P. vaillantii* showed antiproliferative effects on these cancer cells (Vallejo Rodríguez, 2016). Later, Llumiquinga (2017) evaluated the antibacterial potential of skin secretions of eight amphibian species. Specimens of *Atelopus* sp. (San Carlos de Limón, Provincia Morona Santiago), *Gastrotheca riobambae*, *Rana* sp., *B. pellucens*, *B. rosenbergi*,

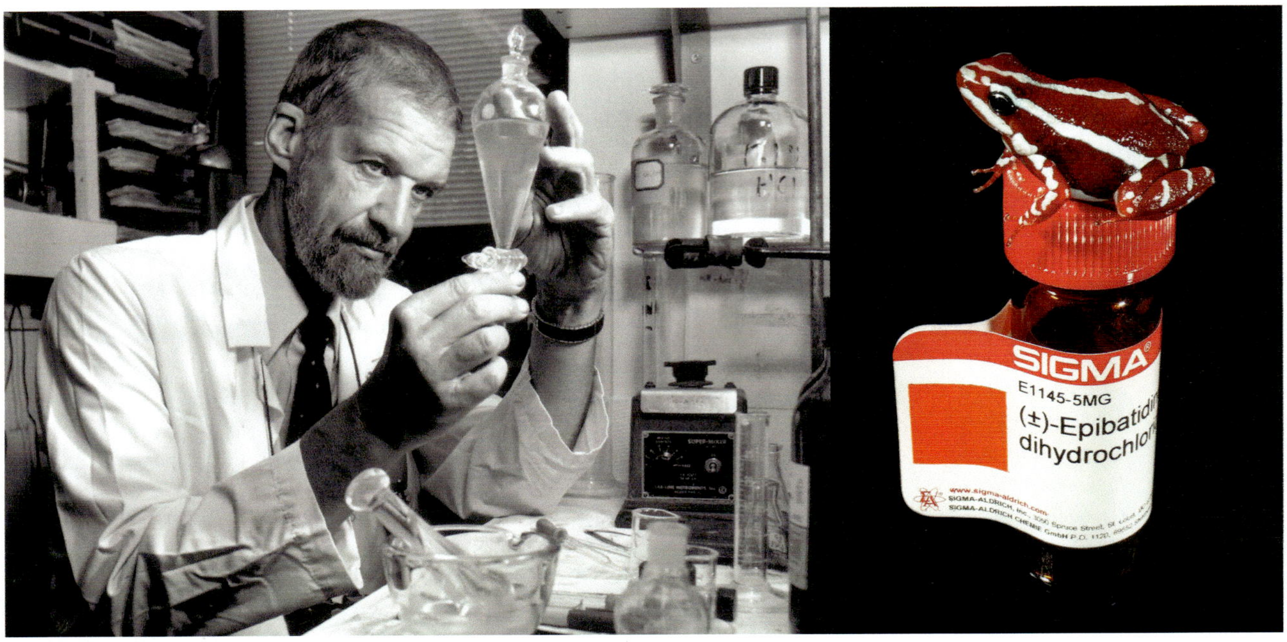

FIGURE 3.1 John W. Daly at the National Institutes of Health (USA). Photo courtesy of Office of NIH History and Stetten Museum. Right: a derivative of epibatidine commercialized by Sigma-Aldrich for research purposes and *Epipedobates anthonyi*, in which it was originally found. LAC.

Oophaga sylvatica, *Smilisca phaeota*, and *Trachycephalus jordani* were used in this study. The antibacterial screening demonstrated that skin secretions of *B. rosenbergi* and *Lithobates* sp. were active against *S. aureus* and *Serratia marcescens* (Llumiquinga, 2017). Likewise, the cutaneous secretion of *Rana* sp. inhibited the growth of extended-spectrum beta-lactamase-producing *E. coli* and *S. aureus* CORSA (Díaz Escobar, 2017). Recently, Viteri Dávila (2020) analyzed the antimicrobial effects of skin secretions of *Hyloscirtus lindae* and *H. pantostictus* against *E. coli*, *S. marscences*, *K. pneumoniae*, *S. aureus*, and *C. albicans*. The antimicrobial assay revealed that the skin secretion of *H. lindae* only inhibited the growth of *S. aureus*, while that of *H. pantostictus* showed no antimicrobial activity (Viteri Dávila, 2020). High-performance liquid chromatography, gel filtration chromatography, and mass spectrometry were used to obtain chemical insight on the molecules present in the secretion with antimicrobial effects. These tools identified proteinaceous compounds in the chromatographic fractions (Dillon, 2012; Ramírez Jiménez, 2012; Cajiao Garcés, 2018; Yambay, 2022; López-Tacoaman, 2022).

Benefit Sharing of Genetic Resources: The Case of Epibatidine and Derivatives

The studies led by John W. Daly at the National Institutes of Health (USA), especially those on natural products and the discovery of epibatidine, not only launched a plethora of research but also sparked global and national discussions about genetic resources, traditional knowledge, bioprospection, intellectual property, biopiracy, patents, benefit sharing, biocommerce, and related issues (e.g., Bravo, 1997; Angerer, 2011, 2013, and citations therein).

John William Daly (1933–2008) was an American biochemist who conducted research for nearly 50 years at the National Institutes of Health in Bethesda, Maryland (Figure 3.1). Some aspects of his life and achievements are highlighted by Creveling (2003), Garraffo (2009), and Jacobson and Kirk (2009). A central focus of his research included the discovery, structure elucidation, synthesis, and pharmacology of alkaloids and other biologically active natural products. He held the position of the foremost global authority on amphibian alkaloids and demonstrated expertise in numerous domains of natural products. Daly was a prolific author, producing approximately 609 papers and book chapter publications, including books and chapters. In 1997, he was elected to the US National Academy of Sciences. In 2002, he was named among the 200 most-cited pharmacologists in the world (NIH Record, 2004).

Daly's research focused on complex alkaloids and the molecular pharmacology of acetylcholine receptors and sodium channel levels, mainly in the Dendrobatidae family, resulting in the identification of hundreds of different alkaloids. Among them is the iconic alkaloid epibatidine. Epibatidine is a natural alkaloid molecule discovered in the skin of the Ecuadorian frog *Epipedobates anthonyi* (considered *E. tricolor* at the time of epibatidine discovery). After isolating it in 1974 in a sub-milligram quantity, Daly showed that this neurotoxic chemical has extraordinary analgesic properties in mice; that is, it is capable of blocking pain. This alkaloid is widely considered an analgesic agent, with a potency ranging from 100- to 200-fold higher than that of the analgesic opioid morphine and 30 times higher than that of nicotine. Its structure was not fully elucidated until 1992 (Spande *et al.*, 1992). It took 18 years for Daly to uncover the secrets of this molecule to elucidate the chemical structure and its antinociceptive mechanism. Surprisingly, epibatidine works by activating nicotinic receptors (cholinergic ion channels), not receptors for opioids. This compound is now a key lead in the quest for future pain relief medicines that might have fewer side effects and not produce tolerance. Epibatidine is toxic. Its toxicity stems from its ability to interact with nicotinic and muscarine acetylcholine receptors. These receptors are involved in the transmission of painful sensations and in movement, among other functions. Epibatidine causes numbness and, eventually, paralysis. Doses are lethal when the paralysis causes respiratory arrest. According to Salehi *et al.* (2019), because of the high toxicity of epibatidine, its therapeutic use is hampered. Due to severe gastrointestinal side effects, the first analog of epibatidine, ABT-594, is not included in current pain therapies in humans. Epiboxidine and ABT-418 are less toxic analogs. Despite its decreased potency and toxicity compared to epibatidine, epiboxidine itself is still too toxic to be developed as a drug for use in humans. It is used in scientific research and as a parent compound to derive newer analogs, which may be safer and have greater potential for clinical development. ABT-418 is used in the treatment of less severe attention deficit hyperactivity disorder in adult patients. ABT-418 has been well tolerated by patients with minor side effects, such us nausea, dizziness, headaches, or skin irritations. Thus, derivatives of epibatidine and its pharmacological effects open new perspectives in drug therapies for anxiety, Parkinson's disease, Tourette's syndrome, epilepsy, inflammatory bowel disorder, schizophrenia, depression, early biomarkers of Alzheimer's disease, appetite suppression, smoking cessation, and many others, and they also represent an important research tool to investigate nicotinic acetylcholine receptor activity (Salehi et al., 2019, and citations therein).

Daly, his group, and collaborators also astonished the world of herpetology in the early 1990s by demonstrating that the alkaloids found in frog skin were not synthesized but were sequestered from an arthropod diet, chiefly ants, small beetles, millipedes, and, most importantly, mites. During a face-to-face meeting in Washington with Daly, Garrafo, and the Ecuadorians Coloma and Acosta-Buenaño, Daly emphasized the significance of persisting in this type of research and the importance of tracking down the arthropod origin and ultimately the potentially fungal origin of poison frog alkaloids. This endeavor was initiated in 2016 by a team of US and Ecuadorian scientists (McGugan *et al.*, 2016) led by Lauren A. O'Connell.

The case of epibatidine is not only an interesting chapter in the history of chemistry and pharmacology (Daly *et al.*, 1999; Daly *et al.*, 2000) but also one of the icons in the struggle of countries rich in diversity to obtain economic benefits from products derived from their genetic resources. The Ecuadorian government has a pending royalties claim to countries that are party to the Convention on Biological Diversity with regards to the commercial use of epibatidine or its derivatives.

This discovery by the scientists in the United States of America, unprecedented in the history of medicine, renewed the search for non-narcotic analgesic agents that can be used therapeutically, especially to treat pain and support smoking cessation (Sudo *et al.*, 2018). Because epibatidine in its initial structure cannot be used due to its great toxicity, the production of epibatidine derivatives was triggered. Its potential market handles an estimated billions of dollars. Animal and human tests are currently underway. It should be noted that the commercialization of derivatives of epibatidine is growing. Many companies are selling epibatidine for research uses, for example, Sigma-Aldrich (see Figure 3.1), Amersham Biosciences, Biotrend, etc. In this current scenario, several questions arise: What is the current situation of the Ecuadorian government's claim on epibatidine? What measures should be taken in relation to other products that derive from their biological and cultural diversity, considering the ineffective application of the Convention on Biological Diversity, Decision 391 of the Andean Community of Nations (CAN), and the pending laws and regulations on the genetic resources of the Ecuadorian government? For instance, a bylaw needed for the commercial use of genetic resources is not yet available. It is well worth it, then, that we review some historical milestones related directly and indirectly to this topic (Box 3.1). Additional information can be found in original references cited by Wikimedia (2023) and under Remarks on the *Epipedobates anthonyi* account.

BOX 3.1

1823–1824. British Captain Charles Stuart Cochranae reports for the first time that "poison frogs" (of the Dendrobatidae family) from the Colombian Chocó are used by indigenous people to poison darts used in hunting (Myers and Daly, 1983).

1932. Scientists discover that pain can be blocked. Nicotine is used and posteriorly opioid compounds like morphine.

1969. Ecuador issues Decree CLP No. 170 and the corresponding regulations to control activities related to the hunting and capture of wild animals, as well as those related to their export (Duarte Valverde, 1969).

1970. Ecuador issues the Law for the Protection of Wild Fauna and Ichthyological Resources (Espinosa R. and Durán Arias, 1970).

1974. John W. Daly (National Institutes of Health, National Institute of Diabetes and Digestive and Kidney Diseases, USA) discovers an analgesic substance (the current epibatidine) based on the observation that it produces the upright tail reaction in mice. This discovery was made in frogs of the Dendrobatidae family from southwestern Ecuador (until 2003 called *Epipedobates tricolor*, today known as *Epipedobates anthonyi* after taxonomic changes).

1975. Ecuador ratifies its adherence to the Convention on International Trade in Endangered Species of Wild Fauna and Flora (CITES).

1976. Daly extracts 1 mg of epibatidine from extracts of the skin of 750 little frogs from the province of Azuay on the Ecuadorian coast.

1984. The Convention on International Trade in Endangered Species of Wild Fauna and Flora (CITES) includes Dendrobatidae species in Appendix II and restricts their export from the countries of origin.

1990. Several scientists in Daly's laboratory decipher the chemical structure of epibatidine.

1992. Spande, Garrafo, Edwards, Yeh, Panel, and Daly (National Institutes of Health, National Institute of Diabetes and Digestive and Kidney diseases, USA) report to science, in the *Journal of the American Chemical Society*, the finding of epibatidine (Spande *et al.*, 1992).

1993. Ecuador becomes a party in the Convention on Biological Diversity, which recognizes the sovereignty of the signatory countries over their genetic resources. The US government does not ratify its adherence to this Convention.

1993. Elias J. Corey, from the Department of Chemistry and Biochemistry at Harvard University (USA), synthesizes three derivatives of epibatidine for the first time in the laboratory (Corey *et al.*, 1993).

1994. ABT-418 (a derivative of epibatidine) is a drug developed by Abbott.

1995 (October 31). Epibatidine and its derivatives, compositions, and methods of pain treatment are patented (US Patent No. 5,462,956). The patent is granted to the American physician John W. Daly (1933–2008) and two of his collaborators: Thomas F. Spande and Hugo M. Garraffo.

1996. Decision 391 of the Andean Community of Nations (CAN) on "Access to genetic resources" enters into force. The sovereignty of the signatory countries over their genetic resources and their derived products is recognized. The Ecuadorian Forest and Natural Areas Institute (INEFAN) restricts the use of biodiversity as a source of genetic resources, based on the CAN decision.

1997. Epiboxidine, a combination of epibatidine and ABT-418, is synthetized (Badio *et al.*, 1997).

1998. Tebanicicline is synthetized (Donnelly-Roberts *et al.*, 1998). The Abbott Laboratories in Chicago (Illinois, USA), which were working with experimental drugs for Alzheimer's disease, discovered that epibatidine is similar to their drugs and launched a derivative of epibatidine, ABT-594 (tebanicline), on the research market. This derivative is produced based on previous research. It does not cause dependency or harmful side effects like morphine (lung damage, constipation) and nicotine (lung cancer, heart disease) do.

1998. Various media around the world and in Ecuador inform the community about tebanicicline and its potential (Strauss, 1998) (*Diario El Comercio*, November 15, 24, and 28, 1998).

1998. The Ecuadorian government, through INEFAN, asks Abbott Laboratories to "voluntarily" recognize the benefits to which it is entitled. Abbott Laboratories explained that their product was derived from previous research, and that claims should be made to the patent owners.

1999. The derivative of epibatidine (ABT-418) began to be used for research on Alzheimer's disease (Potter *et al.*, 1999) and in the treatment of milder attention deficit/hyperactivity disorder in adult patients. ABT-418 has been well tolerated by patients, with minor side effects such as nausea, dizziness, headaches, or skin irritations (Wilens *et al.*, 1999).

2008. Analogs of epiboxidine are synthetized (Rizzi *et al.*, 2008). Epiboxidine is around one-tenth as potent as epibatidine. It has only one-tenth of the analgesic power of epibatidine but is also much less toxic. Despite its decreased potency and toxicity compared to epibatidine, epiboxidine itself is still too toxic to be developed as a drug for use in humans. It is used in scientific research and as a parent compound to derive newer analogs that may be safer and have greater potential for clinical development.

2013 (February). Flubatine, a derivative of epibatidine with potential uses in Alzheimer's disease, is patented in Germany (DE102011052348).

2014 (January). Promising results on the synthesis and biological evaluation of the use of flubatine (a derivative of epibatidine) as an early marker of Alzheimer's disease are made public at the German Research Institute Helmholtz-Zentrum Dresden-Rossendorf.

2023. A search for the word "epibatidine" on Google Scholar (as of February 2024) reveals about 16,800 titles containing the term "epibatidine", most of them research publications.

In Search of Bioactive Peptides from Ecuadorian Frog Skin Secretions

It is estimated that ~200 distinct skin peptides can occur in a single species (Meneses *et al.*, 2011). These peptides are grouped according to their activity such as myotropical, opioids, cortico-tropin-releasing, angiotensins, protease inhibitors, neuropeptides, antioxidants, lectins, insulin-releasing, wound-healing, immunomodulatory, neuronal nitric oxide synthase inhibitors, pheromone peptides, and antimicrobial peptides (Xu and Lai, 2015).

Antimicrobial peptides (AMPs), which are composed of between 10 to 50 amino acid residues, possess a broad spectrum of activities against bacteria, fungi, cancer, and parasitic protozoa, including multidrug-resistant pathogens (Huan *et al.*, 2020). Furthermore, most AMPs act on the membrane of the microorganisms, leading to a low development of resistance compared to conventional drugs. These characteristics make these peptides an ideal scaffold for the development of new drugs (Sarkar *et al.*, 2021). Similarly, frog skin-derived protease inhibitors could be considered potential anticancer agents (Lyu *et al.*, 2018; Chen *et al.*, 2020).

Amphibian skin secretion is a well-recognized source of molecules with therapeutic and biotechnology potential. In this sense, a great effort is being made to investigate bioactive peptides present in the skin secretion of Ecuadorian frogs.

In this chapter, we detail the bioactive peptides isolated in the cutaneous secretions of Ecuadorian frogs with an overview of their biological activity. Most of this work is carried out in the bioprospecting facility led by Carolina Proaño-Bolaños at Ikiam and in cooperation with several research groups.

The use of a proteomic and transcriptomic approach has allowed for the identification and characterization of 90 bioactive peptides in five frog skin secretions. These peptides belong to existing peptide families, while others are first-reported families, thus representing molecular biodiversity. Furthermore, functional assays have demonstrated that some of these compounds have antibacterial, antifungal, and antiparasitic properties, and other proteinaceous compounds are protease inhibitors. Finally, several analogs' peptides are being generated as scaffolds, and those peptides with greater bioactivity are used for the development of new antibiotic candidates.

Peptide Families from Ecuadorian Frogs

Peptidomic analysis has been performed on the skin secretions of several amphibian species of Ecuador, such as *Cruziohyla calcarifer* (Proaño-Bolaños *et al.*, 2016; Proaño-Bolaños *et al.*, 2017; Cuesta *et al.*, 2021b), *Agalychnis spurrelli* (Proaño-Bolaños *et al.*, 2019), *Boana picturata* (Morán-Marcillo *et al.*, 2022), *Callimedusa ecuatoriana*, and *Lithodytes lineatus*, using a proteomic and transcriptomic strategy (Figure 3.2). These analyses included pools of skin secretions from lab-raised frogs and others collected in the field.

Ninety bioactive peptides have been identified in *Agalychnis spurrelli*, *Cruziohyla calcarifer*, *Boana picturata*, *Callimedusa ecuatoriana*, and *Lithodytes lineatus* (Figure 3.3). Based on their primary sequences, 42 peptides were classified into previously reported families, including dermaseptins, phylloseptins, plasticins, medusins, phyllokinin, tryptophyllin, kazal-type protease inhibitors, proline–arginine-rich peptides, and long peptides with α-helical structures (see Figure 3.3).

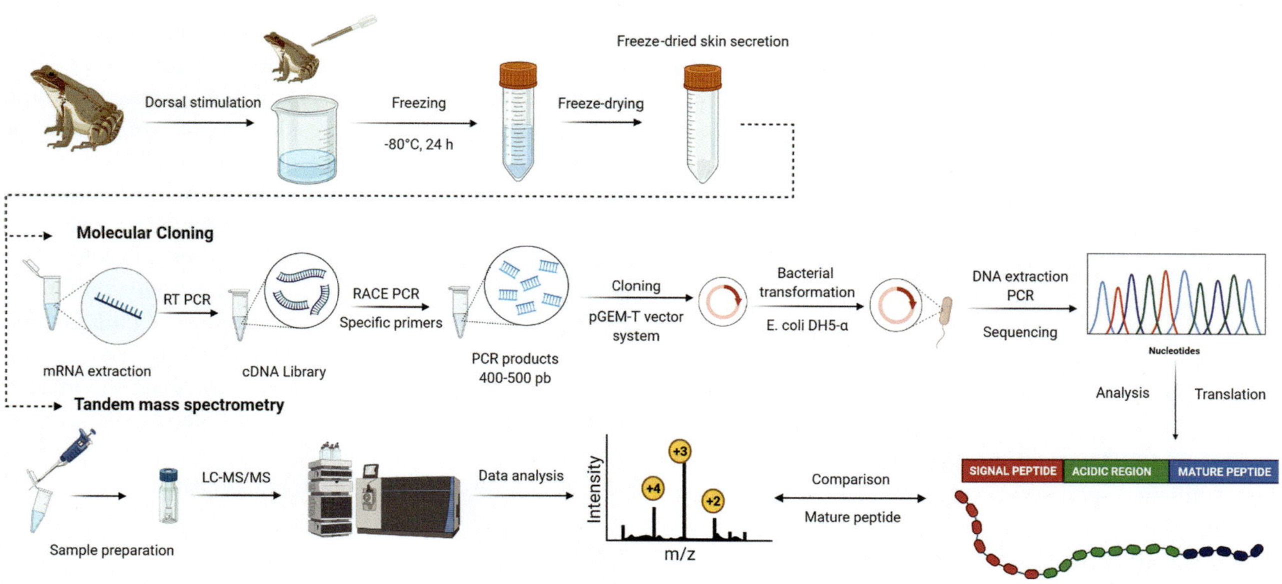

FIGURE 3.2 Molecular cloning and tandem MS/MS strategy to elucidate the primary sequence of bioactive peptides from frog skin secretions. Created with biorender.com.

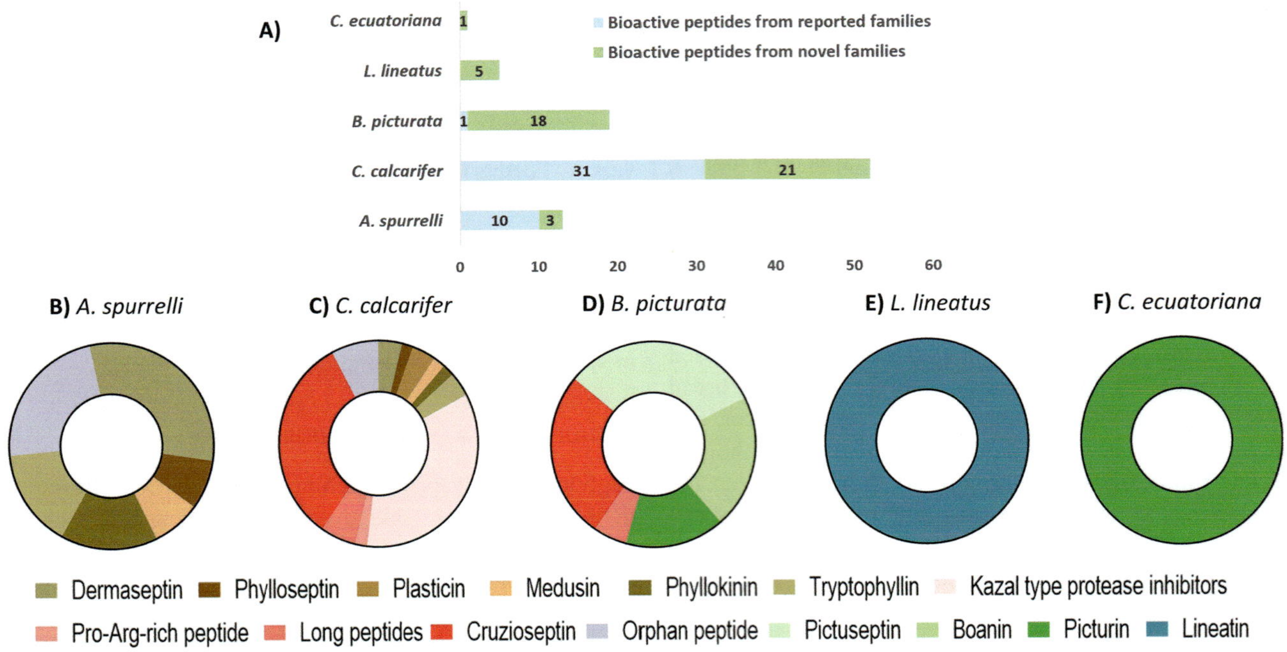

FIGURE 3.3 Peptides present in the skin secretions of frogs collected or lab bred in Ecuador. A: Number of bioactive peptides belonging to previously reported or novel families. Peptide profile of B: *Agalychnis spurrelli*, C: *Cruziohyla calcarifer*, D: *Boana picturata*, E: *Lithodytes lineatus*, and F: *Callimedusa ecuatoriana*.

Dermaseptins

Dermaseptins are a large family of AMPs with more than 70 members reported in different frog species, including *Phyllomedusa* spp., *Agalychnis* spp., and *Phasmahyla* spp. These peptides have a length of 21–35 amino acids, a conserved tryptophan residue in the third position, and a motif sequence AA(A/G)KAAL(G/N)A in the middle region (Amiche *et al.*, 2008; Nicolas and El Amri, 2009). Dermaseptins have shown a broad spectrum of activity against bacteria, fungi, protozoa, yeast, and enveloped viruses. Despite the high similarity in amino acid sequences, the antimicrobial potency of dermaseptins to kill etiologic agents varies considerably (Nicolas and El Amri, 2009). Four novel dermaseptins have been reported in the skin secretion of *Agalychnis spurrelli* (Hylidae). The mature sequences of dermaseptins showed a length of 27–28 residues and were named as DMS-SP2, DMS-SP3, DMS-SP4, and DMS-SP5. The synthetic dermaseptins were active against yeast, parasites, Gram-positive bacteria, and Gram-negative bacteria, including antibiotic-resistant bacteria (Table 3.2). DMS-SP2 displayed the most potent inhibitory activity against *Escherichia coli*, *Staphylococcus aureus*, oxacillin-resistant *S. aureus* (ORSA), *Klebsiella pneumoniae*, and *Candida albicans*. Furthermore, it was observed that none of the dermaseptins induced hemolysis at minimal inhibitory concentrations (MICs) (Cuesta *et al.*, 2019; Proaño-Bolaños *et al.*, 2019). In a more recent study, DMS-SP2 activity was assessed against *Trypanosoma cruzi* (trypomastigote and amastigote stages), *Leishmania mexicana* (promastigote form), and *Plasmodium falciparum* (the drug-sensitive and the multiple-drug-resistant strain). In general, this peptide presented an antiparasitic effect on all parasites in a micromolar range. However, DMS-SP2 showed cytotoxic activity in macrophages, with a low selectivity index (unpublished data).

On the other hand, two dermaseptins were identified in the amphibian skin of *Cruziohyla calcarifer*. These novel peptides were named dermaseptin-CC-1 and dermaseptin-CC-2 because they are the first members of that family reported in this species. With 35 residues, dermaseptin-CC-1 is the longest peptide, whereas dermaseptin-CC-2 shows a length of 33 residues. Both peptides share characteristics with other members of the dermaseptin family, such as the tryptophan residue in position 3 from the N-terminus and the sequence -AAGKAALN- in the central region (data not published). The antimicrobial activity of synthetic dermaseptin-CC1 was evaluated against yeast and bacteria, including drug-resistant clinical isolates. This peptide showed a broad spectrum of activity, inhibiting *Escherichia coli*, *Staphylococcus aureus*, *Candida albicans*, *Klebsiella pneumoniae*, and *Pseudomonas aeruginosa* at micromolar concentrations. Moreover, the toxicity assay of dermaseptin-CC1 against red blood cells revealed that this peptide induces hemolysis by less than 1% at concentrations lower than 64 mg/L (unpublished data).

Phylloseptins

Phylloseptins are a family of AMPs that were first reported in the skin secretions of *Pithecopus hypochondrialis* and *P.*

TABLE 3.2
Biological Activity of Peptides from Ecuadorian Frogs

Peptide	Primary Sequence	Isolated from	Bioactivity	References
Dermaseptin-SP2	ASWKVFLKNIGKAAGKAVLNSV-TDMVNQ-NH_2	*A. spurrelli*	*E. coli, S. aureus*, oxacillin-resistant *S. aureus, K. pneumoniae, C. albicans*	(Cuesta *et al.* (2019), Proaño-Bolaños *et al.* (2019)
Dermaseptin-SP3	SLWSSIKDMAAAAGRAALNAVN-GIVNP-NH_2	*A. spurrelli*	*E. coli, S. aureus*	(Cuesta *et al.* (2019), Proaño-Bolaños *et al.* (2019)
Dermaseptin-SP4	SLWSSIKDMAAAAGRAALNAVNGI-LNP-NH_2	*A. spurrelli*	*E. coli, S. aureus, K. pneumoniae*	(Cuesta *et al.* (2019), Proaño-Bolaños *et al.* (2019)
Dermaseptin-SP5	SLRSSIKDMAAAAGRAALNAVNGI-VNP-NH_2	*A. spurrelli*	*E. coli, C. albicans*	(Cuesta *et al.* (2019), Proaño-Bolaños *et al.* (2019)
Dermapseptin-CC1	GLWSIIKTVGKLAAKEAAKAAG-KAALNTVAEKVNE-NH_2	*C. calcarifer*	*E. coli, S. aureus, C. albicans, K. pneumoniae, P. aeruginosa*	
Phyloseptin-SP1	FLSLIPHVISAIPHVVNALSNL-NH_2	*A. spurrelli*	*L. mexicana, T. cruzi, P. falciparum*	(Proaño-Bolaños *et al.* (2019)
Cruzioseptin 1	GFLDIVKGVGKVALGAVSKLF-NH_2	*C. calcarifer*	*E. coli, S. aureus, C. albicans, L. braziliensis, L. amazonensis, L. mexicana, T. cruzi, P. falciparum*	(Proaño-Bolaños *et al.* (2016) Mendes *et al.* (2020)
Cruzioseptin 2	GFLDVIKHVGKAALGVVTHLINQ-NH_2	*C. calcarifer*	*E. coli, S. aureus, C. albicans, L. braziliensis, L. amazonensis*	(Proaño-Bolaños *et al.* (2016) Mendes *et al.* (2020)
Cruzioseptin 3	GFLDVVKHIGKAALGAVTHLINQ-NH_2	*C. calcarifer*	*E. coli, S. aureus, C. albicans, L. braziliensis, L. amazonensis*	(Proaño-Bolaños *et al.* (2016) Mendes *et al.* (2020)
Cruzioseptin 4	GFLDVIKHVGKAALSVVSHLINE-NH_2	*C. calcarifer*	*E. coli, S. aureus, C. albicans, L. mexicana, T. cruzi, P. falciparum*	
Cruzioseptin 16	GFLDVLKGVGKAALGAVTHLINQ-NH_2	*C. calcarifer*	*E. coli, S. aureus, C. albicans, L. mexicana, T. cruzi, P. falciparum*	(Cuesta *et al.* (2021b)
Cruzioseptin 17	GFLDVVKGVGKAALGAVTHLINQ-NH_2	*C. calcarifer*	*E. coli, S. aureus, C. albicans*	(Cuesta *et al.* (2021b)
Picturin-1	GVFKDALKQLGAALLDKAANALKPK	*B. picturata*	*E. coli, S. aureus*	(Morán-Marcillo *et al.* (2022)
Picturin-2	GVFKDALKQFGAALLDKAANALKPK	*B. picturata*	*E. coli*	(Morán-Marcillo *et al.* (2022)
Picturin-3	GVFKDALKQFGAALLDQAANALKPK	*B. picturata*	*E. coli, S. aureus*	(Morán-Marcillo *et al.* (2022)
Picturin-Ce	GVFKDALKQFGAALPDKAANALKPK	*C. ecuatoriana*	Not active	
Pictuseptin-1	GFLDTLKNIGKTVGRIALNVLT	*B. picturata*	*E. coli, S. aureus, C. albicans*	(Morán-Marcillo *et al.* (2022)
Pictuseptin-2	GFLDTLKNIGKTVGGIALNVLT-NH_2	*B. picturata*	*E. coli, S. aureus, C. albicans*	(Morán-Marcillo *et al.* (2022)
Pictuseptin-3	GFLDTLKNIGKTVGKVALDVAKNVLT	*B. picturata*	*E. coli, S. aureus, C. albicans*	(Morán-Marcillo *et al.* (2022)
Lineatin-2	ILPLLITAAKVIPKIHSFFKNLG-NH_2	*L. lineatus*	*E. coli, S. aureus, C. albicans*	(Yambay, 2022)
CCKP1	AVSAECARYGLACNKMLAPVCGTD-GTTYSNQCMLCYYNRKNKKNIEIRS-RGRC	*C. calcarifer*	Trypsin inhibitory activity	(Proaño-Bolaños *et al.* (2017)

oreades (Leite *et al.*, 2005). To date, 48 phylloseptins have been isolated and identified from several phyllomedusine frogs. The mature sequence of these peptides usually comprises 15–22 residues, a conserved tripeptide (FLS-) in the N-terminal domain, and a C-terminal amidation. Phylloseptins have shown antibacterial, antifungal, antiprotozoan, antibiofilm, and antiproliferative properties (Leite *et al.*, 2005; Wang *et al.*, 2015; Wu *et al.*, 2017). Recently, a novel phylloseptin of 22 amino acids, named phylloseptin-SP1, was found in *Agalychnis spurrelli*. This peptide was not active against *Escherichia coli, Staphylococcus aureus, and Candida albicans* at concentrations of up to 512 mg/L (217.69 μM). In addition, phylloseptin-SP1 induced hemolysis $> 80\%$ at a concentration of 256 mg/L (Proaño-Bolaños *et al.*, 2019). A recent study identified the first phylloseptin, phylloseptin-CC1, in *Cruziohyla calcarifer*. The 19-mer mature sequence of phylloseptin-1 shares a high similarity $(71-76\%)$ with other phylloseptins such as phylloseptin-J6 and phylloseptin-J7 of *Phasmahyla jandaia* and phylloseptin-3 of *P. hypochondrialis* (unpublished data).

Plasticins

Plasticins are dermaseptin-related peptides characterized by being rich in glycine and leucine residues arranged in repetitive 5-mer motifs -GXXXG-, where X is any amino acid. The "plasticin" name is given by their conformational and functional plasticity. These peptides can adopt α-helical, β-sheet, random coil, or β-hairpin structures, possibly owing to their large number of glycine amino acids. With respect to bioactivity, they inhibit several strains of Gram-negative bacteria, Gram-positive bacteria, and fungi. It has been shown that plasticins act via the disruption of prokaryotic and eukaryotic cell membranes or targeting intracellular components. Plasticins have been reported in species belonging to the genera *Agalychnis*, *Leptodactylus*, *Pachymedusa*, and *Phyllomedusa* (El Amri and Nicolas, 2008; Azevedo Calderon *et al.*, 2011; Carlier *et al.*, 2015). Recently, two novel plasticins, plasticin-CC1 and plasticin-CC2, with lengths of 23–29 residues, were identified in *Cruziohyla calcarifer.* The mature peptide sequences were 45–60% similar to plasticins PTC-A1 of *Agalychnis annae*, PTC-C1 and PTC-C2 of *A. callidryas*, PTC-DA1 of *A. dacnicolor*, and PTC-S1 of *Phyllomedusa sauvagii*. In addition, plasticin-CC1 and plasticin-CC2 have two and three -GXXXG- domains in their sequences, respectively. Plasticin-CC1 and plasticin-CC2 were chemically synthesized by SPPS and tested against *Escherichia coli*, *Staphylococcus aureus*, and *Candida albicans*, showing no antimicrobial activity, so their functions remain unknown (unpublished data).

Medusins

Medusins are 18-mer antimicrobial peptides with highly conserved mature sequences: LLGMIP(L/V)AISAIS(A/S) LSKL and C-terminal amidation. Medusins were able to inhibit *Staphylococcus aureus*, *Batrachochytrium dendrobatidis*, and *Candida albicans*. However, these peptides were not effective against *Escherichia coli*, methicillin-resistant *S. aureus* (MRSA), *Enterococcus faecalis*, or *Pseudomonas aeruginosa*. Furthermore, medusin slightly inhibited biofilm formation by *S. aureus*. Four members of this family have been found in the skin secretions of *Agalychnis callidryas*, *Pithecopus hypochondrialis*, *Phyllomedusa dacnicolor*, and *P. tarsius*. The names of the medusins have been designated according to the initials of the species from which they were isolated (medusin-AC for *A. callidryas*, medusin-PH for *P. hypochondrialis*, medusin-PD for *P. dacnicolor*, and medusin-PT for *P. tarsius*) (Xi *et al.*, 2013; Gao *et al.*, 2017). Recently, a medusin of 18 amino acids was discovered in *A. spurrelli* and *Cruziohyla calcarifer*. Medusin-AS and medusin-CC share the same primary structure as medusin-AC and phylloseptin-L1 (isolated from *A. lemur*). Therefore, this is the first AMP identified in four species of two different genera (Proaño-Bolaños *et al.*, 2019).

PHYLLOKININS

Phyllokinins are bradykinin-like peptides with two additional residues at their C-termini: isoleucine and tyrosine. Furthermore, these peptides often display post-translational modifications such as hydroxyproline and tyrosine sulfate. It has been shown that phyllokinins show more potent activity compared to bradykinins. To date, several phyllokinins have been identified in *Pithecopus rohdei*, *P. hypochondrialis*, *P. azureus*, *Phyllomedusa bicolor*, *P. sauvagii*, and, unexpectedly, in *Physalaemus signifier* (Anastasi *et al.*, 1966; Chen and Shaw, 2003). Recently, a 11-mer phyllokinin was identified in *Cruziohyla calcarifer.* The primary sequence was 100% similar to phyllokinin sequences isolated in *P. bicolor*, *P. rodhei*, and *P. hypochondrialis.* Furthermore, two phyllokinins, named [Ser^6, Val^{10}, Asp^{11}]-phyllokinin and [Thr^6, Val^{10}, Asp^{11}]-phyllokinin, were found in the skin secretion of *Agalychnis spurrelli.* The peptide [Thr^6, Val^{10}, Asp^{11}]-phyllokinin showed 100% identity with Thr^6-BK of *A. dacnicolor* and Hyp^3-Thr^6-BK of *A. callidryas*, whereas the amino acid sequence of [Ser^6, Val^{10}, Asp^{11}]-phyllokinin was 100% identical to the bradykinin-like peptide RD-11 of *Ascaphus truei* (Proaño-Bolaños *et al.*, 2019).

Tryptophyllins

Tryptophyllins, a family of small (4–14 residues), heterogeneous peptides, have been isolated in species of Phyllomedusidae. These peptides have tryptophan and proline residues in their sequences. Based on their primary structures, tryptophyllins have been classified into three groups: (1) type 1 tryptophyllins have 7–8 residues and an N-terminal Lys-Pro-(KP-) sequence as well as an internal -Pro-Trp- (-PW-) or -Pro-Pro-Trp- (-PPW-); (2) type 2 is the most heterogeneous group of peptides, with 4–7 residues, an internal -Pro-Trp- (-PW-), and a C-terminal amide; and (3) type 3 tryptophyllins are the most conserved, containing 13 residues with an internal -Pro-Pro-Pro-Ile-Tyr-Pro- (-PPPIYP-) motif. Functional studies of this peptide family have shown a broad range of biological activity such as relaxing or contracting mammalian smooth muscle, bradykinin antagonist activity, antiproliferative activity, and antimicrobial activity. However, certain tryptophillin members have not shown such bioactivity (Chen *et al.*, 2004; Wang *et al.*, 2009, 2013; Ge *et al.*, 2014). In recent years, two tryptophillins, CcT-2 and CcT2-2, were found in *Cruziohyla calcarifer*, which also belongs to Phyllomedusidae. CctT2-2 is a 7-mer mature peptide with a -PW- motif in the central region and a C-terminal amide, whereas the CcT-2 peptide contains two post-translational modifications (PTMs): a pyroglutamic acid (pGlu) instead of a glutamine (Q) at the N-terminus and an amidated valine. The bioactivity assay revealed that CcT-2 was not a bradykinin antagonist and did not inhibit the growth of several cancer cell lines (data not published). Furthermore, tryptophyllin-AsT1 and tryptophyllin-AsT3 are peptides identified in the skin secretion of *Agalychnis spurrelli.* It has been shown that the mature peptide sequence of AsT-1 was 100% identical to tryptophyllin PdT-1 of *A. dacnicolor*, and the mature peptide of AsT3 shared only 62.5% similarity with PhT3 of *Pithecopus rohdei*. The bioactivity of these peptides has not been investigated (unpublished data).

Kazal-Type Protease Inhibitors

Protease inhibitors are another class of proteins present in the secretions of amphibian skin. Protease inhibitors have been classified according to the presence of structural motifs in their sequences such as the Kunitz, Bowman–Birk, and Kazal types. To date, Kazal-type protein inhibitors have only been reported in Phyllomedusidae. As for the biological activity, these molecules exhibit high homology at their active sites, allowing them to inhibit different peptidases at a rate of 1:5. Furthermore, the mechanism of these inhibitors is the interaction with the active site of the peptidase in a substrate-like manner. Possibly, its presence in the skin of amphibians is (1) to confer protection against peptidases produced by pathogenic microorganisms, (2) to regulate the processing reactions of bioactive peptides, and (3) to prevent the degradation of bioactive peptides (Lai *et al.*, 2002; Song *et al.*, 2008; Wang *et al.*, 2012; Li *et al.*, 2012; Proaño-Bolaños *et al.*, 2017). In recent years, seven Kazal-type proteinase inhibitors (CCKP) (with 18 variants) from *Cruziohyla calcarifer* were described. These proteins contain 51–73 amino acids and all share the canonical Kazal motif (CX$_{(6-7)}$)-CX$_{(6-7)}$)-Y-X$_{(2-3)}$)-CX$_{(2)}$)-C). CCKP-1 showed trypsin inhibitory activity. Furthermore, primary and secondary structure analyses revealed that CCKP-4 may possibly be a trypsin inhibitor, while CCKP-2, CCKP-5, and CCKP-7 possibly show chymotrypsin inhibitor activity (Proaño-Bolaños *et al.*, 2017).

Proline–Arginine-Rich Peptides

Rhinophrynin-27 and rhinophrynin-33, proline–arginine-rich peptides, were isolated from the Mexican burrowing toad (*Rhinophrynus dorsalis*). Rhinophrynin-27 showed no antimicrobial or cytotoxic properties (Conlon *et al.*, 2017). On the other hand, this peptide stimulates (1) the production of the anti-inflammatory cytokine IL-10 and (2) the release of insulin from BRIN-BD11 rat clonal β-cells and isolated mouse islets (Scorciapino *et al.*, 2019). A 13-residue peptide, RK-13, was identified in *Cruziohyla calcarifer.* RK-13 also showed insulin-releasing activity and did not inhibit the growth of *Escherichia coli*, *Staphylococcus aureus*, or *Candida albicans* (Abdel-Wahab *et al.*, 2005). Recently, a novel proline–arginine-rich peptide, PR-CC1, was discovered in the skin secretion of *C. calcarifer.* The mature sequence of PR-CC1 has a length of 13 amino acids, 62% of which are arginine (Arg) and proline (Pro). Furthermore, this peptide is 92.3% similar to KR-13, showing a substitution of arginine by lysine in the third position (unpublished data).

Long Peptides with α-Helical Structures

Three novel peptides with a length of 34–52 residues were identified in the skin secretion of *Cruziohyla calcarifer.* They were named as α-long peptide-CC1, α-long peptide-CC2, and α-short peptide-CC3. These peptides showed 34.62–47.05% similarity to dermadistinctin-K from *Phyllomedusa distincta*. Furthermore, secondary structure predictions revealed that these peptides form α-helices. On the other hand, a novel peptide of 55 amino acids, α-long peptide-HP1, was found in *Boana picturata*. The mature sequence of α-long peptide-HP1 shares 78.18% identity with α-long peptide-CC2 from *C. calcarifer* (unpublished data).

Novel Peptide Families from Ecuadorian Amphibians

The comparison of the primary structures of 48 bioactive peptides identified in several species of Ecuadorian frogs (Figure 3.4) led to them being classified into new peptide families (see Figure 3.3), such as cruzioseptins, picturin, pictuseptins, boanins, and lineatins. Several members of these peptide families have shown activity against bacteria, yeast, and parasites (Table 3.2).

Cruzioseptins

Cruzioseptins are a family of AMPs isolated from *Cruziohyla calcarifer* and *Boana picturata* (Proaño-Bolaños *et al.*, 2016; Cuesta *et al.*, 2021b; Morán-Marcillo *et al.*, 2022). To date, 17 cruzioseptins with highly conserved physicochemical properties have been reported. These peptides have a length of 20–32 residues and are characterized by the mid-region motifs -GALVAVSK- or -GKAAL(N/G/S) (V/A)V- and the N-terminal sequence GFLD. Cruzioseptins have a broad spectrum of action, including antibacterial, antifungal, and antiparasitic properties (Table 3.2). For instance, CZS-1, CZS-2, CZS-3, CZS-16, and CZS-17 inhibited *Staphylococcus aureus*, *Escherichia coli*, and *Candida albicans* at micromolar concentrations. The minimum inhibitory concentration (MIC) values are different for each peptide, where CZS-1 displays the highest antimicrobial activity (Proaño-Bolaños *et al.*, 2016; Cuesta *et al.*, 2021b). Similar to other cationic antimicrobial peptides, these peptides also exert their antibacterial action through a membrane disruption mechanism (Valdivieso-Rivera *et al.*, 2022). In regard to the antiparasitic activity of cruzioseptins, their leishmanicidal effect was investigated. CZS-1, CZS-2, and CZS-3 showed activity against promastigote forms of *Leishmania amazonensis* (EC_{50} = 51.4–74.8 μM) and *L. (V) braziliensis* (EC_{50} = 10.9–40.3 μM). Furthermore, the three peptides also displayed inhibitory activity against lesion-isolated amastigotes of *L. amazonensis* (EC_{50} = 9.1–28.5 μM) and *L. (V) braziliensis* (EC_{50} = 13.1–34 μM). The cytotoxicity values of cruzioseptins in macrophages (CC_{50} = 112.34–127.19 μM) were above all EC_{50} values determined, showing that CZS-1 had a selectivity index >10 for both promastigotes and amastigotes of *L. braziliensis*. Insight into the leishmanicidal mechanism in promastigotes was obtained. Cruzioseptins induced the permeabilization of the parasite's cell membrane (Mendes *et al.*, 2020). Recently, the effects of CZS-1, CZS-4, and CZS-16 against *Leishmania mexicana*, *Trypanosoma*

FIGURE 3.4 Species of Ecuadorian frogs for which new peptides families have been described. Top left: *Cruziohyla calcarifer* (cruzioseptins). Top right: *Callimedusa ecuatoriana* (picturins, pictuseptins). Bottom left: *Boana picturata* (cruzioseptins, picturins, boanins). Bottom right: *Lithodytes lineatus* (lineatins). LAC, DAL (top right).

cruzi, and *Plasmodium falciparum* were determined. The study revealed that the three cruzioseptins were active against all parasites tested. CZS-4 showed the most potent and selective antiparasitic activity compared to CZS-1 and CZS-16 (Proaño-Bolaños et al., 2024).

Low stability under physiological conditions is one of the challenges that AMPs must overcome. Indeed, it has been evidenced that the peptides may lose their bioactivity in the presence of salts or serum (Knappe *et al.*, 2010; Ying *et al.*, 2019; Liscano *et al.*, 2020). The influence of salts and serum concentrations in the cruzioseptin MICs were evaluated. The antimicrobial assay revealed that incubated cruzioseptins with different salt concentrations led to an increase of up to two times the MIC. Similarly, the antibacterial effect of cruzioseptins was reduced in the presence of serum with a three-fold increase in the MIC (Valdivieso-Rivera *et al.*, 2022).

Picturins

Picturins are a new family of AMPs identified in the skin secretions of *Boana picturata* and *Callimedusa ecuatoriana*. Picturin 1, picturin 2, and picturin 3 present highly conserved mature sequences with a length of 25 amino acids and two divergence sites: Leu/Phe in position 10 and Lys/Gln in position 17. In silico tools revealed that these peptides adopt an α-helical conformation with a hydrophobic face. In order to evaluate the antimicrobial activity of picturins, the three peptides were chemically synthesized and tested against *Escherichia coli*, *Staphylococcus aureus*, and *Candida albicans*. Picturin 1, picturin 2, and picturin 3 inhibited the growth of *E. coli*. Picturin 3 was slightly active against *S. aureus*. However, picturins were not effective against *C. albicans*. Furthermore, hemolysis caused at the concentration of 128 mg/L for the three picturins was less than 1.4% (Morán-Marcillo *et al.*, 2022). In a more recent study, a new 25-mer picturin was isolated from *C. ecuatoriana*. Picturin Ce showed a $90-96\%$ similarity to the three picturins previously reported. The antimicrobial activity assay reported that picturin Ce does not show antimicrobial properties against *S. aureus*, *E. coli*, or *C. albicans*. Additionally, the hemolytic assay confirmed that picturins are non-hemolytic peptides. Picturin Ce does not induce hemolysis at concentrations up to 2.048 mg/L (unpublished data).

Pictuseptins

Six new peptides found in the cutaneous secretions of *Boana picturata* were assigned to a new family of AMPs: pictuseptins. The mature peptides contain 22–26 amino acid residues and a conserved N-terminal domain (GFLDTLKNIGK). Bioinformatic tools demonstrated that these peptides have a cationic charge and an α-helix structure with a hydrophobic face. Pictuseptins show a broad spectrum of antimicrobial activity. For instance, pictuseptin 1, pictuseptin 2, and pictuseptin 3 were able to inhibit the growth of *Escherichia coli, Staphylococcus aureus, and Candida albicans* in the micromolar range. However, the use of high concentrations of these peptides causes cell lysis in human erythrocytes (Morán-Marcillo *et al.*, 2022).

On the other hand, the antiparasitic activity of pictuseptin-1 (PTS-1) was evaluated. Trypomastigotes and amastigotes of *Trypanosoma cruzi*, promastigotes of *Leishmania mexicana*, and drug-sensitive and multiple-drug-resistant strains of *Plasmodium falciparum* were susceptible to PTS-1. However, this peptide showed great toxicity in macrophages, showing low selectivity (data not published).

Boanins

Boanins are a new peptide family from *Boana picturata*. Four members with 28–37 residues were reported with highly preserved mature sequences. For instance, boanins presented only two divergent changes: Arg/Thr in the seventh position and Leu/Phe in the tenth position. Boanins share 61% of their identity with figainin-05. Furthermore, it was observed through in silico tools that these peptides are cationic and form α-helix structures. The authors argued that future work will focus on evaluating bioactivity against human pathogens (Morán-Marcillo *et al.*, 2022).

Lineatins

Five novel peptides were identified in the cutaneous secretions of *Lithodytes lineatus* and were named as lineatins. The lengths of mature sequences of lineatin 1a, lineatin 1b, lineatin 2, lineatin 3a, and lineatin 3b vary from 13 to 30 amino acid residues. Physicochemical properties showed that these peptides have a hydrophobicity between 0.141 and 0.802 and a net charge in the range of –3 to +4. Furthermore, bioinformatics tools predicted an α-helix as the secondary structure of these peptides. Lineatin 2 was chemically synthesized, and its bioactivity was evaluated against *Escherichia coli*, *Staphylococcus aureus*, and *Candida albicans*. This peptide showed antimicrobial properties in micromolar concentrations against all tested pathogens. Additionally, lineatin-2 induced hemolysis of less than 15% at MIC concentrations (Yambay, 2022).

Orphan Peptides

Orphan peptides are those that have unique primary sequences and cannot be assigned to any of the previously reported families. Several peptides are included in this list, such as DRS-9 from *Phyllomedusa sauvagii* and DRP-PD-3–7 from *Agalychnis dacnicolor* (Nicolas and El Amri, 2009). Recently, three peptides with a length of 11–15 residues were found in *A. spurrelli*. Their mature peptide sequences had no relevant results when compared with NCBI databases using BLAST-p (Proaño-Bolaños *et al.*, 2019). Likewise, four novel peptides were identified in the skin secretion of *Cruziohyla calcarifer*. The mature sequence showed a length of 8–27 amino acids. As with the other three unknown peptides from *A. spurrelli*, the mature sequence does not produce any significant hits. Therefore, the seven peptides were assigned to the orphan peptide family until further information was obtained (data not published).

Amphibian Skin AMPs as a Source of Inspiration for the Development of Therapeutic Antimicrobial Candidates

The great potency and multifunctional activity of antimicrobial peptides (AMPs) have led to the interest in using these molecules in the agricultural, livestock, and pharmaceutical industries (Holaskova *et al.*, 2015; Meneguetti *et al.*, 2016). However, toxicity toward host cells, bioavailability, loss of bioactivity under environmental conditions, and stability are some of the challenges to overcome in order to offer peptide-based biosolutions (Shanmugaraj *et al.*, 2021; Valdivieso-Rivera *et al.*, 2022). Peptide engineering enables us to obtain a new generation of molecules with enhanced desired effects and fewer unwanted properties. Helicity, hydrophobicity, net charge, and amphipathicity are critical features in the bioactivity and toxicity of AMPs (Mohammad *et al.*, 2022). Therefore, the design of new peptide analogs has been based on tuning these parameters using several approaches such as substitution, truncation, hybridization, N-terminal or C-terminal modifications, cyclization, and in silico drug design (Ting *et al.*, 2020; Torres *et al.*, 2021). For instance, specific residue substitution with positively charged amino acids enhanced the antimicrobial potency (Ajingi and Jongruja, 2020; Sarkar *et al.*, 2021). After shortening the length of the chain of natural AMPs, the truncated analogs showed less toxicity to red blood cells without affecting the antimicrobial activity (Zhou *et al.*, 2019; Ying *et al.*, 2019). The cyclization or incorporation of unnatural amino acids has been widely used to reduce degradation by proteases (Barreto-Santamaría *et al.*, 2019). Peptide hybridization, which is the combination of two bioactive peptide fragments, is an approach that improves the antimicrobial activity and reduces the toxicity of AMPs (Yang *et al.*, 2020; Silva *et al.*, 2020; Mohammad *et al.*, 2022). Indeed, some of these strategies have been used in amphibian antimicrobial peptides (B. Guimarães *et al.*, 2016), including dermaseptins (Mor and Nicolas, 1994; Savoia *et al.*, 2008; Jiang *et al.*, 2014; Tan *et al.*, 2018; Li *et al.*, 2019; Ying *et al.*, 2019; Gong *et al.*, 2020; Qin *et al.*, 2021), caerin (Madanchi *et al.*, 2020; Cuesta *et al.*, 2021a), medusin (Gao *et al.*, 2017), ranatuerin (Sonnevend *et al.*, 2004; Chen *et al.*, 2018b; Zhou *et al.*, 2019), and brevinine (Chen *et al.*, 2018a; He *et al.*, 2020; Pei *et al.*, 2021). In this section, we will describe the current status of analogs derived from native Ecuadorian peptides.

TABLE 3.3
Peptide Analogs Derived from Antimicrobial Peptides of Ecuadorian Frogs

		MIC μM (mg/L)			
Peptide	**Primary Sequence**	***E. coli***	***S. aureus***	***C. albicans***	**Reference**
CZS-1	GFLDIVKGVGKVALGAVSKLF-NH2	15.11 (32)	3.77 (8)	3.77 (8)	(Proaño-Bolaños *et al.* (2016)
K8K10	VKGVGKVKLKAVSKL-NH2	>659.36 (>1024)	>659.36 (>1024)	>659.36 (>1024)	(Bermúdez-Puga *et al.* (2023)
K8K10F	VKGVGKVKLKAVSKLF-NH2	150.57 (256)	>301.14 (>512)	>301.14 (>512)	(Bermúdez-Puga *et al.* (2023)
[K4K15]CZS-1	GFLKIVKGVGKVALKAVSKLF-NH2	3.63 (8)	1.81 (4)	14.53 (32)	(Bermúdez-Puga *et al.* (2023)
Picturin-Ce	GVFKDALKQFGAALPDKAANALKPK	>196.99 (>512)	>196.99 (>512)	>196.99 (>512)	
Picturin-CE1a	GVFKKALKQFGAALLRLAANALKPK-NH 2	3.02 (8)	6.03 (16)	12.06 (32)	
Picturin-CE1b	GVFKDALKQFGAALDKAANALKK-NH	106.5 (256)	>212.99 (>512)	212.99 (512)	

Cruzioseptin Analogs

Cruzioseptin 1 (CZS-1), the most potent antimicrobial peptide of the cruzioseptin family (Proaño-Bolaños *et al.*, 2016; Mendes *et al.*, 2020; Cuesta *et al.*, 2021b), was used as an archetype to design three derivatives (Table 3.3). K8K10 (VKGVGKVKLKAVSKL-NH_2) and K8K10F (VKGVG-KVKLKAVSKLF-NH_2) were designed to evaluate the effects of length reduction and increased net charge on bioactivity. K8K10 did not inhibit the growth of tested pathogens, while K8K10F only showed antimicrobial activity against *Escherichia coli*. Both peptides have a net charge of +6 and differ from phenylalanine in the C-terminal. These findings highlight the importance of C-terminus phenylalanine in peptide bioactivity. On the other hand, [K4K15] CZS-1, a 21-mer analog with increased cationicity and reduced hydrophobicity, showed antibacterial properties against *E. coli, Staphylococcus aureus, and Candida albicans*. The substitutions in the analog led to an improvement in selectivity toward bacteria. Furthermore, scanning electron micrographs revealed that the peptide acts on the bacteria through a membranolytic effect. Additionally, [K4K15] CZS-1 exhibited non-selective antiproliferative activity on both non-cancer and cancer cell lines (Bermúdez-Puga *et al.*, 2023).

Picturin Analogs

The amino acid sequence of the inactive AMP picturin-CE was used to design two analogs with antimicrobial properties (Table 3.2). Picturin-CE1a has a continuous α-helix and four substitutions that increase cationicity and hydrophobicity compared to the parent peptide. This peptide was strongly active against the opportunistic yeast *Candida albicans*, Gram-positive bacteria, and Gram-negative bacteria, including ampicillin-resistant isolates. Additionally, picturin-CE1a induced hemolysis <23.59% at the MIC range. On the other hand, picturin-CE1b also presents a well-defined α-helix and the elimination of two prolines compared to the native peptide. This peptide inhibited the growth of *Escherichia coli*, *C. albicans*, and several ampicillin-resistant isolates, such as *Klebsiella pneumoniae*, *Pseudomonas aeruginosa*, and *Bacillus subtilis*. Furthermore, this peptide was more hemolytic than picturin-CE1a at each peptide concentration tested (data not published).

PROGRESS IN AMPHIBIAN SYSTEMATICS AND TAXONOMY IN ECUADOR

Luis A. Coloma and Carolina Reyes-Puig

The current state of progress in amphibian systematics and taxonomy in Ecuador requires the clarification of several terms, such as "systematics" and "taxonomy", which are often used interchangeably, along with a couple of related definitions. Systematics refers to the study of biological diversity and the relationships among organisms, while taxonomy comprises the theory and practice of describing, naming, and ordering groups of organisms, known as taxa. Phylogenetics is a vital aspect of systematics and involves the study of the evolutionary history and relationships among groups of organisms. Species refers to an ancestor-descendant lineage of a population that maintains its identity from other such lineages over time and space (Felsenstein, 1985; Wägele, 2005; Wiley and Lieberman, 2011).

In the past three decades, biological research has seen significant changes in taxonomy and systematics, leading to conceptual improvements in the definition of species and better methods of species delimitation and evolution understanding. This shift is driven by the availability of new characters beyond morphology and acoustics, improvements in molecular analysis, and computer technology progress. The traditional use of phenetics (numerical taxonomy) for the classification of organisms has been replaced by phylogenetic systematics. This approach has made the biodiversity classification process more robust, enabling the use of integrative approaches using multiple lines of evidence at the genetic and phenotypic levels, which have become popular in the last two decades (Pace, 2009; Padial *et al.*, 2010).

What is the purpose of taxonomy, particularly in a diverse group such as amphibians, and why is it crucial in a country like Ecuador? At its simplest, taxonomy is a tool for naming, classifying, and organizing species. It is especially useful in a country with a high number of species in a small area.

In Ecuador, the taxonomy and systematics of amphibians have historically relied on external morphological features, such as osteology, vocalizations, and tadpole characteristics, to describe and group species based on similarity and/or putative synapomorphies. For instance, many alpha taxonomic works from the 18th century to the end of the 20th century relied on such features, as demonstrated by species descriptions and classical taxonomic reviews of caecilians by Taylor (1968), *Atelopus* by Peters (1973), centrolenids by Lynch and Duellman (1973), *Pristimantis* (under *Eleutherodactylus*) by Lynch (1974) (1979) (1981) and Lynch and Duellman (1980) (1997), dendrobatids by Silverstone (1975) (1976) and Coloma (1995), and *Osteocephalus* by Jungfer *et al.* (2000) and Jungfer and Hödl (2002), among others. In the 1970s, anuran phenetic species groups were also recognized using phenograms to represent percentages of similarity (e.g., Duellman, 1972, 1973; Heyer, 1973).

From the 1970s to the 1990s, systematists relied on discrete morphological features to infer evolutionary and phylogenetic relationships, with searching for putative synapomorphic characters being a common practice. While most studies in this period focused on identifying monophyletic groups, the testing and verification of these hypotheses were often lacking. Additionally, there was a constant search for new ways to analyze and interpret the evolutionary histories of organisms, as well as for new kinds of characters. For example, Linda Trueb from KU pioneered the use of osteological characters in this context, as exemplified by her taxonomic treatments in Trueb (1971, 1974) and Trueb and Duellman (1971).

During this time, various groups of species had their phylogenetic relationships proposed, such as the *parviceps* group (Duellman and Crump, 1974), the *surdus* assembly (Lynch, 1980), *Ceratophrys* (Lynch, 1982), the *Lithobates* (under *Rana*) *palmipes* group (Hillis and de Sá, 1988), centrolenidae (Ruiz-Carranza and Lynch, 1991), *Hyloxalus* and related dendrobatidae (Coloma, 1995), and *Pristimantis* (under *Eleutherodactylus*) (Lynch and Duellman, 1997).

In the 1990s, developments in cladistic theory and methodology provided a basis for testing previous mostly intuitive hypotheses, and computer software such as PAUP and MacClade strengthened these efforts. However, these attempts to ascertain phylogenetic relationships based on external morphological and osteological datasets using computer software were of limited use given the relatively few discrete characters available compared to molecular characters. The coding of morphometric data and the use of continuous and polymorphic characters in phylogenetic inference, as shown by Coloma (1997), represented an advance but remained of limited use.

In the late 1980s and early 1990s, advancements in computing technology and the availability of molecular characters, such as serum albumins (Duellman *et al.*, 1988) and allozymes (Duellman and Hillis, 1987, 1990; Wiens and Coloma, 1992), greatly benefited phylogenetics and also species delimitation. However, in the first half of the 2000s, and with the formation of the QCAZ biobank in Ecuador, there was a shift toward the use of DNA sequencing tools in systematics and taxonomy, with genetic, genomics, and phylogenetics becoming predominant. Pioneering studies in this area include those on dendrobatids by Santos *et al.* (2003) and Darst *et al.* (2005). The focus of these studies was to understand the evolution of aposematism, diet specialization, and chemical defense in poison frogs. However, they also uncovered the presence of abundant cryptic diversity. Additionally, Ron *et al.* (2005) proposed a molecular phylogeny for *Engystomops* taxonomy, which identified cryptic species.

Since then, the use of molecular characters to generate phylogenetic hypotheses has spread in systematic and taxonomic studies of Ecuadorian amphibians. For instance, since about the 2010s, the research programs led by Juan Carlos Santos, Lauren A. O'Connell, and Kim Hoke are using phylogenetic approaches to biogeography and the evolution of behavior, physiology, and audition in frogs, among many other biology fields (see Chapter 2 and citations therein, under the subtitle "Centro Jambatu, a Fresh Venture"). Nowadays, phylogenies are commonly used as prerequisites for species delimitation and as part of integrative approaches to taxonomy, typically using genetic, external morphological, and bioacoustic data sets (e.g., Caminer and Ron, 2014; Caminer *et al.*, 2017; Guayasamin *et al.*, 2017; Yánez-Muñoz *et al.*, 2021) but also sometimes osteology, tadpoles (e.g., Coloma *et al.*, 2012; Carvajal-Endara *et al.*, 2019), geographic records (e.g., Arteaga-Navarro *et al.*, 2016), and environmental characters (e.g., Caminer and Ron, 2020).

The emphasis on species identification and delimitation began to rise in the 2000s due to growing concerns regarding threats to biodiversity and the desire to describe as many species as possible as quickly and accurately as possible before they disappear (Wiens, 2007). Despite this progress, there are also some caveats to surpass, for instance, taxonomic inflation. The term "taxonomic inflation" describes the situation where taxa are being described as new based on minor variations, leading to a proliferation of taxonomic names without a commensurate increase in our understanding of the diversity and relationships of these organisms. This taxonomic inflation may also influence other fields such as macroecology and conservation (Isaac *et al.*, 2004). There are recent calls for targeted, thorough geographic sampling across contact zones and into parapatric ranges to test for reproductive isolation and draw inferences about the evolutionary independence (or lack thereof) of populations that exchange genes (Hillis *et al.*, 2021). The danger of taxonomic inflation in the face of limited geographic sampling and a lack of clear diagnostic characters has been shown in *Osornophryne bufoniformis* (Páez-Moscoso and Guayasamin, 2012) and in *Dendropsophus molitor* from the Cordillera Oriental of Colombia (Arias-Cárdenas *et al.*, 2023). It is important to perform a close examination of geographic variation in species and contact zones between potential species to determine if they are isolated from one another or not and to show that two lineages do (or do not) remain distinct through time and space. In this sense, the detailed sampling and studies in Ecuador by Roland *et al.* (2017) and Páez-Vacas *et al.* (2022) on the dendrobatids *Oophaga sylvatica* and *Epipedobates anthonyi* have shed light not only on the lack of evolutionary independence but also on the mechanisms for the evolution and diversification of lineages. Similar approaches are still lacking for putative complexes in the large genus *Pristimantis* (previously *Eleutherodactylus*), which is one of the most diverse groups of vertebrates on the planet and the most diverse in Ecuador.

Pristimantis serves as a prime example of the significant changes that have occurred in species delimitation and phylogenetics over time. According to Reyes-Puig and Mancero (2022), until 2005, *Pristimantis* frogs were primarily described based on morphological characteristics, with 72% of their species being described in this manner. However, even today, many species of Ecuadorian *Pristimantis* are still being described, with the highest rate of description occurring between 2010 and 2021. Some of these new species are described without prior knowledge of their phylogeny (e.g., Yánez-Muñoz *et al.*, 2011, 2014, 2016; Rámirez-Jaramillo *et al.*, 2014; Reyes-Puig *et al.*, 2015; Brito-Zapata and Reyes-Puig, 2021; Reyes-Puig *et al.*, 2022).

Nevertheless, numerous cryptic species of *Pristimantis* are being discovered and described using phylogenetic molecular analyses (e.g., Páez and Ron, 2019; Reyes-Puig *et al.*, 2020b, Ron *et al.*, 2020; Ortega *et al.*, 2022; Zumel *et al.*, 2021), some of which lack morphological diagnostic features but exhibit call differences (Hutter and Guayasamin, 2015). For some groups of species of *Pristimantis* for which molecular phylogenies are available, limited specimens and geographic sampling remain problematic, as well as the lack of other kinds of data which may help to delimit taxa (e.g., calls, larvae, osteology) (Kieswetter and Schneider, 2013; Guayasamin *et al.*, 2018; Franco-Mena *et al.*, 2023).

Systematics and taxonomy are essential, especially when dealing with rare and cryptic species. In such cases, there are two options: (1) the traditional use of morphological characters, which can be extended to differences in calls (Hutter and Guayasamin, 2015) and behavior (Hurd and Enquist, 2005) (which is much less explored), and (2) molecular characters. Both options offer critical phylogenetic evidence to support taxonomic classification. Morphological characters are particularly valuable, and it is recommended to explore them in greater detail to identify "invisible differences" that may only become evident in a broader context. For instance, the genus *Oedipina* comprises tropical salamanders mainly distributed from Mexico to central Ecuador, with its highest diversity peaks in Central America (Frost, 2023). These elusive, fossorial salamanders have not been well studied in Ecuador due to the scarcity of specimens collected and documented in the field (Reyes-Puig *et al.*, 2020a), as well as the lack of tissues for molecular analyses. Historically, the species in this genus found in Ecuador had been identified as *Oedipina complex*. Three specimens are known from Ecuador, one collected in the 19th century and the other two in the 21st century. However, in 2020, Reyes-Puig *et al.* (2020a), revealed the existence of at least two lineages in Ecuador that were distinct from *O. complex* through a rigorous taxonomic process that included a detailed morphological description. This information could only have been identified through the use of morphological characters. It is expected that in the near future, DNA extraction and analysis from formalin-preserved specimens will help confirm or refute the current phylogenetic hypotheses based on morphology.

Taxonomy and Other Fields

The value of taxonomy goes beyond classification and it is evident in various other fields, from ecology to conservation and wildlife management (Agapow *et al.*, 2004; Marsh *et al.*, 2007; Guerra García *et al.*, 2008), as it provides an essential ecosystem service for the benefit of human beings. Taxonomy plays a crucial role in the interpretation of research results, as shown by numerous studies (e.g., Paquin *et al.*, 2008; Sastre and Lobo, 2009). Its absence can have adverse effects on the effective management of biodiversity (Gippoliti *et al.*, 2018). The chosen capacity, mode, and method of classification for a particular living group can lead to different responses, highlighting the critical nature of taxonomy in biological research.

Taxonomic revisions are fundamental for implementing species distribution and ecological niche models (Mori *et al.*, 2019). To understand the biogeographic processes underlying Ecuador's high rates of amphibian endemism and diversification, it is essential to know its phylogeny and recognize macroevolutionary patterns accurately. Including errors in taxonomic entities can result in the overestimation or underestimation of species distribution areas, which is particularly significant for species with conservation interest. Errors in the taxonomic process can change the category and conservation status of a species, which can have a direct effect on conservation actions and subsequent conservation efforts, whether local or regional. Therefore, this preliminary step should not be taken lightly. The preceding is very important if we think of species with conservation interest, as the error in the taxonomic process prior to the generation of one of these models can result in a radical change in the category and conservation status of a species. For example, *Noblella coloma* is categorized as DD on the IUCN Red List vs. CR in the study by Ortega-Andrade *et al.* (2021), which includes data curation processes such as the revision of the coordinates, localities, distribution, and taxonomic identification of populations conducted by researchers specialized in these groups. Similarly, the discovery of cryptic species among many species of *Pristimantis* previously considered more widespread has changed their conservation status dramatically. Considering that species of Andean *Pristimantis* reveal a striking cryptic diversity and small distribution ranges, anthropogenic alterations are likely to have a larger impact on their conservation than previously expected. For instance, this is the case for *Pristimantis trachyblepharis*, *P. phoxocephalus*, *P. calcarulatus*, *P. riveti*, *P. walkeri*, and *P. myersi*, among others, as discussed by Zumel *et al.* (2021). In other words, misidentification can have a direct effect on conservation actions for a given species and subsequent conservation efforts, whether local or regional. An illustrative example of the significance of taxonomy is the detection of a cryptic species of dendrobatid rocket frog endemic to the Yurimagua mining site in the Intag region. The frog played a crucial role in a compelling case presented by the local community, supported by a team of scientists and lawyers, to halt mining operations until conservation measures are implemented (Re:wild, 2022).

Moreover, the interest and motivation in the taxonomy of certain groups can generate biases toward geographic regions or ecosystems, leading to a misinterpretation of biodiversity patterns caused by oversampling (Sastre and Lobo, 2009). Comprehensive taxonomic knowledge of the number of amphibian species can directly impact practical conservation, not only for the species themselves but also for the ecosystems and habitats where they reside. Naming acts that include funding donations in some part of the process for long-term conservation (Reyes-Puig *et al.*, 2022) can further incentivize the importance of taxonomy in conservation efforts (see also Chapter 2, under the subtitle "The INABIO and the Coining of Species Names for Amphibian Conservation").

Discovery and Rates of Descriptions

According to Womack *et al.* (2022), around 150 new amphibian species are described globally each year, and Ecuador has the highest density of new amphibian species described between 2016 and 2020. Ortega *et al.* (2022) attribute this to the efforts of a large group of Ecuadorian amphibian taxonomists working in a hotspot of amphibian diversity (see Chapter 2 for a detailed overview). Despite significant progress in discovering and describing Ecuador's amphibian fauna, there is still much work to be done. Following the Amphibian Conservation Summit in 2005, there has been a call to action to complete inventories of amphibian species and taxonomic work due to the high extinction rates that amphibians face. Some have even referred to this work as "forensic taxonomy" (Mendelson III and Mulcahy, 2010), as new species descriptions can also highlight possibly extinct species. For example, while the number of *Atelopus* species in the Neotropics has increased from 70 to 99 since 2000, there are also 12 formally described species that may be extinct in Ecuador alone, with at least eight more awaiting description (some of them also possibly extinct).

Regrettably, the response to the crisis has not been sufficient to prevent further amphibian species decline and extinction (Bishop *et al.*, 2012). Systematics and taxonomy are key to accelerating discovery and description rates and providing a solid taxonomic basis for further studies and conservation efforts. This requires increased inventories, supported biobanks, and more work on systematics and taxonomy.

As of now, Ecuador is known to have 655 species of amphibians included in 78 genera and 19 families. Since 2000, about 9 new species have been described per year on average, and this number has increased to 12 since 2010. Figures 3.5 and 3.6 illustrates the numbers of species and the cumulative numbers described by year, from Linnaeus' first amphibian descriptions in 1758 up to December 2022. Over the last 23 years, this rate of description has added 206 species to the known list. Among these species, 114 (56%) belong to the single genus *Pristimantis* in the Craugastoridae family, followed by Hylidae (26 species), Centrolenidae

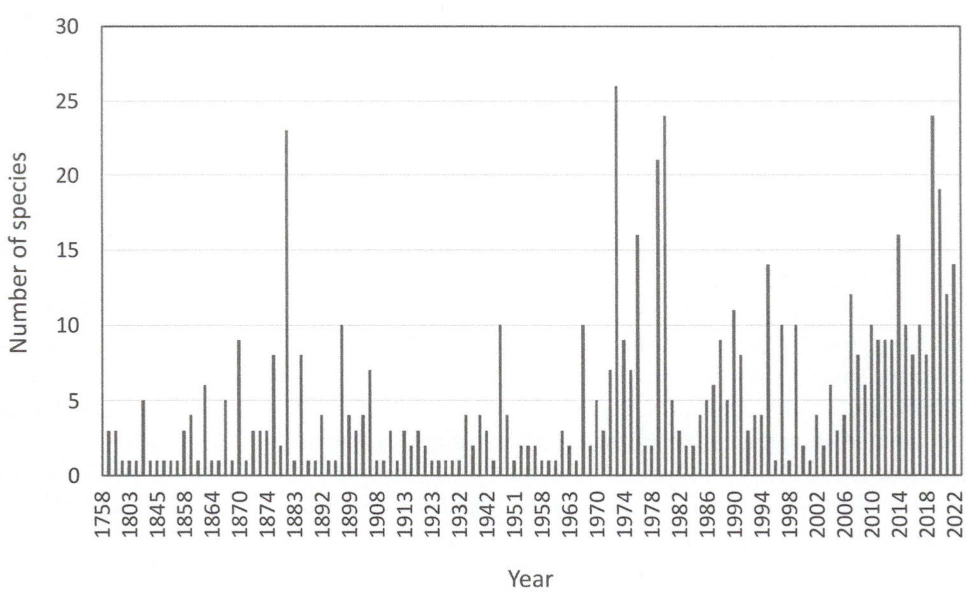

FIGURE 3.5 Number of species of amphibians in Ecuador by year of description (up to 31 December 2022).

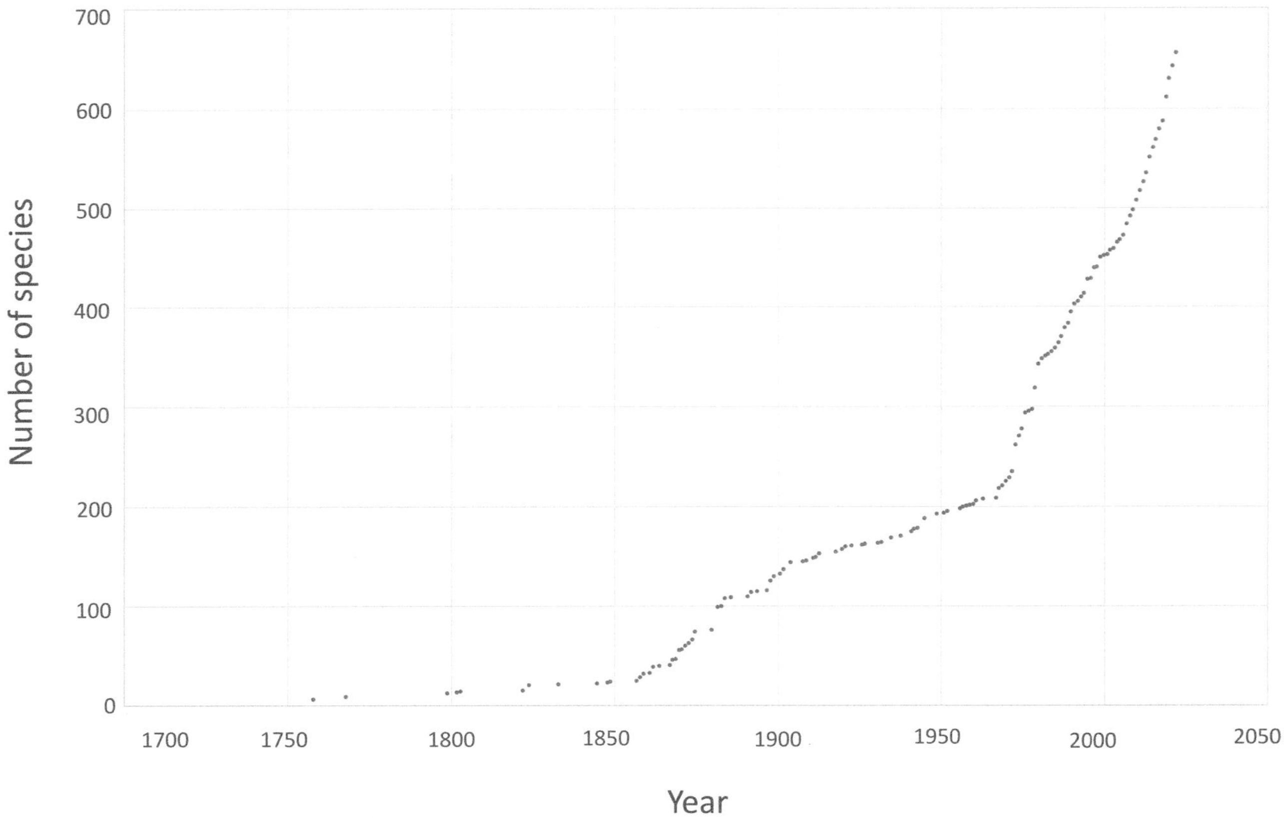

FIGURE 3.6 Cumulative number of species of amphibians in Ecuador by year of description (up to 31 December 2022).

(22), Bufonidae (14), and Dendrobatidae (7), among others (refer to Figure 3.7).

Coloma *et al.* (2007) suggested that the real number of amphibian species in Ecuador may be between 677 and 728. Based on the rates of descriptions in recent years, Reyes-Puig and Mancero (2022) estimated that during the next 10 years, presumably 45 new species of *Pristimantis* may be added to the amphibian inventory of Ecuador. It is now clear that the current number included in this encyclopedia (655), plus the species awaiting discovery and descriptions, is even higher than the upper number previously estimated. For instance, the molecular phylogenetic analyses of Kieswetter and Schneider (2013), Szĕkely *et al.* (2018), Páez and Ron (2019), and Ortega *et al.* (2022) found many candidate species of *Pristimantis* that remain undescribed. Additionally, Franco-Mena *et al.* (2023) found eight candidate species among their *Pristimantis myersi* group (which has 10 species) and suggested that their estimates, although restricted to a relatively small group, may apply to other Andean anurans with similar characteristics (e.g., Terrarana). If we use a conservative increase number (half of the findings of Franco-Mena *et al.*, 2023) for about 200 species of Andean *Pristimantis* in Ecuador, there will be 80 cryptic species to be added. Guayasamin *et al.* (2020) found a 15% increase in the family Centrolenidae. Applying a similar increase for centrolenids in Ecuador, 10 species should be added. If we also consider a speculative, conservative number of 50 species (20 *Pristimantis*, 10 centrolenids, 20 other taxa) that are likely to be awaiting collections in unexplored areas, as well as roughly 90 candidate species of other taxa that have already been identified based on preliminary analyses and/or publications (e.g., Gymnophiona (Fletcher-Lazo, 2002), *Lithobates* (Hillis and Wilcox, 2005), *Hyloxalus* (Páez-Vacas *et al.*, 2010), *Engystomops* (Funk *et al.*, 2011), *Osornophryne* (Páez-Moscoso and Guayasamin, 2012), *Bolitoglossa* (Elmer *et al.*, 2013), Dendrobatidae (Santos *et al.*, 2014), *Dendropsophus* (Gehara *et al.*, 2014), *Adenomera* (Fouquet *et al.*, 2014), *Hyloscirtus* (Ron *et al.*, 2018a; Yánez-Muñoz *et al.*, 2021), *Boana* (Caminer and Ron, 2020), *Pristimantis* (Yánez-Muñoz *et al.*, 2020; Reyes-Puig *et al.*, 2022)), then there are at least 230 new species awaiting description, and the total number of amphibian species predicted in Ecuador becomes 884.

Assuming no significant changes in the rates of discovery and descriptions, it may take another two decades (19 years) to document the additional 230 species at the current rate of species descriptions (12 per year over the most recent 13 years).

In summary, the number of amphibian species in Ecuador is likely to be much higher than previously estimated, with many more species awaiting formal description.

Considering that at least 396 (60%) of the Ecuadorian amphibian species are under categories of extinction risk, a rapid response and fundamental changes in policies and

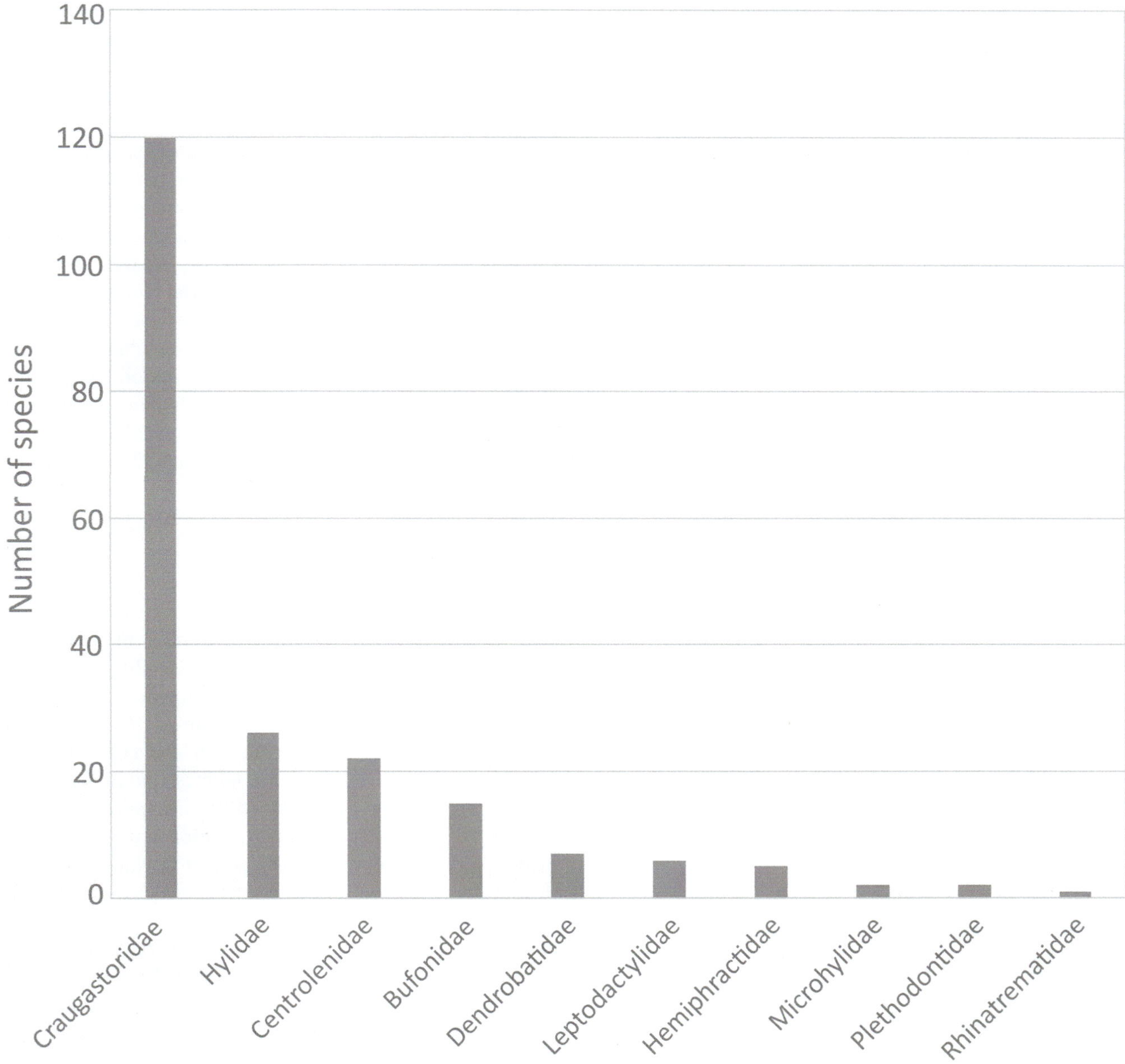

FIGURE 3.7 Families of Ecuadorian amphibians and number of species described and/or added in the period 2000–2022. Total = 206 species.

actions are urgently needed to address the amphibian crisis. This has been emphasized elsewhere (Bishop *et al.*, 2012) due to threats mainly from global warming, pathogens, habitat destruction, and invasive species (Pounds *et al.*, 2006; Lips *et al.*, 2006; Stuart *et al.*, 2004). If the goal established in 2022 by the Convention on Biological Diversity is to reduce extinction rates significantly within the next 20–30 years, then massive investments are needed (Lötters *et al.*, 2023). This task includes, among other actions, funding the exploration of under-sampled areas, supporting amphibian biobanks, and at least doubling the current rate of species descriptions. For this purpose, a major breakthrough is also necessary in the regulations and bureaucratic burden that is impeding biological research and the inventory of biological diversity worldwide and specially in megadiverse countries, as has been discussed elsewhere (Gilbert, 2010; Fernandez, 2011; Bockmann *et al.*, 2018; Prathapan *et al.*, 2018). One of the major challenges is to implement fair, reliable, simple, unambiguous, and transparent regulations. Ecuadorian and allied foreign scientists need juridical security. Conducting systematic research is problematic, and, ironically, in the meantime, the biodiversity crisis and mass extinctions are dramatically underway.

In Ecuador, we urgently need more projects and researchers working in systematics and taxonomy. We need scientists who can identify classification errors, establish phylogenetic relationships, and clarify the taxonomy of amphibians. It is also necessary to have people working in the field and the lab; those committed to learning in scientific collections, museums, and biobanks; and individuals who are willing to use stereoscopic microscopes to examine specimens in detail and improve their specialist skills. Although among the challenges in systematics and taxonomy is also a lack of interest among new biologists, who view it as a science of the past (Guerra García *et al.*, 2008; Wägele *et al.*, 2011), this seems not to be the case of amphibian systematists in Ecuador, given the unique history of the development of systematics in Ecuador through a lineage of Ecuadorian systematists closely tied to the legacy of William E. Duellman (see Chapter 2 under the subtitle "The Duellman KU Era: The Big Bang in Amphibian Systematics and Ecology"). Systematics and taxonomy are still relevant, and it is the duty of senior researchers to mentor young scientists and encourage them to embrace the best practices for the classification and organization of organisms, particularly amphibians. Thus, there is ample opportunity for new generations of researchers to contribute. Moreover, we need more women in taxonomy. Historically, this field has been dominated by men, and it needs active participation from female researchers. For instance, only 20% researchers participating in the descriptions of the hyperdiverse genus *Pristimantis* have been women, and only 2% have done so as main authors (e.g., first author or corresponding author) (Reyes-Puig and Mancero, 2022). A section that includes data and discussion about the participation of women in the Ecuadorian systematics and taxonomy is given in Chapter 2, under the subtitle "Women in Amphibian Biology: An Ongoing Trend of Underrepresentation in Academia."

SPECIES RICHNESS AND PATTERNS OF DISTRIBUTION

Luis A. Coloma, Andrea Terán-Valdez, and Camilo Rosero-Gómez

Ecuador is home to an impressive number of amphibian species, with over 655 species currently included in 19 families and 78 genera, in a relatively small country (276, 370 km²); we use the corrected number of species for Ecuador (sensu this encyclopedia, 655 species as of 2022). In comparison, other countries of similar size have much lower diversity; for example, Burkina Faso in Africa has 23 species, Italy in Europe has 48 species, and the Philippines in Asia have 115 species (AmphibiaWeb, 2022). When compared to the top 17 megadiverse countries listed by Conservation International, Ecuador ranks nineth by the global biodiversity index but fourth in terms of amphibian species diversity (numbers from AmphibiaWeb, 2022), behind larger neighboring countries such as Brazil (1159 species), Colombia (821 species), and Peru (665 species) (Figure 3.8). While the number of species in Ecuador and Peru is very similar, considering Peru's larger size and the fact that the taxonomy of Peruvian amphibians is not as thoroughly documented as that of Ecuador, it is anticipated that Peru harbors a greater diversity of species than Ecuador.

When it comes to amphibian diversity relative to its size, Ecuador emerges as the world's most diverse country, with roughly 2.3 species per 1000 square kilometers (see Figure 3.8). To put this in perspective, Ecuador's amphibian species density is 69 times higher than the United States', 17 times higher than Brazil's, 11 times higher than Mexico's, 5 times higher than Peru's, and 3 times higher than Colombia's. This tremendous diversity is not only numerical but also encompasses genomic, proteomic, chemical, morphological, ecological, and ethological diversity, as well as ancestral knowledge, aesthetic beauty, potential biomedical uses, economic values, and countless other aspects.

Notably, the current Ecuadorian amphibian richness numbers are incomplete, and ongoing work on the systematics and taxonomy of Ecuadorian amphibians is expected to increase the known diversity. Paradoxically, the numbers are also simultaneously decreasing due to species extinction.

Understanding the Richness of Ecuadorian Amphibians

The incredible richness and uniqueness of Ecuador's biodiversity, comprising around 16,500–20,000 vascular plant species (Ulloa Ulloa and Jørgensen, 2004 and onwards), 465 mammal species (Brito *et al.*, 2023), and 655 amphibian species, featuring high levels of endemism (4500 plant species, 60 mammal species, and 319 amphibian species), can be attributed to the interplay of factors such as the Andean uplift and its location at the crossroads of the Andean region, the Amazonian lowlands, and the Pacific Coast. Each of these regions provides distinct habitats, landscapes, climatic and altitudinal gradients, and environmental conditions, resulting in a variety of ecological niches that favors the evolution and diversification of lineages. Also, it cannot be ruled out that genetic drift on near or neutral alleles in a small number of genes might play a role on diversification (Roland *et al.*, 2017; Páez-Vacas *et al.*, 2022).

Hence, to thoroughly explore important forces molding Ecuador's diversity, we provide a brief description of key elements spanning geography and climate.

Geography

Ecuador is located in the northwestern part of South America, sharing borders with Colombia to the north, Peru to the east and south, and the Pacific Ocean to the west. Positioned strategically on the equatorial line, it spans approximately latitudes of 1° 28' N to 5° S.

The geomorphology of Ecuador has been shaped by the emergence of the Cordillera of the Andes, influenced by geological faults, volcanic activity, and soil processes. The Andes result in three major natural regions—the Coast, the Andes, and Amazonia. Additionally, this mountain range separates Ecuador's hydrographical systems into two major watersheds, one flowing westward into the Pacific Ocean and the other eastward into the Amazon River. The first one

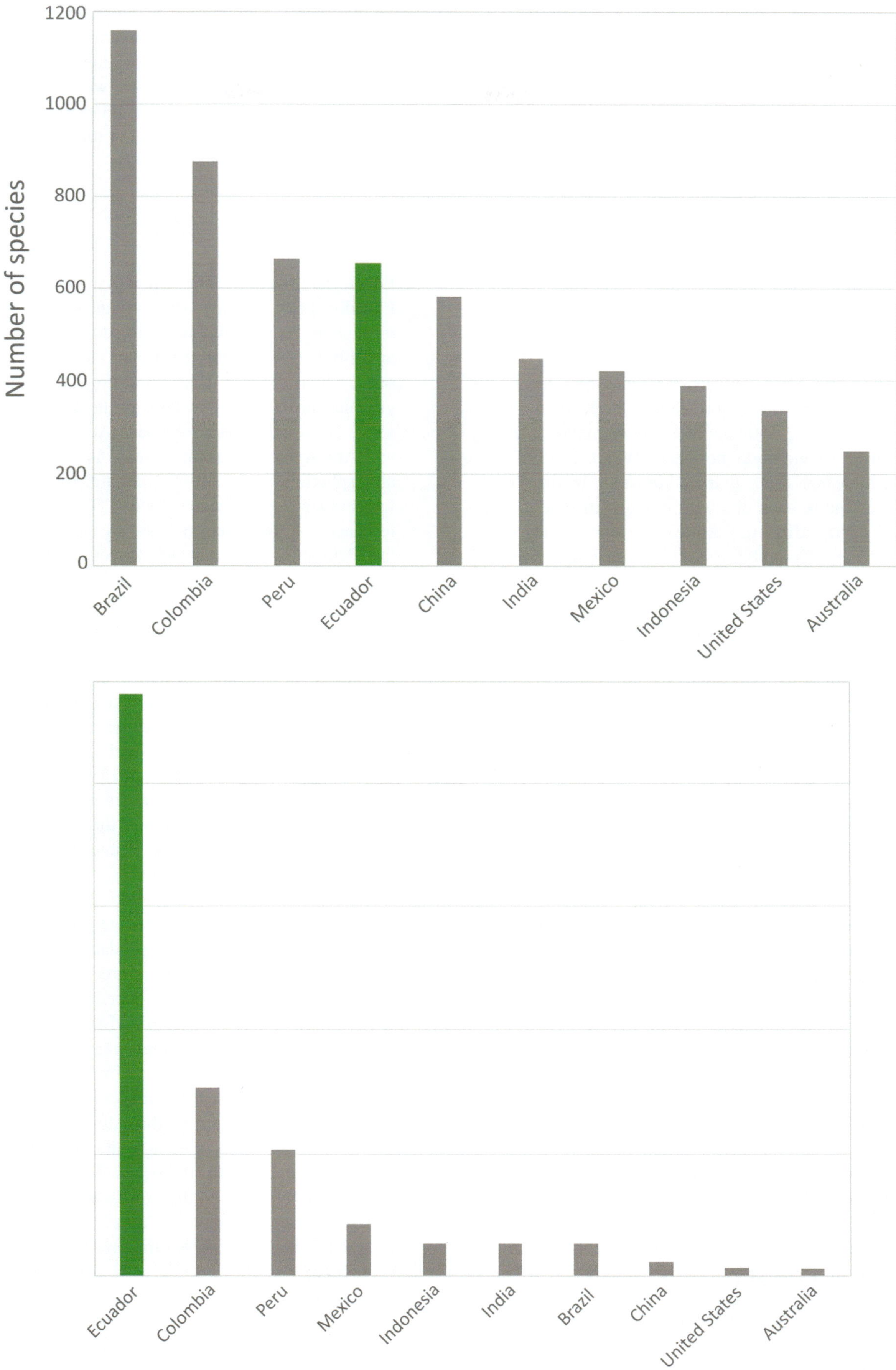

FIGURE 3.8 Comparison of the diversity of amphibians in 10 megadiverse countries. Top: Ecuador ranks fourth in terms of amphibian species diversity. Bottom: Ecuador ranks first; it has a high number of amphibian species relative to its size.

is composed of six major watersheds (Mira, Esmeraldas, Guayas, Manabí, Jubones, and Puyango-Catamayo), and the second one comprises three large watersheds (Napo, Pastaza, and Santiago) (Ministerio del Ambiente del Ecuador, 2014).

The Cordillera of the Andes is composed of three cordilleras, central, west, and east, creating the inter-Andean region between them. This mountainous barrier has an east–west width ranging from 100 to 200 km. The central part is the narrowest area, and the southernmost part is the widest (Winckell, 1997).

The Cordillera is composed of three successive branches: in the north, between the Colombian frontier and the equatorial line, the cordillera has a dominant south–southwest and southwest–northeast (at its northernmost part) orientation. In the central area, between the equatorial line and the 2° 30 ' parallel, the orientation is almost perfectly meridian. To the south of the 2° 30 ' parallel, the orientation is mostly south–southwest and north–northeast (Winckell, 1997).

Also, there are three marked divisions in the Cordillera of the Andes, each one with unique attributes. At the northern part, called the Andes Septentrionales, from the border with Colombia to 2°15 ' S in Palmira-Alausí, the cordillera has the most common and known aspect, the one that Humboldt called the "Avenue of Volcanoes." It is characterized by very steep external flanks with elevations varying by 3000 m, reaching elevations of 4000 and 4500 m in both cordilleras. This zone is composed of a chain of volcanoes, most of them inactive, and the Volcán Chimborazo stands out as the highest, at 6310 m. Between the two cordilleras, the inter-Andean valleys are formed. Along the valley, there are depressions that can be more or less extensive (e.g., Quito) or narrow (e.g., Chota). Each depression is separated from each other by knots, which are formed by coalescent volcanic edifices. At the central part, known as the Central Andes, from Riobamba (Palmira-Alausí) to the south to Zaruma-Saraguro, the cordillera changes significantly, there are no recent volcanic buildings, and the relief is more monotonous, constituted by large volcanic spills in the shape of overlaid crackers. The width between the cordilleras is larger, and the elevation diminishes while approaching to the south, with a maximum height of 3800 m. The arrangement of depression and knots is not visible here, and the cordilleras are connected by an orthogonal branch. Finally, in the region known as the Meridional Andes, from Zaruma-Saraguro to the Peru border, the cordillera loses any kind of arrangement and is more chaotic. The eastern Cordillera extends to the south in a fragmented way and ends at the border with Peru at 3900 m. With the exception of that extension, the relief is characterized by relatively low altitudes, with the highest peaks at 2500 m, hence, without a glacier print. The western Cordillera is replaced by a series of complex slopes that transition gradually to the western coast reliefs (Winckell, 1997).

Regarding the coastal region, it extends to the foot of the western cordillera. At the northern part, it has a width of about 100 km and reduces to a narrow band of 20–40 km at the south of Guayaquil. The highest point in this relief is 830 m at the central–south part. The Amazonian reliefs are located west of the eastern cordillera, where altitudes slowly decrease from west to east, having elevations less than 300 m and a very marked zoning of shapes, which are located as submeridian stripes. Three types of relief are identified: the sub-Andean reliefs, the peri-Andean foothills, and the fluvial hills and plains (Winckell, 1997).

Climate

Factors Shaping Ecuador's Climate

1. Equatorial position: The equatorial position of Ecuador allows for a fairly constant temperature throughout the year, without marked seasons; there are only two "seasons," the dry and the rainy seasons, which differ in time according to the geographical position in Ecuador (i.e., the rainy season varies in time between the coast, Andean region, and Amazonia). Also, the climate is affected by atmospheric systems of low pressure, such as the Intertropical Convergence Zone (ITCZ), the disturbances of the Amazon basin to the west, and the Peruvian Trough, which directly influences the pluviosity regimen during the rainy season. Other regional systems of high atmospheric pressure, such as the South Pacific High, the ridge of the Caribbean High, and the influence of the Tropospheric High of Bolivia during the dry season, play a role. The temporal occurrence of these atmospheric systems characterizes the rainfall and temperature regimes of the natural regions of Ecuador (INAMHI 2011, in Melo *et al.*, 2015).
2. Cordillera of the Andes: Apart from shaping the geomorphology, the Cordillera of the Andes plays a pivotal role in influencing Ecuador's climate. This cordillera acts like a barrier, modifying the formation, movement, and isolation of local or regional masses of air throughout the country. The altitude of the cordillera produces masses of cold air, modifies the regimen of precipitation, and shapes dry depressions. It also impedes the contact between the masses of air coming from the Pacific and those coming from Amazonia. In addition, the impact of the masses of air from the Pacific and Amazonia against the cordillera generates an increase in precipitation on the external slopes of the cordilleras. Also, the high-altitude areas of mountains, over 3200 m, receive less abundant rain of low intensity. In the inter-Andean valleys, there is an accumulation of dense, cold, and dry air, contributing to a stable climate with little rain (Pourrut *et al.*, 1995).
3. Pacific Ocean and marine currents: The Pacific Ocean has a dynamic of tropical oceanic air masses that mobilize into the continent. This normal regimen is modified by the presence of marine currents. Deep-water currents coming from the Indian and Antarctic oceans compensate

> for the disequilibrium between evaporation and precipitation in the Pacific Ocean. These currents transform when arriving at the Peruvian and south Ecuadorian coasts into warmer and less dense waters, forming the south equatorial current. Also, the cold current of Humboldt along the Peruvian and south Ecuadorian coasts comes from the south towards the northwest and mixes with the south equatorial current. This mixing of waters plays an important role in the development of oceanic fauna in Ecuador and Peru (Pourrut *et al.*, 1995). Ecuador's weather is influenced permanently by the Humboldt and the south equatorial currents and exposure to the sporadic action of a flux of warm water coming from the Gulf of Panama, known as the El Niño current, that can be dramatic regarding the change in weather (Pourrut *et al.*, 1995).

Besides these macro factors that characterize Ecuador's climate, several other processes interact, such as small- and large-scale convective cloud systems, local and regional valley/mountain breeze systems, and terrain lines of preferred moisture transport on various time scales. This leads to complex patterns of rainfall in space and time at a more local level (Rollenbeck and Bendix, 2011).

Climate Classification

The characterization of the climate in Ecuador is very complex due to all the factors influencing the climate and creating many microclimates locally. The first approach to classify climate is using the three macro regions, the Coast, the Andes, and Amazonia; each of them have general patterns of rain and temperature. Nevertheless, the influence of other factors creates large heterogeneity along the landscape, with areas receiving less than 150 mm of annual pluviosity and other areas with more than 6000 mm; the areas with the highest pluviosity are the Amazonian region and the northwest of Provincia Esmeraldas. In this context, the study of climate in Ecuador is complex and extends to the study of the bioclimate of the country, which tries to understand the link between numeric data of climate and the distribution of plants and vegetal formations. There have been several attempts to classify Ecuador's bioclimate (e.g., Köppen, 1918, 1931; Walter, 1970, 1976), nevertheless, many of them do not reflect the country's environmental, geographical, and climatic complexity. We present here the classification standardized by the Ministerio del Ambiente del Ecuador (2013) to create a map of the ecosystems of Ecuador. They, at the same time, base their classification on the one proposed by Rivas-Martínez (2004).

This classification is hierarchical, starting at a macro-bioclimate level that gathers climatic factors: temperature, precipitation, and solar radiation. The next unit is bioclimate, which is the interaction between temperature, precipitation, evaporation at the regional scale, and the correspondence with different types of vegetation. The basic units of the classification system are the ombrotypes and thermotypes, which predict the bioclimate at a latitudinal or altitudinal level and at the landscape or local level and relate the precipitation and temperature to vegetal communities. According to this classification, Ecuador is within the tropical macro-bioclimate. As for bioclimate, Ecuador is divided in four large bioclimate regions: desertic, pluvial, seasonal pluviosity, and xeric. At a finer scale, the continental Ecuador presents 11 ombrotypes and 12 thermotypes (Figure 3.9) (Ministerio del Ambiente del Ecuador, 2013).

Based on the climatic characterization and vegetation data, the Ministerio del Ambiente del Ecuador (2013) proposed a classification for this diversity, organizing it into 87 ecosystems. This classification employs a hierarchy of biogeographic units, including biogeographic regions (continental scale), biogeographic provinces (regional scale), and biogeographic sectors (landscape scale) (Table 3.4, Figures 3.10–3.17).

Patterns of Distribution of Amphibians in Ecuador

The distribution patterns of amphibians in South America have been documented by Duellman (1999) and Vasconcelos *et al.* (2019). Duellman delineated 12 natural biogeographic regions for continental South America's amphibians, including the Chocó, Caribbean coastal forest, Amazonia-Guiana, Atlantic forest domain, Llanos, Cerrado-Caatinga-Chaco, Pampean-Monte, Patagonia, Austral temperate forest, Atacama, Andes, and Guiana highlands. Among these, the Andes stand out as the most biodiverse, followed by the Amazonia-Guiana and the Atlantic forest domain. Three of these regions intersect with Ecuador: Chocó, Andes, and Amazonia-Guiana.

In 2019, Vasconcelos proposed a different classification, defining six biogeographic regions for South American anurans: Amazon, DIAGONAL-AF (encompassing the Caatinga-Cerrado-Chaco region and the eastern forest formations of the Atlantic Forest Domain), MID-ANDES, NORTH/SOUTH-ANDES, SUB-TROPICAL (encompassing the Pampa biome, part of the temperate grasslands biome, and forest formations of the Atlantic forest domain and Chaco/dry forest biome), and TEMP-GRASS (representing the southern temperate grasslands). Ecuador is traversed by the North/South Andes and Amazon regions. Vasconcelos *et al.* (2019) highlight the Andes as the region with the highest diversity, where anurans exhibit the smallest ranges on the continent.

At the national level, Ecuador's richness distribution of amphibians can be characterized in various ways. First, there's the administrative delineation considering Ecuador's provinces (Figure 3.18). Morona Santiago and Napo provinces emerge as the richest, hosting 209 and 207 species, respectively, while Santa Elena exhibits the lowest species count at 14. Morona Santiago and Napo are situated within the Amazonian region, whereas Santa Elena is a peninsula on the Pacific coast. However, this political division lacks the ability to reveal any discernible pattern in amphibian distribution; rather, it serves a practical purpose for conservation management at the political level.

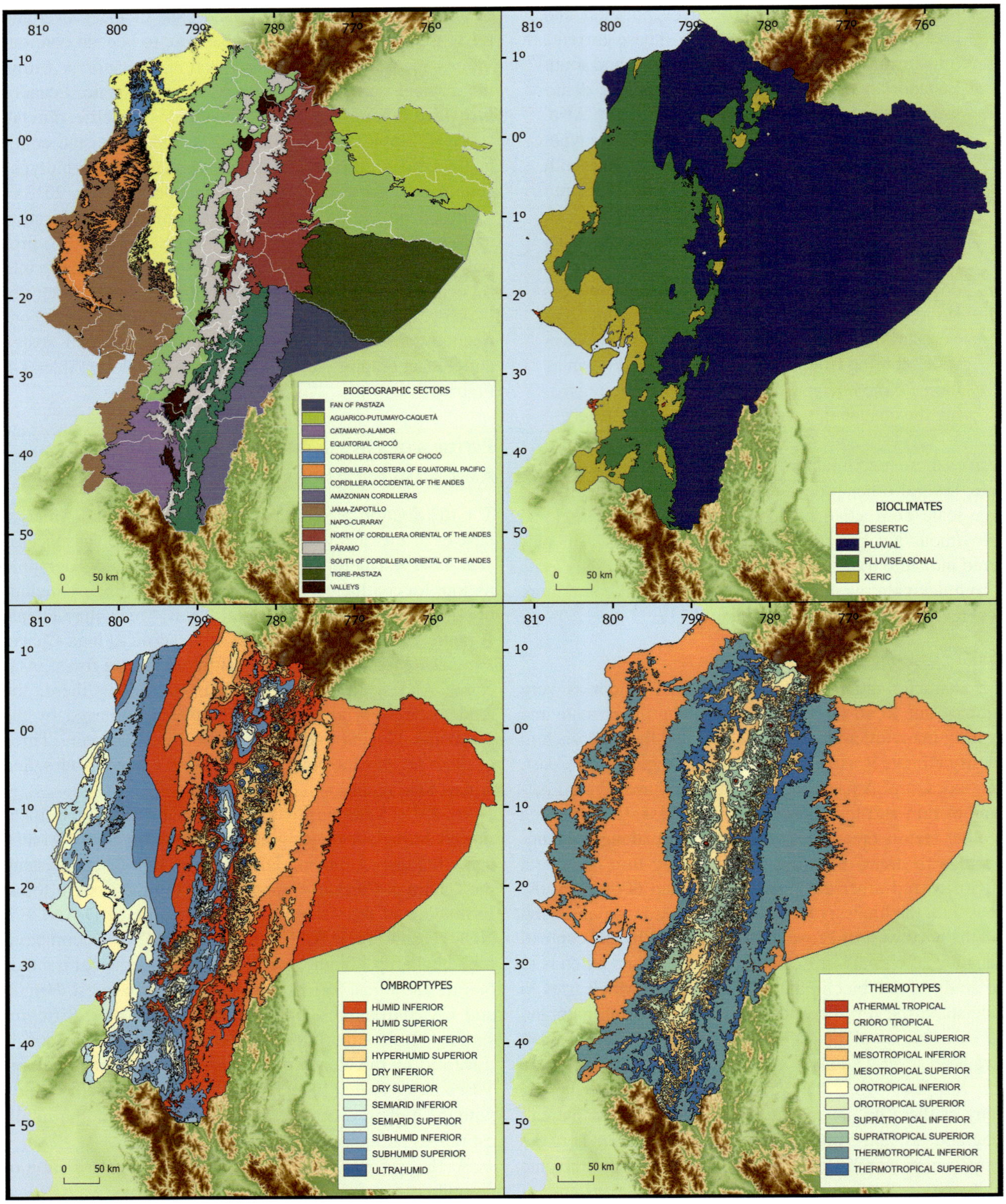

FIGURE 3.9 Climate attributes of continental Ecuador and its biogeographic sectors. Top left: interrelationship between temperature, precipitation, regional evaporation, and their correspondence with different types of vegetation. As a result of this model, Ecuador presents four bioclimates. Top right: relationship between precipitation and average temperature, which gives us an approximation of the predominant humidity in the area. As a result of this model, Equator presents 12 ombrotypes. Bottom left: relationships of temperatures, both maximum, minimum, and mean, which show an approximation to what is known as bioclimatic zones. As a result of this model, Ecuador presents 11 thermotypes. Bottom right: biogeography of continental Ecuador. This map allows us to explain the distribution areas of species according to patterns that identify their centers of origin and dispersion, as well as to understand the spatial patterns of biological diversity. These units are based on vegetation physiognomy, landscape criteria, climatic conditions, and floristic components.

TABLE 3.4
Hierarchical Classification of Biogeographic Areas in Continental Ecuador. (Ministerio del Ambiente del Ecuador [2013])

Biogeographic Scale			Ecosystem
Region	**Province**	**Sector**	**No. Ecosystems**
Coast	Chocó	Equatorial Chocó	7
		Cordillera Costera of Chocó	2
	Equatorial Pacific	Jama-Zapotillo	8
		Coastal Cordillera of Equatorial Pacific	4
Andes	Northern Andes	Cordillera Occidental of the Andes	5
		Catamayo-Alamor	10
		North of Cordillera Oriental of the Andes	4
		South of Cordillera Oriental of the Andes	7
		Páramo	11
		Valleys	4
Amazonia	Northwest Amazonia	Aguarico-Putumayo-Caquetá	1
		Napo-Curaray	1
		Tigre-Pastaza	1
		Fan of Pastaza	1
		Amazonian Cordilleras	11

We have found that the biogeographic sectors outlined by the Ministerio del Ambiente del Ecuador (2013) offer a valuable framework for describing Ecuador's amphibian diversity and the challenges it faces (Figure 3.19, Table 3.5). Notably, the Andean Cordillera emerges as the richest region in Ecuador, with the north of the Cordillera Oriental of the Andes along with the Cordillera Occidental of the Andes biogeographic sectors harboring the highest number of species, totaling 264 and 228, respectively. Particularly noteworthy is the remarkable richness of the Páramo sector despite its harsh climatic conditions, boasting a species count (113) comparable to that of Amazonian areas like the Tigre-Pastaza (129) and Fan of Pastaza (100) sectors, and even surpassing the Equatorial Chocó (87). Furthermore, the Páramo region boasts the highest rate of endemism, with approximately 73% of its species endemic to Ecuador. Other biogeographic sectors, such as the Cordillera Occidental of the Andes, the south of the Cordillera Oriental of the Andes, and the valleys, also stand out for their high levels of endemism, each hosting 50 – 60% endemic species.

This concentration of species in the Andes is unsurprising, given that mountains often serve as barriers to dispersal, thus facilitating speciation (Janzen, 1967; Ruggiero and Hawkins, 2008; Antonelli *et al.*, 2009). Moreover, mountainous regions can promote adaptive speciation through parapatric and sympatric processes, driven by differential selection along environmental gradients (e.g., Rahbek, 1997; Moritz, 2000; Jetz *et al.*, 2004). Indeed, Rangel *et al.* (2008) define the Andean Mountain range as a crucial source of South American diversity from an evolutionary standpoint.

While the Amazon region is globally renowned for its extraordinary biodiversity, the Andean region surpasses it in terms of amphibian richness, evident in the species counts of Amazonian biogeographic sectors such as the Fan of Pastaza, Aguarico-Putumayo-Caquetá, Napo-Curaray, and Tigre-Pastaza, without implying a disregard for the importance of the Amazon in terms of diversity.

Elevation plays a crucial role in shaping species distribution patterns in the Andes, acting as a barrier to dispersal for many species and delineating the distributional ranges of various phylogenetic groups (Figures 3.20, 3.21, 3.22). For example, Telmatobiidae species are predominantly found in high-altitude zones above 2000 m and extending up to 4000 m. Conversely, lower elevations, below 1000 m, exhibit higher species abundance for families such as Hylidae and Dentrobatidae, with species richness decreasing at higher altitudes and a notable decline in hylid presence beyond 1500 m. Bufonidae distribution appears more evenly distributed along the elevation gradient, with species present from sea level up to approximately 4400 m. Centrolenidae species are most abundant at mid-elevations, between 1000 and 2000 m, with the highest recorded occurrence observed at 3400 m. Craugastoridae, boasting the highest species diversity, exhibits a widespread distribution across the entire gradient, with a peak in abundance above 1000 m and around 2500 m.

Additionally, there is a clear pattern of distribution among endemic species (political endemism), with around the 50% of them occupying elevations ranging from 1500 to 2500 m. Conversely, more than the 50% of non-endemic species are distributed below the 1500 m threshold (Figure 3.23).

Several studies have attempted to explain the causes of tropical and Andean amphibian richness and diversity in various taxa such as hylids, dendrobatids, and glassfrogs (e.g., Wiens *et al.*, 2006; Santos *et al.*, 2009; Hutter *et al.*,

FIGURE 3.10 Equatorial Chocó and Chocó Coastal Cordillera (Chocó, Coast). Top: San Antonio, Provincia Carchi. Elicio E. Tapia. Middle: La Concordia, Puerto Quito, Provincia Pichincha, LAC. Bottom: Rocafuerte, Provincia Esmeraldas. LAC.

2013). Overall, they found that the time for speciation effect is important. When analyzing dendrobatids, Santos *et al.* (2009) found that the complex pattern of diversification is strongly intertwined with paleogeographic events. They provided data supporting that the Andes have undergone extended in situ diversification since the late Eocene (38–33.9 Myr), whereas Amazonian and Chocoan dendrobatid diversity is primarily derived from Late Miocene (<10 Myr)

FIGURE 3.11 Jama-Zapotillo and Equatorial Pacific Coastal Cordillera. Top: Arenillas, Provincia El Oro. LAC. Middle: Salango, Parque Nacional Machalilla. LAC. Bottom: Rocafuerte, Provincia Manabí. Fernando Coloma Román.

Andean lineages. Also, Hutter *et al.* (2013) found evidence for the montane museum hypothesis that suggests that the formation of montane cloud forests created isolated habitats that promoted speciation in glassfrogs since ~35 Myr, resulting in high levels of endemism in these regions. The Andes Mountains have acted as a stable refuge for plant and animal species during climatic fluctuations throughout the region's history.

FIGURE 3.12 Cordillera Occidental of the Andes and Catamayo-Alamor (Northern Andes, Andes). Cordillera de Intag, Provincia Imbabura., Gustavo Pazmiño-Otamendi. Middle: Pacto, Provincia Pichinhca. SGS. Bottom: Macará-Catacocha, Provincia Loja. LAC.

UNDESCRIBED SPECIES AND *INCERTAE SEDIS*

Taxonomic stability has improved in most Ecuadorian anuran taxa. However, there are also many species awaiting a formal description and entire groups such as Caudata and Gymnophiona that require further studies. Although both of the latter have been analyzed at the molecular and/or morphological levels (at least partially), species descriptions are pending due to their relatively high crypticity

FIGURE 3.13 North and south of Cordillera Oriental of the Andes (Northern Andes, Andes). Top: Sumaco Volcano, Provincia Napo., DAL. Middle: Sumaco Lake, Provincia Napo. DAL. Bottom: Río Upano, Provincia Morona Santiago. LAC.

(e.g., Franco *et al.*, 2023) and/or publication delays (e g., Fletcher-Lazo, 2002). Many species, especially of the genera *Atelopus*, *Telmatobius*, *Hyloxalus*, *Gastrotheca*, and *Strabomantis* have not been observed for over 30 years, and their phylogeny and taxonomic comparisons are crucial for their conservation. In such species, standardizing, improving, and integrating phenotypic characteristics with sequencing techniques to analyze phylogenetic and evolutionary relationships is essential. The recent progress in extracting DNA from formalin-preserved specimens is

FIGURE 3.14 Valleys and paramos (Northern Andes, Andes). Top: Hoya de Chimbo, Provincia Bolívar. Middle: Laguna de Yambo, Provincia Tungurahua. Bottom: Páramo del Angel, Provincia Carchi. LAC.

the holy grail of systematics, and a project led by the University of Kansas in cooperation with CJ and the USFQ is currently underway, given that many types of Ecuadorian taxa are deposited in their collections. This project will enable the sequencing of species for which no ethanol tissues are yet available.

There are several species well known to occur in Ecuador that belong to species complexes and for which their

FIGURE 3.15 Páramo (Northern Andes, Andes). Top: Mojanda, Provincia Imbabura. SGS. Bottom: Quilotoa Lake, Provincia Cotopaxi. SGS. Bottom: Atillo Lake, Provincia Chimborazo. LAC.

taxonomy remains unresolved and their description and assignment of a specific epithet are pending. Thus, we have not included them in the species accounts and instead we provide herein comments regarding the taxonomy of the names that have been used for them.

Atelopus spp. (*spumarius* Complex) (Anura: Bufonidae) (Figure 3.24)

Frost (2023) summarizes chronologically (under comments) the contents of papers dealing with the taxonomy

FIGURE 3.16 Aguarico-Putumayo-Caquetá and Napo-Curaray (Northwestern Amazonia, Amazonia): Cuyabeno lake, Provincia Sucumbíos, LAC. Middle and bottom: Misahuallí, Provincia Napo. Fernando Coloma Román.

and other aspects of this complex of species currently under the name *Atelopus spumarius* (Cope, 1871) and distributed in the Amazonian lowlands and foothills of Ecuador and eastern Peru (Loreto), into extreme southeastern Colombia in the vicinity of Leticia. Herein, we recount aspects that pertain to the Ecuadorian populations.

The neotype of *Atelopus spumarius* is from Río Ampiyacu, Pebas, Departamento Loreto, Peru. Duellman and Lynch

FIGURE 3.17 Tigre-Pastaza, Fan of Pastaza, and Amazonian Cordilleras (Northwestern Amazonia, Amazonia): Pucayacu River, Provincia Pastaza. LAC. Middle: Reserva Selva Eterna, Provincia Morona Santiago. Jaime Culebras. Bottom: Tepuy in Valley of Río Nangaritza, Cordillera del Cóndor, Provincia Zamora Chinchipe. Morley Read.

(1969) described the tadpole of an Ecuadorian population from Río Puyo, 3 km south of Puyo, Provincia Pastaza. According to Lötters *et al.* (2011), a well-supported phylogenetic branch of Amazonian harlequin frogs is *flavescens–spumarius*. Frogs in this group have been subject to several systematic studies that include the

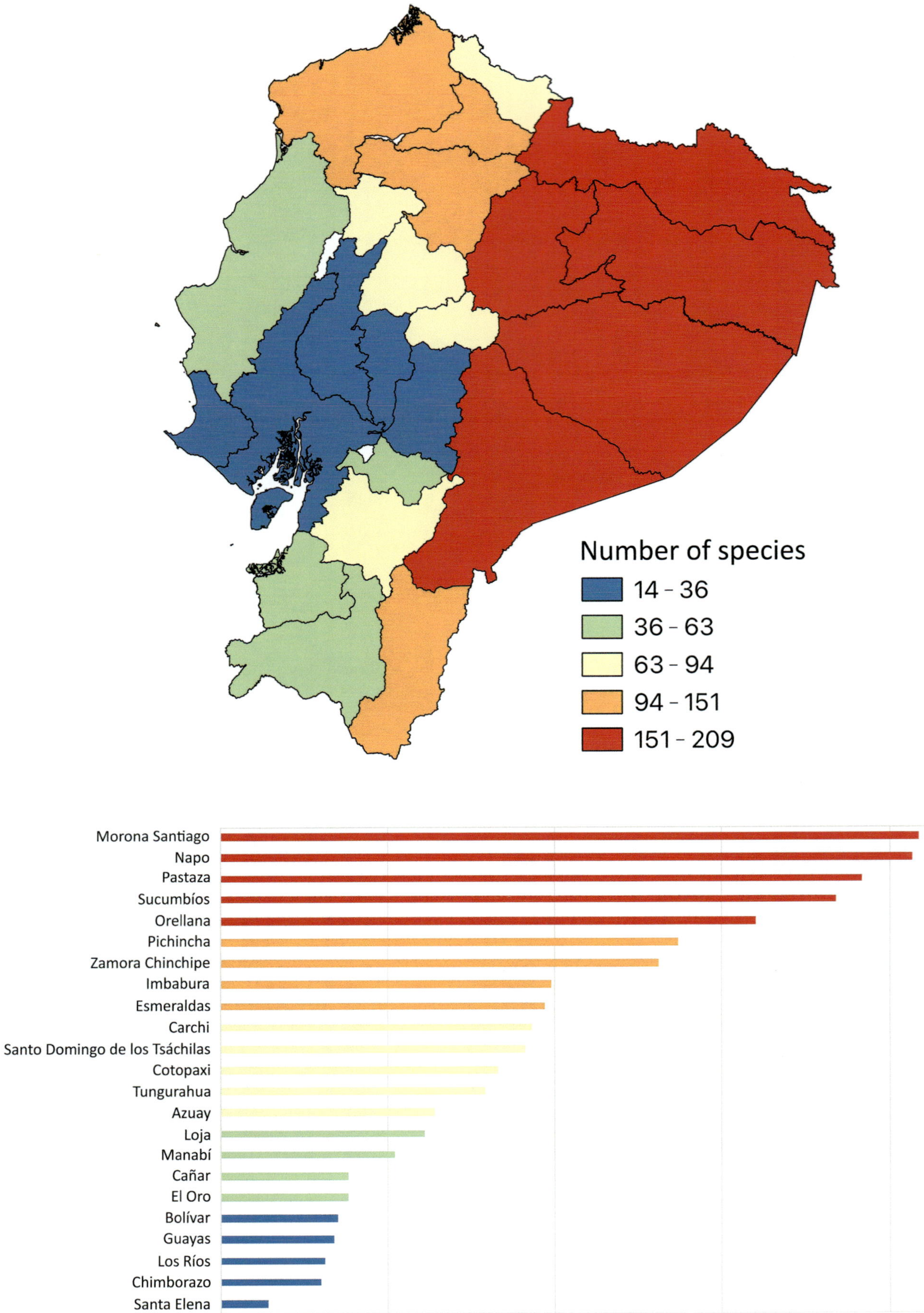

FIGURE 3.18 Total number of species for each province of Ecuador.

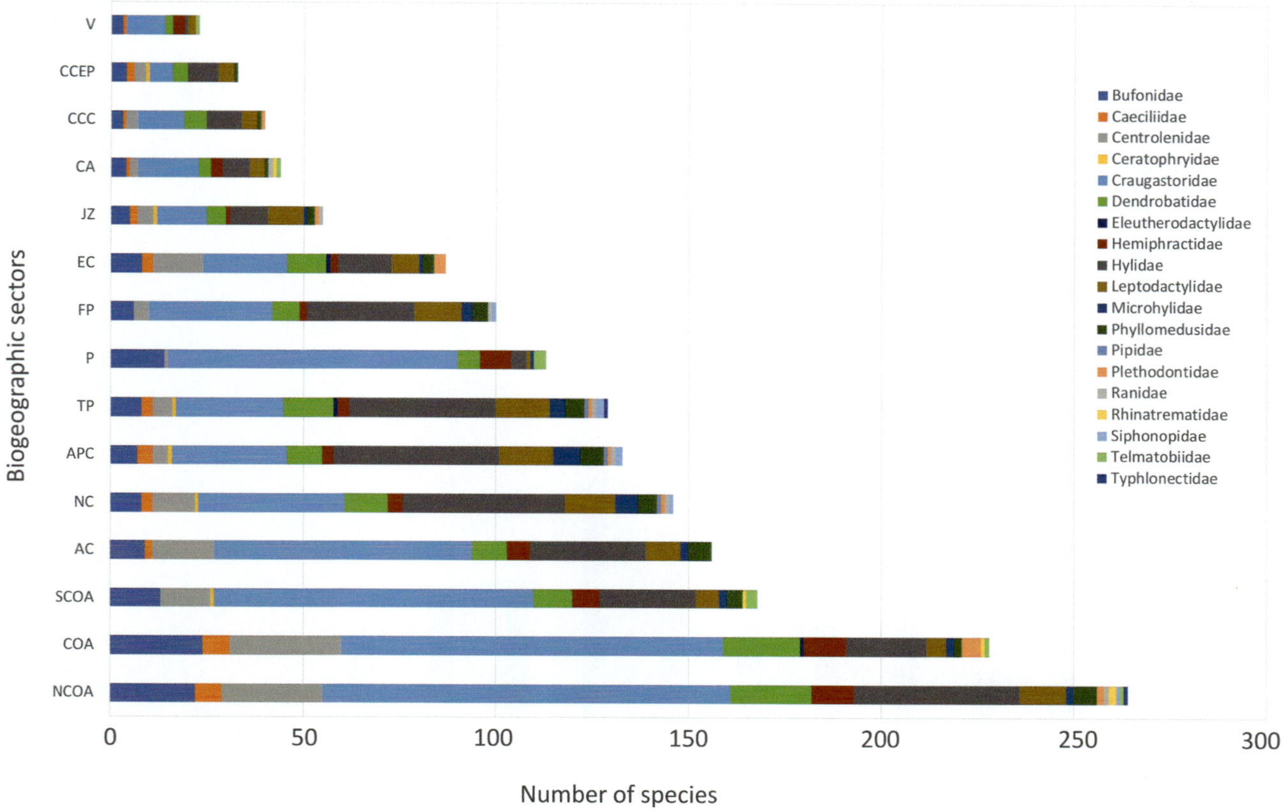

FIGURE 3.19 Number of species, classified by families, for each biogeographic sector of continental Ecuador.

TABLE 3.5
Summary of the Number of Species, Endemism, and Conservation Status of the Species within Each Biogeographic Sector of Continental Ecuador

Biogeographic Sector	Total Species	Endemic Species	Threatened Species
North of the Cordillera Oriental of the Andes	264	113	108 (36.5 %)
Cordillera Occidental of the Andes	228	115	157 (68.9 %)
South of the Cordillera Oriental of the Andes	168	86	86 (51.2 %)
Amazonian Cordilleras	156	58	57 (36.5 %)
Napo-Curaray	146	17	11 (7.5 %)
Aguarico-Putumayo-Caquetá	133	12	6 (4.5%)
Tigre-Pastaza	129	15	5 (3.9 %)
Páramo	113	82	91 (80.5 %)
Fan of Pastaza	100	10	11 (11 %)
Equatorial Chocó	87	19	49 (56.3 %)
Jama-Zapotillo	55	19	22 (40 %)
Catamayo-Alamor	44	17	21 (47.7 %)
Cordillera Costera of Chocó	40	9	15 (37.5 %)
Cordillera Costera of Equatorial Pacific	33	6	5 (15.2 %)
Valleys	23	14	11 (47.8 %)

redescription of *A. pulcher* from Peru (Lötters *et al.*, 2002), a phylogeographic analysis of Guianan populations (Noonan and Gaucher, 2005), and the phylogenetic analyses of several populations from the Amazon Basin and the Guiana Shield (Lötters *et al.*, 2011). A reexamination of old collections deposited at museums, plus molecular phylogenetic analyzes of several populations in Amazonian Ecuador and the neotype locality of

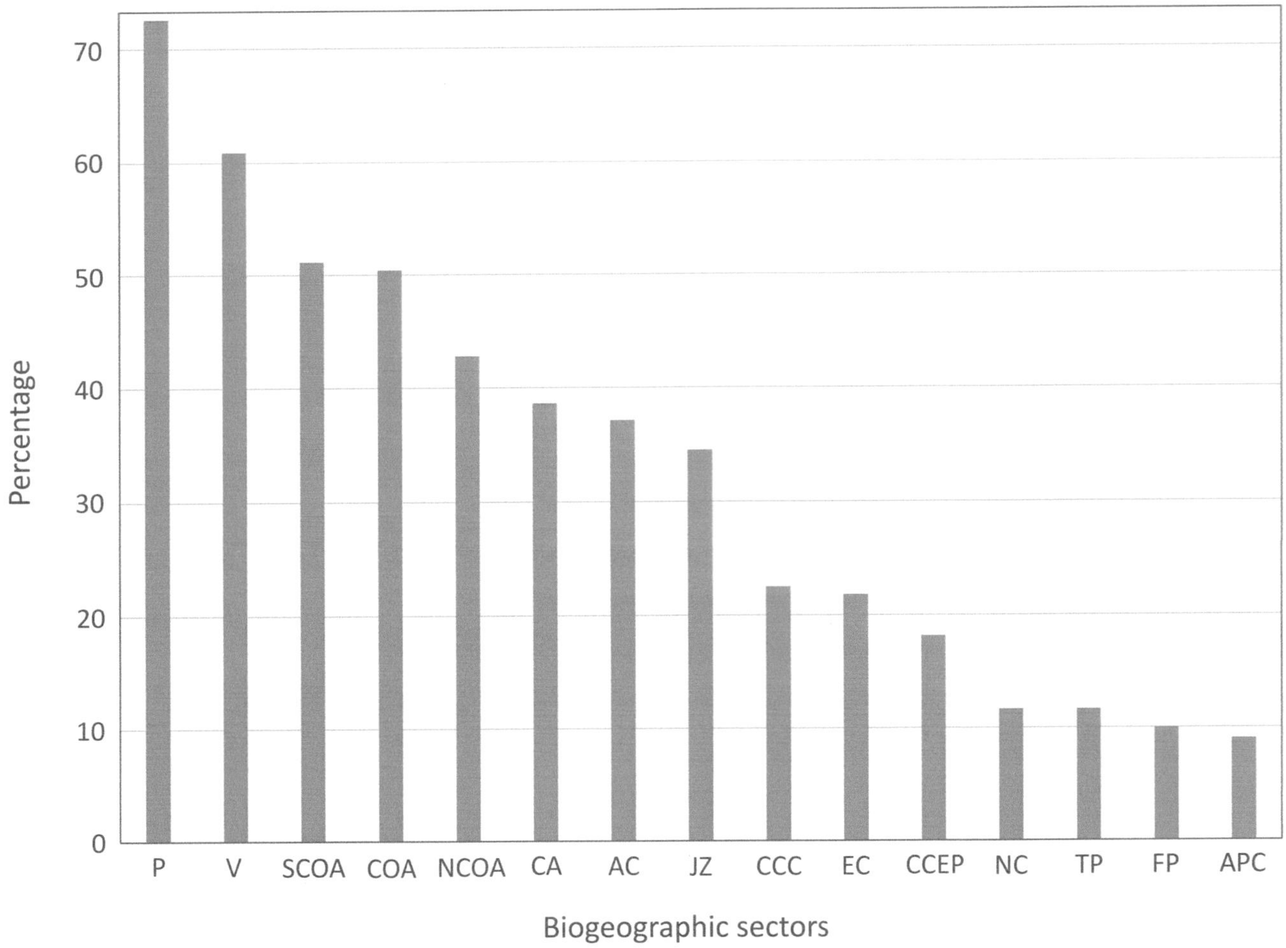

FIGURE 3.20 Percentages of endemism within each biogeographic sector of continental Ecuador.

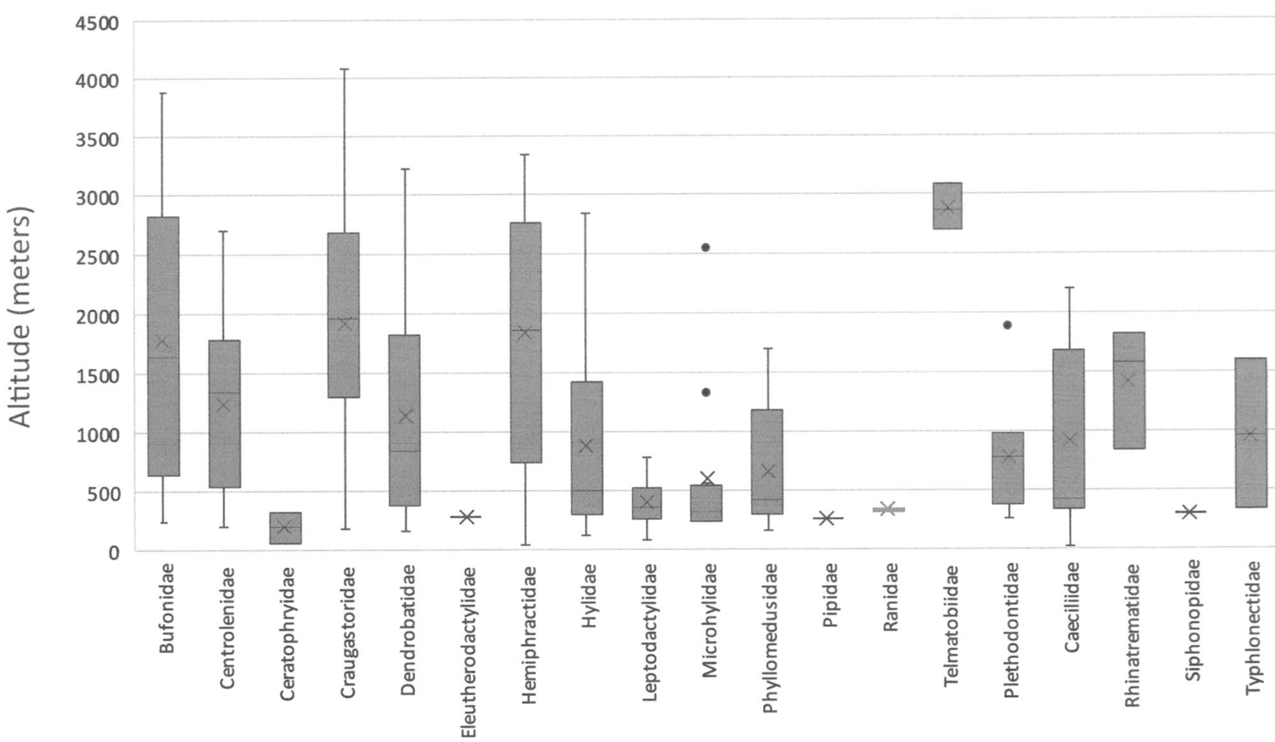

FIGURE 3.21 Distribution of richness of families along the altitudinal gradient.

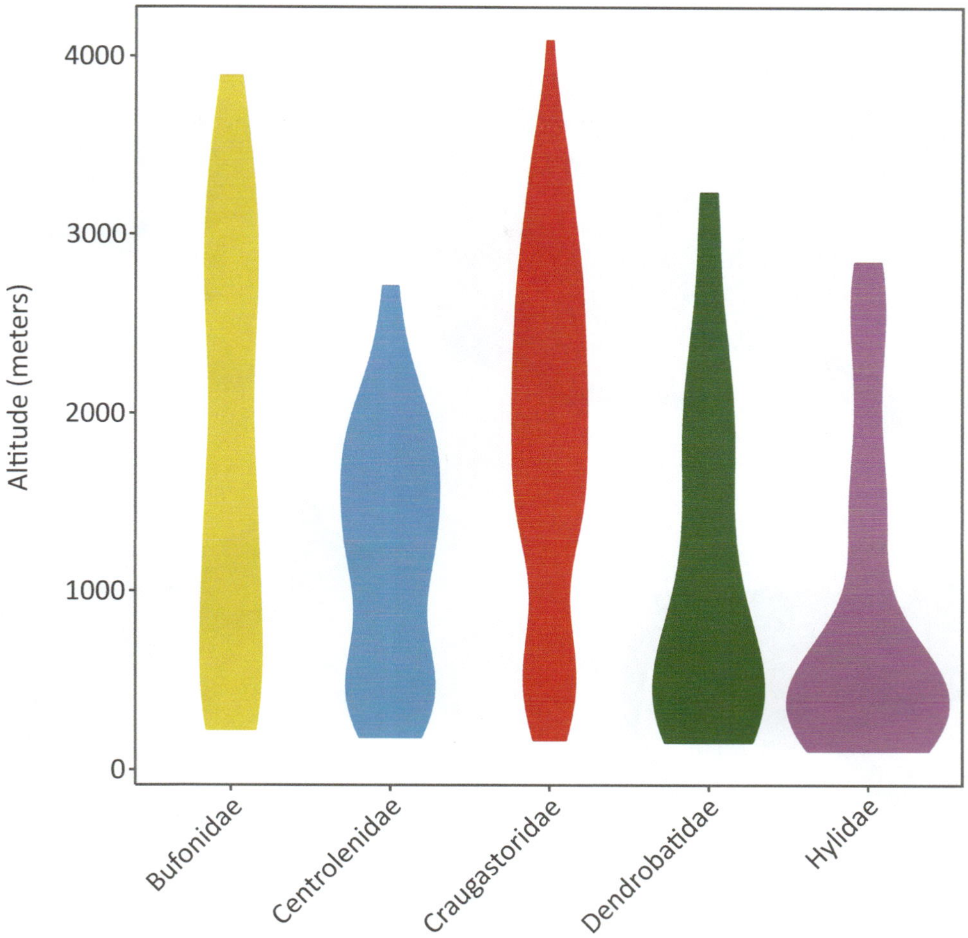

FIGURE 3.22 Distribution of the most abundant families (in terms of number of species) along the altitudinal gradient. These five families account for approximately 82% of the total amphibian richness in Ecuador.

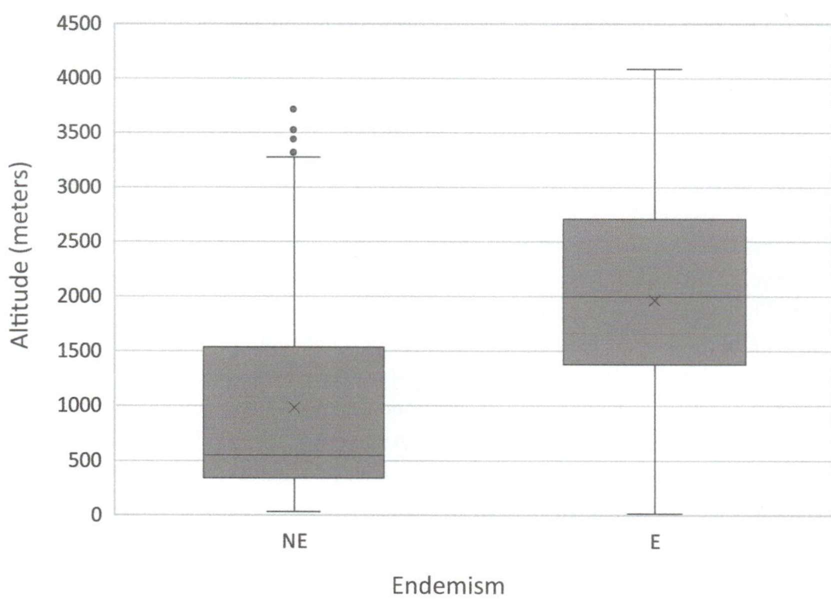

FIGURE 3.23 Political endemism of Ecuadorian amphibians along the featured altitudinal gradient.

FIGURE 3.24 Morphological variation in *Atelopus* spp. (*spumarius* species complex) from Ecuador. Top left: ca Limón, Provincia Morona Santiago. Top right: Parque Nacional Yasuní, Provincia Orellana. Bottom left: CJ 13037, female, Puyo, Pastaza. Bottom right: QCAZ 1925, male, Lorocachi, Provincia Pastaza. LAC.

A. spumarius, as well as morphological data, reveal the presence of at least three undescribed species from Ecuador referable to the *A. spumarius* sensu lato (not to *A. spumarius* sensu stricto), and they are being described elsewhere. Arbeláez Ortiz (2005) provides a brief account of a population from southeastern Ecuador. Its call (from San Carlos de Limón, Provincia Morona Santiago) was described by Womack *et al.* (2018), who tested the hearing of *A.* sp. 1 (*spumarius* complex, from Pucayacu, Provincia Pastaza) and *A.* sp. 2 (*spumarius* complex, from San Carlos de Limón) from Ecuador to assess their hearing sensitivity and to better understand mechanisms of hearing without a complete tympanic middle ear. They described their ear structures. Next, they assessed whether *Atelopus* hearing differs from hearing in other *Atelopus* and bufonids.

Rhinella spp. (*margaritifera* Complex) (Anura: Bufonidae) (Figures 3.25, 3.26)

Frost (2023) provides a complete list of synonyms and summarizes chronologically (under comments) the contents of papers dealing with the taxonomy and other aspects of this complex of species currently under the name *Rhinella margaritifera* (Laurenti, 1768) and distributed in Amazonian and Guianan Brazil, Colombia, Ecuador, French Guiana, Guyana, Panama, Peru, Suriname, and Venezuela at elevations from sea level to 2400 m. Herein, we recount aspects that pertain to the Ecuadorian populations.

Lavilla *et al.* (2017) rediscovered two syntypes of *Rhinella margaritifera* from "Brasilia" (= Brazil), and the name *margaritifera* was restricted to the type locality. However, after 2017, some authors continued using the name *R. margaritifera* for Ecuadorian populations (e.g., Camper *et al.*, 2021; Frost, 2023).

One of the first authors to address the taxonomy of these species is Hoogmoed (1977), who indicated that there are a number of sibling species confused under this name. A year later, Duellman (1978) provided an account (as *Bufo typhonius*) based on specimens from Santa Cecilia and nearby localities of Provincia Sucumbíos in northern Ecuadorian Amazonia. He included data about morphology, occurrence, life history, tadpoles, mating calls, and food. He also made taxonomic comments. Almendáriz (1987) added some natural history and diet data from several sites in Provincia Pastaza in central–south Ecuadorian Amazonia. Hoogmoed (1990) noted a form from west of the Andes of Ecuador and Colombia (which has subsequently been assigned to *Rhinella alata* by dos Santos *et al.,* 2015); and at least one

FIGURE 3.25 Morphological variation in *Rhinella* spp. (*margaritifera* species complex) from Ecuador. Top left: QCAZ 48638. Top right: QCAZ 48637. Morona Santiago, San Luis, Bomboiza, Provincia Morona Santiago. Middle left: QCAZ 31343, Km 38, road Pompeya Sur-Iro, Parque Nacional Yasuní, Perovincia Orellana. Middle right: QCAZ 47314, La Selva Amazon Ecolodge, Limoncocha, Provincia Sucumbíos. Bottom left: QCAZ 37216, La Selva Lodge, Garzacocha, Provincia Sucumbíos. Bottom right: QCAZ 37801, Laguna Grande, Reserva de Producción de Fauna Cuyabeno. LAC.

other in Amazonian Ecuador (among others in the Amazonian region). Fouquet *et al.* (2007) provided molecular evidence that this is a species complex by including samples from Yasuní and western Ecuador in their molecular analyses. Valencia *et al.* (2008) and Beirne and Whitworth (2011) provide brief accounts in field guides.

FIGURE 3.26 Morphological variation in *Rhinella* spp. (*margaritifera* species complex) from Ecuador. Top left: CJ 12740, Río Negro, Provincia Tungurahua. Top right: CJ 8231, female, 52.2 mm SVL, Río Cugusha Chico, 9 de Octubre, Provincia Morona Santiago. SGS. Middle left: CJ 7287. Middle right: Río Blanco, Paquisha Alto, Provincia Zamora Chinchipe. DAL. Bottom left: CJ 1195, female, 66.0 mm SVL. Bottom right: CJ 1175, male, 46.5 mm SVL. 12 km north of El Chaco, on road to Santa María, ca Río Oyacachi, Provincia Napo. LAC.

Dendropsophus sp. (*minutus* Complex) (Anura: Hylidae) (Figure 3.27)

Frost (2023) provides a complete list of synonyms and summarizes chronologically (under comments) the contents of papers dealing with the taxonomy and other aspects of this complex of species currently under the name *Dendropsophus minutus* (Peters, 1872). The complex is distributed in lowlands east of the Andes from Colombia, Venezuela, and Trinidad southward through Ecuador, Peru, and Brazil

FIGURE 3.27 Morphological variation in *Dendropsophus* spp. (*minutus* species complex) from Ecuador. Top left: SC 11342. Top right: SC 11338. Quebrada Río Napinaza, Provincia Morona Santiago. SGS. Bottom left: CJ 4411, male, 21.6 mm SVL. Río Blanco, Cordillera del Cóndor. Provincia Zamora Chinchipe. Bottom right: CJ 868, Río Pucayacu, Reserva Otoyacu, Provincia Pastaza. LAC.

to Bolivia, eastern Paraguay, northeastern Uruguay, and Argentina, up to 2000 m elevation. Herein, we recount aspects that pertain to the Ecuadorian populations.

Duellman (1978) provided an account (as *Hyla minuta*) based on specimens from Santa Cecilia, Provincia Sucumbíos, in the northern Ecuadorian Amazonia. He included data about morphology, occurrence, life history, tadpoles, mating calls, and food. Kaplan (1994) suggested that *Hyla minuta* is a composite of several species. Gehara *et al.* (2014) found on the basis of mitochondrial genes that this species is likely a complex of species. They identified 43 lineages within the single "species," many of which may represent cryptic species, including at least two in Ecuador, one of them from Limón Indanza-Plan de Milagro, Provincia Morona Santiago, and another one from Sacha Yacu, Comunidad Bolívar, Provincia Pastaza. Given that the type locality is Neu-Freiburg [= Nova Friburgo] and the vicinity of Rio de Janeiro, Brazil, the Ecuadorian populations are clearly differentiated from the nominal *Dendropsophus minutus*, which belongs to the Atlantic group of species given the molecular data and tree provided by Gehara *et al.* (2014).

Bolitoglossa spp. (*peruviana* Complex) (Caudata: Plethodontidae)

Elmer *et al.* (2013) did a molecular analysis of mitochondrial DNA and nuclear sequence data of South American *Bolitoglossa*. Given the highly localized nature of genetic differentiation in what has been considered *Bolitoglossa peruviana*, Elmer *et al.* (2013) suggest that none of the Ecuadorian samples should be assigned to that taxon. Instead, *B. peruviana* should be considered to be a Peruvian endemic that so far is known only from the unique holotype. Additional sampling, particularly in southern Ecuador and northern Peru, is required to further resolve phylogenetic relationships. The four major subgroups of *B. peruviana* (sensu lato) are highly divergent: from 9.5 to 11.9% at Cytb and from 1.1 to 2.8% at Rag1. Divergence time estimates among populations that have been assigned to *B. peruviana* average ca. 10.8 Myr.

According to Elmer *et al.* (2013) within *Bolitoglossa peruviana* (sensu lato) four major subgroups are broadly distributed by elevation and river basin. They informally identify these subgroups as follows: *altamazonica*, Upper Napo, Upper Aguarico, and Lower Napo/Lower Aguarico.

The *altamazonica* subgroup spans the greatest latitude: from the lower Napo to the lower Pastaza river basins. This subgroup, which includes the most lowland representative in their study (*B. altamazonica* Loreto, Peru), is not the sister taxon of the other Amazonian basin lowland species from Brazil, tentatively assigned to *B. paraensis*. Instead, the specimens assigned to *B. paraensis* (from Ituxi and Juruá) are more closely related to the southeastern Ecuadorian specimens from higher elevations. The upper Napo subgroup occurs in multiple localities in the vicinity of the headwaters of the Napo River. Although the southeastward distribution of this subgroup remains to be determined, it is replaced to the north at the headwaters of the Aguarico River with the divergent sister subgroup Upper Aguarico. Specimens in the Lower Napo/Lower Aguarico subgroup were found to span the lower Napo and lower Aguarico river basins. Additional sampling is required to determine the eastward distribution of this subgroup.

INTRODUCED SPECIES

The impact of non-native amphibians on Ecuador's ecosystems can vary widely, ranging from invasive species with significant negative effects to those with no negative impact. Ecuador has seen imports of at least nine species of non-native amphibians, most of which apparently have had a non-negative impact, except for one species, the American bullfrog *Aquarana catesbeiana*, introduced in the 1980s for the edible frog industry, which has now established feral populations, is naturalized, and is considered invasive. Chapter 1 provides further details on the bullfrog industry. Eight other species have been imported for the pet trade but have only been kept in aquariums and terrariums, with no reports of feral populations. These species include the African clawed frog *Xenopus laevis* (normal and albino forms (Figure 3.28)), which have been popular pets in aquarium shops since at least the 1970s, and the salamanders *Cynops* sp. and *Notophthalmus* sp., which were imported in the late 1970s and have since disappeared from aquarium shops. The axolotl *Ambystoma mexicanum* (see Figure 3.28) has been available in aquarium shops since the 2000s and is still found in some pet shops. In 2005, *Dendrobates auratus*, *D. azureus*, and *Phyllobates terribilis* were imported under CITES permits for the amphibian exhibit Sapari in Quito, with *Dendrobates auratus* also being used for embryological studies at PUCE. Finally, *Telmatobius* spp. aquatic frogs were brought from Peru by aquarium dealers in 2009, and a few individuals were donated to the QCAZ, where they were kept alive.

Additionally, there are two notable cases of human-mediated amphibian introductions within Ecuador. First, *Scinax quinquefasciatus*, originally from the coastal region of Ecuador, has been accidentally introduced to the Galápagos Islands (see Figure 3.28). It was first collected on the Island of Santa Cruz in 1973, and in 1998, it was also observed in Isla Isabela at Poza Las Diablas, near Puerto Villamil. This species has invaded various parts of the Galápagos Islands, including towns, farms, and lakes in the south of Isla Isabela and Isla Santa Cruz, although there have been no records reported for San Cristóbal since 2001. Despite attempts at eradication, including hand capture, caffeine spraying, and changes in lagoon salinity, they have been unsuccessful (Lever, 2003; Zug, 2013; Cisneros-Heredia, 2018; Ron *et al.*, 2018b).

Second, *Pristimantis unistrigatus*, a species commonly found even in urban areas and widespread across the northern Andes of Ecuador (south to Provincia Chimborazo), has recently been reported in Cuenca, Provincia Azuay, located in southern Ecuador. According to Salazar *et al.* (2022), the presence of *P. unistrigatus* in the urban and peri-urban areas of Cuenca has been reported since around 2019, particularly at the Universidad Católica de Cuenca (Miracielos and Gañansol). Cuenca is densely populated, and the authors speculate that the species may have been accidentally introduced through nursery plants.

To summarize, many of the amphibian species that were imported to Ecuador have vanished. However, two species, *Xenopus laevis* and *Ambystoma mexicanum*, are still available in pet shops. Meanwhile, *Aquarana catesbeiana* is now a part of a thriving frog farming industry and has even established feral populations. Also, the hylid *Scinax quinquefasciatus* from the Pacific lowlands of Ecuador has invaded the Galápagos Islands, whereas *Pristimantis unistrigatus* from the northern highlands of Ecuador is invading the southern highlands.

EXTINCTION AND CONSERVATION

Ecuador is renowned for being a hot spot of biodiversity, and this is particularly true of its amphibian fauna. With 655 recorded species, this group is both megadiverse and unique, with about half of species (319; 49%) being endemic to the country. However, over the four most recent decades, Ecuador has experienced a large-scale, unexpected, and apocalyptic decline in amphibian populations. Currently, 399 species (61%) are under imminent threat of extinction, with at least 43 species (7%) possibly already extinct. Although this decline is part of a global trend, it has had a particularly severe impact on Ecuador given that it is home to 7.6% of the world's amphibian diversity, and the decline has affected most of its endemic amphibian fauna. And if that weren't enough, to make matters worse, our knowledge of amphibians is still precarious. For instance, it is predicted that at least 230 species are still awaiting description. Also, the biology of most species is unknown, and population surveys only exists for a handful of species. The widely recognized causes of decline, such as habitat destruction and contaminants, remain largely uncontrolled. Conversely, issues like climate change and pathogens, which impact even well-protected natural areas, are poorly understood and beyond current control measures. Although

FIGURE 3.28 Top: the invasive species *Scinax quinquefasciatus* introduced to the Galápagos Islands, Provincia Galápagos. Top left: near Laguna Posada de los Flamingos at Isla Isabela, on 25 January 2009. LAC. Right: Salasaca (455 m, 0.63274 S, 90.4178 W) at Isla Santa Cruz, on August 2023. Camilo Rosero-Gómez. Introduced species in Ecuador. Middle left: the African clawed frog *Xenopus laevis* (albino form). Middle right: the invasive American bullfrog *Aquarana catesbeiana*. LAC. Bottom left: the axolotl *Ambystoma mexicanum* (leucistic form). Camilo Ospina. Bottom right: the axolotl. SGS.

amphibian declines and extinctions in Ecuador are a significant problem that has required immediate action and a paradigm shift (Mendelson III *et al.*, 2006), only a few conservation efforts are already in place, and they have not yet been effective in mitigating this phenomenon.

The response to the amphibian crisis has been extremely slow, with actions deemed both too little and too late, as has been discussed elsewhere (Stuart, 2012; Bishop *et al.*, 2012). Furthermore, conservation efforts, in general, have adhered to traditional approaches like habitat conservation, often neglecting the reality that numerous species have vanished from pristine and protected natural areas. There has been a tendency to overlook the significance of promptly implementing ex situ conservation breeding measures. Advocacy for the conservation breeding has been present since the 2000s (Mendelson III *et al.*, 2006; Lötters, 2007; Zippel and Mendelson, 2008) but is only recently gaining substantial attention (Lewis *et al.*, 2019; Luedtke *et al.*, 2023; Lötters *et al.*, 2023).

The high rates of amphibian extinctions and increasing losses can have far-reaching biological, ecological, cultural, and economic impacts and are now a clear indicator of the sixth mass extinction that is ultimately affecting the well-being of human beings. Addressing the causes of amphibian declines (in protected and non-protected areas) and implementing effective research and conservation strategies has become a monumental task that is far from being managed to ensure that Ecuador's amphibian populations continue to thrive for generations to come. Herein, we provide updated data and also explore the causes and consequences of this decline as well as review the actions to solve this critical problem.

The Red List of Ecuadorian Amphibians

The first Red List of Amphibians in Ecuador was compiled by Luis A. Coloma under the request of the Nature Conservancy, but it remained unpublished (Coloma, 1992). Subsequently, Coloma, Santiago R. Ron, and Pablo A. Menéndez-Guerrero prepared the baseline data on Ecuadorian amphibians for the Global Amphibian Assessment (GAA) in 2003. This data was reviewed by a group of additional experts during a meeting in Tandayapa, Provincia Pichincha, Ecuador, held on 18–22 August 2003. The GAA was a comprehensive assessment of the conservation status of the world's amphibian species, leaded by Simon Stuart and published by the International Union for Conservation of Nature (IUCN) between 2004 and 2008 (Stuart *et al.*, 2008). The goal of the assessment was to determine the conservation status of all known species of amphibians and to identify those species most in need of conservation attention. The GAA identified Ecuador as a global hotspot for amphibian conservation. Since then, there have been several general updates for Ecuadorian amphibians (Ron *et al.*, 2008, 2011), as well as species-specific updates on the IUCN Red List website. The latest updates to the website include information on some species accessed on 15 March 2021, which was published in the IUCN Red List of 2022.

The latest national-level Red List assessments of Ecuadorian amphibians were conducted until approximately April 2021 and published on 6 May 2021 by Ortega-Andrade *et al.* (2021). Another recently published assessment is found in Luedtke *et al.* (2023), covering the period from 2016 to 2021. This assessment resulted from a GAA workshop held in Cosanga, Provincia Napo, on 16–19 July 2016. The Ortega-Andrade *et al.* (2021) assessment was part of the project "Conservation of Amphibians and Genetic Resources Project (PARG)," which was Centro Jambatu's initiative supported by the Ministry of the Environment (MAE), the United Nations Development Programme (UNDP), the Global Environment Facility (GEF), and Universidad Regional Amazónica Ikiam. H. Mauricio Ortega-Andrade from Ikiam led a team of amphibian biologists who evaluated 635 species, revealing that a significant 57% of Ecuador's amphibian species were classified as threatened. This finding stands in stark contrast to the assessment conducted by Ron *et al.* (2011), which identified only 30.5% of species as threatened. Ortega *et al.*'s Red List assessment and the Red List of Amphibians of Ecuador were officially validated by the Ministry of Environment of Ecuador (Ministerial Agreement 069). Nonetheless, the assessment by Ortega-Andrade *et al.* (2021) has not yet been incorporated to the IUCN Red List of Threatened species web page (IUCN, 2022), which currently (31 December 2022) remains mostly outdated. In general, the IUCN Red List has failed to be updated, despite being widely used for research (e.g., Jetz and Pyron, 2018; Scheele *et al.*, 2019) and conservation purposes, especially by funding organizations, most of which only rely on the information provided by the IUCN.

Herein, we provide an updated overview of the Red List data on Ecuador's amphibians, covering the year 2022. Our assessment only considers the distribution of Ecuadorian populations, and therefore, categories could differ from IUCN evaluations that consider the species' entire distributional ranges. In the latter case fall about 337 species (51%) of Ecuadorian species that are non-endemic. Given the scope of this publication, we only have refined the data available from Ortega-Andrade *et al.*'s assessment, and we refrain from performing more detailed analyses for the Red List. Nonetheless, it is clear that further Red List analyses are urgently needed. For instance, Ortega-Andrade *et al.* incorporated in their analyses the estimation of environmental contractions on climate change scenarios as a threat. However, more complete analyses of climate change and the synergic effects of climate change and emergent diseases (apparently responsible for the sudden declines) are needed, given that they have particular importance among the threats sustained by Ecuadorian amphibians, as well as climate change indexes (e.g., climate stability index, climate vulnerability index, temperature anomaly index) and the pathogen susceptibility index. Our update differs from that by Ortega-Andrade *et al.* (2021) because we incorporated newly described taxa and taxonomic changes; thus, we evaluated 655 species. Also, we have fixed some geographic and identification errors and made refinements to the level of precision of distribution records. In general terms, we interpreted more carefully the raw taxonomic locality information from collections and literature. This information deserved to be reviewed since for some species, it significantly affected the calculation of the minimum polygon of the species and the variables used for their vulnerability evaluation. In several species, there were georeferencing errors that drastically altered the distribution range (modifying the extent of their occurrence area or their altitudinal range). A recurrent issue in the analyses by Ortega-Andrade *et al.* (2021) was that in many of the species, duplicate and triplicate records were included resulting from independent georeferencing efforts for the same locality. Therefore, what appeared as distinct points on the maps were a single record. Furthermore, we have noted that Ortega-Andrade *et al.* (2021) made an error in categorizing 46 species as near threatened, while they actually provided criteria that meet justifications to be included in a threatened category. The locality data used to build the

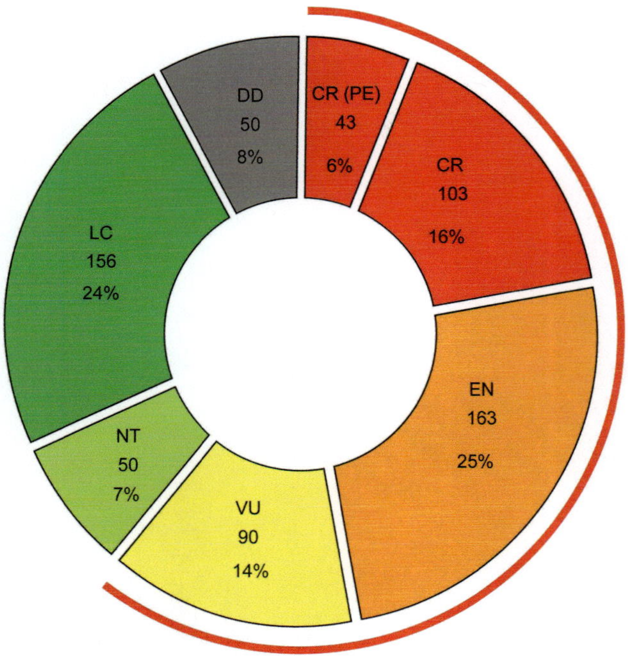

FIGURE 3.29 IUCN Red List of 655 amphibians from Ecuador. Categories: CR (PE) = critically endangered (possibly extinct), CR = critically endangered, EN = endangered, VU = vulnerable, NT = near threatened, LC = least concern, DD = data deficient. Number of species and percentage are indicated below the category. Threatened species (61%) are in red, orange, and yellow bordered by a red line.

maps is available at CJ. When reclassifying a species, we adhered to the criteria and guidelines outlined in version 3.1 of the IUCN (IUCN, 2001), along with the accompanying guidelines for its application (IUCN, 2012).

Our Red List assessment of 655 species shows that 61% (399) of Ecuador's amphibian species are categorized as threatened, with 8% as near threatened, 24% as least concern, and 8% as data deficient (Figure 3.29). See also the categories for each species in Table 2 of Volumes II–IV and additional details in the conservation section of each species account.

The relative number of threatened species (61%) in Ecuador is well above the worldwide number of threatened amphibians according to the IUCN (41%), and these numbers reafirm the findings of the first global amphibian assessment that Ecuador is a global hotspot for amphibian conservation. Regarding the most threatened lineages, the family Craugastoridae represents the highest number of threatened species, followed by Centrolenidae, Bufonidae, Dendrobatidae, and Hylidae (Figure 3.30); this trend is expected since they are the families with the highest number of species. However, if the proportional value of threatened species is considered by removing the factor of number of species, the trend is different, with Telmatobiidae in first place, with 100% of its species in the possibly extinct category. It is followed by Plethodontidae with almost 86% of its species threatened, Hemiphractidae with 76%, and Bufonidae with 72% of its species threatened. These last two have 7 and 13 species, respectively, categorized as possibly extinct. Among various diverse genera, it is noteworthy that several have a significant proportion of threatened species. *Atelopus* and *Telmatobius* have a striking 100% of their species classified as threatened. An analysis conducted by Scheele *et al.* (2019) observed that the 45% of the declines and extinctions at a global level occurred in the Neotropical genera *Atelopus*, *Craugastor*, and *Telmatobius*. At the local level, in Ecuador, three other genera have been severely affected with high rates of threatened species: *Strabomantis* with 86%, *Gastrotheca* with 85%, and *Hyloxalus* with 80%.

Threatened Species in the Ecuadorian Landscape

Geographically, the Andean region harbors a significant proportion of species facing a risk of extinction. For instance, 88% of the species inhabiting the Páramo biogeographic sector are categorized as being at risk of extinction, with nine species identified as possibly extinct (see Figure 3.32). The Cordillera Occidental of the Andes biogeographic sector has 77% of its species in a threatened category. These areas are particularly vulnerable to threats such as habitat destruction and deterioration, as well as to the effects of climate change and pathogens. The Equatorial Chocó also stands out for its high presence of threatened species, with 60% of its species threatened. The main threat and major factor driving the loss of species in this sector is habitat loss (Figure 3.31). On the other hand, the Amazonian sectors have less than 11% of species threatened. When observing the distribution of threatened species along the altitudinal gradient, there is a clear pattern that coincides with the biogeographic sectors with the highest number of threatened species (Figure 3.33). In general, more than 50% of threatened and possibly extinct species are located within an altitudinal range above 1000 m up to approximately 2500 m, while non-threatened species are predominantly found below 1000 m. This elevational pattern matches the perceived lower occurrence of amphibian declines in the Neotropical lowlands (Young *et al.*, 2001). However, at a finer level of analysis, 50% of the possibly extinct species are found between approximately 1500 m and 2500 m; critically endangered species are located between approximately 1500 m and 3000 m. Species classified as endangered and vulnerable occupy lower ranges, from approximately 500 m to 1000 m. This difference in altitudinal ranges also reflects the various threats facing amphibian populations, such as climate change, emerging pathogens, and habitat loss and degradation. Thus, possibly extinct and critically endangered species are responding to the massive decline in amphibians during the 1980s and 1990s, resulting from increased temperatures in high-altitude locations, which create optimal conditions for *Batrachochytrium dendrobatidis* growth and vulnerability of frogs due to reduced thermal tolerances, leading to outbreaks of this pathogen, as proposed by Pounds *et al.* (2006). Additionally, Scheele *et al.* (2019) argue that species at higher elevations have a lower probability of recovery due

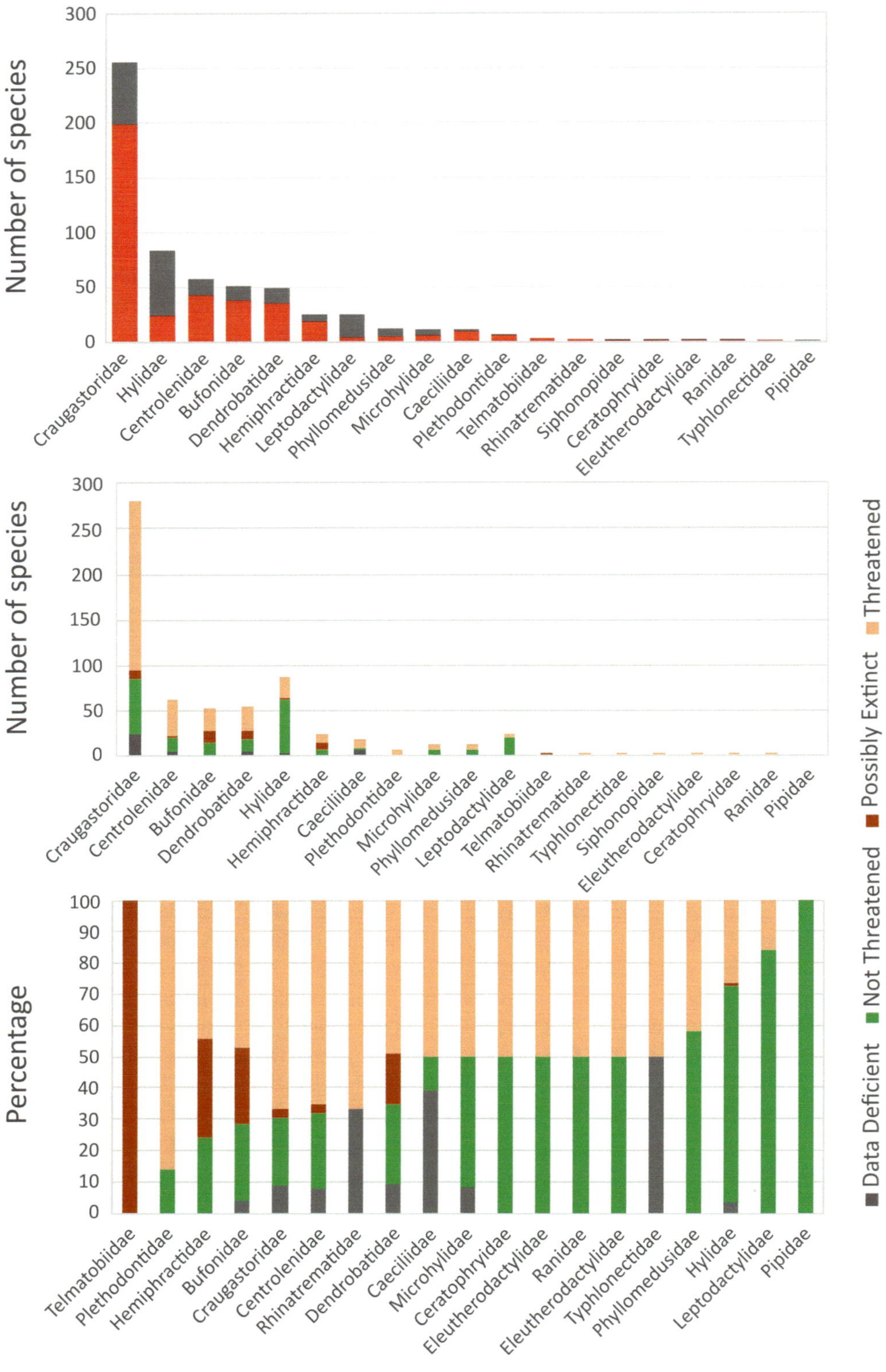

FIGURE 3.30 IUCN Red List of 655 amphibians from Ecuador by families. Top: ordered by the number of threatened species (in red) from highest to lowest. Non-threatened species are in gray. Data-deficient species are not included. Middle: ordered by the number of threatened species (in four categories) from highest to lowest. Bottom: ordered by the percentage of threatened species (in four categories) from highest to lowest.

to ideal climatic conditions for the persistence of chytrid fungus, as well as narrow species distribution ranges and a lack of connectivity with other populations of the same species. It is important to note that many of the possibly extinct and critically endangered species, because many of their populations have disappeared, were located in protected areas where habitat loss does not pose a risk. On the other hand, species classified as endangered and vulnerable are

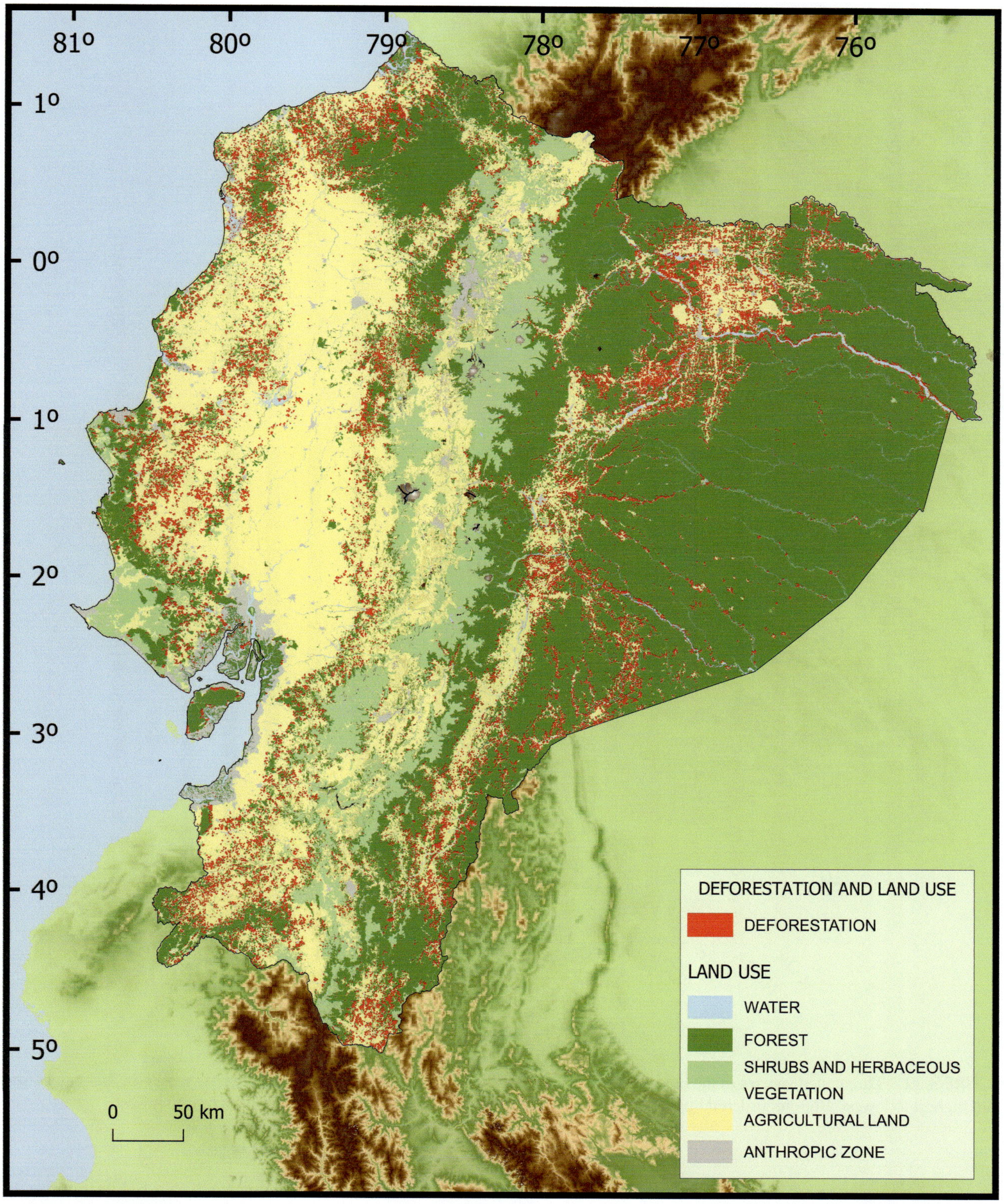

FIGURE 3.31 Map of Ecuador showing land use and land cover (updated in 2022) and deforestation (2020–2022 period). Taken from the Interactive Map of the Ministerio de Ambiente, Agua y Transición Ecológica, 2023. http://ide.ambiente.gob.ec:8080/mapainteractivo/.

exposed to risks associated with land use changes that lead to habitat loss and degradation, without excluding the possibility that some of these species are also exposed to pressures imposed by climate change and emerging diseases. Additionally, many species at higher altitudinal ranges are also facing risks associated with land use change. In conclusion, this altitudinal pattern of threatened species reflects the complexity of amphibian conservation, as these

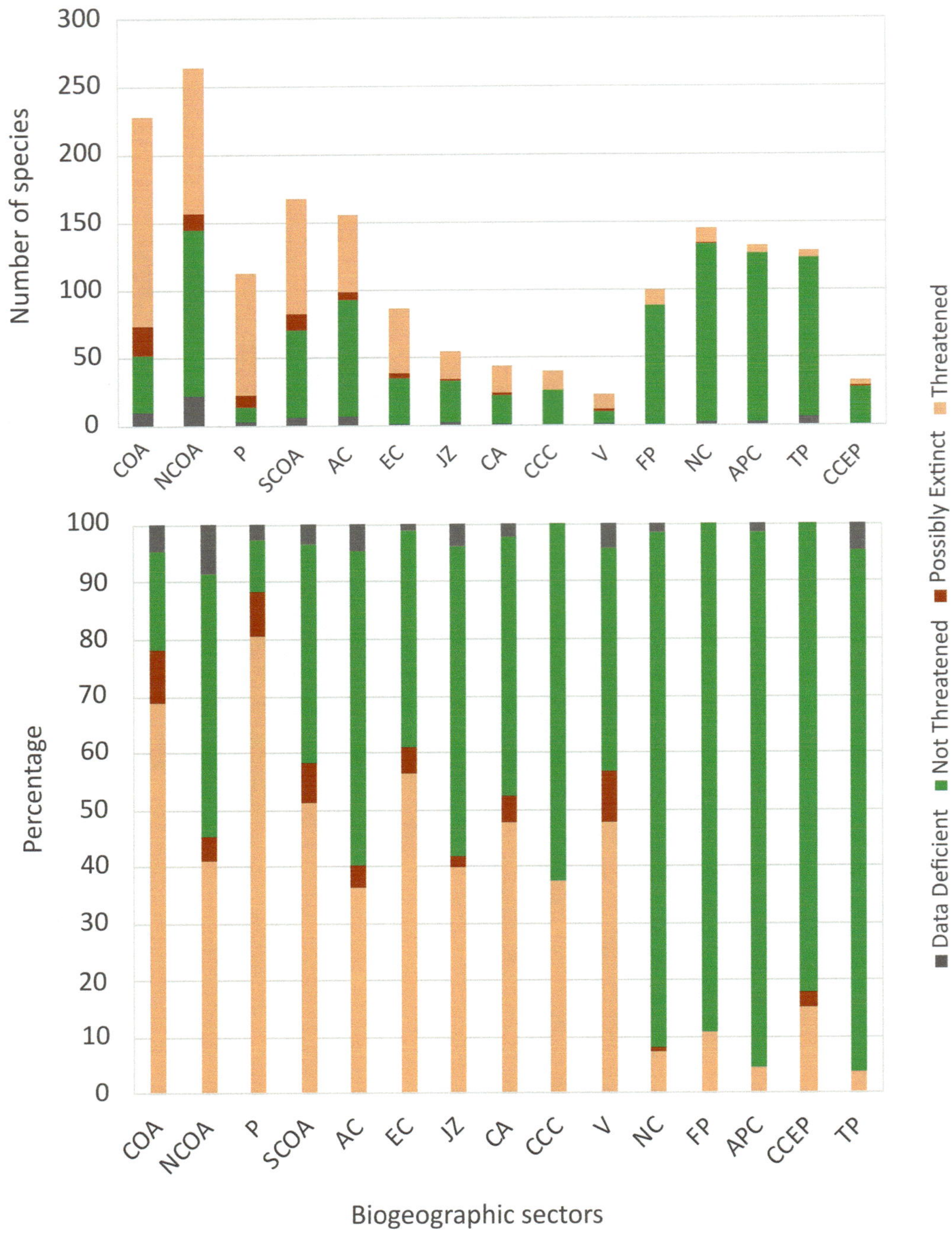

FIGURE 3.32 Threatened species in the biogeographic sectors. Top: ordered by the number of threatened species (in four categories) from highest to lowest. Bottom: ordered by percentage of threatened species (in four categories) from highest to lowest.

species are exposed to multiple risk factors that cannot be addressed in the same way but require specific and creative solutions to the threats.

The Vanishing Legacy: Possibly Extinct Species

In 2004, the IUCN categorized two Ecuadorian species as extinct in the wild, *Atelopus ignescens* and *A. longirostris* (Ron *et al.*, 2004; Bustamante *et al.*, 2004). Listing a species as extinct requires that "exhaustive surveys" have been undertaken in all known or likely habitats throughout its historical range at appropriate times (diurnal, seasonal, annual) and over a timeframe appropriate to its life cycle and life form. The rediscovery of both species in 2016 indicates that this categorization was premature and that a more conservative categorization and or labeling method was needed for species that has not been sighted for decades despite survey efforts. The IUCN provides such an option in its guidelines (IUCN Standards and Petitions Committee, 2022), and the tag of possibly extinct

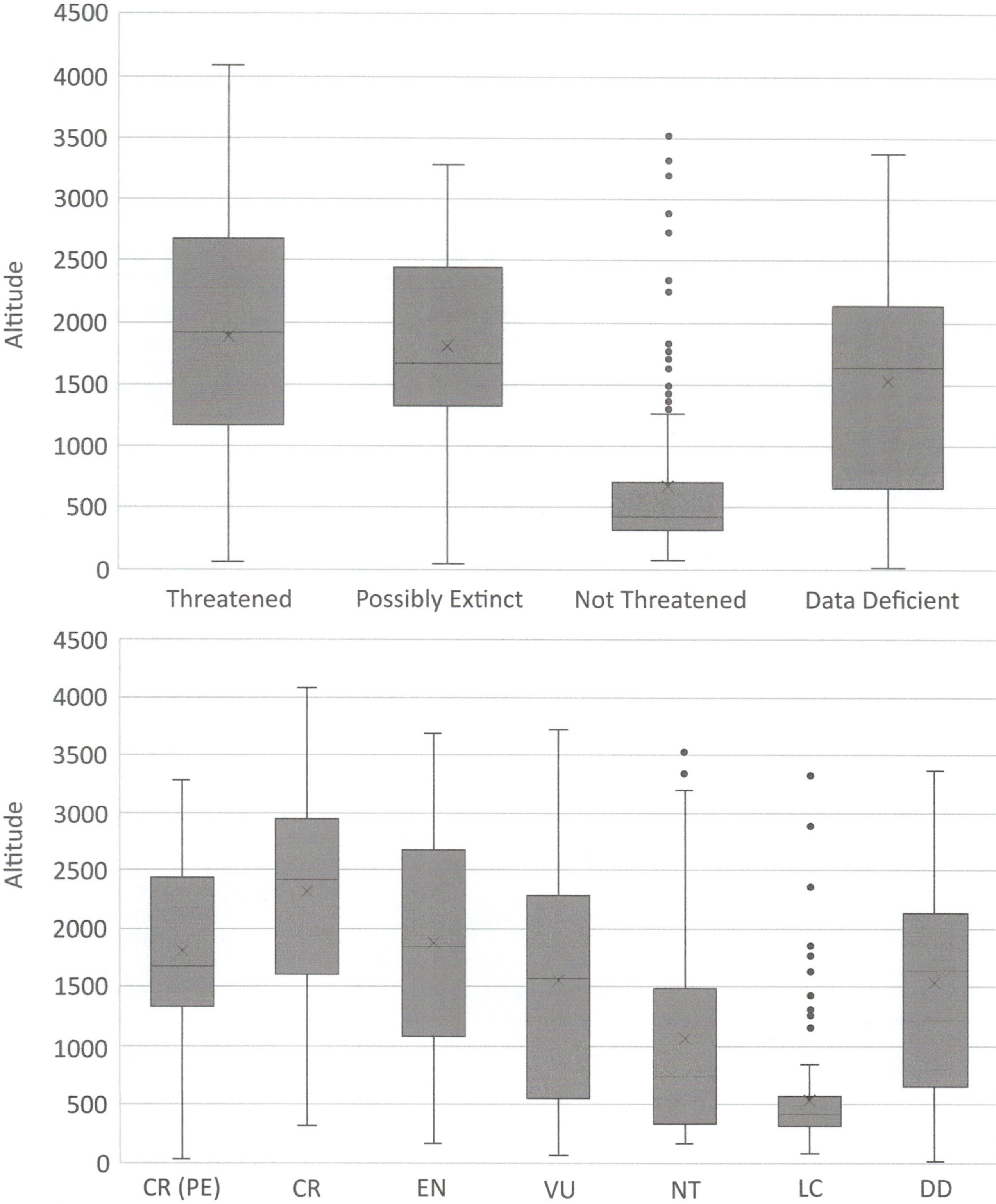

FIGURE 3.33 Distribution of threatened and non-threatened species along the altitudinal range.

has been already incorporated for several species on the IUCN website (e.g., *A. arthuri* by IUCN SSC Amphibian Specialist Group, 2022a). The tag of "possibly extinct" has therefore been developed to identify those critically endangered species that are, on the basis of evidence, likely to be extinct, but for which there is a small chance that they may be extant.

Thus, we have not categorized any species as extinct but tagged them as possibly extinct when survey efforts have not revealed their presence for three decades or more.

There are 43 species of Ecuadorian amphibians that are considered critically endangered (possibly extinct) in Ecuador until further surveys confirm otherwise (Table 3.6), applying IUCN Red List criteria (version 3.1 from 2001). Half of them are endemic to Ecuador. It is noteworthy that all of them are anurans, and no caecilians or salamanders are tagged as possibly extinct.

Three species (*Hemiphractus helioi*, *Paruwrobates whymperi*, *Gastrotheca angustifrons*) have not been sighted in Ecuador for more than a century, when they

TABLE 3.6

Forty-Three Species of Ecuadorian Amphibians That Are Tagged As Possibly Extinct under the IUCN Category Critically Endangered. Dates of last sightings and codes of voucher specimens are indicated. E = endemic

Species	Last Sighting in Ecuador	Museum Code
Atelopus boulengeri (E)	July 1984	ANSP HRP
Atelopus halihelos (E)	4 March 1984	KU 201040
Atelopus planispina (E)	1 September 1988	MHNG 2526.069
Atelopus arthuri (E)	1 February 1988	QCAZ 1870–1874
Atelopus orcesi (E)	May 1988	MHNG 2559.067, 2684.075
Atelopus nepiozomus (E)	1989	Photo Morley Read
Atelopus onorei (E)	21 April 1990	QCAZ 3440–42
Atelopus lynchi	30 May 1977	KU 178416
Atelopus pachydermus	September 1985	MHNG 2259.008–009
Atelopus angelito	22 July 1988	MECN 649–650
Atelopus pastuso	29 June 1993	QCAZ 4798–4802
Atelopus podocarpus	1 December 1994	QCAZ 6801
Rhaebo caeruleostictus (E)	8 April 1990	QCAZ 21719
Centrolene medemi	19 March 1975	KU 164493–94
Centrolene geckoidea	January-May 1999	Notes Diego F. Cisneros-Heredia
Ectopoglossus confusus (E)	8 April 1975	KU 166157–58
Hyloxalus peculiaris (E)	6 August 1962	USNM 282664–65, 282667
Hyloxalus pumilus (E)	25 August 1962	USNM 282812–16
Hyloxalus fuliginosus (E)	25 June 1980	KU 189934
Hyloxalus elachyhistus (E)	21 March 1984	UMMZ 177959
Hyloxalus maquipucuna (E)	2 April 1984	KU 202880–82
Hyloxalus breviquartus	September 1984	MECN
Hyloxalus lehmanni	1 October 1990	QCAZ 1502
Paruwrobates whymperi (E)	1880	BMNH 1947.2.14.25
Gastrotheca psychrophila (E)	8 March 1975	KU 164233–34
Gastrotheca angustifrons	1898	BMNH 1947.2.22.38–39
Gastrotheca andaquiensis	17 November 1984	MCZ 113382–84
Gastrotheca weinlandii	22 February 1985	USNM 284311
Gastrotheca guentheri	Early 1991	QCAZ 2691
Hemiphractus helioi	1880	BMNH 1880.2.5.192
Hemiphractus fasciatus	February 1949	CAS 122205
Hyloscirtus torrenticola (E)	24 February 1979	MCZ 97990
Niceforonia babax	18 February 1979	USNM 285970
Niceforonia dolops	23 February 1979	MCZ 97475
Pristimantis loustes	30 May 1977	KU 179247–50
Pristimantis caprifer	15 September 1984	QCAZ 577–79
Strabomantis helonotus (E)	24–31 May 1959	USNM 195784
Strabomantis necerus (E)	1 May 1985	QCAZ 563–564
Strabomantis anatipes	30 May 1977	KU 177625–26
Strabomantis cerastes	26 April 1977	MCZ 92882
Telmatobius cirrhacelis (E)	14 November 1981	EPN 1749–50
Telmatobius vellardi (E)	1 July 1987	QCAZ 388, 653–54
Telmatobius niger (E)	1 December 1994	QCAZ 160–162

were collected in the 1800s. However, *G. angustifrons* was last seen in Colombia in 29 May 1983. The other 40 species have not been seen in Ecuador in the period between 1949 (*H. fasciatus*) and 1994 (*Telmatobius niger*, *Atelopus podocarpus*). One species, *Centrolene geckoidea*, was last sighted in 1999.

The number of possibly extinct species by families and genera are indicated in Figure 3.34.

The 43 species that are possibly extinct belong to seven families and 13 genera. Thirty-one species of 5 diverse genera (*Atelopus*, *Hyloxalus*, *Gastrotheca*, *Strabomantis*, *Telmatobius*) stand out due to a high percentage of species disappearance. The genus *Telmatobius* is an extreme case, as all 3 Ecuadorian species have vanished. Additionally, there are 12 species from 8 genera that have not been sighted for over 2–3 decades: 2 species of *Centrolene*, 2 Dendrobatids (*Ectopoglossus*, *Paruwrobates*), 2 *Hemiphractus*, 1 *Hyloscirtus*, 5 Craugastoridae (2 *Niceforonia*, 3 *Pristimantis*), and 1 *Rhaebo*. It is noteworthy that, with the exception of *Hemiphractus* and *Gastrotheca*, these species are primarily riparian species, predominantly inhabiting the Andean region.

Most of the possibly extinct species have been affected by sudden and catastrophic declines occurred during the 1980s through their ranges. Climate change and pathogens might be implicated as the likely causes of most of their decline in natural protected areas. Additional threats to many of these species are severe habitat destruction and degradation due to agriculture and clear cutting, logging, grazing, cattle ranching, timber extraction, pollution, human settlements and urbanization, introduced species, fires, volcanic activity, landslides, oil activities, mining, and other factors; however, these threats do not account for the observed decline, given that most of them disappeared even from protected areas.

A dramatic example of sudden declines in protected areas can be seen in the amphibian species that previously inhabited the Cajas National Park, the adjacent Bosque Protector Mazán (Figure 3.35), and surrounding areas in Provincia Azuay. Among these species, five vanished from these protected montane forests and paramo habitats (*Telmatobius niger*, *Hyloxalus anthracinus*, *H. vertebralis*, *Centrolene buckleyi*, *Pristimantis w-nigrum*). Three species (*Atelopus nanay*, *A. exiguus*, *Gastrotheca cuencana*) experienced significant declines, with *A. nanay* and *A. exiguus* currently facing imminent extinction (Cáceres, 2022), while *G. pseustes* underwent a significant decline. Two additional species, *A. onorei* and *Nelsonophryne aequatorialis*, documented in historical records from the natural areas surrounding these protected regions, have not been sighted within or near these areas for approximately three decades. Only three species (*P. morlaco*, *P. philipi*, and *P. cryophilius*) seem to have maintained apparent stability (Arbeláez Ortiz and Vega Toral, 2008).

As was discussed by Ortega-Andrade *et al.* (2021), we agree that amid a general trend of loss of biodiversity, some amphibian taxa show a phylogenetic vulnerability to causes of sudden declines (anthropogenetically driven climate change and/or emerging diseases), most likely as a result of a combination of their distinctive life history traits and immune constraints. These taxa include most species of *Atelopus*, *Hyloxalus*, monophasic species of *Gastrotheca*, *Strabomantis*, and *Telmatobius*. Among their traits, running water habitats are a factor in common for all these lineages, except for the monophasic species of *Gastrotheca* that live in trees and seem do not depend on streams or rivers for their breeding, although some are associated with river margins.

Because they contribute uniquely to the functioning of their communities, the loss of such species is especially worrisome as it is expected to have a disproportionate impact on the stability of local ecosystems, beyond their taxonomic loss. This is of particular importance since most of them are endemic to specific habitats. Herein, we elaborate further on the disappearance of Ecuadorian species in these five most affected genera.

Tragic Loss: Extinction of Ecuador's Highland Aquatic Frogs, *Telmatobius*

The loss of the *Telmatobius* genus, a group of aquatic frogs from the Andean highlands of southern Ecuador, is a tragic event with far-reaching implications for taxonomy, ecology, culture, and beyond. These remarkable frogs, known for their aquatic lifestyle, have long been a prominent presence among the diverse anuran species inhabiting the highlands, thriving at elevations ranging from 2103 to 4110 meters. Their cultural significance is equally notable, having been historically used for sustenance and medicinal purposes by local communities.

Regrettably, the three endemic species within this genus are now believed to be on the brink of extinction, as they have not been sighted for over three decades, despite extensive survey efforts. Although hope remains that they might still exist in unexplored areas, this possibility remains slim. Ironically, *Telmatobius niger* holds a unique historical distinction as the first amphibian from Ecuador to be collected in the 1700s and preserved in a museum, specifically the Museum of Comparative Zoology at Harvard University.

Although the diversity of *Telmatobius* species in Ecuador is not extensive, with only three recorded species, one particular species, *Telmatobius niger*, has historically stood out due to its abundance. This is evident from the existence of over a hundred specimens deposited in collections, despite the challenges associated with collecting this species due to its aquatic habits. The earliest known specimens, discovered by an unknown collector, date back to the 1700s (MCZ 3035–39, *T. niger*) and were found in the Palmira Desert of Provincia Chimborazo. Additional collections took place in the early 20th century (1903, 1907) and later from 1959 to 1994.

Approximately a decade after the most recent record, a publication by Merino-Viteri *et al.* (2005) summarized all available information on the three endemic *Telmatobius*

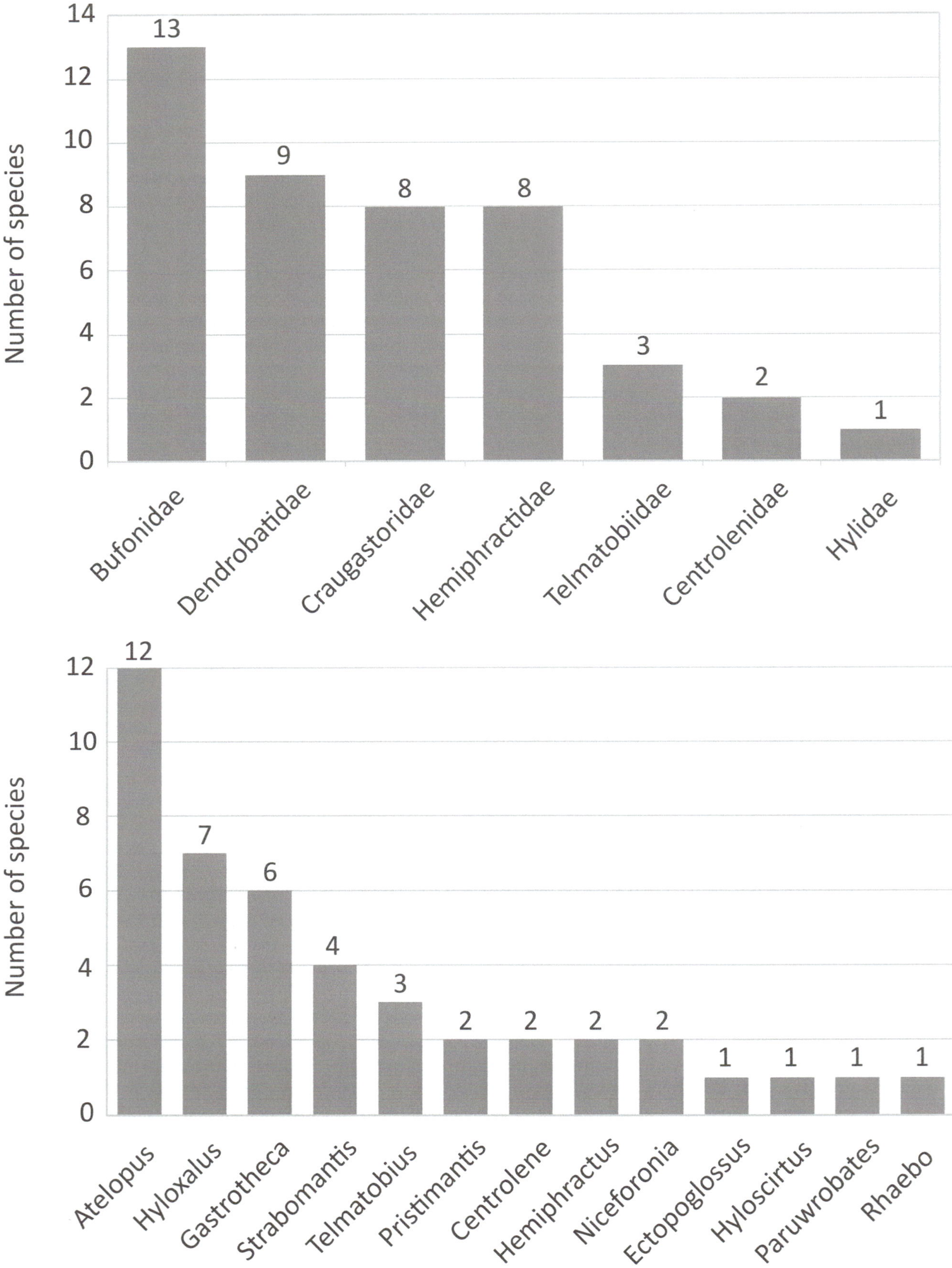

FIGURE 3.34 Top: number of critically endangered (possibly extinct) species from Ecuador according to families, ordered from highest to lowest. Bottom: 45 critically endangered (possibly extinct) species from Ecuador according to genera, ordered from highest to lowest number of species.

FIGURE 3.35 Extinction in protected areas. Six species photographed in 1986 by Morley Read at Bosque Protector Mazán, Provincia Azuay, Ecuador. From top left to right bottom: *Telmatobius niger*, *Hyloxalus anthracinus*, *Atelopus exiguus*, *Hyloxalus vertebralis*, *Gastrotheca cuencana*, and *Centrolene buckleyi*. Four of them disappeared from Mazán, and *Atelopus exiguus* and *G. cuencana* still survive in very low numbers. *Atelopus exiguus* is facing its imminent extinction.

species of Ecuador's central and southern Andes: *cirrhacelis*, *T. niger*, and *T. vellardi* (see also their species accounts in Volume II). The authors provided distributional, ecological, natural history, and presence/absence data, which indicated the absence of *Telmatobius* frogs since 1994. The populations experienced a sudden and steep decline in the late 1980s and early 1990s, even within protected areas such as Parque Nacional Sangay and Parque Nacional Cajas (for *T. niger*) and Parque Nacional Podocarpus (for *T. cirrhacelis* and *T. vellardi*). Notably, *T. cirrhacelis* has not been collected since March 1987, *T. vellardi* since July 1987, and *T. niger* since December 1994.

Merino-Viteri *et al.* (2005) also conducted an analysis of potential causes behind the decline. Between 1985 and 1990, 31% of the collected specimens (juveniles, subadults, and adults) exhibited some form of disease or abnormality. Furthermore, all tadpoles collected between 1989 and 1994 displayed anomalies, including chytridiomycosis, extra limbs and limb malformations, epidermal disorders, and other unidentified illnesses. The examination of climatic data from three meteorological stations in the southern Ecuadorian Andes revealed occurrences of extreme conditions characterized by high temperatures and low precipitation, coinciding with the reported instances of diseases and disappearances. The researchers concluded that these diseases and/or climatic abnormalities may have played a role in the population declines and apparent extinctions of *Telmatobius* frogs in Ecuador during the late 1980s and early 1990s.

From Abundance to Obscurity: The Painful Story of Harlequin Frogs' Decline and Extinction

The species most heavily impacted by population declines in the Neotropics and in Ecuador are harlequin frogs of the genus *Atelopus*. In Ecuador, there are 24 described species (18 endemic) and at least nine awaiting descriptions (Figure 3.36). About one-third of the Ecuadorian species seem to be extinct, while the other two-thirds are critically endangered (Figure 3.37). These colorful and diurnal frogs, terrestrial and semiarboreal, are closely associated with rivers and rocky streams, where females lay eggs and tadpoles develop. Their habitats encompass lowland tropical rainforests, cloud forests, and paramos at an altitudinal range between 14 and 4385 m. Among the 33 species (described and undescribed), two species live in the Pacific lowlands, two in the Amazonian lowlands, and one in the Andean foothills, and the remainder are restricted to the Andean region, usually exceeding 900 meters. Most *Atelopus* species exhibit limited geographic ranges and high levels of regional endemism.

The declines of *Atelopus* in Ecuador are part of a general process that has affected this genus throughout most of its distributional range, including Costa Rica (Lips, 1998), Panama (Lips, 1999), Venezuela (La Marca and Lötters, 1997; Bonaccorso *et al.*, 2003), Colombia (Rueda-Almonacid *et al.*, 2004), and Peru (Vial and Saylor, 1993). Comprehensive reviews of the population status of the genus and possible causes for declines have been published by La Marca *et al.* (2005), Pounds *et al.* (2006), Jaynes *et al.* (2022), and Lötters *et al.* (2023).

Several studies have examined the status and conservation of Ecuadorian *Atelopus* species, focusing on specific species such as *Atelopus ignescens* (Ron *et al.*, 2003), *A. longirostris* (Tapia *et al.*, 2017), and *A. mindoensis* (Barrio-Amorós *et al.*, 2020). Additionally, sets of species

FIGURE 3.36 Species of Ecuadorian harlequin frogs *Atelopus*. The 24 described species are in full color; the eight species awaiting descriptions and one to be reported from Ecuador are faded.

FIGURE 3.37 Conservation status of the 33 species of Ecuadorian *Atelopus*. Critically endangered (possibly extinct) species are in black (12 species), and critically endangered species are in red (20 species). The status of one undescribed species from the Amazonian lowlands is unknown (?).

have been analyzed by Bustamante *et al.* (2005), while comprehensive information on most species can be found in La Marca *et al.*, 2005; Pounds *et al.*, 2006; Ron *et al.*, 2011; Tapia *et al.*, 2017; and Jaynes *et al.*, 2022. Further discussion on the *Atelopus* declines has been provided by Pounds and Coloma (2008). Herein, we aim to provide a general update on their status and conservation, while specific details are available in each species' individual accounts and in Lötters *et al.* (2023).

Our Red List assessment of the 33 species (24 described, nine pending descriptions) of Ecuadorian *Atelopus* shows that all them are threatened except for the unknown status of one undescribed species (see Figures 3.36, 3.37).

Historical accounts and field surveys indicate the former abundance of Ecuadorian *Atelopus* species, but their populations have significantly declined since the mid-1980s. In general terms, Giovanni Onore, the Tapia family, Luis A. Coloma, and Ana Almendáriz-Cabezas were the first-hand witnesses of the sudden transition from abundance to absence of Ecuadorian frogs because in the 1980s and 1990s, they were performing collections in Ecuador and especially in the Andean region (see Chapter 2). The absence of records and reports from local people indicates widespread declines and extinctions, particularly notable in the case of at least 12 species that are probably extinct and the remainder highly threatened.

There is anecdotal evidence of the historical abundance of Ecuadorian *Atelopus* species, such as *Atelopus ignescens*, up to 1985. Reports from the 1800s and field surveys in the 1960s and 1980s documented high densities of *Atelopus* individuals in various locations. They are detailed in Jiménez de la Espada (1875), Peters (1973), Black (1982), Ron *et al.* (2003), and Ron *et al.* (2011). For instance, in December 1864, the Spanish naturalist Marcos Jiménez de la Espada witnessed the occurrence of thousands of individuals of *A. ignescens* in the lush and damp meadows near streams, pools, and lakes at Páramo del Antisana (Laguna de la Mica). Scientific teams from the University of Kansas recorded high densities of *Atelopus* in the 1960s and 1970s (e.g., 194 individuals of *A. ignescens* [0.81 individuals/person/min] at Ingaloma, Provincia de Pichincha, 3780 m, 16 August 1968; 53 individuals of *A.* sp. [0.88/pers/min) 20 km southwest of Santa Rosa, Provincia del Chimborazo, 3700 m, 23 July 1970; 53 individuals of *A.* sp. [0.59 ind/pers/min) 11.2 km northwest of San Juan, Provincia del Chimborazo, 3950 m, 12 July 1971). In 1981, at Páramo del Antisana, densities reached as high as 50 individuals/m^2 (Black, 1982). At Laguna de Limpiopungo in Parque Nacional Cotopaxi, Carlos A. Valle recorded 12–60 individuals in 1984 in an area of about 420 m^2 bordering the stream that feeds the lake (Valle, 1984). Furthermore, there are reports of mass migrations in the mid-20th century that resulted in large numbers of individuals fatally struck across 1–8 km of highways in the Cotopaxi, Tungurahua, and Bolívar provinces in 1958, 1959, and 1985. However, since the mid-1980s, efforts to find *Atelopus* species above 1000 meters in the Ecuadorian Andes have been largely unsuccessful. Numerous surveys conducted from the second half of the 1980s to 2022 failed to record *Atelopus* in most localities, except for a few individuals belonging to the critically endangered species and the ones not yet described, which are also threatened (see Figure 3.37).

The most well-documented case of population decline is that of *Atelopus ignescens*, a historically abundant species endemic to Ecuador. It inhabited various habitats in the Andes but was absent since 1988, except for a dying tadpole in 1989. Despite search efforts in accessible regions with undisturbed vegetation, no records were found during 28 years. Native Americans and local people, who were familiar with *Atelopus* due to its abundance, have also reported its absence for many years. In 2016, a relictual population of this species was rediscovered in Angamarca, Provincia Cotopaxi, and is currently being monitored, while genetic studies are being conducted and ex situ breeding attempts are made.

Despite their former abundance, efforts to find *Atelopus* in Ecuador since the mid-1990s have been generally unsuccessful. Bustamante *et al.* (2005) and Ron *et al.* (2003) reported the absence of records of *Atelopus* in nearly 100 surveys at 21 localities in Ecuador between 1998 and 2002. Except for the isolated population of *Atelopus* sp. (Cordillera del Cóndor), the number of individuals in surviving populations of *Atelopus* are extremely low (see each species account). For instance, only a handful of *A. exiguus* and *A. nanay* frogs have been found despite being monitored during about two decades. Other more recently monitored species such as *A. longirostris*, *A ignescens*, *A. bomolochos*, *A. coynei*, *A. balios*, *A.* sp[1], *A.* sp[2] (*spumarius* complex), and *A.* sp. (*aff. longirostris*) reveal larger numbers than previously recorded, but still extremely low numbers compared to what could be considered healthy populations.

The Extinction and Disappearance of the Rocket Frogs *Hyloxalus*

Hyloxalus frogs represent the most diverse genus (60 species) among dendrobatid frogs. *Hyloxalus* are primarily diurnal frogs that generally live associated with rocky streams; they are characterized by being mostly cryptically colored frogs with complex parental behaviors. They are distributed predominantly in the northern and central Andes of South America and used to be conspicuous components of Andean ecosystems. Coloma (1995) reviewed the population status of *Hyloxalus* (then *Colostethus*). Páez-Vacas *et al.* (2010) discussed the conservation of six species in the *Hyloxalus bocagei* complex. The status of *Hyloxalus* was also discussed by Ron *et al.* (2011). Currently, and notably, out of the 26 recorded species in Ecuador, 21 (80%) are threatened, all of which are native to the Andean region. The remaining three non-threatened species inhabit the Amazonian lowlands (*H. yasuni* and *H. sauli*) and the Pacific lowlands (*H. awa*). Among the recorded species, seven may be extinct, another seven are classified as critically endangered, and seven more are either endangered or vulnerable (Figure 3.38).

Eight species of *Hyloxalus* have not been sighted since 1990. Seven are categorized as critically endangered (possibly extinct) and one as critically endangered. The most dramatic losses and possible extinctions are those of *Hyloxalus fuliginosus*, *H. elachyhistus* (sensu this encyclopedia, see account), and *H. lehmanni*, which were last sighted in 1980, 1984, and 1990, respectively. These species were once abundant in the Ecuadorian Andes during historical times. Survey efforts for them have been extensive, without positive results. It is interesting to note that surveys between 1988 and 2004 in the vicinities of El Reventador and the Quijos-Topo depression on the eastern slopes of the Cordillera Oriental revealed the absence of *H. fuliginosus* but the presence of the sympatric *H. bocagei* and *H. pulchellus* (Bustamante *et al.*, 2005).

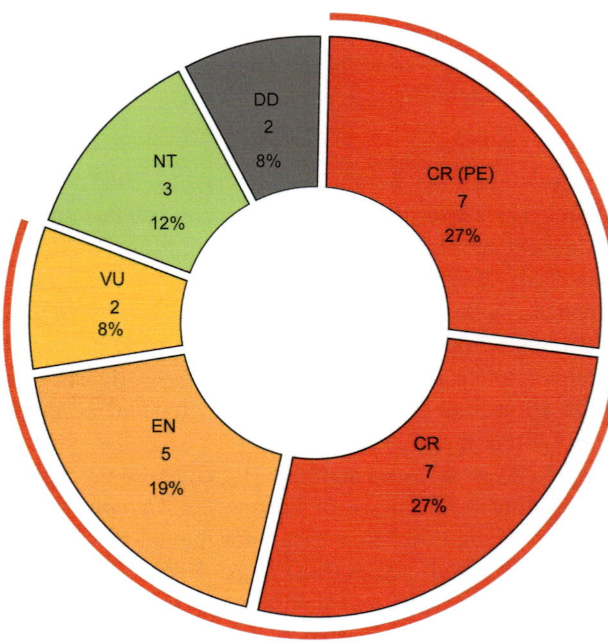

FIGURE 3.38 IUCN Red List of 26 species of *Hyloxalus* from Ecuador. Categories: CR (PE) = critically endangered (possibly extinct), CR = critically endangered, EN = endangered, VU = vulnerable, NT = near threatened, DD = data deficient. Number of species and percentage are indicated below the category. Threatened species (80%) are in red, orange, and yellow bordered by a red line.

Additionally, several rare species of rocket frogs, known only from few individuals and one or two specific locations, were last observed in the 1980s: *Hyloxalus breviquartus* (1984), *H. maquipucuna* (1984), and *H. fallax* (1987). On the other hand, *H. peculiaris* and *H. pumilus* were collected and last sighted in 1962. Except for *H. fallax*, survey efforts for all of them have been conducted and revealed no specimens. More details regarding their status and conservation are provided in each species account.

Another 13 threatened species of *Hyloxalus* have experienced significant population declines and face an imminent risk of extinction unless urgent and effective conservation measures are implemented. For instance, iconic cases are those of the historically abundant and widespread *Hyloxalus jacobuspetersi*, *H. pulchellus*, and *H. vertebralis*. *Hyloxalus jacobuspetersi* has not been recorded since 1989, despite its historically wide distribution in the higher parts of the Cordillera Occidental and inter-Andean valleys. Bustamante *et al.* (2005) provided absence data for *H. jacobuspetersi* at Cashca-Totoras, Provincia Bolívar. Nonetheless, a small relictual population was found in 2008 by Morley Read on a private property and reserve in the region of the Río Pita, near Quito, Provincia Pichincha. *Hyloxalus pulchellus* has not been recorded from the Papallacta area, Provincia Napo, where it was notably abundant. The most recent record at Papallacta is from September 1994. Fortunately, small relictual populations have been found at Yanayacu Reserve, Baeza, and Río Azuela, Provincia Napo. *Hyloxalus vertebralis* disappeared from Parque Nacional Cajas and the adjacent protected Reserva Mazán in the Cordillera Occidental, but isolated populations have been recorded inside abandoned properties in the urban parts of the City of Cuenca and toward the Cordillera Oriental at El Jordán, Paguancay, Provincia Azuay. The habitat at El Jordán comprises pasturelands and secondary vegetation near human habitations.

Less-known *Hyloxalus* are *Hyloxalus delatorreae*, *H. anthracinus*, *H. infraguttatus*, *H. mystax*, and *H. nexipus*. *Hyloxalus delatorreae* was monitored between September 2005 and August 2006 at Cañón del Morán, Provincia Carchi, near its type locality (Yánez-Muñoz and Meza-Ramos, 2006). Researchers estimated a population size of 25 individuals. Subsequent searches have revealed no specimens. *Hyloxalus anthracinus* is a very rare frog that disappeared from most of its range, except in Reserva Mazar, Provincia Cañar, where its most recent sighting is from 9 January 2010. *Hyloxalus infraguttatus* (sensu this encyclopedia), *H. mystax*, and *H. nexipus* are still present in some of the localities of their ranges but are threatened by habitat destruction.

The Tragedy with Direct-Developing Marsupial Frogs and Their Congeners

Twenty species of marsupial frogs of the genus *Gastrotheca* are known from Ecuador. Duellman (2015) and Carvajal-Endara *et al.* (2019) provide detailed information of the Ecuadorian species, and the latter also discuss their conservation. Marsupial frogs stand out because of their evolutionarily unique reproductive behavior and cultural significancy. Sadly, nearly all of them (85%) are at risk of extinction, except for three species that are categorized as near threatened (Figure 3.39).

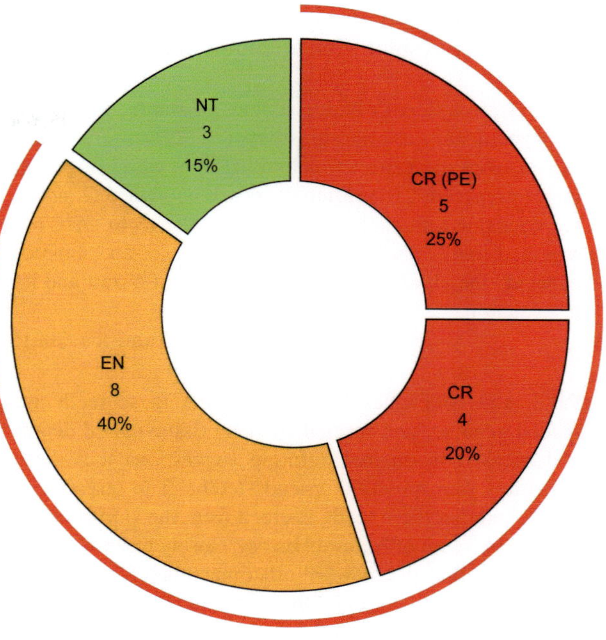

FIGURE 3.39 IUCN Red List of 19 species of *Gastrotheca* from Ecuador. Categories: CR (PE) = critically endangered (possibly extinct), CR = critically endangered, EN = endangered, NT = near threatened. Number of species and percentage are indicated below the category. Threatened species (85%) are in red and orange, bordered by a red line.

Among the threatened species, a major tragedy is the disappearance and possible extinction of half (five) of the monophasic species. The monophasic species are not Ecuadorian endemics and have larger distributions that include Colombia, where no recent records are available either.

One of the monophasic possibly extinct species (*Gastrotheca angustifrons*) comes from the Pacific lowlands and has not been seen in Ecuador since its collection in 1898, although more recent records are from Colombia in 1983. The other three monophasic species that disappeared are the cloud forest species *G. guentheri* (last seen in 1991), *G. andaquiensis* (last seen in 1984), and *G. weinlandii* (last seen in 1985) from the Cordillera Oriental. The species *G. dendronastes*, last observed in Ecuador in 1979 within the Cordillera Occidental, was rediscovered in Colombia in 2004. This suggests that this critically endangered species may still persist. These species were nearly unknown and poorly studied, except *G. guentheri*, for which studies have been conducted on preserved specimens (Wiens *et al.*, 2007; Paluh *et al.*, 2019, 2021). *Gastrotheca guentheri* has true teeth on its lower jaw and is the only one species to possess them, out of the more than 7600 known anurans. Since its finding in 1882, scientists have been perplexed by the uniqueness of this species of marsupial frogs that only recently garnered significant attention worldwide.

> In 2011, evolutionary biologist John Wiens reconstructed the evolutionary relationships between 170 different species of frogs to create a timeline between when frogs lost their lower teeth 230 million years ago and when *G. guentheri* regained its teeth. He found that the teeth were not regained until around 20 million years ago, an "unprecedented" length of time between the loss of a trait and its re-evolution.
>
> (Tamisiea, 2021)

> Paluh and colleagues have shown that these teeth are indeed genuine, and formed by bony tissue called dentin and encased in enamel, refuting suggestions that these structures were possibly "pseudo-teeth." The researchers also found that these teeth, sourced from the University of Kansas Museum of Natural History, closely resemble teeth those along the upper jaws of other marsupial frogs—making them more likely to be the real deal. . . . Paluh hopes that Guenther's marsupial frog likewise reappears—not least because living samples of this amphibian will be crucial for learning more about their teeth.
>
> (Tamisiea, 2021)

Regarding biphasic species, only 1 species out of 10 is possibly extinct, *Gastrotheca psychrophila*. Failure to find it is intriguing given that other biphasic species, although threatened, are still found. It was originally collected in 1968–1971, but many subsequent searc efforts at the type locality have been unsuccessful (Székely *et al.*, 2020). According to Carvajal-Endara *et al.* (2019), the highland biphasic species have survived the catastrophic events that led to the disappearance of many other sympatric anurans. In general, although there have been no studies on their population dynamics, except for *G. riobambae* (Jiménez-Casalombo *et al.*, 2023 and citations therein), other lines of evidence such as a reduction in the area of occupation point to their crises. All biphasic *Gastrotheca* species included seem to tolerate some degree of habitat modification; most of them have been observed in pasture lands where lentic water is available for tadpoles. However, their habitats are in heavily human-populated areas, where the synergistic effects of agriculture, mining, cattle raising, fires, introduced species, pesticide use, pollution, road construction, and urban development are increasing, and the reduction in their areas of occupation suggests that their population size may be decreasing. The distribution of most of these species is indeed narrow and fragmented (as reported for *G. cuencana*, *G. elicioi*, *G. lojana*, *G. espeletia*, *G. turnerorum*, and *G. yacuri*); thus, it is a cause of concern, as geographically restricted species are often at risk of extinction because of demographic stochasticity. Additional significant threats in their areas of occurrence are extreme climatic abnormalities and pathogens, key factors implicated in abrupt anuran declines, and extinctions in the highlands of Ecuador (Ron *et al.*, 2003; Merino-Viteri *et al.*, 2005; Pounds *et al.*, 2006). It is noticeable that the chytrid fungus *Batrachochytrium dendrobatidis*, which has been implicated in anuran declines worldwide (Lips *et al.*, 2008), has been present in most species of *Gastrotheca* since at least 1974 (when no sudden declines were noticed) (Manzano Pasquel, 2014) and is present contemporaneously in most Andean frogs (Guayasamin *et al.*, 2014; Manzano Pasquel, 2014; Tapia *et al.*, 2017).

The Unknown Extinction of the Broad-Headed *Strabomantis* Frogs

There are currently seven known species of *Strabomantis* in Ecuador, but very little is known about their ecology and natural history, and they have been rarely observed or collected. Unfortunately, with the exception of *Strabomantis sulcatus*, the other six species in Ecuador are at risk of extinction. It is disheartening to note that four species with limited distribution areas have not been recorded for over four decades. *Strabomantis helonotus* has not been collected since May 1959, *S. cerastes* since April 1977, *S. anatipes* since May 1977, and *S. necerus* since May 1985. Consequently, these species may be classified as "possibly extinct." The remaining two species (*S. anomalus* and *S. cornutus*) are highly threatened. The most recent record of *S. anomalus* in Ecuador dates back to March 1985, although it was found in Colombia in 2021.

With the exception of *Strabomantis sulcatus*, which is active during the day and night and inhabits the leaf litter on the ground in lowland Amazonian areas, all the other species are riparian and found in lowland and montane forests. Unfortunately, it appears that their dependence on riparian habitats has contributed to their decline and imminent extinction. Their absence from general herpetological surveys in Ecuador and Colombia is perplexing, suggesting that they have experienced catastrophic declines, possibly due to climate change and pathogens, similar to many other stream-dwelling frog species. Additionally, these species face many potential significant threats such as habitat destruction and fragmentation because of deforestation, agriculture, livestock rearing, wood extraction, pollution, mining, landslides, human settlement, roads, and wildfires.

The Data-Deficient Species

Fifty species of Ecuadorian amphibians, accounting for 8% of the total 655 species, are categorized as data deficient. Among these, half (25 species) belong to the genus *Pristimantis*, followed by 9 species of Gymnophiona, 5 Centrolenidae, and 11 in a variety of other genera. Taxonomic uncertainties surround a dozen of the data-deficient species, including species such as *Osornophryne talipes*, *Rhinella proboscidea*, *Allobates trilineatus*, *Epipedobates espinosai*, *Hyloxalus marmoreoventris*, and *Ranitomeya amazonica*, among others. The deficiency of data for *Pristimantis* species primarily stems from their recent description, as they are typically known only from the type locality or a few surrounding sites, with few additional surveys conducted. On the other hand, the lack of data for Gymnophiona species is attributed to their fossorial habits and the scarcity of collections and taxonomic studies focusing on this group. Some other species migh also be poorly known (e.g., *Elachistocleis araios*, *Amazops amazops*) given their low numbers; thus they might be threatened with extinction.

Which Are the Causes of Amphibian Declines in Ecuador?

A comprehensive overview of the factors contributing to amphibian declines worldwide, along with relevant citations, can be found in the Conservation section of AmphibiaWeb (2022). When the recognition of precipitous global amphibian declines first emerged during the First World Congress of Herpetology in 1989, research into the causes of these sudden as well as of the gradual declines began.

Ten years later, a significant effort was made to compile data and summarize knowledge about declines and their causes in Latin America. This initiative involved a series of three regional workshops attended by 88 herpetologists, one of which took place in Quito, Ecuador, in November 1999 (Lips *et al.*, 2000). At that time, it was believed that 15 species from Ecuador were already extinct and eight were experiencing declines (Young *et al.*, 2001).

These workshops served as a catalyst for the initial studies conducted in Ecuador, primarily carried out at the Museo de Zoología of Pontificia Universidad Católica del Ecuador. The findings from these studies began to be published in the 2000s, leading to several influential publications that focused on different aspects of amphibian declines. Among them, Ron and Merino (2000), Ron *et al.* (2003), Merino-Viteri *et al.* (2005), and Bustamante *et al.* (2005) addressed and discussed the potential causes of declines, providing baseline data for subsequent discussions by Pounds *et al.* (2006), Pounds *et al.* (2007), Pounds and Coloma (2008). Ron *et al.* (2011) summarized the knowledge of the factors behind amphibian declines and extinctions in Ecuador until that time. In 2013, Menéndez-Guerrero and Graham (2013) published their evaluation of three major threats implicated in precipitous declines of Ecuadorian amphibians: chytridiomycosis, climate change, and habitat loss. Herein, we provide an update as of 2023. The IUCN provides a Unified Classification of Direct Threats (www.iucnredlist.org/technical-documents/classification-schemes/threats-classification-scheme), where they describe 12 threats. Herein, we discuss some of the most relevant to Ecuadorian amphibians as discussed by the authors earlier mentioned and by Ortega-Andrade *et al.* (2021). The amphibian extinctions and declines, both sudden and gradual, are attributed to a confluence of multiple stressors, often interacting with one another.

Primary factors contributing to the sudden extinction and decline of numerous species, notably in the latter part of the 1980s and the early 1990s, include climate change and diseases. These have significantly impacted both protected and altered areas in the Andean region. Furthermore, amphibians in the Andes, Amazonian, and Pacific lowlands face ongoing and future threats from a spectrum of challenges, including climate change, diseases, habitat destruction, fragmentation, pollution, introduced species, UV radiation, and the compounded effects resulting from their interactions. Habitat destruction is being significantly exacerbated by activities such as logging, agriculture, livestock farming, the oil industry, and urban development. Moreover, both legal and illegal mining have emerged as major threats contributing to this degradation in protected and non-protected areas.

Climate Change

Climate change can be a significant cause for the population declines in amphibians (Winter *et al.*, 2016; Luedtke *et al.*, 2023). Steigerwald (2021) provides an overview of the effects of climate change on amphibians. According to him and citations therein, it has been shown that the anthropogenic climate change crisis has diverse and often profound and devastating repercussions for Earth's biodiversity, especially on amphibians that serve as sensitive indicators. While certain amphibians may endure changes, some might genetically adapt, others will disperse, and, unfortunately, some species may face extinction (see also "Climate Change" under "Genetics of Wild Populations"). The most recent Global Amphibian Assessment (Luedtke *et al.*, 2023) evaluated 8011 species for the IUCN Red List of Threatened Species, including Ecuadorian species. They found that amphibians are the most threatened vertebrate class (40.7% of species are globally threatened). According to this study, ongoing and projected climate change effects are now of increasing concern, driving 39% of status deterioration since 2004, followed by habitat loss (37%). Thus, it became clear that climate change is driving and will drive many amphibians toward extinction. However, it is paradoxical that Luedtke *et al.* (2023), as well as other studies involving assessments of Ecuadorian amphibians (e.g., Scheele *et al.*, 2019; Ortega-Andrade *et al.*, 2021), have predominantly overlooked the threat of climate change in the 1980s up to 1990s but place an emphasis on the current and future scenarios. Instead, these studies have focused on habitat destruction and chytrid fungi, considering them main drivers of extinction. The chytrid fungi are assumed to have acted alone in protected areas, following the novel pathogen hypothesis (Berger *et al.*, 1999). This emphasis persisted due to the challenges associated with incorporating climate change and its interaction with pathogens and other factors into their analyses. Nonetheless, Ron *et al.* (2011) concluded that "undoubtedly, habitat loss has contributed to amphibian population declines in Ecuador but it cannot explain declines in seemingly pristine regions like those reported by Ron *et al.* (2003)." Similarly, Menéndez-Guerrero and Graham (2013) conclude that "climate change, chytridiomicosis, and their synergistic effects, are likely currently exerting the greatest impact on amphibians in Ecuador, while habitat loss does not seem to be causing precipitous declines."

Amphibians are notably sensitive to climate warming, and numerous studies have established a connection between climate change and abrupt declines in amphibian populations worldwide. Discussions about whether climate change was already driving amphibian declines launched in earnest in the early 1990s. Baffled herpetologists were watching amphibian populations disappear in well-protected habitats, where the usual major menace of habitat loss could not be blamed (Blaustein and Wake, 1990; Pounds and Crump, 1994; Lips, 1998; Pounds *et al.*, 1999; Kiesecker *et al.*, 2001; Lips *et al.*, 2005; La Marca *et al.*, 2005). The climate crisis may be a key factor responsible for driving these precipitous declines through direct or indirect mechanisms. A seminal paper by Pounds *et al.* (2006) suggested that global warming has driven *Atelopus* population declines in the Neotropics by favoring disease outbreaks. This publication triggered controversy (e.g., Lips *et al.*, 2008; Pounds *et al.*, 2007; Pounds and Coloma, 2008) because of the difficulties in establishing that climate change and diseases are causally linked, and it was not universally accepted, leading to further discussions and ongoing research on this topic. More recently, experimental work on *Atelopus zeteki* (a species that suffered abrupt declines) by Cohen *et al.* (2019) established this link.

According to the latest IPCC report (IPCC, 2023), the likely range of total human-caused global surface temperature increase from 1850–1900 to 2010–2019 is 0.8°C to 1.3°C, with a best estimate of 1.07°C. In South America, in 2022, the annual temperature was 0.89°C above average—the 12th warmest year since regional records began in 1910. South America's annual temperature has increased at an average rate of 0.14°C per decade since 1910; however, the average rate of increase has been nearly double (0.22°C) that value since 1981 (NOAA National Centers for Environmental Information, 2023). Nevertheless, the extent of this temperature rise (1.07°C) exhibits regional disparities. In the Ecuadorian Andes specifically, the temperature increase has significantly outpaced the global average (Ron *et al.*, 2003; Merino-Viteri *et al.*, 2005). For instance, within the Quito Valley alone, temperatures have surged by 2.01°C over the course of the last century (Ron *et al.*, 2003). In various other Andean locations in Ecuador, both urban and rural, the temperature rise mirrors that observed in Quito, and in some cases, it dramatically surpasses it. An example of such dramatic temperature increases can be seen in Biblián, Provincia Cañar (Merino-Viteri *et al.*, 2005), where the mean temperature was much higher (in 2003) than in historical records, marking an astonishing mean temperature increase of 5.6°C. Regarding precipitation, data between 1972 and 1998 show that there has been a significant decrease in total annual precipitation in the Ecuadorian Andes (Menéndez-Guerrero and Graham, 2013).

Studies carried out in Ecuador by Ron *et al.* (2003) on *Atelopus ignescens*, Merino-Viteri *et al.* (2005) focusing on *Telmatobius*, and Menéndez-Guerrero and Graham (2013) delving into endemic frogs all supported the hypothesis by Pounds *et al.* (2006) regarding the interaction between climate change and pathogens. Remarkably, the two former studies revealed that extreme combinations of dry and warm climates prevailed within the distribution range of these species during the late 1980s, precisely when the last individuals of these species were observed alive throughout most of their ranges. One population of *Atelopus ignescens* (rediscovered in 2016) is known to have survived in Angamarca, Provincia Cotopaxi. The simultaneous occurrence of extreme climate patterns and population declines during the late 1980s and 1990s aligns with the hypothesis that unusual weather conditions might have significantly contributed to the catastrophic decline of the endemic jambato harlequin frog, *A. ignescens*, and led to the possible extinction of the three species of Ecuadorian aquatic species of *Telmatobius*.

A study by Menéndez-Guerrero and Graham (2013) used geographical quantitative analyses, and their results showed that amphibians display a non-random pattern of extinction risk, both geographically and taxonomically.

Rising temperatures due to global warming are anticipated to cause shifts in species ranges, potentially altering the makeup of species communities. These changes are expected to vary based on the topography of the region where extinctions of either species or populations will occur. Projected climate change scenarios have been developed for Ecuadorian amphibians as part of a larger data set of Neotropical amphibians by Menéndez-Guerrero *et al.* (2019). This study predicts that within Neotropical anuran amphibians, a trend toward biotic homogenization may be observed in mountainous areas, whereas a trend toward biotic heterogenization may be more likely in lowland and plains landscapes characterized by only modest variation in altitude. Their models indicate that biotic homogenization (e.g., in the Andes) is mainly driven by an increase in generalist species and, to a lesser extent, a loss of specialists, resulting in greater local species richness overall but lower phylogenetic diversity. In contrast, increasing biotic heterogeneity (e.g., in the Amazonian lowlands) is explained by range contraction and local extinctions of generalists, resulting in reduced species richness but higher phylogenetic dispersion of remaining species.

Pathogens

Emerging infectious diseases are a major proximal cause of amphibian declines worldwide. Three important diseases are among the causes invoked to explain abrupt amphibian declines worldwide (see also "Disease" under "Genetics of Wild Populations"). These include amphibian chytridiomycosis caused by the chytrid fungus *Batrachochytrium dendrobatidis* (*Bd*) (see a review by Berger *et al.* (2016); Scheele *et al.* (2019)), viral diseases caused by different ranaviruses (Rv) (Ruggeri *et al.*, 2023, and citations therein), and the protist Perkinsea (Chambouvet *et al.*, 2015). Parasites such as nematodes also have been implicated in amphibian declines (Johnson *et al.*, 2010).

Nevertheless, although the amphibian declines have generated a huge research effort on pathogens, and especially on chytridiomycosis since it was first reported in 1998, studies of pathogens in Ecuador are relatively limited. Among the Ecuadorian investigations, the focus has predominantly been on chytrid fungus, with only two exceptions: Merino-Viteri *et al.* (2005) surveyed the aquatic *Telmatobius* for multiple pathogens during a period of abrupt declines using preserved specimens, while Urgiles *et al.* (2021) explored the prevalence of multiple pathogens—*Bd*, Rv, and Perkinsea—in terrestrial frogs, specifically *Pristimantis*, during more recent times.

Chytridiomycosis is a disease that disrupts the normal function of the amphibian skin. Chytrids infect the keratinized parts of larval mouths and the epithelial cells of post-metamorphic frogs. In post-metamorphs, *Bd* might cause electrolyte imbalance, dehydration, and eventually heart failure and death by interfering with cutaneous respiration and osmoregulation. Infected amphibians may exhibit lethargy, abnormal posture, and skin discoloration. In amphibian communities, some species thrive even as others around them perish. Around the world, some amphibian species appear to have evolved immunity to chytridiomycosis.

There are two hypotheses that explain the presence and impact of the chytrid on abrupt amphibian declines: the emerging or novel pathogen hypothesis (NP) (Berger *et al.*, 1999; Daszak *et al.*, 1999; Lips *et al.*, 2008) and the endemic pathogen hypothesis (EnP) (Rachowicz *et al.*, 2005; Pounds *et al.*, 2006). According to the NP hypothesis, factors such as global trade, human movement, and the intercontinental movement of amphibians have facilitated the introduction of *Bd* to new regions. In contrast, the EnP hypothesis proposes that *Bd* has been present in amphibian populations for a long time but that climate change and stressors have recently triggered its increased virulence or the onset of disease outbreaks.

Several studies worldwide and in Ecuador contribute to the growing body of evidence supporting the idea that *Bd* has been present in amphibian populations for a longer period than initially thought, consistent with the EnP hypothesis. For instance, Bataille *et al.* (2013) used historical data to assess *Bd* prevalence in amphibians over time, suggesting that *Bd* was present in some populations before the observed disease-related declines. As remarked by Carvalho *et al.* (2024), retrospective studies using museum specimens to track historical *Bd* infections have complicated the original story of *Bd* as a fully novel pathogen that spread quickly and caused epizootics after point introductions in the late 20th century. *Bd* was detected in museum specimens from Costa Rica at least 30 years before the presumed wave-like epizootic of the 1990s, as well as Baja California in the 1890s and southern California in the 1910s, around a century before *Bd* epizootics. In Ecuador, an unpublished study by Manzano Pasquel (2014) shows that the earliest records (1974) are dated to at least a decade before declines were first noticed, after the mid-1980s.

The first reports of the presence of *Bd* in Ecuador were made by Ron and Merino (2000), Merino-Viteri (2001), Merino-Viteri *et al.* (2005), and Ron (2005) in the Andean region. Subsequently, there have been a handful of studies reporting chytrid fungus throughout most of Ecuador, in the Pacific and Amazonian lowlands, and in the Andes (McCracken *et al.*, 2009; Vizcaíno *et al.*, 2013; Tarvin *et al.*, 2014; Guayasamin *et al.*, 2014; Tapia *et al.*, 2017; Jervis *et al.*, 2020; Narváez-Narváez *et al.*, 2021; Urgiles *et al.*, 2021). Most of these studies recorded prevalence data.

Interestingly, despite positive diagnoses of chytridiomycosis in more than 50 species of amphibians in Ecuador since 1974 (including a species of salamander, *Bolitoglossa* sp., Vizcaíno *et al.*, 2013), massive die-offs only seem to have occurred from the mid-1980s to the mid-1990s. The presence of chytrid fungus after the mid-1990s to the present is apparently not causing precipitous declines (Guayasamin *et al.*, 2014; Tapia *et al.*, 2017). The studies published from 2009 to date found no evidence of population declines or that *Bd* is negatively affecting host populations.

The examination of individuals from various species collected during the period of massive disappearances revealed both negative and positive results for the chytrid fungus. For example, *Atelopus ignescens* individuals tested negative, as reported by Ron *et al.* (2003). In the years 1983 (1), 1984 (7), 1985 (3), 1986 (29), 1987 (11), and 1988 (2), 89 individuals of *A. ignescens* collected during the period when the chytrid fungus was already present in Ecuador showed negative test results, including three females of *A. ignescens* found dead in November 1987 20 km southeast of Latacunga (Provincia del Cotopaxi) (Merino-Viteri, 2001). Other specimens of five species that declined or disappeared in the Andes tested positive for chytrid fungus, including *Telmatobius niger* from Azuay, *A. bomolochos* (Provincia Cañar, collected in 1980), *A. pastuso* from Provincia Carchi and Imbabura, *A.* sp. from Otongoro in Provincia Pichincha, and *A.* sp. from Atillo, Provincia Chimborazo. The Atillo frogs were examined because when found in the field, they were either dead or evidently unhealthy and died within a few hours after collection in January 1991 in a massive die-off observed by Luis A. Coloma and Giovanni Onore (Ron and Merino-Viteri, 2000; Merino-Viteri, 2001; Ron *et al.*, 2003).

It is noteworthy that in the studies conducted by Andrés Merino-Viteri and collaborators on *Telmatobius* in Ecuador in the period of its decline (1985–1994), chytridiomycosis was one of several diseases encountered (Merino-Viteri *et al.*, 2005). This study points to a cocktail of diseases, such as chytridiomycosis and parasites. They found extra limbs and limb malformations, epidermal disorders, and other yet-unidentified illnesses in adults and in tadpoles (see under "Tragic Loss: Extinction of Ecuador's Highland Aquatic Frogs, *Telmatobius*"). The pathogen outbreaks during the late 1980s to the mid-1990s could have been mediated by favorable environmental conditions (e.g., dry and/or warm climate). The recent study by Urgiles *et al.* (2021) on terrestrial frogs (genus *Pristimantis*) in 17 localities in the provinces of Azuay and Cañar in the southern Ecuadorian Andes confirm the first detection of Perkinsea in terrestrial-breeding amphibians and in the Andean region, present the first report of Ranavirus in Ecuador, and add to the handful of studies finding *Bd* in Ecuadorian frogs.

Additional pathological diagnoses of preserved specimens during the abrupt decline period are needed to understand the geographic and temporal distribution of chytridiomycosis, ranaviruses, Perkinsea, and other pathogens. As emphasized by Urgiles *et al.* (2021) and Ruggeri *et al.* (2023), it is important to incorporate pathogen surveillance (particularly for *Bd*, ranaviruses, Perkinsea) into biodiversity monitoring. This will serve as a starting point to understand pathogen dynamics, transmission, and impacts, ultimately aiding in the prevention of mortality events.

Habitat Destruction and Fragmentation

Ron *et al.* (2003), Merino-Viteri *et al.* (2005), Ron *et al.* (2011), Menéndez-Guerrero and Graham (2013), and Ortega-Andrade *et al.* (2021) have provided data on, summarized, and discussed the habitat destruction and fragmentation in Ecuador and their relation to the amphibian declines. According to Ortega-Andrade *et al.* (2021), agriculture, transport, infrastructure (i.e., roads, oil pipelines, etc.), production areas (mining, oil camps, etc.), and deforestation are important threats for Ecuadorian amphibians, with $70-98\%$ of taxa associated with each of these categories. In the future, nearly 21–36% of assessed species will exhibit a reduction in more than half of the areas that represent their ecological niches (loss of environmental conditions) in future scenarios.

Human activities in Ecuador have led to the modification and destruction of most of the natural habitats crucial for the survival of amphibians. Currently, the threat of habitat destruction and subsequent fragmentation is the main threat to those species that have managed to endure precipitous declines occurred during the mid-1980s to mid-1990s. The natural habitats that remain in Ecuador are indicated in Figure 3.31.

In Ecuador, deforestation rates have been the highest in South America for over 20 years, with annual rates ranging from 1.5% for 1990–2000 to 1.8% for 2001–2010 (FAO, 2011). Activities such as the expansion of agricultural frontiers, commercial logging, infrastructure development, palm oil production, banana and cocoa plantations, and mining and oil concessions are the reasons for deforestation and forest degradation in Ecuador (Torres *et al.*, 2020 and citations therein).

The gross deforestation in continental Ecuador for the period of 2020–2022 is as follows: an average annual gross deforestation of 95,570 ha/year and an annual gross deforestation rate of -0.78%. The results for net deforestation show an average annual net deforestation of 93,023 ha/year and an annual net deforestation rate of -0.76%. The average annual regeneration result is 2547 ha/year (MAAE, 2023).

Habitat destruction and subsequent fragmentation pose a substantial threat to species with low tolerance for habitat changes. Chronologically, habitat destruction has been especially dramatic in the inter-Andean valleys, followed by destruction in the Pacific lowlands, the Amazonian lowlands, and the Andean versants.

Human impact on the inter-Andean valleys has been significant since precolonial times, but it has been further exacerbated by sustained population growth. Brandbyge and Holm-Nielsen (1991) estimated that only 3.5% of the native forest remains in the inter-Andean region. Despite a relatively low species diversity (23) in these valleys compared to that in other Ecuadorian biogeographic sectors, almost all species are endemic. The surviving populations, which endured drastic declines from the mid-1980s to the mid-1990s, are now under severe threat due to factors such as urban development, intensive agriculture, and water course contamination. Only a handful of species have managed to adapt and thrive in these highly modified habitats, exemplified by *Pristimantis unistrigatus*, a species widely distributed in the northern Andes and introduced in the southern Andes (e.g., Cuenca), where it is spreading. A striking illustration of the depletion of amphibian fauna is observed in Quito, the capital of Ecuador. It becomes evident that the environmental vision of the city of Quito needs to incorporate these animals and place efforts to reintroduce populations of these species to green areas within the city. As advocated by Cuvi (2022), a conscientious citizenship could, instead of exerting pressure on resources from a negative perspective, contribute to the care of water sources, forested areas, and biodiversity. By the end of the 1970s, at least six frog species existed in relatively large numbers (*Gastrotheca riobambae*, *G. pseustes*, *Hyloxalus jacobuspetersi*, *Atelopus ignescens*, *Centrolene buckleyi*, and *P. unistrigatus*, among others). Currently, *P. unistrigatus* is the only abundant species, while the others have either disappeared (*A. ignescens*, *C. buckleyi*, *H. jacobuspetersi*) (Cuvi, 2022) or survive in scarce and fragmented populations (*G. riobambae*, *G. psesustes*) (Jiménez-Casalombo *et al.*, 2023).

Carvajal-Endara *et al.* (2019) discussed the conservation of numerous endemic species of biphasic marsupial frogs in the southern Andes of Ecuador. According to them, it is becoming increasingly evident that, despite their ability to adapt to altered environments, even in the vicinity of cities where some standing water is available (e.g., irrigation ditches, water reservoirs), amphibians are under threat due to the synergistic and cumulative impacts of numerous threats affecting fragmented populations with small distribution ranges (10–1600 km^2). Furthermore, certain species inhabit relatively large cities (e.g., Cuenca, Loja), where the availability of lentic water is not guaranteed. In the green spaces of cities, their in situ conservation could be significantly improved through regulations aimed at preserving and creating water bodies of various sizes surrounded by natural vegetation, particularly incorporating plants such as bromeliads, *Agave* spp., and *Schoenoplectus* sp. (totora).

These policies should be complemented by educational campaigns to shift public perception about frogs and foster a sense of conservation as a moral responsibility. Therefore, there is an urgent need for integrated species conservation plans, encompassing both in situ and ex situ conservation, along with studies on population dynamics for all biphasic *Gastrotheca* species, as consistently advocated for amphibians.

The human impact on the Pacific lowlands has been significant, and it is widely acknowledged that the forests of western Ecuador rank among the most severely threatened biogeographic regions globally (Dodson and Gentry, 1991). In recent years, this impact has been particularly pronounced in the Ecuadorian Chocoan Tropical Rainforest. For instance, in Provincia Esmeraldas, only 6% of the natural vegetation remains (Tamayo *et al.*, 2020). Lynch and Duellman (1997) provided a synopsis of the habitat transformations in the Pacific region in historical times in the context of their review of the frogs of the genus *Pristimantis*, *Strabomantis*, and *Craugastor* (then *Eleutherodactylus*). They finalize their review, regretting that their publication is a requiem for extirpated and surely threatened species of *Prisitmantis*. According to them (and citations therein), prior to the early 1960s, the human population on the Pacific lowlands and on the Pacific slopes of the Andes in Ecuador was relatively low. Indigenous peoples, such as the Chachis (Cayapas) and Tsáchilas (Colorados), were primarily hunter-gatherers, and the few colonists who inhabited the region were mostly subsistence farmers. For decades, the only town of any size on the coastal lowlands was Guayaquil, estimated to have about 28,000 inhabitants in 1879. In the vicinity of Guayaquil and less so in smaller towns and villages, adjacent forests were cut for firewood and construction materials, and palm trees were cut for their fronds for thatched roofs. Most rural inhabitants were located along the rivers that provided transit to towns. As Guayaquil became a port more frequently visited by freighters, timber cruisers began removing economically important trees, such as caoba (*Swietenia macrophylla*), balsa (*Ochroma pyramidale*), and Colorado (*Pouteria* sapota) that were near rivers and easily accessible; during the period of World War II, rubber collectors worked the native rubber trees (*Castilla elastica*). Narrow "avenues" of colonization emerged along the two railroads that crossed the Andes and penetrated the lowlands to the coast; the northern railroad extended from Ibarra to San Lorenzo and the southern from Alausí to Guayaquil. Although these railroads provided important trade routes, they had only limited impact on the natural resources of the region. The first commercial crops of any significance were banana and cacao plantations at lower elevations and coffee plantations in the shade of native trees at higher elevations; most of these plantations were within a short traveling distance to the railroad. Prior to the mid-1900s, few roads penetrated the region; the most important of these was the "old road" from Quito via Chiriboga to Santo Domingo de los Tsáchilas. Aside from subsistence farming and some dairy cattle, coffee was the major agricultural product. Two coincident events in the early 1960s dramatically affected the Pacific lowlands. The first was the initiation of a large road-building program—a new paved road from Quito to Santo Domingo de los Tsáchilas, paved roads from there to Esmeraldas and to Guayaquil, and, later, paved roads from Cuenca to Guayaquil. These roads provided the routes for commerce and colonization, which were bolstered by opportunities for the second event, commercial agriculture involving cacao, sugar cane, oil palms, and especially bananas (the "oro verde" or green gold) and forest extraction for commercial purposes. Within a decade, hundreds of thousands of hectares of forest were cut, burned, and planted. Chemical pesticides (DDT and Bordeaux) were used with abandon; the latter sprayed from planes gave banana plants a bluish hue. During Duellman and Lynch's various trips to the lowlands between Esmeraldas and Babahoyo between 1967 and 1984, they witnessed this wanton destruction and pollution of the lowland tropical forest. The natural resources have been pressured further by the great increase in population by burgeoning colonization and birth rates. In 1995, Guayaquil had more than 2.5 million inhabitants and Provincia Manabí more than 1 million. Small towns or villages by the 1970s were major metropolises, some with more than 100,000 people. According to Dodson and Gentry (1991), the forests of western Ecuador are among the most severely threatened of the world's ecosystems; more than 90% of the forests in the lowlands of the foothills have been converted to agriculture. Although there is notably less destruction of natural forests on the steep slopes, an increasing number of roads (and the accompanying colonists) resulted in ribbons of destruction, even in protected areas, such as the Bosque Protector de Molleturo. Two large natural reserves on the western versant in Ecuador, the Reserva Ecológica Cotacachi-Cayapas and the Reserva Étnica y Forestal Awa, face destruction as squatters establish themselves within the boundaries of the reserves and economically valuable trees are removed. At higher elevations, monocultures of *Eucalyptus* and *Pinus* form a barren, inhospitable habitat for native animals. Over a short period of time, the changes brought by humans equal or exceed those by natural causes during the Pleistocene.

In the Amazonian region of Ecuador, human impact has become significant, driven by petroleum extraction since 1972. The history of oil extraction in Ecuador is summarized by Dayot (2023). According to Finer *et al.* (2008), the history of oil development in the western Amazon is one of massive ecological and social disruption. Since the 1970s, 65% of the Ecuadorian Amazon has been "zoned" for oil activities. The erosion of land and dispersion of wildlife followed hundreds of seismic detonations, and the two billion barrels extracted between 1972 and 2004 provoked ecosystem destruction and deforestation (Dayot, 2023 and citations therein).

Habitat destruction in the Amazonian lowlands was further accelerated by road constructions, colonization, logging, and the establishment of oil palm plantations in the 1980s. Nonetheless, it stands out that 26.8% of the

Ecuadorian Amazonia has been designated for protection (Vergara *et al.*, 2022).

Presently, major threats include agriculture, both legal and illegal mining activities, urban expansion, and fires, among others. According to a recent report (Paz Cardona, 2022) based on data by MapBiomas (2021), between 2001 and 2020, the Ecuadorian Amazon region lost over 623,000 hectares in two decades. About 77% of the deforestation in the Amazon is concentrated in four provinces, with Morona Santiago being the most affected, followed by Sucumbíos, Orellana, and Zamora Chinchipe. A particularly alarming finding is that only two out of the six Ecuadorian Amazon provinces accounted for 46% (287,000 hectares) of all deforestation detected between 2001 and 2020. Morona Santiago witnessed the loss of over 158,000 hectares of forest (25%), while Sucumbíos saw nearly 129,000 hectares affected (21%). Both provinces have a significant presence of extractive activities such as mining and oil exploration.

In line with Vergara *et al.* (2022), Ecuador is the world's leading exporter of balsa wood, supplying 95% of global demand, which accounts for 66% of the Ecuadorian forestry industry. Ecuadorian balsa exports doubled between 2019 and 2020 to a total of US$550 million. The amount of balsa in certain Amazonian regions of Ecuador decreased by as much as 75% in that same period, with logging activities taking place within, or very close to, protected areas, increasing pressure on the natural resources they were designed to protect.

Mining is emerging as a significant risk factor in areas that are not part of Ecuador's National System of Protected Areas. Mining is affecting communities, biodiversity, and conservation efforts, with concessions overlying significant areas of protected forests (Bosques Protectores), Indigenous territories, and unprotected ecosystems rich in biodiversity (Peck *et al.*, 2024). According to Peck *et al.*, 7813 mining concessions over 22,812 km^2 overlay 9.2% of the Ecuadorian mainland, and 2323 concessions (29.7%) overlap 16,081 km^2 of protected forest (4781 km^2, 20%), Indigenous territory (6473 km^2, 8%), and native vegetation outside protected areas and Indigenous territories (13,390 km^2, 9%). With 80% of their protected forests at risk from large-scale mining, the most impacted Indigenous communities are the Shuar. According to Roy *et al.* (2018) and citations therein, during 2016 and 2017, the Ecuadorian Ministry of Mining, with authorization of the Ministry of Environment, increased exploratory mining concessions across the country from roughly 3% to around 13% of the country's continental land area. The Ministry of Environment's authorizations appear paradoxical in light of their disregard for the application of the precautionary principle and the rights of nature, which were constitutionally recognized in Ecuador in 2008. If exploration or exploitation occurs, these concessions will significantly decrease forest protected areas, given that more than 30% of the total land area included protected forests is included in new exploratory mining concessions. The majority of the concessions are located in the hyperdiverse Andean and Amazonian Cordilleras biogeographic sectors and ecosystems, which are unique in holding the highest concentration of diversity of amphibians and high levels of endemism (Table 3.4) and by being one of the most threatened eco-regions on the planet. The exploration and exploitation phases of metal mining decrease biodiversity primarily through deforestation, disturbance from road construction, associated river sedimentation, and water contamination. Deforestation reduces inputs of leaf litter into streams, changing energy inputs to the streams and shifting the trophic structure toward algal-based autotroph systems in montane Chocó-Andean streams. Moreover, after deforestation, mercury mobilization from the soil is the main source of methylated mercury in aquatic systems in the northern Amazon basin, with enormous negative effects on aquatic life and human health. Land slides, soil loss, increased stream sediments, and changes in stream flows are additional problems that result from deforestation. Other impacts from mining activities include the contamination of rivers with heavy metals, acoustic pollution, and social conflicts. Figure 3.40 shows the presence of threatened, as well as many endemic and non-endemic species of amphibians in mining concesions, although inventories in many of these areas are incomplete, as exemplified by the iconic case of two species of amphibians in the mining consecion of Llurimagua (see Chapter 2 under the subtitle "A Victory for Nature's Rights over Mining: Cooperative Work, Two Frogs, and the Science Crusade").

Pollution

Environmental pollutants are also contributing to global amphibian declines worldwide and in Ecuador. According to Hayes *et al.* (2010), environmental pollutants arise as likely important factors in amphibian decline because they have realized potential to affect recruitment. Further, many studies have documented immunosuppressive effects of pesticides, suggesting a role for environmental contaminants in increased pathogen virulence and disease rates. Hayes *et al.* (2010) recommend increased attention to recruitment and ultimate factors that interact with pathogens. Agricultural chemicals, such as pesticides and fertilizers, can exert a toxic effect on amphibians and their habitats. Moreover, the discharge of pollutants in urban areas and from oil and mining activities can have devastating effects on local amphibian populations. However, no studies have been conducted on amphibians and pollutants in Ecuador.

As outlined by Ron *et al.* (2011), the majority of the Ecuadorian economy relies on oil production, agricultural products (e.g., bananas, coffee, flowers, potatoes), and mining. These activities elevate the levels of contaminants in the environment, potentially impacting amphibians (along with other organisms). The list of contaminants encompasses heavy metals, pesticides, herbicides, fungicides, and fertilizers.

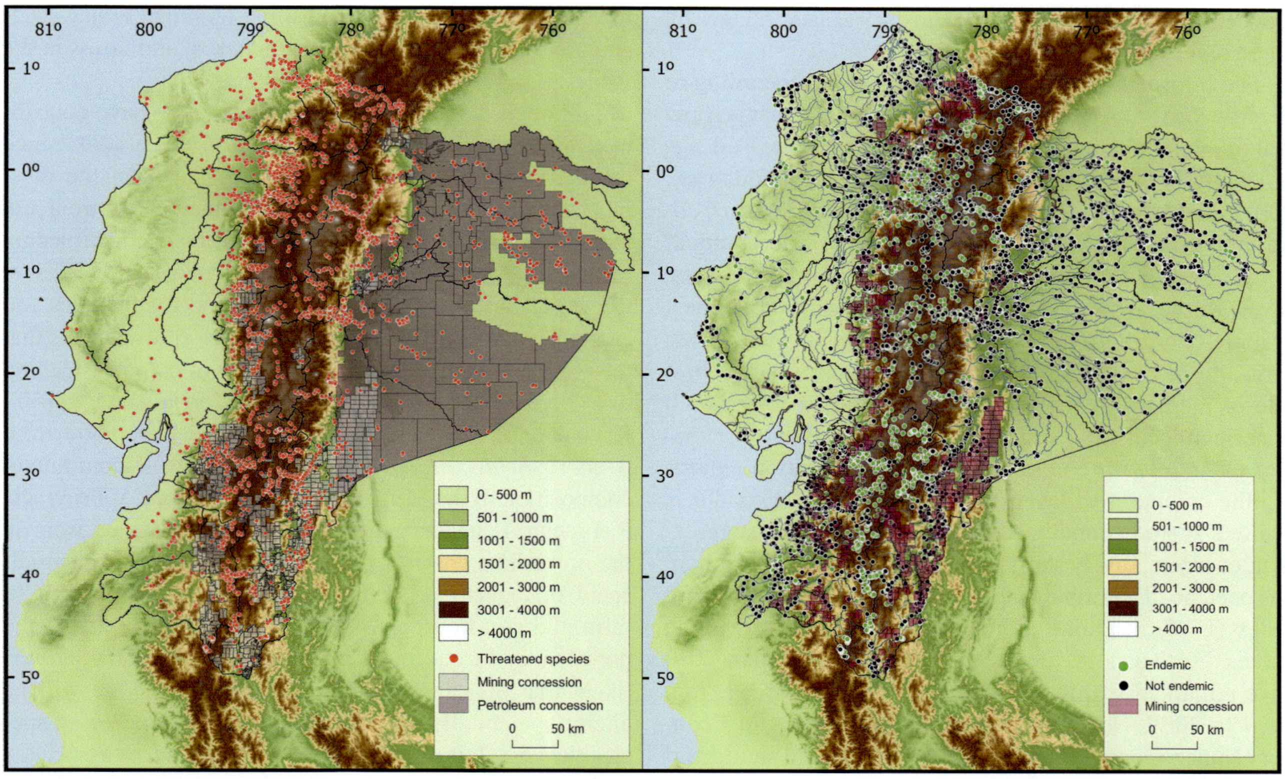

FIGURE 3.40 Maps of Ecuador showing museum records and overlap with mining and petroleum concessions. Left: threatened species. Right: endemic and non-endemic species.

The oil industry in the Amazonian lowlands has not only driven deforestation but also led to numerous oil spills since the early 1970s. According to Hurtig and San Sebastián (2002), extractive technologies in the 1970s resulted in large-scale contamination, and more than 30 billion gallons of toxic wastes and crude oil were released into the Ecuadorian Amazon between 1972 and 1993. Texaco, the primary foreign company to produce oil in Ecuador between 1964 and 1992, is said to have spilled more than the Exxon Valdez spill in Alaska in 1989, as the comparison has become famous in Ecuador. Texaco "did not dismantle many old wells and did not rehabilitate areas surrounding production sites" and used highly contaminating technologies which had long been illegal in the US, notably a mechanism called toxic ghost acres, which turned the Ecuadorian Amazon into a cheap hazardous waste sink of the oil industry (Dayot, 2023 and citations therein). In Ecuador, a total of 10,000 metric tons of crude oil were spilled in 464 events between 2001 and 2011 (Vergara *et al.*, 2022). Oil contaminates soil, water, and air. A 2023 popular consultation in Ecuador decided to leave the oil permanently underground in the ITT fields of the Yasuní National Park, and this probably marks a new era for the conservation of Amazonian biodiversity.

Alternative hypotheses for the decline in amphibian populations include the suggestion that pesticides and their degradation products have induced gonadal changes (Stebbins and Cohen, 1995; Hayes *et al.*, 2006) as well as the assertion that estrogen contamination leads to immunosuppression (Salla *et al.*, 2023). For instance, the herbicide atrazine, a chemical commonly used in Ecuadorian agriculture, has been shown to induce hermaphroditism and demasculinize the larynges of exposed African clawed frog males (*Xenopus laevis*) (Hayes *et al.*, 2002). Salla *et al.* (2023) further observed that exposure to estrogen modulated significant toxicodynamic pathways, evident in the immunosuppression of exposed animals and the initiation of feminization in males, thereby confirming synthetic estrogen's role as an endocrine disruptor. However, no studies on pesticides and estrogens have been conducted in Ecuador.

Ultraviolet Radiation

According to Hird *et al.* (2023) and references therein, increases in ultraviolet radiation (UVR) correlate spatially and temporally with global amphibian population declines and interact with other stressors such as disease and temperature. Declines have largely occurred in high-altitude areas associated with greater UVR and cooler temperatures. UVR is a powerful mutagenic agent, harming organisms largely by damaging DNA. When acutely exposed to UVR at cool temperatures, amphibian larvae exhibit increased levels of DNA damage. Amphibians are particularly vulnerable to UVR exposure, showing increased mortality, morphological malformations in embryos, slowed developmental and growth rates, physiological and behavioral changes, and reduced locomotor performance in larvae (see review by Alton and Franklin (2017)).

The information on UVR increases varies according to specific localities. In general, satellite measurements show that annually averaged UV-B radiation has increased significantly in South America since 1979 (Middleton *et al.*, 2001). According to these measurments, of 20 sites in Central and South America (none in Ecuador) where annual UV-B levels were measured between 1979 and 1998, the site with the highest values was the one with the highest altitude (Cordillera de Mérida at 4000 m in Venezuela). This site in Cordillera de Mérida also shows the highest rate of increase in UV-B when compared to other South American localities (Middleton, 2001). Therein, four species of *Atelopus* are possibly extinct. Information regarding the comparative levels of UV radiation in Ecuador is limited. A study conducted by Parra *et al.* (2019) suggests that, over a relatively short period from 1979 to 2018, there has been no discernible increase in the ultraviolet radiation index (UVI) in Quito, although the UVI shows extreme values. Furthermore, the measured UVI values since 2010 have remained within the historical range recorded from 1979 to 2009. Nonetheless, a study on humans in Ecuador (Núñez-González *et al.*, 2020) documented significant increases in malignant melanoma—the main risk factor for the development of malignant melanoma is increased UV exposure.

No studies on UVR and its impact on amphibians have been conducted in Ecuador, and such research is urgently required. This need is particularly pronounced in the equatorial paramos and highlands, where increased UV-B radiation may have surpassed the tolerance limits of certain highland species. In the Ecuadorian Andes, species potentially vulnerable to this factor are diurnal frogs such as *Atelopus* and *Hyloxalus* and species of marsupial frogs such as *Gastrotheca*, which expose themselves to the sun while incubating embryos. These genera also have aquatic embryos and larvae exposed to direct sunlight. On the other hand, nocturnal species on the paramos and the ones that exhibit direct development (deposit eggs amid soil and/or vegetation, hidden from direct or indirect sunlight) such as *Pristimantis*, *Lynchius*, *Niceforonia*, and *Osornophryne* may not be affected.

Introduced Animals

Humans have introduced animals voluntarily or accidentally all over the world. Some introductions have had a negative impact on local amphibian populations due to competition for food, space, and resources; direct predation on adults and larvae; and serving as vectors or reservoirs for diseases and parasites (Bishop *et al.*, 2012). In Ecuador, hundreds of species of exotic animals have been introduced since colonial times, but mainly three of them cause amphibian declines: the rainbow trout *Oncorhynchus mykiss*, the brown trout *Salmo trutta*, and the bullfrog *Aquarana catesbeiana*, all of them originally from North America. The salmonids arrived to Ecuador in 1932 for farming and recreational fishing (Aguirre *et al.*, 2021) and are widely distributed in rivers, lakes, and farms in the Andes.

Although it is likely that salmonids may have contributed to amphibian declines in Andean ecosystems since their introduction in 1932, they have not been the driving cause of precipitous declines. Nonetheless, trout are blamed for acting as vectors for diseases like saprolegnia (Martín-Torrijos *et al.*, 2016) and iridoviruses (Miller *et al.*, 2011) and for presumably preying on the tadpoles and eggs (Krynak *et al.*, 2020). A laboratory study reports that the mere presence of trout causes stress in tadpoles, leading to a decrease in their growth (Martín-Torrijos *et al.*, 2016). Krynak *et al.* (2020) studied the influence of rainbow trout (*Oncorhynchus mykiss*) farms on glassfrog community composition in the Mindo and Alambi regions of Provincia Pichincha, Ecuador, across 13 sites, six of which were in the immediate proximity of trout farms. Their results indicated that glassfrog communities differed between trout farm and non-trout farm sites. They concluded that if the prevalence of trout farms increases across this region, it is likely that the glassfrog community composition will be altered, potentially resulting in a pattern of decreased species richness. It is also likely that habitat changes associated with trout farming practices, including deforestation, water chemistry changes, and predation pressures by escaped trout, will influence glassfrog species persistence. Mitigation strategies, including improved barriers to decrease trout escape, the incorporation of settling ponds to decrease stream contamination, and the preservation of habitats in areas of high amphibian species richness, are warranted. Studies in the Pyreneans by Miró *et al.* (2020) show that whole amphibian communities from high mountain lakes recover rapidly after eliminating or reducing non-native fish such as the rainbow trout, proving that this is a powerful tool to improve the conservation status of endangered amphibians.

The bullfrog arrived in the mid-1980s for the bullfrog industry, and feral populations are mainly distributed in the coastal region and in the southern Andes of Ecuador. *Aquarana catesbeiana* can affect native amphibians mostly via predation, competition, and as a vector for disease. However, neither has the impact of its presence been studied nor mitigation strategies implemented in Ecuador. See more information in Chapter 1 under the subtitle "The Bullfrog Industry."

CONSERVATION

Considering that at least 399 (61%) of the Ecuadorian amphibian species are under categories of extinction risk, a rapid response and fundamental changes in policies and actions have been urgently needed to address the amphibian crisis. This has been emphasized elsewhere (Bishop *et al.*, 2012) due to threats mainly from global warming, pathogens, habitat destruction, environmental pollutants and invasive species (Pounds *et al.*, 2006; Lips *et al.*, 2006;

Stuart *et al.*, 2004; Hayes *et al.*, 2010). If the goal stablished in 2022 by the Convention on Biological Diversity is to reduce extinction rates significantly within the next 20–30 years, massive investments are needed (Lötters *et al.*, 2023). This task includes, among other actions, funding the exploration of under-sampled areas, supporting amphibian biobanks, and at least doubling the current rate of species descriptions. For this purpose, a major breakthrough is also necessary in the regulations and bureaucratic burden that are impeding biological research and an inventory of biological diversity worldwide and especially in megadiverse countries, as has been discussed elsewhere (Gilbert, 2010; Fernández, 2011; Bockmann *et al.*, 2018; Prathapan *et al.*, 2018). One of the major challenges is to implement fair, reliable, simple, unambiguous, and transparent regulations. Ecuadorian and allied foreign scientists need juridical security. Conducting systematic research is problematic, and, ironically, in the meantime, the biodiversity crisis and mass extinctions are dramatically underway.

What Are the Solutions?

Conservation In Situ

Roy *et al.* (2018, and citations therein) and Mestanza-Ramón *et al.* (2023) provide an overview of the protected lands in Ecuador. The former focus on the major role that areas of forest and protected vegetation are playing in preservation of biodiversity, including amphibians in the northwestern montane cloud forests in the Cordillera Occidental. Ecuador currently has 74 protected areas, with state, decentralized autonomous, community, and private subdivisions.

The protected lands include the following categories:

1. Heritage natural areas (PANE, including national parks, are set aside and funded by the Ecuadorian national government and run as public institutions).
2. Areas of forest and protected vegetation (Áreas de Bosque y Vegetación Protectora = BP) are recognized by the national government but not funded by it. Recognition by the government of BPs enables legal support when conflicts in land use occur, including help with illegal logging and squatters.
3. Private reserves are not regulated by the national government, nor funded by it. However, some national programs exist to promote the conservation of forests by private landowners, such as the successful Socio Bosque program.
4. Community reserves are private reserves that are owned and managed by the community, a legal entity that is particularly common in Indigenous areas.
5. Local government reserves are owned and managed by provincial, municipal, or parish governments. In northwestern Ecuador, forested watershed reserves owned by parishes have become increasingly common, but management practices vary due to the lack of oversight from the national government.

In each of the species' accounts, we provide information on the occurrence of amphibians in public and protected areas. The collection sites of Ecuadorian amphibians are indicated in a map in Figure 3.36. This map shows the regions where collection gaps still exist (as inferred by the absence or lower density of dots) and where additional inventories are needed. The region is notorious for the relatively poor sampling in the central and south eastern Amazonian lowlands, especially at the border with Peru and in the Cordillera de La Costa in Provincia Manabí in the coastal region. Furthermore, the National System of Protected Areas (SNAP) suffers from significant gaps in data collection, particularly in crucial areas, primarily due to challenges related to accessibility. These challenges encompass issues such as inadequate funding, logistical constraints, and difficulties in obtaining collection permits, among others. As a result, numerous regions within the SNAP remain largely unexplored, impeding our understanding and management of these protected areas. Addressing these obstacles is crucial to ensuring comprehensive data collection and enhancing conservation efforts within the SNAP. For instance, in the Amazonian lowlands and foothills of Ecuador, several areas remain largely unexplored, such as the Reserva Ecológica Cofán Bermejo, the southern region of Parque Nacional Yasuní, and a substantial portion of Parque Nacional Sumaco-Napo Galeras. In the Andes, there are insufficient records for extensive areas within reserves like Parque Nacional Cayambe Coca, Reserva Biológica Colonso Chalupas, Parque Nacional Llanganates, Parque Nacional Sangay, and Parque Nacional Podocarpus. Along the coast, the majority of the Cotacachi Cayapas Reserve is under-sampled, and there is also a lack of data for Reserva Ecológica Mache Chindul and Refugio de Vida Silvestre Manglar del Estuario de Río Esmeraldas. Also, there are no documented amphibian records from Reserva de Producción de Fauna Manglares el Salado.

Conservation Ex Situ

Conservation Breeding and Assurance Colonies

Confronting the decline and extinction of amphibians poses one of humanity's most formidable conservation challenges, particularly as numerous species have vanished in recent decades due to climate change and pathogens, placing them at the forefront of the biodiversity crisis. In Ecuador alone, a minimum of 43 species are considered possibly extinct, with hundreds more potentially following the same trajectory. The acknowledgment of these sudden declines globally has spurred research and conservation initiatives, accompanied by a widespread call to prevent further species loss (Mendelson III *et al.*, 2006). Traditional programs, laws, and policies have proven insufficient to address the overarching threats that extend beyond the boundaries of reserves and nations. Effectively responding to these

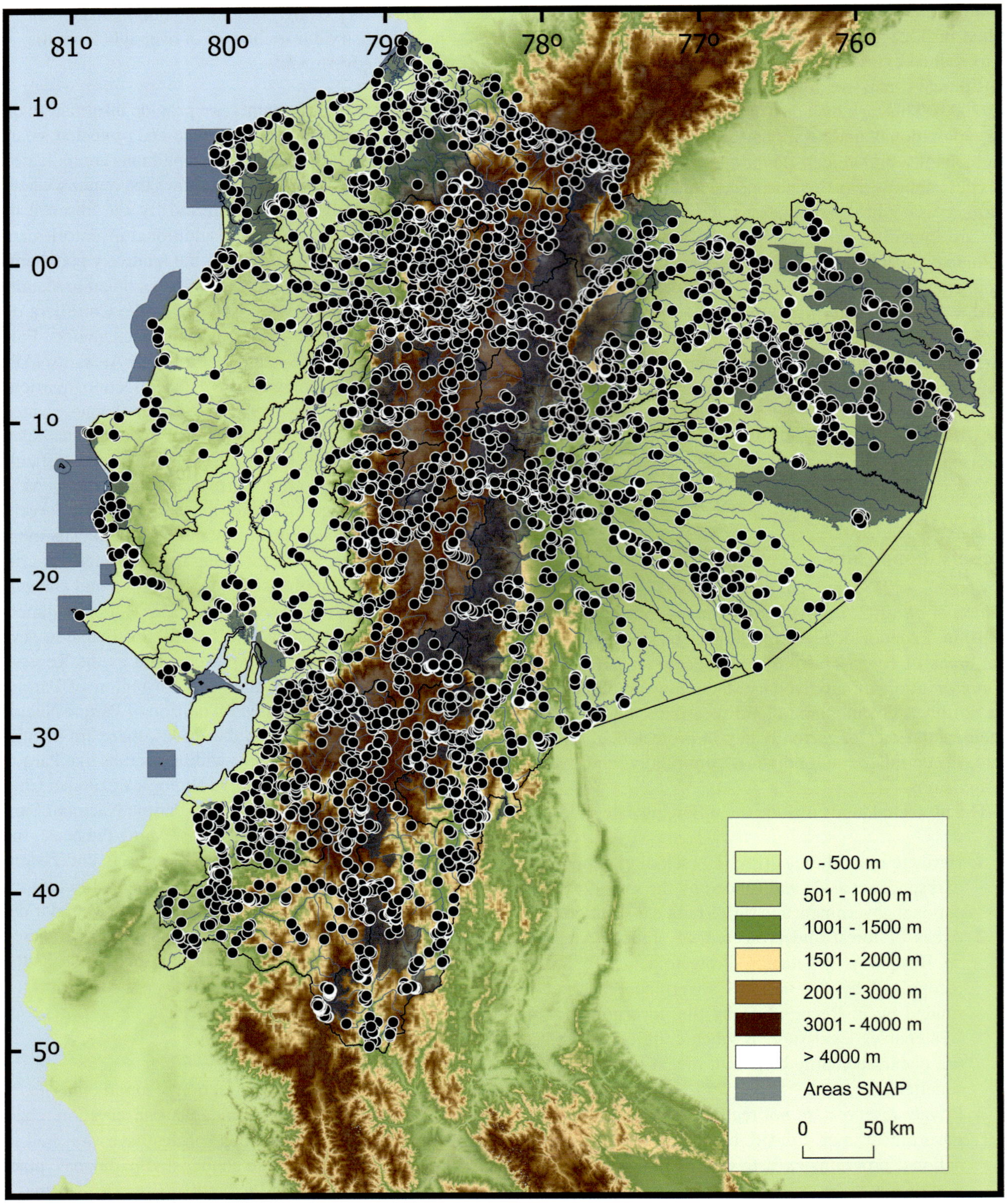

FIGURE 3.41 Map of Ecuador showing collection localities for 655 species of Ecuadorian amphibians. Each dot represents any number of species at a given locality. Note collection gaps in areas lacking dots or with low dot density as a gross approximation of under-sampling efforts.

challenges necessitates innovative, cross-disciplinary initiatives. One such initiative on a global scale is the Amphibian Conservation Action Plan (ACAP) (Gascon *et al.*, 2007; Wren *et al.*, 2015; IUCN SSC Amphibian Specialist Group, 2022b). As part of this plan, the Amphibian Ark was established in 2007 to spearhead ex situ conservation efforts (e.g., Gagliardo *et al.*, 2008; Zippel *et al.*, 2011). In the latest iteration of the plan, Tapley *et al.* (2022) provide a comprehensive overview of the principles underpinning conservation breeding and assurance colonies. They emphasize the following:

> In the face of overwhelming and sometimes acute threats to many amphibians, such as disease or habitat destruction, the only hope in the short-term for populations and species at imminent risk of extinction is immediate rescue for the establishment and management of captive survival-assurance colonies (CSCs). Such programmes are not the final solution for conservation of any species, but in some circumstances may be the only chance to preserve the potential for eventual recovery of a species or population to threat-ameliorated habitat. Captive-assurance strategy should always be implemented as part of an integrated conservation plan that includes research on amphibian biology, advances in husbandry and veterinary care, training and capacity-building in range countries, mitigation of threats in the wild, and ongoing habitat and species protection and, where appropriate, disease risk analysis. The existence of captive assurance colonies also facilitates many of the goals of other ACAP branches, including research on amphibians and their diseases as well as the development and validation of methods that may be later used in the field. Captive programmes do not replace important programmes related to, *inter alia*, habitat preservation, control of harvesting, climate change, and ecotoxicology, but instead provide options and resources to enable survival of some species while these research programmes proceed, and to directly or indirectly support such programmes.

In Ecuador, laboratory breeding efforts with conservation and research purposes commenced as early as 1991. The primary objectives were to develop technologies for the maintenance and breeding of amphibians, establish assurance colonies, and provide animals for research and potential future reintroduction programs. A detailed account of these actions is provided in Chapter 2, titled "The Life Raft for Frogs at QCAZ" or the "Balsa de los Sapos" program (refer also to Zippel and Mendelson, 2008; Coloma *et al.*, 2010). Although this program remained active until 2010, its activities have since declined, and it has shifted its original focus under new direction.

In the early 2000s, the Centro de Conservación de Anfibios Amaru was established as part of the Amaru Zoo. Ernesto Arbeláez-Ortiz, the director, played an enthusiastic role in amphibian ex situ conservation. Currently, CCA-Amaru maintains 24 species of amphibians on its premises (refer to Table 2.7). The program's history and details are narrated by its main actors in Chapter 2, under the subtitle "Amphibian Conservation in Southern Ecuador: The Amaru Initiative."

After departing from Balsa de los Sapos in 2011, Coloma founded Centro Jambatu, which continued ex situ conservation under the name "Arca de los Sapos." As a private non-governmental organization deeply involved in ex situ conservation and other integral conservation actions, CJ's ex situ activities are detailed in Chapter 2, titled "Centro Jambatu, a Fresh Venture." CJ maintains colonies of approximately 75 species, engaging in biobanking and the development of assisted reproductive technologies. Its assurance colonies have created numerous opportunities for integrated research and are utilized in education and reintroduction efforts.

Additional initiatives for frog conservation include those by the Vivarium and the Zoológico de Quito. The latter is particularly engaged in the reintroduction of *Gastrotheca riobambae* in urban green areas of the city of Quito.

Reproductive Technologies and Biobanking

Renato E. Naranjo and Luis A. Coloma

The ongoing challenges in amphibian conservation highlight the need to incorporate innovative techniques such as assisted reproductive technologies (ARTs) and biobanking. This is especially crucial in regions like Ecuador, known to house 655 described species, with almost half of the species being endemic, including endangered amphibian fauna. These initiatives play a pivotal role in conserving genetic diversity and facilitating ex situ management, translocations, and reintroductions, especially for endangered species. Hormonal therapies, sperm cryopreservation, and artificial fertilization constitute essential components of ARTs. They have shown considerable success in inducing breeding behavior and obtaining viable gametes and eventually offspring. Biobanking holds immense significance in preserving amphibian genetic material to ensure the conservation of genetic diversity. Ecuador has successfully established noteworthy biobanks to safeguard genetic resources, encompassing tissue, organs, skin extracts, DNA and sperm. These efforts significantly contribute to the conservation of amphibian diversity within the region.

The number of newly described amphibian species increases annually, as does the number of species facing population declines and disappearances (Luedtke *et al.*, 2023). For instance, in Ecuador alone, about 12 species new to science are described each year, but also 399 species are currently threatened. Conservation breeding programs (CBPs) play a vital role in the sustainable management of amphibian species threatened with extinction. These programs provide an opportunity to ensure the survival of species through long-term commitment, accompanied by management strategies aligned with goals of perpetuation, repopulation, and conservation research (Browne *et al.*, 2019). Despite the critical role of CBPs in conserving numerous endangered species, the effectiveness of laboratory breeding for amphibians is generally limited. Many programs struggle to consistently induce breeding behavior, achieve high fertilization rates, and produce viable offspring (Silla and Byrne, 2019). The incorporation of transformative technologies such as ARTs and biobanking can help CBPs by maintaining genetic diversity and producing individuals for translocations and reintroductions (Clulow and Clulow, 2016; Naranjo *et al.*, 2022; Silla and Byrne, 2019). Although these transformative technologies are a relatively new field of study and are often developed in moderate to high-income industrialized countries (Browne *et al.*, 2024), Ecuador have already initiated these studies (Naranjo *et al.*, 2022).

Assisted Reproductive Technologies ARTs encompass various techniques aimed at understanding and improving reproductive outcomes, as well as enhancing the genetic management of threatened species. These methods include the use of hormonal therapies to induce spermiation and spawning or ovulation, sperm cryopreservation, and artificial fertilization, among others (Silla and Kouba, 2022). The application of reproductive technologies offers several benefits, including: (a) increased efficiency of laboratory breeding; (b) reduction in costs associated with laboratory assurance colonies; (c) decrease in the rate of genetic diversity loss in laboratory and small wild populations; (d) genetic management through the control of mate pairing; and (e) restoration of extinct species (Clulow *et al.*, 2014). ARTs may also be adopted to maintain the health and wellbeing of amphibians under laboratory conditions (Silla *et al.*, 2021). In Ecuador, Centro Jambatu de Investigación y Conservación de Anfibios (CJ) has developed these technologies for critically endangered or endangered anuran species, including harlequin frogs of the genus *Atelopus* and the Blomberg's toad *Rhaebo blombergi*.

Hormonal Therapies The release of gonadotropin-releasing hormone (GnRH) from the hypothalamus triggers the production and release of luteinizing hormone (LH) and follicle-stimulating hormone (FSH) from the pituitary gland. These hormones prompt the gonads to secrete sex steroids (estrogen, progesterone, testosterone), stimulating oocyte and sperm production in adult females and males, respectively (Bronson and Vance, 2019). The efficient and reliable collection of viable gametes (oocytes and spermatozoa) is fundamental to the implementation of cryopreservation and artificial fertilization. In anurans, exogenous hormones like chorionic gonadotropin (hCG), synthetic GnRH agonists (GnRHas), and pituitary homogenates/extracts are commonly used to stimulate spermiation and ovulation (Clulow *et al.*, 2014). These hormones are typically administered to anurans via intravenous, intraperitoneal, or subcutaneous injection (Silla *et al.*, 2021).

In males housed at CJ, the administration of 10 IU hCG induced the release of high sperm concentrations ($1–4 \times 10^6$ cells/mL) in individuals belonging to *Atelopus balios*, *A. elegans*, *A. guanujo*, *A. longirostris*, *Atelopus* sp. 1 (*aff. halihelos* from El Cóndor), *Atelopus* sp. 2 (*aff. halihelos* from Loja-Zamora), *Atelopus* sp. 3 (*aff. nepiozomus*), *Atelopus* sp. 4 (*aff. longirostris*), *Atelopus* sp. 5 (*aff. spumarius* from Pucayacu), and *Atelopus* sp. 6. (*aff. spumarius* from Limón). A male of *Rhaebo blombergi* released an extremely high sperm concentration ($>1 \times 10^7$ cells/mL) with the same dose. Conversely, the combination of 1.2 μg GnRHa + 10 U hCG was optimal to obtain sperm from *Atelopus ignescens.* In females, *A.* sp. 6 and *A. balios* released a properly high number of eggs with 2.5 IU hCG, whereas a dose of 0.4 μg GnRHa is considered the best for *A. longirostris*, *A.* sp 1, and *A. bomolochos* to achieve spawning, promoting natural reproduction within tanks. Figure 3.42B–C depicts the obtainment of both sperm and oocytes from stimulated individuals of *Atelopus* sp. 6. Pituitary homogenates are prepared by crushing the pituitary glands of sacrificed amphibians into a suitable medium. These crushed glands are then given to another individual (recipient) to aid in breeding or to obtain reproductive cells (gametes) (Clulow *et al.*, 2014). It is interesting to note that in another study of an Ecuadorian species, *Rhinella marina*, ARTs induced ovulation using fresh pituitary homogenate from females of the same species (Proaño and Pérez, 2017).

Sperm Cryopreservation Sperm cryopreservation is a valuable conservation tool because spermatozoa can remain frozen for decades, enabling a male to contribute genetically to a population long after its demise (Kouba *et al.*, 2013). Cryopreserved sperm in CBPs maintains male genetic variation, reduces expenses, reduces captive animal numbers, allows for the fertilization of a single female's spawn with spermatozoa from numerous genetically diverse males, and reduces the necessity for animal transport (Browne *et al.*, 2019). Amphibian sperm are typically stored at ultra-low temperatures a −80°C deep freezer or liquid nitrogen −196°C.

In Ecuador, CJ has established the first collection of frozen sperm of endangered species, storing roughly 80 sperm samples from four species of harlequin frogs (*Atelopus guanujo*, *A. ignescens*, *A. longirostris*, and *Atelopus* sp. 6), with more than half belonging to *A. ignescens*, and roughly 20 samples from the giant toad *Rhaebo blombergi.* They are stored in liquid nitrogen. Various assessments on frozen sperm for each species have shown that around 30% of cells recover movement. Base on sperm concentrations and prior studies by Kouba and Vance (Kouba and Vance, 2009), this percent is sufficient to fertilize roughly 100 to 200 eggs. Interestingly, cryopreserved sperm from *A. ignescens* has demonstrated a movement recovery rate over 50% . A possible assumption is that jambato toad populations often inhabit cold locations such as tundra where temperatures are usually below 0°C. Sperm samples from *Rhinella bella*, on the other hand, were stored in a −80°C freezer and proven to have fertilization capacity after thawing (Proaño and Pérez, 2017).

Artificial Fertilization Given that most amphibians undergo external fertilization, artificial fertilization (also known as in vitro fertilization) is relatively uncomplicated, involving gamete combination in an external fertilization environment without requiring subsequent embryo transfer or implantation, as in mammals (Kouba *et al.*, 2013). CJ has successfully conducted artificial fertilization processes in three amphibian species. With both fresh and frozen sperm from *Atelopus* sp. 6, offspring were successfully obtained (Figure 3.42). Similarly, combining fresh gametes from *Atelopus balios* and *Rhaebo blombergi* resulted in the obtainment

FIGURE 3.42 Assisted reproductive technologies in the harlequin frog *Atelopus* sp. (*spumarius* complex from Limón) (top and middle) and embryos of the giant toad *Rhaebo blombergi* (bottom). Top left: female and male are carefully selected. Top right: sperm is obtained by collecting urine (spermatic urine) using a glass micro-capillary tube approximately three hours after hormonal stimulation. Middle left: approximately eight hours after hormonal stimulation, the female undergoes spawning, releasing eggs for fertilization. Middle right: artificial fertilization, spermatic urine is carefully placed on the collected eggs that were spawned just minutes before. Bottom left: embryos 2 days post-fertilization. Bottom right: 5 days post-fertilization. DAL.

of embryos. Figure 3.42 shows the embryos after a few days of in vitro fertilization. In *Rhinella horribilis*, a considerable percentage of embryos were obtained using cryopreserved sperm and fresh oocytes (Proaño and Pérez, 2017). Although gamete function was successful in these four Ecuadorian studies, consequential embryo abnormalities and mortality were observed. Laboratory rearing or cryopreservation conditions could be potential causes; however, these issues should be studied in depth in the future.

Biobanking Around the world, museums, zoos, aquariums, governments, and conservation NGOs are beginning to establish amphibian biobanks to conserve, in perpetuity, the remaining extant genetic diversity for many critically endangered species. A suite of biomaterials has been targeted for cryoconservation, including blood, cell cultures, tissues, spermatozoa, eggs, and embryos (Kouba *et al.*, 2013). The goal of amphibian biobanks is to preserve the genetic diversity of amphibians, which can be used for research, conservation, and reintroduction programs. Almost all the museums that hold amphibians from Ecuador are collecting material and storing it in ultrafreezers and cryogenic tanks, ensuring the preservation of genetic diversity.

Biobanks (also known as genome resource banks) are essentially the storehouses for amphibian genetic material, which can be deposited or withdrawn to support applied reproductive technologies. The preservation of germplasm, embryos, spermatozoa, eggs, tissues, DNA, and cell cultures is essential to offer a diverse range of options for future species management (Kouba and Vance, 2009). Biobanking establishment is a fundamental component of the toolbox needed to address the extinction crisis. The primary goal is to preserve genetic diversity, which can be utilized for research, conservation, and reintroduction programs.

The first institution to hold tissues stored in ethanol is the KU museum. These collections were assembled mostly during the 1970s and 1980s and contain 831 samples. The Ecuadorian museums are collecting material and storing it in ultrafreezers and cryogenic tanks, ensuring the preservation of genetic diversity. Currently, the most prominent biobanks are those of the QCAZ and CJ (see also Chapter 2) in Quito, Ecuador. The QCAZ initiated systematic genome collection in the mid-1990s, leading to a rapidly growing genomic bank. The QCAZ houses the largest collection of tissues (40,068 samples as of 4 April 2023). Preservation is ensured through storage at −80°C, encompassing liver and muscle tissues, along with DNA extracts. Each genome is linked to a specimen in the QCAZ collection, with detailed taxonomic and origin data in a digital database. The bank holds invaluable genomic samples from endangered or extinct wild populations, safeguarding Ecuador's amphibian genetic diversity. This resource promotes genetic-based scientific research and secures future access to genomic information despite ongoing natural erosion. Similarly, CJ has focused its efforts on establishing an organized repository, the Ecuadorian Amphibian Genome Bank, which offers systematic management to preserve biomaterials such as tissues, organs, skin extracts, and sperm. These initiatives also encompass enhancing technological capabilities, including infrastructure, equipment, and training, to expand their role as custodians of the country's amphibian genetic resources. Preservation technologies such as lyophilization and cryopreservation are being employed, along with training staff and researchers for proper storage and utilization. At present (as of 5 October 2023), CJ stores about 13,343 tissues from about 443 species, along with the largest collection of skin extracts (1692 samples of 381 species), which are being lyophilized for molecular analyses. Additionally, it maintains the collection of frozen sperm of endangered species (four species). The future goal of CJ is to enhance and expand the existing biobank, incorporating a wider range of biomaterials, including blood products, RNA, and DNA. Another more recent initiative in Ecuador focused on amphibian biobanking is the establishment of the National Genetic Resources Bank. This initiative is led by the Instituto Nacional de Biodiversidad (INABIO) with the economic support of the Korea International Cooperation Agency (KOICA). The project's primary objectives include creating a database of genetic resources and upgrading the equipment of biocenters that will store genetic and biological material (INABIO, 2021). This initiative is intended to propel the advancement of bioknowledge and the promotion of biotechnological research and development.

CONSERVATION GENETICS

Mónica I. Páez-Vacas

Conservation genetics (see Box 3.2. for a conservation genetics glossary), as a discipline, has been around for more than 50 years (Frankham *et al.*, 2002). It is the application of genetic data to preserve species and their evolutionary potential. Conservation genetics mainly focuses on resolving taxonomic uncertainties, defining conservation units, managing wild and laboratory-reared populations, and understanding population dynamics in the face of specific threats, such as habitat fragmentation, emerging diseases, or climate change. Recently, advances in the field of genetics and genomics have allowed for the study of many more species, as well as the improvement and addition of novel insights into conservation genetics (Allendorf *et al.*, 2022; Shaffer *et al.*, 2015). Nonetheless, its applications have been somewhat restricted mainly to species at higher latitudes (Torres-Florez *et al.*, 2017). Thus, this genetic and genomic revolution has not yet reached the conservation of amphibians in Ecuador, with only a few studies using genetics in a conservation framework (e.g., Burgos *et al.*, 2019; Jaynes *et al.*, 2022). One of main gaps in conservation genetics in Ecuador (not only in amphibians) is the study of genetics at the population level (Torres-Florez *et al.*, 2017) (Campbell Grant *et al.*, 2023). Herein, we review the application of genetics to conservation and highlight recommendations for potential research opportunities in amphibians of Ecuador.

Why Should We Focus More on Populations?

One of the main debates in conservation focuses on what we should conserve. The Convention on Biological Diversity (CBD) has named three levels of biodiversity: genes, species, and ecosystems (CBD; www.cbd.int). The focus on conservation should be on all three of those levels, although considering them separately can facilitate the implementation of specific strategies for each. However, it has been argued that genetic diversity within species (i.e., at the population level) is "the most important level for focusing conservation efforts", as most of the ecological and evolutionary processes, as well as extinction, occur at the population level (Allendorf *et al.*, 2022; Hoban *et al.*, 2020).

Evolutionary change is the result of an interplay between genetic drift, selection, gene flow, and mutation at the population level. Genetic drift refers to random changes in allele frequencies in a population from one generation to the next due to sampling error. Although all these mechanisms are responsible for evolution in both large and small populations, in small populations, genetic drift is predominant and can overpower the effect of the other evolutionary mechanisms. The loss of genetic diversity due to drift is stochastic, and deleterious alleles could become fixated in a population (Wright, 1931). Selection is the differential success of different genotypes in contributing to the next generation. This success refers to higher rates of survival and reproduction in those individuals carrying alleles that confer them some advantages relative to individuals not carrying those alleles. Selection is the only force that causes adaptive evolution, as it will adapt a population to its local environmental conditions. Contrary to genetic drift, selection is less efficient in smaller populations. Gene flow—the genetic exchange between populations through movement of gametes, individuals, or entire populations—reduces genetic differentiation among populations and may restore lost genetic diversity. It can also spread new alleles or new combinations of alleles throughout a species' range (Slatkin, 1987; Frankham *et al.*, 2002). Mutation is the ultimate source of genetic variation, which generates from errors in the transmission of genetic information from parents to offspring. Mutations occurs at a very low rate, and many are neutral, meaning that they do not have effects on phenotypes and, therefore, are not affected by selection. These mutations are usually lost by genetic drift. Neutral genetic variation is normally used to study genetic diversity within and differentiation among populations, which has several applications in conservation (Frankham *et al.*, 2002; Kardos *et al.*, 2021). The most important mutations for evolutionary change and for population persistence are those at loci affecting fitness traits, which can be advantageous or deleterious, and their frequencies will depend on selection and drift, again, as a function of population size (Fisher, 1930; Wright, 1931, 1949). In the last decade, research implementing genomic tools to study adaptive divergence has become more common and allowed for novel insights for conservation (Allendorf *et al.*, 2022; Shaffer *et al.*, 2015).

Extinction is a demographic process, too. Population size reduction is primarily caused by some human activity, usually acting synergistically and related but not limited to habitat degradation and overexploitation. Population isolation can also contribute to a reduction in population size. Population size reduction results in reduced genetic diversity, which is a symptom of an endangerment, not a cause (Holsinger *et al.*, 1999). In small populations, deleterious alleles can become fixed due to genetic drift, reducing fitness and evolutionary potential. Mating with relatives, or inbreeding, is also an issue in small populations. Inbreeding has an important role in conservation, as it results in a loss of heterozygosity, leading to a reduction in fitness known as inbreeding depression (Allendorf *et al.*, 2022). The loss of genetic variation can also reduce populations' evolutionary potential to adapt to future environmental changes such as invasive species, novel pathogens, or climate change (Ceballos *et al.*, 2017). Moreover, small populations are also more vulnerable to stochastic processes. All these events may further decrease survival and/or reproduction rates, thus increasing risk of extinction.

How Can We Apply Genetics to the Conservation of Amphibians in Ecuador?

The threats underlying amphibian declines and extinctions include habitat degradation and fragmentation, climate change, pathogens, introduced species, contaminants, and others. These causes are not mutually exclusive but are complex and might be acting simultaneously and synergistically (Hayes *et al.*, 2010), and most likely their effects vary from one population to another (Briggs *et al.*, 2010). Well-documented threats, namely habitat degradation and fragmentation, are some of the primary causes of many losses of amphibian biodiversity, and they affect genetic diversity, population size, demography, and connectivity between populations (Catenazzi, 2015). Reductions in genetic variation and population size make populations more susceptible to stochastic demographic and environmental events, as well as other threats. Two important causes of amphibian decline are climate change and emerging pathogens, particularly the chytrid fungus *Batrachochytrium dendrobatidis* (Lips *et al.*, 2008; Lips, 2016; Pounds *et al.*, 2006; Rohr *et al.*, 2008; Scheele *et al.*, 2019). Substantial research on the effects and interactions of threats has shed light on how they affect genetic variation and, ultimately, population persistence (Catenazzi, 2015). Genetic information can have several applications for conservation that will be discussed herein: performing the genetic identification of endangered, introduced, or pathogenic species; genetic monitoring; resolving taxonomic uncertainties; identifying conservation units; estimating inbreeding depression; measuring demography and extinction; assessing population connectivity; and managing populations, either wild or bred in the laboratory.

Main Techniques to Assess Genetic Variation in Populations

The study of the genetic variation in natural populations has advanced immensely in the last few decades. The most traditional molecular markers used for conservation genetics are the mitochondrial DNA, a few nuclear genes (e.g., *POMC* and *RAG-1*), and microsatellites. However, the use of genomic approaches has increased since 2000, and nowadays, they are widely used techniques allowing for the detection of polymorphisms, usually single nucleotide polymorphisms (SNPs; Allendorf *et al.*, 2022). There are two main advantages of population genomics over traditional population genetics. First, the use of a larger number of loci (usually 1–20 with genetics) improves traditional applications by providing more precise and accurate estimates of population genetic variation, population differentiation, inbreeding, or demographic history (Funk *et al.*, 2021; Hoban *et al.*, 2016). Second, it allows for the detection and identification of loci that are potentially under the effects of selection, providing information on the genetic mechanisms of adaptation (Allendorf *et al.*, 2022; Shaffer *et al.*, 2015). "Omics" refers to a wide variety of techniques, which include whole-genome sequencing, reduced representation sequencing (e.g., RADseq), transcriptomics, exomics, metagenomics, proteomics, and metabolomics. One of the difficulties of applying genomics to amphibians is the large genome size; therefore, reduced representation sequencing, transcriptomics, and exomics provide more feasible options for amphibians (Funk *et al.*, 2021; see https://www.amphibiagen.org).

Genetic Detection, DNA Barcoding, and Environmental DNA

Genetic analyses can be useful to identify species, populations, and individuals from a sample. This genetic identification can be applied in conservation genetics to identify the origin of a sample or detect the presence of species, whether it is a threatened species or population, an introduced species, a hybrid, or a pathogen (e.g., Farrington *et al.*, 2015). Species detection allows for the assessment of their distribution, a first step of biodiversity research, allowing researchers to identify and protect critical habitats to enhance persistence probability. Furthermore, uses may include the assessment of biodiversity, community composition, or the establishment of ecological relationships (Ficetola *et al.*, 2019). Pomerantz *et al.* (2018) demonstrated the feasibility of genetic sequencing for the species identification of amphibians and reptiles *in situ* using the portable DNA amplicon sequencing with the MinION device in the Ecuadorian Chocó Rainforest, a remote and diverse tropical forest.

DNA barcoding provides a system for species identification that involves the use of a short, standardized, and well-characterized DNA sequence for species identification, usually mitochondrial DNA (Hebert *et al.*, 2003; Lyra *et al.*, 2017). DNA barcoding is an effective and widely used tool for conservation genetics and other potential applications. This includes research on the identification of invasive species (e.g., Dejean *et al.*, 2012; Melo *et al.*, 2014), biodiversity inventories (e.g., Paz and Crawford, 2012), or even the effects of mass-mortality events due to the pathogen *Bd* at the amphibian community level (Crawford *et al.*, 2010). In Neotropical amphibians, the use of DNA barcodes has also revealed great amounts of cryptic diversity (e.g., Funk *et al.*, 2011 in the Amazon basin with emphasis on Ecuador; De Oliveira *et al.*, 2019 in eastern Amazonian Brazil), and helped in species distinction, especially in speciose groups (e.g., Moravec *et al.*, 2020 in *Pristimantis* from Peru; Castroviejo-Fisher *et al.*, 2011 in *Hyalinobatrachium* in Bolivia). In Ecuador, Pomerantz *et al.* (2018) used DNA barcoding to rapidly assess biodiversity *in situ* in a biodiversity hotspot, the Chocó rainforest.

The application of DNA barcoding to identify traces of organisms from environmental samples as soil or water, known as environmental DNA (eDNA), is another approach that is growing rapidly. Several species are difficult to detect, especially if rare or during certain developmental stages. Organisms shed biological materials containing DNA into their environments, which can persist in aquatic environments for extended periods (Ficetola *et al.*, 2008; Barnes and Turner, 2016; Pilliod *et al.*, 2013b). In recent years, the use of eDNA has grown as a powerful and non-invasive tool for species detection, showing itself to be not only comparable but even more efficient for detecting target species than active searching under certain circumstances (e.g., Goldberg *et al.*, 2011; Pilliod *et al.*, 2013a; Lopes *et al.*, 2021; Plante *et al.*, 2021; Valentini *et al.*, 2016). Moreover, the detection of pathogens such as ranaviruses (e.g., Kaganer *et al.*, 2021) or the chytrid fungus *Bd* (e.g., Kamoroff and Goldberg, 2017), can aid prevention of epidemics or the mitigation of negative impacts on host populations. In Ecuador, eDNA has been applied to detect species and quantify amphibian diversity (Quilumbaquin *et al.*, 2023), as well as invasive (e.g., Van der Heyden *et al.*, 2021) and pathogenic (e.g., Riascos-Flores *et al.*, 2021) species. Other studies underway by researchers from different institutions in Ecuador and abroad are applying eDNA mostly to rediscover new populations of species from the most vulnerable genera, such as *Atelopus* and *Telmatobius*. For example, Centro Jambatu, along with the St. Louis Zoo and the Missouri Botanical Garden (USA), is searching for *Atelopus* species that have not been seen for decades. The Jambato Alliance is searching for new populations of the emblematic *Atelopus ignescens* while exploring the habitat use of the species. The Amphibian Conservation Center at Bioparque Amaru, along with the Trier University (Germany) and Universidad Católica de Cuenca, is searching for species of *Telmatobius*.

Resolving Taxonomic Uncertainties

Describing biodiversity is the first step in most biological disciplines, and conservation is not an exception. The exuberant diversity of amphibians in the Neotropics is partially unknown, and estimates mention that we might still have $200-350\%$ to discover and describe, especially due to cryptic diversity (Arteaga-Navarro *et al.*, 2016; Catenazzi and von May, 2021; Funk *et al.*, 2011). As noted in Chapter 3, subtitle "Discovery and Rates of Descriptions", we presume that 230 species are still waiting to be described. Systematic and taxonomic studies are essential to know what we need to preserve and to design efficient strategies for management and conservation. Taxonomic uncertainties can result in the extinction of unknown and unprotected species (Ortega-Andrade *et al.*, 2021). Moreover, describing patterns of species richness allows us to move forward in understanding the evolutionary processes underlying such diversity. This is the area where most (if not nearly all) studies applying genetics for amphibian conservation in Ecuador have focused. Significant effort in the last decades to assess amphibian biodiversity has resulted in an increase in known diversity from 450 species to 565 species in only 22 years (Chapter 3, subtitle "Discovery and Rates of Descriptions"). Some groups have been exhaustively studied, such as poison frogs, glassfrogs, and some groups of terrestrial frogs (e.g., Padial *et al.*, 2014; Grant *et al.*, 2017; Guillory *et al.*, 2019; Guayasamin *et al.*, 2020). However, there is still much work ahead, and many groups remain relatively unattended or awaiting species descriptions (Chapter 3, subtitle "Discovery and Rates of Descriptions").

Conservation Units

Although species are clearly separate units that require separate management, populations within species might also require separate management due to significant genetic and/or adaptive differentiation (Frankham *et al.*, 2002). One of the most important first steps for managing populations is to identify and delineate the boundaries of this intraspecific units, known as conservation units (CUs). Conservation units are defined as populations within species identified to design and implement conservation efforts (Fraser and Bernatchez, 2001; Ryder, 1986; Moritz, 1994). The two most widely known CUs are evolutionarily significant units (ESUs) and management units (MUs). ESUs are populations or groups of populations that merit separate management or priority due to isolation and ecological differences that result in evolutionary distinctiveness (Moritz, 1994; Waples, 1991; Crandall *et al.*, 2000). MUs, on the other hand, are populations that merit separate management because of demographic independence (Moritz, 1994). Traditionally, mitochondrial DNA and microsatellites have been used to define both of these conservation units. However, advances in genomics allow for new approaches to delimit these units, including neutral and adaptive genetic information (see Funk *et al.*, 2011). Understandably, in a region where most efforts are still directed toward inventorying species richness as Ecuador, little has been done to delimit conservation units within species using genetic information. Given the lack of studies using genetic or genomic methods to delineate CUs in Ecuadorian amphibians, the use of mitochondrial DNA would be a relatively feasible approach for most endangered taxa. On the other hand, until genomic tools become widely accessible for research in Ecuador, fostering international collaborations to study adaptive genetic variation is a first step forward. In the meantime, efforts for capacity building are being applied in Ecuador. For example, the Instituto Nacional de Biodiversidad (INABIO), along with the Universidad Regional Amazónica IKIAM and the Korean Cooperation are implementing the project "The Establishment of the National Data Bank of Genetic Resources for Conservation and Sustainable Use of Biodiversity in Ecuador".

Genetics of Wild Populations

Population Size and Demography

Population genetic methods have been widely implemented in several systems to efficiently quantify population sizes and dynamics in the wild for more than 20 years now (Hedrick, 2001; Shaffer *et al.*, 2015). One of the most traditional uses of genetic data in conservation is describing the effects of small population size on population demography and genetic diversity. Population size is captured by the effective population size (*Ne*), defined as the size of the ideal, panmictic population that will result in the same loss of genetic variation due to genetic drift as that in the actual population being studied. Estimating *Ne* is essential, as it affects genetic diversity, the likelihood of accumulation of deleterious alleles, inbreeding depression, and the evolutionary potential of a population, particularly in the face of substantial and diverse threats (Frankham *et al.*, 2011). As noted earlier, genetic diversity is the basis for evolutionary change, and low genetic diversity will increase extinction risk (Spielman *et al.*, 2004; Kardos *et al.*, 2021). However, not all species with low genetic variation are more vulnerable to extinction (Hohenlohe and Rajora, 2021).

Reconstructing the demographic history of a wild population can help explain current levels of genetic diversity and understand consequences for population fitness, as well as proposed conservation management (e.g., Wang *et al.*, 2018 in the Tibetan frog). With the development of microsatellites, estimates of past bottlenecks or population expansion also became possible. Nowadays, with genomic techniques, it is possible to make more accurate and precise inferences that are less limited by sample size (Allendorf *et al.*, 2022). Historical demographic reconstruction can elucidate the effect of genetic erosion on population persistence in comparison to extrinsic factors (e.g., Casas-Marce *et al.*, 2017 in the Iberian lynx) and the consequences of population isolation at the genome level (e.g., Saremi *et al.*, 2019 in pumas) to link population changes to environmental shifts (e.g., Guo *et al.*, 2016 in Andrew's toad, Wang *et al.*, 2018), or estimate the loss of genetic variation post-decline (Jaynes *et al.*, 2022 in harlequin frogs in Ecuador). Harlequin frogs (genus *Atelopus*, Bufonidae) are among the most vulnerable taxa, with about 70% of species considered critically endangered or extinct (IUCN, 2022). Jaynes *et al.* (2022) provided genomic data on six species that experienced a significant population decline, which included a comparison of two historical samples collected prior to the decline. This study showed a correlation between the duration of a species' absence and a reduction in genetic variation. Notably, the decline in heterozygosity was distinct in the two species compared to that in the historical samples, *Atelopus ignescens* and *A. nanay*. This observation underscores how genetic consequences vary from population declines within closely related species.

Perhaps one of the most relevant uses of genetics in conservation is measuring inbreeding, where closely related individuals reproduce, and detecting inbreeding depression, the reduction in fitness due to inbreeding (Kardos *et al.*, 2013, 2016). Inbreeding will occur in both large and small populations; nonetheless, considerable inbreeding will occur in smaller populations even with random mating because individuals will most likely all be related (Allendorf *et al.*, 2022). Inbred individuals usually have reduced fitness compared to non-inbred individuals (e.g., Willis, 1993). Nowadays, genomic techniques offer informative approaches to detect and understand inbreeding and inbreeding depression, such us demographic reconstruction and runs of homozygosity (ROHs). ROHs are chromosomal regions that have few or no heterozygous nucleotide sites because both chromosome copies derive from a single copy in a relatively common ancestor (Ceballos *et al.*, 2018; Kirin *et al.*, 2010). ROHs offer robust estimates of individual inbreeding and candidate loci underlying inbreeding depression. Moreover, the length of the ROH unveils information about the timescale of inbreeding (Kardos *et al.*, 2015). To our knowledge, explicit measures of inbreeding are only known for *Pristimantis kichwarum* (formerly *Eleutherodactylus ockendeni*; Elmer *et al.*, 2007) and *Epipedobates anthonyi* (F_{IS} estimates; Páez-Vacas *et al.*, 2022), but these estimates are not framed in a conservation context.

Population Genetic Structure and Connectivity

One of the most common application of genetics to conservation is to identify populations and their relationships. Assessing genetic variation among populations allows us to understand evolutionary mechanisms such as genetic drift, selection, and gene flow. It is also fundamental to define conservation units, delineate populations, estimate contemporary migration and gene flow, identify which populations are more vulnerable, and assess connectivity and barriers for dispersal between populations. Therefore, population connectivity can contribute significantly to population persistence. The development of genomics allows to expand adaptive landscape genomics beyond the traditional landscape analyses on neutral variation (Storfer *et al.*, 2018; Richter-Voix *et al.*, 2011; Guo *et al.*, 2016). In Ecuadorian amphibians, population genetic structure, estimates of genetic differentiation between population recorded as F_{ST} or analogous metrics, are only known for a few species, mostly in an eco-evolutionary context. Notably, Funk *et al.* (2009) determined that genetic divergence at microsatellite loci in populations of *Engystomops petersi* is related to call variation instead of landscape features. Páez-Vacas *et al.* (2022) explored the environmental drivers of genetic divergence at microsatellite loci and phenotypic divergence in size and color in populations of the epibatidine poison frog *Epipedobates anthonyi*. Roland *et al.* (2017) evaluated the microevolutionary processes behind color diversity by exploring the phylogeographic pattern among color-divergent populations of the little devil poison frog, *Oophaga sylvatica*, using three mitochondrial and three nuclear markers, as well as genome-wide markers from a double-digested RAD (ddRAD) approach. However, two studies have used intraspecific molecular variation in a conservation context. Fuller *et al.* (2013) showed that the distribution of threatened species can be predicted by a correlation with regions of high genetic and morphological divergence of common species of birds, bats, and the frog *Pristimantis w-nigrum* (Fuller *et al.*, 2013). Moreover, Mittan-Moreau *et al.* (2022) tracked the origin of the invasive species *Rhinella marina* and its past demographic history using SNPs obtained by reduced representation sequencing (ddRAD) from native populations in Ecuador, French Guiana, Guyana, Panama, and Texas and from introduced populations in the United States and Caribbean islands. Their findings corroborate that *R. marina* is a species complex, showing that the introduced populations exhibit only slightly lower genetic diversity than native populations and that there are limited differences in putatively adaptive loci between native and introduced populations. The latter implies that selection in the studied introduced populations is based upon existing genetic diversity found in native populations.

Disease

One of the leading causes of amphibian decline is emerging infectious pathogens. Three major infectious disease agents in amphibians are chytrid fungi (mostly *Batrachochytrium dendrobatidis*, *Bd*), ranaviruses, and an undescribed Perkinsea protist (Chambouvet *et al.*, 2015; Hoverman *et al.*, 2011; Scheele *et al.*, 2019). Several advances have been made to unveil the origins, mechanisms of infection, and consequences, mostly for *Bd* (Azat *et al.*, 2022; Lips, 2016). Conservation genetics offers an enormous potential to investigate the biology, epidemiology, evolution, and host dynamics of infectious disease (Allendorf *et al.*, 2022). The detection and quantification of a disease agent allows for the determination of its prevalence, origins, genetic diversity of pathogens, infection dynamics, consequences, and host defense mechanisms (Fisher *et al.*, 2021; Lips, 2016; Van Rooij *et al.*, 2015). Molecular markers are increasingly used to detect pathogens in adults, tadpoles, and the environment. These pathogens are usually detected with quantitative polymerase chain reaction (qPCR), but protocols exist that rely on PCR (e.g., Peñafiel *et al.*, 2019). These methods have high sensitivity, repeatability, and comparability and are widely used (Scheele *et al.*, 2019; Azat *et al.*, 2022). Using whole-genome sequencing and population genomic analyses, at least four major *Bd* lineages are recognized, including the hypervirulent global panzootic lineage (*Bd*GPL), and their origins have been traced to one lineage, BdASIA-1, on the Korean peninsula (Farrer *et al.*, 2011; O'Hanlon *et al.*, 2018). Susceptible species to chytridiomycosis, the disease caused by *Bd*, can be extirpated, while less susceptible species can persist and act as reservoirs, alongside non-amphibian carriers such as crayfish, lizards, and midges (McMahon *et al.*, 2013; Reeder *et al.*, 2012; Neely *et al.*, 2020).

At the host level, susceptibility to chytridiomycosis shows phylogenetic clustering among amphibians, with some lineages more susceptible than others (Azat *et al.*, 2022; Scheele *et al.*, 2019). Consequences at the genome level due to *Bd*-related declines include low genetic diversity, a reduction in heterozygosity, and increased allele fixation in Central America (Byrne *et al.*, 2020; Velo-Antón *et al.*, 2012). In terms of defense mechanisms, heterozygosity in major histocompatibility complex (MHC) class IIB alleles is a significant predictor of host survival following *Bd* infection in laboratory experiments (Savage and Zamudio, 2011). Moreover, selective pressure on MHC-II for *Bd* survival drives rapid adaptation in some natural populations (Kosch *et al.*, 2016; Savage and Zamudio, 2011). Furthermore, antimicrobial defenses, such as the skin microbiome or antimicrobial skin secretions, have been widely studied using genetics in the context of disease resistance (Flechas *et al.*, 2012; Rebollar *et al.*, 2020; Voyles *et al.*, 2018; Woodhams *et al.*, 2007).

In Ecuador, most of the investigations that have applied genetic analyses in the context of amphibian diseases have focused on pathogen detection to determine the presence and prevalence of *Bd* in biodiversity hotspots or vulnerable species. For instance, these studies have been conducted in an Andean frog community in the cloud forest (Guayasamin *et al.*, 2014), in the Ecuadorian Amazon lowland forest (McCracken *et al.*, 2009), in localities where endangered or nearly extinct species of *Atelopus* have been rediscovered or monitored (Jervis *et al.*, 2020; Tapia *et al.*, 2017; Tarvin *et al.*, 2014), and in an urban park in Quito throughout one year (Narváez-Narváez *et al.*, 2021). The only study (Urgiles *et al.*, 2021) on the prevalence of multiple pathogens, *Bd*, Ranavirus, and Perkinsea, in terrestrial frogs (genus *Pristimantis*) underscores the need to drastically expand pathogen studies, which can readily be implemented by adding straightforward swabbing and PCR techniques to current field surveys. Two studies have developed cost-effective molecular protocols for *Bd* detection in amphibians from Ecuador (Burgos *et al.*, 2019; Peñafiel *et al.*, 2019). One study explored the genetic structure of *Bd* from several locations situated mostly throughout the highlands and Andean slopes in Ecuador and found seven haplotypes from a total of 62 samples (Vizcaíno *et al.*, 2013). Other studies have assessed the skin microbiome across different species of amphibians and communities (Bresciano *et al.*, 2015; Jervis *et al.*, 2021) and in laboratory populations of *Atelopus* (Flechas *et al.*, 2017). However, many questions remain in terms of the pathogen diversity, prevalence, modes of transmission, effects at the genome level, and defense mechanisms. *Bd* and other pathogens are unlikely to ever disappear from the regions they have already invaded or to exhibit attenuated virulence; therefore, the recovery of amphibian communities will depend on the traits that allow host–pathogen coexistence in the long term.

Climate Change

Anthropogenic climate change is already influencing the distribution and biology of species and the composition of biological communities (Anjos and de Toledo, 2018; Walther *et al.*, 2002). Populations whose climatic niche is moving rapidly have four not mutually exclusive options: (1) to tolerate new conditions through phenotypic plasticity, the ability of the same genotype to produce different phenotypes in different environments; (2) to adapt through natural selection, which involves changes at the genetic level; (3) to disperse and follow their climate niche; and (4) to become extirpated (Allendorf *et al.*, 2022; Catullo *et al.*, 2019). Molecular approaches can help to distinguish between phenotypic plasticity and adaptation responses to climate change, which can be difficult (Merilä and Hendry, 2014). Adaptation to climate change might be limited by a lack of genetic variation, inbreeding, and gene flow (Allendorf *et al.*, 2022). Thus, obtaining estimates of these factors can give us a better idea of the potential for adaptation in species. Genomic approaches for predicting adaptation to climate change usually involve finding climate-associated loci by techniques such as genotype–environment associations (GEAs), genome-wide association studies (GWAS), and F_{ST} outlier tests (Capblancq *et al.*, 2020). Sets of candidate loci for climate adaptation can then be used to predict potential adaptation, or maladaptation, to future climates using global climate models (GCMs). Also, distribution range shifts are now known for several species (Parmesan, 2006). Amphibians, for example, are following their climate niche in tropical mountains (Seimon *et al.*, 2007). Endemic species with smaller ranges, particularly amphibians from South America, are predicted to face greater extinction risks (Urban, 2015). Moreover, climate change has been predicted to enhance the effects of other threats, such as invasive species and disease dynamics (Ellison *et al.*, 2020; Hellmann *et al.*, 2008; Rohr *et al.*, 2008). Ultimately, if populations cannot respond to climate change, they will become extirpated.

The impacts of climate change will depend on the magnitude of environmental changes and on the biology of the organisms (Tewksbury *et al.*, 2008). Tropical ectotherms, such as amphibians, are considered more vulnerable to environmental variation than other groups because they are expected to have narrower thermal tolerances and limited acclimation capabilities to changes in temperature due to the relative thermal uniformity throughout the year in these regions (Ghalambor, 2006; Janzen, 1967; Polato *et al.*, 2018). Studies of amphibians in Ecuador show that elevation and local microclimate affect thermal limits in *Pristimantis* frogs (Pintanel *et al.*, 2019) and in tadpoles of several species (Pintanel *et al.*, 2022). Also, precipitation is also known to affect phylogenetic diversity in community assemblages in Ecuador (Amador *et al.*, 2019). Furthermore, climate change will likely cause biotic homogenization in Neotropical anuran communities, particularly in the highlands (Menéndez-Guerrero *et al.*, 2020). Nonetheless, studies on the effects of climate change on genetic diversity, population connectivity, plastic or (mal)adaptive responses, range shifts, disease dynamics, communities, and how this knowledge can be translated to management, are still very limited.

Genetic Management of Wild Populations

Genetic information not only provides key information about causes, mechanisms, and consequences of population declines or potential mechanisms of persistence and recovery, but it can also guide specific actions to manage endangered populations. Some possible actions include assisted gene flow, genetic rescue, and deciding on conservation priorities to maximize evolutionary potential (Flanagan *et al.*, 2017). Assisted gene flow is the managed movement of individuals (or gametes) between populations of a species within the species range to facilitate adaptation to future environmental conditions (Aitken and Whitlock, 2013). Genetic rescue refers to the positive effect of fitness on a population due to an introduction of a beneficial genetic variation from another population (Fitzpatrick and Funk, 2021). Genetic rescue typically reduces inbreeding depression and also facilitates adaptation to changing environments (Allendorf *et al.*, 2022). Concerns have existed with regard to outbreeding depression, which is the reduction in offspring fitness due to genetically divergent parents (Frankham *et al.*, 2011). However, evidence shows that the risk of outbreeding depression is low and that genetic rescue benefits usually offset the risks when taking into consideration how distinct populations are (Frankham *et al.*, 2011; Hamilton and Miller, 2016; Ralls *et al.*, 2020). Gathering genetic information on endangered amphibian species can inform potential genetic rescue between populations (e.g., *Atelopus varius* in Panama, Byrne *et al.*, 2020). However, a significant challenge in the genetic management of endangered species is that only one or, at most, two populations are known for several species.

Conservation Breeding

Conservation breeding refers to the managed breeding of a species with conservation purposes and represents a valuable effort to prevent extinction of species. Mostly it involves breeding in controlled environments within laboratory facilities, that is, *ex situ* breeding. But it can also apply to the managed breeding of populations in the wild (Allendorf *et al.*, 2022). Amphibians exhibit attributes that make them well-suited for laboratory breeding, including their small body size and associated low space requirements, high fecundity, applicability of reproductive technologies, short generation time, and relatively low maintenance requirements (Tapley *et al.*, 2015). However, many threatened species do not meet these criteria due to specific habitat requirements for their survival and reproduction (Bradfield *et al.*, 2022). Consequently, this strategy needs to be integrated with a strong research program that leads to a successful breeding program and subsequent reintroductions (Harding *et al.*, 2016). Moreover, as populations kept *ex situ* also provide key information about the biology of endangered species and offer an opportunity for research, ultimately enhancing the prospects for successful reintroduction (Lewis *et al.*, 2019).

Conservation genetics offers various opportunities to strengthen conservation breeding programs. Small *ex situ* populations suffer the same issues as small populations in the wild, namely, a loss of genetic variation, inbreeding, inbreeding depression, and a loss of evolutionary potential. Thus, the same principles and techniques will be applicable to both, and conservation biologists increasingly recognize the blurred distinction between *ex situ* and *in situ* approaches (Allendorf *et al.*, 2022; Tapley *et al.*, 2015). The first step for conservation breeding is to select and/or characterize the founding population. For this, confirming taxonomic identification, maximizing genetic variation, and reducing inbreeding are relatively easy actions where traditional or recently developed molecular markers come in handy. Once in the laboratory, increasing the effective population size (*Ne*) and minimizing genetic drift can be achieved by equalizing the reproductive success among individuals (Allendorf *et al.*, 2022). Also, assessing relatedness among individuals can guide mating pairs to reduce inbreeding (Funk *et al.*, 2021). Another major concern specific to *ex situ* conservation is adaptation to laboratory conditions in the facilities, which can occur in a single generation (Christie *et al.*, 2012). Proposed mitigation strategies include mimicking natural conditions, equalizing reproductive success, maintaining subpopulations in laboratory, mixing them before reintroduction, and performing supportive breeding (Frankham, 2008; Ryman and Laikre, 1991). However, some of these actions might be impractical or have undesirable effects on *Ne* and therefore need to be carefully designed, but they have not been evaluated in most amphibians yet (Allendorf *et al.*, 2022; Pabijan *et al.*, 2020).

One of the main goals of *ex situ* conservation is to reintroduce individuals bred in the laboratory to the wild. Genetics can help in choosing appropriate populations for reintroduction, so they harbor enough genetic variation, promote gene flow among populations, and provide sources of adaptive genetic variation (Byrne *et al.*, 2020; Frankham *et al.*, 2011). Furthermore, other aspects of reintroductions can be tackled with the aid of genetics; for example, metabarcoding can be used to assess amphibian microbiomes for disease mitigation or reintroduction essays, as shown for *Atelopus limosus* in Panama (Estrada *et al.*, 2022; Jiménez and Sommer, 2017), or preserving genetic variation using assisted reproductive technologies (ARTs), cryopreservation, and biobanking (Silla and Byrne, 2019).

BOX 3.2. CONSERVATION GENETICS GLOSSARY

(Modified from Allendorf *et al.*, 2022)

Adaptation: An evolutionary change resulting from natural selection that increases fitness.
Conservation breeding: The breeding of species for conservation.
Conservation genetics: The use of genetics to solve problems in conservation biology.
Conservation unit: A population or group of populations that is considered distinct for conservation purposes.
Evolution: Changes in allele frequencies in a population from one generation to the next.
Evolutionarily significant unit: A classification of conservation unit that refers to populations that have substantial reproductive isolation that has led to adaptive divergence.
Exomics: The study of the exome, or all exon sequences, using a technique called exome capture.
Extinction: The loss of a species so it no longer exists anywhere.
Fixation: One allele becoming the only one present in a population ($p = 1.0$), while others are lost ($p = 0$).
Gene flow: Exchange of genetic material between populations through migration.
Genetic drift: Random changes in allele frequencies from one generation to the next.
Genetics: The study of genes, their transmission from one generation to the next, and their effect on phenotypes.
Genomics: The study of genomes through the analysis of many loci.
Heterozygosity: A measure of genetic variation that estimates the proportion of individuals in a population that are heterozygotes.
Inbreeding depression: A reduction in fitness due to inbreeding.
Inbreeding: Mating between relatives resulting in increased homozygosity.
Locus (pl. loci): The location of a gene or a marker on a chromosome.
Management unit: A classification of conservation unit that refers to a local population that is managed as a distinct unit due to demographic independence.
Metabolomics: The study of the metabolome, the entire set of small-molecule metabolites in a cell, tissue, or species.
Metagenomics: The study of DNA from many species in an environmental sample using high-throughput sequencing.
Microsatellites: Short sequences of DNA repeated in tandem, consisting of 1–6 nucleotides repeated ~5–100 times. Also known as SSR or VNTR.
Mutation: Change in DNA sequences or chromosomes from parent to offspring.
Neutral (mutation, locus): A locus that has no effect on selection because all genotypes have the same fitness.
Outbreeding depression: A reduction in fitness of hybrids compared to parental types.
Phenotypic plasticity: Phenotype differences in similar genotypes due to environmental differences, often resulting from changes in gene expression during development.
Proteomics: The study of the proteome, the complete set of proteins within a cell, tissue, or species.
Reduced representation sequencing: Partial sequencing of genomes using a variety of methods available.
Selection: Differential contribution of genotypes to the next generation due to differences in survival and reproduction.
Single nucleotide polymorphism (SNP): A nucleotide (base pair) in a DNA sequence that is polymorphic in a population.
Supportive breeding: The practice of removing a subset of individuals from a wild population for breeding in the laboratory and then releasing the lab-born offspring back into their native habitat to intermix with the wild-born individuals and increase population size or persistence.
Transcriptomics: The study of the transcriptome, the set of all messenger RNA transcripts of a cell, tissue, or species.
Whole-genome sequencing: Sequencing of nearly all base pairs in the genome.

REFERENCES

Abdel-Wahab, Y. H. A., L. Marenah, D. F. Orr, C. Shaw, and P. R. Flatt. 2005. Isolation and structural characterisation of a novel 13-amino acid insulin-releasing peptide from the skin secretion of *Agalychnis calcarifer*. *Biological Chemistry* 386: 581–587.

Agapow, P. M., O. R. P. Bininda-Emonds, K. A. Crandall, J. L. Gittleman, G. M. Mace, J. C. Marshall, and A. Purvis. 2004. The impact of species concept on biodiversity studies. *Quarterly Review of Biology* 79: 161–179.

Aguirre, W. E., G. Alvarez-Mieles, F. Anaguano-Yancha, R. Burgos Morán, R. V. Cucalón, D. Escobar-Camacho, I. Jácome-Negrete, P. Jiménez Prado, E. Laaz, and K.

Miranda-Troya. 2021. Conservation threats and future prospects for the freshwater fishes of Ecuador: A hotspot of Neotropical fish diversity. *Journal of Fish Biology* 99: 1158–1189.

Aitken, S. N., and M. C. Whitlock. 2013. Assisted gene flow to facilitate local adaptation to climate Change. *Annual Review of Ecology, Evolution, and Systematics* 44: 367–388.

Ajingi, Y., and N. Jongruja. 2020. Antimicrobial peptide engineering: Rational design, synthesis, and synergistic effect. *Russian Journal of Bioorganic Chemistry* 46: 463–479.

Almendáriz, A. 1987. Contribución al conocimiento de la herpetofauna centroriental Ecuatoriana. *Politécnica* 12: 77–133.

Allendorf, F. W., W. C. Funk, S. N. Aitken, M. Byrne, and G. Luikart. 2022. Conservation and the genomics of populations, 3rd edition. *Evolutionary Applications* 15: 1966.

Alton, L. A., and C. E. Franklin. 2017. Drivers of amphibian declines: Effects of ultraviolet radiation and interactions with other environmental factors. *Climate Change Responses* 4: 1–26.

Amador, L., M. Soto-Gamboa, and J. M Guayasamin. 2019. Integrating alpha, beta, and phylogenetic diversity to understand anuran fauna along environmental gradients of tropical forests in western Ecuador. *Ecology and Evolution* 9: 11040–11052.

Amiche, M., A. Ladram, and P. Nicolas. 2008. A consistent nomenclature of antimicrobial peptides isolated from frogs of the subfamily Phyllomedusinae. *Peptides* 29: 2074–2082.

AmphibiaWeb. 2022. *Information on amphibian biology and conservation*. http://amphibiaweb.org/ (accessed September 29, 2022).

Anastasi, A., G. Bertaccini, and V. Erspamer. 1966. Pharmacological data on phyllokinin (bradykinyl-isoleucyl-tyrosine o-sulphate) and bradykinyl-isoleucyl-tyrosine. *British Journal of Pharmacology and Chemotherapy* 27: 479–485.

Angerer, K. 2011. Frog tales—on poison dart frogs, epibatidine, and the sharing of biodiversity. *Innovation: The European Journal of Social Science Research* 24: 353–369.

Angerer, K. 2013. "There is a Frog in South America/Whose Venom is a Cure": Poison alkaloids and drug discovery. *Biologics. A History of Agents Made From Living Organisms in the Twentieth Century* 3: 173–191.

Anjos, L. J. S., and P. M. de Toledo. 2018. Measuring resilience and assessing vulnerability of terrestrial ecosystems to climate change in South America. *PloS One* 13: e0194654–e0194654.

Antonelli, A., J. A. Nylander, C. Persson, and I. Sanmartín. 2009. Tracing the impact of the Andean uplift on Neotropical plant evolution. *Proceedings of the National Academy of Sciences* 106: 9749–9754.

Arbeláez Ortiz, E. 2005. *Atelopus* sp. 13. In *Ranas Arlequines*, eds. J. V. Rueda-Almonacid, J. V. Rodríguez-Mahecha, S. Lötters, E. La Marca, T. R. Kahn, and A. Angulo, 149. Bogotá: Conservación Internacional.

Arbeláez Ortiz, E., and A. Vega Toral. 2008. *Guía de Anfibios, Reptiles y Peces del Parque Nacional Cajas/Cajas National Park amphibian, reptile and fish guide*. Cuenca, Ecuador: ETAPA, Municipalidad Cuenca. Grafisum.

Arias-Cárdenas, A., L. S. Barrientos, C. Pardo-Diaz, A. Paz, A. J. Crawford, and C. Salazar. 2023. Taxonomic inflation and a reconsideration of speciation in the Andes: The case of the high-elevation tree frog *Dendropsophus molitor* (Anura: Hylidae). *Zoological Journal of the Linnean Society*: zlad085.

Arteaga-Navarro, A. F., R. A. Pyron, N. Peñafiel, P. Romero-Barreto, J. Culebras, L. Bustamante, and J. M. Guayasamin. 2016. Comparative phylogeography reveals cryptic diversity and repeated patterns of cladogenesis for amphibians and reptiles in northwestern Ecuador. *PLoS One* 11: 1–40.

Azat, C., M. Alvarado-Rybak, J. J. Solano-Iguaran, A. Velasco, A. Valenzuela-Sánchez, S. V. Flechas, A. Peñafiel-Ricaurte, A. A. Cunningham, and L. D. Bacigalupe. 2022. Synthesis of *Batrachochytrium dendrobatidis* infection in South America: Amphibian species under risk and areas to focus research and disease mitigation. *Ecography* 2022: e05977.

Azevedo Calderon, L., A. de A. E. Silva, P. Ciancaglini, and R. G Stábeli. 2011. Antimicrobial peptides from *Phyllomedusa* frogs: From biomolecular diversity to potential nanotechnologic medical applications. *Amino Acids* 40: 29–49.

Badio, B., H. M. Garraffo, C. V. Plummer, W. L. Padgett, and J. W. Daly. 1997. Synthesis and nicotinic activity of epiboxidine: An isoxazole analogue of epibatidine. *European Journal of Pharmacology* 321: 189–194.

Barnes, M. A., and C. R. Turner. 2016. The ecology of environmental DNA and implications for conservation genetics. *Conservation Genetics* 17: 1–17.

Barreto-Santamaría, A., M. E. Patarroyo, and H. Curtidor. 2019. Designing and optimizing new antimicrobial peptides: All targets are not the same. *Critical Reviews in Clinical Laboratory Sciences* 56: 351–373.

Barrio-Amorós, C. L., M. Costales, J. Vieira, E. Osterman, H. Kaiser, and A. Arteaga. 2020. Back from extinction: Rediscovery of the harlequin toad *Atelopus mindoensis* Peters, 1973 in Ecuador. *Herpetology Notes* 13: 325–328.

Bataille, A., J. J. Fong, M. Cha, G. O. Wogan, H. J. Baek, H. Lee, M. Mi-Sook, and B. Waldman. 2013. Genetic evidence for a high diversity and wide distribution of endemic strains of the pathogenic chytrid fungus *Batrachochytrium dendrobatidis* in wild Asian amphibians. *Molecular Ecology* 22: 4196–4209.

Beirne, C., and A. Whitworth. 2011. *Frogs of the Yachana Reserve*. Global Vision International. https://www.researchgate.net/publication/236504364_Frogs_of_the_Yachana_Reserve.

Berger, L., A. A. Roberts, J. Voyles, J. E. Longcore, K. A. Murray, and L. F. Skerratt. 2016. History and recent progress on chytridiomycosis in amphibians. *Fungal Ecology* 19: 89–99.

Berger, L., R. Speare, and A. Hyatt. 1999. Chytrid fungi and amphibian declines: Overview, implications and future directions. *Declines and Disappearances of Australian frogs. Environment Australia, Canberra* 1999: 23–33.

Bermúdez-Puga, S., G. Morán-Marcillo, N. Espinosa de los Monteros-Silva, R. E. Naranjo, F. Toscano, K. Vizuete, M. Torres Arias, J. R. Almeida, and C. Proaño-Bolaños. 2023. Inspiration from cruzioseptin-1: Membranolytic analogue with improved antibacterial properties. *Amino Acids* 55: 113–124.

Bishop, P. J., A. Angulo, J. P. Lewis, R. D. Moore, G. B. Rabb, and J. Garcia Moreno. 2012. The Amphibian extinction crisis—what will it take to put the action into the Amphibian Conservation Action Plan? *Surveys and Perspectives Integrating Environment and Society* 5: 97–111.

Black, J. 1982. Los páramos de Antisana. *Instituto Geográfico Militar, Revista Geográfica* 17: 25–52.

Blaustein, A. R., and D. B. Wake. 1990. Declining amphibian populations: A global phenomenon. *Trends in Ecology & Evolution* 5: 203–204.

Bockmann, F. A., M. T. Rodrigues, T. Kohsldorf, L. C. Straker, T. Grant, M. C. C. de Pinna, F. L. M. Mantelatto, A. Datovo, J. P. Pombal Jr., and J. C. McNamara. 2018. Brazil's government attacks biodiversity. *Science* 360: 865.

Bonaccorso, E., J. M. Guayasamin, D. Méndez, and R. Speare. 2003. Chytridiomycosis as a possible cause of population declines in *Atelopus cruciger* (Anura: Bufonidae). *Herpetological Review* 34: 331–333.

Bonilla-Jiménez, M. S. 2021. Ocurrencia de actividad antimicrobiana en la secreción de la piel de *Callimedusa ecuatoriana* y caracterización estructural del péptido Ikiam 2599. Tesis de Licenciatura. Universidad Regional Amazónica Ikiam.

Bradfield, K. S., B. Tapley, and K. Johnson. 2022. Amphibians and conservation breeding programmes: How do we determine who should be on the ark? *Biodiversity and Conservation* 32: 885–898.

Brandbyge, J., and L. Holm-Nielsen. 1991. *Reforestación de los andes ecuatorianos con especies nativas*. Quito: Central Ecuatoriana de Servicios Agrícolas (CESA).

Bravo, E. 1997. *La bioprospección en el Ecuador. Biodiversidad, Bioprospección y Bioseguridad*. Quito: Abya-Yala.

Bresciano, J. C., C. A. Salvador, C. Paz-y-Miño, A. M. Parody-Merino, J. Bosch y D. C. Woodhams. 2015. Variation in the presence of anti-*Batrachochytrium dendrobatidis* bacteria of amphibians across life stages and elevations in Ecuador. *EcoHealth* 12: 310–319.

Briggs, C. J., R. A. Knapp, and V. T. Vredenburg. 2010. Enzootic and epizootic dynamics of the chytrid fungal pathogen of amphibians. *Proceedings of the National Academy of Sciences of the United States of America* 107: 9695–9700.

Brito, J., M. A. Camacho, V. Romero, and A. F. Vallejo. 2023. *Mamíferos del Ecuador*. Versión 2023.0. Museo de Zoología, Pontificia Universidad Católica del Ecuador. https://bioweb.bio/faunaweb/mammaliaweb/

Brito-Zapata, D., and C. Reyes-Puig. 2021. A new species of terrestrial-breeding frog *Pristimantis* (Anura: Strabomantidae) from the Cordillera del Cóndor, Zamora Chinchipe, Ecuador. *Neotropical Biodiversity* 7: 213–222.

Bronson E. and C. K. Vance. 2019. Amphibian Reproduction. Chapter 53 in: Fowler's Zoo and Wild Animal Medicine, Vol 9. Editors R E. Miller and M. E. Fowler. Pg. 371–379.

Browne, R. K., Q. Luo, P. Wang, N. Mansour, S.A. Kaurova, E. N. Gakhova, N. V. Shishova, V. K. Uteshev, L. I. Kramarova, G. Venu, S. Vaissi, Z. Taheri-Khas, P. Heshmatzad, M. F. Bagaturov , P. Janzen, R. E. Naranjo, A. Swegen, J. Strand, D. McGinnity, and I. Dunce. 2024. Ecological civilisation and amphibian sustainability through reproduction biotechnologies, biobanking, and conservation breeding programs (RBCs). *Animals* 14: 1455.

Browne, R. K., A. J. Silla, R. Upton, G. Della-Togna, R. Marcec-Greaves, N. V. Shishova, V. K. Uteshev, B. Proaño, O. D. Pérez, N. Mansour, S. A. Kaurova, E. N. Gakhova, J. Cosson, B. Dyzuba, L. I. Kramarova, D. McGinnity, M. Gonzalez, J. Clulow, and S. Clulow. 2019. Sperm collection and storage for the sustainable management of amphibian biodiversity. *Theriogenology* 133: 187–200.

Burgos, G., D. A. Narváez-Narváez, B. Freire-Paspuel, A. Merino-Viteri, C. Muslin, and A. Genoy-Puerto. 2019. Development of a multiplex real-time PCR surveillance assay for monitoring the health status of Ecuadorian amphibians at risk of extinction. *Forensic Science International: Genetics Supplement Series* 7: 674–676.

Bustamante, M. R., S. R. Ron, and L. A. Coloma. 2005. Cambios en la diversidad en siete comunidades de anuros en los Andes de Ecuador. *Biotropica* 37: 180–189.

Bustamante, M. R., W. Bolívar, L. A. Coloma, S. Ron, D. Cisneros-Heredia, F. Castro, J. V. Rueda, S. Lötters, and A. R. Acosta-Galvis. 2004. *Atelopus longirostris*. The IUCN Red List of Threatened Species 2004: e.T54522A11158637. https://dx.doi.org/10.2305/IUCN.UK.2004.RLTS.T54522A11158637.en (accessed April 12, 2017).

Byrne, A. Q., C. L. Richards-Zawacki, J. Voyles, K. Bi, R. Ibáñez, and E. B. Rosenblum. 2020. Whole exome sequencing identifies the potential for genetic rescue in iconic and critically endangered Panamanian harlequin frogs. *Global Change Biology* 27: 50–70.

Cáceres Andrade, J. F. 2022. Informe anual 2022. Parque Nacional Cajas. Subgerencia de Gestión Ambiental, Etapa. Cuenca.

Caicedo, A. 2007. Análisis de secreciones peptídicas de anfibios ecuatorianos con pruebas de susceptibilidad en bacterias patógenas. Tesis de Licenciatura. Pontificia Universidad Católica del Ecuador.

Cajiao Garcés, V. L. 2018. Determinación de la bioactividad de fracciones peptídicas de *Agalychnis spurrelli* y *Phyllomedusa tomopterna* (Anura: Hylidae). Tesis Licenciatura. Pontificia Universidad Católica del Ecuador.

Caminer, M. A., B. Milá, M. Jansen, A. Fouquet, P. J. Venegas, G. Chávez, S. C. Lougheed, and S. R. Ron. 2017. Systematics of the *Dendropsophus leucophyllatus* species complex (Anura: Hylidae): Cryptic diversity and the description of two new species. *PLoS One* 12: e0171785

Caminer, M. A., and S. R. Ron. 2014. Systematics of treefrogs of the *Hypsiboas calcaratus* and *Hypsiboas fasciatus* species complex (Anura, Hylidae) with the description of four new species. *ZooKeys* 370: 1–68.

Caminer, M. A., and S. R. Ron. 2020. Systematics of the *Boana semilineata* species group (Anura: Hylidae), with a description of two new species from Amazonian Ecuador. *Zoological Journal of the Linnean Society* 190: 149–180.

Campbell Grant, E. H., S. M. Amburgey, B. Gratwicke, V. A. Chaves, A. M. Belasen, D. Bickford, C. A. Brühl, N. E. Calatayud, *et al*. 2023. Priority research needs to inform amphibian conservation in the Anthropocene. *Conservation Science and Practice* 5: e12988.

Camper, J. D., O. Torres-Carvajal, S. R. Ron, J. Nilsson, A. Arteaga, T. W. Knowles, and B. S. Arbogast. 2021. Amphibians and reptiles of Wildsumaco Wildlife Sanctuary, Napo Province, Ecuador. *Check List* 17: 729–751.

Capblancq, T., M. C. Fitzpatrick, R. A. Bay, M. Exposito-Alonso, and S. R. Keller. 2020. Genomic prediction of (mal) adaptation across current and future climatic landscapes. *Annual Review of Ecology, Evolution, and Systematics* 51: 245–269.

Carlier, L., P. Joanne, L. Khemtémourian, C. Lacombe, P. Nicolas, C. El Amri, and O. Lequin. 2015. Investigating the role of GXXXG motifs in helical folding and self-association of plasticins, Gly/Leu-rich antimicrobial peptides. *Biophysical Chemistry* 196: 40–52.

Carvajal-Endara, S., L. A. Coloma, M. A. Morales-Mite, J. M. Guayasamin, P. Székely, and W. E. Duellman. 2019. Phylogenetic systematics, ecology, and conservation of marsupial frogs (anura: Hemiphractidae) from the Andes of southern Ecuador, with descriptions of four new biphasic species. *Zootaxa* 4562: 1–102.

Carvalho, T., A. M Belasen, L. F. Toledo, T. Y. James. 2024. Coevolution of a generalist pathogen with many hosts: The case of the amphibian chytrid *Batrachochytrium dendrobatidis*. *Current Opinion in Microbiology* 78: 102435.

Casas-Marce, M., E. Marmesat, L. Soriano, B. Martínez-Cruz, M. Lucena-Perez, F. Nocete, and A. Rodríguez-Hidalgo, *et al*. 2017. Spatiotemporal dynamics of genetic variation in the Iberian Lynx along its path to extinction reconstructed with ancient DNA. *Molecular Biology and Evolution* 34: 2893–2907.

Castroviejo-fisher, S., J. Moravec, J. Aparicio, M. Guerrero-Reinhard, and G. Calderón. 2011. DNA taxonomy reveals two new species records of *Hyalinobatrachium* (Anura: Centrolenidae) for Bolivia. *Zootaxa* 2798: 64.

Catenazzi, A. 2015. State of the world's amphibians. *Annual Review of Environment and Resources* 40: 91–119.

Catenazzi, A., and R. von May. 2021. Systematics and conservation of Neotropical amphibians and reptiles. *Diversity* 13: 45.

Catullo, R. A., J. Llewelyn, B. L. Phillips, and C. C. Moritz. 2019. The potential for rapid evolution under Anthropogenic climate change. *Current Biology* 29: R996–1007.

Ceballos, F. C., P. K. Joshi, D. W. Clark, M. Ramsay, and J. F. Wilson. 2018. Runs of homozygosity: Windows into population history and trait architecture. *Nature Reviews Genetics* 19: 220–234.

Ceballos, G., P. R. Ehrlich, and R. Dirzo. 2017. Biological annihilation via the ongoing sixth mass extinction signaled by vertebrate population losses and declines. *Proceedings of the National Academy of Sciences of the United States of America* 114: E6089–6096.

Chambouvet, A., D. J. Gower, M. Jirků, M. J. Yabsley, A. K. Davis, G. Leonard, F. Maguire, T. M. Doherty-Bone, G. B. Bittencourt-Silva, M. Wilkinson, and T. A. Richards. 2015. Cryptic infection of a broad taxonomic and geographic diversity of tadpoles by Perkinsea protists. *Proceedings of the National Academy of Sciences of the United States of America* 112: E4743–4751.

Chen, Q., P. Cheng, C. Ma, X. Xi, L. Wang, M. Zhou, H. Bian, and T. Chen. 2018a. Evaluating the bioactivity of a novel broad-spectrum antimicrobial peptide Brevinin-1GHa from the frog skin secretion of *Hylarana guentheri* and its analogues. *Toxins* 10.

Chen, T., D. F. Orr, M. O'Rourke, C. McLynn, A. J. Bjourson, S. McClean, D. Hirst, P. Rao, and C. Shaw. 2004. *Pachymedusa dacnicolor* tryptophyllin-1: Structural characterization, pharmacological activity and cloning of precursor cDNA. *Regulatory Peptides* 117: 25–32.

Chen, T., and C. Shaw. 2003. Cloning of the (Thr6)-phyllokinin precursor from *Phyllomedusa sauvagei* skin confirms a non-consensus tyrosine O-sulfation motif. *Peptides* 24: 1123–1130.

Chen, X., D. Chen, L. Huang, X. Chen, M. Zhou, X. Xi, C. Ma, T. Chen, and L. Wang. 2020. Identification and target-modification of SL-BBI: A novel bowman-Birk type trypsin inhibitor from *Sylvirana latouchii*. *Biomolecules* 10: 1254.

Chen, X., L. Zhang, C. Ma, Y. Zhang, X. Xi, L. Wang, M. Zhou, J. F. Burrows, and T. Chen. 2018b. A novel antimicrobial peptide, Ranatuerin-2PLx, showing therapeutic potential in inhibiting proliferation of cancer cells. *Bioscience Reports* 38: BSR20180710.

Christie, M. R., M. L. Marine, R. A. French, and M. S. Blouin. 2012. Genetic adaptation to captivity can occur in a single generation. *Proceedings of the National Academy of Sciences of the United States of America* 109: 238–242.

Chuang, P. H. 2012. Evaluación de la actividad anticancerígena del extracto peptídico crudo de *Agalychnis spurrelli*. Tesis de Licenciatura. Pontificia Universidad Católica del Ecuador.

Cilveti, C. 2013. Genotipaje de la resistencia a antimicrobianos en bacterias entéricas productoras de carbapenemasas y pruebas de inhibición con secreciones peptídicas de anfibios nativos. Tesis de Licenciatura. Pontificia Universidad Católica del Ecuador.

Cisneros-Heredia, D. F. 2018. The hitchhiker wave: Non-native small terrestrial vertebrates in the Galápagos. In *Understanding invasive species in the Galapagos Islands: From the molecular to the landscape*, eds. M. Torres and C. Mena, 95–139. Cham: Springer.

Clarke, B. T. 1997. The natural history of amphibian skin secretions, their normal functioning and potential medical applications. *Biological Reviews of the Cambridge Philosophical Society* 72: 365–379.

Clulow, J., and S. Clulow. 2016. Cryopreservation and other assisted reproductive technologies for the conservation of threatened amphibians and reptiles: Bringing the ARTs up to speed. *Reproduction, Fertility, and Development* 28: 1116–1132.

Clulow, J., V. L. Trudeau, and A. J. Kouba. 2014. Amphibian declines in the twenty-first century: Why we need assisted reproductive technologies. In *Reproductive sciences in animal conservation: Progress and prospects*, eds. W. V. Holt, J. L. Brown, and P. Comizzoli, 275–316. New York: Springer.

Cohen, J. M., D. J. Civitello, M. D. Venesky, T. A. McMahon, and J. R. Rohr. 2019. An interaction between climate change and infectious disease drove widespread amphibian declines. *Global Change Biology* 25: 927–937.

Coloma, L. A. 1992. *Anfibios de Ecuador: Estatus poblacional y de conservación*. Nature Conservancy. Technical Report.

Coloma, L. A. 1995. Ecuadorian frogs of the genus Colostethus (Anura: Dendrobatidae). Miscellaneous Publication. Museum of Natural History, University of Kansas 87: 1–72.

Coloma, L. A. 1997. Morphology, systematics, and phylogenetic relationships among frogs of the genus *Atelopus* (Anura: Bufonidae). PhD dissertation, University of Kansas.

Coloma, L. A., V. Cano, and M. A. Rodríguez. 2010. A "life raft" for amphibians in Ecuador. In *Building a future for wildlife: Zoos and aquariums committed to biodiversity conservation*, eds. G. Dick and M. Gusset, 105–110. Gland: Waza Executive Office.

Coloma, L. A., S. Carvajal-Endara, J. F. Dueñas, A. Paredes-Recalde, M. Morales-Mite, D. Almeida-Reinoso, E. E. Tapia, C. R. Hutter, E. Toral, and J. M. Guayasamin. 2012. Molecular phylogenetics of stream treefrogs of the *Hyloscirtus larinopygion* group (Anura: Hylidae), and description of two new species from Ecuador. *Zootaxa* 78: 1–78.

Coloma, L. A., S. Lötters, W. E. Duellman, and A. Miranda-Leiva. 2007. A taxonomic revision of *Atelopus pachydermus*, and description of two new (extinct?) species of *Atelopus* from Ecuador (Anura: Bufonidae). *Zootaxa* 1557: 1–32.

Conlon, J. M. 2011. Structural diversity and species distribution of host-defense peptides in frog skin secretions. *Cellular and Molecular Life Sciences* 68: 2303–2315.

Conlon, J. M., L. Guilhaudis, J. Leprince, L. Coquet, M. L. Mangoni, S. Attoub, T. Jouenne, and J. D. King. 2017. Peptidomic analysis of skin secretions of the Mexican burrowing toad *Rhinophrynus dorsalis* (Rhinophrynidae): Insight into the origin of host-defense peptides within the Pipidae and characterization of a proline-arginine-rich peptide. *Peptides* 97: 22–28.

Conlon, J. M., and M. Mechkarska. 2014. Host-defense peptides with therapeutic potential from skin secretions of frogs from the family pipidae. *Pharmaceuticals* 7: 58–77.

Corey, E. J., T. P. Loh, S. AchyuthaRao, D. C. Daley, and S. Sarshar. 1993. Stereocontrolled total synthesis of (+)-and (-)-epibatidine. *The Journal of Organic Chemistry* 58: 5600–5602.

Costa-Neto, E. M. 2005. Animal-based medicines: Biological prospection and the sustainable use of zootherapeutic resources. *Anais da Academia Brasileira de Ciências* 77: 33–43.

Crandall, K. A., O. R. P. Bininda-Emonds, G. M. Mace, and R. K. Wayne. 2000. Considering evolutionary processes in conservation biology. *Trends in Ecology & Evolution* 15: 290–295.

Crawford, A. J., K. R. Lips, and E. Bermingham. 2010. Epidemic disease decimates amphibian abundance, species diversity, and evolutionary history in the highlands of central Panama. *Proceedings of the National Academy of Sciences of the United States of America* 107: 13777–13782.

Creveling, C. R. 2003. Foreword: Honoring John William Daly. *Cellular and Molecular Neurobiology* 23: 285–287.

Cuesta, A., M. Vila, A. Molina-Marquez, C. Carbajo, J. A. Lebron, M. López-López, P. López-Cornejo, M. L. Moyá, M. López-Sanmartín, J. R. López, J. I. Navas, J. Vigara, and R. León. 2021a. Influence of adding terminal tags on the structural and antimicrobial properties of the peptide caerin 1.1. *Aquaculture* 532: 736035.

Cuesta, S. A., F. Gallegos, J. Arias, F. Pilaquinga, A. Blasco-Zúñiga, C. Proaño-Bolaños, M. Rivera, and L. Meneses. 2019. Molecular modeling of four dermaseptin-related peptides of the gliding tree frog *Agalychnis spurrelli*. *Journal of Molecular Modeling* 25: 260.

Cuesta, S. A., C. Reinoso, F. Morales, F. Pilaquinga, G. Morán-Marcillo, C. Proaño-Bolaños, A. Blasco-Zúñiga, M. Rivera, and L. Meneses. 2021b. Novel antimicrobial cruzioseptin peptides extracted from the splendid leaf frog, *Cruziohyla calcarifer*. *Amino Acids* 53: 853–868.

Cuvi, N. 2022. *Historia ambiental y ecología urbana para Quito*. Quito: Flacso.

Daly, J. W., H. M. Garraffo, and T. F. Spande. 1999. Alkaloids from amphibian skins. In *Alkaloids: Chemical and biological perspective*, ed. S. W. Pelletier, 1–161. New York: Pergamon.

Daly, J. W., H. M. Garraffo, T. F. Spande, M. W. Decker, J. P. Sullivan, and M. Williams. 2000. Alkaloids from frog skin: The discovery of epibatidine and the potential for developing novel non-opioid analgesics. *Natural Product Reports* 17: 131–135.

Darst, C. R., P. A. Menéndez-Guerrero, L. A. Coloma, and D. C. Cannatella. 2005. Evolution of dietary specialization and chemical defense in poison frogs (Dendrobatidae): A comparative analysis. *American Naturalist* 165: 56–69.

Daszak, P., L. Berger, A. A. Cunningham, A. D. Hyatt, D. E. Green, and R. Speare. 1999. Emerging infectious diseases and amphibian population declines. *Emerging Infectious Diseases* 5: 735–748.

Dayot, J. 2023. Oil conflict and compromises in the Ecuadorian Amazon: The relationships between oil and indigenous people in historical perspective. *Alternautas* 10: 34–62.

de Oliveira, E. A., M. Penhacek, K. L. A. Guimarães, G. A. do Nascimento, L. R. R. Rodrigues, and E. J. Hernández-Ruz. 2019. *Pristimantis* in the Eastern Brazilian Amazon: DNA barcoding reveals underestimated diversity in a megadiverse genus. *Mitochondrial DNA Part A* 30: 731–738.

Dejean, T., A. Valentini, C. Miquel, P. Taberlet, E. Bellemain, and C. Miaud. 2012. Improved detection of an alien invasive species through environmental DNA barcoding: The example of the American bullfrog *Lithobates catesbeianus*. *Journal of Applied Ecology* 49: 953–959.

Díaz Escobar, V. D. 2017. Caracterización molecular de genes de resistencia a β-lactámicos en aislados clínicos de *Escherichia coli* provenientes de urocultivos y pruebas de inhibición con péptidos de *Boana rosenbergi* y *Rana* sp. Tesis de Licenciatura. Pontificia Universidad Católica del Ecuador.

Dillon, X. 2012. Identificación de fracciones con actividad antimicrobiana obtenidas del exudado de la piel de *Agalychnis spurrelli* (Anura: Hylidae). Tesis de Licenciatura. Pontificia Universidad Católica del Ecuador.

Dodson, C. H., and A. H. Gentry. 1991. Biological extinction in western Ecuador. *Annals of the Missouri Botanical Garden* 78: 273–295.

Donnelly-Roberts, D. L., P. S. Puttfarcken, T. A. Kuntzweiler, C. A. Briggs, D. J. Anderson, J. E. Campbell, M. Piattoni-Kaplan, D. G. McKenna, *et al.* 1998. ABT-594 [(R)-5-(2-Azetidinylmethoxy)-2-chloropyridine]: A novel, orally effective analgesic acting via neuronal nicotinic acetylcholine receptors: I. In vitro characterization. *Journal of Pharmacology and Experimental Therapeutics* 285: 777–786.

dos Santos, S. P., R. Ibáñez, and S. R. Ron. 2015. Systematics of the *Rhinella margaritifera* complex (Anura, Bufonidae) from western Ecuador and Panama with insights in the biogeography of *Rhinella alata*. *ZooKeys* 145: 109–145.

Duarte Valverde, A. 1969. *Reglamento para regular las acitividades de caza y exportación de animales vivos de nuestra fauna, así como todo lo relativo a la preservación de esos importantes recursos naturales*. Registro Oficial No 339, del 31 de Diciembre de 1969, Quito, Ecuador.

Duellman, W. E. 1972. A review of the neotropical frogs of the *Hyla bogotensis* group. *Museum of Natural History, University of Kansas* 11: 1–31.

Duellman, W. E. 1973. Frogs of the *Hyla geographica* Group. *Copeia* 1973: 515–533.

Duellman, W. E. 1978. *The biology of an equatorial herpetofauna in Amazonian Ecuador*. University of Kansas Museum of Natural History, Miscellaneous Publications 65. Lawrence: University of Kansas.

Duellman, W. E. 1999. *Patterns of distribution of amphibians: A global perspective*. Baltimore: JHU Press.

Duellman, W. E. 2015. *Marsupial frogs: Gastrotheca and allied genera*. Baltimore: JHU Press.

Duellman, W. E., and M. L. Crump. 1974. Speciation in frogs of the *Hyla parviceps* group in the upper Amazon Basin. *Occasional Papers of the Museum of Natural History, the University of Kansas* 23: 1–40.

Duellman, W. E., and D. M. Hillis. 1987. Marsupial frogs (Anura: Hylidae: *Gastrotheca*) of the Ecuadorian Andes: Resolution of taxonomic problems and phylogenetic relationships. *Herpetologica* 43: 141–173.

Duellman, W. E., and D. M. Hillis. 1990. Systematics of frogs of the *Hyla larinopygion* group. *Occasional Papers of the Museum of Natural History University of Kansas* 134: 1–23.

Duellman, W. E., and J. D. Lynch. 1969. Descriptions of *Atelopus* tadpoles and their relevance to atelopodid classification. *Herpetologica* 25: 231–240.

Duellman, W. E., L. R. Maxson, and C. A. Jesiolowski. 1988. Evolution of marsupial frogs (Hylidae: Hemiphactinae): Immunological evidence. *Copeia* 1988: 527–545.

El Amri, C., and P. Nicolas. 2008. Plasticins: Membrane-damaging peptides with "chameleon-like" properties. *Cellular and Molecular Life Sciences* 65: 895–909.

Ellison, A., K. Zamudio, K. Lips, and C. Muletz-Wolz. 2020. Temperature-mediated shifts in salamander transcriptomic responses to the amphibian-killing fungus. *Molecular Ecology* 29: 325–343.

Elmer, K. R., R. M. Bonett, D. B. Wake, and S. C. Lougheed. 2013. Early Miocene origin and cryptic diversification of South American salamanders. *BMC Evolutionary Biology* 13: 1–16.

Elmer, K. R., J. A. Dávila, and S. C. Lougheed. 2007. Applying new inter-individual approaches to assess fine-scale population genetic diversity in a neotropical frog, *Eleutherodactylus ockendeni*. *Heredity* 99: 506–515.

Espinosa R, R., and J. Durán Arias. 1970. *Decreto No. 099. Reforma al acuerdo ministerial No. 800*. Ecuador: Registro oficial No. 393, del 18 de Marzo 1970.

Estrada, A., D. Medina, B. Gratwicke, R. Ibáñez, and L. K. Belden. 2022. Body condition, skin bacterial communities and disease status: Insights from the first release trial of the limosa harlequin frog, *Atelopus limosus*. *Proceedings. Biological Sciences* 289: 20220586.

FAO. 2011. State of the World's forests 2011. Roma. http://www.fao.org/3/i2000e/i2000e00.htm.

Farrer, R. A., L. A. Weinert, J. Bielby, T. W. J. Garner, F. Balloux, F. Clare, J. Bosch, A. A. Cunningham, C. Weldon, L. H. du Preez, L. Anderson, S. L. K. Pond, R. Shahar-Golan, D. A. Henk, and M. C. Fisher. 2011. Multiple emergences of genetically diverse amphibian-infecting chytrids include a globalized hypervirulent recombinant lineage. *Proceedings of the National Academy of Sciences of the United States of America* 108: 18732–18736.

Farrington, H. L., C. E. Edwards, X. Guan, M. R. Carr, K. Baerwaldt, and R. F. Lance. 2015. Mitochondrial genome sequencing and development of genetic markers for the detection of DNA of invasive bighead and silver carp (*Hypophthalmichthys nobilis* and *H. molitrix*) in environmental water samples from the United States. *PloS One* 10: e0117803–0117803.

Felsenstein, J. 1985. Phylogenies and the comparative method. *American Naturalist* 125: 1–15.

Fernández, F. 2011. The greatest impediment to the study of biodiversity in Colombia. *Caldasia* 33: iii–v.

Ficetola, G. F., R. Manenti, and P. Taberlet. 2019. Environmental DNA and metabarcoding for the study of amphibians and reptiles: Species distribution, the microbiome, and much more. *Amphibia-Reptilia* 40: 129–148.

Ficetola, G. F., C. Miaud, F. Pompanon, and P. Taberlet. 2008. Species detection using environmental DNA from water samples. *Biology Letters* 4: 423–425.

Finer, M., C. N. Jenkins, S. L. Pimm, B. Keane, and C. Ross. 2008. Oil and gas projects in the western Amazon: Threats to wilderness, biodiversity, and indigenous peoples. *PloS One* 3: e2932.

Fisher, M. C., F. Pasmans, and A. Martel. 2021. Virulence and pathogenicity of chytrid fungi causing amphibian extinctions. *Annual Review of Microbiology* 75: 673–693.

Fisher, R. A. 1930. *The genetical theory of natural selection*. Oxford: Clarendon Press.

Fitzpatrick, S. W., and W. C. Funk. 2021. Genomics for genetic rescue. In *Population genomics wildlife*, eds. P. A. Hohenlohe and O. P. Rajora, 437–471. Cham: Springer.

Flanagan, S. P., B. R. Forester, E. K. Latch, S. N. Aitken, and S. Hoban. 2017. Guidelines for planning genomic assessment and monitoring of locally adaptive variation to inform species conservation. *Evolutionary Applications* 11: 1035–1052.

Flechas, S. V., A. Blasco-Zúñiga, A. Merino-Viteri, V. Ramírez-Castañeda, M. Rivera, and A. Amézquita. 2017. The effect of captivity on the skin microbial symbionts in three *Atelopus* species from the lowlands of Colombia and Ecuador. *PeerJ* 5: e3594.

Flechas, S. V., C. Sarmiento, M. E. Cárdenas, E. M. Medina, S. Restrepo, and A. Amézquita. 2012. Surviving chytridiomycosis: Differential anti-*Batrachochytrium dendrobatidis* activity in bacterial isolates from three lowland species of *Atelopus*. *PloS One* 7: e44832–44832.

Fletcher-Lazo, G. 2002. Taxonomía, distribución e historia natural de los ápodos (Amphibia: Gymnophiona) del occidente ecuatoriano. Tesis de Licenciatura. Pontificia Universidad Católica del Ecuador.

Fouquet, A., P. Gaucher, M. Blanc, and C. M. Vélez-Rodríguez. 2007. Description of two new species of *Rhinella* (Anura: Bufonidae) from the lowlands of the Guiana shield. *Zootaxa* 32: 17–32.

Fouquet, A., C. Santana Cassini, C. Fernando Baptista Haddad, N. Pech, and M. Trefaut Rodrigues. 2014. Species delimitation, patterns of diversification and historical biogeography of the Neotropical frog genus *Adenomera* (Anura, Leptodactylidae). *Journal of Biogeography* 41: 855–870.

Franco-Mena, D., J. M. Guayasamin, D. Andrade-Brito, M. H. Yánez-Muñoz, and F. J. M. Rojas-Runjaic. 2023. Unveiling the evolutionary relationships and the high cryptic diversity in Andean rainfrogs (Craugastoridae: *Pristimantis myersi* group). *PeerJ* 11: e14715.

Frankham, R. 2008. Genetic adaptation to captivity in species conservation programs. *Molecular Ecology* 17: 325–333.

Frankham, R., J. D. Ballou, D. A. Briscoe, and K. H. McInnes. 2002. *Introduction to conservation genetics*. New York: Cambridge University Press.

Frankham, R., J. D. Ballou, M. D. B. Eldridge, R. C. Lacy, K. Ralls, M. R. Dudash, and C. B. Fenster. 2011. Predicting the probability of outbreeding depression. *Conservation Biology* 25: 465–475.

Fraser, D. J., and L. Bernatchez. 2001. Adaptive evolutionary conservation: Towards a unified concept for defining conservation units. *Molecular Ecology* 10: 2741–2752.

Frost, D. R. 2023. *Amphibian species of the world: An online reference*. Version 6.1. American Museum of Natural History, New York, USA. http://research.amnh.org/herpetology/amphibia/index.html.

Fuller, T. L., H. A. Thomassen, M. Peralvo, W. Buermann, B. Milá, C. M. Kieswetter, P. Jarrín-V. *et al.* 2013. Intraspecific morphological and genetic variation of common species predicts ranges of threatened ones. *Proceedings. Biological Sciences* 280: 20130423.

Funk, W. C., M. Caminer, and S. R. Ron. 2011. High levels of cryptic species diversity uncovered in Amazonian frogs. *Proceedings of the Royal Society B: Biological Sciences* 279: 1806–1814.

Funk, W. C., D. C. Cannatella, and M. J. Ryan. 2009. Genetic divergence is more tightly related to call variation than landscape features in the Amazonian frogs *Physalaemus petersi* and *P. freibergi*. *Journal of Evolutionary Biology* 22: 1839–1853.

Funk, W. C., K. R. Zamudio, and A. J. Crawford. 2021. Advancing understanding of amphibian evolution, ecology, behavior, and conservation with massively parallel sequencing. In *Population genomics wildlife*, eds. P. A. Hohenlohe, and O. P. Rajora, 211–254. Cham: Springer.

Gagliardo, R., P. Crump, E. Griffith, J. Mendelson, H. Ross, and K. Zippel. 2008. The principles of rapid response for amphibian conservation, using the programmes in Panama as an example. *International Zoo Yearbook* 42: 125–135.

Gao, Y., D. Wu, L. Wang, C. Lin, C. Ma, X. Xi, M. Zhou, J. Duan, O. R. P. Bininda-Emonds, and T. Chen. 2017. Targeted modification of a novel amphibian antimicrobial peptide from *Phyllomedusa tarsius* to enhance its activity against MRSA and microbial biofilm. *Frontiers in Microbiology* 8: 628.

Garraffo, H. M. 2009. John William Daly, 1933–2008. *Cellular and Molecular Neurobiology* 29: 439–440.

Gascon, C., J. P. Collins, R. D. Moore, D. R. Church, J. E. McKay, and J. R. Mendelson III. 2007. *Amphibian conservation action plan*. Gland: IUCN/SSC Amphibian Specialist Group.

Ge, L., P. Lyu, M. Zhou, H. Zhang, Y. Wan, B. Li, R. Li, L. Wang, T. Chen, and C. Shaw. 2014. AcT-2: A novel myotropic and antimicrobial type 2 tryptophyllin from the skin secretion of the Central American red-eyed leaf frog, *Agalychnis callidryas*. *The Scientific World Journal* 2014: 158546.

Gehara, M., A. J. Crawford, V. G. D. Orrico, A. Rodríguez, S. Lötters, A. Fouquet, and L. S. Barrientos, *et al*. 2014. High levels of diversity uncovered in a widespread nominal taxon: Continental phylogeography of the Neotropical tree frog *Dendropsophus minutus*. *PLoS One* 9: e103958.

Ghalambor, C. K. 2006. Are mountain passes higher in the tropics? Janzen's hypothesis revisited. *Integrative and Comparative Biology* 46: 5–17.

Gilbert, N. 2010. Biodiversity law could stymie research. *Nature* 463: 598.

Gippoliti, S., F. P. D. Cotterill, D. Zinner, and C. P. Groves. 2018. Impacts of taxonomic inertia for the conservation of African ungulate diversity: An overview. *Biological Reviews* 93: 115–130.

Goldberg, C. S., D. S. Pilliod, R. S. Arkle, and L. P. Waits. 2011. Molecular detection of vertebrates in stream water: A demonstration using Rocky Mountain tailed frogs and Idaho giant salamanders. *PloS One* 6: e22746–e22746.

Gong, Z., X. Pei, S. Ren, X. Chen, L. Wang, C. Ma, X. Xi, T. Chen, C. Shaw, and M. Zhou. 2020. Identification and rational design of a novel antibacterial peptide dermaseptin-AC from the skin secretion of the red-eyed tree frog *Agalychnis callidryas*. *Antibiotics* 9: 243.

Grant, T., M. Rada, M. Anganoy-Criollo, A. Batista, P. H. Dias, A. M. Jeckel, D. J. Machado, and J. V. Rueda-Almonacid. 2017. Phylogenetic systematics of dart-poison frogs and their relatives revisited (Anura: Dendrobatoidea). *South American Journal of Herpetology* 12: S1–90.

Guayasamin, J. M., A. Arteaga, and C. R. Hutter. 2018. A new (singleton) rainfrog of the *Pristimantis myersi* Group (Amphibia: Craugastoridae) from the northern Andes of Ecuador. *Zootaxa* 4527: 323–334.

Guayasamin, J. M., D. F. Cisneros-Heredia, R. J. Maynard, R. L. Lynch, J. Culebras, and P. S. Hamilton. 2017. A marvelous new glassfrog (Centrolenidae, *Hyalinobatrachium*) from Amazonian Ecuador. *ZooKeys* 2017: 1–20.

Guayasamin, J. M., D. F. Cisneros-Heredia, R. W. McDiarmid, P. Peña, and C. R. Hutter. 2020. Glassfrogs of Ecuador: Diversity, evolution, and conservation. *Diversity* 12: 222.

Guayasamin, J. M., Á. M. Mendoza, A. V. Longo, K. R. Zamudio, and E. Bonaccorso. 2014. High prevalence of *Batrachochytrium dendrobatidis* in an Andean frog community (Reserva Las Gralarias, Ecuador). *Amphibian & Reptile Conservation* 8: 33–44.

Guerra García, J. M., F. Espinosa Torre, and J. C. García Gómez. 2008. Trends in taxonomy today: An overview about the main topics in taxonomy. *Zoológica baetica* 19: 15–49.

Guillory, W. X., M. R. Muell, K. Summers, and J. L. Brown. 2019. Phylogenomic reconstruction of the Neotropical poison frogs (Dendrobatidae) and their conservation. *Diversity* 11: 126.

Guimarães, A., F. Costa, F., O. Júnior, W. Fontes, and M. Castro. 2016. The amazing world of peptide engineering: The example of antimicrobial peptides from frogs and their analogues. *Protein & Peptide Letters* 23: 722–737.

Guo, B., D. Lu, W. B. Liao, J. Merilä. 2016. Genomewide scan for adaptive differentiation along altitudinal gradient in the Andrew's toad *Bufo andrewsi*. *Molecular Ecology* 25: 3884–3900.

Hamilton, J. A., and J M. Miller. 2016. Adaptive introgression as a resource for management and genetic conservation in a changing climate. *Conservation Biology* 30: 33–41.

Harding, G., R. A. Griffiths, and L. Pavajeau. 2016. Developments in amphibian captive breeding and reintroduction programs. *Conservation Biology* 30: 340–349.

Hayes, T., A. Collins, M. Lee, M. Mendoza, N. Noriega, A. A. Stuart, and A. Vonk. 2002. Hermaphroditic, demasculinized frogs after exposure to the herbicide atrazine at low ecologically relevant doses. *Proceedings of the National Academy of Sciences of the United States of America* 99: 5476–5480.

Hayes, T. B., P. Falso, S. Gallipeau, and M. Stice. 2010. The cause of global amphibian declines: A developmental endocrinologist's perspective. *Journal of Experimental Biology* 213: 921–933.

Hayes, T. B., A. A. Stuart, M. Mendoza, A. Collins, N. Noriega, A. Vonk, G. Johnston, L. Roger, and D. Kpodzo. 2006. Characterization of atrazine-induced gonadal malformations in African clawed frogs (*Xenopus laevis*) and comparisons with effects of an androgen antagonist (cyproterone acetate) and exogenous estrogen (17β-estradiol): Support for the demasculinization/feminization hypothesis. *Environmental Health Perspectives* 114 (Suppl 1): 134–141.

He, H., Y. Chen, Z. Ye, X. Chen, C. Ma, M. Zhou, X. Xi, J. F. Burrows, T. Chen, and L. Wang. 2020. Modification and targeted design of N-terminal truncates derived from brevinin with improved therapeutic efficacy. *Biology* 9: 209.

Hebert, P. D. N., A. Cywinska, S. L. Ball, and J. R. DeWaard. 2003. Biological identifications through DNA barcodes. *Proceedings of the Royal Society of London. Series B: Biological Sciences* 270: 313–321.

Hedrick, P. W. 2001. Conservation genetics: Where are we now? *Trends in Ecology & Evolution* 16: 629–636.

Hellmann, J. J., J. E. Byers, B. G. Bierwagen, and J. S. Dukes. 2008. Five potential consequences of climate change for invasive species. *Conservation Biology* 22: 534–543.

Heyer, W. R. 1973. Systematics of the *marmoratus* group of the frog genus *Leptodactylus* (Amphibia, Leptodactylidae). *Natural History Museum of Los Angeles County Contributions in Science* 251: 1–50.

Hillis, D. M., E. A. Chambers, and T. J. Devitt. 2021. Contemporary methods and evidence for species delimitation. *Ichthyology & Herpetology* 109: 895–903.

Hillis, D. M., and R. O. de Sá. 1988. Phylogeny and taxonomy of the *Rana palmipes* group (Salientia: Ranidae). *Herpetological Monographs* 2: 1–26.

Hillis, D. M., and T. P. Wilcox. 2005. Phylogeny of New World true frogs (*Rana*). *Molecular Phylogenetics and Evolution* 34: 299–314.

Hird, C., R. L. Cramp, and C. E. Franklin. 2023. Thermal compensation reduces DNA damage from UV radiation. *Journal of Thermal Biology* 117: 103711.

Hoban, S., M. Bruford, J. D'Urban Jackson, M. Lopes-Fernandes, M. Heuertz, P. A. Hohenlohe, I. Paz-Vinas, *et al*. 2020. Genetic diversity targets and indicators in the CBD post-2020 Global Biodiversity Framework must be improved. *Biological Conservation* 248: 108654.

Hoban, S., J. L. Kelley, K. E. Lotterhos, M. F. Antolin, G. Bradburd, D. B. Lowry, M. L. Poss, L. K. Reed, A. Storfer, and M. C. Whitlock. 2016. Finding the genomic basis of local adaptation: Pitfalls, practical solutions, and future directions. *The American Naturalist* 188: 379–397.

Hohenlohe, P. A., and O. P. Rajora. 2021. *Population genomics: Wildlife*. Cham: Springer.

Holaskova, E., P. Galuszka, I. Frebort, and M. T. Oz. 2015. Antimicrobial peptide production and plant-based expression systems for medical and agricultural biotechnology. *Biotechnology Advances* 33: 1005–1023.

Holsinger, K. E., R. J. Mason-Gamer, and J. Whitton. 1999. Genes, demes, and plant conservation. In *Genetics and the extinction of species: DNA and the conservation of biodiversity*, eds. L. Landweber and A Dobson, 23–46. Princeton, NJ: Princeton University Press.

Hoogmoed, M. S. 1977. On the presence of *Bufo nasicus* Werner in Guiana, with a redescription of the species on the basis of recently collected material. *Zoologische Mededelingen. Leiden* 51: 265–275.

Hoogmoed, M. S. 1990. Biosystematics of South American Bufonidae, with special reference to the *Bufo "typhonius"* group. In *Vertebrates in the tropics*, eds. G. Peters and R. Hutterer, 113–123. Bon: Museum Alexander Koenig.

Hoverman, J. T., M. J. Gray, N. A. Haislip, and D. L. Miller. 2011. Phylogeny, life history, and ecology contribute to differences in amphibian susceptibility to ranaviruses. *EcoHealth* 8: 301–319.

Huan, Y., Q. Kong, H. Mou, and H. Yi. 2020. Antimicrobial peptides: Classification, design, application and research progress in multiple fields. *Frontiers in Microbiology* 11: 582779.

Hurd, P. L., and M. Enquist. 2005. A strategic taxonomy of biological communication. *Animal Behaviour* 70: 1155–1170.

Hurtig, A.-K., and M. San Sebastián. 2002. Geographical differences in cancer incidence in the Amazon basin of Ecuador in relation to residence near oil fields. *International Journal of Epidemiology* 31: 1021–1027.

Hutter, C. R., and J. M. Guayasamin. 2015. Cryptic diversity concealed in the Andean cloud forests: Two new species of rainfrogs (*Pristimantis*) uncovered by molecular and bioacoustic data. *Neotropical Biodiversity* 1: 36–59.

Hutter, C. R., J. M. Guayasamin, and J. J. Wiens. 2013. Explaining Andean megadiversity: The evolutionary and ecological causes of glassfrog elevational richness patterns. *Ecology Letters* 16: 1135–1144.

INABIO. 2021. Autoridades coreanas y ecuatorianas conocen más de la propuesta del Banco Nacional de Recursos Genéticos del INABIO. *INABIO* (August 11). http://inabio.biodiversidad.gob.ec/2021/08/11/autoridades-coreanas-y-ecuatorianas-conocen-mas-de-la-propuesta-del-banco-nacional-de-recursos-geneticos-del-inabio/.

IPCC. 2023. Summary for policymakers. In *Climate change 2023: Synthesis report*, eds. Core Writing Team, H. Lee, and J. Romero, 1–34. Geneva: IPCC.

Isaac, N. J. B., J. Mallet, and G. M. Mace. 2004. Taxonomic inflation: Its influence on macroecology and conservation. *Trends in Ecology & Evolution* 19: 464–469.

IUCN. 2001. *Red list categories*. Version 3.1. Gland, Switzerland: UICN Species Survival Commission.

IUCN. 2022. *The IUCN red list of threatened species*. Version 2022–2. https://www.iucnredlist.org.

IUCN Species Survival Commission. 2012. *Guidelines for application of IUCN Red List Criteria at regional and national levels*. Version 4.0. Gland, Switzerland and Cambridge: IUCN.

IUCN SSC Amphibian Specialist Group 2022a. *Atelopus arthuri*. The IUCN Red List of Threatened Species 2022: e.T54490A98640562. https://dx.doi.org/10.2305/IUCN.UK.2022-1.RLTS.T54490A98640562.en.

IUCN SSC Amphibian Specialist Group 2022b. *The Amphibian Conservation Action Plan (ACAP): A status review and roadmap for global amphibian conservation*. https://ecoevorxiv.org/brfas/ (accessed May 9, 2019).

IUCN Standards and Petitions Committee. 2022. *Guidelines for using the IUCN Red List Categories and Criteria*. Version 15.1. Prepared by the Standards and Petitions Committee. https://www.iucnredlist.org/documents/RedListGuidelines.pdf

Jacobson, K. A., and K. L. Kirk. 2009. John W. Daly-An appreciation. *Heterocycles* 79: 61.

Janzen, D. H. 1967. Why mountain passes are higher in the tropics. *The American Naturalist* 101: 233–249.

Jaynes, K. E., M. I. Páez-Vacas, D. Salazar-Valenzuela, J. M. Guayasamin, A. Terán-Valdez, F. R. Siavichay, S. W. Fitzpatrick, and L. A. Coloma. 2022. Harlequin frog rediscoveries provide insights into species persistence in the face of drastic amphibian declines. *Biological Conservation* 276: 109784.

Jervis, P., B. Karlsdottir, R. Jehle, D. Almeida-Reinoso, F. Almeida-Reinoso, S. Ron, M. C. Fisher, and A. Merino-Viteri. 2020. Disease reservoirs threaten the recently rediscovered Podocarpus Stubfoot Toad (*Atelopus podocarpus*). *Amphibian & Reptile Conservation* 14: 157–164.

Jervis, P., P. Pintanel, K. Hopkins, C. Wierzbicki, J. M. G. Shelton, E. Skelly, G. M. Rosa, D. Almeida-Reinoso, M. Eugenia-Ordoñez, S. Ron, X. Harrison, A. Merino-Viteri, and M. C. Fisher. 2021. Post-epizootic microbiome associations across communities of neotropical amphibians. *Molecular Ecology* 30: 1322–1335.

Jetz, W., C. Carbone, J. Fulford, and J. H. Brown. 2004. The scaling of animal space use. *Science* 306: 266–268.

Jetz, W., and R. A. Pyron. 2018. The interplay of past diversification and evolutionary isolation with present imperilment across the amphibian tree of life. *Nature Ecology & Evolution* 2: 850–858.

Jiang, Z., A. I. Vasil, M. L. Vasil, and R. S. Hodges. 2014. Specificity determinants improve therapeutic indices of two antimicrobial peptides piscidin 1 and dermaseptin S4 against the gram-negative pathogens *Acinetobacter baumannii* and *Pseudomonas aeruginosa*. *Pharmaceuticals* 7: 366–391.

Jiménez-Casalombo, G. B., A. P. Guarderas-Valverde, and L. A. Coloma. 2023. Abundancia de la rana marsupial *Gastrotheca riobambae* (Fowler, 1913) (Hemiphractidae, Anura) en un gradiente de urbanización en la Provincia Pichincha, Ecuador. *Revista Latinoamericana de Herpetología* 6: 45–70.

Jiménez de la Espada, M. 1875. *Vertebrados del viaje al Pacífico verificado de 1862 a 1865 por una comisión de naturalistas enviada por el Gobierno Español*. Batracios. Madrid: Miguel Ginesta.

Jiménez, R. R., and S. Sommer. 2017. The amphibian microbiome: Natural range of variation, pathogenic dysbiosis, and role in conservation. *Biodiversity and Conservation* 26: 763–786.

Johnson, P. T. J., M. K. Reeves, S. K. Krest, and A. E. Pinkney. 2010. A decade of deformities: Advances in our understanding of amphibian malformations and their implications. *Ecotoxicology of Amphibians and Reptiles* 2: 511–536.

Jungfer, K.-H., and W. Hödl. 2002. A new species of *Osteocephalus* from Ecuador and a redescription of *O. leprieurii* (Duméril & Bibron, 1841) (Anura: Hylidae). *Amphibia-Reptilia* 23: 21–46.

Jungfer, K.-H., S. Ron, R. Seipp, and A. Almendáriz. 2000. Two new species of hylid frogs, genus *Osteocephalus*, from Amazonian Ecuador. *Amphibia-Reptilia* 21: 327–340.

Kaganer, A. W., L. D. Nagel, T. E. Youker-Smith, E. M. Bunting, and M. P. Hare. 2021. Environmental DNA-derived pathogen gene sequences can expand surveillance when pathogen titers are decoupled in eDNA and hosts. *Environmental DNA* 3: 1192–1207.

Kamoroff, C., and C. S. Goldberg. 2017. Using environmental DNA for early detection of amphibian chytrid fungus *Batrachochytrium dendrobatidis* prior to a ranid die-off. *Diseases of Aquatic Organisms* 127: 75–79.

Kaplan, M. 1994. A new species of frog of the genus *Hyla* from the Cordillera Oriental in northern Columbia with comments on the taxonomy of *Hyla minuta*. *Journal of Herpetology* 28: 79–87.

Kardos, M., F. W. Allendorf, and G. Luikart. 2013. Evaluating the role of inbreeding depression in heterozygosity-fitness correlations: How useful are tests for identity disequilibrium? *Molecular Ecology Resources* 14: 519–530.

Kardos, M., E. E. Armstrong, S. W. Fitzpatrick, S. Hauser, P. W. Hedrick, J. M. Miller, D. A. Tallmon, and W. C. Funk. 2021. The crucial role of genome-wide genetic variation in conservation. *Proceedings of the National Academy of Sciences* 118: e2104642118.

Kardos, M., G. Luikart, and F. W. Allendorf. 2015. Measuring individual inbreeding in the age of genomics: Marker-based measures are better than pedigrees. *Heredity* 115: 63–72.

Kardos, M., H. R. Taylor, H. Ellegren, G. Luikart, and F. W. Allendorf. 2016. Genomics advances the study of inbreeding depression in the wild. *Evolutionary Applications* 9: 1205–1218.

Kiesecker, J. M., A. R. Blaustein, and L. K. Belden. 2001. Complex causes of amphibian population declines. *Nature* 410: 681–684.

Kieswetter, C. M., and C. J. Schneider. 2013. Phylogeography in the northern Andes: Complex history and cryptic diversity in a cloud forest frog, *Pristimantis w-nigrum* (Craugastoridae). *Molecular Phylogenetics and Evolution* 69: 417–429.

Kirin, M., R. McQuillan, C. S. Franklin, H. Campbell, P. M. McKeigue, and J. F. Wilson. 2010. Genomic runs of homozygosity record population history and consanguinity. *PloS One* 5: e13996–13996.

Knappe, D., P. Henklein, R. Hoffmann, and K. Hilpert. 2010. Easy strategy to protect antimicrobial peptides from fast degradation in serum. *Antimicrobial Agents and Chemotherapy* 54: 4003–4005.

Köppen, W. 1918. Klassifikation der Klimate nach Temperatur, Niederschlag. *Jahreslauf. Pet. Mitt* 64: 243–248.

Köppen, W. P. 1931. *Grundriss der klimakunde*. Berlin und Leipzig: Walter de Gruyter & Co.

Kosch, T. A., A. Bataille, C. Didinger, J. A. Eimes, S. Rodríguez-Brenes, M. J. Ryan, and B. Waldman. 2016. Major histocompatibility complex selection dynamics in pathogen-infected túngara frog (*Physalaemus pustulosus*) populations. *Biology Letters* 12: 20160345.

Kouba, A. J., R. E. Lloyd, M. L. Houck, A. J. Silla, N. Calatayud, V. L. Trudeau, J. Clulow. *et al*. 2013. Emerging trends for biobanking amphibian genetic resources: The hope, reality and challenges for the next decade. *Biological Conservation* 164: 10–21.

Kouba, A. J., and C. K. Vance. 2009. Applied reproductive technologies and genetic resource banking for amphibian conservation. *Reproduction, Fertility and Development* 21: 719–737.

Krynak, K. L., D. G. Wessels, S. M. Imba, T. J. Krynak, E. B. Snyder, J. A. Lyons, and J. M. Guayasamin. 2020. Call survey indicates rainbow trout farming alters glassfrog community composition in the Andes of Ecuador. *Amphibian & Reptile Conservation* 14: 1–11.

La Marca, E., K. R. Lips, S. Lötters, R. Puschendorf, R. Ibañez, J. V Rueda-Almonacid, R. Schulte, C. Marty, *et al*. 2005. Catastrophic population declines and extinctions in Neotropical harlequin frogs (Bufonidae: *Atelopus*). *Biotropica* 37: 190–201.

La Marca, E., and S. Lötters. 1997. Monitoring of declines in Venezuelan *Atelopus* (Amphibia: Anura: Bufonidae). *Herpetologia Bonnensis* 1997: 207–231.

Lai, R., H. Liu, W. H. Lee, and Y. Zhang. 2002. Identification and cloning of a trypsin inhibitor from skin secretions of Chinese red-belly toad *Bombina maxima. Comparative Biochemistry and Physiology Part B: Biochemistry and Molecular Biology* 131: 47–53.

Lavilla, E. O., J. A. Langone, U. Caramaschi, J. P. Pombal Jr, R. O. de Sá. 2017. Comments on the neotype of *Rana margaritifera* Laurenti, 1768 (Anura, Bufonidae). *Zootaxa* 4286: 113–114.

Leite, J. R. S. A., L. P. Silva, M. I. S. Rodrigues, M. V Prates, G. D. Brand, B. M. Lacava, R. B. Azevedo, A. L. Bocca, S. Albuquerque, and C. Bloch. 2005. Phylloseptins: A novel class of anti-bacterial and anti-protozoan peptides from the *Phyllomedusa* genus. *Peptides* 26: 565–573.

Lever, C. 2003. *Naturalized reptiles and amphibians of the world*. New York: Oxford University Press.

Lewis, C. H. R., C. L. Richards-Zawacki, R. Ibáñez, J. Luedtke, J. Voyles, P. Houser, and B. Gratwicke. 2019. Conserving Panamanian harlequin frogs by integrating captive-breeding and research programs. *Biological Conservation* 236: 180–187.

Li, M., X. Xi, C. Ma, X. Chen, M. Zhou, J. F. Burrows, T. Chen, and L. Wang. 2019. A novel dermaseptin isolated from the skin secretion of *Phyllomedusa tarsius* and its cationicity-enhanced analogue exhibiting effective antimicrobial and anti-proliferative activities. *Biomolecules* 9: 628.

Li, R., H. Wang, Y. Jiang, Y. Yu, L. Wang, M. Zhou, Y. Zhang, T. Chen, and C. Shaw. 2012. A novel Kazal-type trypsin inhibitor from the skin secretion of the Central American red-eyed leaf frog, *Agalychnis callidryas*. *Biochimie* 94: 1376–1381.

Lips, K. R. 1998. Decline of a tropical montane amphibian fauna. *Conservation Biology* 12: 106–117.

Lips, K. R. 1999. Mass mortality and population declines of anurans at an upland site in Western Panama. *Conservation Biology* 13: 117–125.

Lips, K. R. 2016. Overview of chytrid emergence and impacts on amphibians. *Philosophical Transactions of the Royal Society of London. Series B, Biological Sciences* 371: 20150465.

Lips, K. R., F. Brem, R. Brenes, J. D. Reeve, R. A. Alford, J. Voyles, C. Carey, L. Livo, A. P. Pessier, and J. P. Collins. 2006. Emerging infectious disease and the loss of biodiversity in a Neotropical amphibian community. *Proceedings of the National Academy of Sciences of the United States of America* 103: 3165–3170.

Lips, K. R., P. A. Burrowes, J. R. Mendelson, and G. Parra-Olea. 2005. Amphibian declines in Latin America: Widespread population declines, extinctions, and impacts. *Biotropica* 37: 163–165.

Lips, K. R., J. Diffendorfer, J. R. Mendelson III, and M. W. Sears. 2008. Riding the wave: Reconciling the roles of disease and climate change in amphibian declines. *PLoS Biology* 6: 441–454.

Lips, K. R., B. Young, J. Reaser, R. Ibáñez, and A. Salas. 2000. Amphibian declines in Latin America: Workshops to design a monitoring protocol and database. *Froglog* 37: 1–4.

Liscano, Y., J. Oñate-Garzón, and J. P. Delgado. 2020. Peptides with dual antimicrobial-anticancer activity: Strategies to overcome peptide limitations and rational design of anticancer peptides. *Molecules* 25: 4245.

Lopes, C. M., D. Baêta, A. Valentini, M. L. Lyra, A. F. Sabbag, J. L. Gasparini, T. Dejean, C. F. B. Haddad, and K. R. Zamudio. 2021. Lost and found: Frogs in a biodiversity hotspot rediscovered with environmental DNA. *Molecular Ecology* 30: 3289–3298.

López-Tacoaman, Y. F. 2022. Estudio del potencial antimicrobiano de la secreción cutánea de *Atelopus nanay* (Bufonidae). Tesis de Licenciatura. Universidad Regional Amazónica Ikiam.

Lötters, S. 2007. The fate of the harlequin toads—help through a synchronous multidisciplinary approach and the IUCN. Amphibian Conservation Action Plan. *Mitteilungen Museum Für Naturkunde Berlin, Zoologische Reihe* 83: 69–73.

Lötters, S., W. Haas, S. Schick, and W. Böhme. 2002. On the systematics of the harlequin frogs (Amphibia: Bufonidae: *Atelopus*) from Amazonia. II: Redescription of *Atelopus pulcher* (Boulenger, 1882) from the eastern Andean versant in Peru. *Salamandra* 38: 165–184.

Lötters, S., A. Plewnia, A. Catenazzi, K. Neam, A. R. Acosta-Galvis, Y. Alarcón Vela, J. P. Allen, *et al*. 2023. Ongoing harlequin toad declines suggest the amphibian extinction crisis is still an emergency. *Communications Earth & Environment* 4: 412.

Lötters, S., A. Van Der Meijden, L. A. Coloma, R. Boistel, P. Cloetens, R. Ernst, E. Lehr, and M. Veith. 2011. Assessing the molecular phylogeny of a near extinct group of vertebrates: The Neotropical harlequin frogs (Bufonidae; *Atelopus*). *Systematics and Biodiversity* 9: 45–57.

Luedtke, J. A., J. Chanson, K. Neam, L. Hobin, A. O. Maciel, A. Catenazzi, A. Borzée, A. Hamidy, A. Aowphol, and A. Jean. 2023. Ongoing declines for the world's amphibians in the face of emerging threats. *Nature* 622: 308–314.

Lynch, J. D. 1974. New species of frogs (Leptodactylidae: *Eleutherodactylus*) from the Amazonian lowlands of Ecuador. *Occasional Papers of the Museum of Natural History, the University of Kansas* 31: 1–22.

Lynch, J. D. 1979. Leptodactylid frogs of the genus *Eleutherodactylus* from the Andes of southern Ecuador. *The University of Kansas, Museum of Natural History, Miscellaneous Publications* 66: 1–62.

Lynch, J. D. 1980. Two new species of earless frogs allied to *Eleutherodactylus surdus* (Leptodactylidae) from the Pacific slopes of the Ecuadorian Andes. *Proceedings of the Biological Society of Washington* 93: 327–338.

Lynch, J. D. 1981. Leptodactylid frogs of the genus *Eleutherodactylus* in the Andes of northern Ecuador and adjacent Colombia. Miscellaneous Publication. Museum of Natural History, University of Kansas 72: 1–46.

Lynch, J. D. 1982. Relationships of the frogs of the genus *Ceratophrys* (Leptodactylidae) and their bearing on hypothesis of Pleistocene Forest refugia in South America and punctuated equilibria. *Systematic Zoology* 31: 166–179.

Lynch, J. D., and W. E. Duellman. 1973. A review of the centrolenid frogs of Ecuador, with descriptions of new species. *Occasional Papers of the Museum of Natural History, University of Kansas* 16: 1–66.

Lynch, J. D., and W. E. Duellman. 1980. The *Eleutherodactylus* of the Amazonian slopes of the Ecuadorian Andes (Anura: Leptodactylidae). Museum of Natural History, University of Kansas 69: 1–86.

Lynch, J. D., and W. E. Duellman. 1997. Frogs of the genus *Eleutherodactylus* (Leptodactylidae) in western Ecuador: Systematics, ecology and biogeography. *University of Kansas, Natural History Museum, Special Publication* 23: 1–236.

Lyra, M. L., C. F. B. Haddad, and A. M. L. de Azeredo-Espin. 2017. Meeting the challenge of DNA barcoding Neotropical amphibians: Polymerase chain reaction optimization and new COI primers. *Molecular Ecology Resources* 17: 966–980.

Lyu, P., L. Ge, R. Ma, R. Wei, C. M. McCrudden, T. Chen, C. Shaw, and H. F. Kwok. 2018. Identification and pharmaceutical evaluation of novel frog skin-derived serine proteinase inhibitor peptide-PE-BBI (*Pelophylax esculentus* Bowman-Birk inhibitor) for the potential treatment of cancer. *Scientific Reports* 8: 14502.

Llumiquinga, M. 2017. Potencial antimicrobiano de la secreción cutánea total de anuros ecuatorianos sobre cepas bacterianas tipo y aislados clínicos. Tesis de Licenciatura. Pontificia Universidad Católica del Ecuador, Quito, Ecuador.

Ministerio de Ambiente, Agua y Transición Ecológica. 2023. *Deforestación y regeneración a nivel provincial del período 2020–2022 del Ecuador continental*. Quito: MAATE.

Madanchi, H., S. Sardari, H. Shajiee, S. Taherian, M. Ashkar, B. Johari, A. A. Shabani, and S. Sharafi. 2020. Design of new truncated derivatives based on direct and reverse mirror repeats of first six residues of Caerin 4 antimicrobial peptide and evaluation of their activity and cytotoxicity. *Chemical Biology & Drug Design* 96: 801–811.

Manzano Pasquel, A. L. 2014. *Revisiting the enigmatic: Emergence of Bd coincide with amphibian decline in Los Andes of Ecuador*. CA: San Francisco State University.

MapBiomas. 2021. *Proyecto MapBiomas Amazonía. Colección 4 de los mapas anuales de cobertura y uso del suelo*. https://amazonia.mapbiomas.org/ (accessed May 9, 2019).

Marsh, H., A. Dennis, H. Hines, A. Kutt, K. McDonald, E. Weber, S. Williams, and J. Winter. 2007. Optimizing allocation of management resources for wildlife. *Conservation Biology* 21: 387–399.

Martín-Torrijos, L., J. V. Sandoval-Sierra, J. Muñoz, J. Diéguez-Uribeondo, J. Bosch, and J. M. Guayasamin. 2016. Rainbow trout (*Oncorhynchus mykiss*) threaten Andean amphibians. *Neotropical Biodiversity* 2: 26–36.

McCracken, S. F., J. P. Gaertner, M. R. Forstner, and D. Hahn. 2009. Detection of *Batrachochytrium dendrobatidis* in amphibians from the forest floor to the upper canopy of an Ecuadorian Amazon lowland rainforest. *Herpetological Review* 40: 190.

McGugan, J. R., G. D. Byrd, A. B. Roland, S. N. Caty, N. Kabir, E. E. Tapia, S. A. Trauger, L. A. Coloma, and L. A. O'Connell. 2016. Ant and mite diversity drives toxin variation in the little devil poison frog. *Journal of Chemical Ecology* 42: 537–551.

McMahon, T. A., L. A. Brannelly, M. W. H. Chatfield, P. T. J. Johnson, M. B. Joseph, V. J. McKenzie, C. L. Richards-Zawacki, M. D. Venesky, and J. R. Rohr. 2013. Chytrid fungus *Batrachochytrium dendrobatidis* has nonamphibian hosts and releases chemicals that cause pathology in the absence of infection. *Proceedings of the National Academy of Sciences of the United States of America* 110: 210–215.

Melo, M. A., M. L. Lyra, A. M. Brischi, V. C. Geraldi, and C. F. B. Haddad. 2014. First record of the invasive frog *Eleutherodactylus johnstonei* (Anura: Eleutherodactylidae) in São Paulo, Brazil. *Salamandra* 50: 177–180.

Melo, P., X. Herrera, and R. Galeas. 2015. *Modelo bioclimático para la reparación cartográfica de ecosistemas de Ecuador Continental*. Quito: Ministerio de Ambiente del Ecuador.

Mendelson III, J. R., K. R. Lips, R. W. Gagliardo, G. B. Rabb, J. P. Collins, J. E. Diffendorfer, P. Daszak, *et al*. 2006. Confronting amphibian declines and extinctions. *Science* 313: 48–48.

Mendelson III, J. R., and D. G. Mulcahy. 2010. A new species of toad (Bufonidae: *Incilius*) from central Panama. *Zootaxa* 2396: 61–68.

Mendes, B., C. Proaño-Bolaños, F. R. Gadelha, J. R. Almeida, and D. C. Miguel. 2020. Cruzioseptins, antibacterial peptides from *Cruziohyla calcarifer* skin, as promising leishmanicidal agents. *Pathogens and Disease* 78: ftaa053.

Meneguetti, B. T., L. D. S. Machado, K. G. N. Oshiro, M. L. Nogueira, C. M. E. Carvalho, and O. L. Franco. 2016. Antimicrobial peptides from fruits and their potential use as biotechnological tools-a review and outlook. *Frontiers in Microbiology* 7: 2136.

Menéndez-Guerrero, P. A., and C. H. Graham. 2013. Evaluating multiple causes of amphibian declines of Ecuador using geographical quantitative analyses. *Ecography* 36: 756–769.

Menéndez-Guerrero, P. A., D. M. Green, and T. J. Davies. 2019. Climate change and the future restructuring of Neotropical anuran biodiversity. *Ecography* 43: 222–235.

Menéndez-Guerrero, P. A., D. M. Green, and T. J. Davies. 2020. Climate change and the future restructuring of Neotropical anuran biodiversity. *Ecography* 43: 222–235.

Meneses, E. P., O. Villa-Hernández, L. Hernández-Orihuela, R. Castro-Franco, V. Pando, M. B. Aguilar, and C. V. F. Batista. 2011. Peptidomic analysis of the skin secretions of the frog *Pachymedusa dacnicolor*. *Amino Acids* 40: 113–122.

Merilä, J., and A. P. Hendry. 2014. Climate change, adaptation, and phenotypic plasticity: The problem and the evidence. *Evolutionary Applications* 7: 1–14.

Merino-Viteri, A. 2001. Análisis de posibles causas de las disminuciones de poblaciones de anfibios en los Andes del Ecuador. Tesis de Licenciatura. Pontificia Universidad Católica del Ecuador.

Merino-Viteri, A., L. A. Coloma, and A. Almendáriz. 2005. Los *Telmatobius* (Leptodactylidae) de los Andes del Ecuador y su declive poblacional. In *Estudios sobre las ranas andinas de los géneros Telmatobius y Batrachophrynus (Anura: Leptodactylidae)*, eds. E. O. Lavilla and I. De la Riva, 9–37. Valencia: Asociación Herpetológica Española.

Mestanza-Ramón, C., J. Monar-Nuñez, P. Guala-Alulema, Y. Montenegro-Zambrano, R. Herrera-Chávez, C. B. Milanes, C. Arguello-Guadalupe, P. Buñay-Guisñan, and M. Toledo-Villacís. 2023. A review to update the protected areas in Ecuador and an analysis of their main impacts and conservation strategies. *Environments* 10.

Middleton, E. M., J. R. Herman, E. A. Celarier, J. W. Wilkinson, C. Carey, and R. J. Rusin. 2001. Evaluating ultraviolet radiation exposure with satellite data at sites of amphibian declines in Central and South America. *Conservation Biology* 15: 914–929.

Miller, D., M. Gray, and A. Storfer. 2011. Ecopathology of ranaviruses infecting amphibians. *Viruses* 3: 2351–2373.

Ministerio del Ambiente del Ecuador. 2013. *Metodología para la representación cartográfica de los ecosistemas del Ecuador continental*. Quito: Subsecretaría de Patrimonio Natural.

Ministerio del Ambiente del Ecuador. 2014. *Demarcaciones Hidrográficas 2014*. http://ide.ambiente.gob.ec:8080/mapainteractivo/

Miró, A., D. O'Brien, J. Tomàs, T. Buchaca, I. Sabás, V. Osorio, F. Lucati, Q. Pou-Rovira, and M. Ventura. 2020. Rapid amphibian community recovery following removal of non-native fish from high mountain lakes. *Biological Conservation* 251: 108783.

Mittan-Moreau, C. S., C. Kelehear, L. F. Toledo, J. Bacon, J. M. Guayasamin, A. Snyder, and K. R. Zamudio. 2022. Cryptic lineages and standing genetic variation across independent cane toad introductions. *Molecular Ecology* 31: 6440–6456.

Mohammad, A., A.-T. Alaa, A. T. Yara, and A.-H. Mohammad. 2022. SAMA peptide, a rationally designed antimicrobial peptide. *Journal of Applied Pharmaceutical Science* 12: 182–189.

Mor, A., and P. Nicolas. 1994. The NH2-terminal alpha-helical domain 1–18 of dermaseptin is responsible for antimicrobial activity. *The Journal of Biological Chemistry* 269: 1934–1939.

Morán-Marcillo, G., V. S. Hinojosa, N. E. de Los Monteros-Silva, A. Blasco-Zúñiga, M. Rivera, R. E. Naranjo, J. R. Almeida, L. Wang, M. Zhou, T. Chen, C. Shaw, and C. Proaño-Bolaños. 2022. Picturins and pictuseptins, two novel antimicrobial peptide families from the skin secretions of the Chachi treefrog, *Boana picturata*. *Journal of Proteomics* 264: 104633.

Moravec, J., E. Lehr, and K. Kodejš. 2020. A new species of *Pristimantis* (Amphibia, Anura, Strabomantidae) from the Pui Pui Protected Forest (central Peru), with comments on *Pristimantis albertus* Duellman & Hedges, 2007. *ZooKeys* 994: 125–148.

Mori, E., M. Menchetti, R. Zozzoli, and P. Milanesi. 2019. The importance of taxonomy in species distribution models at a global scale: The case of an overlooked alien squirrel facing taxonomic revision. *Journal of Zoology* 307: 43–52.

Moritz, C. 1994. Defining "Evolutionarily Significant Units" for conservation. *Trends in Ecology & Evolution* 9: 373–375.

Moritz, H. 2000. Geodetic reference system 1980. *Journal of Geodesy* 74: 128–133.

Moskowitz, N. A., A. Alvarez-Buylla, C. R. Morrison, A. Chamba, J. Rentería, E. E. Tapia, L. A. Coloma, D. A. Donoso, and L. A. O'Connell. 2022. Poison frog diet and chemical defense are influenced by availability and selectivity for ants. *bioRxiv* 6.

Myers, C. W., and J. W. Daly. 1983. Dart-poison frogs. *Scientific American* 248: 120–133.

Naranjo, R. E., E. Naydenova, C. Proaño-Bolaños, K. Vizuete, A. Debut, M. T. Arias, and L. A. Coloma. 2022. Development of assisted reproductive technologies for the conservation of *Atelopus* sp. (*spumarius* complex). *Cryobiology* 105: 20–31.

Narváez-Narváez, D. A., A. Cabrera-Andrade, A. Merino-Viteri, C. Paz-y-Miño, G. Burgos, and A. Genoy-Puerto. 2021. Infection dynamics of *Batrachochytrium dendrobatidis* in two frog species inhabiting Quito's metropolitan Guangüiltagua park, Ecuador. *Journal of Wildlife Diseases* 57: 749–760.

Neely, W. J., S. E. Greenspan, L. P. Ribeiro, T. Carvalho, R. A. Martins, D. Rodriguez, J. R. Rohr, C. F. B. Haddad, L. F. Toledo, and C. G. Becker. 2020. Synergistic effects of warming and disease linked to high mortality in cool-adapted terrestrial frogs. *Biological Conservation* 245: 108521.

NIH Record. 2004. NIDDK's Daly Delivers NIH Director's Lecture, Jan. 2. U. S. Departement of Health and Human Services. National Institutes of Health. LVI: 1, 4–5.

Nicolas, P., and C. El Amri. 2009. The dermaseptin superfamily: A gene-based combinatorial library of antimicrobial peptides. *Biochimica et Biophysica Acta* 1788: 1537–1550.

NOAA National Centers for Environmental Information. 2023. *Monthly global climate report for annual 2022*. https://www.ncei.noaa.gov/access/monitoring/monthly-report/global/202213 (accessed May 9, 2019).

Noonan, B. P., and P. Gaucher. 2005. Phylogeography and demography of Guianan harlequin toads (*Atelopus*): Diversification within a refuge. *Molecular Ecology* 14: 3017–3031.

Núñez-González, S., E. Bedoya, D. Simancas-Racines, and C. Gault. 2020. Spatial clusters and temporal trends of malignant melanoma mortality in Ecuador. *SAGE Open Medicine* 8: 2050312120918285.

O'Hanlon, S. J., A. Rieux, R. A. Farrer, G. M. Rosa, B. Waldman, A. Bataille, T. A. Kosch, K. A. Murray, B. Brankovics, and M. Fumagalli. 2018. Recent Asian origin of chytrid fungi causing global amphibian declines. *Science* 360: 621–627.

Ortega, J. A., J. Brito, and S. R. Ron. 2022. Six new species of *Pristimantis* (Anura: Strabomantidae) from Llanganates

National Park and Sangay National Park in Amazonian cloud forests of Ecuador. *PeerJ* 10: e13761.

Ortega-Andrade, H. M., M. Rodes Blanco, D. F. Cisneros-Heredia, N. Guerra Arévalo, K. G. López de Vargas-Machuca, y J. C. Sánchez-Nivicela, *et al*. 2021. Red List assessment of amphibian species of Ecuador: A multidimensional approach for their conservation. *PloS One* 16: e0251027.

Pabijan, M., G. Palomar, B. Antunes, W. Antoł, P. Zieliński, and W. Babik. 2020. Evolutionary principles guiding amphibian conservation. *Evolutionary Applications* 13: 857–878.

Pace, N. R. 2009. Mapping the tree of life: Progress and prospects. *Microbiology and Molecular Biology Reviews* 73: 565–576.

Padial, J. M., T. Grant, and D. R. Frost. 2014. Molecular systematics of terraranas (Anura: Brachycephaloidea) with an assessment of the effects of alignment and optimality criteria. *Zootaxa* 3825: 1–132.

Padial, J. M., A. Miralles, I. de la Riva, and M. Vences. 2010. The integrative future of taxonomy. *Frontiers in Zoology* 7: 16.

Páez, N. B., and S. R. Ron. 2019. Systematics of *Huicundomantis*, a new subgenus of *Pristimantis* (Anura, Strabomantidae) with extraordinary cryptic diversity and eleven new species. *ZooKeys* 868: 1–112.

Páez-Moscoso, D. J., and J. M. Guayasamin. 2012. Species limits in the Andean toad genus *Osornophryne* (Bufonidae). *Molecular Phylogenetics and Evolution* 65: 805–822.

Páez-Vacas, M., L. A. Coloma, and J. C. Santos. 2010. Systematics of the *Hyloxalus bocagei* complex (Anura: Dendrobatidae), description of two new cryptic species, and recognition of *H. maculosus*. *Zootaxa* 2711: 1–75.

Páez-Vacas, M. I., D. R. Trumbo, and W. C. Funk. 2022. Contrasting environmental drivers of genetic and phenotypic divergence in an Andean poison frog (*Epipedobates anthonyi*). *Heredity* 128: 33–44.

Paluh, D. J., W. A. Dillard, E. L. Stanley, G. J. Fraser, and D. C. Blackburn. 2021. Re-evaluating the morphological evidence for the re-evolution of lost mandibular teeth in frogs. *Evolution* 75: 3203–3213.

Paluh, D. J., E. L. Stanley, and D. C. Blackburn. 2019. First dietary record of *Gastrotheca guentheri* (Boulenger, 1882), the lone anuran with true mandibular teeth. *Herpetology Notes* 12: 699–700.

Paquin, P., N. Dupérré, J. C. Cokendolpher, K. White, and M. Hedin. 2008. The fundamental importance of taxonomy in conservation biology: The case of the eyeless *Cicurina bandida* (Araneae: Dictynidae) of central Texas, including new synonyms and the description of the male of the species. *Invertebrate Systematics* 22: 139–149.

Parmesan, C. 2006. Ecological and evolutionary responses to recent climate change. *Annual Review of Ecology, Evolution, and Systematics* 37: 637–669.

Parra, R., E. Cadena, and C. Flores. 2019. Maximum UV index records (2010–2014) in Quito (Ecuador) and its trend inferred from remote sensing data (1979–2018). *Atmosphere* 10: 787.

Paz, A., and A. J. Crawford. 2012. Molecular-based rapid inventories of sympatric diversity: A comparison of DNA barcode clustering methods applied to geography-based vs clade-based sampling of amphibians. *Journal of Biosciences* 37: 887–896.

Paz Cardona, A. J. 2022. La Amazonía ecuatoriana ha perdido más de 623 mil hectáreas en dos décadas. *Mongabay* (November 2). https://es.mongabay.com/2022/11/amazonia-ecuatoriana-ha-perdido-mas-de-623-mil-hectareas-en-dos-decadas/.

Peck, M. R., M. Desselas, S. Bonilla-Bedoya, G. Redín, and J. Durango-Cordero. 2024. The conflict between rights of nature and mining in Ecuador: Implications of the Los Cedros Cloud Forest case for biodiversity conservation. *People and Nature* 00: 1–20.

Pei, X., Z. Gong, Q. Wu, X. Chen, L. Wang, C. Ma, X. Xi, T. Chen, C. Shaw, and M. Zhou. 2021. Characterisation of a novel peptide, Brevinin-1H, from the skin secretion of *Amolops hainanensis* and rational design of several analogues. *Chemical Biology & Drug Design* 97: 273–282.

Peñafiel, N., D. M. Flores, J. Rivero De Aguilar, J. M. Guayasamin, and E. Bonaccorso. 2019. A cost-effective protocol for total DNA isolation from animal tissue. *Neotropical Biodiversity* 5: 69–74.

Peters, J. A. 1973. The frog genus *Atelopus* in Ecuador (Anura: Bufonidae). *Smithsonian Contributions to Zoology* 145: 1–49.

Peters, W. C. H. 1872. Über eine Sammlung von Batrachiern aus Neu Freiburg in Brasilien. Monatsberichte der Königlichen Preussische Akademie des Wissenschaften zu Berlin 1872: 680–684.

Pilliod, D. S., C. S. Goldberg, R. S. Arkle, and L. P. Waits. 2013a. Estimating occupancy and abundance of stream amphibians using environmental DNA from filtered water samples. *Canadian Journal of Fisheries and Aquatic Sciences* 70: 1123–1130.

Pilliod, D. S., C. S. Goldberg, R. S. Arkle, and L. P. Waits. 2013b. Factors influencing detection of eDNA from a stream-dwelling amphibian. *Molecular Ecology Resources* 14: 109–116.

Pintanel, P., M. Tejedo, A. Merino-Viteri, F. Almeida-Reinoso, S. Salinas-Ivanenko, A. C. López-Rosero, G. A. Llorente, and L. M. Gutiérrez-Pesquera. 2022. Elevational and local climate variability predicts thermal breadth of mountain tropical tadpoles. *Ecography* 2022: e05906.

Pintanel, P., M. Tejedo, S. R. Ron, G. A. Llorente, and A. Merino-Viteri. 2019. Elevational and microclimatic drivers of thermal tolerance in Andean *Pristimantis* frogs. *Journal of Biogeography* 46: 1664–1675.

Plante, F., P. Bourgault, Y. Dubois, and L. Bernatchez. 2021. Environmental DNA as a detection and quantitative tool for stream-dwelling salamanders: A comparison with the traditional active search method. *Environmental DNA* 3: 1128–1141.

Polato, N., B. Gill, A. Shah, M. Gray, K. Casner, A. Barthelet, P. Messer, M. Simmons, *et al*. 2018. Narrow thermal tolerance and low dispersal drive higher speciation in tropical mountains. *Proceedings of the National Academy of Sciences* 115: 12471–12476.

Pomerantz, A., N. Peñafiel, A. Arteaga, L. Bustamante, F. Pichardo, L. A. Coloma, C. L. Barrio-Amorós, D. Salazar-Valenzuela, and S. Prost. 2018. Real-time DNA barcoding in a rainforest using nanopore sequencing: Opportunities for rapid biodiversity assessments and local capacity building. *GigaScience* 7: 1–14.

Potter, A., J. Corwin, J. Lang, M. Piasecki, R. Lenox, and P. A. Newhouse. 1999. Acute effects of the selective cholinergic channel activator (nicotinic agonist) ABT-418 in Alzheimer's disease. *Psychopharmacology* 142: 334–342.

Pounds, J. A., M. R. Bustamante, L. A. Coloma, J. A. Consuegra, M. P. Fogden, P. N. Foster, E. La Marca, K. L. Masters, *et al*. 2006. Widespread amphibian extinctions from epidemic disease driven by global warming. *Nature* 439: 161–167.

Pounds, J. A., M. R. Bustamante, L. A. Coloma, J. A. Consuegra, M. P. L. Fogden, P. N. Foster, E. La Marca, K. L. Masters, *et al*.

2007. Global warming and amphibian losses; The proximate cause of frog declines? (Reply). *Nature* 447: E5–6.

Pounds, J. A., and L. A. Coloma. 2008. Beware the lone killer: Why are harlequin frogs disappearing across the American tropics? A resifiting of the evidence backs up the conclusion that global warming is a key conspirator in the losses. *Nature Reports Climate Change* 2: 57–59.

Pounds, J. A., and M. L. Crump. 1994. Amphibian declines and climate disturbance: The case of the golden toad and the harlequin frog. *Conservation Biology* 8: 72–85.

Pounds, J. A., M. P. L. Fogden, and J. H. Campbell. 1999. Biological response to climate change on a tropical mountain. *Nature* 398: 611–615.

Pourrut, P., G. Gómez, A. Bermeo, and A. Segovia. 1995. Factores condicionantes de los regímenes climáticos e hidrológicos. In *El agua en el Ecuador: clima, precipitaciones, escorrentía*, ed. P. Pourrut, 7–12. Quito: Corporación Editora Nacional/Colegio de Geógrafos del Ecuador/ORSTOM.

Prathapan, K. D., R. Pethiyagoda, K. S. Bawa, P. H. Raven, P. D. Rajan, and 172 co-signatories from 35 countries. 2018. When the cure kills—CBD limits biodiversity research. *Science* 360: 1405–1406.

Proaño, B., and O. D. Pérez. 2017. In vitro fertilizations with cryopreserved sperm of *Rhinella marina* (Anura: Bufonidae) in Ecuador. *Amphibian and Reptile Conservation* 11: 1–6.

Proaño-Bolaños, C. 2004. Ocurrencia de compuestos antimicrobiales secretados por la piel de algunos anfibios de las familias: Hylidae, Dendrobatidae, Ranidae, y purificación parcial de 5 péptidos antimicrobianos de *Agalychnis spurrelli* (Hylidae: Anura). Tesis de Licenciatura. Pontificia Universidad Católica del Ecuador.

Proaño-Bolaños, C., A. Blasco-Zúñiga, J. R. Almeida, L. Wang, M. A. Llumiquinga, M. Rivera, M. Zhou, T. Chen, and C. Shaw. 2019. Unravelling the skin secretion peptides of the Gliding leaf frog, *Agalychnis spurrelli* (Hylidae). *Biomolecules* 9: 667.

Proaño-Bolaños, C., R. Li, M. Zhou, L. Wang, X. Xi, E. E. Tapia, L. A. Coloma, T. Chen, and C. Shaw. 2017. Novel Kazal-type proteinase inhibitors from the skin secretion of the Splendid leaf frog, *Cruziohyla calcarifer*. *EuPA Open Proteomics* 15: 1–13.

Proaño-Bolaños, C., M. Zhou, L. Wang, L. A. Coloma, T. Chen, and C. Shaw. 2016. Peptidomic approach identifies cruzioseptins, a new family of potent antimicrobial peptides in the splendid leaf frog, *Cruziohyla calcarifer*. *Journal of Proteomics* 146: 1–13.

Proaño-Bolaños, C., G. Morán-Marcillo, N. Espinosa de los Monteros-Silva, S. Bermúdez-Puga, M. A. Salazar, A. Blasco-Zúñiga, ... and J. A. Costales. 2024. Bioactivity of synthetic peptides from Ecuadorian frog skin secretions against *Leishmania mexicana, Plasmodium falciparum, and Trypanosoma cruzi*. *Microbiology Spectrum*, 12: e03339-23.

Qin, H., H. Fang, X. Chen, L. Wang, C. Ma, X. Xi, T. Chen, C. Shaw, and M. Zhou. 2021. Exploration of the structure-function relationships of a novel frog skin secretion-derived bioactive peptide, t-DPH1, through use of rational design, cationicity enhancement and in vitro studies. *Antibiotics* 10: 1529.

Quilumbaquin, W., A. Carrera-Gonzalez, C. Van der Heyden, and H. M. Ortega-Andrade. 2023. Environmental DNA and visual encounter surveys for amphibian biomonitoring in aquatic environments of the Ecuadorian Amazon. *PeerJ* 11: e15455–e15455.

Rachowicz, L. J., J. M. Hero, R. A. Alford, J. W. Taylor, J. A. T. Morgan, V. T. Vredenburg, J. P. Collins, and C. J. Briggs. 2005. The novel and endemic pathogen hypotheses: Competing explanations for the origin of emerging infectious diseases of wildlife. *Conservation Biology* 19: 1441–1448.

Rahbek, C. 1997. The relationship among area, elevation, and regional species richness in Neotropical birds. *The American Naturalist* 149: 875–902.

Ralls, K., P. Sunnucks, R. C. Lacy, and R. Frankham. 2020. Genetic rescue: A critique of the evidence supports maximizing genetic diversity rather than minimizing the introduction of putatively harmful genetic variation. *Biological Conservation* 251: 108784.

Rámirez-Jaramillo, S. R., M. B. Pérez, J. P. Reyes-Puig, C. Reyes-Puig, and L. M. H. Yánez-Muñoz. 2014. Tres nuevas especies de ranas terrestres *Pristimantis* (Anura: Craugastoridae) de la cuenca alta del Río Pastaza, Ecuador. *ACI Avances en Ciencias e Ingenierías* 6: 5–25.

Ramírez Jiménez, M. G. 2012. Detección de fracciones peptídicas con actividad antimicrobiana extraídas de secreciones cutáneas de *Agalychnis spurrelli* (Anura: Hylidae). Tesis Licenciatura. Pontificia Universidad Católica del Ecuador.

Re: wild. 2022. In a vote of defiance, newly discovered rocket frog in Ecuador is named "resistance rocket frog" during Amphibian Week. *Re: wild* (May 4). https://www.rewild.org/press/in-a-vote-of-defiance-newly-discovered-rocket-frog-in-ecuador-is-named.

Rebollar, E. A., E. Martínez-Ugalde, and A. H. Orta. 2020. The amphibian skin microbiome and its protective role against chytridiomycosis. *Herpetologica* 76: 167.

Reeder, N. M. M., A. P. Pessier, and V. T. Vredenburg. 2012. A reservoir species for the emerging Amphibian pathogen *Batrachochytrium dendrobatidis* thrives in a landscape decimated by disease. *PloS One* 7: e33567–33567.

Reyes-Puig, C., W. D. B. Ake, R. Kotharambath, J. W. Streicher, C. Koch, D. F. Cisneros-Heredia, M. H. Yánez-Muñoz, and S. Ron. 2020a. Two extremely rare new species of fossorial salamanders of the genus *Oedipina* (Plethodontidae) from northwestern Ecuador. *PeerJ* 8: e9934.

Reyes-Puig, C., and E. Mancero. 2022. Beyond the species name: An analysis of publication trends and biases in taxonomic descriptions of rainfrogs (Amphibia, Strabomantidae, *Pristimantis*). *ZooKeys* 1134: 73.

Reyes-Puig, J. P., C. Reyes-Puig, D. Franco-Mena, L. Jost, and M. H. Yánez-Muñoz. 2022. Strong differentiation between amphibian communities on two adjacent mountains in the Upper Rio Pastaza watershed of Ecuador, with descriptions of two new species of terrestrial frogs. *ZooKeys* 2022: 35–87.

Reyes-Puig, J., C. Reyes-Puig, M. B. Pérez L., and M. H. Yánez-Muñoz. 2015. Dos nuevas especies de ranas *Pristimantis* (Craugastoridae) de la cordillera de los Sacha Llanganatis, vertiente oriental de los Andes de Ecuador. *ACI Avances en Ciencias e Ingenierías* 7: 61–74.

Reyes-Puig, J. P., M. H. Yánez-Muñoz, D. F. Cisneros-Heredia, and S. Ramírez. 2011. "2010". Una nueva especie de rana *Pristimantis* (Terrarana: Strabomantidae) de los bosques nublados de la cuenca alta del río Pastaza, Ecuador. *Avances en Ciencias e Ingenierías* 2: 78–82.

Reyes-Puig, C., M. H. Yánez-Muñoz, J. A. Ortega, and S. R. Ron. 2020b. Relaciones filogenéticas del subgénero *Hypodictyon* (Anura: Strabomantidae: *Pristimantis*) con la descripción de tres especies nuevas de la región del Chocó. *Revista Mexicana de Biodiversidad* 91: 18.

Riascos-Flores, L., A. Carrera, L. Naranjo, J. Yanez, P. Goethals, J. Bonilla, J. Celi, C. Van der Heyden, and M. Ortega. 2021. DNA-based biomonitoring in the tropics: Detection and control of *Batrachochytrium dendrobatidis* in Ecuadorian ecosystem. *ARPHA Conference Abstracts* 4: e65304.

Richter-Voix, A., M. Quintela, G. Segelbacher, and A. Laurila. 2011. Genetic analysis of differentiation among breeding ponds reveals a candidate gene for local adaptation in *Rana arvalis*. *Molecular Ecology* 20: 1582–1600.

Rivas-Martínez, S. 2004. *Global bioclimatics. Clasificación bioclimática de la Tierra*. Versión 27–08–04. Phytosociological Research Center. https://www.biourb.net/app/download/5782396363/Clasif.+Bioclim.+de+la+Tierra.pdf.

Rizzi, L., C. Dallanoce, C. Matera, P. Magrone, L. Pucci, C. Gotti, F. Clementi, and M. De Amici. 2008. Epiboxidine and novel-related analogues: A convenient synthetic approach and estimation of their affinity at neuronal nicotinic acetylcholine receptor subtypes. *Bioorganic & Medicinal Chemistry Letters* 18: 4651–4654.

Rohr, J. R., T. R. Raffel, J. M. Romansic, H. McCallum, and P. J. Hudson. 2008. Evaluating the links between climate, disease spread, and amphibian declines. *Proceedings of the National Academy of Sciences of the United States of America* 105: 17436–17441.

Roland, A. B., J. C. Santos, B. C. Carriker, S. N. Caty, E. E. Tapia, L. A. Coloma, and L. A. O'Connell. 2017a. Radiation of the polymorphic Little Devil poison frog (*Oophaga sylvatica*) in Ecuador. *Ecology and Evolution* 7: 9750–9762.

Rollenbeck, R., and J. Bendix. 2011. Rainfall distribution in the Andes of southern Ecuador derived from blending weather radar data and meteorological field observations. *Atmospheric Research* 99: 277–289.

Ron, S. R. 2005. Predicting the distribution of the amphibian pathogen *Batrachochytrium dendrobatidis* in the New World. *Biotropica* 37: 209–221.

Ron, S. R., M. A. Caminer, A. Varela-Jaramillo, and D. Almeida-Reinoso. 2018a. A new treefrog from Cordillera del Cóndor with comments on the biogeographic affinity between Cordillera del Cóndor and the Guianan Tepuis. *ZooKeys* 124: 97–124.

Ron, S. R., J. Carrión, M. A. Caminer, Y. Sagredo, M. J. Navarrete, J. A. Ortega, A. Varela-Jaramillo, G. A. Maldonado-Castro, and C. Terán. 2020. Three new species of frogs of the genus *Pristimantis* (Anura, Strabomantidae) with a redefinition of the *P. lacrimosus* species group. *ZooKeys* 2020: 121–155.

Ron, S. R., L. A. Coloma, and D. C. Cannatella. 2005. A new, cryptic species of *Physalaemus* (Anura: Leptodactylidae) from western Ecuador with comments on the call structure of the *P. pustulosus* species group. *Herpetologica* 61: 178–198.

Ron, S. R., L. A. Coloma, S. Lötters, W. Duellman, M. R. Bustamante, W. Bolívar, and E. La Marca. 2004. *Atelopus ignescens*. The IUCN Red List of Threatened Species 2004: e.T54518A11157432. https://dx.doi.org/10.2305/IUCN.UK.2004.RLTS.T54518A11157432.en

Ron, S. R., W. E. Duellman, M. A. Caminer, and D. Pazmiño. 2018b. Advertisement calls and DNA sequences reveal a new species of *Scinax* (Anura: Hylidae) on the Pacific lowlands of Ecuador. *PloS One* 13: 1–26.

Ron, S. R., W. E. Duellman, L. A. Coloma, and M. R. Bustamante. 2003b. Population decline of the Jambato toad *Atelopus ignescens* (Anura: Bufonidae) in the Andes of Ecuador. *Journal of Herpetology* 37: 116–126.

Ron, S. R., J. M. Guayasamin, L. A. Coloma, and P. A. Menéndez-Guerrero. 2008. *Lista roja de los anfibios de Ecuador*. Version 1.0 Museo de Zoología, Pontificia Universidad Católica del Ecuador (May 8). https://web.archive.org/web/20090917184123/http://www.puce.edu.ec/zoologia/sron/roja%0D405/index.htm. (accessed April 12, 2017).

Ron, S. R., J. M. Guayasamin, and P. Menéndez-Guerrero. 2011. Biodiversity and conservation status of Ecuadorian amphibians. In *Amphibian biology*, eds. H. Heatwole, C. L. Barrio-Amorós, and W. H. Wilkinson, 129–170. Baulkham Hills: Surrey Beatty & Soons PTY Limited.

Ron, S. R., and A. Merino-Viteri. 2000. Amphibian declines in Ecuador: Overview and first report of chytridiomycosis from South America. *Froglog* 42: 2–3.

Roy, B. A., M. Zorrilla, L. Endara, D. C. Thomas, R. Vandegrift, J. M. Rubenstein, T. Policha, B. Ríos-Touma, and M. Read. 2018. New mining concessions could severely decrease biodiversity and ecosystem services in Ecuador. *Tropical Conservation Science* 11: 1940082918780427.

Rueda-Almonacid, J. V., J. D. Lynch, and A. Amézquita. Eds. 2004. *Libro rojo de los anfibios de Colombia*. Bogotá: Conservación Internacional Colombia, Instituto de Ciencias Naturales–Universidad Nacional de Colombia, Ministerio del Medio Ambiente.

Ruggeri, J., M. R. Pontes, L. P. Ribeiro, K. L. Gendreau, R. L. M. Sousa, and L. F. Toledo. 2023. Predominant prevalence of ranavirus in southern Brazil, a region with widespread occurrence of the amphibian chytrid. *Animal Conservation* 3: 338–349.

Ruggiero, A., and B. A. Hawkins. 2008. Why do mountains support so many species of birds?. *Ecography* 31: 306–315.

Ruiz-Carranza, P. M., and J. D. Lynch. 1991. Ranas Centrolenidae de Colombia I. Propuesta de nueva clasificación genérica. *Lozania* 57: 1–30.

Ryder, O. A. 1986. Species conservation and systematics: The dilemma of subspecies. *Trends in Ecology and Evolution* 1: 9–10.

Ryman, N., and L. Laikre. 1991. Effects of supportive breeding on the genetically effective population size. *Conservation Biology* 5: 325–329.

Salazar, J., F. Siavichay, and G. Maldonado. 2022. *Haciendo visible lo invisible: Fauna urbana de la Universidad Católica de Cuenca*. Cuenca: Editorial Don Bosco.

Salehi, B., S. Sestito, S. Rapposelli, G. Peron, D. Calina, M. Sharifi-Rad, F. Sharopov, N. Martins, and J. Sharifi-Rad. 2019. Epibatidine: A promising natural alkaloid in health. *Biomolecules* 9: 6.

Salla, R. F., M. J. Costa, F. C. Abdalla, C. R. Oliveira, E. Tsukada, G. A. N. S. Boeing, J. Prado, *et al.* 2024. Estrogen contamination increases vulnerability of amphibians to the deadly chytrid fungus. *Science of the Total Environment* 917: 170337.

Santos, J. C., M. Baquero, C. L. Barrio Amorós, L. A. Coloma, L. K. Erdtmann, A. P. Lima, and D. C. Cannatella. 2014. Aposematism increases acoustic diversification and speciation in poison frogs. *Proceedings of the Royal Society B: Biological Sciences* 281: 20141761.

Santos, J. C., L. A. Coloma, and D. C. Cannatella. 2003. Multiple, recurring origins of aposematism and diet specialization in poison frogs. *Proceedings of the National Academy of Sciences* 100: 12792–12797.

Santos, J. C., L. A. Coloma, K. Summers, J. P. Caldwell, R. Ree, and D. C. Cannatella. 2009. Amazonian amphibian diversity is primarily derived from Late Miocene Andean lineages. *PLOS Biology* 7: 1–14.

Santos, J. C., R. D. Tarvin, and L. A. O'Connell. 2016. A review of chemical defense in poison frogs (Dendrobatidae): Ecology, pharmacokinetics, and autoresistance. In *Chemical signals in vertebrates 13*, eds. B. A. Schulte, T. E. Goodwin, and M. H. Ferkin, 305–337. New York: Springer Science/Business Media.

Saremi, N. F., M. A. Supple, A. Byrne, J. A. Cahill, L. L. Coutinho, L. Dalén, H. V. Figueiró, *et al*. 2019. Puma genomes from North and South America provide insights into the genomic consequences of inbreeding. *Nature Communications* 10: 4769.

Sarkar, T., M. Chetia, and S. Chatterjee. 2021. Antimicrobial peptides and proteins: From nature's reservoir to the laboratory and beyond. *Frontiers in Chemistry* 9: 691532.

Sastre, P., and J. M. Lobo. 2009. Taxonomist survey biases and the unveiling of biodiversity patterns. *Biological Conservation* 142: 462–467.

Savage, A. E., and K. R. Zamudio. 2011. MHC genotypes associate with resistance to a frog-killing fungus. *Proceedings of the National Academy of Sciences of the United States of America* 108: 16705–16710.

Savoia, D., R. Guerrini, E. Marzola, and S. Salvadori. 2008. Synthesis and antimicrobial activity of dermaseptin S1 analogues. *Bioorganic & Medicinal Chemistry* 16: 8205–8209.

Scheele, B. C., F. Pasmans, L. F. Skerratt, L. Berger, A. Martel, W. Beukema, A. A. Acevedo, P. A. Burrowes, T. Carvalho, A. Catenazzi, I. De Riva, M. C. Fisher, and S. V. Flechas. 2019. Amphibian fungal panzootic causes catastrophic and ongoing loss of biodiversity. *Science* 1463: 1459–1463.

Scorciapino, M. A., P. Carta, J. Pantic, M. L. Lukic, A. Lukic, V. Musale, Y. H. A. Abdel-Wahab, and J. M. Conlon. 2019. Conformational analysis and in vitro immunomodulatory and insulinotropic properties of the frog skin host-defense peptide rhinophrynin-27 and selected analogs. *Biochimie* 167: 198–206.

Seimon, T. A., A. Seimon, P. Daszak, S. R. P. Halloy, L. M. Schloegel, C. A. Aguilar, P. Sowell, A. D. Hyatt, B. Konecky, and J. E. Simmons. 2007. Upward range extension of Andean anurans and chytridiomycosis to extreme elevations in response to tropical deglatiation. *Global Change Biology* 13: 288–299.

Shaffer, H. B., M. Gidiş, E. McCartney-Melstad, K. M. Neal, H. M. Oyamaguchi, M. Tellez, and E. M. Toffelmier. 2015. Conservation genetics and genomics of amphibians and reptiles. *Annual Review of Animal Biosciences* 3: 113–138.

Shanmugaraj, B., C. J. I. Bulaon, A. Malla, and W. Phoolcharoen. 2021. Biotechnological insights on the expression and production of antimicrobial peptides in plants. *Molecules* 26: 4032.

Silla, A. J., and P. G. Byrne. 2019. The role of reproductive technologies in amphibian conservation breeding programs. *Annual Review of Animal Biosciences* 7: 499–519.

Silla, A. J., N. E. Calatayud, and V. L. Trudeau. 2021. Amphibian reproductive technologies: Approaches and welfare considerations. *Conservation Physiology* 9: coab011.

Silla, A., and A. Kouba. 2022. Integrating reproductive technologies into the conservation toolbox for the recovery of amphibian species. In *Reproductive technologies and biobanking for the conservation of amphibians*, eds. A. Silla, A. Kouba, and H. Heatwole. Clayton: CSIRO Publishing.

Silva, O. N., M. D. T. Torres, J. Cao, E. S. F. Alves, L. V. Rodrigues, J. M. Resende, L. M. Lião, W. F. Porto, *et al*. 2020. Repurposing a peptide toxin from wasp venom into antiinfectives with dual antimicrobial and immunomodulatory properties. *Proceedings of the National Academy of Sciences of the United States of America* 117: 26936–26945.

Silverstone, P. A. 1975. A revision of the poison-arrow frogs of the genus *Dendrobates* Wagler. Science Bulletin. Natural History Museum of Los Angeles County 21: 1–55.

Silverstone, P. A. 1976. A revision of the poison-arrow frogs of the genus *Phyllobates* Bibron in Sagra (Family Dendrobatidae). Science Bulletin. Natural History Museum of Los Angeles County 27: 1–53.

Slatkin, M. 1987. Gene flow and the geographic structure of natural populations. *Science* 236: 787–792.

Song, G., M. Zhou, W. Chen, T. Chen, B. Walker, and C. Shaw. 2008. HV-BBI—a novel amphibian skin Bowman-Birk-like trypsin inhibitor. *Biochemical and Biophysical Research Communications* 372: 191–196.

Sonnevend, A., F. C. Knoop, M. Patel, T. Pál, A. M. Soto, and J. M. Conlon. 2004. Antimicrobial properties of the frog skin peptide, ranatuerin-1 and its [Lys-8]-substituted analog. *Peptides* 25: 29–36.

Spande, T. F., H. M. Garraffo, M. W. Edwards, H. J. C. Yeh, L. Pannell, and J. W. Daly. 1992. Epibatidine: A novel (chloropyridyl) azabicycloheptane with potent analgesic activity from an Ecuadoran poison frog. *Journal of the American Chemical Society* 114: 3475–3478.

Spielman, D., B. W. Brook, and R. Frankham. 2004. Most species are not driven to extinction before genetic factors impact them. *Proceedings of the National Academy of Sciences of the United States of America* 101: 15261–15264.

Stebbins, R. C., and N. W. Cohen. 1995. *A natural history of amphibians*. New Yersey: Princeton University Press.

Steigerwald, E. 2021. Impacts of climate change on amphibians. *AmphibiaWeb* (October 11). https://amphibiaweb.org/declines/climatechange.html.

Storfer, A., A. Patton, and A. K. Fraik. 2018. Navigating the interface between landscape genetics and landscape genomics. *Frontiers in Genetics* 9: 68.

Strauss, E. 1998. New nonopioid painkiller shows promise in animal tests. *Science* 279: 32–33.

Stuart, S. N. 2012. Responding to the amphibian crisis: Too little, too late. *Alytes* 29: 9–12.

Stuart, S. N., J. S. Chanson, N. A. Cox, B. E. Young, A. S. L. Rodrigues, D. L. Fischman, and R. W. Waller. 2004. Status and trends of amphibian declines and extinctions worldwide. *Science* 306: 1783–1786.

Stuart, S. N., M. Hoffman, J. S. Chanson, N. A. Cox, R. J. Berridge, P. Ramani, and B. E. Young. 2008. *Threatened amphibians of the world*. Barcelona: Lynx editions; Gland: IUCN, Gland; Arlington: Conservation International.

Sudo, R. T., K. Hayashida, A. N. Santos, M. Kawatani, C. E. S. Monteiro, R. D. Moreira, M. M. Trachez, G. C. Montes, and G. Zapata-Sudo. 2018. Novel agonist of $\alpha4\beta2^*$ neuronal nicotinic receptor with antinociceptive efficacy in rodent models of acute and chronic pain. *Journal of Pain Research* 2018: 2453–2462.

Székely, P., J. S. Eguiguren, L. Ordóñez-Delgado, D. Armijos-Ojeda, and D. Székely. 2020. Fifty years after: A taxonomic revision of the amphibian species from the Ecuadorian biodiversity hotspot Abra de Zamora, with description of two new *Pristimantis* species. *PLoS One* 15: 1–59.

Székely, P., J. S. Eguiguren, D. Székely, L. Ordóñez-Delgado, D. Armijos-Ojeda, M. L. Riofrío-Guamán, and D. Cogălniceanu. 2018. A new minute *Pristimantis* (Amphibia: Anura: Strabomantidae) from the Andes of southern Ecuador. *PLoS One* 13: e0202332.

Tamayo, F., B. Torres, R. Fischer, A. Lajones, R. Cervantes, C. Corozo, R. Ferrer Velasco, and S. Günter. 2020. Caracterización de paisajes forestales en el Noroccidente Ecuatoriano:

Deforestación y aspectos socioculturales. In *Deforestación en paisajes forestales tropicales del Ecuador: Bases científicas para perspectivas políticas*, eds. R. Fischer, J. C. Vargas, and S. Günter, 71–92. Quito: INABIO.

Tamisiea, J. 2021. This mysterious frog re-evolved a full set of teeth. *National Geographic* (November 10). https://www.nationalgeographic.com/animals/article/marsupial-frog-teeth-evolutionary-mystery.

Tan, Y., X. Chen, C. Ma, X. Xi, L. Wang, M. Zhou, J. F. Burrows, H. F. Kwok, and T. Chen. 2018. Biological activities of cationicity-enhanced and hydrophobicity-optimized analogues of an antimicrobial peptide, dermaseptin-PS3, from the skin secretion of *Phyllomedusa sauvagii*. *Toxins* 10: 320.

Tapia, E. E., L. A. Coloma, G. Pazmiño-Otamendi, and N. Peñafiel. 2017. Rediscovery of the nearly extinct longnose harlequin frog *Atelopus longirostris* (Bufonidae) in Junín, Imbabura, Ecuador. *Neotropical Biodiversity* 3: 157–167.

Tapley, B., K. S. Bradfield, C. Michaels, and M. Bungard. 2015. Amphibians and conservation breeding programmes: Do all threatened amphibians belong on the ark? *Biodiversity and Conservation* 24: 2625–2646.

Tapley, B., K. Johnson, C. Michaels, K. Bradfield, D. Barber, L. Randall, K. Vasudevan, C. Paszkowski, R. A. Griffiths, N. Calatayud, E. Griffith, and F. Kacoliris. 2022. Conservation breeding. In *The Amphibian Conservation Action Plan (ACAP): A status review and roadmap for global amphibian conservation*, eds. IUCN SSC Amphibian Specialist Group. https://ecoevorxiv.org/brfas/.

Tarvin, R. D., C. M. Borghese, W. Sachs, J. C. Santos, Y. Lu, L. A. O'Connell, D. C. Cannatella, R. A. Harris, and H. H. Zakon. 2017. Interacting amino acid replacements allow poison frogs to evolve epibatidine resistance. *Science* 357: 1261–1266.

Tarvin, R. D., P. Peña, and S. R. Ron. 2014. Changes in population size and survival in *Atelopus spumarius* (Anura: Bufonidae) are not correlated with chytrid prevalence. *Journal of Herpetology* 48: 291–297.

Tarvin, R. D., J. C. Santos, L. A. O'Connell, H. H. Zakon, and D. C. Cannatella. 2016. Convergent substitutions in a sodium channel suggest multiple origins of toxin resistance in poison frogs. *Molecular Biology and Evolution* 33: 1068–1081.

Taylor, E. H. 1968. *The caecilians of the world: A taxonomic review*. Lawrence: University of Kansas Press.

Tewksbury, J. J., R. B. Huey, and C. A. Deutsch. 2008. Putting the heat on tropical animals. *Science* 320: 1296–1297.

Ting, D. S. J., R. W. Beuerman, H. S. Dua, R. Lakshminarayanan, and I. Mohammed. 2020. Strategies in translating the therapeutic potentials of host defense peptides. *Frontiers in Immunology* 11: 983.

Torres, B., R. Fischer, J. C. Vargas, and S. Günter. 2020. *Deforestación en paisajes forestales tropicales del Ecuador: Bases científicas para perspectivas políticas*. Quito: Universidad Estatal Amazónica—Instituto Johann Heinrich von Thünen. Puyo, Ecuador/Serie de publicaciones misceláneas del INABIO.

Torres, D. 2009. Actividad antifúngica de los péptidos de la piel de *Agalychnis spurrelli* sobre hongos fitopatógenos. Tesis de Licenciatura. Pontificia Universidad Católica del Ecuador.

Torres, M. D. T., J. Cao, O. L. Franco, T. K. Lu y C. de la Fuente-Nunez. 2021. Synthetic biology and computer-based frameworks for antimicrobial peptide discovery. *ACS Nano* 15: 2143–2164.

Torres-Florez, J. P., W. E. Johnson, M. F. Nery, E. Eizirik, M. A. Oliveira-Miranda, and P. M. Galetti. 2017. The coming of age of conservation genetics in Latin America: What has been achieved and what needs to be done. *Conservation Genetics* 19: 1–15.

Trueb, L. 1971. Phylogenetic relationships of certain Neotropical toads with the description of a new genus (Anura: Bufonidae). Contributions in Science. Natural History Museum of Los Angeles County 216: 1–40.

Trueb, L. 1974. Systematic relationships of neotropical horned frogs, genus *Hemiphractus* (Anura: Hylidae). *Occasional papers of the Museum of Natural History, the University of Kansas* 29: 1–60.

Trueb, L., and W. E. Duellman. 1971. A synopsis of neotropical hylid frogs, genus *Osteocephalus*. *Occasional papers of the Museum of Natural History, the University of Kansas* 1: 1–47.

Ulloa-Ulloa, C., and P. M. Jørgensen. 2004. *Arboles y arbustos de los Andes del Ecuador*. Quito: Abya-Yala.

Urban, M. C. 2015. Accelerating extinction risk from climate change. *Science* 348: 571–573.

Urgiles, V. L., E. R. Ramírez, C. I. Villalta, D. C. Siddons, and A. E. Savage. 2021. Three pathogens impact terrestrial frogs from a high-elevation tropical hotspot. *EcoHealth* 18: 451–464.

Valdivieso-Rivera, F., S. Bermúdez-Puga, C. Proaño-Bolaños, and J. R. Almeida. 2022. Deciphering the limitations and antibacterial mechanism of cruzioseptins. *International Journal of Peptide Research and Therapeutics* 28: 73.

Valentini, A., P. Taberlet, C. Miaud, R. Civade, J. Herder, P. F. Thomsen, E. Bellemain, *et al*. 2016. Next-generation monitoring of aquatic biodiversity using environmental DNA metabarcoding. *Molecular Ecology* 25: 929–942.

Valencia, J., E. Toral, M. Morales, R. Betancourt, and A. Barahona. 2008. *Guía de campo de anfibios del Ecuador*. Quito Fundación Herpetológica Gustavo Orcés, Simbioe.

Valle, C. 1984. Aspectos ecológicos del *Atelopus ignescens*. Pontificia Universidad Católica del Ecuador. Proyecto de Ecología.

Vallejo Rodríguez, G. M. 2016. Identificación de secreciones peptídicas de 15 especies de anuros amazónicos con actividad antimicrobiana y anticancerígena, y su aporte como servicio ecosistémico de tipo provisión. Tesis de Licenciatura. Pontificia Universidad Católica del Ecuador.

Van der Heyden, C., L. Riascos, A. Carrera-Gonzalez, K. E. Apunte Ramos, M. Carbrera, R. Espinosa, P. Boets, *et al*. 2021. DNA-based monitoring for assessing the effect of invasive species on aquatic communities in the Amazon basin of Ecuador. *ARPHA Conference Abstracts* 4: e65376.

Van Rooij, P., A. Martel, F. Haesebrouck, and F. Pasmans. 2015. Amphibian chytridiomycosis: A review with focus on fungus-host interactions. *Veterinary Research* 46: 137.

van Zoggel, H., Y. Hamma-Kourbali, C. Galanth, A. Ladram, P. Nicolas, J. Courty, M. Amiche, and J. Delbé. 2012. Antitumor and angiostatic peptides from frog skin secretions. *Amino Acids* 42: 385–395.

Vargas Hurtado, A. P. 2012. Pruebas antimicrobianas de secreciones cutáneas de *Agalychnis spurrelli* (Anura: Hylidae) en cinco especies de levaduras patógenas del género *Candida*. Tesis de Licenciatura. Pontificia Universidad Católica del Ecuador.

Vasconcelos, T. S., F. R. da Silva, T. G. Dos Santos, V. H. Prado, and D. B. Provete. 2019. *Biogeographic patterns of South American anurans*. Cham, Switzerland: Springer International Publishing.

Velo-Antón, G.,. Rodríguez, A. E. Savage, G. Parra-Olea, K. R. Lips, and K. R. Zamudio. 2012. Amphibian-killing fungus loses genetic diversity as it spreads across the New World. *Biological Conservation* 146: 213–218.

Vergara, A., M. Arias, B. Gachet, L. G. Naranjo, L. Román, J. Surkin, and V. Tamayo. 2022. *Living Amazon report 2022*. Quito: WWF.

Vial, J. L., and L. Saylor. 1993. *The status of amphibian populations: A compilation and analysis*. Gland: IUCN Species Survival Commission.

Viteri Dávila, C. I. 2020. Actividad antimicrobiana de las secreciones cutáneas de *Hyloscirtus lindae* e *Hyloscirtus pantostictus* (Anura: Hylidae) sobre cepas tipo de bacterias Gram positivas, Gram negativas y levaduras. Tesis de Licenciatura. Pontificia Universidad Católica del Ecuador.

Vizcaíno, C., C. W. Barnes, and E. Ordóñez. 2013. Estructuración genética de muestras del hongo *Batrachochytrium dendrobatidis* en Ecuador. *Revista Ecuatoriana de Medicina y Ciencias Biologicas* 34: 1–2.

Voyles, J., D. C. Woodhams, V. Saenz, A. Q. Byrne, R. Perez, G. Rios-Sotelo, M. J. Ryan, *et al.* 2018. Shifts in disease dynamics in a tropical amphibian assemblage are not due to pathogen attenuation. *Science* 359: 1517–1519.

Wägele, H., A. Klussmann-Kolb, M. Kuhlmann, G. Haszprunar, D. Lindberg, A. Koch, and J. W. Wägele. 2011. The taxonomist-an endangered race. A practical proposal for its survival. *Frontiers in Zoology* 8: 1–7.

Wägele, J.-W. 2005. Phylogenetic systematics. *Verlag Dr. Friedrich Pfeil, München, Germany* 3: 27–29.

Walter, H. 1970. *Vegetationszonen und Klima*. Stuttgart: Verlag Eugen Ulmer, 382 pp. [abridged version of *Die Vegetation der Erde*; 2nd edition, 1973; 3rd edition, 1977; 4th edition, 1979, renamed *Vegetation und Klimazonen*; 5th edition, 1984; 6th edition, 1990; 7th edition, 1999, with S.-W. Breckle].

Walter, H. 1976. *Die ökologischen Systeme der Kontinente (Biogeosphäre). Prinzipien ihrer Gliederung mit Beispielen*. Stuttgart, New York: Fischer.

Walther, G.-R., E. Post, P. Convey, A. Menzel, C. Parmesan, T. J. C. Beebee, J.-M. Fromentin, O. Hoegh-Guldberg, and F. Bairlein. 2002. Ecological responses to recent climate change. *Nature* 416: 389–395.

Wang, G. 2020. Bioinformatic analysis of 1000 amphibian antimicrobial peptides uncovers multiple length-dependent correlations for peptide design and prediction. *Antibiotics* 9: 491.

Wang, G.-D., B.-L. Zhang, W.-W. Zhou, Y.-X. Li, J.-Q. Jin, Y. Shao, H.-C. Yang, *et al.* 2018. Selection and environmental adaptation along a path to speciation in the Tibetan frog *Nanorana parkeri*. *Proceedings of the National Academy of Sciences of the United States of America* 115: E5056–5065.

Wang, H., L. Wang, M. Zhou, M. Yang, C. Ma, T. Chen, Y. Zhang, M. Zeller, M. Hornshaw, and C. Shaw. 2012. Functional peptidomics of amphibian skin secretion: A novel kunitz-type chymotrypsin inhibitor from the African hyperoliid frog, *Kassina senegalensis*. *Biochimie* 94: 891–899.

Wang, L., M. Zhou, T. Chen, B. Walker, C. Shaw. 2009. PdT-2: A novel myotropic type-2 tryptophyllin from the skin secretion of the Mexican giant leaf frog, *Pachymedusa dacnicolor*. *Peptides* 30: 1557–1561.

Wang, R., T. Chen, M. Zhou, L. Wang, and C. Shaw. 2013. PsT-1: A new tryptophyllin peptide from the skin secretion of Waxy Monkey Leaf Frog, *Phyllomedusa sauvagei*. *Regulatory Peptides* 184: 14–21.

Wang, Y.-W., J.-M. Tan, C.-W. Du, N. Luan, X.-W. Yan, R. Lai, and Q.-M. Lu. 2015. A novel trypsin inhibitor-like cysteine-rich peptide from the frog *Lepidobatrachus laevis* containing proteinase-inhibiting activity. *Natural Products and Bioprospecting* 5: 209–214.

Waples, R. S. 1991. Pacific salmon, *Oncorhynchus* spp., and the definition of "species" under the Endangered Species Act. *Marine Fisheries Review* 53: 11–22.

Wiens, J. J. 2007. Species delimitation: New approaches for discovering diversity. *Systematic Biology* 56: 875–878.

Wiens, J. J., and L. A. Coloma. 1992. A new species of the *Eleutherodactylus myersi* (Anura: Leptodactylidae) assembly from Ecuador. *Journal of Herpetology* 26: 196–207.

Wiens, J. J., C. H. Graham, D. S. Moen, S. A. Smith, and T. W. Reeder. 2006. Evolutionary and ecological causes of the latitudinal diversity gradient in hylid frogs: Treefrog trees Unearth the roots of high tropical diversity. *The American Naturalist* 168: 579–596.

Wiens, J. J., C. A. Kuczynski, W. E. Duellman, and T. W. Reeder. 2007. Loss and re-evolution of complex life cycles in marsupial frogs: Does ancestral trait reconstruction mislead? *Evolution* 61: 1886–1899.

Wikimedia. 2023. *Epibatidine*. https://es.wikipedia.org/wiki/Epibatidina (accessed May 10, 2023).

Wilens, T. E., J. Biederman, T. J. Spencer, J. Bostic, J. Prince, M. C. Monuteaux, J. Soriano, C. Fine, A. Abrams, and M. Rater. 1999. A pilot controlled clinical trial of ABT-418, a cholinergic agonist, in the treatment of adults with attention deficit hyperactivity disorder. *American Journal of Psychiatry* 156: 1931–1937.

Wiley, O., and B. S. Lieberman. 2011. *Phylogenetics: The theory and practice of phylogenetic systematics*. New York: John Wiley & Sons.

Willis, J. H. 1993. Partial self-fertilization and inbreeding depression in two populations of *Mimulus guttatus*. *Heredity* 71: 145–154.

Winckell, A. 1997. Los grandes rasgos del relieve en el Ecuador *En las condiciones generales del medio natural Geografía básica del Ecuador, Tomo IV Geografía Física, Volumen 1*. Quito: CEDIG, IPGM, IGM, IRD, PUCE, IFEA.

Winter, M., W. Fiedler, W. M. Hochachka, A. Koehncke, S. Meiri, and I. de la Riva. 2016. Patterns and biases in climate change research on amphibians and reptiles: A systematic review. *Royal Society Open Science* 3: 160158.

Womack, M. C., J. Christensen-Dalsgaard, L. A. Coloma, and K. L. Hoke. 2018. Sensitive high-frequency hearing in earless and partially eared harlequin frogs (*Atelopus*). *The Journal of Experimental Biology* 221: 1–8.

Womack, M. C., E. Steigerwald, D. C. Blackburn, D. C. Cannatella, A. Catenazzi, J. Che, M. S. Koo, J. A. McGuire, S. R. Ron, and C. L. Spencer. 2022. State of the Amphibia 2020: A review of five years of amphibian research and existing resources. *Ichthyology & Herpetology* 110: 638–661.

Woodhams, D. C., K. Ardipradja, R. A. Alford, G. Marantelli, L. K. Reinert y L. A. Rollins-Smith. 2007. Resistance to chytridiomycosis varies among amphibian species and is correlated with skin peptide defenses. *Animal Conservation* 10: 409–417.

Wren, S., A. Angulo, H. Meredith, J. Kielgast, M. Dos Santos, and P. Bishop. 2015. *Amphibian conservation action plan*. IUCN SSC Amphibian Specialist Group. https://www.iucn-amphibians.org/resources/acap/

Wright, S. 1931. Evolution in mendelian populations. *Genetics* 16: 97–159.

Wright, S. 1949. The genetical structure of populations. *Annals of Eugenics* 15: 323–354.

Wu, X., J. Pan, Y. Wu, X. Xi, C. Ma, L. Wang, M. Zhou, and T. Xen. 2017. PSN-PC: A novel antimicrobial and anti-biofilm peptide from the skin secretion of *Phyllomedusa-camba* with cytotoxicity on human lung cancer cell. *Molecules* 22: 1896.

Xi, X., R. Li, Y. Jiang, Y. Lin, Y. Wu, M. Zhou, J. Xu, L. Wang, T. Chen, and C. Shaw. 2013. Medusins: A new class of antimicrobial peptides from the skin secretions of phyllomedusine frogs. *Biochimie* 95: 1288–1296.

Xu, X., and R. Lai. 2015. The chemistry and biological activities of peptides from amphibian skin secretions. *Chemical Reviews* 115: 1760–1846.

Yambay, E. 2022. Caracterización molecular de péptidos antimicrobianos a partir de la secreción cutánea de *Lithodytes lineatus* (Anura: Leptodactylidae). Tesis de Licenciatura. Universidad Regional Amazónica Ikiam.

Yánez-Muñoz, M. H., D. Batallas, D. Franco-Mena, P. A. Meza-Ramos, L. A. Oyagata, D. Padilla, C. Paucar, J. P. Reyes-Puig, *et al*. 2020. *Anfibios en los Ecosistemas Andino-Tropicales de la Provincia del Carchi*. Quito: Serie de Publicaciones del Gobierno Autónomo Descentralizado Provincial del Carchi/Instituto Nacional de Biodiversidad.

Yánez-Muñoz, M. H., P. Bejarano-Muñoz, J. Brito-M., and D. Batallas. 2014. Ranas terrestres de los Andes surorientales de Ecuador II: Una nueva especie de *Pristimantis* verde espinosa de los bosques montanos del Parque Nacional Sangay (Anura: Craugastoridae). *Avances en Ciencias e Ingenierías* 6: 63–77.

Yánez-Muñoz, M. H., and P. Meza-Ramos. 2006. Generando información para conservar a *Colostethus delatorreae* (Anura: Dendrobatidae): una especie de rana nodriza críticamente amenazada en los Andes Ecuatorianos. *Boletín Especies Amenazadas UICN* 6: 1–3.

Yánez-Muñoz, M. H., P. A. Meza-Ramos, D. F. Cisneros-Heredia, and J. P. Reyes-Puig. 2011. "2010". Descripción de tres nuevas especies de ranas del género *Pristimantis* (Anura: Terrarana: Strabomantidae) de los bosques nublados del Distrito Metropolitano de Quito, Ecuador. *Avances en Ciencias e Ingenierías* 2: 16–27.

Yánez-Muñoz, M. H., J. P. Reyes-Puig, D. Batallas-Revelo, C. Broaddus, M. Urgilés-Merchán, D. F. Cisneros-Heredia, and J. M. Guayasamin. 2021. A new Andean treefrog (Amphibia: *Hyloscirtus bogotensis* group) from Ecuador: An example of community involvement for conservation. *PeerJ* 9: e11914.

Yánez-Muñoz, M. H., J. C. Sánchez-Nivicela, and C. Reyes-Puig. 2016. Tres nuevas especies de ranas terrestres *Pristimantis* (Anura: Craugastoridae) de la Provincia de El Oro, Ecuador. *Avances en Ciencias e Ingenierías*, 8: 5–25.

Yang, Y., D. Wu, C. Wang, A. Shan, C. Bi, Y. Li, and W. Gan. 2020. Hybridization with insect Cecropin A (1–8) improve the stability and selectivity of naturally occurring peptides. *International Journal of Molecular Sciences* 21: 1470.

Ying, Y., H. Wang, X. Xi, C. Ma, Y. Liu, M. Zhou, Q. Du, J. F. Burrows, M. Wei, T. Chen, and L. Wang. 2019. Design of N-terminal derivatives from a novel dermaseptin exhibiting broad-spectrum antimicrobial activity against isolates from cystic fibrosis patients. *Biomolecules* 9: 646.

Young, B. E., K. R. Lips, J. K. Reaser, R. Ibáñez, A. W. Salas, J. R. Cedeño, L. A. Coloma, S. Ron, *et al*. 2001. Population declines and priorities for amphibian conservation in Latin America. *Conservation Biology* 15: 1213–1223.

Zhou, X., D. Shi, R. Zhong, Z. Ye, C. Ma, M. Zhou, X. Xi, L. Wang, T. Chen, and H. F. Kwok. 2019. Bioevaluation of ranatuerin-2Pb from the frog skin secretion of *Rana pipiens* and its truncated analogues. *Biomolecules* 9: 249.

Zippel, K. C., and J. R. Mendelson III. 2008. The amphibian extinction crisis: A call to action. *Herpetological Review* 39: 23–24.

Zippel, K., K. Johnson, R. Gagliardo, R. Gibson, M. McFadden, R. Browne, C. Martinez, and E. Townsend. 2011. The Amphibian Ark: A global community for ex situ conservation of amphibians. *Herpetological Conservation and Biology* 6: 340–352.

Zug, G. R. 2013. *Reptiles and amphibians of the Pacific Islands*. Berkeley: University of California Press.

Zumel, D., D. Buckley, and S. R. Ron. 2021. The *Pristimantis trachyblepharis* species group, a clade of miniaturized frogs: Description of four new species and insights into the evolution of body size in the genus. *Zoological Journal of the Linnean Society* 195: 315–354.

Index

Note: Page numbers in *italics* indicate a figure, page numbers in bold-italic shows authorship, and page numbers in **bold** indicate a table on the corresponding page.

A

Acosta-Buenaño, N. A., xxiii, xxiv, **147**, *152*, *153*, 154–157, 177
Acosta López, D. (DAL), xxv, 185
adaptation, 201, 299, 300
Adler, K., 98, 99, 104–106, 108
African clawed frog, *see Xenopus laevis* (African clawed frog)
Agalychnis buckleyi, 97, 124
Agalychnis dacnicolor, 232, 235
Agalychnis spurrelli (Spurrell's gliding leaf frog), *69*, *72*, 78, 200, 224, *230*
  exports, *70*
  domestic trade, *77*
  skin secretion of, 232
Aguarico-Putumayo-Caquetá, 247, *254*
Aguiar, G., xxv, *178*
Augustine, L., xxv, 181
Albuja Viteri, L., 110, 111, *114*, 146
Allobates kingsburyi, 75, 155
Allobates zaparo (napi piripri), 1
  exports, *70*
  domestic trade, *77*
Almeida Reinoso, D., xxv, xxvi, *152*, *153*, 155, 157, 177
Almeida Reinoso, F., xxv, *152*, 155, 177
Almendáriz-Cabezas, A., xxiii, xxiv, ***106–108***, ***111–115***, *113–115*, 121, *123*, 130, *131*, 139, 141, 146, 164, 191, 193, 195, *196*, 197, 201, 202, 277
Altamirano-Benavides, M., 164
Alto Napo, 34, *41*
Amaru Zoo, 179, 184–186, 291, *see* Centro de Conservación de Anfibios Amaru (CCA-Amaru)
Amazonian Cordilleras, *255*, 286
Ambatiug/Ambatyu/Ambatio/Ambato/Ampatu/Ambatukuna, 5, 6
Ameerega bilinguis (wikiri), 1, 10, 65, 74
  exports, *70*
  domestic trade, *77*
American bullfrog, *see Aquarana catesbeiana* (American bullfrog)
Amoroso, E. C., 48, *61*
amphibian biology, women inclusion, 189–194, *191*, **192**, *194*
Amphibian Conservation Action Plan (ACAP), 290, 291
amphibian genome bank, 142, 151–155
amphibian iconography, Ecuadorian archaeology, 10–11
  integration period, 20–27
  late formative period, 11–13
  rain rituals, 27–34
  regional development period, 14–20
Amphibian Species of the World (AMNH), 48, 120
AmphibiaWeb, 48, 158, 159, 281
AmphibiaWebEcuador, 156–158, 168
Andersson, L. G., 100, 104, *105*, *123*
Anfibios de/del Ecuador, encyclopedia, 156–159
anthropozoomorphic seated figurine, *23*
antimicrobial peptides (AMPs), 229, 230, 232–236
  amphibian skin, 235
  peptide analogs derived from, **236**
appliqués, 11, 14, 15, 20, 22
  bottle/container with anurans, *17*
  cocina de brujo pot, 34
  lime pot/container with anurans, *16*
  pedestal cup with figurines, *18*
  plate decorated with amphibians, *26*
  pot decorated with anurans, *26*, 36
  ritual table decorated with figurines, *38*
  vessel decorated with anurans, *36*
Aquarana catesbeiana (American bullfrog), 57, 59, 264, *265*, 288
  feral populations of, *58*
Arbeláez-Ortiz, E., xxiii, 6, ***183–188***, *184*, 260
Arcos Terán, L., 135, 142, 144, *145*
Ardila-Robayo, M. C., 191
Argo, A., 48
Arias, C., 5
Armijos-Ojeda, D., *188*, 189
Arteaga-Navarro, A. F., xxiv, xxvi, *123*, 141, 195, *200*, 203
artificial fertilization, 291, 292, *293*
assisted reproductive technologies (ARTs), 291, 292, 300
Astroblepus (catfish), 5
Atelopus balios, *73*, 74, 75, 130, 155, 186, 292
  exports, *70*
  domestic trade, *77*
Atelopus bomolochos, 49, 283
Atelopus boulengeri, 100, 105
Atelopus cocha, 22
  Pasto culture, *35*
Atelopus coynei, 186, 278
Atelopus elegans, *73*, 95, 224
  exports, 70
  domestic trade, *77*
Atelopus flavescens-spumarius, 255
Atelopus exiguus (small harlequin frog), 10, *275*
Atelopus guanujo (puca-sapo), 7, 59
Atelopus ignescens (jambato harlequin frog), 5, 6, 8, 10, 44, 47, 48, *49*, *50*, 53, 59, 87, 88, *93*, 96, 105, 109, 110, 113, 124, 125, 145, 146, 270, 283, 284, 292, 296, 297
  decline and extinction, 276–282
  rediscovery, 179
Atelopus longirostris (longnose harlequin frog), 49, 89, 270, 292
  decline and extinction, 276–278
  the science crusade, 179–181
Atelopus onorei (Onore's harlequin frog), 10
Atelopus pastuso, 105, 283
  Pasto culture, 22, *35*
  Tulcán, *35*
Atelopus petersi, *93*, *110*
Atelopus planispina, 94
Atelopus sp. (*aff. longirostris*), 186, 278
Atelopus spp. (*spumarius* complex), *73*, 124, 164, 185, 186, 253–260, *260*, 278, 292, *293*
  exports, 70
  domestic trade, 77
  morphological variation in, *26*
Atelopus spumarius, 254, 257, 260
Aubér-Thomay, M., 61, 62, *63*, 150
Avilés, A., *163*
Avilés Chevalier, L., *163*

B

Bahía culture, 17, 18
  anthropozoomorphic seated figurine, *23*
  Ceratophrys stolzmanni, 21
Bahía Monster, 17
Banana Republic frogs outflow, 60–61
Banisteriopsis caaspi (Yagé plant), 43
Baquero Rivadeneira, M, M., **147**, *153*
Barragán, A. R., 142
Barragán, M. E., 177
Barriga, R., 110, 111, *114*
Barrio-Amorós, C. L., 10
Bastidas Andrade, D., *163*
Bataille, A., 283
Batallas-Revelo, D. A., xxiv, xxvi, 113, *123*, *162*, 169, *200*, 202
Batékes tribe, *143*, 144
Batrachochytrium dendrobatidis (*Bd*), 179, 193, 232, 267, 280, 282–284, 296, 298–299
BBC Wildlife, 160, *161*, 203
Beirne, C., 261
Bejarano Muñoz, P., *162*, 165, *166*, *167*, 169, 191
Bermúdez-Puga, S., xxiii, ***224–236***
Betancourt, R., *162*, *167*, 169, 177, 191
bifurcated swampfrog, *see Dendropsophus bifurcus* (bifurcated swampfrog)
bioactive peptides, 229, 233, 235
biobanks, 138, 294–295
biogeographic areas, **247**
biogeographic sectors, *see* climate
bioprospection, amphibian skin, 224–226
biotrade, amphibians
  Banana Republic frogs outflow, 60–61
  bans to exportation of Ecuadorian amphibians, 65–66
  domestic use of live Amphibians, 71–79, *72–74*, *75–76*
  origins and global exports, 67–71
  poison dart frogs, 63–65
  sustainable, 66–67
  value of live amphibian trade, 78–79
Bioweb, 155–157
Bish, R., 48
Blomberg, R., 105

Blomberg's toad, *see Rhaebo blombergi* (Blomberg's toad)
Boana appendiculata, 55, *102*
Boana boans, 55, 87
  syntype of, *88*
Boana calcarata, 87
Boana cinerascens, 87
Boana fasciata, 98
Boana picturata (Chachi frog/Imbambura treefrog/colorful aroboreal frog/ painted tree frog), 9, *72*, 229, *230*, 233, *234*, 235
  exports, *70*
  domestic trade, *77*
Boana punctata, 87
Boana rosenbergi (gladiator frog), 96, *102*
boanins, 233, 235
Böhme, W., 65
Bokermann, W. C. A., 108
Bolitoglossa palmata, 88
Bolitoglossa peruviana, 263–264
Bonaccorso, E., xxiii, 154, 156, 157, 168, 174, 191, 193
Bonner, J., xxv, 181
Boonman, J., 61, 62
Bosque Protector Mazán, 273, *275*
Bosque Protector Siete Cascadas, *12*
Botasso, J., 10, 39, 55, 59
Boul, K., 135
Boulenger, G. A., 47, 88, 95–100, *99*, *101–103*, 104, 105, *123*
Brandbyge, J., 284
Brenner, W., 61, 62
Bringsøe, H., 66
British Museum (BMNH), 94–95, 98–104
Brito Molina, J., xxiv, xxvi, 113, *123*, *162*, *166*, 168, 195, *200*, 201–202
Brito-Zapata, D., xxiii, ***170–177***, *176*, 177
broad-headed *Strabomantis* frogs, 280
Brown, J. L., 66
Brunner, R., 193
Buckley, C., 95, *96*, 97, 100
Buckley's glass frog, *see Centrolene buckleyi* (Buckley's glass frog)
bullfrog industry, 55–59
Burrowes, P. A., ***xiii***, xxviii, *123*, 128, *138*, 141, 191, 197
Bustamante, M. R., **7**, 88, *131*, 137, 141, **147**, *151*, *152*, 154, 155, 164
Bustamante, T., 88, 90

C

Caecilia buckleyi, 97
caecilian, *see Epichrionops petersi* (caecilian)
Caecilia tentaculata, 87
Calle, F., *152*
Callimedusa ecuatoriana, 229, 230, *234*
  exports, *70*
Camacho-Badani, T., 193
Caminer Rodríguez, M. A., xxiv, 98, *123*, 158, 195, *200*, 202–203
Campbell, J. R., 66
Campbell Grant, E. H., 189, 294
Campos Yánez, F., 144, *148*, 171, 172
Cannatella, D. C., 117, *123*, 128, 130, 131, 134–137 (*134*, **136**), 142, 147, 158, 163, *196*, 197, 198
Caramaschi, U., 10
Caranqui culture, 20, 22
  vessel decorated with anurans, *36*
Carr, J., *112*, 113, *114*
Carrión Mora, C., 100, *104*, 108
Carvajal Campos, A., *152*
Carvajal-Endara, S., xxiv, xxv, 44, 141, **147**, 157, 177, 178, 191, 197, *200*, 201, 284
Carvalho, T., 283
casque-headed frog, *see Osteocephalus festae* (casque-headed frog)
Castillo-Trenn, P., 75, **147**, 193
Castro, C., *166*
Catamayo-Alamor, *250*
Catequil, 6, 39–44
Center of Biodiversity and Environment (CBA), 142, 145, 146
Centro Científico Río Palenque, 64, 111, 117, 118, 119, 150, 173
Centro de Conservación de Anfibios Amaru (CCA-Amaru), 183–184
  amphibians at CCA-Amaru, **186**
  birth of, 184–186
  communities through education, 187
  ex situ conservation program and education, **186**
  inspiring students, 187
  preventing, loss of natural habitat, 187–188
Centro Jambatu (CJ), 48, 177–179, 182, *183*, 202, 291
  foreign input to research, 181–183
Centrolene buckleyi (Buckley's glass frog), 6
ceramic vessels, 11, 14, 20, 22
Ceratophrys cornuta, 87
  type of, *88*
Ceratophrys stolzmanni (Stolzmann's pac-man frog), 11, 17, 18, 29, *68*, 70, *72*
  amplectant pair of, *40*
  Bahía culture, *21*
  Correos del Ecuador, *50*
  domestic trade, *77*
  exports, *70*
  Guangala culture, *24*
  Jama Coaque culture, *40*
  from Reserva Ecológica Arenillas, *22*, *40*
Ceratrophrys stolzamnni, 29
Chachi frog/Imbabura treefrog/colorful arboreal frog/painted tree frog, *see Boana picturata*
Chapman, F. M., 107
Charles III, king, Prince Charles, 181, *182*
Chimbirico/Jimbirico/Guiltre, 9
Chocó Coastal Cordillera, *248*
Chorrera culture, 11, 17, 18
  iconographic representations, 14
  Rhinella horribilis, *12*, *13*
Chorreran styles, 20
Christmann, S. P., 65
Chuglli/Shugshi, 9
Cicala, M. M., 46, 59
Cilveti, C., 224
Cisneros-Heredia, D. F., xxiv, 57, *123*, 141, 164, 165, 169, *172*, ***170–177***, 170–174, *176*, 177, 193, 195, *196*, 199, 201, 264
Clarke-Macintyre, W., 105
climate
  biogeographic sector, 245, *246*, **247**, *257*, **257**, *258*, 267, *270*, 284, 286
  classification, 245
  factors, 244–245
climate change, 281–283, 289, 291, 299–300
Cochran, D. M.., 190
cocina de brujo pot, 22, *34*
Cofán mythology, 43–44
Cohen, J. M., 282
Collins, D., 117
Coloma, L. A. (LAC), ix, xvii, xx, xxi, xxiii, xxiv, xxv, 7, 10, 47, 48, 62, *64*, 66, 74, 75, 107, *112*, 117, 121, *123*, 127, 130, *131*, 134, 135, 137–139, 141, 142, 144, 163, 168, 177, *180*, 185, 191, 193, 195, *196*, 197, 198, 201, 227, 236, 237, 240, 242, 266, 277, 278, 281, 283, 291
  at CJ, 181–183, 182
  at EPN, 110–113, **112**, 113
  at QCAZ, **147**, 146–157, *151*, *152*, *153*
colorful arboreal frog, 9
conservation, 74, 149–150, 288–289
  amphibian research, 142–146
  CCA-Amaru, 183–184
  dendrobatid frogs, 65
  EcoCiencia, 130
  ex situ, **186**, 289–294
  genetics, 294–300
  INABIO, 169–170
  priority areas for, 48
  in situ, 289
conservation breeding, 156, 265, 289, 291, 300
conservation breeding programs (CBPs), 291, 292, 300
conservation units (CUs), 297
Convention on Biological Diversity (CBD), 294
Convention on International Trade in Endangered Species of Wild Fauna and Flora (CITES), 60, 65–68, 70, 77, *78*, 264
Cope, E. D., 89, 100, *123*
Cordillera Occidental of the Andes, 247, *250*, 267
Cordillera Oriental of the Andes, 247, *251*
Cornalia, E., 87–88, *90*
Cornejo, J., 52
Correa Delgado, R. 113, 142, 168, 188
Correa Monge, G., 63, 75, **147**, 148, 155
Cosanga ceramics, 20
Cosanga culture, *26*
Cosanga-Píllaro culture, 20
Costales, P., 5, 39, 40, see also Peñaherrera de Costales, P.
Costales Samaniego, A., **7**, 9
Costales Peñaherrera, J., 39
Cotán, 6, 39, 40, 41
crafts, 1, 34–35
  from Kichwa, *41*
  from tagua, *41*, *42*
Creveling, C. R., 226
Crump, M. L., ***xiii***, ***xxiv***, 124, 127, 128–130, *129*, *137*, 142, 155, 191, 193
Cruz-Cordovez, C., 57
Cruziohyla calcarifer, *100*, 201, 229–235, *234*
  exports, 70
  domestic trade, *77*
cruzioseptin 1 (CZS-1), 233, 234, 236
cruzioseptins, 201, 233–234
cryptic dendrobatids, 120–121
Ctenophryne aequatorialis, 9, 105, **186**
Ctenophryne aterrima, **70**, 99
  domestic trade, *77*
cuico rainfrog, *see Pristimantis unistrigatus* (cuico rainfrog)
Culebras, J., xxiv, xxvi, 48, *123*, *180*, 195, *200*, 203

Cutín, 2, 6, 10
Cuvi, N., 9, 284

D

DAL, *see* Acosta López, D.
Daly, J. W., 65, 127, 224, *226*, 226–228
Darst, C. R., 135, 237
Darwin, C., 90, 95, 97, 146
data-deficient species, 280–281
Dávila-Játiva, M., xxiii, ***170–177***, *176*
Dayot, J., 285
Deckert, K., 61, 62
Declining Amphibian Populations Task Force (DAPTF), 120
del Pino, E. M., 61, 71, 74, *116*, 135, 142, 146, 193, 195, *196*, 197
  and marsupial frogs, 115–117, *116*
Dendrobates auratus, 74, 112, 264
Dendropsophus bifurcus (bifurcated swamp frog), 45, 55
Dendropsophus carnifex, 124, 174
Dendropsophus marmoratus, 87
Dendropsophus minutus, 262–263
Dendropsophus triangulum, 99, 203
dermaseptins, 229, 230, 235
Deutsche Gesellschaft fur Herpetologie und Terrarienkunde (DGHT), 64–65
de la Torre, S., 144, *148*, 174
De Velasco, Juan, 5, 6, 9, 46, 54, 59
de Vries, 122, *148*
DHMECN, 160–164
  Universidad Central and taxonomy, 164–169
  Mario H. Yánez-Muñoz at, 164–169
Díaz-Proaño, M., **147**
De Rossi, M., xxv, *180*
distribution, Ecuadorian amphibians 245–250
DNA barcoding, 296
Dodoka/Kodoka, 9
Dodson, C. H., 285
Donoso, D., 182, *183*
dos Santos, S. P., 193, 260
doubledigested RAD (ddRAD), 298
Duellman, W. E., ix, xxiii, 11, 12, 48, 62, 106, 110, 112, *114*, 117, *123*, 142, 146, 148, 149, *152*, 156, 157, 158, 160, 163, 170, 173, 198, 242, 245, 285
  in memoriam, xiii
  KU era, 123–141 (*126*, *131*, *132*, *138*, *140*)
  Dueñas, J. F., 152
Dunn, E. R., 99, 100, 104, *105*, 106

E

Ecociencia (Fundación Ecuatoriana de Estudios Ecológicos, 156
Ecuador
  altitudinal gradient, *258*
  amphibian iconography in, 10–34
  amphibian richness, *259*
  amphibian taxonomy, 87–89
  bioactivity skin secretions, **225**
  biogeographic sector of, *257*, *258*
  cited publications of amphibians, **136**
  educational system, 45–48
  edible amphibians, 54–59, **56**–57
  endemism, 114, 238, **257**, *258*, 286
  gender diversity in, 195–204
  names for amphibians in, **2–4**
  non-threatened species, *271*
  political endemism of, *259*
  possibly extinct species, 267, 268, 270–274, **272**, *274*, 289
  primary collectors, *131*
  species for each province, *256*
  tadpoles in, **7**, *8*
  threatened species in, 267–270
Edalhorina perezi, *93*
Edwards, S. R., ***xiv-xv***, 66, 121, 125, 128, *137*, 137–138, 148
egg-brooding frog, *see Gastrotheca riobambae* (marsupial frogs)
Eleutherodactylus frogs, 121–123
Elmer, K. R., 191, 193, 263, 298
English names, 9–10
Encalada, A., *148*
Engystomops petersi, 91, 135, 298
Ensinck, F. H., 61, 62
Enterococcus faecalis, 232
environmental DNA (eDNA), 296
environmental pollutants, 286–287
epibatidine, discovery of, 224, *226*, 226–228, 298
epibatidine poison frog, *see Epipedobates anthonyi* (epibatidine poison frog)
Epichrionops petersi (caecilian), 110
Epipedobates anthonyi (epibatidine poison frog), 9, 10, 63–66, 69, *72*, 74, 78, 117, 226–228, 238, 298
  exports, 70
  domestic trade, 77
  Sigma-Aldrich for research purposes, *226*
Epipedobates boulengeri (poison frog), 1, 67, 100
Epipedobates darwinwallacei, 67
  exports, *70*
Epipedobates machalilla, 74, 75, 117, 155
Epipedobates tricolor (tricolor poison frog), 69, *72*, 117, *178*, 181, **186**, 198, 226
  exports, *70*
  domestic trade, 77
  type of, *178*
Equatorial Chocó, 247, *248*, 267
Equatorial Pacific Coastal Cordillera, *249*
Escherichia coli, 224, 230, 232
Escobar, B., 116
Escuela Politécnica Nacional (EPN) (museum), 106–108, 110–115
  Gustavo Orcés Villagómez at, 106–108
  Ana Almendáriz-Cabezas at, 111–115
Espin, C., 5
Espinosa, G., 5
Espinosa Cordero, S., 54
Espinosa R, R., 227
Europe, 6, 47, 60, 67, 70
  bans to exportation of Ecuadorian amphibians, 65–66
  museums, 94–95
  poison dart frogs, 63–65
  scientists and frog lovers in, 61–62
evolutionarily significant units (ESUs), 297
ex situ conservation, 74, 155, 171, **186**
  breeding and assurance colonies, 289–291, *290*
  reproductive technologies and biobanking, 291–294
  exomics, 296
  extinction, 1, 53, 67, 128, 144, 152, 154, 155, 156, 173, 178, 179, 180, 181, 185, 239, 240, 241, 242, 264–294, 295, 296, 297, 299, 300
  broad-headed *Strabomantis*, 280
  harlequin frogs, 276–278, *276*, *277*
  highland aquatic frogs, *Telmatobius*, 273–276
  marsupial frogs and congeners, 279–280
  rocket frogs *Hyloxalus*, 278–279

F

Fan of Pastaza, 247, *255*
Félix-Novoa, C., **147**, *152*, 156, 157
Fernández, C., 193
Festa, E. L., 95, 98, 105
films, Ecuadorian Amphibians, 48
Finer, M., 285
Fischer, E. K., 193
Fitzpatrick, S. W., 179
five-lined agile frog, *see Scinax quinquefasciatus* (five-lined agile frog)
fixation, 299
Flectonotus pygmaeus (Venezuelan marsupial frog), 115, 116
Fletcher-Laso, G., **147**
Flores, G., xxiv, 120–123, *122*
FonoZoo.com, 48
food, humans
  bullfrog industry, 55–59
  native species, 54–55, **56–57**
Fouquet, A., 240, 261
fragmentation, 280, 281, 284–286, 294, 295
Franco-Mena, D., xxviii, 176, 191, 240
Frank, N., 10
Fraser, L., 95, 97, 98, 99, 100
Freire, D., 193
Fritts, T. H., 62, 126, *127*, 128
Fuentes Roldán, A., 6, 45, *46*
Fugler, C. M., 124, *131*, 170
frog skin secretions
  bioactive peptides from, 229
  bioactivity, **225**
Frost, D. R., xx, 10, 66, 92, 99, 139, 147, 156, 253, 260, 262
Fundación Ecominga, 160, 168, 169, 170, 202
Fundación Ecuatoriana de Estudios Ecológicos, see Ecociencia
Fundación Herpetológica Gustavo Orcés (FHGO), 57, 108, 164, 169, 171, 172, 173, 177, 191, 203
Fundación Jambatu, 178
Fundación Jatun Sacha, 120
Fundación Jocotoco, 164, 203
Fundación Natura, 160
Fundación Naturaleza Kakaram, 202
Fundación Oscar Efrén Reyes, 202, 203
Fundación Otonga, *xxiii*, xxv, 142, 145, 146, 178, 182
Funk, C. W., 137, *152*, 154, 201, 298

G

Gagliardo, R., xxiii, *151*, 154, 156
Galápagos Islands, 176, 197, 264, *265*
Galarza, L., *116*, 147
Gallardo, G. G., 10, 55, 57, 59, **147**
García, F., 45
García Moreno, G., 90
García Saltos, M., 93, *148*
Gardner, A. L., 97, 98
Garraffo, H. M., 227, 228

Garzón, K., 9
Gastrotheca cornuta, *101*
Gastrotheca guntheri (lower-toothed marsupial frog), 99, *102*
Gastrotheca plumbea, 62, 74, 150
Gastrotheca pseustes, 9, *37*, 185, 201, 273
  exports, *70*
  domestic trade, *77*
Gastrotheca riobambae (gualataco marsupial frog), 2, 5–9, 47, 48, *68*, *72*, 74, 279–280
  distribution, 201
  domestic trade, *77*
  exports, *70*
  gravid females of, *37*
  mating pair of, *61*
  pot decorated with anurans, *36*
  in Quito, 74
Gehara, M., 263
gene flow, 295, 298–300
generalized linear models (GLMs), 190, **192**
genetic detection, 296
genetic drift, 242, 295, 297, 298, 300
genetics, wild populations, 297–300
genetic variation, 295–298, 300
genome resource banks, *see* biobanks
genome-wide association studies (GWAS), 299
genomics, 295, 296, 297, 298
genotype–environment associations (GEAs), 299
Gentry, A. H., 285
geography, 242–244
Gill, B. J., 60
Goin, C. J, *123*
gold-spotted glassfrog, *see Hyalinobatrachium aureoguttatum* (gold-spotted glassfrog)
Goldewicht, A., *129*
gladiator frog, *see Boana rosenbergi* (gladiator frog)
glassfrogs, *51*, 106, 117, 130–132, 173, 164, 174, 179, 198, 247, 249, 288, 297
CITES appendix II, 66
Global Amphibian Assessment (GAA), 266
global climate models (GCMs), 299
gold discs, 22, *35*
gonadotropinreleasing hormone (GnRH), 292
Gorzula, S., 66
graffiti, 53–54, *55*
Graham, C. H., 193, 281, 282, 284
Grenard, S., 61
Guerra, M. A., 115, 135, 137, **147**, *152*
Guevara Salvador, S. (SGS), xxv
Guangala culture, 10, 18, 29
  lime pots, *24*, *39*
  small bowl, *25*
Guarderas, M. D., xxiii, xxv, ***66–79***, 67, *69*, 157, 177, 178
Guayaquil frogs, 46
Guayasamin, J. M., ***xv***, xxiii, xxiv, 10, 53, *123*, 134, 135, 139, 140, 141, 142, **147**, 149, *151*, *152*, 154, 156, 157, 164, 168, 169, ***170–177***, *172*, 174–177, *176*, 178, 180, 193, 195, *196*, 198, 240
Guerrero, V., *114*
Günther, A. K. L. G., 95, 97, 98–99, *99*, *123*
Gutiérrez Usillos, A., 11, 14, 17, 18, 22, 29, 34

H

habitat destruction, 241, 264, 284–286, 288, 291
Hämäläinen, M., 96, 97
Hambatu/Hampatu/Hamppatu/Hambatug/Hamp'atu/Hambato, 2, 5, 39, 40, 56
harlequin frogs, 68, 108–110, 140, 184, 185, 186, 297
  markets, *70*, *73*, *77*
  declines and extinction, 276–278, *276*, *277*
  assisted reproductive technologies, 292–293, *293*
Harrison Matthews, L., 61
Hayes, T., 286
Hemiphractus fasciatus, 88, 273
Hemiphractus scutatus, 87
herpetological communities, tropical rain forests, **124**
Herrera, G., *107*, 108, 110, *114*
Herrera, I., 54
heterozygosity, 295, 297, 299
Heyer, W. R., 120–123, *122*
Hidalgo Alzamora, L., 52
Hillis, D. M., 128, 137–*138*, 140, 147
Hird, C., 287
Hirtz, A., xxiv, 11, 28–33, 41
Hödl, W., 237
Hoffstetter, R. J., 106, 107
Hoke, K., xxiii, xxv, 137, 181, 182, *183*, 193, 237
Holm-Nielsen, L., 284
Hoogmoed, M. S., 61, *62*, 66, 110–112, 163, 260
Hualaj/ Huartaj/Ty'ungil/Gualag/Gualataco, 2, 6
Huamoni, N., xxv, 159, 160
Hurtig, A.-K., 287
Hutter, C. R., *123*, 197, 249
Hyalinobatrachium aureoguttatum (gold-spotted glassfrog), *72*
  exports, *70*
  domestic trade, *77*
hylid frog, *see Pseudacris crucifer* (hylid frog)
Hyloscirtus conscientia, 169, 170
Hyloscirtus larinopygion, 138
Hyloscirtus palmeri, 98
Hyloxalus awa, 63, 75, 155
Hyloxalus bocagei, 6, 137, 198, 278
Hyloxalus delatorreae, 164, 164, 279
Hyloxalus frogs, 278–279
Hyloxalus jacobuspetersi (Quito rocket frog), 110, 279
Hyloxalus vertebralis, 9, 98, 159, 185, 186, *275*, 279

I

Icaza, J., 5, 9, 47
Ilulo, Ilulo, 6, 10, 48
Imbabura treefrog, 9
inbreeding, 295–300
inbreeding depression, 295, 297, 298, 300
Incertae sedis, 250–253
indigenous names, 1–7
in situ conservation, 145, 160, 180, 185, 289
Instituto de Ecología Aplicada (ECOLAP), 171
Instituto Ecuatoriano Forestal, de Áreas Naturales y Vida Silvestre (INEFAN), 57, 59, 67, 67, 228
Instituto Nacional de Biodiversidad (INABIO), 75, *162*, *165*, 168, 199, 294
  species names for amphibian conservation, 169–170
  Mario H. Yánez Muñoz at, 169–170
International Union for Conservation of Nature (IUCN), 266–267
introduced animals, 288
Isla Isabela, 264, *265*
Isla Santa Cruz, 264, *265*
IUCN Red List of Threatened Species, 48, 174, 179, *267*, *268*, *278*, *279*

J

Jacobson, K. A., 226
Jácome, C. E., 142
Jácome, G. A., 50
Jama Coaque culture, 10, 14, 29
  Ceratophrys stolzmanni, *40*
  Leptodactylus ventrimaculatus, *16–19*
  lime pot, *16*, *20*
Jama-Zapotillo, *249*
jambato harlequin frog, *see Atelopus ignescens* (Jambato harlequin frog)
Jambatu/Jambato/Jampatok/Jamp'atu/Jampatu/Janpatu/Jambatiu/Jambatio/Jambatyu/Jambatyug/Jambatiug/Jambacho, 5, 6, 10, 44
Jaynes, K. E., 179, 276, 277, 297
Jiménez de la Espada, M., 5, 89–94, *91*, **92**, 105, *123*, 277
Jiménez Sánchez Mariscal, J. D., 5
jokes, 53–54, *55*
Jones, R. E., 61, 62
Jordan's bone-headed treefrog, *see Trachycephalus jordani* (Jordan's bone-headed treefrog)
Jungfer, K.-H., 65, 158, 237

K

Kaila/K'ayra/Kayra/Cayla, 6
Kaká frogs, 41
Kaplan, M., 263
Kästle, W., 61
Kazal-type protein inhibitors, 233
Kichwa, 8–10, 34, 71, 94, 171
  in Alto Napo, *41*
Kieswetter, C. M., 240
Kingdom of Quito, 46
Kirk, K. L., 226
Kirkland, B., 122
Kizirian, D. A., 128, 147, *148*
Klebsiella pneumoniae, 224, 230, 236
Koch, C., 169, 191
Kohn, E., 6, 10, 55, 56
Krynak, K. L., 193, 288
Kuartank myth, 41–43
Kulu, 9

L

LAC, *see* Coloma, L. A.
La Chimba culture, 20, 22, *26*
Lamar, W. W., *172*
Lambert, C., 117
Las Orquídeas site, 11, 20, *26*
Latin American Association of Zoos and Aquariums (ALPZA), 187, 188
La Tolita–Tumaco culture, 14

Laurent, R., 117
Lavilla, E. O., 260
León-Reyes, A., xxiii, ***170–177***, 170–172
Leptodactylus labrosus, 17, 94
Leptodactylus mystaceus, 87
Leptodactylus pentadactylus, 5, 6, 55, 87, 87, 94
Leptodactylus peritoaktites, 11, 12, *13*, 55
Leptodactylus rhodomerus, 6, *13*
Leptodactylus stenodema, 94
Leptodactylus ventrimaculatus, *16–19*
Leptodactylus wagneri, 55, 88
Leptodactylus frogs, 121
Lessmann, J., 193
Letellier, F., 61–62
Licenciatura students, 116, 117, 146, **147**, 155, 158
lime pots, 11
  Chorrera culture, *12*
  Guangala culture, *24*, 39
  Jama Coaque culture, *16*, *20*
  Rhinella horribilis, 17
lineatins, 233, 235
Linnaeus, C., 1, 87, 95, 239
Lips, K. R., 154
Lithobates palmipes, 55, 59, 87, 138
Lithodytes lineatus, 87
loci, 295, 296, 298, 299
Lombeida, D., 39, 44, 135, 146, **147**
long peptides with α-helicalstructures, 229, 233
long-snouted toad, *see Rhinella festae* (long-snouted toad)
Loor-Vela, S., 116
López, L. E., 151
Lötters, S., 65, 142, 149, *151*, 255, 276, 277
lower-toothed marsupial frog, *see Gastrotheca guntheri* (lower-toothed marsupial frog)
Luna, A., *112*, 113
luteinizing hormone (LH), 292
Lutz, B., 190
Lynch, J. D., 117, 122, *123*, 127, 128, 130–132, *131*, *132*, 135, 138, 142, 158, 163, 168, 170, 173, *196*, 237, 254, 285

M

major histocompatibility complex (MHC), 299
Mancero, E., 238, 240
Manteño Huancavilca culture, 20
  miniature art, *26*, *27*, *29–33*
Manzano Pasquel, A. L., 283
management unit, 297
MapBiomas, 286
Martín-Torrijos, L., 193
Matheus Pozo, J. C., 111, 160, *163*
marsupial frogs, *see Gastrotheca riobambae* (gualataco marsupial frog)
McDiarmid, R. W., 118, *172*, 173
medusins, 229, 232, 235
Meggers, B., 11
Mena Valenzuela, P., *114*
Mendelson III, J. R., xxiii, xxiv, ***xvii***, 123, 141, 154, 156
Menéndez-Guerrero, P. A., 135, *152*, 154, 156, 157, 168, 266, 281, 282, 284
Merino-Viteri, A., *151*, *152*, 153, 154, 155, 156, *157*, 158, 185, 273, 275, 281, 282, 283, 284
Mesías-Tapia, V., **147**
Mestanza-Ramón, C., 289
mestizo, 1, 2, 6, 39, 44–54
metabolomics, 296
metagenomics, 296
methicillinresistant *S. aureus* (MRSA), 232
Meza-Ramos, P. A., *112*, *162*, *166*, 167, 168, 111, 114, 164
microsatellites, 297
Milagro Quevedo culture, 21, *34*
Miller, E., xxv, 181
Ministerio de Agricultura y Ganadería (MAG), 57, 65–66
Ministerio del Ambiente del Ecuador, 78, 245, 247
Miró, A., 288
Miyata, K., 117–120, *118*, 122, 131, 142, 156, 173
Mochica/Moche culture, 11, 12
molecular diversity, 224–226
Molineros, J., **147**
Montalvo-Puente, C., xxiii, ***10–35***
Moreno Espinosa, M., 160, *163*, 164
Morales, P., 55
Morales-Mite, M., xxiv, xxv, *162*, 168, 177, *178*, 197
Moskowitz, N., 193
Mossman, K., 116
Muñoz B., I., *126*, 128, 130
Museo de Arte Precolombino Casa del Alabado (MAPCA), 11
Museo de Zoología of UTPL (MUTPL), 189
Museo Ecuatoriano de Ciencias Naturales (MECN), 160
  Mario H. Yánez Muñoz at, 164–170
Museo Nacional del Ecuador (MuNa), 11
Muséum d'histoire naturelle de la Ville de Genève (MHNG), 144, 145, 177, 131
Museum of Comparative Zoology (MCZ), 104, 118, 121–122, 144, 170
music, 1, 47, 50–53, 96
mutation, 295
Muysken, P. C., 6
mythology, 27, 35–44

N

Names, *see* vernacular names, tapdole names
Napo-Curaray, 247, *254*
Naranjo, R. E., ***291–294***
National Biodiversity Strategy 2015–2030, 67
National System of Protected Areas (SNAP), 289
Naturhistorisches Museum Wien (NHMW), 88
Navarrete, M. J., xxiv, *123*, 158, 169, 191, 193, 195, *204*
Naydenova, E., xxv, *180*
Nototrema marsupiatum, 47, 95, 96
novel peptide families from Ecuadorian amphibians, 233
Nyctymantis rugiceps, *102*

O

O'Connell, L. A., xxiii, xxiv, xxv, 137, 181–*183*, 193, 195, 197, *200*, 202, 227, 237
Oeser, R., 61, 62
oil industry, 287
Olalla family, 107, 108, 121
Olalla, J., 107, *114*
Olalla, M., 107, 110
Olalla, R., *107*, 108
Onore, G., xxiii, xxiv, 44, 62, *64*, 122, *131*, 142–146, *142*, *145*, 149, 150, 178, 277, 283
Onore's harlequin frog, *see Atelopus onorei* (Onore's harlequin frog)
Oña-Alava, I., **147**
Oophaga sylvatica (little devil frog), 1, 7, *50*, 64, 67, *72*, 298
  domestic trade, *77*
  exports, imported from Ecuador, *70*, *77*
  intrapopulation color variation, *75*
  morphs of, *76*
  total sales, 79
Orcés-Villagómez, G., 106–108, 110–113, *111*, *114*, 146, *148*, 160, *163*
Ordoñez-Alvarez, M. P., 10
orphan peptides, 235
Ortega, J., xxiv, *123*, 158, 169, 191, *204*, 239, 240
Ortega-Andrade, H. M., xxiv, 141, *162*, 164, *166*, 168, 169, 170, 197, 238, 266, 273, 281, 284
Ortiz Crespo, F., 142, 160, *163*
Osornophryne bufoniformis, 105, 237
Osteocephalus buckleyi, 6, 97, 159, 164, 273, *275*
Osteocephalus festae (casque-headed frog), 98, 105
outbreeding depression, 300
Oxford, P., 48
Oyagata C, L., *162*, 165, *166*, *167*, *168*

P

Páez, N. B., xxiv, *123*, 158, 191, 195, 203–*204*
Páez-Vacas, M. I., xxiii, xxiv, 141, **147**, 150, *151*, *152*, ***189–195***, 191, 193, 195, 197, *200*, 201, 238, 278, 294, 298
Padilla, A., *145*
Padilla, S., **147**
painted tree frogs, *see Boana picturata* (Chachi frog/Imbabura treefrog/ colorful arboreal frog/painted tree frog)
Palmer, M. G., 95, 98, 100
Palta language, 9
Paluh, D. J., xi, xxiii, 280
Páramo, 247, *253*, 267, 277
Paredes Recalde, A., **147**
Parker, H. W., 100, *104*
Parmelee, J., *133*
Parque Nacional Yasuní (PNY), 160
Parra R. J. H., 39, 43
Parra, R., 288
Paruwrobates erythromos (rocket frog), 119
Paruwrobates whymperi, 95
Pasto culture, 22
  gold discs with embossed anurans, *35*
pathogens, 282–284, 286, 288, 289
patterns of distribution, amphibians in Ecuador, 245–250
Patzelt, E., xxvi, 6, 47, 48, 106, *114*
Paucar Guerra, D., **147**
Paucar Veintimilla, C. R., *162*, 164, 165, *167*
Paymal, N., 38
Pazmiño-Otamendi, G., xxiv, **147**
Pellizzaro, S., 39, 41
Peña Loyola, P., *152*
Peñaherrera-Romero, E., xxiii, ***170–177***, 174, *176*
Peñaherrera de Costales, P., 39, *see* Costales, P.
peptides

biological activity, **231**
frog skin secretions, 229–230, *230*
peptide analogs derived from, **226**
Pepper, M., *151*
Pepper, J., *151*
Peracca, M. G., 95, 98, 100, 104–*105*
Pérez, J. D., 48
Pérez Lara, M., 165, *166*, 168, 191
Pessier, A. P., *151*
Peters, J. A., 89, 97, 106, 107, 108–110, *109*, **110**, 121, *123*, 237, 277
Peters, W. K. H., 88, 91
petroglyphs, 22, 27
depicting frogs, *38*
Sapo Rumi rock, 27, *37*
phenotypic plasticity, 299
Philadelphia Zoo, 184, 185, 188
Phryniscus laevis, *see Atelopus ignescens*
phyllokinins, 229, 232
Phyllomedusa sauvagii, 232, 235
phylloseptins, 229–231
Phylogenetic Analyses Using Parsimony (PAUP) software, 149, 237
Phytelephas aequatorialis, 34
Pickett, J., 66
picturins, 201, 234
analogs, 236
pictuseptins, 201, 233–235
Pímbalo, 9
Pipa pipa, 87
syntypes of, *89*
plasticins, 229, 231–232
poetry, 1, 50–53
poison frogs (poison dart frogs), 10, 54, 63–65, 125, 135, 137, 182, 183, 202, 227, 237, 297
poison frog, *see Epipedobates boulengeri* (poison frog)
pollution, 286–287
polymerase chain reaction (PCR), 299–300
Ponce, P., *112*
Pontificia Universidad Católica del Ecuador, *see* QCAZ
Porras, P., 22, 27, 37
Pounds, J. A., *151*, 154, 267, 276, 277, 281, 282
Pramuk, J. B., *123*, 191, 197
Preñadillas, 5
Pristimantis buckleyi, 97
Pristimantis carvalhoi, 190, 193
Pristimantis chalceus, 88
Pristimantis conspicillatus, 98, *103*
Pristimantis festae (rainfrog), 98, 105
Pristimantis palmeri, 98
Pristimantis petersi, 110
Pristimantis petersioides, 110
Pristimantis surdus, 97, *103*
Pristimantis unistrigatus (cuico rainfrog), 45, 47, 53, 116, 154, 284
Proaño, B., 193
Proaño-Bolaños, C., xxiii, xxiv, *178*, 193, 197, 199–201, *200*, ***224–236***, 229
proline–arginine-rich peptides, 229, 233
proteomics, 296
Proyecto Arqueológico Huataviro (TH), 11
Pseudacris crucifer (hylid frog), 121
Pseudomonas aeruginosa, 224, 230, 232, 236
puca-sapo, *see Atelopus guanujo* (puca-sapo)
Puruhá culture, 22
pot decorated with anurans, *36*
Puruhá people, 6
Pyles, R. A., 190

Q

QCAZ, 144–146
amphibian collection growth, *150*
amphibians at, 141–142
Anfibios de Ecuador, 156–157
establishment of, 149–150
life raft for frogs at, 155–156
Luis A. Coloma at, 146–149
Martin R. Bustamante at, *152*
Néstor Acosta Buenaño at, *153*
Santiago R. Ron at, 157–159
quantitative polymerase chain reaction (qPCR), 298
Quelal, P., 10
Quiguango-Ubillús, A., xxv, 75, **147**, 155, 156, 193
Quito rocket frog, *see Hyloxalus jacobuspetersi* (Quito rocket frog)
Quitu-Cara, 6, 9
myth of the Catequil, 39–44

R

rainfrog, *see Pristimantis festae* (rainfrog)
rain rituals, 22, 27–34, *38*
Ramírez J., S., *166*
Ramus, E., 10
rana/sapo (Spanish for frog/toad) 6–7
Read, M., xxiv, xxv, xxvi, 150, 157, *131*, *159*, 159–160, *161*, 180, 195, *196*, 197, 275, 279
Real Academia Española, 6, 7, 54
Red List of Amphibians in Ecuador, 266–267
reduced representation sequencing (ddRAD), 296
Reece Dousdebés, R., *163*
Regalado, J, *112*
Rehabilitation Center of Manabí, 57
Renjifo, A., 156
Reserva Biológica Bilsa, *13*, *19*, 120
Reserva Ecológica Arenillas, *22*, *40*
Reserva Integral Otonga, 145
Reserva Otongachi, 145, 146
Reserva Otokiki, *39*, 71, 145, 146, 182
Reserva Otoyacu, 145, 146, 263
Reserva Quitahuayco, 186, 188
Reyes-Puig, J. P., xxiv, xxvi, 27, *123*, *162*, 164, 165, *166*, 168, *169*, 170, 195, *200*, 202, 236
Reyes-Puig, C., xxiii, xxiv, xxvi, *123*, 141, *162*, 165, *166*, 168, 169, ***170–177***, *176*, 177, 191, 193, 195, *200*, 203, ***236–242***, 238, 240
Reyes-Puig, M., 164, *167*, 168
Rhaebo blombergi (Blomberg's toad), *69*, 292, *293*
Rhaebo caerulesotictus, *102*
Rhaebo guttatus, 87
Rhaebo haematiticus, 14
Tolita culture, *14*
Rhinella festae (long-snouted toad), 98, 105, 134
Rhinella horribilis, 7, 11, 14, 29, 29, 46, 60
from Bosque Protector Siete Cascadas, *12*
in Chorrera culture, *12*, *13*
in Guangala culture, *24*, *39*
in Jama Coaque culture, *20*
from Reserva Otokiki, *39*
stylized representation of, *19*
in Tolita culture, *15*, 18
Rhinella spp. (*margaritifera* complex), 94, 260–262, *261*, *262*
Rhinella margaritifera, 260
Rhinella marina, 55, 87, 292, 293, 298
type of, *88*
Rhinella proboscidea, 87, 280
rhinophrynin-27, 233
rhinophrynin-33, 233
Rijksmuseum van Naturlijke Historie in Leiden (RMNHG), 62
Riobamba marsupial frog, 2
Riobamba pouched frog, 2
Río Palenque, *see* Centro Científico Río Palenque
Río Guayas basin, 21, 57
Rivas-Martínez, S., 245
Rivera, M., 191, 193, 224
Rivero Quintero, J. A., 120–121, *122*, *123*, 138
rocket frog, *see Paruwrobates erythromos* (rocket frog)
Rodríguez, M. A., *152*, 155, 156, 157, 185
Roland, A. B., 238, 298
Román Carrión, A. *163*
Román Jiménez, G., *163*
Román Jiménez, H., *163*
Romo, D., xxiii, ***170–177***, 170–172, *172*, 199
Ron, S. R., 48, 98, 117, *123*, *131*, 135, 137, 139, 140, 142, 149, 151, 154, 156, *157*, 157–159, 168, 169, 193, 195, *196*, 197–198, 199, 202, 203, 204, 237, 240, 266, 277, 278, 281, 282, 283, 284, 286
Ron, M., *163*
Rosenberg, W. F. H., 95, 96, 100
Rosero-Gómez, C. A., xxiii, xxiv, ***242–259***
Ruggeri, J., 284
Ruiz-Carranza, P. M., 117, *123*, 149
runs of homozygosity (OHs), 298
Ryan, M. J., 135, 137

S

Sagredo, Y., 191
Salazar, A., 48
Salazar, J., 264
Salazar-Nicholls, M. J., 193
Salazar-Valenzuela, D., *152*, 185
Salehi, B., 227
Salla, R. F., 287
Sánchez, M., *151*
Sánchez-Nivicela, J. C., xxiii, xxiv, xxvi, *123*, *162*, 168, ***170–177***, 177, *200*, 201
Sánchez-Carvajal, M. J., 193
Sandoval, S., 164
San Sebastián, M., 287
Santos, J. C., xxiii, xxiv, *134*, 135, 137, 141, **147**, *153*, 182, 195, *196*, 197, 198–199, 237, 248
Santos Ortiz de Villalba, J., 5
sapo game, *see* toad game
Sapo Rumi, 27
rock petroglyphs, 37
Sarmiento Rodríguez, F., *163*
Savage, J. M., 90, 106
Sayings, 53–54, *55*
Scheele, B. C., 267, 282
Schjellerup, I., 55
Schmidt, H., 63
Schmidt, W., 65
Schneider, C. J., 240
Schneider, J. G. T., 87

School of Biology of Pontificia Universidad Católica del Ecuador (PUCE), 67, 71, 74, 107, 115, *116*
Licenciatura at, **147**
Scinax quinquefasciatus (five-lined agile frog), 45, 46, 60, 264, *265*
Scinax ruber, 87
Seba, A., 87, *88*, *89*
Secretaría Nacional de Ciencia y Tecnología (SENACYT), 150, 154, 156, 157
selection, 158, 183, 202, 295, 296, 298, 299
SGS, *see* Guevara Salvador, S.
Shreve, B., 100, 104
Shuar, 10, 39, 43, 55
mythology, 41
protected forests at, 286
Shugshi, 9
Siavichay-Pesántez, F., ***183–188***, *184*, 186
Silva Chiriboga, D., 156
Silva del Pozo, X., 62
Silverstone, P. A., 237
Simmons, J. E., ***xv–xvi***, xxiv, xxvi, 5, 44, 59, 94, 95, 113, *114*, 117, 123, 127, 128, 129, 138, 139, 146, 147
Simons, P. O., 95, 96, 97–98, 100
single nucleotid polymorphisms (SNPs), 295
Siona/Secoyas mythology, 43–44
Siphonops annulatus, 6, 108
Siu-Ting, K. 191
small harlequin frog, *see Atelopus exiguus* (small harlequin frog)
Smith, E. N., *162*, 164, 165, *167*
Sornoza-Molina, F., 164, *166*, 168
Sosa, C., 38
south Cordillera Oriental of the Andes, 247, *251*
Spanish names, 9–10
Spannhof, I., 61, 62
Spannhof, L., 61, 62
species richness, 242
climate, 244–245
geography, 242–244
patterns of distribution, 245–250
sperm cryopreservation, 292
Sphaenorhynchus lacteus, 87, 87, 108
Spurrell's gliding leaf frog, *see Agalychnis spurrelli* (Spurrell's gliding leaf frog)
stamps, 44, 48, *49*, *50*, 149
Stănescu, F., 193
Stark, L. R., 6
Stolzmann's pac-man frog, *see Ceratophrys stolzmanni* (Stolzmann's pac-man)
Strabomantis anomalus, *103*
Stuster, R., 67
superstitions, 53–54, *55*
supportive breeding, 300
sustainable bioethical trade, *see* biotrade, amphibians
synthetic GnRH agonists (GnRHas), 292
Székely, D., *188*, xxiv, 189, 191, 193, 197, 198, *200*, 201
Székely, P., *188*, xxiv, xxvi, *123*, 189, 195, *196*, 198, 240

T

Tachina, 14
tadpoles, 2, 5, 28, 34, 59, 128, 298
names, 7–9, **7**, *8*
Tapia, C., *145*, 146, 150
Tapia, E. E., xxiv, xxv, xxvi, *131*, *145*, 146, *151*, 157, 177, 179, 182, 183, *184*, *196*, 197, 198
Tapia, I. G., xxiii, xxv, *131*, *145*, 146, *151*, *152*, 155, 157, 195, *196*
Tapia, M. G., *145*
Tapia Caisaguano family, *131*, 142, 144, *145*, 277
Tapley, B., 290
Tarvin, B., 137, 193, 204
taxonomic uncertainties, 296–297
Taylor, E. H., 100, 104, *105*, 106, *123*, 237
Telmatobius, 6, 10, 134, 251, 264, 267, 273, 282, 283, 296
Extinction, 273–276
Telmatobius cirrhacelis, 6, 134, 191, 193, 275
Telmatobius niger, 6, 9, 39, 55, 60, 74, 100, 155, 159, 188, 273–276, *275*
medicinal uses of, 60
Tepuihyla tuberculosa, *102*
Terán-Valdez, A., xxiii, xxiv, xxv, 179, *180*, ***189–195***, 191, 193, 197, *200*, 202, ***242–259***
Tigre-Pastaza, 247, *255*
Timbul, 9
trade, see biotrade, amphibians
toad game, 45
Tobar Suárez, C., *162*, 164, *167*, 193
Tolita culture, 10
Rhaebo haematiticus, *14*
Rhinella horribilis, *15*
Torres Carvajal, O., 134, 139, 140, 149, 168
Touzet, J.-M., 163, 164, 171, *172*, 173, 174, 177
Trachycephalus jordani (Jordan's bone-headed treefrog), *60*, 226
Trachycephalus quadrangulum, *102*
traditional medicine, 59–60
transcriptomics, 296
tricolor poison frog, *see Epipedobates tricolor* (tricolor poison frog)
Trueb, L., xxiii, 12, *114*, 124, 125, *126*, 127, 128, 131, *132*, *133*, 132–134, 135, 138, 139–141, *140*, 146, 149, 158, 174, 191, 198, 237
Trujilllo, F., *114*
tryptophyllins, 229, 232
Tsáchila culture, 12
Tulcán, *35*, 89, 105, 170, 202

U

Ubidia, A., 50, 52
Ucug, 6, 60
Ugalde, M. F., 14, 22
Uilli-Uilli, 8–9
Ultio, 9
ultraviolet radiation (UVR), 287–288
ultraviolet radiation index (UVI), 288
undescribed species, 250–253, 260, 277
Universidad Central del Ecuador, 106, 107, 109, 146, 155, 177, 201
biologists from, 164, 199, 201, 202, 203
Universidad del Azuay, 168, 184, 185, 201, 203
Universidad Técnica Particular de Loja, 115, 188–189, 193, 198
Universidad San Francisco de Quito (USFQ), 166
amphibian biologists, 176
amphibians described by, 175–176
Cisneros-Heredia at, 172–174
Guayasamin at, 174–175
Reyes Puig, C. at, 176–177
first researchers and collections, 171–172
initial steps, 170
species described, **175–176**
University of Kansas, 106, 113, 117, 121, 123, **124**, 127, 130, 133, 134, 137, 138, 139, 141, 146, 148, 154, 156, 157, 174, 191, 198, 252, 277, 280
Urgilés Merchán, M. A., *162*, 165, *166*, *167*
Urgilés, V. L., xxvi, *123*, 168, 191, 193, *200*, 203, 283, 284
USA, 59, 60, 70, 75
bans to exportation of Ecuadorian amphibians, 65–66
scientists and frog lovers in, 61–62
Uzendoski, M., 27

V

Vaca Guerrero, J., *112*, 113, 114
Valarezo Delgado, S., *163*
Valdivia period, 11
Valencia-Aguilar, A., 193
Valencia, J. H., xxiv, 9, *123*, 177, 195, *196*, 199, 261
Valencia Rojas, B., *163*
Vallejo Rodríguez, G. M., 224
valleys and paramos, 8, *252*
Valdivieso, S., *148*
Vane-Wright, R. I., *96*
Vanzolini, P. E., 117
Varela-Jaramillo, A., 191
Vargas Mena, P. xxiv, xxv, *178*
Vasconcelos, T. S., 245
Vega Toral, A., 6, 59–60, 183, *184*, 185
Vela, C., *163*
Velasco, A. M., 130, 144, 171, *172*, 177
Velasco Ibarra, J. M., 54, 65, 106
Venezuelan marsupial frog, *see Flectonotus pygmaeus* (Venezuelan marsupial frog)
Vergara, A., 286
Verkade family, 67
Verkade, J. C., xxiv, xxvi, 67, *69*, *178*, 181
Verkade, J., xxiv, *69*, 181
Verkade, T., xxiv, xxv, 181
vernacular names, 1–7, **2–4**, **56**
Vigle, G. O., xxiii, xxiv, xxvi, 111, ***117–120***, 119–120, *120*, ***121–123***, 164, 197
Viteri Dávila, C. I., 226
Vitreorana ritae, 190, 193
Vivarium de Quito, 177
Vizcaíno-Barba, M., xxiii, ***189–195***, 193

W

Walker, C. F., 61, 125, *126*, 128
Walker, J., 61
Walls, J. G., 65
Wampuch'/Wampucha, 9
Wanner, M., xxiii, xxiv, xxv, 181, *183*
water toad fountains, 45, *46*
Weiss, H., 61, 62
Weygoldt, P., 64
Whitworth, A., 261
whole-genome sequencing, 296, 298
Whymper, E., 47, *95*, 95–96, 100
Wiens, J. J., 121, 128, 147, *148*, 280
Wikiri, 65, *69*, 71, 178

Ameerega bilinguis, 1
Ceratophrys stolzmanni, 68
frogs trade by year, *78*
harlequin frogs, *73*
long-term growth and success, 67
Oophaga sylvatica, 76, *183*
Wilbert, J., 20, *26*, 34, 35
Wild, E. R., *133*
wild populations, genetics
climate change, 299–300
disease, 298–299
genetic management of, 300
size and demography, 297–298
structure and connectivity, 298
Wilson, B., *151*
Williams, C., *69*
Williams, E. E., 117–119, 121, 122, 125
Wolf, F. T., 47
Womack, M. C., xxv, 142, 158, *183*, 193, 230, 239, 260
WWF, 156, 160

X

Xenopus laevis (African clawed frog), 115, 184, 197, 264, *265*

Y

Yagé plant, *see Banisteriopsis caaspi* (Yagé plant)
Yanayaku river, 27
Yánez-Muñoz, M. H., xxiii, xxiv, xxvi, 111, *112*, *123*, *131*, 141, *162*, ***160–170***, 160–164, 165, *166*, 168, *169*, 174, 193, 195, *196*, 199
Yépez, A., *111*
Yantsaush frog, 41
Yeager, J., 67, 71

Z

Zambrano, R., 54
Zambrano Aveiga, R., *152*
Zamora Chinchipe, 57, 100, *166*, 185, *255*, 286
Aquarana catesbeiana, *58*
Zimmermann, E., 64, *64*
Zimmermann, H., 64, *64*
Zoologische Staatssammlung Mûnchen (ZSM) museum, 88, 89
zoomorphic pipe, 17, *19*
zoomorphic vase, 17, *21*
Zippel, K., xxiii, 156, *182*
Zumel, D., 239

For Product Safety Concerns and Information please contact our
EU representative GPSR@taylorandfrancis.com Taylor & Francis
Verlag GmbH, Kaufingerstraße 24, 80331 München, Germany